A Concrete Approach to Abstract Algebra

A Concrete Approach to Abstract Algebra

From the Integers to the Insolvability of the Quintic

Jeffrey Bergen

DePaul University
Chicago, Illinois

ELSEVIER

AMSTERDAM • BOSTON • HEIDELBERG • LONDON
NEW YORK • OXFORD • PARIS • SAN DIEGO
SAN FRANCISCO • SINGAPORE • SYDNEY • TOKYO

Academic Press is an imprint of Elsevier

Academic Press is an imprint of Elsevier
30 Corporate Drive, Suite 400, Burlington, MA 01803, USA
525 B Street, Suite 1900, San Diego, California 92101-4495, USA
84 Theobald's Road, London WC1X 8RR, UK

Notices

Knowledge and best practice in this field are constantly changing. As new research and experience broaden our understanding, changes in research methods, professional practices, or medical treatment may become necessary.

Practitioners and researchers must always rely on their own experience and knowledge in evaluating and using any information, methods, compounds, or experiments described herein. In using such information or methods they should be mindful of their own safety and the safety of others, including parties for whom they have a professional responsibility.

To the fullest extent of the law, neither the Publisher nor the authors, contributors, or editors, assume any liability for any injury and/or damage to persons or property as a matter of products liability, negligence or otherwise, or from any use or operation of any methods, products, instructions, or ideas contained in the material herein.

Library of Congress Cataloging-in-Publication Data
Bergen, Jeffrey, 1955.
   A concrete approach to abstract algebra : from the integers to the insolvability of the quintic / Jeffrey Bergen.
      p. cm.
   Includes bibliographical references and index.
   ISBN 978-0-12-374941-3 (hard cover : alk. paper) 1. Algebra, Abstract. I. Title.
   QA162.B45 2010
   512'.02–dc22

                                                                                    2009035349

British Library Cataloguing-in-Publication Data
A catalogue record for this book is available from the British Library.

ISBN: 978-0-12-374941-3

For information on all Academic Press publications
visit our Web site at *www.elsevierdirect.com*

Typeset by: diacriTech, India

Printed in the United States
10 11 12   9 8 7 6 5 4 3 2 1

To Donna

Contents

Preface

Abstract algebra, perhaps more than any other subject studied in college, has strong ties to the mathematics courses students have taken in high school. A course in abstract algebra can provide answers to many questions that are posed but not answered in high school mathematics courses. This is one reason that all mathematics majors, especially those hoping to teach at the high school or college level, can benefit from a course in abstract algebra.

Many instructors have witnessed students who, despite having had success in courses up through multivariable calculus and linear algebra, struggle in abstract algebra. Some of these instructors wonder if abstract algebra should even be required for math majors with a secondary education concentration. While writing this book, I was keenly aware of these issues.

This book was written because of my conviction that all mathematics majors should take abstract algebra, and, more importantly, all mathematics majors can learn abstract algebra. Some of the features that I believe will assist students in learning this subject are:

1. *Links to previous mathematics courses*: This book uses abstract algebra to answer basic questions that arise in courses in algebra, geometry, trigonometry, precalculus, and calculus. Concepts in abstract algebra are introduced as the tools needed to solve these basic questions.

2. *Exercises*: Courses up through multivariable calculus and linear algebra provide students with many exercises that allow them to practice and master new concepts. This book has 1996 exercises, many of which give students lots of practice working with concrete examples of new concepts. For example, in Chapter 8, students will have many chances to look at a multiplication table of a group and then compute cyclic subgroups, cosets, and centralizers.

    At various points, the exercises may appear to be somewhat repetitive. This is deliberate. In many books, instructors find an interesting exercise and then are faced with the choice of whether to include it in the lecture or in the homework. Perhaps the exercise's solution is in the solutions manual and the instructor prefers to assign problems where the solution is not readily available. Sometimes the exact opposite situation occurs. To avoid these problems, this book often includes many similar-looking exercises. This also gives the student more chances to practice and master concepts than is typically found in abstract algebra texts.

3. *Examples before definitions*: Students in abstract algebra courses are often overwhelmed or intimidated by the sheer volume of definitions and new objects. Whenever possible, this book attempts to provide examples before definitions so that definitions reflect the collecting of properties common to several concrete examples. For example, the integers, rational numbers, real numbers, and complex numbers are introduced before the definitions of commutative rings and fields are given. Similarly, concrete objects such as the invertible elements of the integers modulo n and the bijections of a set are studied before our formal discussion of groups. When new concepts are introduced, such as automorphisms in Chapter 5 and ring homomorphisms in Chapter 9, they are immediately applied to familiar problems such as finding roots of polynomials and determining when polynomials are irreducible.

4. *Fundamental Theorem of Algebra*: Virtually every abstract algebra textbook mentions the Fundamental Theorem of Algebra, but very few contain a proof. The reason is that a primarily algebraic proof requires so many new ideas that it would take most books too far off course. However, in Chapter 6, we present a proof based on some familiar ideas from one and two variable calculus. We have chosen this direction for both philosophical and practical reasons.

    One of the goals of this book is to help students develop a deep understanding of the roots and factoring of polynomials over different number systems. Occasionally, this requires examining topics that are not traditionally part of an algebra course, such as the Intermediate Value Theorem and the Fundamental Theorem of Algebra. However, these topics are essential for an understanding of the differences in the behavior of polynomials over the rational numbers, the real numbers, and the complex numbers.

    As a practical matter, having the Fundamental Theorem of Algebra at our disposal makes it much easier to introduce Galois theory and then prove the insolvability of the quintic. Students often struggle with the level of abstraction in Galois theory. However, when examining the roots of polynomials with rational coefficients, the Fundamental Theorem of Algebra allows us to always work with fields that lie between the rational numbers and complex numbers. This more concrete approach makes the key ideas of Galois theory easier to understand and greatly simplifies the proof of the insolvability of the quintic.

5. *Theorems with proofs*: With the occasional exception of results from courses below abstract algebra, if a theorem appears in this book, so will its proof. A philosophy underlying this book is that reading proofs is an essential part of abstract algebra. Sometimes textbooks will state powerful theorems, without proof, and then use them to obtain other important results. For example, abstract algebra books often state, without

proof, the Fundamental Theorem of Algebra or the Fundamental Theorem of Galois theory and then use them to prove other results. I believe this approach can stand in the way of students gaining a deep understanding and appreciation of algebra.

There will be times in this book when the theorems we state, prove, and apply are not the most general results known. However, as opposed to applying stronger results whose proofs they have never seen, I believe students will learn more applying results whose proofs they have worked through.

Please feel free to e-mail your thoughts, comments, and corrections to me at jbergen@depaul.edu. You can find a list of corrections at www.depaul.edu/~jbergen.

A User's Guide

A yearlong course in abstract algebra can cover this entire book with sufficient time for a thorough treatment of each section. However, it can easily be adapted to courses that meet for only one quarter, one semester, or two quarters. For courses that run less than a year, the chapter summaries following should help instructors decide which sections to skip and how to sequence the sections that are covered.

Chapter 1—This introductory chapter points out that many questions that arose and were left unanswered in a student's previous courses in algebra, geometry, trigonometry, precalculus, and calculus can now be answered using abstract algebra. It previews many of the results that will be proven in this text, such as the insolvability of the quintic, the Fundamental Theorem of Algebra, the impossibility of trisecting angles, rational values of trigonometric functions, partial fraction decomposition, and multiple roots of polynomials. This chapter can either be covered in class or left as a reading assignment. It is not a prerequisite for any of the later chapters.

Chapter 2—Sections 2.1, 2.2, and 2.3 begin by discussing the importance of both intuition and rigor in mathematics. They then focus on proofs by contradiction, the Well Ordering Principle, and Mathematical Induction. Throughout this book, it will be very important for your students to have a solid understanding of these sections. However, if your students are already adept at writing proofs, these sections can be left as a reading assignment.

Section 2.4 introduces functions and binary operations. This section will be the foundation for much of the material in this book. To make our detailed examination of groups in Chapter 8 more accessible to students, examples of groups will appear at various points before then. In particular, groups are briefly discussed in Section 2.4 when we look at injective, surjective, and bijective functions.

Chapter 3—This chapter focuses on properties of the integers, such as prime numbers and the Euclidean Algorithm. The most important result in this chapter is the existence and uniqueness of prime factorization. Exercises 31–37, immediately after Section 3.3, might be particularly helpful to students who wonder why a concept as intuitive as unique factorization requires proof. The ideas presented in this chapter are used throughout this book. In particular, our discussion of polynomial rings in Chapter 12 follows the pattern set forth in this chapter.

Chapter 4—Sections 4.1 and 4.2 contain topics that are not required for later chapters. Section 4.1 examines rational numbers and the relationship between fractions and repeating decimals. Section 4.2 compares the rational numbers and the real numbers and focuses on the least upper bound property, the Intermediate Value Theorem, and roots of polynomials with real coefficients. Some instructors may choose to skip these sections, as they cover topics that rarely appear in abstract algebra courses. However, if a student has not seen these topics in previous courses, they have the opportunity to see them here.

Equivalence relations and equivalence classes are introduced in Section 4.3. These topics will reappear many times throughout this book. Since students often struggle with quotient groups and quotient rings, many examples and exercises are provided that examine the addition and multiplication of equivalence classes and when these operations are well defined.

Chapter 5—This chapter introduces the complex numbers and uses them, along with the integers, rational numbers, and real numbers, to motivate the definitions of commutative rings and fields. Complex conjugation and its relationship to roots of polynomials are then used to motivate the definitions of automorphisms and Galois groups. Chapters 8, 15, and 17 contain a more detailed and theoretical treatment of groups and automorphisms. However, it is helpful for students to gain experience, at this stage, working with concrete examples of these objects.

Chapter 6—One of the themes of Chapters 5 and 6 is to demystify the complex numbers and to show that they are as real as the real numbers. In Chapter 5, we show that the construction of the complex numbers from the real numbers is simpler and more straightforward than either the construction of the rational numbers from the integers or the real numbers from the rational numbers. In Sections 6.1, 6.2, and 6.3, polar form and DeMoivre's Theorem are introduced and are used to help show that the addition and multiplication of complex numbers can be viewed in a very concrete and geometric manner.

Section 6.4 contains a proof of the Fundamental Theorem of Algebra. This allows us to deal with fields, Galois groups, and the insolvability of the quintic more concretely in Chapters 15 and 17, as we only need to work with fields that are contained in the complex numbers. Abstract algebra courses that do not run for a full year might need to omit Chapters 15 and 17. In this case, Section 6.4 can also be omitted.

Chapter 7—Sections 7.1, 7.2, and 7.4 examine the integers modulo n and provide many examples of commutative rings, fields, and groups. The ideas in these sections are needed when we examine polynomials with integer and rational coefficients in Chapter 9 and also for the proof of Kronecker's Theorem in Chapter 17. Section 7.3 looks at the Euler ϕ function and is not a prerequisite for any of the later chapters.

Chapter 8—This chapter, which examines the structure of finite groups, can be covered in many different ways depending on how the instructor structures the course. Since students will have already worked with examples of groups in Chapters 2, 5, and 7, they should be well prepared for the more formal and detailed treatment in Chapter 8. If a course proceeds sequentially through this text, Sections 8.1 and 8.2 will be covered toward the end of the first semester. Therefore, even if students only take one semester of abstract algebra, they can still see a proof of Sylow's Theorem.

Sections 8.3 and 8.4 deal with solvable and symmetric groups and are only needed for Chapters 15 and 17. Since Chapter 8 is quite long, instructors may decide to take a short break from group theory after Section 8.2, as Section 8.3 can be covered at any point before Section 15.3 and Section 8.4 at any point before Chapter 17.

If an abstract algebra course runs for only one quarter, one semester, or two quarters, the instructor may determine that the brief introduction to groups in Chapters 2, 5, and 7 is sufficient and then skip Chapter 8 entirely. This would allow time to cover some of the links between abstract algebra and the high school curriculum in Chapters 9, 11, and 13 that do not require group theory.

Chapter 9—This chapter helps to illustrate the importance of ring homomorphisms and the integers modulo p by using them to prove the Rational Root Test, Gauss' Lemma, and Eisenstein's Criterion. Since this chapter examines the roots and irreducibility of polynomials over the integers, rationals, reals, and complex numbers, it should be particularly useful for students planning to teach algebra at the high school or community college level.

Chapter 10—Section 10.1 shows how to find the roots of polynomials of degrees less than 5, and Section 10.2 informally discusses some consequences of Galois' work. Section 10.1 can be covered at any point in the course, and Section 10.2 only requires an understanding of Eisenstein's Criterion. The material in this chapter is not a prerequisite for any of the later chapters.

Chapter 11—This chapter examines rational values of trigonometric functions and explains why the 30°–60°–90° and 45°–45°–90° triangles tend to be the only right triangles studied in trigonometry classes. The only background material needed for this chapter is Mathematical Induction and the Rational Root Test. This is another chapter that should be particularly useful for future teachers. The material in this chapter is not a prerequisite for any of the later chapters.

Chapter 12—In this chapter, it is shown that polynomials over fields satisfy analogs of many properties satisfied by the integers. The proofs in Sections 12.1–12.4 are very similar to those in Chapter 3. In Section 12.5, the relationship between multiple roots of polynomials and derivatives is examined. The results in this chapter will be used repeatedly throughout the remainder of the book.

Chapter 13—This chapter contains material that should be of particular interest to teachers of precalculus and calculus. In Section 13.1, difference functions are used to find the polynomial of smallest degree that can produce a collection of data. As an application, Section 13.2 shows how to derive many of the formulas that students merely verify when first learning about Mathematical Induction. Section 13.3 shows why the partial fraction decomposition algorithm in calculus courses actually works. This section relies heavily on the division algorithm and Euclidean Algorithm for polynomial rings in Chapter 12. The material in this chapter is not a prerequisite for any of the later chapters.

Chapter 14—This chapter examines some of the key concepts in linear algebra: basis, dimension, spanning set, and linear independence. The material in Sections 14.1, 14.2, and 14.3 is essential for the final three chapters of this book. However, instructors may choose to skip this chapter if the students have already taken a course in linear algebra.

Chapter 15—Section 15.1 examines degrees of field extensions, and Sections 15.2 and 15.3 look at splitting fields and Galois groups. The material in this chapter is the foundation for the work in Chapter 16 on ruler and compass constructions and in Chapter 17 on the insolvability of the quintic. If a course does not allow time for a proof of the insolvability of the quintic, instructors can go directly from Section 15.1 to Chapter 16 and can also skip Sections 8.3 and 8.4.

Chapter 16—This chapter contains the proof that angles cannot be trisected with ruler and compass. It relies very heavily on Section 15.1. Although this result appears near the end of the book, by carefully choosing which sections to skip, it can be covered in a one-semester course. The results in this chapter are not used in Chapter 17.

Chapter 17—Sections 17.1 and 17.2 contain the proof of the insolvability of the quintic and also show how to produce infinite families of fifth- and seventh-degree polynomials that are not solvable by radicals. Section 17.3 contains additional material, such as Kronecker's Theorem and the Isomorphism Theorem for Rings, that should be of particular interest for students planning to pursue graduate study. This section exploits one of the recurring themes of this book: the similarities between the integers and polynomials rings.

Acknowledgments

I would like to thank the many people who supported me as my class notes became a book. First, my thanks to my DePaul colleagues Allan Berele and Stefan Catoiu for teaching from preliminary drafts and to Susanna Epp and Lynn Narasimhan for encouraging me to take on this project. Second, I thank Ken Price of the University of Wisconsin-Oshkosh for providing useful feedback after using a preliminary draft. Third, my thanks to Glenn Olson of Maine East High School for providing me with information about the connection between complex numbers and electrical circuits.

I owe a debt of gratitude to Dan Tripamer of St. Viator High School for all his work producing the diagrams. Lauren Schultz Yuhasz of Elsevier has been enormously helpful, and the comments by the reviewers she found helped shape the final product. My thanks to Phil Bugeau of Elsevier for his help in the final stages of this project. I would also like to thank the University Research Council at DePaul University for their support.

Finally, a special thank you to my wife Donna and children Renee, Sabrina, Mark, and Melisa for their continuous love and support.

Jeffrey Bergen
July 2009

CHAPTER 1

What This Book Is about and Who This Book Is for

You are about to embark on a journey. Often this journey is referred to as abstract algebra. Others call it modern algebra, and still others simply call it algebra. But it is probably very different from any type of algebra you have ever studied before.

When they are first introduced to this subject, many students feel quite intimidated. They feel as if they are drowning in an unending sea of meaningless definitions. Terms like *group, ring, field, vector space, basis, dimension, homomorphism, isomorphism*, and *automorphism* appear, often for no apparent reason.

Almost all of us, at some point, are intimidated by a new project. Many home repair projects have that effect on me. A walk through the aisles of a home improvement store can intimidate me to the point where it becomes difficult to even formulate an intelligent question for a sales clerk. The aisles and aisles of bizarre-looking devices and gadgets overwhelm me. However, every item is there for a reason. Each one is a tool needed to solve a problem. Suddenly one odd-looking device is exactly what I need to unclog my bathtub. Yet another is precisely what I need to make my vacuum cleaner work again.

Abstract algebra is a subject that arose in an attempt to solve some very concrete problems. It is likely that you have already come across many of these problems in your previous courses as they occur very naturally in algebra, geometry, trigonometry, and calculus. However, in those courses, these problems are usually dismissed with the comment that they are beyond the scope of the course.

It may seem like an odd analogy, but reading through a book in abstract algebra is not all that different from walking through the aisles of a home improvement store. All those intimidating new terms you come across in an abstract algebra book are actually tools. They are precisely the tools needed to finally solve many of the problems that arose but remained unsolved in your previous courses.

In this book, you will be introduced to the basic terms, ideas, and concepts of abstract algebra. Each of these new ideas will be presented as concretely as possible. New terms and concepts will be introduced as the tools needed to solve well-known problems. Each time we come across a new abstract object, we will be equipped with both the knowledge of the problem it is being used to solve, as well as multiple concrete examples of the object. This should help eliminate the intimidating aspects of this subject and will allow us to understand and appreciate both the beauty and the importance of the subject.

Let us now look at some of the problems that we will use abstract algebra to solve. We will list them according to the course where you may have first seen them.

1.1 Algebra

1.1.1 Finding Roots of Polynomials

Long ago, you learned that in order to find the root of the polynomial $2x + 1$, we first subtract 1 from both sides of the equation

$$2x + 1 = 0$$

to obtain the equation

$$2x = -1,$$

and then divide both sides by 2 to obtain the root

$$x = -\frac{1}{2}.$$

More generally, if a and b are real numbers, with $a \neq 0$, then to find the root of the polynomial $ax + b$, we first subtract b from both sides of the equation

$$ax + b = 0$$

to obtain the equation

$$ax = -b,$$

and then divide both sides by a to obtain the root

$$x = -\frac{b}{a}.$$

Thus, we know how to find the root of any polynomial of degree 1. Moving on to polynomials of degree 2, any such polynomial can be written as

$$ax^2 + bx + c,$$

where a, b, and c are real numbers, with $a \neq 0$. In high school, we derived the quadratic formula that told us that the roots of $ax^2 + bx + c$ are

$$x = \frac{-b \pm \sqrt{b^2 - 4ac}}{2a}.$$

Therefore, for polynomials of degrees 1 and 2, it was not too difficult to find formulas for their roots. Note that these formulas were expressions involving only the coefficients of the polynomials and that the coefficients were combined in various ways via addition, subtraction, multiplication, division, and taking square roots. The next natural step is to look for a formula for the roots of polynomials of degree 3. We would like to find a formula that once again involves the coefficients. However, we would expect that, at this point, we might not only need to take square roots but to take cube roots as well.

More generally, our goal is to find formulas for the roots of polynomials of all possible degrees where these formulas involve only the coefficients and the coefficients are combined in various ways via addition, subtraction, multiplication, division, and taking roots. By taking roots, we mean square roots, cube roots, fourth roots, and so on.

In Chapter 10, we will show that such formulas do indeed exist for polynomials of degrees 3 and 4. The quadratic formula $x = \frac{-b \pm \sqrt{b^2 - 4ac}}{2a}$ is significantly more complicated than the formula $x = -\frac{b}{a}$ for the root of polynomials of degree 1. In light of this, it is not surprising that the formula for the roots of polynomials of degree 3 is significantly more complicated than the quadratic formula. Again, it is no surprise that the formula for the roots of polynomials of degree 4 is significantly more complicated than its predecessors.

The logical next step is to move on to polynomials of degree 5. Unfortunately, if one tries to generalize or adapt the techniques used to find the roots of polynomials of degrees 1, 2, 3, and 4, nothing seems to work. There are two possible reasons why nothing seems to work for polynomials of degree 5. The first possible reason is that the formula is so complicated that we just haven't hit upon the approach needed to find it. Since the formula for the roots of polynomials of degree 4 is so much more complicated than its predecessors, it is logical to assume that finding a formula for the roots of polynomials of degree 5 should be an extremely difficult task. However, there is another possible reason why we have been unsuccessful. Perhaps there is no formula for the roots of polynomials of degree 5. This seems to be a disturbing possibility. Not only would it be disappointing to not have a formula available for finding the roots of polynomials of degree 5, but we also need to ask ourselves how can one possibly prove that no such formula exists. After all, how do we prove that something can't be done or doesn't exist?

In one of the greatest achievements in abstract algebra, it was shown by Galois that no formula exists for finding the roots of polynomials of degree 5. In fact, Galois showed that for any integer $n \geq 5$, there is no formula for finding the roots of polynomials of degree n. Once again, by a "formula" we mean an expression involving only the coefficients of the polynomial where the coefficients are combined in various ways via addition, subtraction, multiplication, division, and taking roots. This famous problem is known as the **insolvability of the quintic**. Its solution will require an enormous amount of mathematical machinery and appears in Chapter 17. The main tools needed to solve it will be group theory and Galois theory. In fact, many of the terms and concepts appearing in this book are included because they are the tools needed to solve this famous problem.

Before leaving this particular topic, we must remember that there are other approaches to finding the roots of polynomials. Essentially, the insolvability of the quintic tells us that a purely algebraic approach comes up short in trying to find the roots of some polynomials. However, depending on the application you have in mind, you may not need a formula for the roots of a polynomial that involves various combinations of the coefficients. Instead, you may need the roots computed to a certain number of decimal places. There are many numerical algorithms that can give you the roots of polynomials to as many decimal places of accuracy as you desire (or at least as many decimal places as the machine you are using can handle). Many of these algorithms are built into or can be easily programmed into a graphing calculator. Although this does not technically give you the exact answer, having the answer correct to a large number of decimal places may well be sufficient for the application you have in mind.

1.1.2 Existence of Roots of Polynomials

As mentioned in the preceding paragraph, the phrase "finding a root" can have slightly different meanings, depending on the context. If we are looking for the largest root of the polynomial $x^4 - 14x^2 + 9$, then in an algebra course, you would probably write the answer in the form $\sqrt{2} + \sqrt{5}$. (At this point, you should take a moment to check that $\sqrt{2} + \sqrt{5}$ is indeed a root of $x^4 - 14x^2 + 9$.) However, depending on the application you had in mind, you might want the answer to 5 decimal places, and, in this case, 3.65028 would be your answer. If you wanted the answer to 10 decimal places, then 3.6502815398 would be the answer. On the other hand, in the unlikely event that you needed the answer to 32 decimal places, then the answer would be

$$3.65028153987288474521086239294097.$$

Similarly, the phrase "existence of a root" can mean different things depending on the context. We begin by considering the polynomial $x + 5$; if we restrict ourselves to dealing only with positive integers, then this polynomial has no roots. Once we expand our horizons to the set of integers, we see that this polynomial certainly has a root and the root is -5. In a similar vein, if we restrict ourselves to dealing only with integers, then the polynomial $2x - 7$ has no roots. By once again expanding our horizons, this time to the set of rational numbers, then our polynomial certainly has a root and the root is $\frac{7}{2}$.

At various points in this book, including Chapters 2 and 3, we will show that $\sqrt{2}$ is not a rational number. Therefore, in order for the polynomial $x^2 - 2$ to have a root, we must look beyond the rational numbers. In calculus, one proves that there is indeed a positive real number whose square is 2. This is an issue that we will reexamine in Chapter 4. Therefore, by looking at yet another larger set of numbers—the real numbers—our polynomial has the roots $\pm\sqrt{2}$. At this point, we can begin to wonder if we must continually expand the set of numbers we are using in order to guarantee the existence of roots of all polynomials. After all, even the real numbers do not suffice, as they do not contain a root of the polynomial $x^2 + 1$.

The complex numbers contain an element, denoted as i, with the property that $i^2 = -1$. Therefore both i and $-i$ are roots of the polynomial $x^2 + 1$. Every complex number can be written in the form $a + bi$, where a and b are real numbers. It turns out that the complex numbers are "big enough" that they contain the roots of all polynomials. More precisely, we mean that any polynomial of degree at least one, whose coefficients are real numbers, has a root in the complex numbers. Therefore the polynomials $x + 5$, $2x - 7$, $x^2 - 2$, and $x^2 + 1$, as well as more complicated polynomials like $x^5 - 6x + 2$ and $\pi x^3 - 3x^2 + \sqrt{7}x - \frac{11}{219}$, all have a root in the complex numbers. This beautiful and important result is known as the **Fundamental Theorem of Algebra**. Interestingly enough, proofs of this result relying almost entirely on algebra are extremely difficult, whereas fairly elementary proofs exist that use only some of the basic ideas of complex numbers and multivariable calculus. We will present one of these relatively elementary proofs in Chapter 6.

Think about the old story of how hard it is to find a needle in a haystack. Imagine how much more difficult the situation would be if you weren't entirely sure there even *was* a needle in the haystack. If you were unable to find the needle, you would never know if the problem was that you hadn't searched well enough or that the needle wasn't there in the first place. There is a clear parallel with trying to find the roots of polynomials. Finding roots can be a difficult task, but imagine how much more difficult it would be if you didn't know whether a root was there to be found. But thanks to the Fundamental Theorem of Algebra, we are guaranteed that there will always be a root in the complex numbers. Therefore, although it may be difficult to find a root of a polynomial, we know there is always a root in the complex numbers waiting to be found.

1.1.3 Solving Linear Equations

Let us consider the following three similar-appearing systems of linear equations:

$$(\mathrm{I}) \quad 2x + 5y = 7 \quad \& \quad 2x + 3y = 1,$$

$$(\mathrm{II}) \quad 2x + 5y = 7 \quad \& \quad 4x + 10y = 1,$$

$$(\mathrm{III}) \quad 2x + 5y = 7 \quad \& \quad 4x + 10y = 14.$$

Despite looking somewhat similar, when we look at their solutions, we see that these systems of linear equations differ greatly from one another. Note that system (I) has the unique solution $x = -4$ and $y = 3$, whereas system (II) has no solutions, and system (III) has an infinite number of solutions.

In your earlier algebra courses, you probably noticed that every system of linear equations had one solution, no solutions, or an infinite number of solutions. Perhaps you have wondered if this is always the case. Or is it possible for a system of linear equations to have exactly two solutions or three solutions or some other number of solutions?

Chapter 14 will include an investigation of systems of linear equations. At that point, we will show that when dealing with familiar number systems like the rational numbers and real numbers, it is indeed the case that every system of linear equations has one solution, no solutions, or an infinite number of solutions. However, there are other types of number systems, which will be introduced to in Chapter 7, where there are other possibilities for the number of solutions.

1.2 Geometry

1.2.1 Ruler and Compass Constructions

In a course in geometry, we construct various geometric objects using a ruler and a compass. One of the first constructions is to take a line segment and divide it into two equal pieces. It is not much harder to take a line segment and divide it into three, four, or any number of equal pieces. At that point, we might turn our attention to angles. Just as it was not difficult to bisect a line segment, it is also not hard to divide an angle into two equal angles. However, when we try to divide an angle into three equal angles, difficulties seem to arise. Certainly some angles can be trisected. For example, we can trisect a 90° angle and obtain a 30° angle. However, all attempts at finding a procedure that will work for all possible angles seem to fail. This raises the type of question we dealt with earlier when we discussed looking for a formula for the roots of polynomials of degree 5. Are we unable to find a technique for trisecting angles because we simply haven't hit upon the right idea, or is it impossible to trisect angles with only a ruler and a compass? Once again, we are confronted with the difficult question of how to show that something is impossible.

Using a ruler and compass, it is not difficult to construct equilateral triangles. Since all three angles of an equilateral triangle are equal, that means that we have succeeded in constructing a 60° angle. If indeed it were possible to trisect all angles, then we could trisect our 60° angle to obtain a 20° angle. In Chapter 16, using tools on field extensions developed in Chapter 15, we will show that it is impossible to construct a 20° angle. Thus, it is indeed impossible to use a ruler and compass to trisect all possible angles.

It is very common to think of positive real numbers as representing distances between points on a number line. We can therefore think of a positive real number a as being constructible, meaning that with a ruler and compass, we can construct a line segment whose length is a. What we will really be showing in Chapter 16 is that many, many real numbers such as $2^{\frac{1}{3}}$, $29^{\frac{2}{3}}$, and $11^{\frac{1}{7}}$ are not constructible. The key to proving that $20°$ angles cannot be constructed will be to first show that $\cos(20°)$ is not a constructible real number.

1.3 Trigonometry

1.3.1 Rational Values of Trigonometric Functions

When we study trigonometry, we pay special attention to the $30°-60°-90°$ and $45°-45°$ $-90°$ right triangles. There are two reasons for this. The first is that it is fairly easy to compute the values of the sine, cosine, and tangent functions for these two types of triangles. The second is that for these triangles, the values of some of the trigonometric functions are so nice—in particular, $\sin(30°) = \cos(60°) = \frac{1}{2}$ and $\tan(45°) = 1$. This raises some questions. In a course in trigonometry, why do we not examine other triangles that also produce particularly nice values of the trig functions? Why do we not look for angles θ_1, θ_2 such that $\sin(\theta_1) = \frac{2}{3}$ or $\tan(\theta_2) = 5$? We know that there exist acute angles θ_1, θ_2 with these properties, so why do we not study them?

In trigonometry courses, we tend to look at angles using radian measure. Recall that $30° = \pi/6$, $45° = \pi/4$, and $60° = \pi/3$. Each of these angles is a rational number times π. By using other geometric and trigonometric facts, such as the double-angle formula, we can compute the values of the trigonometric functions at several other angles that are a rational number times π. In fact, in the exercises after Chapter 11, we will show how to compute the exact values of $\cos\left(\frac{\pi}{8}\right)$ and $\cos\left(\frac{\pi}{5}\right)$. Although these two values of the cosine do not turn out to be rational, it certainly seems that in our trigonometry courses, we should be able to study additional angles that are a rational number times π such that the values of sine, cosine, or tangent are rational numbers.

We know that there are rational multiples of π that can make the sine and cosine functions take on the rational values $0, \pm\frac{1}{2}, \pm 1$ and the tangent function take on the rational values $0, \pm 1$. However, it is a surprising fact that these are the *only* rational values of the sine, cosine, and tangent functions that can be obtained by plugging in an angle that is a rational number times π. In Chapter 11, we will prove the somewhat stronger result that when plugging in angles that are a rational number times π, the only values of the sine and cosine functions whose squares are rational belong to the set $\left\{0, \pm\frac{1}{2}, \pm\frac{\sqrt{2}}{2}, \pm\frac{\sqrt{3}}{2}, \pm 1\right\}$. In addition, we will see in Chapter 11 that the only values of the tangent functions whose squares are rational that are obtained by plugging in angles that are a rational number times π belong to the set $\left\{0, \pm\frac{\sqrt{3}}{3}, \pm 1, \pm\sqrt{3}\right\}$. Observe that these are all values of the sine, cosine, and tangent functions that were examined in our previous courses in trigonometry. Thus, Chapter 11 will explain why, in our previous courses, we never examined any other values of the trigonometric functions that are rational or whose square is rational.

1.4 Precalculus

1.4.1 Recognizing Polynomials Using Data

In courses in algebra, precalculus, and calculus, we are often presented with a table of values of a function and asked to determine if the function that produced those values was linear, quadratic, or some other type of polynomial. Therefore, it is natural to look for a way to recognize if these values were indeed produced by a polynomial. Going one step further, if the values were produced by a polynomial, can we find the polynomial? Shortly, we will need to be a bit more precise about what we are asking, but first let us look at an example.

x:	1	4	7	10	13	16	19	22	25
$f(x)$:	7	13	19	25	31	37	43	49	55

In the table, x continues to increase by 3 as we move from left to right. As you may recall, a function $f(x)$ is linear precisely if a fixed change in x always results in a fixed change in $f(x)$. To test whether $f(x)$ is linear, we will introduce a new function $f_{(1)}(x)$, which we define as $f_{(1)}(x) = f(x+3) - f(x)$. So, for example,

$$f_{(1)}(4) = f(7) - f(4) = 19 - 13 = 6 \quad \text{and} \quad f_{(1)}(22) = f(25) - f(22) = 55 - 49 = 6.$$

We can see that the function $f_{(1)}(x)$ measures the change in $f(x)$ as x increases by 3. Next, we place the values of $f_{(1)}(x)$ alongside those of x and $f(x)$ on our table.

x:	1	4	7	10	13	16	19	22	25
$f(x)$:	7	13	19	25	31	37	43	49	55
$f_1(x)$:	6	6	6	6	6	6	6	6	

Note that the table is left blank in one position, as we cannot compute $f_{(1)}(25)$ because $f_{(1)}(25) = f(28) - f(25)$, and we were not given the value of $f(28)$. Looking at the values of $f_{(1)}(x)$, we see that $f_{(1)}(x)$ always gives us a value of 6. This tells us that whenever x increases by 3, then $f(x)$ increases by 6. Thus, the values on the table can indeed be produced by a linear function.

At this point we need to be a little careful about our wording. Note that we did not say that $f(x)$ was a "linear" function. The reason is that we were only given 9 values of $f(x)$. It is possible that if we were given additional values of $f(x)$, then the new values of $f_{(1)}(x)$ might not continue to always be 6.

It turns out that given any finite number of data points, there are an infinite number of polynomials that could have produced that data. However, if we are given exactly n data points, where n is an arbitrary positive integer, then there will always be exactly one polynomial of degree at most $n - 1$ that could produce that data. This generalized the fact—which we have seen in our previous algebra courses—that given two data points, there is only one linear function that could produce that data. These are facts that we will prove in

Chapter 13. In light of this, our goal will be to find the polynomial of the smallest possible degree that could produce a collection of data points. In the preceding example, the data for $f(x)$ is produced by a linear function, and using the point-slope formula (or various other techniques from your previous algebra courses), we see that the values on the table could be produced by the function $2x + 5$.

Let us look at a second example.

x:	1	2	3	4	5	6	7	8	9
$g(x)$:	2	7	14	23	34	47	62	79	98

In this table, x is increasing by 1, so to see if $g(x)$ could be linear, we look at the function $g_{(1)}(x) = g(x+1) - g(x)$, which measures the change in $g(x)$ as x increases by 1. Placing the values of $g_{(1)}(x)$ on the preceding table, we obtain

x:	1	2	3	4	5	6	7	8	9
$g(x)$:	2	7	14	23	34	47	62	79	98
$g_{(1)}(x)$:	5	7	9	11	13	15	17	19	

Since the values of $g_{(1)}(x)$ are not constant, the values of $g(x)$ cannot be produced by a linear function. However, we can consider the function $g_{(1)}(x)$ as measuring the "first differences" of $g(x)$. We can now define a new function $g_{(2)}(x)$, as $g_{(2)}(x) = g_{(1)}(x+1) - g_{(1)}(x)$. Thus, we can consider $g_{(2)}(x)$ as measuring the "second differences" of $g(x)$. We will now place the values of $g_{(2)}(x)$ alongside those of $g(x)$ and $g_{(1)}(x)$, noting that we do not have enough information to compute either $g_{(2)}(9)$ or $g_{(2)}(8)$.

x:	1	2	3	4	5	6	7	8	9
$g(x)$:	2	7	14	23	34	47	62	79	98
$g_{(1)}(x)$:	5	7	9	11	13	15	17	19	
$g_{(2)}(x)$:	2	2	2	2	2	2	2		

We can see that the values of $g_{(2)}(x)$, the "second differences" of $g(x)$, are constant. It turns out that although the table of values of $g(x)$ cannot be produced by a linear function, the values can be produced by a quadratic function. In fact, the function $x^2 + 2x - 1$ does the trick.

Both of our examples are actually examples of a more general phenomenon. Suppose we are given a table of values for a function $F(x)$, where, throughout the table, the change in x is the fixed number a. We can define a new collection of functions as follows:

$$F_{(1)}(x) = F(x+a) - F(x) \quad \text{and} \quad F_{(n+1)}(x) = F_{(n)}(x+a) - F_{(n)}(x),$$

where n is any positive integer. We call the function $F_{(n)}(x)$ the nth difference function of $F(x)$. Note that computing the various difference functions of $F(x)$ is not unlike computing the higher derivatives of a function in calculus. In order to find the third derivative of a function, you must have already found its second derivative, which in turn requires having already computed the first derivative. Similarly, to compute $F_{(3)}(x)$, one must have already found $F_{(2)}(x)$, which means that you need to have already found $F_{(1)}(x)$.

In Chapter 13, we will show that if the nth differences of $F(x)$ are constant while the $n-1$st differences are not constant, then the values of $F(x)$ are produced by a unique polynomial of degree n. Observe that this is quite similar to a consequence of the Mean Value Theorem in calculus, which states that if the nth derivative of a function is constant and the $n-1$st derivative is not constant, then the function must be a polynomial of degree n. We will also show how to find the polynomial of smallest possible degree that can produce a given table of values.

When a student is first introduced to Mathematical Induction, they are usually asked to check the validity of formulas like

$$2^2 + 4^2 + 6^2 + \cdots + (2n-2)^2 + (2n)^2 = \frac{(2n)(n+1)(2n+1)}{3}.$$

Observe that whereas we are asked to verify such formulas, we do not address the more important and far more interesting question of how one goes about deriving formulas like the preceding one. In Chapter 13, we will apply our results on difference functions to show how to derive formulas like the preceding one.

1.5 Calculus

1.5.1 Partial Fraction Decomposition

Among the problems we confront in a calculus course are

$$\int \frac{1}{x(x+1)}\, dx \quad \text{and} \quad \sum_{n=1}^{\infty} \frac{1}{n(n+1)}.$$

The key to both problems is the fact that we can write

$$\frac{1}{x(x+1)} = \frac{1}{x} - \frac{1}{x+1}.$$

Using this fact, the first problem becomes

$$\int \frac{1}{x(x+1)}\, dx = \int \frac{1}{x} - \frac{1}{x+1}\, dx = \ln|x| - \ln|x+1| + C.$$

Similarly, the second problem becomes

$$\sum_{n=1}^{\infty} \frac{1}{n(n+1)} = \sum_{n=1}^{\infty}\left(\frac{1}{n} - \frac{1}{n+1}\right) = \left(1 - \frac{1}{2}\right) + \left(\frac{1}{2} - \frac{1}{3}\right) + \left(\frac{1}{3} - \frac{1}{4}\right) + \cdots = 1.$$

The rational functions $\frac{1}{x}$ and $\frac{1}{x+1}$ are examples of a special type of rational function known as a **partial fraction**. Recall that there are two types of partial fractions. The first, and simpler type, consists of a real number in the numerator and a linear function raised to a positive integer in the denominator. Examples of the first type of partial fraction are

$$\frac{3}{(2x-5)^7}, \quad \frac{\sqrt{7}}{x^9}, \quad \frac{\pi}{\left(6x+\frac{2}{3}\right)^{20}}.$$

The second type of partial fraction consists of a real number or linear function in the numerator and an irreducible quadratic raised to a positive integer in the denominator. When we refer to an irreducible quadratic, we mean a quadratic polynomial that has no real roots. Therefore, in this context, we do not consider $x^2 - 2$ to be irreducible, since it has the real roots $\pm\sqrt{2}$ and can be factored as

$$x^2 - 2 = \left(x - \sqrt{2}\right)\left(x + \sqrt{2}\right).$$

Examples of the second type of partial fraction are

$$\frac{8}{x^2+1}, \quad \frac{2x-3}{(2x^2+9)^3}, \quad \frac{\pi}{(x^2+2x+73)^{15}}.$$

In calculus courses, partial fractions are occasionally used to compute the sum of an infinite series, as in the preceding example. However, the primary use of partial fractions in calculus courses is to assist us in the integration of rational functions. In this context, when we say that we are integrating a function, what we really mean is that we are finding an antiderivative. Partial fractions of the first type are very easy to integrate. There is also a straightforward algorithm that can be used to integrate partial fractions of the second type, although the computations can get very, very messy when the exponent in the denominator is large. For example, solving $\int \frac{1}{x^2+1}\, dx$ is easy, and the solution is $\arctan(x) + C$. On the other hand, to solve

$$\int \frac{1}{(x^2+1)^{15}}dx,$$

we must first use the trig substitution $x = \tan(y)$. This then reduces the problem to solving

$$\int \cos^{28}(y)\, dy.$$

This integral can be solved in a straightforward way, but the work is incredibly long and tedious and probably takes several pages. However, the bottom line is that all partial fractions can be integrated.

In calculus, one states but does not prove that every rational function can be written as the sum of a polynomial and partial fractions. For example,

$$\frac{2x^4 + 3x^3 + 10x^2 + 11x - 13}{(x+1)(x^2+4)} = 2x + 1 - \frac{3}{x+1} + \frac{4x-5}{x^2+4}.$$

Since we can certainly integrate any polynomial, the fact that we can also integrate any partial fraction now means that we can integrate any rational function. However, all of this is predicated on the fact that we can indeed decompose any rational function into the sum of a polynomial and partial fractions. This takes place, almost magically, in calculus by solving a system of linear equations. No explanation is given as to why this procedure always works. It turns out that an investigation of the greatest common divisors of polynomials holds the key and, in Chapter 13, we will show why partial fraction decomposition is always possible.

1.5.2 Detecting Multiple Roots of Polynomials

In calculus courses, one often uses the derivative to graph polynomials. Along the way, you usually look for the roots of both the original function and its first derivative. Perhaps you have noticed that the points where the original function has a multiple root are precisely the points where the function and derivative have a common root. For example, consider the polynomial

$$f(x) = x^3 - 2x^2 + x = x(x-1)^2.$$

In this case,

$$f'(x) = 3x^2 - 4x + 1 = (3x-1)(x-1).$$

Observe that 1 is a double root of $f(x)$ and 1 is a root of both $f(x)$ and $f'(x)$.

This is no coincidence, and in Chapter 12 we shall show that a root of a polynomial $g(x)$ is a double root if and only if it is also a root of $g'(x)$. We have already discussed how difficult it can be to find the roots of a polynomial. Given the difficulty in finding the roots of both a polynomial and its derivative, you might assume that it would still be quite difficult to test if a polynomial has multiple roots. However, we shall see in Chapter 12 that there is an easy algorithm for determining if a polynomial has multiple roots. This algorithm involves examining the greatest common divisor of a polynomial and its derivative, and we can apply this algorithm even if we have no idea what the roots of our polynomial are. One consequence of this algorithm is that if a polynomial and its derivative have no real roots in common, then the polynomial will have no multiple roots in the real numbers. Furthermore, if a polynomial and its derivative have no complex roots in common, then the polynomial will have no multiple roots in the complex numbers.

Several times in this chapter we used the terms *tools* and *machinery*. These are words you are certainly familiar with, but they may appear out of place in a math book. Perhaps they will look less out of place after considering the following situation: You are given the job of clearing off a field after ten inches of snow has fallen. One approach you might take is to spend the time and money necessary to obtain a snowblower, whereas a second approach is to do the entire job using only a shovel. Notice that there are advantages to each approach. Some advantages of the first approach are that you would spend far less time out in the field and you would be much less tired after completing the job. On the other hand, with the second approach, you don't need to spend either the time or the money obtaining a snowblower. In addition, you might enjoy working outside and might find it very fulfilling to do the job without the help of a machine. In many ways this is analogous to what can go on when attempting to solve a math problem. One approach is to put a good deal of time and effort into developing mathematical tools that can then be used to quickly solve the problem. An alternative approach is to try to solve the problem by doing lots and lots of calculations and computations that require hard work and patience but don't require advanced mathematical ideas. As with the previous situation, there are advantages to each approach. The decision whether to buy the snowblower might be strongly influenced on how often there are heavy snowfalls. If it snows heavily nine or ten times per year, you are much more likely to buy a snowblower than if it only snows heavily once every five years. Applying the same type of thinking to math problems, we are much more likely to invest the time developing sophisticated mathematical tools if we suspect that these tools will be used repeatedly to solve problems. Throughout this book and throughout abstract algebra, a great deal of effort is invested in developing mathematical tools to solve problems. Like the snowblower that will be frequently used, these mathematical tools will be frequently used. Occasionally we will apply one of these tools to a problem that could be done without sophisticated mathematical tools. However, in the long run, we come out far ahead for having developed these tools.

The title of this chapter consists of two parts. Hopefully, the preceding series of examples have answered the first part of the title: "What This Book Is about."

Teachers of mathematics at the high school and college levels will probably spend much of their career teaching courses in algebra, geometry, trigonometry, and calculus. As a result, this book is written, to a great extent, with teachers (both future and present teachers) in mind. An understanding of abstract algebra can help teachers better explain and appreciate many of the topics they teach. Thus, one of the primary goals of this book is to provide teachers with the understanding and appreciation of abstract algebra needed to make them better teachers. This book includes many topics not found in other books on abstract algebra, and they are included here because they relate directly to questions and problems that arise in courses in algebra, geometry, trigonometry, and calculus. As much as I would hope that all math majors, especially teachers, would study abstract algebra for an entire year, I understand that this is often not the case. If you will only be studying abstract algebra for one quarter, one semester, or two quarters, you will still find plenty of topics in this book of interest to teachers.

Fortunately, many of you will study abstract algebra for an entire year. This will give you the opportunity to read the entire book and work through the more advanced material on group theory, field extensions, and Galois Theorem. In particular, you will see the proof of the insolvability of the quintic. The concrete approach throughout this book will give you the experience and confidence needed to master the more theoretical topics in Chapters 8, 15, and 17. Regardless of whether or not you plan to teach, this book will provide you with the background in abstract algebra required for courses leading to advanced degrees in either pure or applied mathematics. Hopefully, this also answers the second part of this chapter's title: "Who This Book Is for."

Enjoy and learn well!

Exercises for Chapter 1

1. Let $P(x) = x^4 - 14x^2 + 9$.

    (a) Find the roots of $P(x)$ by (i) letting $t = x^2$, (ii) using the quadratic formula to find t, and (iii) having found t, solve for x.

    (b) Check, by plugging into $P(x)$, that $\sqrt{2} + \sqrt{5}$, $\sqrt{2} - \sqrt{5}$, $-\sqrt{2} + \sqrt{5}$, and $-\sqrt{2} - \sqrt{5}$ are all roots of $P(x)$.

    (c) The four roots of $P(x)$ in parts (a) and (b) must be the same, yet they look quite different. For each of the four roots in part (a), find the root in part (b) that it is equal to. If necessary, use a calculator to examine decimal equivalents for the roots in parts (a) and (b).

2. Let $G(x) = x^4 - 20x^2 + 16$.

    (a) Find the roots of $G(x)$ by (i) letting $t = x^2$, (ii) using the quadratic formula to find t, and (iii) having found t, solve for x.

    (b) Check, by plugging into $G(x)$, that $\sqrt{3} + \sqrt{7}$, $\sqrt{3} - \sqrt{7}$, $-\sqrt{3} + \sqrt{7}$, and $-\sqrt{3} - \sqrt{7}$ are all roots of $P(x)$.

    (c) The four roots of $G(x)$ in parts (a) and (b) must be the same, yet they look quite different. For each of the four roots in part (a), find the root in part (b) that it is equal to. If necessary, use a calculator to examine decimal equivalents for the roots in parts (a) and (b).

3. Find the four roots of the polynomial $H(x) = x^4 - 26x^2 + 81$. If necessary, use the technique outlined in part (a) of exercises 1 and 2.

For exercises 4–8, please first read the following:

The work of Galois on the insolvability of the quintic tells us that for polynomials of degree ≥ 5, there are no general formulas that will provide us with the roots of the polynomial. Again, in this context, by a *formula* we mean an expression that only involves the coefficients of the polynomial where the coefficients are combined in various ways via addition, subtraction, multiplication, division, and taking roots.

However, there are some polynomials of degree ≥ 5 where we can find the roots using purely algebraic techniques. To assist with exercises 4–6, you may want to refer to the formula

$$(x+y)^5 = x^5 + 5x^4 y + 10x^3 y^2 + 10x^2 y^3 + 5xy^4 + y^5.$$

4. Find the roots of $x^5 + 5x^4 + 10x^3 + 10x^2 + 5x + 33$ and $x^5 + 5x^4 + 10x^3 + 10x^2 + 5x - 20$.

5. Find the roots of $x^5 - 10x^4 + 40x^3 - 80x^2 + 80x - 61$.

6. Find the roots of $x^5 + 15x^4 + 90x^3 + 270x^2 + 405x + 360$.

7. Find the roots of $x^6 + 10x^3 + 21$.

8. Find the roots of $(x^2 - 1)^4 - 23$.

For exercises 9–11, you may assume the following:

(a) If you can construct angles with degree measure n and m, then you can also construct angles with degree measure $n + m$, $n - m$, and $\frac{n}{2}$.

(b) $30°$ and $45°$ angles can be constructed.

(c) $20°$ angles cannot be constructed.

9. Which of the following angles can be constructed: $10°$, $15°$, $40°$, $75°$, $95°$, $105°$? Explain your answers.

10. It can be shown that $36°$ angles can be constructed. In light of this fact, which of the following angles can be constructed: $1°$, $2°$, $3°$, $4°$, $5°$, $6°$, $7°$, $8°$, $9°$? Explain your answers.

11. In light of your answers to exercise 10, determine those positive integers n for which $n°$ angles can be constructed.

12. In this exercise, you may need to use the formula

$$\cos(\theta_1 + \theta_2) = \cos(\theta_1)\cos(\theta_2) - \sin(\theta_1)\sin(\theta_2).$$

(a) Find the exact value of $\cos(75°)$.

(b) Show that $\cos(75°)$ is a root of the polynomial $16x^4 - 16x^2 + 1$.

(c) Find the other three roots of $16x^4 - 16x^2 + 1$.

(d) For each of the other three roots of $16x^4 - 16x^2 + 1$, find an angle such that the root is the value of the cosine function at that angle.

13. Suppose θ is an angle such that $\cos(\theta)$ is a root of the polynomial $x^2 + \alpha x + \beta$, where α, β are real numbers. Use the trigonometric identity $\sin^2(\theta) + \cos^2(\theta) = 1$ to show that $\sin(\theta)$ is a root of the polynomial

$$x^4 + (\alpha^2 - 2\beta - 2)x^2 + (\beta^2 + 2\beta + 1 - \alpha^2).$$

14. Here is a table of some of the values of the function $f(x)$.

x:	-23	-18	-13	-8	-3	2	7	12	17
$f(x)$:	-51	-36	-21	-6	9	24	39	54	69

Explain why this table can be produced by a linear function, and then find the linear function that can produce this table.

15. Here is a table of some of the values of the function $g(x)$.

x:	-5	-3	-1	1	3	5	7	9	11
$g(x)$:	89	31	-3	-13	1	39	101	187	297

(a) Explain why this table cannot be produced by a linear function but can be produced by a quadratic function.

(b) Find the quadratic function that produces this table.

16. Here is a table of some of the values of the function $h(x)$.

x:	-11	-7	-3	1	5	9	13	17	21
$h(x)$:	-214	-102	-22	26	42	26	-22	-102	-214

Find the polynomial of smallest possible degree that can produce this table.

17. Here is a table of some of the values of the function $k(x)$.

x:	-4	-3	-2	-1	0	1	2	3	4
$k(x)$:	-3413	-987	-207	-35	-9	-3	13	213	1155

Find the smallest positive integer n such that this table can be produced by a polynomial of degree n. (You do not need to find the polynomial, merely its degree.)

18. Find real numbers a and b such that

$$\frac{2x - 19}{x^2 + 11x + 28} = \frac{a}{x+7} + \frac{b}{x+4}.$$

19. Find real numbers a, b, and c such that

$$\frac{11x^2 - 2x + 4}{x^3 + x} = \frac{a}{x} + \frac{bx + c}{x^2 + 1}.$$

20. When decomposing a rational function into a sum of partial fractions, all the denominators we use are powers of polynomials that cannot be factored any further. A similar type of decomposition can be done for rational numbers, as we can decompose them into sums of fractions such that all the denominators are powers of prime numbers. To illustrate this, find integers A and B such that

$$\frac{43}{20} = \frac{A}{4} + \frac{B}{5}.$$

As was the case with partial fractions, one should begin by multiplying both sides of the equation by the denominator of the left-hand side of the equation. Note that when solving for A and B, many different answers are possible.

21. Let $F(x) = 12x^3 - 8x^2 - 17x - 5$.
    (a) Find $F'(x)$, and then find all the roots of $F'(x)$.

    (b) Check to see if any of the roots of $F'(x)$ are also roots of $F(x)$.

    (c) Use your results from parts (a) and (b) to write both $F(x)$ and $F'(x)$ as products of linear polynomials.

22. Let $G(x) = x^5 - 8x - 23$.
    (a) Find $G'(x)$.

    (b) Show that any real number that is a root of $G'(x)$ cannot possibly also be a root of $G(x)$.

    (c) Does $G(x)$ have any multiple roots in the real numbers? Explain your answer.

CHAPTER 2

Proof and Intuition

Think about the effect a good novel can have on you. It can move you to laughter or tears. It can terrify you to the point where you feel your heart racing. Pity the poor reader who considers a novel to be nothing more than a collection of letters and punctuation marks that obey various rules of grammar and spelling.

Similarly, consider the effect a symphony can have on you. It can evoke emotions and feelings that were buried inside you for years. If your friends felt that a symphony was merely a long sequence of notes and tones, wouldn't you feel sorry for them?

In order to transcribe or reproduce a novel or piece of music, it sometimes must be reduced to a sequence of symbols on a piece of paper. However, a novel or symphony is so much more than a collection of symbols. In the same way, a mathematical proof is much more than a collection of symbols that obey various rules of syntax and logic. A proof should evoke an appreciation and understanding of the subject. When reading a mathematical proof, one needs to internalize what it is saying. Sadly, many people view mathematics as merely the formal manipulation of symbols. Even worse, they believe that the rules that dictate the manipulation of the symbols are arbitrary or random.

We need to ask ourselves, where do mathematical proofs come from? When trying to prove something in mathematics, you must first spend time developing an understanding of the problem at hand. You need to gain experience by working with special cases and examples. You need to experiment with all types of possible solutions. This requires using your imagination. When you begin, you never know what ideas will eventually be needed. Even if your first 20 ideas do not lead to a solution, perhaps the 21st or 31st or 51st idea you have will lead to a solution. A combination of perseverance and imagination will be required.

Suppose after minutes or hours or days of work, you succeed in finding the proof that you were looking for. When you prepare a written version of your proof, you no longer record all the false attempts or ideas that didn't work. You only write down what *did* work. Whoever reads your proof will probably never be aware of all the different thoughts you had while trying to solve the problem. They will never be aware that the idea that ultimately solved the problem may have been a combination of bits and pieces of ideas from seven or eight failed attempts at a solution. In reality, the path to a solution is rarely a direct route. But this is not what will come across to your reader. They will only see a direct path to the solution. They will not be aware of all the hard work and false starts. Much of the intuition and imagination needed to solve the problem probably never comes across to your reader. The proof your reader will read is likely to be relatively short and extremely formal.

At certain points, formality and precision are crucial in the writing of mathematics. When one transcribes a novel or piece of music for future generations, there is no place for sloppiness. Every word or note must be reproduced correctly; otherwise, we are no longer reflecting the true intentions of the composer or author. Similarly, there are places in mathematical proofs where there is no room for sloppiness. A mathematical proof should be written in such a way that it is convincing to an informed, skeptical reader. The reader must have an understanding of the terms and concepts being discussed. In addition, they should question every step of the proof. No statement should be accepted as true unless it has been logically demonstrated beyond a shadow of a doubt. In light of this, there is no place in a mathematical proof for fuzziness or ambiguities. There can be no loopholes.

Unfortunately, the need for formality and rigor in the writing of proofs often masks the ideas that ultimately led to the proof. Throughout this book, you will often be required to either read or write proofs. In doing so, you will be forced to write in a very formal and rigorous way. The arguments you make must be airtight. But never forget that the proofs you write are actually the by-product of hard work, experimentation, and imagination. It is creativity that drives new discoveries in mathematics. Although the proofs you read and write may be very formal objects, try to look for the creativity and ideas that motivated those proofs. When reading a proof, you will justify statements on a line-by-line basis. But when you are done, go back and try to find the main ideas in the proof. Try to recognize what is special about the proof and how it rises above and beyond the mere manipulation of symbols.

In addition to the many proofs that will appear in this book, we will also include many paragraphs denoted as "Intuition." The goal of these additional paragraphs is to help you see and understand the ideas and intuition that ultimately led to the proofs. In mathematics, we must always try to see the forest for the trees. By necessity, texts in advanced mathematics are filled with formal proofs. But hopefully, you will always remain aware that proofs are merely a record of discoveries that would be impossible without imagination, experimentation, and perseverance.

2.1 The Well Ordering Principle

Let us begin this section by looking at two examples that illustrate the importance of both rigor and imagination in mathematics.

Problem 2.1:   Trisecting Angles with a Ruler and Compass

As mentioned in Chapter 1, we will prove in Chapter 16 that there is no algorithm for trisecting angles with a ruler and compass. In particular, we will show that 60° angles cannot be trisected. However, let us consider the following procedure, where we will choose one of the markings on our ruler and make a special note of it. We will refer to the point on the ruler with this marking as M.

Step 1   Begin with an acute angle, and let O denote its vertex.

Step 2   Place the ruler along one edge of the angle while placing the beginning of the ruler at O. Let A denote the point on the edge where the marking M lands.

Step 3   Use the ruler to extend the line segment OA past the point A on one side and past the point O on the other.

Step 4   Use the compass to draw a circle with center O such that A lies on the circle.

Step 5   Let B denote the intersection of the circle with the second edge of the angle.

Step 6   Place the beginning of the ruler at point O such that the marking M lies on the point B. Then slide the beginning of the ruler along the line OA away from the point A. While doing so, make sure that the ruler continues to pass through point B. Continue doing this until the marking M lands on the circle. Now let C denote the point on the line OA where the beginning of the ruler is and let D denote the point on the circle where the marking M is.

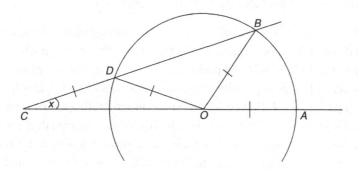

We will now compare $\angle OCD$ and $\angle AOB$. If we draw the line connecting points O and D, it is clear that line segments OB, OD, and CD all have the same length. Therefore, triangles $\triangle OCD$ and $\triangle OBD$ are isosceles. We will let the symbol $m\angle$ denote the measure of an angle, and, for convenience, we let $x = m\angle OCD$. Since $m\angle OCD = m\angle DOC$ and $m\angle ODB$ is an exterior angle of $\triangle OCD$, it follows that

$$m\angle ODB = m\angle OCD + m\angle DOC = 2x.$$

However, since $m\angle ODB = m\angle OBD$, it also follows that $m\angle OBD = 2x$. But $\angle AOB$ is an exterior angle of $\triangle OCB$, so

$$m\angle AOB = m\angle OCD + m\angle OBD = x + 2x = 3x = 3(m\angle OCD).$$

As a result, we see that $m\angle OCD$ is one-third of $m\angle AOB$, and so we have trisected $\angle AOB$.

The preceding algorithm enables us to trisect any acute angle. But we had previously said that it is impossible to trisect 60° angles. How is this possible? How can we resolve this apparent contradiction? The problem is that the mathematical world is very imprecise when using the phrase "ruler and compass." The preceding algorithm indicates that you really can trisect angles using a ruler and compass. However, when mathematicians use the word ruler in the phrase "ruler and compass constructions," they are really referring to a ruler that has no markings on it. If mathematicians were being rigorous and precise, they would not call a ruler without markings a ruler. In fact, a ruler without markings is commonly called a straightedge. As a result, what we will really be proving in Chapter 16 is that it is impossible to trisect angles with only a straightedge and compass. Note that the key step in our trisection algorithm was Step 6, which makes strong use of the marking. As you can see, the imprecise use of the word *ruler* resulted in a great deal of confusion as to whether or not angles can be trisected with a ruler and compass. This example points out how rigorous and precise we must be when using mathematical terminology. If not, all types of apparent contradictions, like the preceding one, can occur.

Problem 2.2:   Moving Knights around an Altered Chessboard

A famous problem asserts that you can place a knight on a standard 8×8 chessboard and move it around so it lands on every other square on the board exactly once. We will not prove that, as we will be looking at a slightly different problem. But before going any further, we should review the moves that a knight is permitted to make on a chessboard. A knight must move two spaces vertically (up or down) followed by one space horizontally (left or right) *or* one space vertically followed by two spaces horizontally. The following diagram denotes with the letter B all the possible places a knight could go with its next move if it was at position A. Note that if a knight was too close to one or more of the edges of the board, then it would no longer have eight places to go on its next move.

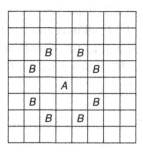

We can now pose our problem. Suppose you removed the bottom/left square as well as the top/right square. This would leave us with an altered chessboard with only 62 squares. Is it still possible to place the knight on the board and move it around so it hits each of the remaining 61 squares exactly once? Before reading further, you should take some time to play with and think about this problem.

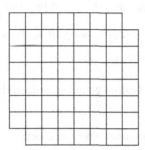

If you play with this problem long enough, you may begin to suspect that it is impossible to move the knight to each square exactly once. But how would we prove such a thing? After all, the inability to solve a problem hardly proves that it cannot be done. In order to prove that it is impossible for the knight to move in the desired way, some type of idea is needed. This is where the creative and imaginative aspects of mathematics come into play. The idea that will ultimately solve this problem is an incredibly easy one, but it will illustrate that mathematics is much more than formally manipulating symbols.

Recall that every square on a chessboard is colored either red or black. Furthermore, every square has the opposite color from the squares that share a side with it. We can now observe that the bottom/left and top/right squares actually have the same color. Therefore, when we remove these two squares, we are left with 32 squares of one color and 30 squares of the other color. Now think about how a knight moves around a chessboard. As it moves, the color of the square it is on alternates. If it started on a red square, then it moves to a black square, then a red square, and so on. Similarly, if it started on a black square, then it moves to a red square, then a black square, and so on. That means that if it were possible to place a knight on one square and to move it so that it touched every other square exactly once, then it would touch 31 red squares and 31 black squares. However, our altered chessboard has 32 squares of one color and 30 of the other. So it is impossible for the knight to touch every square exactly once.

The preceding solution is fairly simple once we had the idea to think about the colors of the remaining 62 squares and how a knight alternates colors as it moves. But without the imagination to think about the colors on the chessboard, it is not obvious how in the world we could possibly solve this problem. In light of this, we see that the solution to a mathematics problem is often the result of a creative or imaginative idea and not the result of boring and formal symbol manipulations.

In mathematics, statements are *always* either true or false but *never* both. Some examples of typical mathematical statements are

1. An even integer plus an odd integer is always odd.

2. An odd integer times an odd integer is always even.

3. Every even integer can be written as the sum of two odd integers.

4. Every even integer greater than 2 can be written as the sum of two prime numbers.

5. Every finite group with an odd number of elements is solvable.

6. Every finite group arises as the Galois group of a finite field extension of the rational numbers.

Some of the preceding statements are obviously true, and some are obviously false. For some of the others, we may not yet understand what they are talking about. In fact, for statements 4 and 6, mathematicians have been unable to determine whether or not these statements are true. However, our inability to determine whether or not a mathematical statement is true does not change the fact that it must be true or false but cannot be both.

Throughout this book, we will be proving whether various mathematical statements are true or false. As Problems 2.1 and 2.2 indicated, finding and writing proofs will involve not only formality and rigor but also creativity and imagination. The first algebraic object we will study in this course is the **positive integers**, which is the infinite set $\{1, 2, 3, 4, 5, 6, \ldots\}$. The positive integers are frequently called the **natural numbers** and are abbreviated by the symbol $\mathbb{N}$. Interestingly enough, proofs about more abstract objects like vector spaces, rings, and groups will frequently rely on basic facts about the positive integers. Such proofs often use a fundamental property of the positive integers known as the **Well Ordering Principle**.

There are various ways to look at the Well Ordering Principle. Suppose you start by listing the days of your life

$$\{1, 2, 3, \ldots, 500, 501, 502, \ldots, 2205, 2206, 2207, \ldots\}.$$

Next, if you have ever gone to a museum, place a line over each day of your life that you went to a museum. Your list would then look something like

$$\left\{1, 2, 3, \ldots 1166, \overline{1167}, 1168, \ldots, 2131, \overline{2132}, 2133, \ldots, 2596, \overline{2597}, 2598, \ldots\right\}.$$

If the preceding list was correct, then it indicates that the first time you went to a museum was on day number 1167 of your life. It seems quite clear that if you have ever gone to a museum, then there must have been a first time. Similarly, if you listed the days of your life and placed bars over each day that you attended school, then there must have been a first day that you attended school. Viewing these examples in a more mathematical mode, the list of the days of your life can be viewed as a listing of the positive integers $\mathbb{N}$. The days that you went to a museum can be viewed as a subset S of $\mathbb{N}$. The fact that there was a day on which you first went to a museum, provided that you have ever been to a museum, means that the set S has a smallest element. Similarly, the fact that there was a first day that you attended school, provided that at some point you have gone to school, indicates that the set of days that you attended school also has a smallest element. This leads us to the formal statement of the Well Ordering Principle.

The Well Ordering Principle. *Every nonempty subset S of the positive integers $\mathbb{N}$ has a smallest element.*

Note that if you have never been to a museum, then the set S of days on which you went to a museum would be the empty set. Therefore, S would not have a smallest element, as it has no elements at all. Thus, in order to guarantee that a subset of $\mathbb{N}$ has a smallest element, all we need to know is that it is not empty.

It is likely that you have used the Well Ordering Principle in the past in subtle ways and have not even realized that you were using it. For example, let us consider the following:

Proposition 2.3. *Every rational number can be written in lowest terms.*

Intuition. Suppose we start with a fraction like $\frac{792}{1008}$. It is not yet in lowest terms, as we can easily find common factors of the numerator and denominator. In particular, 2 is a common factor, and after dividing the numerator and denominator by 2, we see that $\frac{792}{1008} = \frac{396}{504}$. Once again, 2 is a common factor and we reduce further to $\frac{396}{504} = \frac{198}{252}$. There are still common factors. In particular, 2 is still a common factor, but we are free to divide the numerator and denominator by any common factor. Since 3 is also a common factor, we can divide the numerator and denominator by 3 to obtain $\frac{198}{252} = \frac{66}{84}$. The argument we usually give to convince ourselves that we can write our fraction in lowest terms is that we can continue to divide the numerator and denominator by common factors until there are no common factors left.

But how do we know that this procedure ever comes to an end? As we continue to divide the numerator and denominator by common factors, what guarantee is there that we will ultimately end up with a fraction in lowest terms? Since it seems obvious that this procedure comes to an end, we usually don't go any further in proving it. However, when we say that this procedure eventually comes to an end, we are subtly using the Well Ordering Principle, even if we don't realize it. It is now time to see how the Well Ordering Principle is used in a very precise and explicit way to solve this problem.

If we go back and reexamine our original fraction $\frac{792}{1008}$, observe that we can rewrite this fraction using many different denominators. Our previous calculations show that we can rewrite $\frac{792}{1008}$ in equivalent forms with $504, 252$, and 84 as denominators. However, the Well Ordering Principle guarantees that there is a smallest denominator that can be used in rewriting $\frac{792}{1008}$. Even if we do not care to find or compute the smallest possible denominator, the Well Ordering Principle guarantees that there is one. But notice: Once you have found this smallest possible denominator, it must give you a fraction in lowest terms. For if there still remained a common factor of the numerator and denominator that was greater than 1, then dividing the numerator and denominator by that factor would result in an equivalent fraction with a smaller denominator, which is an impossibility. Observe that for the special case of the fraction $\frac{792}{1008}$, the smallest denominator that works is 14, and when we write $\frac{792}{1008} = \frac{11}{14}$, we have written it in lowest terms.

Proof. If q is a rational number, let

$$S = \left\{ b \in \mathbb{N} \mid \text{ there exist some integer } a \text{ such that } q = \frac{a}{b} \right\}.$$

Since q is rational, it can be written as a quotient of integers; therefore, S is a subset of $\mathbb{N}$ that is not empty. Thus, S contains a smallest element, which we can denote as d. Since we can write q as a fraction with d as the denominator, there is some integer c such that $q = \frac{c}{d}$.

We contend that the fraction $\frac{c}{d}$ is in lowest terms. If not, then there is a common factor e of both c and d such that $e > 1$. Since e is a factor of both c and d, we can write

$$c = c' \cdot e \text{ and } d = d' \cdot e,$$

where c' and d' are integers and $d > d' \geq 1$. As a result, we now have

$$q = \frac{c}{d} = \frac{c' \cdot e}{d' \cdot e} = \frac{c'}{d'}.$$

However, we have now written q as a fraction with a denominator d' that is less than d. This is impossible, so it cannot possibly be the case that c and d have a common factor that exceeds 1. Therefore, the fraction $\frac{c}{d}$ does indeed express q in lowest terms. $\qquad \square$

2.2 Proof by Contradiction

There is an important technique of proof that is very common in all of mathematics, especially in algebra, which is often used along with the Well Ordering Principle and is known as **proof by contradiction**. As we commented earlier, every statement in mathematics must be either true or false but cannot be both. That means that if we are trying to prove that a mathematical statement is true, then it is good enough to show that it cannot possibly be false. There will be many times where it will be difficult to prove in a direct manner that a certain statement is true, but it will be much easier to show that it cannot be false. In some sense, we are taking an indirect route in proving our original statement to be true, so one often refers to a proof by contradiction as an **indirect proof**.

This then raises the important question of how we can prove that a statement cannot be false. To do that we consider what logical consequences would follow if the statement were indeed false. Suppose that one of the logical consequences of our statement being false is the truth of a new statement that we know cannot possibly be true because it contradicts other statements that we already know to be true. In order for this contradiction to have arisen, somewhere in our mathematical argument there must have been a mistake or faulty piece of logic. However, if we were careful and made no mathematical mistakes or logical errors, then the only thing that could have caused our contradiction was the presumption that our original statement was false. Since this tells us that it is impossible for our statement to be false, then our statement must be true.

There are some disadvantages or dangers in using a proof by contradiction. Let us ask ourselves, when doing a proof by contradiction, how can we tell when we are "done"? The answer is we are done when we reach a contradiction. However, if you have made even the slightest computational or logical error, then that could be the real reason why you arrived at a contradiction. Therefore, if we make a mistake while doing the proof, we will incorrectly think that we have proven something when we really haven't. That is the danger in using proofs by contradiction.

It goes without saying that the main purpose of a proof in mathematics is to prove something. However, it is always preferable when a proof also adds to our intuition and understanding. It would be nice if every time we read a proof, we also came away with an understanding of what the key ideas are that were behind the proof. In a proof by contradiction, the proof proceeds until we reach a contradiction, and this type of reasoning often camouflages the key ideas behind the proof. This is occasionally a disadvantage of using a proof by contradiction. On the other hand, if using a proof by contradiction is the only way (or the easiest way) to prove something, we would be foolish to not use one.

The proof of the next proposition is an example of a proof that combines both the Well Ordering Principle and proof by contradiction. On the plus side, this proof shows the power and usefulness of the Well Ordering Principle. On the negative side, it is an example of a proof where the key ideas are, unfortunately, camouflaged, and the proof adds little to our intuition or understanding.

Proposition 2.4. *If n is a positive integer, then $\sqrt{n}$ is either a whole number or is irrational.*

After reading a mathematical statement, one should first get a thorough understanding of what it really means before reading the proof. So let's first think about what Proposition 2.4 says. We are to consider the numbers

$$\sqrt{1}, \sqrt{2}, \sqrt{3}, \sqrt{4}, \sqrt{5}, \ldots, \sqrt{99}, \sqrt{100}, \sqrt{101}, \ldots.$$

Some of them, like $\sqrt{1}, \sqrt{4}, \sqrt{9}, \sqrt{16}, \ldots$, are obviously whole numbers. The proposition then asserts that all the other numbers on the list

$$\sqrt{2}, \sqrt{3}, \sqrt{5}, \sqrt{6}, \sqrt{7}, \sqrt{8}, \sqrt{10}, \ldots, \sqrt{99}, \sqrt{101}, \ldots$$

must be irrational.

Proof. Let n be a positive integer such that $\sqrt{n}$ is not a whole number; we must show that $\sqrt{n}$ is irrational. We will proceed using a proof by contradiction. Therefore, we will assume that $\sqrt{n}$ is a rational number but is still not a whole number, and we will arrive at a contradiction. Since $\sqrt{n}$ is not a whole number, it must lie between two consecutive positive integers. Therefore, there exists a positive integer a such that

$$a < \sqrt{n} < a+1.$$

Next, we let

$$S = \left\{ b \in \mathbb{N} \mid b\sqrt{n} \text{ is an integer} \right\}.$$

Since $\sqrt{n}$ is rational, the set S is not empty, and the Well Ordering Principle asserts that S contains a smallest element t. Thus, t is the smallest positive integer that when multiplied by $\sqrt{n}$, results in an integer.

Subtracting a from the inequalities $a < \sqrt{n} < a+1$ results in

$$0 < \sqrt{n} - a < 1.$$

Multiplying these inequalities by t gives us

$$0 < t(\sqrt{n} - a) < t.$$

Since $t\sqrt{n}$ is an integer, it now follows from the preceding inequalities that $t(\sqrt{n} - a)$ is a positive integer that is now less than t. We already know that t is the smallest positive integer that when multiplied by $\sqrt{n}$ gives us an integer. Since $t(\sqrt{n} - a)$ is a positive integer that is less than t, it cannot give us an integer when multiplied by $\sqrt{n}$. However,

$$(t(\sqrt{n} - a))\sqrt{n} = t(\sqrt{n})^2 - ta\sqrt{n} = tn - a(t\sqrt{n}),$$

which is indeed an integer. Therefore, we have reached a contradiction, and this concludes the proof. $\qquad\square$

In the final exercise following Section 2.3, you will be asked to generalize the argument used in the previous proof and show that if a and n are positive integers such that $a^{1/n}$ is not a whole number, then $a^{1/n}$ must not be rational. We will also prove this in Chapter 3, and you might wonder why we chose to prove this in two consecutive chapters. In this section, our goal is to illustrate that the ability to use the Well Ordering Principle in a proof by contradiction is an enormously useful and powerful mathematical skill. However, some of the proofs that use this skill do not provide much intuition or a deep understanding of what is really going on. In other words, if a goal of a mathematical proof is to be illustrative and enlightening, then the shortest or most clever proof might not be the best proof.

On the other hand, the proofs in Chapter 3 might be longer and require the development of more mathematical tools, but they will be more intuitive and will reveal more about the structure of the positive integers. In particular, these proofs will be based on some fascinating properties that are shared by both integers and polynomials. An important and recurring theme throughout this book, and throughout abstract algebra, is the strong similarity between the properties of integers and polynomials. By developing an understanding of these properties of the integers in Chapter 3, we will be providing a blueprint for our study of polynomials in Chapters 9, 12, and 17.

At this point, we also need to be careful when saying that we have proven that numbers like $2^{1/2}$, $7^{1/2}$, and $21^{1/5}$ are irrational. The proof of Proposition 2.4 does not technically show that $2^{1/2}$ is irrational, but it shows that there is no rational number whose square is 2. Similarly, it does not technically show that $7^{1/2}$ is irrational, but it shows that there is no rational number whose square is 7. Along the same lines, the final exercise after Section 2.3 deals with numbers of the form $a^{1/n}$. It appears to imply that $21^{1/5}$ is irrational, but it really implies that there is no rational number whose fifth power is 21. It might seem like we are quibbling over a minor point. However, in order to prove that a number is irrational, it is not enough to prove that it is not rational. Recall that irrational numbers are those *real numbers* that are not rational. Therefore, in order to prove that $2^{1/2}$, $7^{1/2}$, and $21^{1/5}$ are irrational, it is not enough to prove that they are not rational. We must also prove that they are real numbers. Thus, before saying that $2^{1/2}$, $7^{1/2}$, and $21^{1/5}$ are irrational, we must verify that there is a positive real number whose square is 2, a positive real number whose square is 7, and a positive real number whose fifth power is 21. All of these issues will be dealt with in Chapter 4 in our discussion of the rational numbers and real numbers.

2.3 Mathematical Induction

The Well Ordering Principle seems to lend itself to use in proofs by contradiction. It is reasonable to wonder if it is possible to restate the Well Ordering Principle in a way that would allow us to prove things directly (and hopefully more intuitively) and occasionally avoid proofs by contradiction. To this end, let us a consider a set T that has the following two properties: (a) T contains the number 1, and (b) whenever T contains a number k, it also contains the number $k+1$. What can we say about the set T? For example, does T contain the number 3 or 15 or 1951?

We know from property (a) that T contains 1. At first we didn't know if T contained 2, but now that we know that T contains 1, property (b) tells us that T must contain $1+1=2$. (Think about this!) Similarly, we didn't originally know that T contains 3, but since we showed that T contains 2, property (b) tells us that T must also contain $2+1=3$. Now that we have the ball rolling, it looks like we can continue to apply property (b). Since T contains 3, it must now contain $3+1=4$. But therefore it must also contain 5, and so on. As a result, it certainly looks like T must contain all the positive integers. Unfortunately, the preceding argument is not a complete proof, as the phase "and so on" is not sufficiently clear or rigorous. However, once again the Well Ordering Principle can be used to finish the proof. The new principle that we are obtaining from the Well Ordering Principle is known as Mathematical Induction, and it can be stated several ways.

Mathematical Induction—First Version. *Let T be a subset of $\mathbb{N}$ satisfying the following two properties:*

(a)   *T contains 1.*

(b)   *Whenever T contains a number k, then it also contains the number $k+1$.*

Then T contains all positive integers.

Proof. Once again, we will proceed with a proof by contradiction. Therefore, we will suppose that there are positive integers that are not in T. Now let

$$S = \{n \in \mathbb{N} \mid T \text{ does not contain } n\},$$

so S consists of those positive integers that are *not* in T. Since we are assuming that there are positive integers not in T; we see that S is not empty, and the Well Ordering Principle guarantees that S contains a smallest element, which we will denote as m. Using property (a), we know that T contains 1, so S does not contain 1. As a result $m \neq 1$, so it must be the case that $m > 1$. Now consider $m - 1$. Since $m > 1$, we see that $m - 1 \geq 1$. Thus, $m - 1$ is a positive integer that is less than m. Since m is the smallest positive integer in S, it must follow that $m - 1$ is not in S. But this tells us that $m - 1$ must be in T. We can now apply property (b). Since $m - 1$ belongs to T, property (b) asserts that T contains $(m - 1) + 1$. However, $m = (m - 1) + 1$, so T contains m. But this contradicts the fact that S contains m. Having arrived at a contradiction, the proof is complete. $\square$

There are countless statements about the positive integers that can be proved using either the Well Ordering Principle or Mathematical Induction. The choice of which to use is up to you.

The difference is that proofs using the Well Ordering Principle usually are proofs by contradiction, whereas proofs using Mathematical Induction are usually more direct. This might make it appear that, whenever possible, it is preferable to use Mathematical Induction as opposed to the Well Ordering Principle. However, there is another important factor to consider. The Well Ordering Principle, at first glance, seems much more believable than Mathematical Induction. Students often feel that there is some hocus pocus going on in a Mathematical Induction proof. Mathematical Induction allows us to assert that a certain set T contains all positive integers provided it satisfies properties (a) and (b). Unfortunately, students are often extremely unnerved by the arguments used in proving that T satisfies property (b).

In proving that T satisfies property (b), we are actually showing that *if* T contains a number k, then it also contains the number $k + 1$. Note that we are not assuming that T contains k, merely saying that *if* T contains k, then it also contains $k + 1$. However, at this point many students feel that we are assuming that T contains all positive integers k and they feel that we are making the error of assuming what we are trying to prove. Although we are not assuming what we are trying to prove, it is difficult for many students to rid themselves of that feeling, and, because of this, many students do not trust or believe Mathematical Induction. Hopefully, after several additional readings of the proof of how Mathematical Induction follows logically from the Well Ordering Principle, you will become more comfortable with Mathematical Induction and accept that it is a valid tool to use.

A common way to visualize Mathematical Induction is to think about dominoes. Suppose you are setting up a long row of dominoes. You want to set them up in such a way that after you knock down the first domino, all the dominoes will eventually fall. What do you need to do to guarantee that all the dominoes will fall? In particular, you need to make sure that the 11th domino is positioned so that when the 10th domino falls, it will knock down the 11th. Similarly, you need to make sure that the 8456th domino is positioned so that when the 8455th domino falls, it will knock down the 8456th. More generally, we need to make sure that domino number $k + 1$ is positioned so that when domino number k falls, it will knock down domino number $k + 1$. Suppose we now let U denote the set of dominoes that eventually fall at some point after we knock down the first domino. What properties does the set U have? Since we are knocking down the first domino, U certainly contains the first domino. Furthermore, we have arranged the dominoes such that if U contains domino number k, then U will also contain domino number $k + 1$. Thus, U satisfies properties completely analogous to properties (a) and (b) in Mathematical Induction. Just as all the dominoes will eventually fall, the set T in Mathematical Induction will contain all positive integers.

There are other useful aspects of the domino analogy. When doing a proof by Mathematical Induction, we need to justify that our set T satisfies properties (a) and (b). In most proofs, the harder case is showing that T satisfies property (b). As a result, students often consider property (a) to be relatively unimportant. However, if we think about our row of dominoes, nothing happens until we knock down that first domino. Not a single domino will ever fall if we fail to knock that first one down. Hopefully, you can see that knocking down the first domino is analogous to T satisfying property (a). Indeed, if we do not show that T contains 1, then we can never invoke property (b) and we will be unable to prove that T contains any integers.

For a slight variation, suppose you set the dominoes up as before but instead you ignore the first few dominoes and then knock the 4th domino toward the 5th. In this case, all the dominoes starting with the 4th will eventually fall. Similarly, if you chose to ignore the first 20 dominoes but then knocked the 21st domino toward the 22nd domino, then all the dominoes starting with the 21st will eventually fall. In light of this, it is easy to see that we can restate our first version of Mathematical Induction in a form that looks more general.

Mathematical Induction—First Version Revisited. *Let T be a subset of $\mathbb{N}$ satisfying the following two properties:*

(a)   *T contains the number m.*

(b)   *Whenever T contains a number k, where $k \geq m$, then it also contains the number $k+1$.*

Then T contains all positive integers greater than or equal to m.

At this point, we should give several examples of how Mathematical Induction can be used.

Problem 2.5:   The Two-Color Problem for Planes Divided Up by Lines

A famous question in mathematics asks, "What is the minimum number of colors needed to color any map?"

When coloring a map, our only demand is that any two regions that share a boundary must be colored in different colors. The following picture indicates that, in general, three colors are not sufficient, as the four regions in the picture must all be different colors. In one of the great mathematical achievements of its era, it was shown in the 1970s that four colors can be used to color any map.

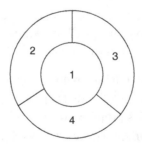

However, we will look at a simpler problem. Suppose the plane is split up into different regions using only straight lines. What is the minimum number of colors needed to color any such map? For an example of the type of situation we are dealing with, consider the following picture. Note that by putting the numbers 1 and 2 into the various regions, we can see that we can color this map with only two colors.

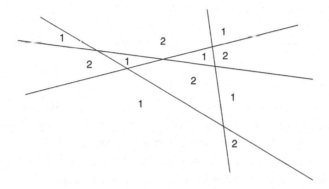

At this point, you should draw several pictures and consider cases that have 4, 5, or 6 lines dividing the plane and see how many colors you need in each case. Presumably, you will observe that in each of the cases you looked at, two colors sufficed. The question is, how would you go about proving that two colors always suffice? The proof will proceed by Mathematical Induction.

Proof. We need to show that no matter how many lines are used to divide up the plane, we can color the plane using only two colors. To this end, we let

$$T = \{n \in \mathbb{N} \mid \text{whenever a plane is divided up by } n \text{ lines,}$$
$$\text{it can be colored with 2 colors}\}.$$

We will be done if we can show that T contains all positive integers, and we will do this by applying Mathematical Induction. Therefore, we need to show that T satisfies properties (a) and (b). To show that T satisfies property (a), we need to consider the situation where there is only one line cutting across the plane. Since the line splits the plane into only two regions, we can use one color for the first region and a second color for the second region. So we see that two colors certainly suffice in this case and T contains 1.

To show that T satisfies property (b), we are in the situation where T contains some positive integer k, and we need to show that T contains $k+1$. So suppose we are given a plane that is divided up by $k+1$ lines. We need to show that we can color this plane using only two colors. The following picture will illustrate the procedure that we are about to discuss.

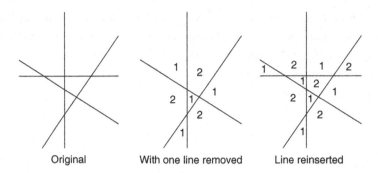

Original   With one line removed   Line reinserted

Focus on one particular line, and, for the sake of simplicity, rotate the plane so this particular line is now horizontal. For the moment, erase that particular line from the plane, thereby leaving only k lines on the plane. Since we now have a plane divided up by only k lines and since T does contain the number k, we can color this plane using only two colors. In doing this, keep in mind that regions that share a border must be different colors, but regions that have only a point in common do not need to be different colors.

The key idea in this proof is that when we place line $k+1$ back on the plane, we reverse the color of all the regions that lie above this line. Let us examine the effect this has on the coloring of the plane. If two regions share a border, then part or all of their border could be below line $k+1$, part or all of their border could be above line $k+1$, or their entire border is along line $k+1$. Any two regions that share a border below line $k+1$ had already been colored in different colors before we reinserted line $k+1$. Since no changes were made below line $k+1$ when it was reinserted, these regions are still different colors. Similarly, any two regions that share a border above line $k+1$ had already been colored in different colors before we reinserted line $k+1$. However, when we reinserted line $k+1$, all the colors above the line were reversed. Therefore, if two regions above line $k+1$ had different colors before line $k+1$ was reinserted, they still have different colors after their colors have been reversed.

The only remaining case to consider is two regions whose border lies along line $k+1$. Before line $k+1$ split these regions into two separate entities, they combined to form a single region and were therefore the same color. However, when line $k+1$ is reinserted, the upper part of the region has its color reversed, so it is a different color from the lower part of the region. As a result, in all three cases we see that after line $k+1$ is reinserted, regions with a common border must be different colors. Therefore, T does contain the number $k+1$, and T is indeed the entire set of positive integers. □

With experience, you will see that Mathematical Induction is a wonderful tool for proving things that you already believe to be true but have not yet proven to be true. However, the more important and exciting part of mathematics is the creativity and experimenting needed in deciding what we think is true. In the previous problem, the really interesting part was realizing, after experimenting with several examples, that two colors did apparently suffice. Mathematical Induction played the important role of enabling us to prove that two colors did indeed suffice, but more importantly, we first needed to understand the problem well enough to suspect that two colors sufficed.

Problem 2.6: For any positive integer n,

$$1+4+7+\cdots+(3n-2) = \frac{3n^2-n}{2}.$$

Intuition. Before trying to solve this problem, we should first try to get a feeling for what the problem is saying. When $n = 1$, the left-hand side of the equation is the sum of all numbers of the form $3n-2$ starting with 1 and ending with $3(1)-2$. In other words, the left-hand side is merely the number 1. On the other hand, the right-hand side is $\frac{3(1)^2-1}{2}$, which is also equal to 1. For a more interesting case, when $n = 5$, the left-hand side of the formula is the sum of all numbers of the form $3n-2$ starting with 1 and ending with $3(5)-2$. Therefore, the left-hand side is

$$1+4+7+10+13.$$

On the other hand, the right-hand side is $\frac{3(5)^2-5}{2}$, and we see that both sides are equal to 35.

For large values of n, the preceding formula tells us that computing the sum of a long list of numbers can be greatly simplified by merely plugging the value of n into the quadratic polynomial $\frac{3n^2-n}{2}$. Many of the most common problems and exercises involving Mathematical Induction will be of this type. We will frequently be asked to prove that computing sums of long lists of numbers often simplify into plugging one number into a polynomial. Although problems of this type serve as excellent practice, they are somewhat misleading as an indicator of how mathematics is really done. In this example, we will be using Mathematical Induction to *verify* that a certain formula is true. However, far more creativity and mathematical thinking takes place in actually *finding* a possible formula.

Mathematical Induction enables us to prove that various formulas work, but the more important and interesting piece of these problems is in finding these formulas. In Chapter 13, we will see how to derive many of the formulas that seem to magically appear in textbooks as exercises on Mathematical Induction.

Proof. We will let

$$T = \left\{ n \in \mathbb{N} \mid 1+4+7+\cdots+(3n-2) = \frac{3n^2-n}{2} \right\}.$$

We will need to show that $T = \mathbb{N}$, and we will do this by using Mathematical Induction. As we just saw, if we plug $n = 1$ into both sides of the formula $1 + 4 + 7 + \cdots + (3n - 2) = \frac{3n^2 - n}{2}$, we obtain the equality $1 = 1$. This tells us that T contains 1, and therefore T satisfies property (a). In order to show that T satisfies property (b), we now need to show that if T contains a positive integer k, then it also contains $k + 1$. Therefore, we may assume that k is a particular positive integer such that

(1)
$$1 + 4 + 7 + \cdots + (3k - 2) = \frac{3k^2 - k}{2}.$$

We need to show that T contains $k + 1$. If we let $n = k + 1$ in the formula

$$1 + 4 + 7 + \cdots + (3n - 2) = \frac{3n^2 - n}{2},$$

we see that we need to show that

(2)
$$1 + 4 + 7 + \cdots + (3(k + 1) - 2) = \frac{3(k + 1)^2 - (k + 1)}{2}.$$

If we observe that the next-to-last summand of the left-hand side of (2) is $3k - 2$, we have

$$1 + 4 + 7 + \cdots + (3(k + 1) - 2) = 1 + 4 + 7 + \cdots + (3k - 2) + (3k + 1).$$

However, if we plug equation (1) into our last equation, we obtain

$$1 + 4 + 7 + \cdots + (3(k + 1) - 2) = (1 + 4 + 7 + \cdots + (3k - 2)) + (3k + 1) =$$
$$\frac{3k^2 - k}{2} + (3k + 1).$$

Since

$$\frac{3k^2 - k}{2} + (3k + 1) = \frac{3k^2 + 5k + 2}{2},$$

we see that the left-hand side of (2) is equal to $\frac{3k^2 + 5k + 2}{2}$. On the other hand, it is easy see that

$$\frac{3(k + 1)^2 - 3(k + 1)}{2} = \frac{3k^2 + 5k + 2}{2},$$

so both sides of (2) are equal to $\frac{3k^2 + 5k + 2}{2}$. Thus, T contains $k + 1$, and we have shown that T satisfies property (b), thereby concluding our proof.   □

There is another version of Mathematical Induction that, depending on the problem you are studying, may be more useful.

Mathematical Induction—Second Version. *Let T be a subset of* $\mathbb{N}$ *satisfying the following two properties:*

(a) *T contains* 1

(b) *whenever T contains the set of numbers* $\{1, 2, \ldots, k\}$, *then it also contains the number* $k + 1$.

Then T contains all positive integers.

If you understood the proof of the first version of Mathematical Induction using the Well Ordering Principle, then it should be fairly easy for you to use the Well Ordering Principle to prove the second version of Mathematical Induction. In fact, there are many slightly modified versions of Mathematical Induction that may be of use to you depending on the problem you are studying. All of these versions follow from the Well Ordering Principle, and some of them will appear as exercises at the end of this section.

When reading a mathematics text, we repeatedly come across the words **lemma, theorem, corollary, and proposition**. Usually the text contains no explanation of what these terms mean. Theorems are important mathematical facts. In mathematics, when you feel that you have succeeded in proving something of substance, you almost always call it a theorem. Often, in order to prove a theorem, you first need to prove a series of smaller mathematical facts, and we call these lemmas. In some sense, lemmas are identical to theorems, as they are both mathematical statements that require proofs. However, the difference is that we usually reserve the name *theorem* for our major results, and we call the smaller facts leading up to them lemmas. Once we have proven a theorem, it is often possible to use the theorem to prove other interesting facts. We call these corollaries. Corollaries are also mathematical statements that require proof. To make an analogy, one could say that a lemma is to a theorem as a theorem is to a corollary as, in both cases, the first is used to prove the second. A proposition is yet another name for a mathematical statement that requires proof. Usually an author will call a statement a proposition if they consider it to be important but not as important as a theorem. Also, a symbol that appears in many textbooks and has appeared several times in this chapter is $\square$. It is used to denote the end of a proof.

Exercises for Sections 2.1, 2.2, and 2.3

1. If r is a real number such that $1 + r > 0$, prove that

$$(1 + r)^n \geq 1 + rn,$$

for all integers $n \in \mathbb{N}$.

2. In precalculus courses we saw that

    $$\ln(r_1 r_2) = \ln(r_1) + \ln(r_2) \quad \text{and} \quad e^{(s_1+s_2)} = e^{s_1} \cdot e^{s_2},$$

    for all positive real numbers r_1, r_2 and real numbers s_1, s_2. Using these facts, prove
    (a) if $n \in \mathbb{N}$ and $r_1, r_2, \ldots, r_n$ are positive real numbers, then

    $$\ln(r_1 \cdot r_2 \cdots r_n) = \ln(r_1) + \ln(r_2) + \cdots + \ln(r_n);$$

    (b) if $n \in \mathbb{N}$ and $r_1, r_2, \ldots, r_n$ are real numbers, then

    $$e^{(r_1 + r_2 + \cdots + r_n)} = e^{r_1} \cdot e^{r_2} \cdots e^{r_n}.$$

3. In calculus we learned that

    $$(x)' = 1 \quad \text{and} \quad (f(x)g(x))' = f'(x)g(x) + f(x)g'(x),$$

    for all differentiable functions $f(x), g(x)$. Using these facts, prove
    (a) $(x^n)' = nx^{n-1}$, for all $n \in \mathbb{N}$;
    (b) $(f(x)^n)' = nf(x)^{n-1}f'(x)$, for all $n \in \mathbb{N}$ and differentiable functions $f(x)$.

4. Suppose F and G are functions both of whose domain is the real numbers such that

    $$F(r_1 + r_2) = F(r_1) + F(r_2) \quad \text{and} \quad G(r_1 \cdot r_2) = G(r_1) \cdot G(r_2),$$

    for all r_1, r_2 in the real numbers.
    (a) Prove that if $n \in \mathbb{N}$ and $r_1, r_2, \ldots, r_n$ belong to the real numbers, then

    $$F(r_1 + r_2 + \cdots + r_n) = F(r_1) + F(r_2) + \cdots + F(r_n).$$

    (b) Prove that if $n \in \mathbb{N}$ and $r_1, r_2, \ldots, r_n$ belong to the real numbers, then

    $$G(r_1 \cdot r_2 \cdots r_n) = G(r_1) \cdot G(r_2) \cdots G(r_n).$$

5. Prove that

    $$2 + 5 + 8 + \cdots + (3n - 1) = \frac{3n^2 + n}{2},$$

    for all $n \in \mathbb{N}$.

6. Prove that

    $$1^2 + 3^2 + 5^2 + \cdots + (2n - 1)^2 = \frac{4n^3 - n}{3},$$

    for all $n \in \mathbb{N}$.

7. Prove that

$$1^3 + 2^3 + 3^3 + \cdots + n^3 = \frac{n^2(n+1)^2}{4},$$

for all $n \in \mathbb{N}$.

For exercises 8–9, please first read the following:

The Fibonacci sequence F_n is defined as

$$F_1 = 1, \quad F_2 = 1, \quad \text{and} \quad F_{n+2} = F_n + F_{n+1}, \quad \text{for all } n \in \mathbb{N}.$$

8. Prove that, for any $n \in \mathbb{N}$, the largest positive integer that divides both F_n and F_{n+1} is 1.

9. Prove that, for any $n \in \mathbb{N}$, $|(F_{n+1})^2 - F_{n+2}F_n| = 1$.

10. If a, r are real numbers, with $r \neq 1$, prove that

$$a + ar + \cdots + ar^{n-1} + ar^n = \frac{ar^{n+1} - a}{r - 1},$$

for all $n \in \mathbb{N}$.

11. The triangle inequality states that if r_1, r_2 are real numbers, then

$$|r_1 + r_2| \leq |r_1| + |r_2|.$$

Use this fact to prove that if $n \in \mathbb{N}$ and $r_1, r_2, \ldots, r_n$ are real numbers, then

$$|r_1 + r_2 + \cdots + r_n| \leq |r_1| + |r_2| + \cdots + |r_n|.$$

12. Prove that if $n \in \mathbb{N}$, then

$$\frac{1}{1 \cdot 2} + \frac{1}{2 \cdot 3} + \frac{1}{3 \cdot 4} + \cdots + \frac{1}{n(n+1)} = \frac{n}{n+1}.$$

13. Prove that if $n \in \mathbb{N}$, then

$$\frac{1}{1 \cdot 3} + \frac{1}{2 \cdot 4} + \frac{1}{3 \cdot 5} + \cdots + \frac{1}{n(n+2)} = \frac{3n^2 + 5n}{4(n+1)(n+2)}.$$

14. Prove that if $n \in \mathbb{N}$, then

$$\frac{1}{1 \cdot 4} + \frac{1}{2 \cdot 5} + \frac{1}{3 \cdot 6} + \cdots + \frac{1}{n(n+3)} = \frac{11n^3 + 48n^2 + 49n}{18(n+1)(n+2)(n+3)}.$$

15. Prove that if a set has n elements, then it has $\frac{n(n-1)}{2}$ subsets with exactly two elements, for every integer $n \geq 2$.

16. Prove that if a set has n elements, then it has $\frac{n(n-1)(n-2)}{6}$ subsets with exactly three elements, for every integer $n \geq 3$.

17. Prove that a set with n elements has 2^n subsets, for every $n \in \mathbb{N}$.

18. Let s_n be the sequence defined as

$$s_1 = 11 \quad \text{and} \quad s_{n+1} = \frac{2}{3}(s_n + 5), \quad \text{for all } n \in \mathbb{N}.$$

   (a) Prove that $s_n > 10$, for all $n \in \mathbb{N}$.

   (b) Prove that $s_{n+1} < s_n$, for all $n \in \mathbb{N}$.

19. Let t_n be the sequence defined as

$$t_1 = 9 \quad \text{and} \quad t_{n+1} = \frac{2}{3}(t_n + 5), \quad \text{for all } n \in \mathbb{N}.$$

   (a) Prove that $t_n < 10$, for all $n \in \mathbb{N}$.

   (b) Prove that $t_{n+1} > t_n$, for all $n \in \mathbb{N}$.

20. Let u_n be the sequence defined as

$$u_1 = 5 \quad \text{and} \quad u_{n+1} = \frac{3}{5}(u_n + 12), \quad \text{for all } n \in \mathbb{N}.$$

   (a) Prove that $u_n < 18$, for all $n \in \mathbb{N}$.

   (b) Prove that $u_{n+1} > u_n$, for all $n \in \mathbb{N}$.

21. Let v_n be the sequence defined as

$$v_1 = 20 \quad \text{and} \quad v_{n+1} = \frac{3}{5}(v_n + 12), \quad \text{for all } n \in \mathbb{N}.$$

   (a) Prove that $v_n > 18$, for all $n \in \mathbb{N}$.

   (b) Prove that $v_{n+1} < v_n$, for all $n \in \mathbb{N}$.

22. In calculus it is shown that if $f(x), g(x)$ are continuous functions, then so is $f(x) + g(x)$. Use this fact to prove that if $n \in \mathbb{N}$ and $f_1(x), f_2(x), \ldots, f_n(x)$ are continuous functions, then so is $f_1(x) + f_2(x) + \cdots + f_n(x)$.

23. Let F be a function whose domain is $\mathbb{N}$, which is defined as

$$F(1) = 5 \quad \text{and} \quad F(n+1) = \frac{(n+1)F(n)}{3}, \quad \text{for all } n \in \mathbb{N}.$$

Prove that $F(n) = 5\left(\frac{n!}{3^{n-1}}\right)$, for all $n \in \mathbb{N}$, where $n! = 1 \cdot 2 \cdot 3 \cdots (n-1) \cdot n$.

24. Let G be a function whose domain is $\mathbb{N}$, which is defined as

$$G(1) = 8 \quad \text{and} \quad G(n+1) = G(n) + 3n^2 + 7n + 3, \quad \text{for all } n \in \mathbb{N}.$$

Prove that $G(n) = n^3 + 2n^2 + 5$, for all $n \in \mathbb{N}$.

25. Prove that $n^2 > 6n - 5$, for all integers $n \geq 6$.

26. Prove that $n^2 > 11n - 24$, for all integers $n \geq 9$.

27. Prove that

$$\frac{1}{1^2} + \frac{1}{2^2} + \cdots + \frac{1}{n^2} < 2 - \frac{1}{n},$$

for all integers $n \geq 2$.

28. Prove that $n^3 + 5n$ is a multiple of 3, for all $n \in \mathbb{N}$.

For exercises 29–34, please first read the following:

You will frequently come across mathematical statements that look like they can be proven with Mathematical Induction, yet all the versions of Mathematical Induction you have already seen may not apply. In light of this, it will often be necessary to use different versions of Mathematical Induction. All of these additional versions of Mathematical Induction can be proved using the Well Ordering Principle. In exercises 29–32, you will use the Well Ordering Principle to prove additional versions of Mathematical Induction. Then, in exercises 33–34, you may need to apply some new versions of Mathematical Induction to prove that various mathematical statements are true.

29. In section 2.3 of this chapter, we stated but did not prove Mathematical Induction— Second Version. Prove this version of Mathematical Induction.

30. Prove the following:

Let T be a subset of the integers satisfying the following two properties:
(a) T contains the number m

(b) whenever T contains a number k, where $k \geq m$, then it also contains the number $k+1$.

Then T contains all integers greater than or equal to m.

31. Prove the following:

    Let T be a subset of $\mathbb{N}$ satisfying the following two properties:
    (a) T contains 2

    (b) whenever T contains a number k, then it also contains the number $k + 2$.

    Then T contains all positive even integers.

32. Prove the following:

    Let T be a subset of $\mathbb{N}$ satisfying the following two properties:
    (a) T contains 3

    (b) whenever T contains a number k, then it also contains the number $k + 3$.

    Then T contains all positive integers that are multiples of 3.

33. (a) Prove that $9^n - 1$ is a multiple of 8, for all $n \in \mathbb{N}$.

    (b) Prove that $9^n - 1$ is a multiple of 80, whenever n is a positive even integer.

34. (a) Prove that $7^n - 1$ is a multiple of 48, whenever n is a positive even integer.

    (b) Prove that $7^n - 1$ is a multiple of 2400, whenever n is a positive integer which is a multiple of 4.

For exercises 35–36, please first read the following:

Throughout these exercises, you have used Mathematical Induction to verify formulas. However, more important and far more exciting than verifying formulas is finding formulas. In the next two exercises, you will be asked to (i) try to find a formula and (ii) prove the formula you found does indeed work. These exercises may shed some light on where the formulas in exercises 5–7 came from and also serve as a preview of some of our work in Chapter 13.

35. Consider the sum

$$1 + 3 + 5 + \cdots + (2n - 1).$$

    Suppose you suspect that the preceding sum is equal to a quadratic polynomial. If so, there would exist real numbers a, b, c such that $1 + 3 + 5 + \cdots + (2n - 1) = an^2 + bn + c$.
    (a) Plug three different values of n into the preceding equation to obtain three linear equations in a, b, c.

    (b) Solve for a, b, c. At this point, you will have derived a formula that you know works for at least three values of n.

    (c) Use Mathematical Induction to prove that the formula you found in part (b) holds for all $n \in \mathbb{N}$.

36. Consider the sum

$$1^2 + 2^2 + 3^2 + \cdots + n^2.$$

Suppose you suspect that the preceding sum is equal to a cubic polynomial. If so, there would exist real numbers a, b, c, d such that $1^2 + 2^2 + 3^2 + \cdots + n^2 = an^3 + bn^2 + cn + d$.

(a) Plug four different values of n into the preceding equation to obtain four linear equations in a, b, c, d.

(b) Solve for a, b, c, d. At this point, you will have derived a formula that you know works for at least four values of n.

(c) Use Mathematical Induction to prove that the formula you found in part (b) holds for all $n \in \mathbb{N}$.

For exercises 37–40, please first read the following:

Although there are many different versions of Mathematical Induction, they all contain two parts. Therefore, anytime you use Mathematical Induction to prove something, you will need to verify that both parts of Mathematical Induction are satisfied. It is often the case that verifying part (a) of Mathematical Induction is quite routine, whereas verifying part (b) can be quite difficult. As a result, students sometimes fall into the trap of thinking that a proof using Mathematical Induction requires only that you verify part (b). However, there are countless statements that are clearly false even though part (b) of Mathematical Induction can be easily verified. A very simple example is the statement:

All positive integers are greater than 1,000,000.

Obviously, this statement is false, but observe that part (b) of Mathematical Induction is satisfied if k is a positive integer that exceeds 1,000,000, then $k + 1$ is certainly also a positive integer that exceeds 1,000,000. In the next four exercises, you will examine other mathematical statements that are false despite the fact that part (b) of Mathematical Induction holds.

37. In doing this exercise, you may want to refer back to exercise 33.

(a) Consider the statement:

$$9^n - 1 \text{ is divisible by } 9, \quad \text{for all } n \in \mathbb{N}.$$

Explain why this statement is false, and then show that part (b) of Mathematical Induction is satisfied.

(b) Consider the statement:

$$9^n - 1 \text{ is divisible by } 81, \text{ whenever } n \text{ is a positive even integer.}$$

Explain why this statement is false, and then show that part (b) of Mathematical Induction is satisfied.

38. In doing this exercise, you may want to refer back to exercise 34.
    (a) Consider the statement:

    $7^n - 1$ is divisible by 49, whenever n is a positive even integer.

    Explain why this statement is false, and then show that part (b) of Mathematical Induction is satisfied.

    (b) Consider the statement:

    $7^n - 1$ is divisible by 2401, whenever n is a positive integer which is a multiple of 4.

    Explain why this statement is false, and then show that part (b) of Mathematical Induction is satisfied.

39. In this exercise, the sequences s_n and t_n are those defined in exercises 18 and 19.
    (a) Consider the statement:

    $$s_n < 10, \quad \text{for all } n \in \mathbb{N}.$$

    Explain why this statement is false, and then show that part (b) of Mathematical Induction is satisfied.

    (b) Consider the statement:

    $$t_n > 10, \quad \text{for all } n \in \mathbb{N}.$$

    Explain why this statement is false, and then show that part (b) of Mathematical Induction is satisfied.

40. In this exercise, the sequences u_n and v_n are those defined in exercises 20 and 21.
    (a) Consider the statement:

    $$u_n > 18, \quad \text{for all } n \in \mathbb{N}.$$

    Explain why this statement is false, and then show that part (b) of Mathematical Induction is satisfied.

    (b) Consider the statement:

    $$v_n < 18, \quad \text{for all } n \in \mathbb{N}.$$

    Explain why this statement is false, and then show that part (b) of Mathematical Induction is satisfied.

41. Let α be a real number, and define the sequence s_n as follows:

$$s_1 = \alpha \quad \text{and} \quad s_{n+1} = \frac{1 + s_n}{2}, \quad \text{for all } n \in \mathbb{N}.$$

(i) Consider the statement:

$$s_n > 1, \quad \text{for all } n \in \mathbb{N}.$$

Explain why part (b) of Mathematical Induction is satisfied.

(ii) Consider the statement:

$$s_n < 1, \quad \text{for all } n \in \mathbb{N}.$$

Compare this statement to the statement in part (i), and then explain why part (b) of Mathematical Induction is again satisfied.

(iii) Analyze whether the statements in parts (i) and (ii) are true.

(iv) Show that if $\alpha > 1$, then the statement in part (i) is true.

(v) Show that if $\alpha < 1$, then the statement in part (ii) is true.

(vi) What can you say about the sequence if $\alpha = 1$?

(vii) What does this tell you about the importance of part (a) of Mathematical Induction?

42. Let a and n be positive integers such that $a^{1/n}$ is a rational number.
(a) Show that $a^{m/n}$ is rational, for every $m \in \mathbb{N}$.

(b) Show that there exists a smallest positive integer t with the property that $t \cdot a^{i/n}$ is an integer, for all $1 \le i \le n - 1$.

(c) Show that there is a positive integer b with the property that $0 \le a^{1/n} - b < 1$.

(d) Show that $t(a^{1/n} - b)$ is a nonnegative integer that is less than t.

(e) Show that $t(a^{1/n} - b)$ also has the property that $(t(a^{1/n} - b)) \cdot a^{i/n}$ is an integer, for all $1 \le i \le n - 1$.

(f) Show that $t(a^{1/n} - b) = 0$.

(g) Conclude that $a^{1/n}$ is a whole number.

(h) Use parts (a)–(g) to prove that if a and n are positive integers, then either $a^{1/n}$ is a whole number or is not rational.

2.4 Functions and Binary Operations

Functions play an important role in virtually every branch of mathematics. In abstract algebra not only do we use functions to examine the structure of objects like fields, rings, and groups, but sets of functions often form important algebraic objects in their own right.

In calculus, you examined the graphs and other properties of functions like

$$f(x) = e^x, \quad g(t) = -16t^2 + 400, \quad \text{and} \quad h(\theta) = \tan(\theta).$$

At that point, almost all the functions you looked at were represented by a fairly straightforward rule or formula. For that reason, we informally defined a function as a "rule" that assigns to each element in a set an element in a second set.

This intuitive approach is a very appropriate way to look at functions in calculus where many functions arise in an attempt to describe or model real-world situations. Therefore, it comes as no surprise that most functions in calculus are represented by a straightforward rule or formula. As we have indicated at various points in this chapter, mathematics usually requires a mix of intuition and rigor. Defining a function as a "rule" works quite well in many contexts such as calculus. However, since we don't really know exactly what constitutes a rule, a more rigorous approach is sometimes needed. As will be the case throughout this book, rigorous definitions will often be preceded by intuitive examples.

When we look at $f(x) = e^x$, we know that

$$f(0) = 1, \quad f(1) = e, \quad \text{and} \quad f(\ln(5)) = 5.$$

Therefore, three of the ordered pairs that belong to the graph of f are $(0, 1)$, $(1, e)$, and $(\ln(5), 5)$. Observe that if we let $\mathbb{R} \times \mathbb{R}$ denote all ordered pairs of real numbers, then every element of the graph of f is an element of the set $\mathbb{R} \times \mathbb{R}$. Thus, we can think of the graph of f as being a particular type of subset of $\mathbb{R} \times \mathbb{R}$.

For an example far removed from calculus, let $V = \{a, b, c\}$ and $W = \{1, 2\}$. Next, suppose $G : V \rightarrow W$ is the function such that

$$G(a) = 2, \quad G(b) = 1, \quad \text{and} \quad G(c) = 1.$$

Then we can think of the graph of G, or simply think of G, as the ordered pairs $\{(a, 2), (b, 1), (c, 1)\}$. If we let $V \times W$ denote all ordered pairs where the first term belongs to V and the second term belongs to W, then we can think of the function G as being a subset of $V \times W$.

However, not all subsets of $V \times W$ fit our intuitive notion of a function. Consider the ordered pairs

$$\{(a, 2), (b, 1), (b, 2), (c, 1)\} \subset V \times W.$$

If this represented some function H, then we can ask ourselves, what is $H(b)$? The ordered pair $(b, 1)$ indicates that $H(b)$ should be 1. On the other hand, the ordered pair $(b, 2)$ indicates that $H(b)$ should be 2. When this type of ambiguity arises, we will not consider the ordered pairs to represent a function.

There is yet another issue to consider. Using the same V and W as in the previous paragraph, let's examine the ordered pairs

$$\{(a, 1), (c, 2)\} \subset V \times W.$$

If these ordered pairs represent a function J, then what is $J(b)$? We will deal with this situation somewhat differently than you might have done in the past. In calculus, you examined the function $h : \mathbb{R} \to \mathbb{R}$ defined as $h(\theta) = \tan(\theta)$. Note that h is not defined whenever θ is of the form $\frac{\pi}{2} + n\pi$, for all integers n. However, we still wrote $h : \mathbb{R} \to \mathbb{R}$, even though there are many elements in $\mathbb{R}$ for which h is not defined. In the more formal approach typically used in abstract algebra, we will handle things differently. When we refer to a function f from a set S to a set T, we will insist that $f(s)$ is defined for every element of S.

Given sets S and T, we can now put all the pieces together and describe exactly which subsets of $S \times T$ can represent a function f from S to T. We first observed that every element of S occurs in *at most* one ordered pair of f. Next, we observed that every element of S must occur in *at least* one ordered pair of f. Together, these observations lead us to

Definition 2.7. *If S and T are sets then a function $f : S \to T$ is a subset of $S \times T$ such that every element of S belongs to exactly one ordered pair of f.*

Given a function $f : S \to T$, we call S the *domain* of f. If an ordered pair (a, b) belongs to f, we write $f(a) = b$ and say that b is the image of a. Furthermore, we call the set $\{b \in T \mid b = f(a), \text{ for some } a \in S\}$ the *range* of f. Thus, the range of f is the subset of T consisting of all values of the function f.

■ Example

Let $S = \{\alpha, \beta\}$ and $T = \{x, y, z\}$ and let $f : S \to T$ be the function $\{(\alpha, y), (\beta, x)\}$. Then, in this case, $f(\alpha) = y$, $f(\beta) = x$, and the range of f is the set $\{x, y\}$.

■

We will now contrast the previous function f to the new function $g : S \to T$ defined as $\{(\alpha, z), (\beta, z)\}$. One big difference between f and g is that g "repeats values." More precisely, g takes on the value of z twice as $g(\alpha) = z = g(\beta)$. We now give a special name to functions that never repeat values.

Definition 2.8. *Let $f : S \rightarrow T$ be a function. We say that f is* injective *or* one-to-one *if f never repeats any values. This means that whenever a and b are different elements of S, then $f(a)$ and $f(b)$ are different elements of T.*

■ Example

Let $f : \mathbb{R} \rightarrow \mathbb{R}$ and $g : \mathbb{R} \rightarrow \mathbb{R}$ be given by the formulas $f(x) = x^2$ and $g(x) = x^3$. Observe that f is not injective, as there are many cases where $a \neq b$, yet $f(a) = f(b)$. For example, $f(5) = f(-5)$. However, g is injective, for if $a, b \in \mathbb{R}$ and $a \neq b$, then $g(a) = a^3 \neq b^3 = g(b)$.

■

If you suspect that a function $h : S \rightarrow T$ is injective, then an approach to take in trying to prove it is injective is to assume that $s_1, s_2 \in S$ such that $h(s_1) = h(s_2)$ and then show that $s_1 = s_2$.

■ Example

Let $h : \mathbb{R} \rightarrow \mathbb{R}$ be given by $h(x) = 3x - 5$. To show that h is injective, suppose $a, b \in \mathbb{R}$ such that $h(a) = h(b)$. Thus, $3a - 5 = 3b - 5$. But this quickly implies that $3a = 3b$, which then yields $a = b$. Thus, h is indeed injective.

■

To illustrate another important property of functions, consider the following:

■ Example

Let $S = \{1, 2, 3, 4\}$ and $T = \{a, b, c\}$. Next, let $m : S \rightarrow T$ be given by the ordered pairs $\{(1, c), (2, c), (3, a), (4, a)\}$ and let $n : S \rightarrow T$ be given by the ordered pairs $\{(1, b), (2, a), (3, a), (4, c)\}$. Observe that the range of m is not all of T, as b is not a value of m. On the other hand, the range of n is all of T, as every element of T is indeed a value of n. This leads to

■

Definition 2.9. *Let $f : S \rightarrow T$ be a function. We say that f is* surjective *or* onto *if the range of f is all of T. This means that for every $t \in T$, there exists some $s \in S$ such that $f(s) = t$.*

To illustrate our next concept, let us look at the following:

■ Example

Let $S = \{1, 2, 3\}$, $T = \{\alpha, \beta\}$, and $W = \{a, b, c\}$. Next, let $g : S \to T$ be the function $\{(1, \beta), (2, \alpha), (3, \alpha)\}$ and $f : T \to W$ be the function $\{(\alpha, c), (\beta, a)\}$.

Observe that we can create a new function from S to W, which we will denote at $f \circ g$, by first "plugging" elements of S into g and then plugging the result into f. More precisely,

$$(f \circ g)(1) = f(g(1)) = f(\beta) = a,$$

$$(f \circ g)(2) = f(g(2)) = f(\alpha) = c,$$

$$(f \circ g)(3) = f(g(3)) = f(\alpha) = c.$$

Therefore, the new function $f \circ g : S \to W$ consists of the ordered pairs $\{(1, a), (2, c), (3, c)\}$.

■

Definition 2.10. *If $g : S \to T$ and $f : T \to W$ are functions, then the composition $f \circ g : S \to W$ is the function consisting of all ordered pairs of the form $(s, f(g(s)))$, where $s \in S$.*

Now let us push the previous example a bit further.

■ Example

Let S, T, W, g, and f be as in the previous example, and now let $V = \{x, y, z, w\}$. Next, let $h : V \to S$ be the function consisting of the ordered pairs $\{(x, 2), (y, 1), (z, 2), (w, 3)\}$. Therefore, in addition to the composition $f \circ g : S \to W$, we can also look at the composition $g \circ h : V \to T$. Some straightforward computations tell us that

$$(g \circ h)(x) = g(h(x)) = g(2) = \alpha,$$

$$(g \circ h)(y) = g(h(y)) = g(1) = \beta,$$

$$(g \circ h)(z) = g(h(z)) = g(2) = \alpha,$$

$$(g \circ h)(w) = g(h(w)) = g(3) = \alpha.$$

Therefore, $g \circ h$ is the subset of $V \times T$ consisting of the four ordered pairs $\{(x, \alpha), (y, \beta), (z, \alpha), (w, \alpha)\}$.

At this point, we can now form two new compositions $(f \circ g) \circ h$ and $f \circ (g \circ h)$, both of which are functions from V to W. Computing the ordered pairs in $(f \circ g) \circ h$, we have

$$((f \circ g) \circ h)(x) = (f \circ g)(h(x)) = (f \circ g)(2) = f(g(2)) = f(\alpha) = c,$$

$$((f \circ g) \circ h)(y) = (f \circ g)(h(y)) = (f \circ g)(1) = f(g(1)) = f(\beta) = a,$$

$$((f \circ g) \circ h)(z) = (f \circ g)(h(z)) = (f \circ g)(2) = f(g(2)) = f(\alpha) = c,$$

$$((f \circ g) \circ h)(w) = (f \circ g)(h(w)) = (f \circ g)(3) = f(g(3)) = f(\alpha) = c.$$

On the other hand, when we compute the ordered pairs in $f \circ (g \circ h)$, we have

$$(f \circ (g \circ h))(x) = f((g \circ h)(x)) = f(g(h(x))) = f(g(2)) = f(\alpha) = c,$$

$$(f \circ (g \circ h))(y) = f((g \circ h)(y)) = f(g(h(y))) = f(g(1)) = f(\beta) = a,$$

$$(f \circ (g \circ h))(z) = f((g \circ h)(z)) = f(g(h(z))) = f(g(2)) = f(\alpha) = c,$$

$$(f \circ (g \circ h))(w) = f((g \circ h)(w)) = f(g(h(w))) = f(g(3)) = f(\alpha) = c.$$

∎

Observe that in the preceding example the ordered pairs for both $(f \circ g) \circ h$ and $f \circ (g \circ h)$ are $\{(x, c), (y, a), (z, c), (w, c)\}$. Therefore, the functions $(f \circ g) \circ h$ and $f \circ (g \circ h)$ are the same. This is no coincidence, and it leads us to

Theorem 2.11. *The composition of functions is associative. This means that if $f : C \to D$, $g : B \to C$, and $h : A \to B$ are functions, then the functions $(f \circ g) \circ h : A \to D$ and $f \circ (g \circ h) : A \to D$ are the same and consist of all ordered pairs in $A \times D$ of the form $(a, f(g(h(a))))$, where $a \in A$.*

Intuition. Before providing a very short formal proof, we provide a more informal discussion of what is taking place. Observe that if $a \in A$, then when we compute where $(f \circ g) \circ h$ sends a, we first plug a into h and then plug $h(a)$ into $f \circ g$. But plugging $h(a)$ into $f \circ g$ means first plugging $h(a)$ into g to obtain $g(h(a))$ and then plugging $g(h(a))$ into f to give us the final answer of $f(g(h(a)))$.

On the other hand, if $a \in A$, then to see where $f \circ (g \circ h)$ sends a, we first plug a into $g \circ h$. This gives us an a value of $g(h(a))$, which next gets plugged into f to give us $f(g(h(a)))$. The bottom line is that regardless of whether $a \in A$ is plugged into $(f \circ g) \circ h$ or $f \circ (g \circ h)$, we end up with a final value of $f(g(h(a)))$.

Proof. If $a \in A$, then

$$((f \circ g) \circ h)(a) = (f \circ g)(h(a)) = f(g(h(a))),$$

whereas

$$(f \circ (g \circ h))(a) = f((g \circ h)(a)) = f(g(h(a))).$$

Thus, both $(f \circ g) \circ h$ and $f \circ (g \circ h)$ send a to $f(g(h(a)))$ and therefore consist of all ordered pairs of the form $(a, f(g(h(a))))$, where $a \in A$. $\qquad\qquad\square$

Let us stop for a moment and see where we are. If S is a set, we will let $F(S) = \{f : S \to S\}$. Thus, $F(S)$ is the set of functions from S to S. In light of Definition 2.10, whenever $f, g \in F(S)$ the composition $f \circ g \in F(S)$. Therefore

(i) $F(S)$ is a set and

(ii) given any two elements $f, g \in F(S)$, there is a way of combining them, denoted as $\circ$, to obtain a third element $f \circ g$ of $F(S)$.

(Note: When we talk about two elements of $F(S)$ or a third element of $F(S)$, we are not making any assumptions about whether they are all different or whether some of them might be the same.)

Therefore, we are in a situation where we can always combine two elements of a set to obtain a third. As we will see in the following examples, this is a very common situation.

■ Examples

1. Let $\mathbb{R}$ be the set of real numbers and let $\circ$ denote ordinary multiplication. Then, whenever $a, b \in \mathbb{R}$, we have $a \circ b = ab \in \mathbb{R}$.

2. Let $\mathbb{R}^+$ be the set of positive real numbers and let $\circ$ denote ordinary addition. Since the sum of two positive real numbers is a positive real number, we see that whenever $a, b \in \mathbb{R}^+$, we have $a \circ b = a + b \in \mathbb{R}^+$.

3. Let $\mathbb{Q}$ be the set of rational numbers and let $\circ$ be defined as $a \circ b = a + b + ab$, for all $a, b \in \mathbb{Q}$. It is easy to see that if $a, b \in \mathbb{Q}$, then $a \circ b \in \mathbb{Q}$.

4. This example will differ from the first three. In this case, we again consider the set $\mathbb{R}^+$ of positive real numbers, but this time we let $\circ$ denote ordinary subtraction. In this situation, there are many positive real numbers a, b such that $a \circ b = a - b \in R^+$. However, there also exist some $a, b \in R^+$ such that $a \circ b \notin R^+$. For one out of many possible examples, we have $4, 7 \in \mathbb{R}^+$, but $4 \circ 7 = 4 - 7 = -3 \notin \mathbb{R}^+$. ■

In our first example, we can think of $\circ$ as a function whose domain is $\mathbb{R} \times \mathbb{R}$ and whose range is contained in $\mathbb{R}$. Similarly, in the second and third examples, $\circ$ can be thought of as a function whose domain is of the form $S \times S$ and whose range is contained in S. But things are different in the fourth example. In that example, $\circ$ is a function whose domain is $\mathbb{R}^+ \times \mathbb{R}^+$, but the range is not contained in $\mathbb{R}^+$. To better describe the difference between the fourth example and the first three, we have

Definition 2.12. *A binary function $\circ$ on a set S is a function whose domain is $S \times S$. If we are in the case where $\circ : S \times S \to S$, then we say that S is* closed *under $\circ$ and say that $\circ$ is a* binary operation *on S.*

Applying the language of Definition 2.12 to our previous examples, we have

(i)   multiplication is a binary operation on the set $\mathbb{R}$,

(ii)   addition is a binary operation on the set $\mathbb{R}^+$,

(iii) $\circ$ defined as $a \circ b = a + b + ab$ is a binary operation on the set $\mathbb{Q}$,

(iv)   subtraction is *not* a binary operation on the set $\mathbb{R}^+$ as $\mathbb{R}^+$ is not closed under subtraction. Observe that subtraction is a binary function on $\mathbb{R}^+$ but it is not a binary operation as the range of $\circ$ contains negative numbers.

Furthermore, Definition 2.12 indicates that composition of functions is a binary operation on the set of functions from a set to itself. Having already defined injective and surjective functions, the next definition should come as no surprise.

Definition 2.13. *Let $f : S \to T$ be a function. We say that f is* bijective *if it is both injective and surjective.*

We can now examine whether the properties of being injective, surjective, and bijective are preserved when we compose functions. Calculus courses have given most students experience, at least on an informal level, with injective and surjective functions. Despite this, students often have great difficulty proving statements about how injective and surjective functions behave under composition. This might be because we are now forced to deal with topics that are much more formal and less intuitive than those we have dealt with previously. Thus, reading and writing proofs about the composition of injective and surjective functions require a thorough understanding of all the relevant definitions.

Theorem 2.14. *Let $f : T \to U$ and $g : S \to T$ be functions.*

(a)   *If f and g are both injections, then the composition $f \circ g$ is also an injection.*

(b)   *If f and g are both surjections, then the composition $f \circ g$ is also a surjection.*

(c)   *If f and g are both bijections, then the composition $f \circ g$ is also a bijection.*

Proof. For part (a), we need to show that if $a, b \in S$ such that $(f \circ g)(a) = (f \circ g)(b)$, then $a = b$. Since $(f \circ g)(a) = f(g(a))$ and $(f \circ g)(b) = f(g(b))$, we see that $f(g(a)) = f(g(b))$. The function f is injective, and $g(a)$ and $g(b)$ give us the same value when plugged into f, so $g(a) = g(b)$. But g is also injective, and a and b give us the same value when plugged into g, so $a = b$, as desired.

For part (b), we need to show that if $c \in U$, then there exists $a \in S$ such that $(f \circ g)(a) = c$. We will do this in two steps. First, since f is surjective, we know that there exists $b \in T$ such that $f(b) = c$. Next, the surjectivity of g tells us that there exists $a \in S$ such that $g(a) = b$. Thus, g sends a to b and then f sends b to c. More formally, we can write this as

$$(f \circ g)(a) = f(g(a)) = f(b) = c,$$

as desired.

For part (c), observe that since f and g are both injective and surjective, parts (a) and (b) tell us that $f \circ g$ is also injective and surjective. Thus, $f \circ g$ is indeed bijective. $\square$

If S is a set, we already saw that composition of functions is a binary operation on the set of functions from S to S. If we let $Inj(S)$, $Sur(S)$, and $Bij(S)$ denote, respectively, the injections, surjections, and bijections from S to S, then using the language of Definition 2.12, it immediately follows from Theorem 2.14 that

Corollary 2.15. *Let S be set and let $\circ$ denote composition of functions.*

(a) $\circ$ *is a binary operation on the set $Inj(S)$.*

(b) $\circ$ *is a binary operation on the set $Sur(S)$.*

(c) $\circ$ *is a binary operation on the set $Bij(S)$.*

Before concluding this chapter, we will make a few more observations about the set $Bij(S)$. Suppose we define $e : S \rightarrow S$ as $e(s)$, for all $s \in S$, We call e the *identity map* and note that it can be considered to be subset of $S \times S$ consisting of all ordered pairs where the first and second terms are the same. It is easy to see that e is both injective and surjective, so $e \in Bij(S)$. Next, if $f : S \rightarrow S$ then, since $f(s) \in S$, for all $s \in S$, it follows that $e(f(s)) = f(s)$. Therefore, if $s \in S$, we have

$$(f \circ e)(s) = f(e(s)) = f(s),$$

and

$$(e \circ f)(s) = e(f(s)) = f(s).$$

Thus, $f \circ e$, $e \circ f$, and f are all the same function; hence,

$$f \circ e = f = e \circ f.$$

Observe that e plays the same role for the binary operation $\circ$ that the number 0 plays for ordinary addition and the number 1 plays for ordinary multiplication. In particular, no matter what function f you compose with e, you always get f back. For this reason, we consider e to be the identity element of the sets $F(S)$, $Inj(S)$, $Sur(S)$, and $Bij(S)$ under $\circ$.

Now suppose $f \in Bij(S)$ and consider the set of ordered pairs $(f(s), s)$, where $s \in S$. First, since f is surjective, we can see every element of S occurs as the first term in at least one ordered pair of this type. Next, since f is injective, no element of S occurs as the first term in two different ordered pairs of this type. Therefore, the ordered pairs we described satisfy the requirements of being a function from S to S. Since the ordered pairs in this new function are the reverse of the ordered pairs of f, we can think of this new function as "undoing" f. If we think of f as pushing points two feet to the right, then this new function would push points two feet to the left. For another example, if f was a function that doubled all the numbers you plugged into it, then this new function would take half of each number you plugged into it.

Observe that if you plugged something into f and then plugged the answer into the new function, you end up back where you started. If we switch the order and first plug something into this new function and then plug the answer into f, we again end up back where we started. Therefore, regardless of the order, whenever we apply f and this new function, the end result is the same as applying the identity map. Since this new function essentially "inverts" f, we denote it as f^{-1} and call it the inverse function of f. Before formalizing things, let's work through

■ Example

Let $S = \{1, 2, 3, 4\}$ and let $f \in Bij(S)$ be described by the ordered pairs $\{(1, 3), (2, 2),$ $(3, 4), (4, 1)\}$. Then when we reverse the order of the elements of each ordered pair, we see that the function f^{-1} is described by the ordered pairs $\{(3, 1), (2, 2), (4, 3), (1, 4)\}$. Observe that f^{-1} is also a bijection. Furthermore, if you compute the composition $f \circ f^{-1}$ and $f^{-1} \circ f$, you will observe that

$$(f \circ f^{-1})(s) = f(f^{-1}(s)) = s = e(s)$$

and

$$(f^{-1} \circ f)(s) = f^{-1}(f(s)) = s = e(s),$$

for all $s \in S$. (You should check this by letting s take on all the values 1, 2, 3, 4 and examining how the two compositions behave.)

■

Recall that under ordinary addition, the additive inverse of an element is the element that after adding gets you back to 0. Similarly, under ordinary multiplication, the multiplicative inverse of an element is the element that after multiplying gets you back to 1. Now, f^{-1} plays that same role for f under composition of functions, as it gets us back to the identity map e. We now collect and formalize many of our observations.

Theorem 2.16. *If S is a set, let $Bij(S)$ denote the bijections from S to S, and let $\circ$ denote composition of functions. Then $\circ$ is a binary operation on $Bij(S)$ satisfying*

(a) *Associative Law: $(f \circ g) \circ h = f \circ (g \circ h)$, for all $f, g, h \in Bij(S)$;*

(b) *Identity Element: There is an element $e \in Bij(S)$ such that $f \circ e = f = e \circ f$, for all $f \in Bij(S)$;*

(c) *Inverses: For every $f \in Bij(S)$, there is an element in $Bij(S)$, which we denote as f^{-1}, such that $f \circ f^{-1} = e = f^{-1} \circ f$.*

Proof. In Theorem 2.14(c), we showed that $\circ$ is a binary operation on $Bij(S)$, and in Theorem 2.11, we showed that the composition of functions is associative. Furthermore, in the discussion leading up to this theorem, we showed that $Bij(S)$ has an identity element under $\circ$. Therefore, it only remains to prove part (c).

Let $g : T \to U$ be a function, and let us analyze, in terms of the ordered pairs corresponding to g inside of $T \times U$, exactly what it means for g to be bijective. Note that g being injective is identical to every element of U appearing as the second term in *at most* one ordered pair of g. In addition, g being surjective is the same as every element of U occurring as the second term in *at least* one ordered pair of g. As a result, g being bijective is equivalent to every element of U occurring as the second term in exactly one ordered pair of g.

Now suppose $f \in Bij(S)$; we now know that not only does every element of S occur exactly once as the first term of an ordered pair of f, but every element of S also occurs exactly once as the second term of an ordered pair of f. Therefore, when we let f^{-1} be the function consisting of all ordered pairs of the form $(f(s), s)$, where $s \in S$, we can see that f^{-1} satisfies the criterion in the previous paragraph for a function to be a bijection. Thus, $f^{-1} \in Bij(S)$.

Since we defined f^{-1} to consist of the "reverse" of all the ordered pairs in f, it follows that the ordered pair $(a, b) \in S \times S$ belongs to f if and only if the ordered pair (b, a) belongs to f^{-1}. Now suppose $a \in S$; then the ordered pair $(a, f(a))$ belongs to f and the ordered pair $(f(a), a)$ belongs to f^{-1}. This means that f sends a to $f(a)$ and f^{-1} send $f(a)$ back to a. Therefore,

$$(f^{-1} \circ f)(a) = f^{-1}(f(a)) = a = e(a),$$

where e is the identity map.

Similarly, if $b \in S$, then when we let $c = f^{-1}(b)$, we know that the ordered pair (b, c) belongs to f^{-1}. But this means that the ordered pair (c, b) belongs to f, which is another way of saying that $f(c) = b$. As a result, f^{-1} sends b to c and f then sends c back to b. More formally, we have

$$(f \circ f^{-1})(b) = f(f^{-1}(b)) = f(c) = b = e(b).$$

Thus, $f \circ f^{-1} = e = f^{-1} \circ f$, and every element of $Bij(S)$ does indeed have an inverse in $Bij(S)$ under $\circ$. □

As we will see throughout this book, **groups** are among the most important objects in abstract algebra and are an essential part of Galois' proof of the insolvability of the quintic. Later, in much greater detail, we will discuss the fact that groups are sets with a binary operation satisfying the three conditions in Theorem 2.16. Thus, $Bij(S)$ with $\circ$, for various sets S, provide us with some of the many, many examples of groups we will come across in this course. In fact, groups of the form $Bij(S)$, where S is the root of a polynomial, are essential to Galois' work.

Exercises for Section 2.4

In exercises 1–3, let $S = \{1, 2, 3, 4, 5\}$ and $T = \{a, b, c\}$ and consider the following six subsets of $S \times T$:

$A = \{(1, b),\ (2, b),\ (3, a),\ (3, c),\ (4, c),\ (5, c)\},$

$B = \{(1, b),\ (2, b),\ (3, c),\ (4, c),\ (5, b)\},$

$C = \{(1, c),\ (2, b),\ (3, b),\ (4, c),\ (5, a)\},$

$D = \{(1, c),\ (2, c),\ (3, a),\ (5, b)\},$

$E = \{(1, c),\ (2, a),\ (3, a),\ (4, c),\ (5, a)\},$

$F = \{(1, b),\ (2, c),\ (3, a),\ (4, c),\ (5, b)\}.$

1. Which of the six subsets represents a function from S to T?

2. Using your answer from exercise 1, which of these subsets represents an injective function?

3. Using your answer from exercise 1, which of these subsets represents a surjective function?

In exercises 4–6, let $U = \{\alpha, \beta, \gamma\}$ and $V = \{x, y, z, w\}$ and consider the following six subsets of $U \times V$:

$A = \{(\alpha, y), (\beta, y), (\gamma, w)\}$,

$B = \{(\alpha, w), (\gamma, x)\}$,

$C = \{(\alpha, w), (\beta, y), (\gamma, x)\}$,

$D = \{(\alpha, y), (\beta, x), (\gamma, z)\}$,

$E = \{(\alpha, x), (\beta, y), (\gamma, z), (\gamma, w)\}$,

$F = \{(\alpha, z), (\beta, y), (\gamma, z)\}$.

4. Which of the six subsets represents a function from U to V?

5. Using your answer from exercise 4, which of these subsets represents an injective function?

6. Using your answer from exercise 4, which of these subsets represents a surjective function?

In exercises 7–24, let $S = \{1, 2, 3\}$ and let $f, g, h \in Bij(S)$ be as follows:

$$f = \{(1, 3), (2, 2), (3, 1)\}, \quad g = \{(1, 3), (2, 1), (3, 2)\}, \quad h = \{(1, 1), (2, 3), (3, 2)\}.$$

In exercises 7–24, you will be doing computations in $Bij(S)$. In your answers, express elements of $Bij(S)$ as subsets of $S \times S$.

7. Find $f \circ g$.

8. Find $g \circ f$.

9. Find f^{-1}.

10. Find g^{-1}.

11. Find $f^{-1} \circ g^{-1}$.

12. Find $g^{-1} \circ f^{-1}$.

13. Find $(f \circ g)^{-1}$.

14. Find $(g \circ f)^{-1}$.

15. Compare your answers from exercises 13 and 14 to your answers from exercises 11 and 12 and briefly discuss what you noticed.

16. Find $g \circ h$.

17. Using your answer from exercise 16, find $f \circ (g \circ h)$.

18. Using your answer to exercise 7, find $(f \circ g) \circ h$. In light of your answer to exercise 17, is this a surprise?

19. Find $g \circ g$. How do $g \circ g$ and g^{-1} compare?

20. Based on your answer to the second part of exercise 19, what do you think $(g \circ g) \circ g$ is equal to? Check your answer by computing $(g \circ g) \circ g$.

21. Find h^{-1}.

22. Find $g^{-1} \circ h^{-1}$.

23. Find $h^{-1} \circ g^{-1}$.

24. In light of the previous exercises, do you expect $(g \circ h)^{-1}$ to equal $g^{-1} \circ h^{-1}$ or $h^{-1} \circ g^{-1}$? Check your answer by computing $(g \circ h)^{-1}$.

In exercises 25–32, $f : S \to T$ is defined using the formula $f(x) = x^2$, for all $x \in S$. S and T will be various subsets of the real numbers $\mathbb{R}$.

25. If $S = [0, 5]$ and $T = \mathbb{R}$, is f injective and is f surjective?

26. If $S = [0, 5]$ and $T = [0, 25]$, is f injective and is f surjective?

27. If $S = [-5, 5]$ and $T = [-25, 25]$, is f injective and is f surjective?

28. If $S = [-5, 5]$ and $T = [0, 25]$, is f injective and is f surjective?

29. If $S = [1, 2]$ and $T = [0, 4]$, is f injective and is f surjective?

30. If $S = [-1, 0] \cup [1, 2]$ and $T = [0, 4]$, is f injective and is f surjective?

31. If $S = (-1, 0] \cup [1, 2]$ and $T = [0, 4]$, is f injective and is f surjective?

32. If $S = [-1, 2]$ and $T = [-4, 4]$, is f injective and is f surjective?

33. Give an example of a function $f : \mathbb{R} \to \mathbb{R}$ that is injective but not surjective.

34. Give an example of a function $g : \mathbb{R} \to \mathbb{R}$ that is surjective but not injective.

35. Show that if S is a finite set, then every injective function $h : S \to S$ is also surjective.

36. Show that if S is a finite set, then every surjective function $j : S \to S$ is also injective.

37. If T is a set and $f_1, f_2, \ldots, f_n$ are injective functions from T to T, show that the composition $f_1 \circ f_2 \circ \cdots \circ f_n$ is also injective. (In light of the associative law, we do not need to use any parentheses when composing these n functions.)

38. If V is a set and $g_1, g_2, \ldots, g_n$ are surjective functions from V to V, show that the composition $g_1 \circ g_2 \circ \cdots \circ g_n$ is also surjective. (In light of the associative law, we do not need to use any parentheses when composing these n functions.)

In exercises 39–53, you are given a set S with binary function $\circ$. Determine if S is closed under $\circ$, thereby making $\circ$ a binary operation. If S is not closed under $\circ$, give an example of $a, b \in S$ such that $a \circ b \notin S$.

39. Let $S = \{a + bx \mid a, b \in \mathbb{R}\}$ and let $\circ$ be the ordinary addition of polynomials.

40. Let $S = \{a + bx \mid a, b \in \mathbb{R}\}$ and let $\circ$ be the ordinary multiplication of polynomials.

41. Let S be the set of polynomials with rational coefficients and let $\circ$ be the ordinary multiplication of polynomials.

42. Let S be the set of polynomials with rational coefficients of odd degree and let $\circ$ be the ordinary multiplication of polynomials.

43. Let S be the set of polynomials with rational coefficients of even degree and let $\circ$ be the ordinary multiplication of polynomials.

44. Let S be the set of odd integers and let $\circ$ be ordinary addition.

45. Let S be the set of odd integers and let $\circ$ be ordinary multiplication.

46. Let S be the set of nonzero real numbers and let $\circ$ be ordinary multiplication.

47. Let S be the set of positive real numbers and let $\circ$ be ordinary multiplication.

48. Let S be the set of negative real numbers and let $\circ$ be ordinary addition.

49. Let S be the set of negative real numbers and let $\circ$ be ordinary multiplication.

50. Let S be the set of polynomials with integer coefficients and let $\circ$ be ordinary multiplication.

51. Let S be the set of real numbers that are roots of the polynomial $x^2 - 1$ and let $\circ$ be ordinary multiplication.

52. Let S be the set of real numbers that are roots of the polynomial $x^2 - 1$ and let $\circ$ be ordinary addition.

53. Let S be the set of polynomials with rational coefficients of degree less than 50 and let $\circ$ be ordinary multiplication.

In exercises 54–59, let S, T, V be sets and let $f : T \to V$ and $g : S \to T$ be functions.

54. Show that if $f \circ g$ is injective, then g must be injective.

55. Show that if $f \circ g$ is injective and g is surjective, then f must be injective.

56. Give an example of S, T, V, f, and g such that $f \circ g$ is injective and f is not injective.

57. Show that if $f \circ g$ is surjective, then f must be surjective.

58. Show that if $f \circ g$ is surjective and f is injective, then g must be surjective.

59. Give an example of S, T, V, f, and g such that $f \circ g$ is surjective and g is not surjective.

In exercises 60–71, let S be a set with 3 elements, T a set with 4 elements, and V a set with 5 elements.

60. How many functions $f : S \to T$ are there?

61. How many functions $f : T \to S$ are there?

62. How many functions $f : S \to T$ are injective?

63. How many functions $f : T \to S$ are injective?

64. How many functions $f : S \to T$ are surjective?

65. How many functions $f : T \to S$ are surjective?

66. How many functions $f : S \to V$ are there?

67. How many functions $f : V \to S$ are there?

68. How many functions $f : S \to V$ are injective?

69. How many functions $f : V \to S$ are injective?

70. How many functions $f : S \to V$ are surjective?

71. How many functions $f : V \to T$ are surjective?

In exercises 72–74, S is a set with $n \geq 1$ elements and T is a set with $m \geq 1$ elements.

72. How many functions $f : S \to T$ are there?

73. If $n \leq m$, how many functions $f : S \to T$ are injective?

74. If $m = n + 1$, how many functions $f : T \to S$ are surjective?

75. If S is a finite set and $f \in Bij(S)$, show that there is a positive integer t, such that the composition of t copies of f, $f \circ f \circ \cdots \circ f$, is equal to the identity map in $Bij(S)$.

CHAPTER 3

The Integers

Throughout this book, we will work with many number systems that you have probably seen before, such as the integers $\mathbb{Z}$, the rational numbers $\mathbb{Q}$, the real numbers $\mathbb{R}$, and the complex numbers $\mathbb{C}$. In addition, we will look at other types of number systems that may be new to you, such as the integers modulo a prime $\mathbb{Z}_p$. The first number system we will examine is the integers which is the infinite set

$$\mathbb{Z} = \{\dots, -5, -4, -3, -2, -1, 0, 1, 2, 3, 4, 5, \dots\}.$$

We begin with the integers for three reasons:

(a) Many other number systems are built up from the integers.

(b) Other algebraic objects, such as polynomials, have important properties in common with the integers.

(c) Proofs of facts about other algebraic objects will often reduce down to questions about the integers.

In previous courses, you have not only added and multiplied integers but you have probably been introduced to concepts such as divisibility, prime numbers, greatest common divisors, and factorization into primes. Not only will similar concepts occur when studying sets of polynomials such as $\mathbb{Q}[x]$, $\mathbb{R}[x]$, $\mathbb{C}[x]$ but they also arise in even more abstract settings. By first developing a deep understanding of these concepts, as they relate to $\mathbb{Z}$, we will be well prepared to apply them to polynomials in Chapters 9, 12, and 17. In fact, a recurring theme of this book is the important similarity between the algebraic structure of the integers and various sets of polynomials.

3.1 Prime Numbers

In our study of the integers, the first concept we will be introduced to is divisibility.

Definition 3.1. *Given integers a and b (with a ≠ 0), we say that a divides b, written a | b, if there exists an integer m such that b = a · m. In this case, we also say that a is a divisor of b or that b is a multiple of a.*

If a is not a divisor of b, we write $a \nmid b$.

■ Examples

2 | 4, 4 ∤ 2, 45 | 0, −7 | 63, 5 | −25, −10 | −10, 13 | 39, 13 ∤ 93.

■

Observe that 1 and −1 are divisors of all integers, and every nonzero integer is a divisor of 0. A useful fact that we'll use throughout this chapter is that if a and b are integers such that $a \mid b$ and b is positive, then $b \geq a$.

The basic building blocks of the integers are the **prime numbers**. A thorough understanding of the properties of prime numbers will be very useful in proving facts about the integers and other algebraic objects.

Definition 3.2. *An integer n is called* prime *if n > 1 and the only positive divisors of n are 1 and n.*

As a result, the list of prime numbers begins

$$2, 3, 5, 7, 11, 13, 17, 19, 23, 29, 31, 37, 41, 43, \ldots$$

At this point, two questions you may have about prime numbers are:

(a) The definition of prime numbers says that all primes must be greater than 1. This seems somewhat arbitrary. Why is 1 not considered to be a prime number? After all, its only positive divisor is 1.

(b) How many prime numbers are there? In particular, is the number of primes finite or infinite?

We will answer both of these questions shortly. Despite the fact that prime numbers are very concrete and have been studied for centuries, there are still some very basic and easily stated questions about them that remain unanswered after all these years. The following are two of these questions.

Goldbach's Conjecture
Can every even integer greater than 2 be written as a sum of two prime numbers?

For example, we have $4 = 2 + 2$, $6 = 3 + 3$, $8 = 3 + 5$, $10 = 3 + 7$, and $10 = 5 + 5$.

Twin Primes Conjecture

We say that a pair of prime numbers are *twin primes* if they differ by 2. For example, 3 and 5 are twin primes, as well as 17 and 19. The twin primes conjecture asks if there are an infinite number of pairs of twin primes.

In some of your previous mathematics courses, you may have gotten the impression that everything that can be known in mathematics is known and has been known for hundreds of years. However, the preceding two questions are just two of the many questions being studied today by researchers in mathematics. At present, there are thousands of people around the world actively engaged in mathematical research. In fact, the last 50 years has witnessed more research in mathematics than any other period in our history.

There are several interesting comments we can make about primes and twin primes that are related to concepts you have probably seen in a calculus course. Recall that when you studied infinite series, you saw that the sum $\sum_{n=1}^{\infty} \frac{1}{n}$ was infinite. Suppose instead of letting n range over the entire set of positive integers, we only let n range through the set of prime numbers. It has also been shown that the sum

$$\sum_{n \text{ is prime}} \frac{1}{n} = \frac{1}{2} + \frac{1}{3} + \frac{1}{5} + \frac{1}{7} + \frac{1}{11} + \frac{1}{13} + \frac{1}{17} + \frac{1}{19} + \frac{1}{23} + \frac{1}{29} + \cdots$$

is infinite. On the other hand, we can look at the sum where we only let n range through the set of primes that are part of a pair of twin primes to obtain the new sum

$$\sum_{n \text{ is a twin prime}} \frac{1}{n} = \frac{1}{3} + \frac{1}{5} + \frac{1}{7} + \frac{1}{11} + \frac{1}{13} + \frac{1}{17} + \frac{1}{19} + \frac{1}{29} + \cdots$$

Even though this sum looks very similar to the previous sum, it turns out that this sum is finite. What does all this tell us? First, it says that the set of prime numbers is certainly infinite. This is a fact we will prove shortly using different ideas and techniques. Second, although it does not tell us whether the set of pairs of twin primes is infinite, it does say that in some sense the number of primes that are part of a pair of twin primes is much smaller than the number of primes. Admittedly, at this point, we are being very informal and imprecise in discussing the relative sizes of infinite sets. However, it is worthwhile to get a taste of some of the things that one can prove about prime numbers in a course in number theory.

We can now begin our examination of **prime factorization**. Let us consider the following examples:

$$6 = 2^1 \cdot 3^1, \quad 48 = 2^4 \cdot 3^1, \quad 125 = 5^3, \quad 363 = 3^1 \cdot 11^2, \quad 360 = 2^3 \cdot 3^2 \cdot 5^1.$$

Each of the numbers 6, 48, 125, 363, and 360 has been written as a product of prime numbers. Observe that the only other way we could write these numbers as a product of primes is to juggle the order of the primes. For example,

$$363 = 3^1 \cdot 11^2 = 11^1 \cdot 3^1 \cdot 11^1 = 11^2 \cdot 3^1,$$

are the *only* ways to write 363 as a product of primes. In light of this, we say that the factorization of 363 as $3^1 \cdot 11^2$ is *unique up to order*. The main result of this chapter will be that every integer greater than 1 can be written uniquely (up to order) as a product of primes. In order to prove this fact, we will need to develop two important pieces of mathematical machinery: the **Division Algorithm** and the **Euclidean Algorithm**.

Before stating our main result on prime factorization, two comments should be made. First, when we use the expression "product of primes," we allow products of length 1. By this we mean that the factorization of the number 5 into primes is simply as the single prime 5^1. In other words, whereas the factorization of 15 into primes is $15 = 3^1 \cdot 5^1$, the way we factor 5 into primes is $5 = 5^1$. Normally, we think of a product as requiring at least two numbers to be multiplied, but we will now consider a single number as a product of length 1.

Second, we can write the number 6 as a product in the following different ways:

$$6 = 2^1 \cdot 3^1 = 1^1 \cdot 2^1 \cdot 3^1 = 1^2 \cdot 2^1 \cdot 3^1 = 1^3 \cdot 2^1 \cdot 3^1 = 1^4 \cdot 2^1 \cdot 3^1.$$

Note that this says if we consider 1 to be a prime number, then in the preceding equation, we have provided five different ways to express 6 as a product of primes. Therefore, in order to have a reasonable concept of unique factorization, it is important that we do not consider 1 to be a prime number.

3.2 Unique Factorization

We can now state the main result of this chapter.

Theorem 3.3—Unique Factorization Theorem. *Every integer $n > 1$ can be written uniquely (up to order) as a product of primes.*

There are actually two parts to this theorem. First, we will need to show that every integer $n > 1$ *can* be written as a product of primes. That will not be a difficult task. The harder job will be to show that this factorization is unique (up to order).

Intuition. In practice, how do we factor numbers into primes? The basic idea is that if a number is prime, then it is already factored into primes as a product of length 1, but if it is not prime, then it can be written as a product of two smaller numbers. We then apply this same procedure to these two smaller numbers and continue repeating this procedure until only prime numbers remain.

Earlier we observed that $360 = 2^3 \cdot 3^2 \cdot 5^1$. We can now apply the preceding procedure to the number 360 to see how this factorization was obtained. At various points in this procedure, we will be expressing a number as a product of two smaller numbers. The way that you choose to do this is entirely up to you. For example, we could begin by saying $360 = 18 \cdot 20$ or $360 = 8 \cdot 45$ or $360 = 12 \cdot 30$. An important aspect of the *uniqueness* of prime factorization is that regardless of the choices you make along the way, you will always end up with the same factorization.

We begin by observing that 360 is not prime, since 10 is a divisor. Therefore

$$360 = 10 \cdot 36.$$

However, 10 is not prime, since 5 is a divisor. Therefore,

$$10 = 2 \cdot 5 \quad \text{and now} \quad 360 = 2 \cdot 5 \cdot 36.$$

But 36 is not prime, since 2 is a divisor. Therefore,

$$36 = 2 \cdot 18 \quad \text{and now} \quad 360 = 2 \cdot 5 \cdot 2 \cdot 18.$$

But 18 is not prime, since 6 is a divisor. Therefore,

$$18 = 6 \cdot 3 \quad \text{and now} \quad 360 = 2 \cdot 5 \cdot 2 \cdot 6 \cdot 3.$$

But 6 is not prime, since 3 is a divisor. Therefore,

$$6 = 3 \cdot 2 \quad \text{and now} \quad 360 = 2 \cdot 5 \cdot 2 \cdot 3 \cdot 2 \cdot 3.$$

Since 2, 3, and 5 are all primes, the factorization of the number 360 into primes is complete and, by juggling the order of the primes, we obtain

$$360 = 2^3 \cdot 3^2 \cdot 5^1.$$

You should try factoring 360 again, but this time, make different choices than the preceding ones. Then, at the end, juggle the order of your primes until you once again obtain the factorization $360 = 2^3 \cdot 3^2 \cdot 5^1$. As we reflect on this procedure, we can ask ourselves, how do we know when we are "done"? The answer is, when the only factors remaining are prime numbers. But how do we know that we ever reach a point where only prime numbers remain? This may sound very similar to the question we posed after Proposition 2.3, when we wondered if the procedure for reducing fractions ever came to an end. In light of this, it may come as no surprise to you that we will need to use the Well Ordering Principle in a proof by contradiction to rigorously show that this procedure does indeed always reach a point where only prime factors remain.

Proof of the first part of Theorem 3.3—the existence of prime factorization. We proceed with a proof by contradiction, so we will suppose that there exists an integer $n > 1$ that cannot be written as a product of primes. The Well Ordering Principle now guarantees that there is a smallest positive integer m such that $m > 1$ and m cannot be written as a product of primes. Let us now examine the nature of m. One possibility is that m is prime, but in this case we can factor m into primes, as $m = m^1$. In this case, we see that, simultaneously, m can and cannot be factored into primes. This is certainly a contradiction, so the case of m being a prime cannot occur.

The only remaining possibility is that m is not prime, so we can write m as a product of two smaller positive integers. As a result, there exist integers a, b such that

$$m = a \cdot b, \quad \text{where} \quad 1 < a < m \quad \text{and} \quad 1 < b < m.$$

Since m is the smallest integer that is bigger than 1 that cannot be written as a product of primes, we see that a and b can be written as a product of primes. Therefore, there exist prime numbers $p_1, p_2, \ldots, p_k, q_1, q_2, \ldots q_l$ such that

$$a = p_1 \cdot p_2 \cdots p_k \quad \text{and} \quad b = q_1 \cdot q_2 \cdots q_l.$$

Note that the list of primes $p_1, p_2, \ldots, p_k, q_1, q_2, \ldots q_l$ is allowed to have the same prime number occurring more than once.

Since $m = a \cdot b$, we now have

$$m = a \cdot b = p_1 \cdot p_2 \cdots p_k \cdot q_1 \cdot q_2 \cdots q_l.$$

However, the preceding equation illustrates that m can be written as a product of primes. This is a contradiction, since m, simultaneously, can and cannot be written as a product of primes. We have now shown that both of our cases, m being prime and m not being prime, lead to a contradiction. Therefore, there does not exist an integer $n > 1$ that cannot be written as a product of primes. We can now conclude that every integer $n > 1$ can indeed be written as a product of primes. $\square$

Earlier we asked if there is an infinite number of primes. We will now see that the answer is yes and that it follows from the existence of prime factorization.

Theorem 3.4. *There is an infinite number of prime numbers.*

Intuition. Suppose you are given a list of prime numbers like 2, 5, 13, 29. What happens if we multiply all of these numbers and then add 1? In this case, we obtain

$$3770 = 2 \cdot 5 \cdot 13 \cdot 29 \quad \text{and} \quad 3771 = 3770 + 1 = 2 \cdot 5 \cdot 13 \cdot 29 + 1.$$

Observe that any number that divides 3771 and 3770 must divide $3771 - 3770 = 1$. As a result, the only positive integer that divides both 3771 and 3770 is 1. In particular, there is no prime number that divides both 3771 and 3770. In light of this, none of the primes 2, 5, 13, 29 can divide 3771.

The existence of prime factorization tells us that there are prime numbers that divide 3771 and our previous argument tells us that none of those primes can be equal to 2, 5, 13, or 29. Therefore, any prime that divides 3771 is a prime that is not on our original list. What this argument really tells us is that given any finite list of prime numbers, there exists a prime number that is not on the list. It will not be difficult to formalize this argument to produce a proof of Theorem 3.4. Returning to our example,

$$3771 = 3^2 \cdot 419$$

and 3 and 419 are primes not on our original list 2, 5, 13, 29. $\qquad\square$

Proof. We will proceed with a proof by contradiction, so suppose there is only a finite number of prime numbers. Then there is a finite list $p_1, p_2, \ldots, p_m$ consisting of all the prime numbers. Next, let

$$n = p_1 \cdot p_2 \cdots p_m + 1.$$

Thus, n is 1 more than the product of all the prime numbers.

Any integer that divides both n and $p_1 \cdot p_2 \cdots p_m$ must also divide

$$n - p_1 \cdot p_2 \cdots p_m = 1.$$

As a result, the only positive integer that divides both n and $p_1 \cdot p_2 \cdots p_m$ is 1. In particular, there is no prime number that divides both n and $p_1 \cdot p_2 \cdots p_m$. However, the existence of prime factorization tells us that there is some prime q that divides n. Since $p_1, p_2, \ldots p_m$ is a complete list of prime numbers, q appears somewhere on the list, and that tells us that q is also a factor of $p_1 \cdot p_2 \cdots p_m$. This is a contradiction, since we already know that no prime can divide both n and $p_1 \cdot p_2 \cdots p_m$. $\qquad\square$

3.3 Division Algorithm

In order to be able to prove the uniqueness of prime factorization, we need to first turn our attention to **division**. For example, if we divide 64 by 7, we obtain

$$64 = 9 \cdot 7 + 1.$$

Recall that we call 9 the **quotient** and 1 the **remainder**. Similarly, if we divide 101 by 8, we obtain

$$101 = 12 \cdot 8 + 5,$$

so 12 is the quotient and 5 is the remainder.

We need to be a little more careful when dividing negative integers. Suppose we divide -64 by 7. We obtain

$$-64 = -9 \cdot 7 - 1 \quad \text{and} \quad -64 = -10 \cdot 7 + 6.$$

Looking at the preceding equations, your first instinct might be to say that -9 is the quotient and -1 is the remainder. On the other hand, perhaps you feel that -10 is the quotient and 6 is the remainder. Which of these answers do we consider to be the "correct" answer?

Recall that when we divided 64 by 7, we said that 9 was the quotient and 1 was the remainder. Suppose someone else said that since $64 = 6 \cdot 7 + 22$, we should then consider 6 to be the quotient and 22 to be the remainder. We would probably say that we do not consider this to be the correct answer, as we demand that the remainder when dividing by 7 must be less than 7 and also not be negative. For this reason, if someone else pointed out that

$$64 = 10 \cdot 7 - 6,$$

we would also not consider 10 and -6 to be the quotient and remainder, as the remainder must be less than 7 and also not be negative. In light of this, when we reexamine the problem of dividing -64 by 7, we consider -10 to be the quotient and 6 to be the remainder, since the remainder must be less than 7 and also not be negative.

In the preceding examples, it is the restrictions we place on the size of the remainder that allow us to obtain a unique quotient and remainder when dividing by positive integers.

Theorem 3.5—The Division Algorithm. *If a is a positive integer and n is any integer, then there exist* unique *integers q and r with the properties that*

$$n = q \cdot a + r \quad and \quad 0 \le r < a.$$

We call q the quotient and r the remainder.

Intuition.   To see what happens when we perform division, let us examine the special cases where we divide the numbers 64 and -64 by 7. When we divide 64 by 7, we can begin subtracting copies of 7 away from 64 to obtain the list

$$64, 57, 50, 43, 36, 29, 22, 15, 8, 1, -6, -13, -20, \ldots$$

The smallest positive number on this list is 1, and it was obtained by subtracting 9 copies of 7 from 64. In other words, this tells us that $64 = 9 \cdot 7 + 1$.

On the other hand, when we divide -64 by 7, we can begin adding copies of 7 to -64 to obtain the list

$$-64, -57, -50, -43, -36, -29, -22, -15, -8, -1, 6, 13, 20, \ldots$$

The smallest positive number on this list is 6, and it was obtained by adding 10 copies of 7 to -64. This tells us that $-64 = -10 \cdot 7 + 6$.

In both of the preceding examples, we either added or subtracted copies of 7 and found the smallest positive integer that occurs in this way. That is how we found the remainder. The quotient then refers to how many copies of 7 we needed to add or subtract to reach the remainder. At this point, it should come as no surprise that when writing the proof of the Division Algorithm, it will be the Well Ordering Principle that will guarantee that there is always a smallest positive integer that can be obtained by either adding or subtracting copies of the number we are dividing by.

Proof. We will first show that given a positive integer a and an integer n, there exist integers q and r with the properties that

$$n = q \cdot a + r \quad \text{and} \quad 0 \le r < a.$$

We will handle the question of the uniqueness of the quotient and remainder immediately thereafter.

The first possibility is that a is a divisor of n. Therefore, in this case, there exists an integer q such that $n = q \cdot a$. As a result, q is our quotient and 0 is our remainder. Note that this will be the only case where the remainder can be 0.

We are now in the case where a is not a divisor of n, and we will let

$$B = \{n - t \cdot a \mid t \in \mathbb{Z}\}.$$

In other words, B is the collection of all numbers that can be obtained from n by adding and subtracting copies of a. Observe that regardless of whether n is positive or negative, the set B contains some positive integers. The Well Ordering Principle guarantees that there exists a smallest positive integer in B, and we will denote this integer as r. In addition, since r belongs to the set B, we know that there exists some integer q such that

$$r = n - q \cdot a.$$

We claim that $1 \le r < a$. To see this, observe that

$$(1) \qquad\qquad r - a = n - (q + 1) \cdot a.$$

Note that if $r - a = 0$, then $n = (q + 1) \cdot a$, which says that a is a divisor of n. Since we are in the case where a is not a divisor of n, that would be a contradiction. As a result, $r - a \neq 0$, which says that $r \neq a$.

Now suppose that $r > a$; this says that $r - a > 0$. But this would say that $r - a$ is a positive integer that is smaller than r and, using (1), is also in the set B. This is also a contradiction. Since we have shown that both of the cases $r = a$ and $r > a$ cannot occur, it must be the case that $r < a$. However, if we combine this with the fact that we already knew r was positive, we have indeed shown that $1 \le r < a$. Therefore, in this case, we see that q and r are integers with the properties that

$$n = q \cdot a + r \quad \text{and} \quad 1 \le r < a.$$

As a result, we have shown that, in all cases, there exist a quotient q and a remainder r with the desired properties.

To complete the proof, we need to show that the quotient q and remainder r are unique. This means that no matter how many times you divide n by a, you will *always* obtain the same quotient and remainder, provided that we insist that the remainder be less than a and not be negative.

Suppose, on two different occasions, we divide n by a and obtain

$$n = q_1 \cdot a + r_1 \quad \text{and} \quad n = q_2 \cdot a + r_2,$$

where r_1 and r_2 are both less than a and not negative. To prove that the quotient and remainder are unique, we need to show that $q_1 = q_2$ and $r_1 = r_2$.

Since

$$q_1 \cdot a + r_1 = n = q_2 \cdot a + r_2,$$

if we subtract both $q_2 \cdot a$ and r_1 from the previous equation, we obtain

$$q_1 \cdot a - q_2 \cdot a = r_2 - r_1.$$

Therefore,

$$(2) \qquad\qquad (q_1 - q_2) \cdot a = r_2 - r_1.$$

Since r_1 and r_2 are both less than a and not negative, it follows that

$$-a < r_2 - r_1 < a.$$

In particular, the absolute value of $r_2 - r_1$ must be less than a. On the other hand, the absolute value of any nonzero multiple of a must be greater than or equal to a. Therefore, using (2), we see that $(q_1 - q_2) \cdot a$ is a multiple of a whose absolute value is less than a. As a result,

$$(3) \qquad\qquad (q_1 - q_2) \cdot a = 0.$$

Since $a \neq 0$, it follows immediately from (3) that $q_1 - q_2 = 0$. Thus, $q_1 = q_2$. Returning to (2), since $q_1 = q_2$, we see that $r_2 - r_1 = 0$. Therefore, $r_1 = r_2$. □

Exercises for Sections 3.1, 3.2, and 3.3

For the exercises from these sections, we need to briefly introduce two pieces of terminology and notation.

(a) Any positive integer n such that $n \geq 2$ and n is *not* prime is called **composite**. As a result, every integer $n \geq 2$ is either prime or composite but not both. The list of composites begins

$$4, 6, 8, 9, 10, 12, 14, 15, 16, 18, 20, 21, 22, 24, 25, 26, 27, 28, \ldots$$

(b) If $n \in \mathbb{N}$, we let $n! = (1) \cdot (2) \cdot (3) \cdots (n-1) \cdot (n)$, and we refer to $n!$ as n factorial. Observe that if $n \in \mathbb{N}$, then $(n+1)! = (n+1) \cdot n!$. As a result, we have

$$1! = 1, \quad 2! = 2, \quad 3! = 6, \quad 4! = 24, \quad 5! = 120, \quad 6! = 720, \quad 7! = 5040, \ldots$$

For convenience, we also let $0! = 1$.

1. Immediately after Definition 3.2, we listed all the primes less than 44. Continue the list so it includes all prime numbers less than 100.

2. Use your list from exercise 1 to show that every even integer n, where $4 \leq n \leq 40$, can be written as a sum of two prime numbers. By doing so, you will have verified all cases of Goldbach's Conjecture up to 40.

3. Use your list from exercise 1 to find all pairs of twin primes where the primes are less than 100.

4. (a) If n is composite, show that there exists a prime number p that divides n and is less than or equal to $\sqrt{n}$.

    (b) Use part (a) to show that in order to check whether a positive integer $m \geq 2$ is prime, you only need to confirm that it fails to be divisible by all primes less than or equal to $\sqrt{m}$.

(c)   In light of part (b), how would you go about determining whether the numbers 323, 353, and 371 are prime?

(d)   Which of 323, 353, and 371 are prime? How do you know?

5.   Find the prime factorization of each of the following integers: 480, 850, 7623.

6.   Find the prime factorization of each of the following integers: 378, 2205, 7007.

7.   Shortly after the statement of Theorem 3.3, we found the prime factorization of 360 by starting with the observation $360 = 10 \cdot 36$. Show that if we begin with the observation $360 = 18 \cdot 20$, then, up to order, we obtain the same prime factorization of 360.

8.   As in exercise 7, show that the observations $360 = 8 \cdot 45$ and $360 = 12 \cdot 30$ both lead to the same factorization of 360, up to order, that we found after the statement of Theorem 3.3.

9.   (a)   What is the prime factorization of 13!?

(b)   What is the prime factorization of 27!?

10.   (a)   What is the prime factorization of 20!?

(b)   What is the prime factorization of 41!?

11.   In this exercise, you may want to refer to your answers in exercise 9.
     (a)   When computing the exact value of 13!, how many zeroes appear at the far right?

(b)   When computing the exact value of 27!, how many zeroes appear at the far right?

12.   (a)   When computing the exact value of 1000!, how many zeroes appear at the far right?

(b)   When computing the exact value of 200,000!, how many zeroes appear at the far right?

(c)   When computing the exact value of 5,000,000!, how many zeroes appear at the far right?

13.   (a)   Find the smallest positive integer n such that the exact value of $n!$ has exactly 1000 zeroes appearing at the far right. (You will probably need some trial and error to solve this.)

(b)   Use your answer to part (a) to find all positive integers t such that the exact value of $t!$ has exactly 1000 zeroes appearing at the far right.

14.   Write a proof of the first part of Theorem 3.3 that uses the Second Version of Mathematical Induction and does not use a proof by contradiction.

15. Suppose today is Monday.

    (a) What day of the week will it be 100 days from now?

    (b) What day of the week will it be 10,000 days from now?

    (c) What day of the week will it be 1,000,000 days from now?

16. An oval track has a circumference of 440 yards. Therefore, after running 440 yards, you are back at the starting point.

    (a) How far from the starting point are you after running 1395 yards?

    (b) How far from the starting point are you after running 4963 yards?

    (c) After running 3440 yards, how far do you need to run to get back to the starting point?

17. Let $a = 2 \cdot 3 \cdot 5 \cdot 7 + 11$, $b = 2 \cdot 3 \cdot 5 + 7 \cdot 11$, and $c = 3 \cdot 5 \cdot 11 + 2 \cdot 7$. Without computing the exact values of a, b, c, explain why none of these three numbers is divisible by a prime less than 13.

18. Compute the numerical values of a, b, c from exercise 17 and find the prime factorization of all three numbers.

19. (a) Use your list from exercise 1 to find all examples of three consecutive odd integers that are less than 100 and are all prime numbers.

    (b) Show that if m is an integer then $m, m + 2, m + 4$ all have different remainders when divided by 3.

    (c) Use part (b) to show that if n is a positive integer, then exactly one of $n, n + 2, n + 4$ is divisible by 3.

    (d) Use part (c) to show that the example you found in part (a) is the *only* time that three consecutive odd integers are all prime.

20. (a) Show that the sum of any three consecutive integers must be divisible by 3.

    (b) Show that the sum of any five consecutive integers must be divisible by 5.

21. Generalize your results from exercise 20 to show that if n is a positive odd integer, then the sum of any n consecutive integers must be divisible by n.

22. (a) Explain why the following ten integers $11! + 2, 11! + 3, 11! + 4, \ldots, 11! + 10$, $11! + 11$ are all composite.

    (b) Explain why the following 50 integers $51! + 2, 51! + 3, 51! + 4, \ldots, 51! + 50$, $51! + 51$ are all composite.

(c) Show that there exist 1,000,000 consecutive positive integers that are all composite.

23. Generalize your results from exercise 22, and show that for any positive integer n, there exist n consecutive positive integers that are all composite.

For exercises 24–29, please first read the following:

In exercises 22–23, we showed that there can exist long gaps between prime numbers. This is quite interesting when contrasted with the fact that there exists an infinite number of primes. In the next six exercises, we will see some additional proofs that there is an infinite number of primes and will also prove that various subsets of $\mathbb{N}$ contain an infinite number of primes.

24. (a) If $n \in \mathbb{N}$, show that any prime number that divides $n! + 1$ must be larger than n.

(b) Use part (a) to show that for any integer n, there exists a prime number p such that $p > n$.

(c) Explain how parts (a) and (b) can be used to give another proof that there is an infinite number of primes.

25. Explain how your work in exercise 24 actually provides a proof of the following statement: For any $n \in \mathbb{N}$, there exists a prime number p such that $n < p \leq n! + 1$.

26. (a) Show that if n is an odd integer, then there exists an integer k such that either $n = 4k + 1$ or $n = 4k + 3$.

(b) Use part (a) to show that if n is an odd integer, then there is an integer l such that $n^2 = 4l + 1$.

27. (a) Let $k_1, k_2, \ldots, k_m$ be integers, and then let

$$t = (4k_1 + 1)(4k_2 + 1) \cdots (4k_m + 1).$$

Show that there exists an integer s such that $t = 4s + 1$.

(b) Suppose n is an integer of the form $n = 4k + 3$, for some integer k. Use part (c) to show there is a prime p that divides n such that $p = 4w + 3$, for some integer w.

28. We know that there is an infinite number of odd prime numbers and that in light of exercise 26(a), each odd prime number is either of the form $4k + 1$ or $4k + 3$, for some integer k. However, at this point, we do not know how many of these odd primes are of the form $4k + 1$ and how many are of the form $4k + 3$. In this exercise, we will apply various portions of exercises 26 and 27 to show that there is indeed an infinite number of primes of the form $4k + 3$. To this end, let $v \geq 3$ be an integer, and let $p_1, p_2, \ldots, p_m$

be a complete list of the odd primes that are less than or equal to v and then let
$n = p_1 \cdot p_2 \cdots p_m$.

(a) Show that there is an integer l such that $n^2 + 2 = 4l + 3$.

(b) By looking at those primes that divide $n^2 + 2$, show that there exists a prime p such that $p > v$ and $p = 4s + 3$, for some integer s.

(c) Use part (b) to show that there is an infinite number of primes of the form $4k + 3$.

29. You can easily use the division algorithm to check that every odd prime number other than 3 is either of the form $6k + 1$ or $6k + 5$, for some integer k. However, at this point, we do not know how many of the infinite number of odd primes are of the form $6k + 1$ and how many are of the form $6k + 5$. In this exercise, we will show that there is an infinite number of primes of the form $6k + 5$. With some modifications, we will follow the plan set out in exercise 28. Let v be a positive integer and let $p_1, p_2, \ldots, p_m$ be all the odd prime numbers that are more than 3 and less than or equal to v. Then let
$n = p_1 \cdot p_2 \cdots p_m$.

(a) Show that there is an integer l such that $n^2 + 4 = 6l + 5$.

(b) By looking at those primes that divide $n^2 + 4$, show that there exists a prime p such that $p > v$ and $p = 6s + 5$, for some integer s.

(c) Use part (b) to show that there is an infinite number of primes of the form $6k + 5$.

30. For this exercise, we need to recall two facts on infinite series. The first is that $\sum_{n=1}^{\infty} \frac{1}{n}$ is infinite. The second is that if $m > 1$, then $\sum_{i=0}^{\infty} \frac{1}{m^i} = \frac{m}{m-1}$.

(a) If $\{p_1, p_2, \ldots, p_t\}$ is a finite set of prime numbers and if

$$B = \left(\sum_{i=0}^{\infty} \frac{1}{p_1^i} \right) \cdot \left(\sum_{i=0}^{\infty} \frac{1}{p_2^i} \right) \cdots \left(\sum_{i=0}^{\infty} \frac{1}{p_t^i} \right),$$

show that B must be finite.

(b) Show that B is equal to the sum of the reciprocals of all positive integers whose prime divisors are a subset of $\{p_1, p_2, \ldots, p_t\}$.

(c) Show that part (b) can be used to provide another proof of the fact that the number of primes is infinite.

For exercises 31–37, please read the following:

When trying to understand a new mathematical concept or property, it is desirable to look at some examples that have this property as well as some examples that do not. For example, when learning about continuity, things make much more sense when we contrast graphs of continuous functions with graphs of functions that are not continuous. Similarly, to understand limits, we should contrast points where limits exist with points where they do not exist.

Students sometimes have trouble understanding unique factorization, or the need to prove the uniqueness of prime factorization, because such a contrast does not seem to exist. We cannot contrast examples of integers greater than 1 that factor uniquely into primes with integers that do not, as there are no integers greater than 1 that do not factor uniquely. It is difficult to internalize a concept or property without also looking at objects that do not possess this property. Therefore, students sometimes have difficulty imagining how something could fail to factor uniquely. Motivated by this problem, we present an example of a set that is very similar to the set of integers greater than 1, where unique factorization does not hold.

We let $\mathbb{E}$ denote the set of positive *even* integers. Thus,

$$\mathbb{E} = \{2, 4, 6, 8, 10, 12, \dots\}.$$

We say that an element $n \in \mathbb{E}$ is $\mathbb{E}$-**prime** is n cannot be written as the product of smaller elements of $\mathbb{E}$. The $\mathbb{E}$-prime elements of $\mathbb{E}$ will play a role in $\mathbb{E}$ analogous to the role played by ordinary primes in $\mathbb{N}$. In particular, in the following exercises, we will show that every element of $\mathbb{E}$ can be written as a product of $\mathbb{E}$-primes. However, we will also show that factorization into $\mathbb{E}$-primes is *not* unique. Hopefully, this example will help put the concept of unique factorization into primes in $\mathbb{N}$ into better perspective.

31. Show that every element of $\mathbb{E}$ can be written as a product of $\mathbb{E}$-primes. (Remember, when necessary, we allow products of length 1.)

32. Show that $2, 6, 18$ are all $\mathbb{E}$-primes.

33. Show that $4, 8, 20$ are not $\mathbb{E}$-primes.

34. Generalize exercises 32 and 33 and show that the $\mathbb{E}$-primes are precisely those positive even integers that are *not* divisible by 4 in $\mathbb{N}$.

35. Show that 36 can be factored more than one way, up to order, into $\mathbb{E}$-primes in $\mathbb{E}$.

36. Show that 60 can be factored more than one way, up to order, into $\mathbb{E}$-primes in $\mathbb{E}$.

37. Determine a simple criteria for when elements of $\mathbb{E}$ factor uniquely into $\mathbb{E}$-primes in $\mathbb{E}$.

3.4 Greatest Common Divisors

The next tool we will need to prove the uniqueness of prime factorization is **greatest common divisors**.

Definition 3.6. *If a and b are nonzero integers, then we let gcd(a, b) denote the* greatest common divisor *of a and b, which is the largest positive integer that is a divisor of both a and b. More generally, if $a_1, a_2, \ldots, a_n$ are integers, with at least one of them nonzero, then $gcd(a_1, a_2, \ldots, a_n)$ is the largest positive integer that is a divisor of every one of the a_i's.*

■ Examples

$gcd(36, 10) = 2$, $gcd(10, 36) = 2$, $gcd(36, -10) = 2$,
$gcd(27, 16) = 1$, $gcd(1000, 144) = 8$, $gcd(42, 57, 30) = 3$, $gcd(6, 10, 15) = 1$

■

In light of the preceding definition, there are several questions you may have:

(a) Given two nonzero integers, is there any guarantee that they actually have a greatest common divisor?

(b) How do we compute greatest common divisors?

(c) How does this concept relate to the uniqueness of prime factorization?

To illustrate what is happening, let us examine $gcd(3476, -948)$. The first observation we can make is that the numbers -948 and 948 have precisely the same divisors. As a result, $gcd(3476, -948) = gcd(3476, 948)$. Therefore, when computing greatest common divisors, we can replace each integer by its absolute value without changing the final answer.

The number 1 is certainly a divisor of both 3476 and 948, so 3476 and 948 have a common divisor. However, do they have a *greatest* common divisor? Observe that 948 is the largest divisor of 948. Therefore, no common divisor of 3476 and 948 can possibly exceed 948. Now consider the finite set $\{1, 2, 3, \ldots, 947, 948\}$. Every positive divisor of 948 belongs to this set. One by one, we can check the 948 elements of this set to see if any of them, other than 1, are divisors of both 3476 and 948. The largest number in $\{1, 2, 3, \ldots, 947, 948\}$ that divides both 3476 and 948 must certainly be the greatest common divisor of 3476 and 948. It turns out that 316 is the largest divisor of both 3476 and 948, so $gcd(3476, -948) = gcd(3476, 948) = 316$.

There was nothing special about the numbers 3476 and 948. The same reasoning just used shows that any two nonzero integers have a greatest common divisor. In fact, our reasoning provides an algorithm for finding the greatest common divisor. Given two positive integers a and b, we simply check every integer from 1 up to the smaller of a and b to see which of them are divisors of both a and b. The largest number we check that turns out to be a divisor of both a and b is indeed the greatest common divisor of a and b. Although this algorithm is particularly tedious and inefficient, it does tell us that greatest common divisors exist and that there is at least one straightforward way to compute them.

If you were introduced to greatest common divisors in elementary or high school, you may have seen a different algorithm for computing them that uses prime factorization. Suppose you are given two positive integers that are already factored into primes. For example, consider

$$106480 = 2^4 \cdot 5^1 \cdot 11^3 \quad \text{and} \quad 6776 = 2^3 \cdot 7^1 \cdot 11^2.$$

The prime numbers that appear in the prime factorization of both 106480 and 6776 are the only possible primes that can appear in the factorization of $gcd(106480, 6776)$. However, to find the exponent of a prime in the factorization of $gcd(106480, 6776)$, we compare its exponents in the factorization of both 106480 and 6776 and choose the *smaller* one. Therefore, the only primes that appear in the factorization of $gcd(106480, 6776)$ are 2 and 11. Furthermore, the smallest exponent of 2 that appears in the factorization of 106480 and 6776 is 3, and the smallest exponent of 11 that appears in the factorization of 106480 and 6776 is 2. As a result, $gcd(106480, 6776) = 2^3 \cdot 11^2 = 968$.

At first glance, the preceding procedure appears to be a fairly quick and easy way to compute greatest common divisors. However, for both theoretical and practical reasons, the preceding procedure is not the best for computing greatest common divisors. When we used it to compute $gcd(106480, 6776)$, we used the prime factorization of 106480 and 6776. However, at this point, we haven't yet proven that factorization into primes is unique. If it were somehow possible to factor 106480 and 6776 in different ways, that could lead to different answers when computing $gcd(106480, 6776)$. Therefore, until we prove the uniqueness of prime factorization, it is not really valid to use this procedure to compute greatest common divisors.

It turns out that if you are given a large number, it is extremely difficult to write it as a product of primes. If you are given a 50-digit number, although you know that it *can* be written as a product of primes, it may take many hours of computing time to figure out what its prime factorization looks like. Therefore, as a practical matter, using prime factorization to find greatest common divisors is a very slow and inefficient algorithm. At this point, we can introduce an algorithm for computing greatest common divisors that is not only much more efficient (when dealing with large numbers) but also has the added benefit that we will be able to use it to prove the uniqueness of prime factorization. Let us begin by making some observations.

$$gcd(36, 10) = 2 \quad \text{and} \quad 2 = 2 \cdot 36 + (-7) \cdot 10,$$

$$gcd(15, 28) = 1 \quad \text{and} \quad 1 = (-13) \cdot 15 + 7 \cdot 28,$$

$$gcd(1000, 144) = 8 \quad \text{and} \quad 8 = (-1) \cdot 1000 + 7 \cdot 144,$$

$$gcd(-36, 10) = 2 \quad \text{and} \quad 2 = (-2) \cdot (-36) + (-7) \cdot 10.$$

Let's think about what this means. First we observed that 2 was the greatest common divisor of 36 and 10, and we found integers m and n such that

$$2 = m \cdot 36 + n \cdot 10.$$

Next we observed that 1 was the greatest common divisor of 15 and 28, and we found integers m and n such that

$$1 = m \cdot 15 + n \cdot 28.$$

Note that we also did the same for $gcd(1000, 144)$ and $gcd(-36, 10)$.

To formalize what we just did, given integers a and b, we say that an *integral combination* of a and b is any number that can be written in the form

$$m \cdot a + n \cdot b,$$

where m and n are also integers. Therefore, we see that $gcd(36, 10)$ can be written as an integral combination of 36 and 10. Similarly, $gcd(15, 28)$ can be written as an integral combination of 15 and 28. It will be very useful for us to show that $gcd(a, b)$ can *always* be written as an integral combination of a and b. To do this we will need a short lemma.

Lemma 3.7. *Let a, b, m, n and c be integers such that c is a divisor of a and b. Then c is also a divisor of $m \cdot a + n \cdot b$.*

Proof. Since c is a divisor of both a and b, there exist integers s and t such that $a = s \cdot c$ and $b = t \cdot c$. We now have

$$m \cdot a + n \cdot b = m \cdot (s \cdot c) + n \cdot (t \cdot c) = (m \cdot s) \cdot c + (n \cdot t) \cdot c = (m \cdot s + n \cdot t) \cdot c.$$

As a result, c is a divisor of $m \cdot a + n \cdot b$. $\square$

3.5 Euclidean Algorithm

A special case of Lemma 3.7, which we will refer to quite often, is that if a, b, q, and r are integers such that $b = q \cdot a + r$, and if c is a divisor of both a and r, then c is also a divisor of b. We can now introduce the **Euclidean Algorithm** for finding greatest common divisors. It will be best to begin with some examples.

■ Example

Let's try to compute $gcd(81, 24)$. If we apply the division algorithm to 81 and 24, we obtain

$$81 = 3 \cdot 24 + 9.$$

Now, apply the division algorithm to 24 and 9 to obtain

$$24 = 2 \cdot 9 + 6.$$

Next, apply the division algorithm to 9 and 6 to obtain

$$9 = 1 \cdot 6 + 3.$$

Finally, apply the division algorithm to 6 and 3 to obtain

$$6 = 2 \cdot 3 + 0.$$

∎

We claim that this collection of calculations can be used to show that $gcd(81, 24) = 3$ and that 3 can be written as an integral combination of 81 and 24. First, observe that our last equation tells us that 3 is a divisor of 6. Moving up by one on our list of equations, since 3 is a divisor of 6 and 3, Lemma 3.7 tells us that 3 is also a divisor of 9. Again moving up by one on our list of equations, since 3 is a divisor of 9 and 6, Lemma 3.7 now tells us that 3 is a divisor of 24. Finally, moving to the top equation, since 3 is a divisor of 24 and 9, Lemma 3.7 tells us that 3 is a divisor of 81. Therefore, 3 is certainly a common divisor of 81 and 24, but we haven't yet shown that it is the greatest common divisor.

Let us go back to our next to last equation and rewrite it as

$$3 = 9 - 1 \cdot 6.$$

Moving up an equation, we can replace 6 by $24 - 2 \cdot 9$ to obtain

$$3 = 9 - 1 \cdot 6 = 9 - 1 \cdot (24 - 2 \cdot 9) = -1 \cdot 24 + 3 \cdot 9.$$

Again moving up an equation, we can replace 9 by $81 - 3 \cdot 24$ to obtain

$$3 = -1 \cdot 24 + 3 \cdot 9 = -1 \cdot 24 + 3 \cdot (81 - 3 \cdot 24) = 3 \cdot 81 + (-10) \cdot 24.$$

As a result, we have expressed 3 as an integral combination of 81 and 24. Now suppose that d is a positive number that is a divisor of both 81 and 24. Then, Lemma 3.7 tells us that d must also be a divisor of any integral combination of 81 and 24. In particular, d must be a divisor of

$$3 = 3 \cdot 81 + (-10) \cdot 24.$$

Since d is a divisor of 3, d cannot be greater than 3. Therefore, 3 is greater than or equal to any divisor of both 81 and 24. Hence, 3 is indeed the greatest common divisor of 81 and 24.

Let us work through another example before formalizing things.

■ Example

To compute $gcd(166, 75)$, we begin by dividing 166 by 75 to obtain

$$166 = 2 \cdot 75 + 16.$$

Now, divide 75 by 16 to obtain

$$75 = 4 \cdot 16 + 11.$$

Next, divide 16 by 11 to obtain

$$16 = 1 \cdot 11 + 5.$$

Next, divide 11 by 5 to obtain

$$11 = 2 \cdot 5 + 1.$$

Finally, divide 5 by 1 to obtain

$$5 = 5 \cdot 1 + 0.$$

■

We can now crawl upward through our list of equations to show that 1 can be written as an integral combination of 166 and 75. The next to last equation can be rewritten as

$$1 = 11 - 2 \cdot 5.$$

Moving up an equation and replacing 5 by $16 - 1 \cdot 11$, we obtain

$$1 = 11 - 2 \cdot 5 = 11 - 2(16 - 1 \cdot 11) = (-2) \cdot 16 + 3 \cdot 11.$$

Moving up an equation, we can replace 11 by $75 - 4 \cdot 16$ to obtain

$$1 = (-2) \cdot 16 + 3 \cdot 11 = (-2) \cdot 16 + 3 \cdot (75 - 4 \cdot 16) = 3 \cdot 75 + (-14) \cdot 16.$$

Finally, we move to the top equation and replace 16 by $166 - 2 \cdot 75$ to obtain

$$1 = 3 \cdot 75 + (-14) \cdot 16 = 3 \cdot 75 + (-14) \cdot (166 - 2 \cdot 75) = (-14) \cdot 166 + 31 \cdot 75.$$

Having expressed 1 as an integral combination of 166 and 75, we see that any common divisor of 166 and 75 must also be a divisor of 1. Since 1 is certainly a common divisor of 166 and 75, we see that $1 = gcd(166, 75)$.

In the preceding examples, we have used the Euclidean Algorithm to express $gcd(a, b)$ as an integral combination of a and b. In our examples, a and b were both positive. We will now see what to do if a or b is negative.

■ Example

Let's look at $gcd(-166, 75)$, $gcd(166, -75)$, and $gcd(-166, -75)$. Using the fact that

$$1 = gcd(166, 75) = (-14) \cdot 166 + 31 \cdot 75,$$

we can multiply in the appropriate places by -1 to obtain

$$1 = gcd(-166, 75) = 14 \cdot (-166) + 31 \cdot 75,$$
$$1 = gcd(166, -75) = (-14) \cdot 166 + (-31) \cdot (-75),$$
$$1 = gcd(-166, -75) = 14 \cdot (-166) + (-31) \cdot (-75).$$

■

In light of these observations, we can write $gcd(a, b)$ as an integral combination of a and b regardless of whether a and b are positive or negative. We can now formalize all of our observations.

Theorem 3.8. *If a and b are nonzero integers, then $gcd(a, b)$ is the* smallest *positive integer that can be written as an integral combination of a and b.*

Proof. The proof consists of describing and analyzing the Euclidean Algorithm in a more formal way. If at any point you find this formal argument hard to follow, you should go back and compare it to the examples we worked through earlier. By the preceding observations, we may assume that a and b are both positive. We begin by applying the division algorithm to a and b to obtain integers q_1 and r_1 such that

$$b = q_1 \cdot a + r_1 \quad \text{and} \quad 0 \le r_1 < a.$$

Next, divide a by r_1 to obtain integers q_2, r_2 such that

$$a = q_2 \cdot r_1 + r_2 \quad \text{and} \quad 0 \le r_2 < r_1.$$

Then divide r_1 by r_2 to obtain integers q_3, r_3 such that

$$r_1 = q_3 \cdot r_2 + r_3 \quad \text{and} \quad 0 \le r_3 < r_2.$$

Observe that every time we apply the division algorithm, we obtain a smaller remainder than in the previous step. In particular, $a > r_1 > r_2 > r_3$. Therefore, if we continue this process of dividing remainder r_i by the next remainder r_{i+1}, we will eventually obtain a remainder of 0. Let us now suppose that n is the positive integer such that r_n is the *last* remainder that is *not* zero. This says that if we continue to apply this procedure, we will eventually obtain the equations

$$r_{n-2} = q_n \cdot r_{n-1} + r_n \quad \text{and} \quad r_{n-1} = q_{n+1} \cdot r_n + 0,$$

where

$$a > r_1 > r_2 > r_3 > \cdots > r_{n-2} > r_{n-1} > r_n > 0.$$

The last equation tells us that r_n is a divisor r_{n-1}. Applying Lemma 3.7 to the next to last equation tells us that r_n is also a divisor of r_{n-2}. We can continue to move upward through our list of equations, and if we apply Lemma 3.7 at every step, we see that

$$r_n \mid r_{n-1}, \quad r_n \mid r_{n-2}, \quad r_n \mid r_{n-3}, \ldots, \quad r_n \mid r_2, \quad r_n \mid r_1, \quad r_n \mid a, \quad r_n \mid b.$$

Therefore, r_n is a common divisor of a and b.

The next to last equation shows that r_n can be written as an integral combination of r_{n-2} and r_{n-1}. Moving up to the next equation enables us to replace r_{n-1} by an integral combination of r_{n-3} and r_{n-2}, which shows that r_n can be written as an integral combination of r_{n-3} and r_{n-2}. Continuing in this way, we see that eventually r_n can be written as an integral combination of a and r_1, and, finally, r_n can be written as an integral combination of b and a.

If d is any common divisor of a and b, then Lemma 3.7 asserts that d is a divisor of any integral combination of a and b. Therefore, d is also a divisor of r_n, so r_n must be greater than or equal to d. Since we have shown that r_n is both a common divisor of a and b and is also greater than or equal to any common divisor of a and b, we see that $r_n = gcd(a, b)$.

Having succeeded in showing that $gcd(a, b)$ can be written as an integral combination of a and b, we now need to show that $gcd(a, b)$ is the smallest positive integer that can be written this way. To this end, suppose f is a positive integer that can be also written as an integral combination of a and b. We need to show that $gcd(a, b)$ is less than or equal to f.

Since $gcd(a, b)$ is a common divisor of a and b, Lemma 3.7 implies that $gcd(a, b)$ is also a divisor of f. However, since $gcd(a, b)$ is a divisor of f, we see that $gcd(a, b)$ is less than or equal to f. Thus, $gcd(a, b)$ is indeed the smallest positive integer that can be written as an integral combination of a and b.   □

The Euclidean Algorithm is a fast and easy algorithm for computing greatest common divisors. It can easily be programmed into a graphing calculator and is far faster than algorithms that involve finding the prime factorization of a number. Not only will we use it to prove the uniqueness of prime factorization, but in Chapter 12 we will use a modified version of it to better understand polynomials.

If a and b are integers such that $gcd(a, b) = 1$, we say that a and b are **relatively prime**. Two easy, but important, observations are

(a)   If a, b are nonzero integers, then a and b are relatively prime if and only if there is no prime number that is a factor of both a and b.

(b)   If p and q are prime numbers, then either they are equal or they are relatively prime.

The following somewhat technical lemma will turn out to be extremely useful in proving the uniqueness of prime factorization.

Lemma 3.9.   *If a, b, n are nonzero integers, where $n \mid (a \cdot b)$ and $gcd(a, n) = 1$, then $n \mid b$.*

Intuition.   Consider the case where $n = 12$, $a = 8$, $b = 18$. In this case, $12 \nmid 8$ and $12 \nmid 18$, yet $12 \mid (8 \cdot 18)$. Therefore, in general, it is possible for an integer n to divide the product $a \cdot b$ even though it divides neither a nor b. However, the point of Lemma 3.9 is that if n divides the product $a \cdot b$ and if n is relatively prime to a, then n must divide b. The key to the proof will be that we can write 1 as an integral combination of a and n. Note that in our example, $gcd(a, n) = gcd(8, 12) = 4$ and $gcd(b, n) = gcd(18, 12) = 6$. Thus, n is relatively prime to neither a nor to b.

Proof.   Since a and n are relatively prime, we can write 1 as an integral combination of a and n. Therefore, there exist integers r and s such that

$$1 = r \cdot a + s \cdot n.$$

Multiplying this equation by b results in

$$b = b \cdot (r \cdot a) + b \cdot (s \cdot n) = r \cdot (a \cdot b) + (b \cdot s) \cdot n.$$

Therefore, b is an integral combination of $a \cdot b$ and n. Since n is a divisor of both $a \cdot b$ and n, Lemma 3.7 implies that $n \mid b$. $\square$

We can now record an important consequence of Lemma 3.9.

Corollary 3.10.   *Let $p, q_1, q_2, q_3, \ldots, q_n$ be prime numbers (which are not necessarily distinct). If $p \mid (q_1 \cdot q_2 \cdot q_3 \cdots q_n)$, then p is equal to one of the q_i's.*

Intuition.   Let us consider the special case where q_1, q_2 are prime numbers and the prime 5 divides the product $q_1 \cdot q_2$. We want to show that either $q_1 = 5$ or $q_2 = 5$. There are two possibilities, either $q_2 = 5$ or $q_2 \neq 5$. In the first case, we are done. In the second case, 5 and q_2

are relatively prime. But then Lemma 3.9 implies that 5 divides q_1. However, since q_1 is prime, this immediately tells us that $q_1 = 5$.

In the preceding argument, there was nothing special about the prime 5. So the argument really tells us that if a prime p divides a product $q_1 \cdot q_2$ of primes, then p is equal to q_1 or q_2. The general result works for a product of n primes, not just two primes, and will require Mathematical Induction. It turns out that this result is a relatively easy consequence of Lemma 3.9 but would be quite difficult to prove without Lemma 3.9. This explains the importance of Lemma 3.9 in the proof of the uniqueness of prime factorization.

Proof. We let T be the set of positive integers n such that whenever a prime p divides the product $q_1 \cdot q_2 \cdot q_3 \cdots q_n$ of n primes, then p is equal to one of the q_i's. We need to show that $T = \mathbb{N}$, and we will proceed by using Mathematical Induction. First we need to show that T contains 1. So let us consider the case where p and q_1 are primes such that $p | q_1$. Note that in this case we are considering q_1 to be a product of primes of length 1. Since $p | q_1$, $p > 1$, and q_1 is prime, it immediately follows that $p = q_1$. Therefore, $1 \in T$.

Next we consider the case where T contains some positive integer k. We need to show that T also contains $k + 1$. Therefore, suppose that we are now in the situation where $p, q_1, q_2, q_3, \ldots,$ q_k, q_{k+1} are prime numbers such that

(4) $$p | (q_1 \cdot q_2 \cdot q_3 \cdots q_k \cdot q_{k+1}).$$

We need to show that p is equal to one of the q_i's.

There are two possibilities: either $p = q_{k+1}$ or $p \neq q_{k+1}$. In the first case, we are done. In the second case, let

$$b = q_1 \cdot q_2 \cdot q_3 \cdots q_k,$$

then (4) becomes $p | (b \cdot q_{k+1})$. However, in this second case, p and q_{k+1} are relatively prime. Therefore, we can apply Lemma 3.9 to assert that $p | b$. But b is a product of k primes and T contains k. Therefore, p is indeed equal to one of the q_i's that appear in b, and we are also done in this case. $\square$

We now have all the mathematical machinery needed to prove the uniqueness of prime factorization. Before writing the formal proof, let us work through an example that should help us understand the proof.

■ Example

How does Corollary 3.10 help us study prime factorization? Consider the number 85; we know it can be factored into primes as $85 = 5^1 \cdot 17^1$. Let us see why this is the *only* way that 85 can be factored into primes. To this end, suppose $q_1, \ldots, q_n$ are prime numbers such that

$$5^1 \cdot 17^1 = 85 = q_1 \cdot q_2 \cdots q_n.$$

Since $5 \mid 85$, Corollary 3.10 asserts that 5 must be equal to one of the q_i's. By reordering the q_i's we may assume that $q_1 = 5$. Dividing both sides of the preceding equation by 5, we obtain

$$17^1 = \frac{85}{5} = q_2 \cdots q_n.$$

Since 17 divides the left-hand side of the preceding equation, 17 must divide $q_2 \cdots q_n$. However, Corollary 3.10 now asserts that 17 must equal one of the remaining q_i's. By reordering the remaining q_i's, we may assume that $q_2 = 17$. Dividing both sides of the previous equation by 17 results in

$$1 = \frac{85}{5^1 \cdot 17^1} = q_3 \cdots q_n.$$

Since all prime numbers exceed 1, no product of primes can equal 1. Therefore, it is impossible to have the product $q_3 \cdots q_n$ equal to 1. As a result, there do not exist primes $q_3, q_4, \ldots, q_n$ in the factorization of 85. Therefore, 85 is the product of only two primes, q_1 and q_2. However, we showed that after reordering, $q_1 = 5$ and $q_2 = 17$. As a result, up to order, $5^1 \cdot 17^1$ is the *only* way to write 85 as a product of primes.

∎

Observe that in the preceding argument, we used that fact that any product of primes is greater than 1. This illustrates the usefulness of *not* considering 1 to be prime.

Oftentimes, the notation we use can simplify the writing of a proof. In the following proof, when we write an integer as a product of primes, we will write the primes in *nondecreasing* order with all exponents equal to one. What does this mean? Using this notation, we write 12 as $12 = 2 \cdot 2 \cdot 3$. Similarly, we write 720 as $720 = 2 \cdot 2 \cdot 2 \cdot 2 \cdot 3 \cdot 3 \cdot 5$. More generally, in this notation, if we factor n into primes as

$$n = p_1 \cdot p_2 \cdot p_3 \cdots p_m$$

then

$$p_1 \leq p_2 \leq p_3 \leq \cdots \leq p_m.$$

The advantage of this notation is that we have *fixed* the order in which the primes can occur, and we will not need to worry about reordering them during the proof.

Proof of the final part of Theorem 3.3—uniqueness of prime factorization. We will proceed using a slightly modified form of the Second Version of Mathematical Induction. More precisely, we will let T denote those integers that are at least 2 and can be factored uniquely into primes. Our goal is to show that T contains all integers that are at least 2. Mathematical Induction asserts that it will be enough for us to show that T contains 2 and that whenever T contains the set of numbers $\{2, 3, \ldots, k\}$, then it also contains the number $k + 1$.

The number 2 can certainly be factored into primes as $2 = 2^1$. Since 2 is the smallest prime number, any other product of primes would exceed 2. Therefore, $2 = 2^1$ is the only way to factor 2 into a product of primes, and we see that T contains 2.

Now suppose that T contains the set of numbers $\{2, 3, \ldots, k\}$; we need to show that T contains $k + 1$. Therefore, we may assume that every integer that is less than $k + 1$ and larger than 1 can be factored uniquely into primes and our job is to show that $k + 1$ can also be factored uniquely into primes. In order to do this, we need to show that any two factorizations of $k + 1$ into primes are the same. To this end, suppose

$$k + 1 = p_1 \cdot p_2 \cdots p_n \quad \text{and} \quad k + 1 = q_1 \cdot q_2 \cdots q_m$$

are two factorizations of $k + 1$ into primes using the special ordering of primes we discussed earlier. To show that these two factorizations of $k + 1$ are the same, we need to show two things. First, we must show that the number of primes in each factorization is the same, which means that we need to show that $n = m$. Then we will need to show that the same exact primes occur in each factorization, which means that we need to show that $p_i = q_i$, for every $i \leq n$.

Since p_1 and q_1 are both divisors of $k + 1$, we have

$$q_1 \mid (p_1 \cdot p_2 \cdots p_n) \quad \text{and} \quad p_1 \mid (q_1 \cdot q_2 \cdots q_m).$$

Applying Corollary 3.10, we see that $q_1 = p_i$, for some $i \leq n$, and $p_1 = q_j$, for some $j \leq m$. In light of the special ordering we are using, we know that $p_1 \leq p_i$ and $q_1 \leq q_j$. Therefore,

$$p_1 \leq p_i = q_1 \leq q_j = p_1.$$

As a result, $p_1 = q_1$.

Since

$$\frac{k + 1}{p_1} = \frac{k + 1}{q_1},$$

we can now consider the factorization of $\frac{k+1}{p_1}$. We now have

(5) $$\frac{k + 1}{p_1} = p_2 \cdots p_n \quad \text{and} \quad \frac{k + 1}{p_1} = q_2 \cdots q_m.$$

There are two cases to consider; either

$$\frac{k+1}{p_1} = 1 \quad \text{or} \quad \frac{k+1}{p_1} > 1.$$

In the first case, there are no primes other than $p_1 = q_1$ in the prime factorization of $k+1$; otherwise, the products $p_2 \cdots p_n$ and $q_2 \cdots q_m$ would exceed 1. As a result, $n = m = 1$. When we combine this with the fact that $p_1 = q_1$, we see that our two factorizations of $k+1$,

$$k+1 = p_1 \quad \text{and} \quad k+1 = q_1$$

are the same. Therefore, in this case, there is only one way to factor $k+1$, and so T contains $k+1$.

The final case to consider is when $\frac{k+1}{p_1} > 1$. However, in this case $\frac{k+1}{p_1}$ is less than $k+1$ and larger than 1. Thus, T contains $\frac{k+1}{p_1}$. Therefore the factorization of $\frac{k+1}{p_1}$ into primes must be unique. This tells us that the number of primes in our two factorizations of $\frac{k+1}{p_1}$ must be the same. Going back to equation (5), we see that $n - 1 = m - 1$, which immediately tells us that $n = m$. The next thing we know after looking back at equation (5) is that the same exact primes appear in each factorization of $\frac{k+1}{p_1}$. Therefore,

$$p_2 = q_2, \ p_3 = q_3, \dots, p_n = q_n.$$

Since we already knew that $p_1 = q_1$, we now see that our two factorizations of $k+1$ are the same. Therefore, in both cases, T contains $k+1$. □

Theorem 3.3 has some very nice applications to the values of familiar functions like $x^{1/2}$, $x^{1/3}$, and $\log_{10}(x)$. To prove these applications, we first need to prove a fact which is a direct consequence of Theorem 3.3.

Lemma 3.11.   *If a and n are positive integers, then the primes that are divisors of a^n are precisely the same primes that are divisors of a.*

Intuition.   Let us consider some examples:

$$36 = 2^2 \cdot 3^2 \quad \text{and} \quad 36^{10} = 2^{20} \cdot 3^{20},$$

$$245 = 5^1 \cdot 7^2 \quad \text{and} \quad 245^8 = 5^8 \cdot 7^{16},$$

$$99 = 3^2 \cdot 11^1 \quad \text{and} \quad 99^{501} = 3^{1002} \cdot 11^{501}.$$

Based on these examples, Lemma 3.11 certainly seems to be true. Its proof will follow easily from Theorem 3.3.

Proof. If $a = 1$, then $a^n = a = 1$, and there are no primes that divide either a or a^n. On the other hand, if $a > 1$, then we can uniquely factor a into primes as

$$a = p_1{}^{t_1} \cdot p_2{}^{t_2} \cdots p_m{}^{t_m}.$$

Therefore,

$$a^n = (p_1{}^{t_1} \cdot p_2{}^{t_2} \cdots p_m{}^{t_m})^n = p_1{}^{n \cdot t_1} \cdot p_2{}^{n \cdot t_2} \cdots p_m{}^{n \cdot t_m}.$$

When we compare the prime factorizations of a and a^n, we can see that the same primes appear in both. The only difference is that the exponent of each prime in the factorization of a^n is n times greater than in the factorization of a. $\qquad\square$

In Proposition 2.4, we showed that every number on the list

$$\sqrt{1}, \sqrt{2}, \sqrt{3}, \sqrt{4}, \sqrt{5}, \ldots, \sqrt{99}, \sqrt{100}, \sqrt{101}, \ldots$$

is either an integer or is irrational. The proof we gave was not very intuitive and used only the Well Ordering Principle. Now consider the following lists.

$$\sqrt{1}, \sqrt{2}, \sqrt{3}, \sqrt{4}, \sqrt{5}, \ldots, \sqrt{99}, \sqrt{100}, \sqrt{101}, \ldots$$

$$\sqrt[3]{1}, \sqrt[3]{2}, \sqrt[3]{3}, \sqrt[3]{4}, \sqrt[3]{5}, \ldots, \sqrt[3]{99}, \sqrt[3]{100}, \sqrt[3]{101}, \ldots$$

$$\sqrt[4]{1}, \sqrt[4]{2}, \sqrt[4]{3}, \sqrt[4]{4}, \sqrt[4]{5}, \ldots, \sqrt[4]{99}, \sqrt[4]{100}, \sqrt[4]{101}, \ldots$$

$$\sqrt[5]{1}, \sqrt[5]{2}, \sqrt[5]{3}, \sqrt[5]{4}, \sqrt[5]{5}, \ldots, \sqrt[5]{99}, \sqrt[5]{100}, \sqrt[5]{101}, \ldots$$

As an application of the tools developed in this chapter, we will now show that every number on these lists either is an integer or is irrational. The proof will be much simpler and more intuitive than the proof of Proposition 2.4.

Corollary 3.12. *If a and n are positive integers, then $a^{1/n}$ is either an integer or is irrational.*

Proof. Suppose $a^{1/n}$ is rational; it will suffice to show that it must be an integer. To this end, if $a^{1/n}$ is rational, then we can write it as a fraction,

$$a^{1/n} = \frac{c}{d},$$

where c, d are positive integers. Furthermore, we may assume that the fraction is in lowest terms. Therefore, to show that $a^{1/n}$ is an integer, it will be enough to show that $d = 1$. We know that every integer greater than 1 is divisible by some prime. In fact, a surprisingly common technique for proving that a positive integer is equal to 1 is to show that it is not divisible by any prime. Indeed, this is how we will show that $d = 1$.

If $a^{1/n} = \frac{c}{d}$, then raising both sides to the nth power results in

$$a = \left(\frac{c}{d}\right)^n,$$

which immediately implies that

$$a \cdot d^n = c^n.$$

If p is a prime number that divides d, then the preceding equation implies that p divides c^n. However, Lemma 3.11 now asserts that p divides c. This says that p is a common factor of c and d, contradicting our assumption that $\frac{c}{d}$ is in lowest terms. In light of this, it is impossible for d to be divisible by *any* prime number. However, Theorem 3.3 tells us that the only positive integer that is not divisible by a prime is 1. Thus, $d = 1$ and $a^{1/n}$ is indeed an integer.   □

We now consider a similar question for the values of the function $\log_{10}(x)$. Consider the list of numbers

$$\log_{10}(1), \log_{10}(2), \log_{10}(3), \log_{10}(4), \ldots, \log_{10}(99), \log_{10}(100), \log_{10}(101), \ldots$$

Some of the numbers like $\log_{10}(1), \log_{10}(10), \log_{10}(100), \ldots$ are obviously integers. We will show that all the other numbers on the list must be irrational.

Corollary 3.13. *If a is a positive integer, then $\log_{10}(a)$ is either an integer or is irrational.*

Proof. Suppose $\log_{10}(a)$ is rational; it will suffice to show that it must be an integer. Since $\log_{10}(1) = 0 \in \mathbb{Z}$, for the remainder of the proof, we may assume that $a > 1$. Observe that if $\log_{10}(a)$ is rational, then we can write it as a fraction,

$$\log_{10}(a) = \frac{c}{d},$$

where c, d are positive integers. We will show that $\frac{c}{d}$ is an integer, and it is worth pointing out that we will not need to assume that $\frac{c}{d}$ is in lowest terms. The preceding equation implies that

$$10^{\frac{c}{d}} = a$$

and raising both sides to the dth power results in

$$10^c = a^d.$$

Using the fact that $10 = 2 \cdot 5$, the preceding equation becomes

$$2^c \cdot 5^c = a^d.$$

As a result, if p is a prime that divides a, then p must divide $2^c \cdot 5^c$, and Corollary 3.10 now implies that $p = 2$ or $p = 5$. On the other hand, since 2 and 5 divide $2^c \cdot 5^c$, they must both divide a^d, and Lemma 3.11 now implies that they both divide a. As a result, the prime factorization of a must be $a = 2^s \cdot 5^t$, where s, t are positive integers. Substituting this into the preceding equation, we obtain

$$2^c \cdot 5^c = (2^s \cdot 5^t)^d = 2^{s \cdot d} \cdot 5^{t \cdot d}.$$

The equation gives two factorizations of the same number, so the uniqueness of prime factorization implies that the exponents of 2 and 5 in both factorizations must be the same. Therefore,

$$c = s \cdot d \quad \text{and} \quad c = t \cdot d.$$

These equations tell us that

$$s = \frac{c}{d} = t.$$

Therefore, $\frac{c}{d}$ is indeed an integer. $\qquad\square$

In light of Corollaries 3.12 and 3.13, we make a remark similar to one in the final paragraph of Section 2.2. These corollaries tell us that $\sqrt[3]{2}$, $\sqrt[5]{6}$, $\log_{10}(3)$, and $\log_{10}(53)$ are not rational numbers. To assert that they are irrational, we would technically first need to prove that they are real numbers. As remarked earlier, this issue will be dealt with in Chapter 4.

The properties of $\mathbb{Z}$ that we have discussed in this chapter will be used throughout this book to solve a variety of problems. The four basic concepts about $\mathbb{Z}$ that you should always keep in mind are **prime numbers**, **division algorithm**, **Euclidean Algorithm**, and the **existence and uniqueness of prime factorization**. Later, when you study sets of polynomials such as $\mathbb{Q}[x]$, $\mathbb{R}[x]$, and $\mathbb{C}[x]$, be sure to notice the similarities among their algebraic structures and that of $\mathbb{Z}$. In Chapter 12, it will also be interesting to examine the surprising fact that the set $\mathbb{Z}[x]$ of polynomials with integer coefficients actually has less in common with $\mathbb{Z}$ than do the sets $\mathbb{Q}[x]$, $\mathbb{R}[x]$, and $\mathbb{C}[x]$.

Exercises for Sections 3.4 and 3.5

1. Show that any two consecutive positive integers must be relatively prime.

2. (a) Show that any two consecutive positive odd integers must be relatively prime.

    (b) Show that the greatest common divisor of any two consecutive positive even integers is 2.

3. If $a, b, n \in \mathbb{N}$, show that $gcd(n \cdot a, n \cdot b) = n \cdot gcd(a, b)$.

4. If $a, b \in \mathbb{N}$ and $c = gcd(a, b)$, show that $\frac{a}{c}$ and $\frac{b}{c}$ are relatively prime.

5. If a, b are distinct positive odd integers and are relatively prime, find $gcd(a+b, a-b)$.

6. If a, b are distinct positive even integers and $gcd(a, b) = 2$, show that either $gcd(a+b, a-b) = 2$ or $gcd(a+b, a-b) = 4$. Give examples of a and b that show that both possibilities can occur.

7. Show that if $a, b, m, n \in \mathbb{N}$ such that a and b are relatively prime, then a^m and b^n are also relatively prime.

8. If $a, b, c, n \in \mathbb{N}$ and $c = gcd(a, b)$, show that $c^n = gcd(a^n, b^n)$.

9. Suppose $a, b, c \in \mathbb{N}$ and let $d = gcd(a, b)$.
    (a) Show that $gcd(a, b, c) = gcd(c, d)$.

    (b) Show that $gcd(a, b, c)$ is the smallest positive integer that can be written in the form $A \cdot a + B \cdot b + C \cdot c$, where $A, B, C \in \mathbb{Z}$.

10. This exercise will generalize exercise 9. Suppose $n, a_1, a_2, \ldots, a_n \in \mathbb{N}$, and let $c = gcd(a_1, a_2, \ldots, a_{n-1})$.
    (a) Show that $gcd(a_1, a_2, \ldots, a_n) = gcd(a_n, c)$.

    (b) Show that $gcd(a_1, a_2, \ldots, a_n)$ is the smallest positive integer that can be written in the form $A_1 \cdot a + A_2 \cdot a_2 + \cdots + A_n \cdot a_n$, where every $A_i \in \mathbb{Z}$.

11. (a) Find integers A, B such that $gcd(1477, 770) = 1477A + 770B$.

    (b) Use your answer from part (a) to find integers C and D such that $42 = 1477C + 770D$.

    (c) Do there exist integers E and F such that $45 = 1477E + 770F$? Briefly explain.

    (d) Describe all integers M such that there exist integers G and H such that $M = 1477G + 770H$.

12. (a) Find integers A, B such that $gcd(207, 348) = 207A + 348B$.

    (b) Find integers C and D such that $33 = 207C + 348D$.

    (c) Do there exist integers E and F such that $20 = 207E + 348F$? Briefly explain.

    (d) Describe all integers M such that there exist integers G and H such that $M = 207G + 348H$.

13. (a) Find integers A, B such that $gcd(4411, 2486) = 4411A + 2486B$.

    (b) Find integers C and D such that $55 = 4411C + 2486D$.

(c)   Do there exist integers E and F such that $23 = 4411E + 2486F$? Briefly explain.

(d)   Describe all integers M such that there exist integers G and H such that $M = 4411G + 2486H$.

14.   (a)   Find integers A, B such that $gcd(2665, 1976) = 2665A + 1976B$.

(b)   Find integers C and D such that $65 = 2665C + 1976D$.

(c)   Do there exist integers E and F such that $30 = 2665E + 1976F$? Briefly explain.

(d)   Describe all integers M such that there exist integers G and H such that $M = 2665G + 1976H$.

15.   Find integers A, B such that $\frac{1}{63} = \frac{A}{9} + \frac{B}{7}$.

16.   Find integers A, B such that $\frac{29}{210} = \frac{A}{10} + \frac{B}{21}$.

17.   (a)   Find integers A, B, C such that $gcd(156, 123, 114) = 156A + 123B + 114C$.

(b)   Find integers D, E, F such that $30 = 156D + 123E + 114F$.

(c)   Do there exist integers G, H, I such that $40 = 156G + 123H + 114I$? Briefly explain.

(d)   Describe all integers J such that there exist integers K, L, M such that $J = 156K + 123L + 114M$.

18.   (a)   Find integers A, B, C such that $gcd(952, 700, 546) = 952A + 700B + 546C$.

(b)   Find integers D, E, F such that $42 = 952D + 700E + 546F$.

(c)   Do there exist integers G, H, I such that $32 = 952G + 700H + 546I$? Briefly explain.

(d)   Describe all integers J such that there exist integers K, L, M such that $J = 952K + 700L + 546M$.

19.   Find integers A, B, C such that $\frac{1}{900} = \frac{A}{4} + \frac{B}{9} + \frac{C}{25}$.

20.   Find integers A, B, C such that $\frac{1}{39200} = \frac{A}{32} + \frac{B}{25} + \frac{C}{49}$.

21.   You are given a scale that works in the following manner: You can place weights on the left and right sides of the scale. The side of the scale with the greater total weight will tilt downward. If both sides of the scale have the same weight, there will be no movement. Suppose you are given a large collection of 25-ounce weights and 15-ounce weights.

(a)   Explain two different ways that you could determine if a pebble weighed 5 ounces.

(b)   Explain why you could not determine if a pebble weighed 1 ounce.

(c)   If you were additionally given a single 2-ounce weight, could you determine if a pebble weighed 1 ounce? In this case, could you determine if a pebble weighed 3 ounces? Explain your answers.

(d)   Suppose instead of being given one 2-ounce weight, you were given two 2-ounce weights. Could you now determine if a pebble weighed 1 ounce? Explain.

22.   Prove the uniqueness portion of Theorem 3.3 with a proof that does not use Mathematical Induction but does use the Well Ordering Principle in a proof by contradiction.

23.   If $n = 2^7 \cdot 3^4 \cdot 7^2 \cdot 13^3$ and $m = 2^5 \cdot 3^5 \cdot 5^2 \cdot 11^4 \cdot 13^1$, find $gcd(n, m)$.

24.   If $a = 3^4 \cdot 5^1 \cdot 11^5 \cdot 17^2 \cdot 19^5$, $b = 2^2 \cdot 3^3 \cdot 5^2 \cdot 11^3 \cdot 17^2 \cdot 23^8$, and $c = 2^1 \cdot 5^3 \cdot 7^3 \cdot 11^4 \cdot 23^3$, find $gcd(a, b, c)$.

25.   If $a, b \in \mathbb{N}$, let the *least common multiple* of a and b, denoted $lcm(a, b)$, be the smallest positive integer that is a multiple of both a and b. Prove that $lcm(a, b) = \frac{a \cdot b}{gcd(a,b)}$.

26.   If n, m are as in exercise 23, find $lcm(n, m)$.

27.   If n is a positive integer and $A, B, C, D, E \in \mathbb{Z}$ such that $n = 2^A \cdot 3^5 \cdot 5^B \cdot 7^3 \cdot 11^C$ and $n = 2^5 \cdot 3^D \cdot 5^8 \cdot 7^E \cdot 11^2$, find A, B, C, D, E.

28.   If n is a positive integer and $A, B, C, D, E \in \mathbb{Z}$ such that $n = 3^A \cdot 11^6 \cdot 17^B \cdot 23^C$ and $n = 2^D \cdot 3^8 \cdot 11^E \cdot 23^9$, find A, B, C, D, E.

29.   If m is a positive integer and $x, y \in \mathbb{Z}$ such that $m = 5^{2x+y} \cdot 29^4$ and $m = 5^{11} \cdot 29^{3x-y}$, find x and y.

30.   If m is a positive integer and $x, y \in \mathbb{Z}$ such that $m = 7^2 \cdot 13^{4x+10y}$ and $m = 7^{x+3y} \cdot 13^{12}$, find x and y.

31.   If a, b, c, d are integers such that $\frac{28}{375} = 2^a \cdot 3^b \cdot 5^c \cdot 7^d$, find a, b, c, d.

32.   If a, b, c, d are integers such that $\frac{63}{80} = 2^a \cdot 3^b \cdot 5^c \cdot 7^d$, find a, b, c, d.

33.   Let $S = \{p_1, p_2, \ldots, p_n, q_1, q_2, \ldots, q_m\}$ be a set consisting of $n + m$ different prime numbers. Prove that the number $(p_1 \cdot p_2 \cdots p_n) + (q_1 \cdot q_2 \cdots q_m)$ is *not* divisible by any of the elements of S.

34.   Use exercise 33 to come up with another proof that there is an infinite number of primes.

35.   One of the main ideas of this chapter was the unique factorization of positive integers into primes. In this exercise, we will see that if we allow negative exponents, then we can

uniquely factor into primes those positive rational numbers that are in lowest terms. Observe this was the idea you applied in exercises 31 and 32. Now suppose $\{p_1 < p_2 < \cdots < p_t\}$ are prime numbers such that

$$p_1{}^{a_1} \cdot p_2{}^{a_2} \cdots p_t{}^{a_t} = p_1{}^{b_1} \cdot p_2{}^{b_2} \cdots p_t{}^{b_t},$$

where every a_i, b_j is an integer. Show that $a_i = b_i$, for every i.

36. (a) For each of the following integers, determine how many positive integers are divisors of the given number: $8, 9, 10, 15, 16, 17, 24, 25, 26$.

    (b) In which of these cases is the number of divisors odd? Do you detect any pattern?

37. (a) For each of the following integers, determine how many positive integers are divisors of the given number: $27, 28, 29, 35, 36, 49, 50, 51, 64$.

    (b) In which of these cases is the number of divisors odd? Do you detect any pattern?

38. (a) Suppose $n > 1$ is a positive integer and $n = p_1{}^{t_1} \cdot p_2{}^{t_2} \cdots p_m{}^{t_m}$ is the prime factorization of n (you may assume that $p_1 < p_2 < \cdots < p_m$). Show that the number of positive integers which are divisors of n is $(t_1 + 1) \cdot (t_2 + 1) \cdots (t_m + 1)$.

    (b) Look at the prime factorizations of the numbers in part (a) of either exercise 36 or 37. Then check that in all cases the formula from part (a) of this problem yields the same number of positive divisors.

    (c) Use part (a) of this problem to show that $n \in \mathbb{N}$ will have an odd number of positive divisors if and only if n is a perfect square. Then examine your answer to part (b) of either exercise 36 or 37.

39. Without using Corollary 3.12, prove that $7^{\frac{2}{3}}$ is not a rational number.

40. Without using Corollary 3.12, prove that $(210)^{\frac{3}{5}}$ is not a rational number.

41. (a) Show that $\log_5(16)$ is not a rational number.

    (b) Find all positive integers m such that $\log_5(m)$ is rational.

    (c) Find all positive integers m such that $\log_{25}(m)$ is rational.

42. (a) Show that $\log_{36}(40)$ is not a rational number.

    (b) Find all positive integers m such that $\log_{36}(m)$ is rational.

    (c) Find all positive integers m such that $\log_6(m)$ is rational.

43. Show that there cannot exist rational numbers a, b such that $\sqrt{3} = a + b\sqrt{2}$.

44. Show that there cannot exist rational numbers a, b such that $\sqrt{15} = a + b\sqrt{6}$.

45.   Generalize the results in exercises 43–44 and show that if $n, m \in \mathbb{N}$ such that none of $\sqrt{n}, \sqrt{m}, \sqrt{nm}$ are rational, then there cannot exist rational numbers a, b such that $\sqrt{n} = a + b\sqrt{m}$.

46.   (a)   If $n, k \in \mathbb{N}$ such that $k \leq n$, show that $\frac{n!}{(n-k+1)! \cdot (k-1)!} + \frac{n!}{(n-k)! \cdot k!} = \frac{n+1}{(n-k+1)! \cdot k!}$.

      (b)   Use part (a) and the Well Ordering Principle to show that if $n \in \mathbb{N}$ and if k is an integer with $0 \leq k \leq n$, then $\frac{n!}{(n-k)! \cdot k!}$ is an integer.

      (c)   Use part (b) to conclude that if $m \in \mathbb{N}$, then the product of any m consecutive positive integers is divisible by $m!$.

47.   If we let $\mathbb{Q}^+$ denote the set of positive rational numbers, define the function

$$f : \mathbb{Q}^+ \to \mathbb{N}$$

as $f(\frac{p}{q}) = 2^p \cdot 3^q$, where the rational number $\frac{p}{q}$ is in lowest terms and $p, q \in \mathbb{N}$. Compute $f(\frac{5}{3})$, $f(\frac{4}{7})$, and $f(9)$.

48.   Prove that the function $f : \mathbb{Q}^+ \to \mathbb{N}$ in exercise 47 is an injection.

49.   If we let $\mathbb{Z}[x]$ denote the set of polynomials with integer coefficient, define the function

$$g : \mathbb{Z}[x] \to \mathbb{Q}^+$$

as

$$g(a_0 + a_1 x + a_2 x^2 + \cdots + a_n x^n) = 2^{a_0} \cdot 3^{a_1} \cdot 5^{a_2} \cdots p_{n+1}^{a_n},$$

where each $a_i \in \mathbb{Z}$ and p_t represents the tth smallest prime number, for all $t \in \mathbb{N}$. Compute $g(3 + 2x)$, $g(-5 + 4x^2)$, and $g(3 - 4x - 6x^2 + 5x^4)$.

50.   Prove that the function $g : \mathbb{Z}[x] \to \mathbb{Q}^+$ in exercise 49 is an injection.

51.   Use exercises 48 and 50 to show that there exists a function $h : \mathbb{Z}[x] \to \mathbb{N}$ which is an injection.

CHAPTER 4

The Rational Numbers and the Real Numbers

Much of this course is focused on the solutions of various types of equations. Let us look at two equations involving polynomials with integer coefficients

$$2x - 1 = 0 \quad \text{and} \quad x^2 - 2 = 0.$$

In the previous chapter, we examined some of the interesting properties of the integers. However, the integers are not a big enough number system to contain the solutions of the preceding equations.

The solution to $2x - 1 = 0$ is $x = \frac{1}{2}$ and $\frac{1}{2}$ is a rational number that is not an integer. Therefore, our investigation of the roots of polynomials will lead us beyond the integers to the set of rational numbers $\mathbb{Q}$. Similarly, the solutions of $x^2 - 2 = 0$ are $x = \pm\sqrt{2}$ and $\pm\sqrt{2}$ are real numbers that are not rational. Thus, additional investigations will take us beyond the rational numbers to the real numbers $\mathbb{R}$. Furthermore, to find the roots of the polynomial $x^2 + 1$, we will need to move beyond the real numbers to the complex numbers $\mathbb{C}$. In this chapter, we will examine the rational numbers and the real numbers. The complex numbers will be examined in Chapters 5 and 6.

4.1 Rational Numbers

There are two approaches we could take when studying the rational and real number systems. The first approach is based on the assumption that our previous courses in calculus have given us an intuitive understanding of the real numbers. This allows us to view the rational numbers as those real numbers that can be written as a quotient of integers. We then show that this corresponds to those real numbers whose decimal expansion eventually repeats.

Later in this chapter, we present the second and more theoretical approach. In this approach, we view the rational numbers as essentially being pairs of integers. We then construct the real numbers by looking at certain sequences of rational numbers. Both approaches will present us with some subtle and important issues that we will need to deal with. Our motivation for studying the rational and real numbers is to understand and find the solutions to various equations. If you are interested in a more complete and detailed look at the construction of the real numbers, it can be found in a book on real analysis. We begin the first approach with

Proposition 4.1. *A real number α can be written as a quotient of integers if and only if its decimal expansion eventually repeats. Real numbers of this form are called* rational numbers, *and we denote the set of rational numbers as $\mathbb{Q}$.*

Intuition. Let's consider the decimal expansion of $\frac{23}{7}$. It is obtained by long division and the expansion begins as

$$\frac{23}{7} = 3.285714\cdots$$

(You should get a paper and pencil and perform the long division.) When we do the long division, the first seven remainders we obtain, in order, are 2, 6, 4, 5, 1, 3, and 2. Note that the first time we obtained a remainder of 2, it was followed by the remainders 6, 4, 5, 1, 3, and 2. Therefore, *every* time we have a remainder of 2, it will be followed by remainders of 6, 4, 5, 1, 3, and 2. As a result, the sequence of remainders 2, 6, 4, 5, 1, and 3 will continue to repeat forever. However, these remainders cause the term 285714 to appear in the decimal expansion of $\frac{23}{7}$. Therefore, the term 285714 repeats forever in the decimal expansion of $\frac{23}{7}$, and we have

$$\frac{23}{7} = 3.285714285714285714285714285714285714285714\cdots$$

Remember, the division algorithm told us that when dividing by 7, the only possible remainders are 0, 1, 2, 3, 4, 5, and 6. Therefore, when performing long division by 7 to find the decimal expansion of a fraction whose denominator is 7, one of the remainders must eventually occur a second time. As we just saw, that will cause the terms in the decimal expansion to repeat forever. Certainly there is nothing special about the number 7. No matter what the denominator of a fraction is, the division algorithm allows only a finite number of possible remainders. Therefore, one of the remainders eventually appears a second time, and this causes the terms in the decimal expansion to repeat forever.

For a second example, consider the decimal expansion of $\frac{73}{13}$. The decimal expansion begins as

$$\frac{73}{13} = 5.615384\cdots$$

Using long division by 13, the first seven remainders obtained, in order, are 8, 2, 7, 5, 11, 6, and 8. Therefore, the sequence of remainders 8, 2, 7, 5, 11, and 6 will repeat forever. This results in the term 615384 repeating forever in the decimal expansion of $\frac{73}{13}$. Thus, we have

$$\frac{73}{13} = 5.615384615384615384615384615384615384\cdots$$

It is also worth remembering that if a remainder of 0 ever occurs, then the decimal expansion terminates. However, in this case, we can still consider this as a decimal expansion that eventually repeats, since 0 repeats forever. For example, we have

$$\frac{59}{8} = 7.375 = 7.3750000000000 \cdots$$

We will now examine an algorithm you were probably first introduced to in elementary or high school. It allows us to rewrite any real number whose decimal expansion eventually repeats as a quotient of integers. Suppose we are given a real number whose decimal expansion eventually repeats, such as

$$\alpha = 54.6491591591591591591591591 \cdots$$

The term 915 repeats forever, and we can multiply α by an appropriate power of 10 so the first appearance of the term 915 is immediately to the right of the decimal point. In particular, we have

(1) $$10^2 \cdot \alpha = 5464.915915915915915915915 \cdots$$

We can also multiply α by a different power of 10 so now the second appearance of the term 915 is immediately to the right of the decimal point. In particular, we also have

(2) $$10^5 \cdot \alpha = 5464915.915915915915915915915 \cdots$$

We have arranged things so all the terms to the right of the decimal point in both (1) and (2) are identical. Therefore, if we subtract equation (1) from equation (2), we obtain

$$10^5 \cdot \alpha - 10^2 \cdot \alpha = 5464915 - 5464,$$

which tells us that

$$\alpha = \frac{5459451}{99900}.$$

There are some subtle issues related to our assertion that all the terms to the right of the decimal point cancel out when we subtract equation (1) from equation (2). We will deal with these issues after the proof of Proposition 4.1. However, even if one is unaware of these subtleties, you can always use this algorithm to rewrite a real number with a decimal expansion that eventually repeats as a quotient of integers.

Before proving Proposition 4.1, we need to briefly address the question "What is a real number?" In other words, what does the expression "decimal expansion" really mean? To help answer this, let us consider

$$\beta = 1021.3437052096436887541987 \cdots$$

The number β refers to the sum or infinite series

$$1021 + \frac{3}{10^1} + \frac{4}{10^2} + \frac{3}{10^3} + \frac{7}{10^4} + \frac{0}{10^5} + \frac{5}{10^6} + \frac{2}{10^7} + \frac{0}{10^8} + \frac{9}{10^9} + \frac{6}{10^{10}} + \frac{4}{10^{11}} +$$

$$\frac{3}{10^{12}} + \frac{6}{10^{13}} + \frac{8}{10^{14}} + \frac{8}{10^{15}} + \frac{7}{10^{16}} + \frac{5}{10^{17}} + \frac{4}{10^{18}} + \frac{1}{10^{19}} + \frac{9}{10^{20}} + \frac{8}{10^{21}} + \frac{7}{10^{22}} + \cdots$$

Therefore, to truly understand the nature of the number β, we need to understand infinite series and convergence. Real numbers can be viewed as convergent infinite series of rational numbers. On the one hand, we can view the integers and rational numbers as purely algebraic objects, since they are defined in terms of addition, subtraction, multiplication, and division. On the other hand, the real numbers are defined in terms of limits of convergent series. Thus, the real numbers are not purely an algebraic object. Therefore, it is not surprising that the proof of Proposition 4.1 will require a basic fact from calculus on the convergence of certain series. We will state this familiar fact without proof.

Lemma 4.2. *Let a and r be real numbers such that $|r| < 1$. Then*

$$a + ar + ar^2 + ar^3 + ar^4 + \cdots ar^n + ar^{n+1} + \cdots = \frac{a}{1-r}.$$

■ Examples

$$5 + \frac{5}{2} + \frac{5}{2^2} + \frac{5}{2^3} + \frac{5}{2^4} + \cdots = \frac{5}{1 - \frac{1}{2}} = 10,$$

$$\frac{3}{10} + \frac{3}{10^2} + \frac{3}{10^3} + \frac{3}{10^4} + \frac{3}{10^5} + \cdots = \frac{\frac{3}{10}}{1 - \frac{1}{10}} = \frac{1}{3}.$$

The types of series referred to in Lemma 4.2 are known as **geometric series**. When viewing real numbers as infinite series, an important observation to make is that those real numbers whose decimal expansions eventually repeat are precisely those where the corresponding infinite series is a sum of a finite number of terms followed by a geometric series. This will become clearer during the proof of Proposition 4.1.

Lemma 4.3. *Let n be a positive integer and let $d_1, d_2, \ldots, d_n$ be elements of the set $\{0, 1, 2, 3, \ldots, 8, 9\}$. Then the repeating decimal*

$$.d_1 d_2 \cdots d_n d_1 d_2 \cdots d_n d_1 d_2 \cdots d_n d_1 d_2 \cdots d_n d_1 d_2 \cdots d_n d_1 d_2 \cdots d_n \cdots$$

is equal to the fraction $\frac{d_1 d_2 \cdots d_n}{10^n - 1}$.

Intuition. It is important to not be confused by the notation. The term $d_1 d_2 \cdots d_n$ refers to a listing of n digits in a row. It represents an integer that is greater than or equal to zero and is less than 10^n. It does not represent the product of the d_i's. In particular, the lemma asserts that

$$.1492149214921492149214921492\cdots = \frac{1492}{10^4 - 1} = \frac{1492}{9999},$$

$$.072072072072072072\cdots = \frac{72}{10^3 - 1} = \frac{72}{999} = \frac{8}{111}.$$

Proof. We can interpret the repeating decimal

$$.d_1 d_2 \cdots d_n d_1 d_2 \cdots d_n d_1 d_2 \cdots d_n d_1 d_2 \cdots d_n d_1 d_2 \cdots d_n \cdots$$

as representing the geometric series

$$\frac{d_1 d_2 \cdots d_n}{10^n} + \frac{d_1 d_2 \cdots d_n}{10^{2n}} + \frac{d_1 d_2 \cdots d_n}{10^{3n}} + \frac{d_1 d_2 \cdots d_n}{10^{4n}} + \frac{d_1 d_2 \cdots d_n}{10^{5n}} + \cdots$$

Applying Lemma 4.2, with $a = \frac{d_1 d_2 \cdots d_n}{10^n}$ and $r = \frac{1}{10^n}$, we see that the sum of this series is

$$\frac{\frac{d_1 d_2 \cdots d_n}{10^n}}{1 - \frac{1}{10^n}} = \frac{d_1 d_2 \cdots d_n}{10^n - 1}.$$

$\square$

We can now prove Proposition 4.1. Not surprisingly, the proof will use the division algorithm. More precisely, we need the fact that when we divide by an integer b, there are only a finite number of possible remainders.

Proof of Proposition 4.1. It is easy to see that α can be written as a quotient of integers if and only if $-\alpha$ can. Similarly, the decimal expansion of α eventually repeats if and only if the same is true for $-\alpha$. Furthermore, the result is certainly true when $\alpha = 0$. Therefore, to prove the proposition, it is enough to consider the case where α is positive. In one direction, let us suppose that the decimal expansion of α eventually repeats. Therefore, there is a block of digits of the form $d_1 d_2 \cdots d_n$ that, at some point to the right of the decimal point, eventually repeats. Thus, there exist digits $a_1, a_2, \ldots, a_m$ such that

$$\alpha = a_1 a_2 \cdots a_s . a_{s+1} \cdots a_m d_1 d_2 \cdots d_n d_1 d_2 \cdots d_n d_1 d_2 \cdots d_n d_1 d_2 \cdots d_n \cdots$$

Note that the decimal point occurs between a_s and a_{s+1}, which is to the left of the block $d_1 d_2 \cdots d_n$. We can multiply α by an appropriate power of 10 to move the decimal point to the right and obtain

$$10^{m-s} \cdot \alpha = a_1 a_2 \cdots a_s a_{s+1} \cdots a_m . d_1 d_2 \cdots d_n d_1 d_2 \cdots d_n d_1 d_2 \cdots d_n \cdots$$

Therefore,

$$10^{m-s} \cdot \alpha = a_1 a_2 \cdots a_m + .d_1 d_2 \cdots d_n d_1 d_2 \cdots d_n d_1 d_2 \cdots d_n d_1 d_2 \cdots d_n \cdots$$

When we apply Lemma 4.3 to the repeating decimal

$$.d_1 d_2 \cdots d_n d_1 d_2 \cdots d_n d_1 d_2 \cdots d_n d_1 d_2 \cdots d_n d_1 d_2 \cdots d_n \cdots$$

the previous equation becomes

$$10^{m-s} \cdot \alpha = a_1 a_2 \cdots a_m + \frac{d_1 d_2 \cdots d_n}{10^n - 1}.$$

Dividing both sides by 10^{m-s} yields

$$\alpha = \frac{a_1 a_2 \cdots a_m}{10^{m-s}} + \frac{d_1 d_2 \cdots d_n}{10^{m-s} \cdot (10^n - 1)}.$$

Adding the two fractions on the right-hand side of the previous equation, we obtain

$$\alpha = \frac{(10^n - 1) \cdot (a_1 a_2 \cdots a_m) + d_1 d_2 \cdots d_n}{10^{m-s} \cdot (10^n - 1)}.$$

Therefore, we have succeeded in writing α as a quotient of integers.

On the other hand, suppose α can be written as $\frac{a}{b}$, where $a, b \in \mathbb{N}$. Let us consider the remainders that are obtained when we divide the numbers $10, 10^2, 10^3, 10^4, \ldots$ by b. Since all of the remainders are nonnegative and less than b, eventually we can find positive integers n, m with $n > m$ such that 10^n and 10^m have the same remainder when divided by b. The division algorithm asserts that there exist integers q_1, q_2, r with $0 \le r < b$ such that

$$10^n = q_1 \cdot b + r \quad \text{and} \quad 10^m = q_2 \cdot b + r.$$

Thus,

$$10^n - 10^m = (q_1 \cdot b + r) - (q_2 \cdot b + r) = (q_1 - q_2) \cdot b,$$

which is certainly a multiple of b. If we multiply both sides of the preceding equation by a and divide by $b \cdot (10^n - 10^m)$, we obtain

$$\alpha = \frac{a}{b} = \frac{(q_1 - q_2) \cdot a}{10^n - 10^m}.$$

We can certainly rewrite the preceding equation as

$$\alpha = \frac{1}{10^m} \cdot \frac{(q_1 - q_2) \cdot a}{10^{n-m} - 1}.$$

Next, we apply the division algorithm and divide $(q_1 - q_2) \cdot a$ by $10^{n-m} - 1$ to obtain

$$(q_1 - q_2) \cdot a = c \cdot (10^{n-m} - 1) + d,$$

where $0 \le d < 10^{n-m} - 1$. Substituting the result of this equation into the previous one, we see that

$$\alpha = \frac{1}{10^m} \cdot \left(c + \frac{d}{10^{n-m} - 1} \right).$$

Since $0 \le d < 10^{n-m} - 1$, there exist digits $d_1, d_2, \cdots d_{n-m}$ such that $d = d_1 d_2 \cdots d_{n-m}$. In addition, we can also write c as a block of digits $a_1 a_2 \cdots a_t$ to obtain

$$\alpha = \frac{1}{10^m} \cdot \left(a_1 a_2 \cdots a_t + \frac{d_1 d_2 \cdots d_{n-m}}{10^{n-m} - 1} \right).$$

Note that we have some control over the number of digits that appear in c as we could attach as many zeroes on the left as we wish. This means that if necessary, we could extend the length of c. In the final step of the proof, we will need $t - m$ to be at least 0. Therefore, let us now extend the length of c and make t large enough that $t - m \ge 0$.

Applying Lemma 4.3 to $\frac{d_1 d_2 \cdots d_{n-m}}{10^{n-m} - 1}$ and substituting that result into the previous equation gives us

$$\alpha = \frac{1}{10^m} \cdot (a_1 a_2 \cdots a_t . d_1 d_2 \cdots d_{n-m} d_1 d_2 \cdots d_{n-m} d_1 d_2 \cdots d_{n-m} \cdots).$$

Since division by 10^m moves the decimal point m places to the left and $t - m \ge 0$, we see that

$$\alpha = a_1 a_2 \cdots a_{t-m} . a_{t-m+1} \cdots a_t d_1 d_2 \cdots d_{n-m} d_1 d_2 \cdots d_{n-m} d_1 d_2 \cdots d_{n-m} \cdots .$$

As a result, the decimal expansion of α does indeed eventually repeat. □

We can now consider some of the subtle points that lie beneath the surface when we rewrite a real number with a decimal expansion that eventually repeats as a quotient of integers. To this end, let's look at two calculations which both lead to incorrect conclusions. Suppose

$$a = 1 + 2 + 4 + 8 + 16 + \cdots .$$

If we multiply this equation by 2, we obtain

$$2a = 2 + 4 + 8 + 16 + \cdots .$$

The only difference between the right-hand sides of these two equations is that the top one has a 1 in it and the bottom one doesn't. Therefore, if we subtract the bottom equation from the top, we obtain

$$-a = 1,$$

which implies that $a = -1$. However, this is clearly false.

Let's look at one more example. Suppose

$$b = 10^1 + 10^2 + 10^3 + 10^4 + 10^5 + \cdots.$$

Multiplying this equation by 10 yields

$$10b = 10^2 + 10^3 + 10^4 + 10^5 + \cdots.$$

Subtracting the bottom equation from the top yields

$$-9b = 10,$$

which implies that $b = -\frac{10}{9}$. This also is clearly false.

These two calculations are very similar to the one where we showed that the real number $54.6491591591591591591591591591 \cdots$ could be rewritten as the quotient $\frac{5459451}{99900}$. In all of these calculations, we canceled out two copies of a sum of an infinite number of terms. Since one of these calculations gave us the correct answer and two gave us incorrect answers, it raises the question, what went wrong in two of these calculations?

In all of these calculations, we have a sum of an infinite number of terms like

$$\frac{915}{10^3} + \frac{915}{10^6} + \frac{915}{10^9} + \frac{915}{10^{12}} + \frac{915}{10^{15}} + \cdots$$

or

$$2 + 4 + 8 + 16 + 32 + \cdots$$

appearing in more than one equation. We then subtracted one equation from another and claimed that all of the identical terms in the two sums canceled out. Now comes the subtle but essential point. It is only valid to cancel out a sum of an infinite number of terms from two equations if the sum you are dealing with *converges*. By Lemma 4.2, the sum

$$\frac{915}{10^3} + \frac{915}{10^6} + \frac{915}{10^9} + \frac{915}{10^{12}} + \frac{915}{10^{15}} + \cdots$$

converges and represents a real number. Therefore, it is valid to cancel out two copies of

$$\frac{915}{10^3} + \frac{915}{10^6} + \frac{915}{10^9} + \frac{915}{10^{12}} + \frac{915}{10^{15}} + \cdots$$

when trying to write

$$54.6491591591591591591591591591 \cdots$$

as a quotient.

However, the sum

$$2+4+8+16+32+\cdots$$

does not converge and does not represent a real number. Therefore,

$$2+4+8+16+32+\cdots$$

cannot be used in calculations and cannot be subtracted from itself to give 0 as answer. This explains why attempts to cancel two copies of

$$2+4+8+16+32+\cdots$$

leads to the contradiction

$$-1=1+2+4+8+16+32+\cdots.$$

To summarize, when we use an algorithm to rewrite a real number with a decimal expansion that eventually repeats as a quotient, we are usually not concerned or interested in the theory surrounding geometric series and convergence. However, as the preceding examples indicate, convergence is indeed necessary if we are to perform a calculation that requires canceling two copies of a sum that involves an infinite number of terms.

4.2 Intermediate Value Theorem

In Chapters 2 and 3, we showed that $\sqrt[3]{2}$, $\sqrt[5]{6}$, $\log_{10}(3)$, and $\log_{10}(53)$ were not rational numbers. Phrased somewhat differently, this means that the equations

$$x^3 = 2, \quad x^5 = 6, \quad 10^x = 3, \quad 10^x = 53$$

do not have any solutions in the rational numbers. This raises the question whether these equations have solutions in the real numbers? If these equations do indeed have real solutions, then $\sqrt[3]{2}$, $\sqrt[5]{6}$, $\log_{10}(3)$, and $\log_{10}(53)$ are real numbers that are not rational, so we can say that they are irrational. The mathematical tool needed to show that the preceding equations have solutions in the real numbers is the Intermediate Value Theorem.

The Intermediate Value Theorem can be found in calculus and real analysis texts but is usually not found in abstract algebra books. However, it is an important tool for finding roots of polynomials and helps illustrate key differences between the rational numbers and the real numbers. The Intermediate Value Theorem will appear in Chapter 6 to help prove the Fundamental Theorem of Algebra and again in Chapters 9 and 17 in our work on the insolvability of the quintic. Since abstract algebra students should be familiar with the Intermediate Value Theorem, we state it now and will prove it later in this section.

Theorem 4.4—The Intermediate Value Theorem. *Suppose the function $f(x)$ is continuous on the closed interval $[a, b]$. If $f(a) < 0$ and $f(b) > 0$, then there exists some real number c in the open interval (a, b) such that $f(c) = 0$.*

In calculus, we deal with functions whose domains are often the entire set of real numbers. We do not look at functions whose domains are merely the rational numbers. The reason for this can be seen in the following example. Consider the polynomial $f(x) = x^2 - 2$. Certainly $f(1) < 0$ and $f(2) > 0$. However, there is no *rational* number c between 1 and 2 such that $f(c) = 0$. This is because $\sqrt{2}$ is not a rational number. This example indicates that if we restrict the domains of our functions to the rational numbers, then important theorems like the Intermediate Value Theorem no longer hold. In light of this, there must be something special about the real numbers that does not hold for the rational numbers that enables us to prove the Intermediate Value Theorem. To proceed, we must first introduce some terminology.

Definition 4.5. *Let S be a subset of $\mathbb{R}$. An element $c \in \mathbb{R}$ is called an upper bound for S if $c \geq s$, for all $s \in S$.*

Observe that if c is an upper bound for S, then any number bigger than c is also an upper bound for S. Therefore, if a set S has one upper bound, then it has an infinite number of upper bounds.

■ Examples

1. Let S be the open interval $(3, 5)$; then some of the upper bounds for S are $5, 6, 2\pi, 1988$.

2. Let S be the closed interval $[3, 5]$: then some of the upper bounds for S are $5, 6, 2\pi, 1988$.

3. Let $S = \left\{ \frac{3}{n} \mid n \in \mathbb{N} \right\}$; then some of the upper bounds for S are $3, \pi, 100, 1025$.

4. Let $S = \left\{ \frac{n}{3} \mid n \in \mathbb{N} \right\}$; then S has no upper bounds.

5. Let $S = \{x \in \mathbb{R} \mid x^2 < 2\}$; then some of the upper bounds for S are $\sqrt{2}, 2, 3, 2001$.

6. Let $S = \{x \in \mathbb{Q} \mid x^2 < 2\}$; then some of the upper bounds for S are $\sqrt{2}, 2, 3, 2001$.

7. Let $S = \{x \in \mathbb{R} \mid x^2 > 2\}$; then S has no upper bounds.

■

In our seven examples, five of the sets have upper bounds. Certainly, none of these sets has a largest upper bound, for if c is an upper bound for a set, then $c + 1$ is a larger upper bound. However, observe that each of our five sets that has an upper bound has a smallest one. In examples (1) and (2), the smallest upper bound is 5; in example (3), the smallest upper bound is 3; and in examples (5) and (6), the smallest upper bound is $\sqrt{2}$. This is no accident, and it is precisely this special property of the real numbers that enables us to prove results like the Intermediate Value Theorem. Before stating this property, we need one more definition.

Definition 4.6. *Let S be a subset of* $\mathbb{R}$. *An element* $c \in \mathbb{R}$ *is called a least upper bound for S if*

(a) *c is an upper bound for S, and*

(b) *if d is any upper bound for S, then* $c \leq d$.

Note that the least upper bound of a set S need not be an element of the set S. Observe that 5 is the least upper bound of $(3, 5)$ and $5 \notin (3, 5)$. Similarly, $\sqrt{2}$ is the least upper bound of both $\{x \in \mathbb{R} \mid x^2 < 2\}$ and $\{x \in \mathbb{Q} \mid x^2 < 2\}$, yet $\sqrt{2}$ does not belong to either set. We can now state the key property of the real numbers that makes the real numbers significantly different from the rational numbers.

Least Upper Bound Property of the Real Numbers *Let S be a nonempty subset of R. If S has an upper bound, then there is a real number c such that c is the least upper bound of S.*

Before proceeding, there are some observations we should make about the Least Upper Bound Property.

(a) If c_1 and c_2 are both least upper bounds for S, then they are both upper bounds for S, and they are both less than or equal to any other upper bound for S. This tells us that $c_1 \leq c_2$ and $c_2 \leq c_1$, so $c_1 = c_2$. As a result, a set can have *at most* one least upper bound. For that reason, we refer to *the* least upper bound as opposed to *a* least upper bound.

(b) If the set S is empty then, technically, every real number is an upper bound for the set. Therefore, the set would not have a smallest upper bound. For this technical reason, when discussing least upper bounds, we only deal with sets that are nonempty.

(c) Suppose S is a nonempty subset of $\mathbb{Q}$ that has an upper bound, then S is certainly a nonempty subset of $\mathbb{R}$ that has an upper bound. Therefore, S must have a least upper bound, although the least upper bound need not be rational. For example, if $S = \{x \in \mathbb{Q} \mid x^2 < 3\}$, then the least upper bound of S is $\sqrt{3}$, which is not rational. In light of this, we can say that $\mathbb{Q}$ **does not satisfy the Least Upper Bound Property**. By this we mean that although nonempty subsets of $\mathbb{Q}$ that have an upper bound must have a least upper bound, the least upper bound does not necessarily belong to $\mathbb{Q}$. As our example indicates, the least upper bound must belong to $\mathbb{R}$ but may not belong to $\mathbb{Q}$.

Thus, the crucial difference between the real numbers and the rational numbers is that the Least Upper Bound Property is satisfied by $\mathbb{R}$ but not by $\mathbb{Q}$. Whereas the Intermediate Value

Theorem does not hold for functions defined only on the rational numbers, the Least Upper Bound Property will enable us to prove the Intermediate Value Theorem for functions defined on the real numbers. To prove the Intermediate Value Theorem, we must first review the definition of continuity.

Definition 4.7. *A function $f : \mathbb{R} \longrightarrow \mathbb{R}$ is said to be continuous at the point $x = c$ if for every $\epsilon > 0$ there exists a $\delta > 0$ such that whenever $|x - c| < \delta$, we have $|f(x) - f(c)| < \epsilon$.*

Intuitively, if a function is continuous at $x = c$, it tells us that we can keep all the values of $f(x)$ as close to $f(c)$ as we desire as long as we only plug in numbers close to c. In particular, if $f(c) > 0$, then there is an open interval I containing c such that whenever a point in I is plugged into $f(x)$, the result is positive. Similarly, if $f(c) < 0$, then there is an open interval I containing c such that whenever a point in I is plugged into $f(x)$, the result is negative. These observations will be needed in the proof of the Intermediate Value Theorem. In calculus, there is a slight difference between the concepts of a function being continuous on the closed interval $[a, b]$ and a function being continuous at every point of the interval $[a, b]$. The differences concern what happens just to the right of the point b or just to the left of the point a. However, for the purposes of this course, we will be interested in using the Intermediate Value Theorem to show that various polynomials have real roots. Since polynomials are continuous everywhere, such technical concerns do not arise in our applications. For this reason, when we use the phrase that a function is continuous on the interval $[a, b]$, we will give that phrase the somewhat nonstandard meaning that the function is continuous at every point in $[a, b]$.

Proof of the Intermediate Value Theorem. Let $S = \{x \in [a, b] \mid f(x) < 0\}$. The set S is certainly nonempty as $a \in S$, and, in addition, b is an upper bound for S. Therefore, the Least Upper Bound Property asserts that S has a least upper bound, and we will denote it as c. Since $a \in S$, we see that $a \leq c$, and since b is an upper bound for S, we see that $c \leq b$. Thus, $c \in [a, b]$. There are now three possibilities; $f(c) > 0$, $f(c) < 0$, or $f(c) = 0$. We will show that each of the first two possibilities leads to a contradiction, so we will be able to conclude that $f(c) = 0$.

Suppose that $f(c) > 0$; if we let $\epsilon = \frac{f(c)}{2}$, then the continuity of $f(x)$ at $x = c$ implies that there exists some $\delta > 0$ such that if x is within δ of c, then $f(x)$ is with $\frac{f(c)}{2}$ of $f(c)$. Since $f(a) < 0$ and $f(c) > 0$, it follows that $c > a$. Therefore, there exists some d_1 with the properties that (i) d_1 is within δ of c, (ii) $d_1 < c$, and (iii) $d_1 > a$.

If x belongs to the closed interval $[d_1, c]$, then x is within δ of c. Therefore, $f(x)$ is within $\frac{f(c)}{2}$ of $f(c)$. As a result, $f(x) > \frac{f(c)}{2} > 0$. Now suppose $e \in S$; since c is an upper bound for S, $e \leq c$. Since $f(e) < 0$ and $f(x) > 0$, for all $x \in [d_1, c]$, it follows that e does not belong to the interval $[d_1, c]$. Thus, $e < d_1$. This tells us that d_1 is both an upper bound for S and is less than c, which contradicts the fact that c is the least upper bound of S. Therefore, it is impossible for $f(c)$ to be positive.

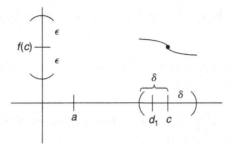

Now suppose that $f(c) < 0$; in this case, we let $\epsilon = -\frac{f(c)}{2}$. The continuity of $f(x)$ at $x = c$ implies that there exists some $\delta > 0$ such that if x is within δ of c, then $\frac{3f(c)}{2} < f(x) < \frac{f(c)}{2} < 0$. Since $f(b) > 0$ and $f(c) < 0$, it follows that $c < b$. Therefore, there exists some d_2 with the properties that (i) d_2 is within δ of c, (ii) $d_2 > c$, and (iii) $d_2 < b$.

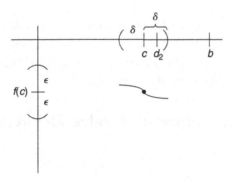

Since $d_2 \in [a, b]$ and $f(d_2) < 0$, we see that $d_2 \in S$. Therefore, d_2 is an element of S that is greater than c, and this contradicts the fact that c is an upper bound for S. Thus, it is also impossible for $f(c)$ to be negative.

As a result, we now know that $f(c) = 0$. Since $f(a) < 0$ and $f(b) > 0$, we see that $c \neq a$ and $c \neq b$. Therefore, c is indeed an element of (a, b) such that $f(c) = 0$, as desired. $\qquad\square$

In mathematics, we often define certain objects in terms of the properties we would like them to have. For example, we define $\sqrt{2}$ to be the positive real number whose square is equal to 2. Similarly, we define $7^{\frac{1}{3}}$ to be the real number whose cube is equal to 7. However, just because we make a definition in no way guarantees that there exists an object that has those properties. For example, we could define an object to be the positive rational number whose square is 2, but, as we know, there is no rational number whose square is 2. Similarly, we could define an object to be a real number whose square is -1, but there is no real number with that property, since the square of every real number is greater than or equal to zero.

We can now use the Intermediate Value Theorem to easily show that real numbers like $\sqrt{2}$ and $7^{\frac{1}{3}}$ do indeed exist.

Corollary 4.8. *Let $a, n \in \mathbb{N}$. Then there exists a positive real number c such that $c^n = a$. In other words, there does exist a real number $a^{\frac{1}{n}}$.*

Proof. If $a = 1$, then $c = 1$ will work, and therefore we may now assume that $a > 1$. Let $f(x)$ be the polynomial $x^n - a$. Observe that

$$f(0) = -a < 0 \quad \text{and} \quad f(a) = a^n - a = a(a^{n-1} - 1) > 0.$$

Since polynomials are continuous everywhere and $f(x)$ goes from negative to positive on the interval $[0, a]$, the Intermediate Value Theorem tells us that there exist some $c \in (0, a)$ such that $f(c) = 0$. As a result, $c^n - a = 0$, which implies that $c^n = a$. $\qquad\square$

At first glance, the Intermediate Value Theorem seems to tell us that roots of certain polynomials exist, but it doesn't appear to help us find those roots. However, if we consider a real number to be a decimal expansion, then the Intermediate Value Theorem does indeed provide an algorithm for finding roots of polynomials.

■ Example: Using the Intermediate Value Theorem to find $\sqrt{2}$

Let $f(x) = x^2 - 2$. Since $f(1) < 0$ and $f(2) > 0$, the Intermediate Value Theorem guarantees that there is a root between 1 and 2, so

$$1 < \sqrt{2} < 2.$$

Next, partition the interval $[1, 2]$ into the ten equal subintervals

$$[1, 1.1], [1.1, 1.2], [1.2, 1.3], [1.3, 1.4], [1.4, 1.5],$$
$$[1.5, 1.6], [1.6, 1.7], [1.7, 1.8], [1.8, 1.9], [1.9, 2.0].$$

Since $f(x)$ has no rational roots, none of the endpoints of the ten subintervals will be a root. However, since $f(x)$ changes from negative to positive in the interval $[1, 2]$, it must change from negative to positive in one of the ten subintervals. Therefore, if we plug the endpoints of our ten subintervals into $f(x)$, we will see where $f(x)$ changes from negative to positive. In particular, $f(1.4) < 0$ and $f(1.5) > 0$, so

$$1.4 < \sqrt{2} < 1.5.$$

We can now iterate the above procedure. Partition the interval $[1.4, 1.5]$ into ten equal subintervals and plug the endpoints of each subinterval into $f(x)$ to determine on which subinterval $f(x)$ changes sign. We will see that $f(1.41) < 0$ and $f(1.42) > 0$, so

$$1.41 < \sqrt{2} < 1.42.$$

We can apply this algorithm as many times as we desire, to determine the decimal expansion to as many decimal places as we wish. For example, if we apply this algorithm ten times, we will see that

$$1.4142135623 < \sqrt{2} < 1.4142135624.$$

If we apply this algorithm twenty times, we will see that

$$1.41421356237309504880 < \sqrt{2} < 1.41421356237309504881.$$

As this example indicates, the Intermediate Value Theorem not only tells us that there exists a positive real number whose square is 2, but it also provides us with an algorithm to compute its decimal expansion to any desired number of decimal places.

∎

In Chapter 6, we will prove the Fundamental Theorem of Algebra. There are many equivalent ways to state the theorem, and one of them is that any polynomial with real coefficients of degree at least one has a root in the complex numbers. Although there are certainly polynomials with real coefficients of even degree that do not have real roots, such as $x^2 + 1$ and $x^4 + 3x^2 + 7$, we can use the Intermediate Value Theorem to see that all polynomials of *odd* degree with real coefficients must have a real root.

Corollary 4.9. *Let $p(x)$ be a polynomial of odd degree with real coefficients. Then $p(x)$ has a root in $\mathbb{R}$.*

Intuition. If the leading term of $p(x)$ is positive, then as $x \to \infty$, the leading term of $p(x)$ dominates and $p(x) \to \infty$. Similarly, as $x \to -\infty$, the leading term of $p(x)$ again dominates. However, since the exponent of x is odd and the leading coefficient is positive, the leading term now approaches $-\infty$. As a result, as $x \to -\infty$, it follows that $p(x) \to -\infty$. Therefore, there exists a negative number a such that $p(a) < 0$ and a positive number b such that $p(b) > 0$. Since polynomials are continuous everywhere, the Intermediate Value Theorem tells us that there is some c between a and b such that $p(c) = 0$.

If the leading term of $p(x)$ is negative, we can consider the polynomial $-p(x)$. The leading term of $-p(x)$ is positive and the preceding argument implies that there is a real number c such that $-p(c) = 0$. However, it is now clear that $p(c) = 0$.

Proof. Let

$$p(x) = a_n x^n + a_{n-1} x^{n-1} + \cdots + a_1 x_1 + a_0,$$

where the $a_i \in \mathbb{R}$ and n is odd. The preceding argument indicates that if $a_n < 0$, we could instead consider the polynomial $-p(x)$. Therefore, without loss of generality, we may assume that $a_n > 0$.

Observe that

$$p(x) = a_n x^n + a_{n-1} x^{n-1} + \cdots + a_1 x + a_0 = x^n \left(a_n + \frac{a_{n-1}}{x} + \cdots + \frac{a_1}{x^{n-1}} + \frac{a_0}{x^n} \right).$$

If x is a real number such that $|x|$ is sufficiently large, then each of the terms

$$\frac{a_{n-1}}{x}, \ \cdots \ , \frac{a_1}{x^{n-1}}, \frac{a_0}{x^n}$$

can be made as close to zero as we desire. Therefore, by choosing x so that $|x|$ is sufficiently large, we can make the sum

$$a_n + \frac{a_{n-1}}{x} + \cdots + \frac{a_1}{x^{n-1}} + \frac{a_0}{x^n}$$

as close to a_n as we wish. Since $a_n > 0$, this tells us that when $|x|$ is large,

(3) $$a_n + \frac{a_{n-1}}{x} + \cdots + \frac{a_1}{x^{n-1}} + \frac{a_0}{x^n} > 0.$$

In particular, we can find a negative number a and a positive number b with sufficiently large absolute values that they both make the expression in (3) positive.

Since $a < 0$ and n is odd, $a^n < 0$. Therefore,

$$p(a) = a_n a^n + a_{n-1} a^{n-1} + \cdots + a_1 a + a_0 = a^n \left(a_n + \frac{a_{n-1}}{a} + \cdots + \frac{a_1}{a^{n-1}} + \frac{a_0}{a^n} \right)$$

is the product of a negative number and a positive number. Thus, $p(a) < 0$.

Similarly, since $b > 0$, we have $b^n > 0$. Therefore,

$$p(b) = a_n b^n + a_{n-1} b^{n-1} + \cdots + a_1 b + a_0 = b^n \left(a_n + \frac{a_{n-1}}{b} + \cdots + \frac{a_1}{b^{n-1}} + \frac{a_0}{b^n} \right)$$

is the product of two positive numbers. Thus, $p(b) > 0$.

Since $p(a) < 0$, $p(b) > 0$, and polynomials are continuous everywhere, we can apply the Intermediate Value Theorem to conclude that there exists a real number c between a and b such that $p(c) = 0$. $\qquad \square$

Note that the Intermediate Value Theorem can be applied to *all* polynomials of odd degree to find real roots, but it can only be applied to some polynomials of even degree to find real roots.

For example, we certainly used the Intermediate Value Theorem to find a real root of $x^2 - 2$. However, we cannot apply the Intermediate Value Theorem to find a real root of $x^2 + 1$, since the values of this polynomial never change from negative to positive.

In the next two chapters, we will examine the complex numbers and provide a proof of the Fundamental Theorem of Algebra. The complex numbers are constructed from the real numbers, so an understanding of the real numbers is very important for an understanding of the complex numbers. Certainly, Corollary 4.9 is an important and useful tool in our study of the roots of polynomials. However, its proof relies on the Intermediate Value Theorem that, in turn, relies on the Least Upper Bound Property. Therefore, Corollary 4.9 is an example of an important theorem with applications to algebra whose proof is not of a purely algebraic nature.

It is also the case that any proof of the Fundamental Theorem of Algebra will require some tools that are not of a purely algebraic nature. The proof we will provide in Chapter 6 is relatively easy and uses several standard facts from one- and two-variable calculus. Proofs of the Fundamental Theorem of Algebra that rely almost entirely on algebraic tools are more difficult, but they also need to use Corollary 4.9.

Exercises for Sections 4.1 and 4.2

In exercises 1–6, write each of the quotients as a decimal expansion that eventually repeats.

1. $\frac{4}{7}$

2. $\frac{3}{11}$

3. $\frac{21}{13}$

4. $\frac{32}{25}$

5. $\frac{241}{99}$

6. $\frac{59}{7}$

In exercises 7–13, write each decimal as a quotient of positive integers.

7. $62.525252525252525252525\cdots$

8. $23.47444444444444444444\cdots$

9. $853.239393939393939393939\cdots$

10. $5676.767676767676767676767\cdots$

11. $2.467856785678568756785\cdots$

12. $78538.53853853853853853853\cdots$

13.   Let a, b be relatively prime positive integers. Show that the decimal expansion of $\frac{a}{b}$ terminates if and only if the set of prime divisors of b is a subset of $\{2, 5\}$.

For exercises 14–17, please first read the following:

Ordinarily, when doing arithmetic, we work in base 10. In fact, the word *decimal* indicates that we are working in base 10, as the word has Latin roots meaning "dealing with tenths." However, it remains the case that regardless of the base, those real numbers that can be written as a quotient of integers correspond to those whose expansion eventually repeats. (Notice that if we are not in base 10, we can no longer refer to the expansion as a decimal expansion.) In the next four exercises, you will work in different bases and convert repeating expansions into quotients of integers.

14.   In Base 2, write $0.111111111111\cdots$ as a quotient of positive integers.

15.   In Base 3, write $0.111111111111\cdots$ as a quotient of positive integers.

16.   In Base 5, write $2.323232323232323232323\cdots$ as a quotient of positive integers.

17.   In Base 7, write $2.323232323232323232323\cdots$ as a quotient of positive integers.

In exercises 18–23, find the sum of the geometric series.

18. $\sum_{n=0}^{\infty} 3\left(\frac{4}{5}\right)^{n}$

19. $\sum_{n=3}^{\infty} 3\left(\frac{4}{5}\right)^{n}$

20. $\sum_{n=0}^{\infty} 3\left(\frac{-4}{5}\right)^{n}$

21. $\sum_{n=0}^{\infty} 6\left(\frac{7}{11}\right)^{n}$

22. $\sum_{n=-2}^{\infty} 6\left(\frac{7}{11}\right)^{n}$

23. $\sum_{n=0}^{\infty} 6\left(\frac{-7}{11}\right)^{n}$

In exercises 24–27, use the Intermediate Value Theorem to show that each polynomial has at least three real roots.

24. $x^5 - 6x + 3$

25. $x^5 - 4x + 2$

26. $x^5 - 8x + 6$

27. $x^5 - 30x + 5$

For exercises 28–31, please first read the following:

In calculus, it follows from Rolle's Theorem that the number of roots of a polynomial is at most one more than the number of roots of its derivative. In exercises 28–31, use this observation along with your work in exercises 24–27 to show that each of the following polynomials has exactly three real roots.

28. $x^5 - 6x + 3$

29. $x^5 - 4x + 2$

30. $x^5 - 8x + 6$

31. $x^5 - 30x + 5$

32. Use the Intermediate Value Theorem and the algorithm after Corollary 4.8 to approximate $\sqrt{13}$ to five decimal places.

33. Use the Intermediate Value Theorem and the algorithm after Corollary 4.8 to approximate $\sqrt{37}$ to five decimal places.

34. Use the Intermediate Value Theorem and the algorithm after Corollary 4.8 to approximate $\sqrt[3]{11}$ to five decimal places.

35. Use the Intermediate Value Theorem and the algorithm after Corollary 4.8 to approximate $\sqrt[5]{43}$ to five decimal places.

36. Use the Intermediate Value Theorem and the algorithm after Corollary 4.8 to approximate the real root of $2x^3 + 5x - 11$ to five decimal places.

37. Use the Intermediate Value Theorem and the algorithm after Corollary 4.8 to approximate the two real roots of $3x^4 - 4x - 9$ to five decimal places.

38. Use Corollary 4.9 to prove the following: if a is a real number and n is a positive odd integer, then there exists a real number c such that $c^n = a$.

39. Prove that if $a > 0$ is a real number and n is a positive even integer, then there exist exactly two real numbers c such that $c^n = a$.

40. For each set, find the least upper bound (if it exists). In each case, if there is a least upper bound, state whether it is an element of the set.
    (a) $\{x \in \mathbb{R} \mid x^2 \leq 5\}$

    (b) $\{x \in \mathbb{R} \mid x^2 < 5\}$

    (c) $\{x \in \mathbb{R} \mid x^2 > 5\}$

    (d) $\{x \in \mathbb{Q} \mid x^2 \leq 5\}$

(e) $\{x \in \mathbb{Q} \mid x^2 < 5\}$

(f) $\{x \in \mathbb{Z} \mid x^2 \leq 5\}$

(g) $\{x \in \mathbb{N} \mid x^2 \leq 5\}$

(h) $\{x \in \mathbb{N} \mid x^2 > 5\}$

41. For each set, find the least upper bound (if it exists). In each case, if there is a least upper bound, state whether it is an element of the set.

    (a) $\{x \in \mathbb{R} \mid x^3 \leq 10\}$

    (b) $\{x \in \mathbb{R} \mid x^3 < 10\}$

    (c) $\{x \in \mathbb{R} \mid x^3 > 10\}$

    (d) $\{x \in \mathbb{Q} \mid x^3 \leq 10\}$

    (e) $\{x \in \mathbb{Q} \mid x^3 < 10\}$

    (f) $\{x \in \mathbb{Z} \mid x^3 \leq 10\}$

    (g) $\{x \in \mathbb{N} \mid x^3 \leq 10\}$

    (h) $\{x \in \mathbb{N} \mid x^3 > 10\}$

42. For each set, find the least upper bound (if it exists). In each case, if there is a least upper bound, state whether it is an element of the set.

    (a) $\{x \in \mathbb{R} \mid x^3 - 3x^2 + 2x \leq 0\}$

    (b) $\{x \in \mathbb{R} \mid x^3 - 3x^2 + 2x < 0\}$

    (c) $\{x \in \mathbb{R} \mid x^3 - 3x^2 + 2x > 0\}$

    (d) $\{x \in \mathbb{Z} \mid x^3 - 3x^2 + 2x \leq 0\}$

    (e) $\{x \in \mathbb{Z} \mid x^3 - 3x^2 + 2x < 0\}$

    (f) $\{x \in \mathbb{Z} \mid x^3 - 3x^2 + 2x > 0\}$

43. For each set, find the least upper bound (if it exists). In each case, if there is a least upper bound, state whether it is an element of the set.

    (a) $(\sqrt{3}, \sqrt{47})$

    (b) $[\sqrt{3}, \sqrt{47}]$

    (c) $(\sqrt{3}, \sqrt{47}]$

    (d) $[\sqrt{3}, \sqrt{47})$

(e) $(-\infty, \sqrt{47})$

(f) $(\sqrt{3}, \infty)$

(g) $\{x \in \mathbb{Q} \mid x \in (\sqrt{3}, \sqrt{47})\}$

(h) $\{x \in \mathbb{Q} \mid x \in [\sqrt{3}, \sqrt{47}]\}$

(i) $\{x \in \mathbb{N} \mid x \in (\sqrt{3}, \sqrt{47})\}$

44. Use the Well Ordering Principle to show that if S is a nonempty subset of $\mathbb{N}$ with an upper bound, then the least upper bound of S must be an element of S.

45. If $M = \{a + b\sqrt{2} \mid a, b \in \mathbb{Q}\}$, show that M does not satisfy the Least Upper Bound Property. In other words, show that there exists a nonempty subset S of M such that S has an upper bound but the least upper bound of S does not belong to M.

For exercise 46, please read the following:

The set $\mathbb{N}$ is certainly a subset of the set of positive rational numbers, $\mathbb{Q}^+$, as well as of the set of polynomials with integer coefficients, $\mathbb{Z}[x]$. Therefore, in some sense, we can think of $\mathbb{Q}^+$ and $\mathbb{Z}[x]$ as being "larger" than $\mathbb{N}$. However, in the exercises that follow Section 3.5, exercises 48 and 51 show that there exist injections $f : \mathbb{Q}^+ \to \mathbb{N}$ and $h : \mathbb{Z}[x] \to \mathbb{N}$. Observe that the range of f is a subset of $\mathbb{N}$ and f is a bijection between $\mathbb{Q}^+$ and the range of f. Therefore, in some sense, we can think of a copy of $\mathbb{Q}^+$ as living inside of $\mathbb{N}$. Therefore, although $\mathbb{N}$ is a subset of $\mathbb{Q}^+$, there is some justification for thinking of $\mathbb{N}$ and $\mathbb{Q}^+$ as having the same size. Since $\mathbb{N}$ is a subset of $\mathbb{Z}[x]$, yet there is an injection from $\mathbb{Z}[x]$ to $\mathbb{N}$, we can also think of $\mathbb{N}$ and $\mathbb{Z}[x]$ as having the same size.

Let us now consider the open interval $(0, 1)$ that consists of all real numbers between 0 and 1. We can ask if there exists an injection $\nu : (0, 1) \to \mathbb{N}$? If the answer is *no*, then we could think of the set $(0, 1)$ as being larger than $\mathbb{N}$. It would then be logical to assert that the set $\mathbb{R}$ is larger than the sets $\mathbb{N}$, $\mathbb{Q}^+$, and $\mathbb{Z}[x]$. The goal of exercise 46 is to prove that no injection exists from $(0, 1)$ to $\mathbb{N}$.

46. Suppose $\nu : (0, 1) \to \mathbb{N}$ is an injection. We may think of every element $\alpha \in (0, 1)$ as a decimal expansion $.\alpha_1\alpha_2 \ldots \alpha_n \ldots$. In order to describe an element of $(0, 1)$, for every $n \in \mathbb{N}$, we need to say which element of the set $\{0, 1, 2, 3, 4, 5, 6, 7, 8, 9\}$ is in the nth decimal place.

We will define the element $\alpha = .\alpha_1\alpha_2 \ldots \alpha_n \cdots \in (0, 1)$ as follows:

(1) if $n \in \mathbb{N}$ and n is *not* in the range of ν, then we let $\alpha_n = 1$;

(2) if $n \in \mathbb{N}$ and $n = \nu(\beta)$, where $\beta \in (0, 1)$ and the nth decimal place of β is *not* 1, then $\alpha_n = 1$;

(3)   if $n \in \mathbb{N}$ and $n = \nu(\beta)$, where $\beta \in (0, 1)$ and the nth decimal place of β is 1, then
$\alpha_n = 2$.

Let $m = \nu(\alpha)$ and then examine the mth decimal place of α. If you can obtain a contradiction, you will have shown that no injection $\nu : (0, 1) \to \mathbb{N}$ can exist.

4.3 Equivalence Relations

Earlier in this chapter, we pointed out that there is a second approach that can be used to examine the rational numbers and real numbers. In this approach, instead of viewing the rational numbers as a subset of the real numbers, we construct the rational numbers directly from the integers and then construct the real numbers directly from the rational numbers. As mentioned earlier, for a more thorough discussion of this approach, one should look at a book on real analysis. However, some of the ideas and techniques used in this approach reappear frequently throughout algebra, and it will be helpful to become acquainted with these ideas at this time.

We are accustomed to viewing rational numbers as quotients like

$$\frac{7}{23}, \frac{11}{22}, \frac{36}{25}, \frac{3}{6}, \frac{-26}{21}.$$

Therefore, it would be natural to try to define the rational numbers as all ordered pairs of integers where the second number is not zero. In other words, we would be looking at

$$\{(a, b) \mid a \in \mathbb{Z}, b \in \mathbb{Z}, \quad \text{and} \quad b \neq 0\}.$$

However, the situation is a little more complicated. We consider the quotients $\frac{11}{22}$ and $\frac{3}{6}$ to represent the same number. More precisely, we consider the quotients $\frac{a}{b}$ and $\frac{c}{d}$ to be the same precisely when $ad = bc$. In light of this, in constructing the rational numbers from the integers, we need to consider the ordered pairs (a, b) and (c, d) to be the same object precisely when $ad = bc$. Therefore, using this approach, we can define the rational numbers as the set

$$\mathbb{Q} = \{(a, b) \mid a \in \mathbb{Z}, b \in \mathbb{Z}, \quad \text{and} \quad b \neq 0\},$$

where (a, b) and (c, d) are considered to be the same object precisely when $ad = bc$.

Having defined the rational numbers in this way, we next need to describe how to add and multiply rational numbers. Recall, when we viewed rational numbers as quotients, we defined addition and multiplication as

(4) $$\frac{a}{b} + \frac{c}{d} = \frac{ad + bc}{bd} \quad \text{and} \quad \frac{a}{b} \cdot \frac{c}{d} = \frac{ac}{bd}.$$

As a result, now that we are viewing elements of $\mathbb{Q}$ as ordered pairs, we should define addition and multiplication as

$$(a, b) + (c, d) = (ad + bc, bd) \quad \text{and} \quad (a, b) \cdot (c, d) = (ac, bd).$$

At this point, there is a very subtle point that is usually glossed over. Let's look at two addition problems. If we apply the rule in (4) for the addition of rational numbers, we see that

$$\frac{2}{3} + \frac{1}{2} = \frac{7}{6} \quad \text{and} \quad \frac{6}{9} + \frac{2}{4} = \frac{42}{36}.$$

In this procedure, we consider $\frac{2}{3}$ and $\frac{6}{9}$ to be two names for the same object. Similarly, we consider $\frac{1}{2}$ and $\frac{2}{4}$ to be two names for the same object. Therefore, when we add $\frac{2}{3}$ to $\frac{1}{2}$, we need to get the same answer as when we add $\frac{6}{9}$ and $\frac{2}{4}$. In the first case, we obtain the answer $\frac{7}{6}$, and in the second case, $\frac{42}{36}$. Fortunately, the two answers $\frac{7}{6}$ and $\frac{42}{36}$ are two names for the same object, since $7 \cdot 36 = 6 \cdot 42$. The concern we now have is, since the same rational number can have many different names, how do we know all of these names yield the same answer when used in addition and multiplication problems?

At first glance, you might consider this concern to be somewhat silly. Your reaction might be that obviously $\frac{1}{2}$ and $\frac{2}{4}$ are the same number, so we will get the same answer regardless of which we use in an addition or multiplication problem. In our previous and more intuitive approach, we viewed the rational numbers as being a subset of the real numbers. In that approach, every rational number corresponds to a decimal that eventually repeats. Therefore, there is no difference between $\frac{1}{2}$ and $\frac{2}{4}$, as they both correspond to the decimal .5. However, in this more rigorous approach, we are constructing the rational numbers from the integers and then constructing the real numbers from the rational numbers. Therefore, we cannot use decimal expansions or properties of the real numbers to justify statements about the rational numbers. As a result, it is necessary to prove that addition and multiplication problems give us the same answer regardless of which ordered pair we use to represent a rational number. To make things clearer, it may be helpful to contrast our situation to the following:

■ Example

Let $T = \{(a, b) \mid a \in \mathbb{Z}, b \in \mathbb{Z}, \text{ and } b \neq 0\}$, where (a, b) and (c, d) are considered to be the same object precisely when $ab = cd$. Therefore, in this example, $(10, 1)$, $(5, 2)$, and $(-2, -5)$ are three different names for the same object. Let us now define addition in T exactly the same way as we did in $\mathbb{Q}$. Thus,

$$(a, b) + (c, d) = (ad + bc, bd) \quad \text{and} \quad (a, b) \cdot (c, d) = (ac, bd).$$

Observe that

$$(10, 1) + (10, 1) = (10 \cdot 1 + 1 \cdot 10, 1 \cdot 1) = (20, 1)$$

and

$$(5, 2) + (5, 2) = (5 \cdot 2 + 2 \cdot 5, 2 \cdot 2) = (20, 4).$$

Also note that since $20 \cdot 1 \neq 20 \cdot 4$, it follows that $(20, 1) \neq (20, 4)$. Therefore, in this example

$$(10, 1) = (5, 2) \text{ but } (10, 1) + (10, 1) \neq (5, 2) + (5, 2).$$

Clearly this is a problem. If $(10, 1)$ and $(5, 2)$ represent the same object, then replacing one by the other should make no difference in an addition problem. However, we showed that replacing $(10, 1)$ by $(5, 2)$ did indeed change the answer. The problem is that we are no longer adding "numbers" but are adding "classes." By this we mean than $(10, 1)$ and $(5, 2)$ are merely two members of a class of ordered pairs that also includes $(2, 5)$, $(1, 10)$, $(-10, -1)$, $(-5, -2)$, $(-2, -5)$, and $(-1, -10)$. When one writes down a rule for addition like $(a, b) + (c, d) = (ad + bc, bd)$, you are giving a formula for the addition of two ordered pairs. But there is no guarantee that if we replace one of the ordered pairs by another member of its class, we will get the same answer. In order for addition and multiplication to be *well defined*, it must be the case that all members of the same class give us the same answer in addition and multiplication problems. As a result, addition in the set T is not well defined. In the exercises, you will be asked to show that multiplication in T is well defined. Furthermore, you will also be asked to show that if we instead define addition as

$$(a, b) + (c, d) = (ab + cd, 1),$$

then addition in T is now well defined.

■

The previous example indicates that when we define something in terms of a collection of classes, it is not automatic that addition and multiplication are well defined. Therefore, we will now prove that addition and multiplication in $\mathbb{Q}$ are well defined.

Proposition 4.10. *Addition and multiplication in $\mathbb{Q}$ are well defined.*

Proof. Let (a_1, b_1) and (a_2, b_2) be two members of one class and let (c_1, d_1) and (c_2, d_2) be two members of a second class. Therefore, $a_1 b_2 = b_1 a_2$ and $c_1 d_2 = d_1 c_2$. We will first show that addition is well defined. Thus, we must show that $(a_1, b_1) + (c_1, d_1)$ belongs to the same class as $(a_2, b_2) + (c_2, d_2)$. The rule for addition in $\mathbb{Q}$ tells us that

$$(a_1, b_1) + (c_1, d_1) = (a_1 d_1 + b_1 c_1, b_1 d_1)$$

and

$$(a_2, b_2) + (c_2, d_2) = (a_2 d_2 + b_2 c_2, b_2 d_2).$$

It is easy to check that if n is a nonzero integer, then the ordered pairs (a, b) and (an, bn) belong to the same class, so we can say that $(a, b) = (an, bn)$. Therefore, if we multiply both terms in $(a_1 d_1 + b_1 c_1, b_1 d_1)$ by $b_2 d_2$ and multiply both terms in $(a_2 d_2 + b_2 c_2, b_2 d_2)$ by $b_1 d_1$, we obtain

$$(a_1, b_1) + (c_1, d_1) = (a_1 d_1 + b_1 c_1, b_1 d_1) = (a_1 b_2 d_1 d_2 + b_1 b_2 c_1 d_2, b_1 d_1 b_2 d_2)$$

and

$$(a_2, b_2) + (c_2, d_2) = (a_2 d_2 + b_2 c_2, b_2 d_2) = (a_2 b_1 d_1 d_2 + b_1 b_2 c_2 d_1, b_1 d_1 b_2 d_2).$$

We can replace $a_2 b_1$ by $a_1 b_2$ and $c_2 d_1$ by $c_1 d_2$ in the equations, and this results in

$$(a_2, b_2) + (c_2, d_2) = (a_2 b_1 d_1 d_2 + b_1 b_2 c_2 d_1, b_1 d_1 b_2 d_2) =$$

$$(a_1 b_2 d_1 d_2 + b_1 b_2 c_1 d_2, b_1 d_1 b_2 d_2) = (a_1, b_1) + (c_1, d_1).$$

Therefore, $(a_1, b_1) + (c_1, d_1)$ and $(a_2, b_2) + (c_2, d_2)$ do indeed belong to the same class, so addition in $\mathbb{Q}$ is well defined.

The algebra is somewhat easier when we show that multiplication is well defined. In this case, we must show that $(a_1, b_1) \cdot (c_1, d_1)$ belongs to the same class as $(a_2, b_2) \cdot (c_2, d_2)$. The rule for multiplication in $\mathbb{Q}$ tells us that

$$(a_1, b_1) \cdot (c_1, d_1) = (a_1 c_1, b_1 d_1) \quad \text{and} \quad (a_2, b_2) \cdot (c_2, d_2) = (a_2 c_2, b_2 d_2).$$

To determine if $(a_1 c_1, b_1 d_1)$ and $(a_2 c_2, b_2 d_2)$ belong to the same class, we need to check if $a_1 c_1 b_2 d_2 = b_1 d_1 a_2 c_2$. Using the facts that $a_1 b_2 = b_1 a_2$ and $c_1 d_2 = d_1 c_2$, we see that

$$a_1 c_1 b_2 d_2 = a_1 b_2 \cdot c_1 d_2 = b_1 a_2 \cdot d_1 c_2 = b_1 d_1 a_2 c_2.$$

Thus, $(a_1, b_1) \cdot (c_1, d_1)$ and $(a_2, b_2) \cdot (c_2, d_2)$ belong to the same class, and multiplication in $\mathbb{Q}$ is well defined. $\qquad \square$

The concept of dealing with a class of objects as opposed to a single object is an important theme which runs throughout mathematics. The idea behind this is that there are objects in mathematics that may not be identical but, in certain circumstances, can be thought of as equivalent. The ordered pairs $(1, 2)$ and $(2, 4)$ are not identical but, in the context of constructing the rational numbers, can be thought of as equivalent. For an everyday example, cash and credit cards are not identical. If you are arriving at a tollbooth, you would rather have cash, whereas if you are trying to rent a car, you would rather have a credit card. However, in many contexts, such as for most purchases in stores or restaurants, having cash is equivalent to having a credit card.

For another example, suppose we consider two days in November to be equivalent if they land on the same day of the week. In this example, the 1st, 8th, 15th, 22nd, and 29th of the month are all equivalent to each other. Similarly, none of these days are equivalent to the 2nd, but the 2nd, 9th, 16th, 23rd, and 30th of the month are all equivalent. When two objects are, in a certain context, considered to be equivalent, we use the symbol $\sim$. Therefore, in this example, we have

$$1 \sim 8, \quad 8 \sim 1, \quad 25 \sim 25, \quad 10 \sim 17, \quad 21 \sim 14, \quad 21 \sim 7.$$

If two objects are not considered to be equivalent, we use the symbol $\nsim$. Therefore, in this example, we have

$$1 \nsim 4, \quad 10 \nsim 20, \quad 25 \nsim 8, \quad 2 \nsim 29.$$

We can now formalize this concept.

Definition 4.11. *Given a set S, we say that $\sim$ is an equivalence relation if it satisfies*

(a) *Reflexive Property: For every $x \in S$, $x \sim x$.*

(b) *Symmetric Property: For every $x, y \in S$, if $x \sim y$, then $y \sim x$.*

(c) *Transitive Property: For every $x, y, z \in S$, if $x \sim y$ and $y \sim z$, then $x \sim z$.*

Equivalence relations can be defined even more formally as certain types of subsets of $S \times S$. However, for our purposes, a more formal approach is not needed.

■ Examples

In each of these examples, you should convince yourself that $\sim$ satisfies the reflexive, symmetric, and transitive properties.

1.   Given the set $\mathbb{Z}$, define $\sim$ as $x \sim y$ precisely when 2 divides $x - y$.

2.   Given the set $\mathbb{Z}$, if n is a fixed positive integer, define $\sim$ as $x \sim y$ precisely when n divides $x - y$.

3.   Given the set $\mathbb{Q}$, define $\sim$ as $x \sim y$ precisely when $x - y$ is an integer.

4. Given the set $\mathbb{R}$, define $\sim$ as $x \sim y$ precisely when $x^2 = y^2$.

5. Given the set $\mathbb{R}[x]$, which is all polynomials with real coefficients, define $f(x) \sim g(x)$ precisely when $f(0) = g(0)$.

∎

We now consider several examples where $\sim$ is not an equivalence relation.

∎ Examples

In each of these examples, you should convince yourself that $\sim$ satisfies exactly two of the three properties needed to be an equivalence relation.

1. Given the set $\mathbb{Z}$, define $\sim$ as $x \sim y$ precisely when $x \leq y$. This example is reflexive and transitive, but it is not symmetric. For example, $1 \sim 2$ but $2 \not\sim 1$.

2. Given the set $\mathbb{R}$, define $\sim$ as $x \sim y$ precisely when $|x - y| \leq 1$. This example is reflexive and symmetric, but it is not transitive. For example, $2 \sim 1$ and $1 \sim 0$ but $2 \not\sim 0$.

3. Given the set $\mathbb{Q}$, define $\sim$ as $x \sim y$ precisely when $xy > 0$. This example is symmetric and transitive, but it is not reflexive. In particular, $0 \not\sim 0$.

∎

Remember, our reason for looking at equivalence relations is that, as was the case when we constructed the rational numbers, we will often need to look at classes of objects instead of individual objects. In light of this, it is now natural to introduce the concept of **equivalence classes**.

Definition 4.12. *Let S be a set with equivalence relation $\sim$. If $x \in S$, then we define the equivalence class of x, denoted as [x], as*

$$[x] = \{y \in S \mid y \sim x\}.$$

In other words, [x] consists of all those elements in S that are equivalent to x.

Let us now return to some of our previous examples.

∎ Examples

1. Let $T = \{(a, b) \mid a \in \mathbb{Z}, b \in \mathbb{Z}, \text{ and } b \neq 0\}$ and define $\sim$ as $(a, b) \sim (c, d)$ precisely when $ad = bc$. This is the equivalence relation used to construct the rational numbers, and before continuing, you should convince yourself that $\sim$ is indeed an equivalence relation. In this case, $[(a, b)]$ consists of all ordered pairs that yield

fractions that are equivalent to $\frac{a}{b}$. For example, $[(1, 2)]$ consists of all ordered pairs that yield fractions that are equal to $\frac{1}{2}$ and some of the elements in $[(1, 2)]$ are $(2, 4)$, $(-3, -6)$, $(501, 1002)$, and $(-231, -462)$.

2. Let $\sim$ be the equivalence relation defined as $x \sim y$ precisely when 2 divides $x - y$. In this case, there are only two equivalence classes:

$$\{0, \pm 2, \pm 4, \pm 6, \pm 8, \ldots\} \quad \text{and} \quad \{\pm 1, \pm 3, \pm 5, \pm 7, \pm 9, \ldots\}.$$

We can certainly think of one class as being the even integers and the other as the odd integers. It is important to note that each class has an infinite number of names. For example, $[0]$, $[2]$, $[1988]$, and $[-9876]$ are four of the infinite number of different names for one of the classes. Similarly, $[1]$, $[1951]$, and $[-1955]$ are three of the infinite number of different names for the other class.

When dealing with the equivalence classes of an equivalence relation, it will always be the case, as it is preceding, that any two classes you choose either have no elements in common or they are the same class, perhaps with different names. In the previous example, the equivalence classes partition $\mathbb{Z}$ into two disjoint pieces: the evens and the odds. It will always be the case that the equivalence classes of an equivalence relation partition a set into disjoint pieces. We record these observations as

Proposition 4.13. *Let S be a set with equivalence relation $\sim$. If $[x]$ and $[y]$ are two equivalence classes, then either $[x] \cap [y] = \emptyset$ or $[x] = [y]$. Since every element of S belongs to an equivalence class, this says that the equivalence classes partition S into disjoint pieces.*

Proof. Suppose $[x]$ and $[y]$ are two equivalence classes such that $[x] \cap [y] \neq \emptyset$; we now need to show that $[x] = [y]$. To this end, let $z \in [x] \cap [y]$; therefore, $z \sim x$ and $z \sim y$. Applying both the symmetric and transitive properties, we first see that $x \sim z$ and $y \sim z$ and then see that $x \sim y$ and $y \sim x$. Now suppose $a \in [x]$; thus, $a \sim x$ and $x \sim y$ and transitivity implies that $a \sim y$. As a result, $a \in [y]$, and we see that $[x] \subseteq [y]$.

In the other direction, suppose $b \in [y]$. Thus, $b \sim y$ and $y \sim x$, so transitivity tells us that $b \sim x$. As a result, $b \in [x]$, and we see that $[y] \subseteq [x]$. Hence, $[x] = [y]$.

In order to conclude the proof, we only need to show that every element of S is in some equivalence class. So far, this proof has only used the symmetric and transitive properties. To finish the proof, we will use the reflexive property. If $x \in S$, we need to show that x belongs to some equivalence class. However, the reflexive property tells us that $x \sim x$, so $x \in [x]$, as desired. □

So far you have seen two approaches to the rational numbers. We first viewed them as a subset of the real numbers but later viewed them as equivalence classes of ordered pairs of integers. You may wonder what is the purpose of the second approach, since it appears to be much more abstract and complicated than the first. The reason we introduced the second approach is that there are some problems beneath the surface with the first approach. Recall that we needed to first assume that we had an intuitive understanding of real numbers as decimal expansions. For example, we can think of

$$\sqrt{2} = 1.41421356237309504880168887242097\cdots$$

and

$$\sqrt{3} = 1.7320508075688772935274463415 0587\cdots.$$

This raises some questions as to how we do basic arithmetic? In particular, how do we express numbers like $\sqrt{2} + \sqrt{3}$, $\sqrt{2} \cdot \sqrt{3}$, or $\frac{1}{\sqrt{2}}$ as decimals? In ordinary arithmetic, we perform operations like addition and multiplication from right to left. By that we mean that if we were computing $315 + 721$, we would (a) first add the 5 and 1 in the one's column, then (b) move to the left and add the 1 and 2 in the ten's column, and then (c) move to the left and add the 3 and 7 in the hundred's column with part of the answer carrying over to the thousands column. The problem is, whereas we usually start basic arithmetic operations at the far right, a decimal expansion can continue on forever to the right. As a result, it is not clear how to add and multiply numbers whose decimal expansions go on forever.

In an attempt to resolve this problem, we can think of a decimal expansion as a sequence of rational numbers. In other words, we can think of $\sqrt{2}$ and $\sqrt{3}$ as the following sequences:

$$\sqrt{2} = 1, 1.4, 1.41, 1.414, 1.4142, 1.14121, 1.414213, 1.4142135, 1.41421356, \ldots$$

$$\sqrt{3} = 1, 1.7, 1.73, 1.732, 1.7320, 1.73205, 1.732050, 1.7320508, 1.73205080, \ldots.$$

We can now easily generate a sequence that represents the sum or product of $\sqrt{2}$ and $\sqrt{3}$ by adding or multiplying on a term-by-term basis. For example, the fifth term of the sequence for $\sqrt{2} + \sqrt{3}$ would be $1.4142 + 1.7320$, since 1.4142 and 1.7320 are the fifth terms in the respective sequences representing $\sqrt{2}$ and $\sqrt{3}$. As a result, we now have

$$\sqrt{2} + \sqrt{3} = 2, 3.1, 3.14, 3.146, 3.1462, 3.14626, 3.146263, 3.1462643, \ldots$$

and

$$\sqrt{2} \cdot \sqrt{3} = 1, 2.38, 2.4393, 2.449048, 2.4493944, 2.449482403, \ldots.$$

Hopefully, it is now becoming clearer why we introduced a second approach to viewing the rational numbers. On an intuitive level, it might have been easier to simply view the rational numbers as a subset of the real numbers. However, this intuitive approach ignored some serious problems regarding how to do basic arithmetic with decimal expansions. As we just illustrated, these problems can be dealt with by viewing real numbers as sequences of rational numbers. Therefore, from a logical viewpoint, it makes sense to construct the rational numbers from the integers and then to construct the real numbers from the rational numbers. Thus, our second approach to the rational numbers, despite being more abstract, is needed. Before concluding this discussion, there are two more points to deal with.

We would like to consider real numbers as sequences of rational numbers. In this way, it becomes possible to perform ordinary arithmetic operations with real numbers. However, how do we handle sequences like

$$0, 1, 0, 1, 0, 1, 0, 1, 0, 1, 0, 1, 0, 1, \ldots ?$$

Certainly this sequence is a legitimate sequence, and it can be added to or multiplied with other sequences. But this sequence cannot possibly be used to represent a real number as it oscillates forever between two different numbers. Similarly, the sequence

$$1, 2, 3, 4, 5, 6, 7, 8, 9, 10, 11, 12, 13, \ldots$$

cannot represent a real number as it increases without bound. Your impulse might be to say that we should only consider those sequences of rational numbers which converge. If so, you are on the right track, but, once again, another technical difficulty arises. Think about a sequence we used to represent $\sqrt{2}$ like

$$1, 1.4, 1.41, 1.414, 1.4142, 1.14121, 1.414213, 1.4142135, 1.41421356, \ldots .$$

We know that this sequence converges as it converges to $\sqrt{2}$. However, we now run the danger of using some circular reasoning. We are at the point in our reasoning where we know that the rational numbers exist, and we are trying to use them to give a formal definition of the real numbers. We want to say that the preceding sequence represents $\sqrt{2}$, since the terms converge to $\sqrt{2}$. However, we cannot say that the preceding sequence converges until we know that the real number $\sqrt{2}$ exists, yet we cannot say that the real number $\sqrt{2}$ exists until we know that the sequence converges. You should certainly spend some time thinking about this problem before continuing.

The bottom line is that we need a way to describe those sequences that we think represent real numbers without using the concept of convergence. Recall from calculus that if a sequence converges, then the limit need not be one of the terms of the sequence. In particular, $\sqrt{2}$ does not appear as one of the terms in the sequence we used to represent it. To resolve our problem, we need a concept that only refers to the terms in the sequence and doesn't refer to an object, like the limit, which may not be one of the terms of the sequence. The concept that resolves this problem is **Cauchy sequences**.

Definition 4.14. *Let*

$$\{a_n\}_{n=1}^{\infty} = a_1, a_2, a_3, a_4, a_5, \ldots$$

be a sequence. We say $\{a_n\}_{n=1}^{\infty}$ is a Cauchy sequence if for every $\epsilon > 0$, there exists a natural number N, such that whenever $n, m > N$, we have $|a_n - a_m| < \epsilon$.

Intuitively, being a Cauchy sequence means that eventually all the terms of the sequence become close to each other and stay as close to each other as you desire. In the exercises, you will be asked to show that every sequence that converges, using the definition of convergence from calculus, is a Cauchy sequence. Therefore, it now appears that we can now consider the real numbers to be the set of all Cauchy sequences of rational numbers.

However, we have one more technical point to deal with. Consider the following sequences:

$$0, 0, 0, 0, 0, 0, 0, 0, 0, 0, 0, 0, 0, \ldots$$

and

$$1, \frac{1}{2}, \frac{1}{3}, \frac{1}{4}, \frac{1}{5}, \frac{1}{6}, \frac{1}{7}, \frac{1}{8}, \frac{1}{9}, \frac{1}{10}, \frac{1}{11}, \frac{1}{12}, \frac{1}{13}, \frac{1}{14}, \ldots$$

Both of these sequences should represent the number 0. Similarly, the sequences

$$2, \frac{3}{2}, \frac{4}{3}, \frac{5}{4}, \frac{6}{5}, \frac{7}{6}, \frac{8}{7}, \frac{9}{8}, \frac{10}{9}, \frac{11}{10}, \frac{12}{11}, \frac{13}{12}, \frac{14}{13}, \ldots$$

and

$$\frac{1}{2}, \frac{2}{3}, \frac{3}{4}, \frac{4}{5}, \frac{5}{6}, \frac{6}{7}, \frac{7}{8}, \frac{8}{9}, \frac{9}{10}, \frac{10}{11}, \frac{11}{12}, \frac{12}{13}, \frac{13}{14}, \ldots$$

should both represent the number 1. In light of this, several different sequences can represent the same real number. This may sound like a familiar situation, as earlier we saw that different ordered pairs can represent the same rational number. The solution, as before, is to define an appropriate equivalence relation and then to look at the equivalence classes. Intuitively, if two sequences have the same limit, then we should consider them to be equivalent. However, as mentioned before, our reasoning would be circular if, in our attempt to construct the real numbers, we defined the equivalence relation in terms of limits. Instead, we need to define an equivalence relation which only refers to the terms of the sequences.

Definition 4.15. *Let T be the set of Cauchy sequences of rational numbers. Define the equivalence relation $\sim_{cs}$ as $\{a_n\}_{n=1}^{\infty} \sim_{cs} \{b_n\}_{n=1}^{\infty}$ precisely when for every $\epsilon > 0$, there exists some natural number N, such that whenever $n > N$, we have $|a_n - b_n| < \epsilon$.*

You should convince yourself that not only is $\sim_{cs}$ an equivalence relation but that if two Cauchy sequences of rational numbers have the same limit, using the ordinary definition of limit from calculus, then they will be equivalent. Therefore, we can now consider the real numbers $\mathbb{R}$ to be the set of equivalence class of Cauchy sequences of rational numbers using the equivalence relation $\sim_{cs}$. Throughout mathematics, we will continue to define certain sets as the equivalence classes under a particular equivalence relation. As we did for the rational numbers, we should always check that if an algebraic object is defined in terms of equivalence classes, then operations like addition and multiplication are well defined. In the exercises, you will be asked to show that addition and multiplication of real numbers are indeed well defined using the definition above.

Exercises for Section 4.3

In exercises 1–8, $\sim$ will be defined on $\mathbb{Q}$. You will need to determine which of the reflexive, symmetric, and transitive properties are satisfied by $\sim$ on $\mathbb{Q}$. Whenever a property fails, provide an example that illustrates its failure.

1. $\sim$ is defined as $x \sim y$ whenever $x - y$ is an integer

2. $\sim$ is defined as $x \sim y$ whenever $x + y$ is an integer

3. $\sim$ is defined as $x \sim y$ whenever xy is an integer

4. $\sim$ is defined as $x \sim y$ whenever $x^2 > y^2$

5. $\sim$ is defined as $x \sim y$ whenever $x^2 \geq y^2$

6. $\sim$ is defined as $x \sim y$ whenever $xy > 0$

7. $\sim$ is defined as $x \sim y$ whenever $xy \geq 0$

8. $\sim$ is defined as $x \sim y$ whenever $xy \leq 0$

In exercises 9–16, $\sim$ will be defined on $\mathbb{Z}[x]$, the set of polynomials with integer coefficients. You will need to determine which of the reflexive, symmetric, and transitive properties are satisfied by $\sim$ on $\mathbb{Z}[x]$. Whenever a property fails, provide an example that illustrates its failure.

9. $\sim$ is defined as $f(x) \sim g(x)$ whenever $f(3) - g(3) = 0$

10. $\sim$ is defined as $f(x) \sim g(x)$ whenever $f(3) - g(3) \in \{-1, 1\}$

11. $\sim$ is defined as $f(x) \sim g(x)$ whenever $f(3) - g(3) \in \{-1, 0, 1\}$

12. $\sim$ is defined as $f(x) \sim g(x)$ whenever $f(3) - g(3) < 0$

13. $\sim$ is defined as $f(x) \sim g(x)$ whenever $f(3) - g(3) \le 0$

14. $\sim$ is defined as $f(x) \sim g(x)$ whenever $f(5) \cdot g(5) = 0$

15. $\sim$ is defined as $f(x) \sim g(x)$ whenever $f(5) \cdot g(5) \ge 0$

16. $\sim$ is defined as $f(x) \sim g(x)$ whenever $f(5) \cdot g(5) \le 0$

In exercises 17–24, $\sim$ will be defined on $\mathbb{R}[x]$, the set of polynomials with real coefficients. You will need to determine which of the reflexive, symmetric, and transitive properties are satisfied by $\sim$ on $\mathbb{R}[x]$. Whenever a property fails, provide an example that illustrates its failure.

17. $\sim$ is defined as $f(x) \sim g(x)$ whenever $f(c) \le g(c)$, for all $c \in \mathbb{R}$.

18. $\sim$ is defined as $f(x) \sim g(x)$ whenever $f(c) \le g(c)$, for some $c \in \mathbb{R}$

19. $\sim$ is defined as $f(x) \sim g(x)$ whenever $f(c) \cdot g(c) \ge 0$, for all $c \in \mathbb{R}$.

20. $\sim$ is defined as $f(x) \sim g(x)$ whenever $f(c) \cdot g(c) > 0$, for all $c \in \mathbb{R}$.

21. $\sim$ is defined as $f(x) \sim g(x)$ whenever $f(c) - g(c) \in \mathbb{Q}$, for all $c \in \mathbb{R}$.

22. $\sim$ is defined as $f(x) \sim g(x)$ whenever $f(c) - g(c) \in \mathbb{Q}$, for some $c \in \mathbb{R}$.

23. $\sim$ is defined as $f(x) \sim g(x)$ whenever $f(c) - g(c) \in \mathbb{Z}$, for all $c \in \mathbb{R}$.

24. $\sim$ is defined as $f(x) \sim g(x)$ whenever $f(c) - g(c) \in \mathbb{N}$, for all $c \in \mathbb{R}$.

25. For the set $\mathbb{Z}$, define $\sim$ as $a \sim b$ if and only if $a - b$ is divisible by 5.
    (a) Show that $\sim$ is an equivalence relation on the set $\mathbb{Z}$.

    (b) How many equivalence classes are there? Briefly explain.

    (c) How many elements are in each equivalence class? Briefly explain.

    (d) List five elements of the equivalence class $[23]$.

    (e) List five elements of the equivalence class $[-19]$.

26. For the set $\mathbb{Z}$, define $\sim$ as $a \sim b$ if and only if $a - b$ is divisible by 7.
    (a) Show that $\sim$ is an equivalence relation on the set $\mathbb{Z}$.

    (b) How many equivalence classes are there? Briefly explain.

    (c) How many elements are in each equivalence class? Briefly explain.

    (d) List five elements of the equivalence class $[11]$.

    (e) List five elements of the equivalence class $[-11]$.

27. Let T be the set of positive rational numbers and then define $\sim$ as $a \sim b$ if and only if $a = b \cdot 2^n$, for some $n \in \mathbb{Z}$. (For example, $5 \sim 10$ as $5 = 10 \cdot 2^{-1}$, whereas $5 \nsim 11$, since $5 \neq 11 \cdot 2^n$, for all $n \in \mathbb{Z}$.)

    (a) Show that $\sim$ is an equivalence relation on the set T.

    (b) How many equivalence classes are there? Briefly explain.

    (c) How many elements are in each equivalence class? Briefly explain.

    (d) List five elements in the equivalence class $\left[\frac{7}{2}\right]$.

    (e) List five elements in the equivalence class $\left[\frac{23}{10}\right]$.

28. Let T be the set of positive rational numbers and then define $\sim$ as $a \sim b$ if and only if $a = b \cdot 3^n$, for some $n \in \mathbb{Z}$. (For example, $21 \sim \frac{7}{3}$ as $21 = \frac{7}{3} \cdot 3^2$, whereas $21 \nsim 42$, since $21 \neq 42 \cdot 3^n$, for all $n \in \mathbb{Z}$.)

    (a) Show that $\sim$ is an equivalence relation on the set T.

    (b) How many equivalence classes are there? Briefly explain.

    (c) How many elements are in each equivalence class? Briefly explain.

    (d) List five elements in the equivalence class $\left[\frac{23}{27}\right]$.

    (e) List five elements in the equivalence class $\left[\frac{19}{6}\right]$.

29. Let $\mathbb{R}[x]$ be the set of polynomials with real coefficients and then define $\sim$ as $f(x) \sim g(x)$ if and only if $f'(x) = g'(x)$.

    (a) Show that $\sim$ is an equivalence relation on the set $\mathbb{R}[x]$.

    (b) How many equivalence classes are there? Briefly explain.

    (c) How many elements are in each equivalence class? Briefly explain.

    (d) List five elements in the equivalence class $[x]$.

    (e) List five elements in the equivalence class $[6x^2 - 2x + 8]$.

30. Let $\mathbb{R}[x]$ be the set of polynomials with real coefficients and then define $\sim$ as $f(x) \sim g(x)$ if and only if $f''(x) = g''(x)$.

    (a) Show that $\sim$ is an equivalence relation on the set $\mathbb{R}[x]$.

    (b) How many equivalence classes are there? Briefly explain.

    (c) How many elements are in each equivalence class? Briefly explain.

    (d) List five elements in the equivalence class $[x]$.

    (e) List five elements in the equivalence class $[3x^8 - 6]$.

31. Let $\mathbb{Q}[x]$ be the set of polynomials with rational coefficients and then define $\sim$ as $f(x) \sim g(x)$ if and only if $f(1) = g(1)$ and $f(-1) = g(-1)$.

    (a) Show that $\sim$ is an equivalence relation on the set $\mathbb{Q}[x]$.

    (b) How many equivalence classes are there? Briefly explain.

    (c) How many elements are in each equivalence class? Briefly explain.

    (d) List five elements in the equivalence class $[1]$.

    (e) List five elements in the equivalence class $[x]$.

32. Let $\mathbb{Q}[x]$ be the set of polynomials with rational coefficients and then define $\sim$ as $f(x) \sim g(x)$ if and only if $f(\sqrt{2}) = g(\sqrt{2})$.

    (a) Show that $\sim$ is an equivalence relation on the set $\mathbb{Q}[x]$.

    (b) How many equivalence classes are there? Briefly explain.

    (c) How many elements are in each equivalence class? Briefly explain.

    (d) List five elements in the equivalence class $[2]$.

    (e) List five elements in the equivalence class $[x]$.

33. In this exercise, we revisit the set T that appeared in the example preceding Proposition 4.10 and show that the multiplication of equivalence classes is well defined. When we were first introduced to the set T, we had not yet defined equivalence classes. Therefore, we will now restate the problem more formally in terms of equivalence relations and equivalence classes. Let $T = \{(a, b) \mid a \in \mathbb{Z}, b \in \mathbb{Z}, b \neq 0\}$ and define $\sim$ on T as $(a, b) \sim (c, d)$ whenever $ab = cd$. Next, we define the multiplication of equivalence classes as

    $$[(a, b)] \cdot [(c, d)] = [(ac, bd)],$$

    for all equivalence classes $[(a, b)]$, $[(c, d)]$. Prove that the multiplication of equivalence classes is well defined.

34. Using the set and equivalence relation in exercise 25, define the addition and multiplication of equivalence classes as

    $$[a] + [b] = [a + b] \quad \text{and} \quad [a] \cdot [b] = [ab],$$

    for all $[a]$, $[b]$.

    (a) Show that the addition of equivalence classes is well defined.

    (b) Show that the multiplication of equivalence classes is well defined.

35. Using the set and equivalence relation in exercise 26, define the addition and multiplication of equivalence classes as

$$[a] + [b] = [a + b] \quad \text{and} \quad [a] \cdot [b] = [ab],$$

for all $[a]$, $[b]$.
    (a) Show that the addition of equivalence classes is well defined.

    (b) Show that the multiplication of equivalence classes is well defined.

36. For the set $\mathbb{R}$, define $\sim$ as $x \sim y$ whenever $|x| = |y|$. You may assume that $\sim$ is an equivalence relation. Next, define the addition and multiplication of equivalence classes as

$$[x] + [y] = [x + y] \quad \text{and} \quad [x] \cdot [y] = [xy],$$

for all $[x]$, $[y]$.
    (a) Show that the multiplication of equivalence classes is well defined.

    (b) Give an example that illustrates that the addition of equivalence classes is not well defined.

37. For the set $\mathbb{Q}[x]$, define $\sim$ as $f(x) \sim g(x)$ whenever $f(x) - g(x)$ is a constant. You may assume that $\sim$ is an equivalence relation. Next, define the addition and multiplication of equivalence classes as

$$[f(x)] + [g(x)] = [f(x) + g(x)] \quad \text{and} \quad [f(x)] \cdot [g(x)] = [f(x)g(x)],$$

for all $[f(x)]$, $[g(x)]$.
    (a) Show that the addition of equivalence classes is well defined.

    (b) Give an example that illustrates that the multiplication of equivalence classes is not well defined.

38. For the set $\mathbb{Z}[x]$, define $\sim$ as $f(x) \sim g(x)$ whenever $f(5) - g(5)$ is a multiple of 3. You may assume that $\sim$ is an equivalence relation. Next, define the addition and multiplication of equivalence classes as

$$[f(x)] + [g(x)] = [f(x) + g(x)] \quad \text{and} \quad [f(x)] \cdot [g(x)] = [f(x)g(x)],$$

for all $[f(x)]$, $[g(x)]$.
    (a) Show that the addition of equivalence classes is well defined.

    (b) Show that the multiplication of equivalence classes is also well defined.

39. For the set $\mathbb{Z}[x]$, define $\sim$ as $f(x) \sim g(x)$ whenever $f(8) - g(8)$ is a multiple of 7. You may assume that $\sim$ is an equivalence relation. Next, define the addition and multiplication of equivalence classes as

$$[f(x)] + [g(x)] = [f(x) + g(x)] \quad \text{and} \quad [f(x)] \cdot [g(x)] = [f(x)g(x)],$$

    for all $[f(x), [g(x)]$.
    (a) Show that the addition of equivalence classes is well defined.

    (b) Show that the multiplication of equivalence classes is also well defined.

40. For the set $\mathbb{Z}[x]$, define $\sim$ as $f(x) \sim g(x)$ whenever the $f(x)$ and $g(x)$ have the same leading coefficient. You may assume that $\sim$ is an equivalence relation. Next, define the addition and multiplication of equivalence classes as

$$[f(x)] + [g(x)] = [f(x) + g(x)] \quad \text{and} \quad [f(x)] \cdot [g(x)] = [f(x)g(x)],$$

    for all $[f(x), [g(x)]$.
    (a) Give an example that illustrates that the addition of equivalence classes is not well defined.

    (b) Show that the multiplication of equivalence classes is well defined.

41. For the set $\mathbb{Z}$, define $\sim$ as $a \sim b$ whenever $a - b$ is divisible by 10. You may assume that $\sim$ is an equivalence relation and may also assume that the addition and multiplication of equivalence classes is well defined where we define addition and multiplication as

$$[a] + [b] = [a + b] \quad \text{and} \quad [a] \cdot [b] = [ab],$$

    for all $[a], [b]$.
    (a) Find a positive integer s such that $[s] + [7] = [0]$.

    (b) Find a positive integer t such that $[t] + [4] = [2]$.

    (c) Find a positive integer u such that $[u] \cdot [7] = [1]$.

    (d) Find a positive integer v such that $[v] \cdot [7] = [9]$.

    (e) Find a positive integer w such that $[w] \cdot [8] = [0]$.

42. For the set $\mathbb{Z}$, define $\sim$ as $a \sim b$ whenever $a - b$ is divisible by 12. You may assume that $\sim$ is an equivalence relation and may also assume that the addition and multiplication of equivalence classes is well defined where we define addition and multiplication as

$$[a] + [b] = [a + b] \quad \text{and} \quad [a] \cdot [b] = [ab],$$

    for all $[a], [b]$.

(a)   Find a positive integer s such that $[s] + [5] = [0]$.

(b)   Find a positive integer t such that $[t] + [8] = [3]$.

(c)   Find a positive integer u such that $[u] \cdot [7] = [1]$.

(d)   Find a positive integer v such that $[v] \cdot [7] = [2]$.

(e)   Find a positive integer w such that $[w] \cdot [9] = [0]$.

43.   For the set $\mathbb{R}$, define $\sim$ as $a \sim b$ whenever $a - b$ is an integer. You may assume that $\sim$ is an equivalence relation. Next, define the addition and multiplication of equivalence classes as

$$[a] + [b] = [a + b] \quad \text{and} \quad [a] \cdot [b] = [ab],$$

for all $[a]$, $[b]$.

(a)   Show that the addition of equivalence classes is well defined.

(b)   Give an example that illustrates that the multiplication of equivalence classes is not well defined.

(c)   Find a positive real number s such that $[s] + [\sqrt{2}] = [0]$.

(d)   Find a positive real number t such that $[t] + [\pi] = [0]$.

44.   For the set $\mathbb{Q}(\sqrt{2}) = \{u + v\sqrt{2} \mid u, v \in \mathbb{Q}\}$, define $\sim$ as $a \sim b$ whenever $a - b \in \mathbb{Q}$. You may assume that $\sim$ is an equivalence relation. Next, define the addition and multiplication of equivalence classes as

$$[a] + [b] = [a + b] \quad \text{and} \quad [a] \cdot [b] = [ab],$$

for all $[a]$, $[b]$.

(a)   Show that the addition of equivalence classes is well defined.

(b)   Give an example which illustrates that the multiplication of equivalence classes is not well defined.

(c)   Find a positive number $s \in \mathbb{Q}(\sqrt{2})$ such that $[s] + [\sqrt{2}] = [0]$.

(d)   Find a positive number $t \in \mathbb{Q}(\sqrt{2})$ such that $[t] + [2 + 3\sqrt{2}] = [0]$.

45.   Write down the first five terms of three different Cauchy sequences of rational numbers that can be used to represent the number 15.

46.   Write down the first five terms of three different Cauchy sequences of rational numbers that can be used to represent the number -9.

47. Write down the first five terms of three different Cauchy sequences of rational numbers that can be used to represent the number $\sqrt{13}$. You may want to refer to the answer to exercise 32 that appears after Section 4.2.

48. Write down the first five terms of three different Cauchy sequences of rational numbers that can be used to represent the number $\sqrt{37}$. You may want to refer to the answer to exercise 33 that appears after Section 4.2.

    In exercises 49–54, you will need to use the following approximations of $\sqrt{5}$ and $\sqrt{7}$:

    $$\sqrt{5} \approx 2.2360679774997\cdots \quad \text{and} \quad \sqrt{7} \approx 2.6457513110645\cdots.$$

Your answers from exercise 49 will then be used in exercises 50–54.

49. Write down the first ten terms of Cauchy sequences of rational numbers that can be used to represent $\sqrt{5}$ and $\sqrt{7}$.

50. Write down the first five terms of a Cauchy sequence of rational numbers that can be used to represent $\sqrt{5} + \sqrt{7}$.

51. Write down the first five terms of a Cauchy sequence of rational numbers that can be used to represent $\sqrt{5} \cdot \sqrt{7}$.

52. Write down the first five terms of a Cauchy sequence of rational numbers that can be used to represent $\frac{1}{\sqrt{5}}$.

53. Write down the first five terms of Cauchy sequences of rational numbers that can be used to represent $2\sqrt{5}$ and $3\sqrt{7}$.

54. Use your answer from exercise 53 to write down the first five terms of Cauchy sequences of rational numbers that can be used to represent $2\sqrt{5} + 3\sqrt{7}$.

In exercises 55–60, you will need to use the following approximations of $\sqrt{6}$ and $\sqrt{10}$:

$$\sqrt{6} \approx 2.4494897427831\cdots \quad \text{and} \quad \sqrt{10} \approx 3.1622776601683\cdots.$$

Your answers from exercise 55 will then be used in exercises 56–60.

55. Write down the first ten terms of Cauchy sequences of rational numbers that can be used to represent $\sqrt{6}$ and $\sqrt{10}$.

56. Write down the first five terms of a Cauchy sequence of rational numbers that can be used to represent $\sqrt{6} + \sqrt{10}$.

57. Write down the first five terms of a Cauchy sequence of rational numbers that can be used to represent $\sqrt{6} \cdot \sqrt{10}$.

58. Write down the first five terms of a Cauchy sequence of rational numbers that can be used to represent $\frac{1}{\sqrt{6}}$.

59. Write down the first five terms of Cauchy sequences of rational numbers that can be used to represent $3\sqrt{6}$ and $-2\sqrt{10}$.

60. Use your answers from exercise 59 to write down the first five terms of a Cauchy sequence of rational numbers that can be used to represent $3\sqrt{6} - 2\sqrt{10}$.

61. Let $\{a_n\}_{n=1}^{\infty}$ be a convergent sequence of real numbers. Prove that $\{a_n\}_{n=1}^{\infty}$ is a Cauchy sequence.

62. Let $\{b_n\}_{n=1}^{\infty}$ be a Cauchy sequence of real numbers. Prove that $\{b_n\}_{n=1}^{\infty}$ is bounded.

63. Prove that $\sim_{cs}$, as defined in Definition 4.15, is an equivalence relation on the set of Cauchy sequences of rational numbers.

64. Show that the sum of any two Cauchy sequences is also a Cauchy sequence.

65. Show that the product of any two Cauchy sequences is also a Cauchy sequence.

66. Let E be the collection of equivalence classes from exercise 63 and define the addition and multiplication of equivalence classes as

$$\left[\{a_n\}_{n=1}^{\infty}\right] + \left[\{b_n\}_{n=1}^{\infty}\right] = \left[\{a_n + b_n\}_{n=1}^{\infty}\right]$$

and

$$\left[\{a_n\}_{n=1}^{\infty}\right] \cdot \left[\{b_n\}_{n=1}^{\infty}\right] = \left[\{a_n b_n\}_{n=1}^{\infty}\right],$$

for all $\left[\{a_n\}_{n=1}^{\infty}\right], \left[\{b_n\}_{n=1}^{\infty}\right]$. Note that in these definitions we are using the facts from the previous exercise that the sum and product of any two Cauchy sequences are also Cauchy sequences.

(a) Show that the addition of equivalence classes is well defined.

(b) Show that the multiplication of equivalence classes is well defined.

CHAPTER 5

The Complex Numbers

In our examination of the real numbers, we saw that $\mathbb{R}$ is large enough to contain a root of every polynomial of odd degree. However, $\mathbb{R}$ also has the property that the square of each element is greater than or equal to zero. Therefore, there are many polynomials of even degree, such as $x^2 + 1$ and $x^4 + 2x^2 + 9$, which have no roots in $\mathbb{R}$. In light of this, to find the roots of all polynomials with real coefficients, we will need a number system larger than $\mathbb{R}$. This leads us to a study of the complex numbers $\mathbb{C}$.

In this chapter, we will define the complex numbers $\mathbb{C}$ and examine some of its basic properties. This will lead us to examples of the three most important objects of study in abstract algebra: fields, commutative rings, and groups. Obtaining experience with examples of fields, commutative rings, and groups at this early stage will make it much easier for you to develop a deeper understanding of them when you reexamine them in much greater detail in Chapters 8, 15, and 17. In the next chapter, we will take a more concrete and geometric view of the complex numbers. That chapter will culminate in a proof of the Fundamental Theorem of Algebra. Hopefully, the combination of approaches used to examine $\mathbb{C}$ in these two chapters will make the complex numbers as real and important to you as the real numbers.

5.1 Complex Numbers

There are many different ways to introduce the complex numbers. We have chosen the one that makes computations the easiest. To get started, let's look at what happens if we add and multiply polynomials with real coefficients of degree at most one. Applying the associative and commutative laws, we see

$$(a + bx) + (c + dx) = (a + c) + (b + d)x$$

and

$$(a + bx) \cdot (c + dx) = ac + (ad + bc)x + bdx^2,$$

for all $a, b, c, d \in \mathbb{R}$.

In defining the complex numbers $\mathbb{C}$, we will let i be a symbol that behaves the same way as x does in the preceding computations, with one big exception. Whenever the term i^2 appears, we replace it by the number -1. Therefore, if we were to replace x by i in the calculation, the term bdi^2 would become $-bd$. This leads us to

Definition 5.1. *The complex numbers $\mathbb{C}$ are all objects of the form $a + bi$, where $a, b \in \mathbb{R}$. Addition and multiplication in $\mathbb{C}$ are defined as:*

$$(a + bi) + (c + di) = (a + c) + (b + d)i \quad and \quad (a + bi) \cdot (c + di) = (ac - bd) + (ad + bc)i,$$

for all $a, b, c, d \in \mathbb{R}$.

■ Examples

$$(3 + 4i) + (-8 + 7i) = -5 + 11i,$$
$$(-6 + (1 - \sqrt{2})i) + (31 + \sqrt{2}i) = 25 + i,$$
$$(2 + i) \cdot (-3 + 6i) = (2 \cdot (-3) - 1 \cdot 6) + (2 \cdot 6 + 1 \cdot (-3))i = -12 + 9i,$$
$$(4 + 7i) \cdot (4 - 7i) = (4 \cdot 4 - 7 \cdot (-7)) + (4 \cdot (-7) + 7 \cdot 4)i = 65.$$

■

When looking at a complex number $a + bi$, where $a, b \in \mathbb{R}$, we often refer to a as the **real** part and bi as the **complex** part. In fact, whenever we write a complex number α in the form $a + bi$, we will always be assuming, unless explicitly stated otherwise, that $a, b \in \mathbb{R}$. As a shorthand, we will write complex numbers like $2 + 0i$ as 2, $0 - 3i$ as $-3i$, and $-5 + 1i$ as $-5 + i$. In particular, we see that every real number a can be viewed as the complex number $a + 0i$. Observe that in $\mathbb{C}$, the sum and product of $a + 0i$ and $c + 0i$ are, respectively, $(a + c) + 0i$ and $ac + 0i$. Thus, when we view real numbers as living inside of $\mathbb{C}$, they give us the same sums as products as they did in $\mathbb{R}$.

When doing a computation by hand, it is often easier to simply treat i the way we treated the variable x and then replace i^2 by -1 at the end of the problem. For example, we have

$$(5 + 5i)^2 = (5 + 5i) \cdot (5 + 5i) = 25 + 50i + 25i^2 = 25 + 50i - 25 = 50i.$$

Previously we mentioned that the polynomials $x^2 + 1$ and $x^4 + 2x^2 + 9$ have no roots in $\mathbb{R}$. However, it should now be clear to you that i and $-i$ are both roots of $x^2 + 1$. Furthermore, you should now apply the rules for addition and multiplication in $\mathbb{C}$ and check, by hand, that $1 + \sqrt{2}i, -1 + \sqrt{2}i, 1 - \sqrt{2}i, -1 - \sqrt{2}i$ are all roots of the polynomial $x^4 + 2x^2 + 9$.

At some point in your previous courses, you were probably introduced to the complex numbers. Based on your experiences with the complex numbers, some questions you might have are:

(a) The complex numbers are often referred to as the imaginary numbers. Are the complex numbers an imaginary object or are they as real and legitimate as other number systems like $\mathbb{Z}$, $\mathbb{Q}$, and $\mathbb{R}$?

(b) Are there ways to view complex numbers as concretely as we view real numbers?

One of the goals of Chapters 5 and 6 is to convince you that the complex numbers are every bit as real and legitimate an object as the real numbers. In fact, we will see that constructing the complex numbers from the real numbers is much simpler and more straightforward than constructing the real numbers from the rational numbers.

Over the years, an imprecise and misleading use of language has evolved regarding the complex numbers. The set $\mathbb{R}$ of real numbers can be used to represent the points on a number line. Therefore, the distance between any two points can be represented by an element of $\mathbb{R}$. This perspective helps us to view the elements of $\mathbb{R}$ as very concrete and very "real." As a result, the set $\mathbb{R}$ is called the real numbers. This led to the somewhat natural but very inaccurate and misleading impression that the only numbers that can be considered as "real" objects are those that belong to $\mathbb{R}$. Therefore, people began to informally think of any number that did not belong to $\mathbb{R}$ as not being real and to say that it was imaginary. Thus, we began to refer to complex numbers as imaginary numbers.

However, as we will see in Chapter 6, there are also very concrete ways of viewing the complex numbers. In fact, in the exercises after Section 5.2, we will see that complex numbers can be used to study practical problems such as the opposition to the flow of electricity in electrical circuits. Thus, complex numbers can and should be considered as "real" as real numbers. But unfortunately, the term *imaginary number* remains popular. The negative effect of this term is that it leads students to believe that complex numbers don't really exist. As a result, many students have the false impression that complex numbers are merely esoteric, abstract objects invented by mathematicians. Sadly, students often fail to appreciate that complex numbers are as important and as legitimate as integers, rational numbers, and real numbers.

In mathematics and other walks of life, it is often difficult to become comfortable with new ideas. There was a time when people had difficulty accepting the concept of the number zero. The feeling was that adding zero was the same as adding nothing and nothing could not exist. Therefore, they concluded that the number zero could not exist. Not surprisingly, it also took people a long time to accept that negative numbers were as valid a concept as positive numbers.

Today, we have no trouble accepting the fact that $\sqrt{2}$ is a number that cannot be written as a quotient of integers. However, there was a time when people felt that $\sqrt{2}$ did not exist, as they could not accept that there existed numbers that could not be written as a quotient of integers.

In light of this, it is not surprising that many people are initially uncomfortable with the more difficult concept that there exist numbers whose square is -1. However, in time, the complex numbers have been accepted as a number system that is as important and legitimate as $\mathbb{Z}$, $\mathbb{Q}$, and $\mathbb{R}$.

5.2 Fields and Commutative Rings

During our examination of $\mathbb{Z}$, $\mathbb{Q}$, and $\mathbb{R}$, we have been tossing around the term *number system*. Actually, there is no formal definition for *number system*. However, it is reasonable to stop and look at some of the important properties that the sets of numbers $\mathbb{Z}$, $\mathbb{Q}$, and $\mathbb{R}$ all have in common. Then we will check and see how many of these properties are also satisfied by $\mathbb{C}$.

Properties of $\mathbb{Q}$ and $\mathbb{R}$

1. *Associative Law of Addition: For every x, y, z, $(x+y)+z = x+(y+z)$.*

2. *Additive Identity: There is an element, usually denoted as 0, such that $x+0 = x = 0+x$, for every x.*

3. *Additive Inverses: For every x there is an element, usually denoted as $-x$, such that $x+(-x) = 0 = (-x)+x$.*

4. *Commutative Law of Addition: For every x, y, $x+y = y+x$.*

5. *Associative Law of Multiplication: For every x, y, z, $(xy)z = x(yz)$.*

6. *Multiplicative Identity: There is an element, usually denoted as 1, such that $x \cdot 1 = x = 1 \cdot x$, for every x.*

7. *Multiplicative Inverses: For every x other than 0, there is an element, usually denoted as x^{-1}, such that $x \cdot x^{-1} = 1 = x^{-1} \cdot x$.*

8. *Commutative Law of Multiplication: For every x, y, $xy = yx$.*

9. *Distributive Laws: For every x, y, z, $x(y+z) = xy+xz$ and $(x+y)z = xz+yz$.*

As we will soon see, $\mathbb{C}$ also satisfies properties 1–9. Since $\mathbb{C}$ satisfies the same nice algebraic properties as $\mathbb{Q}$ and $\mathbb{R}$, we can consider $\mathbb{C}$ to be an algebraic object as legitimate and worthy of study as $\mathbb{Q}$ and $\mathbb{R}$. A recurring theme throughout mathematics is that when many different sets satisfy a common collection of properties, we give a special name to the sets that satisfy those properties. Since many sets in addition to $\mathbb{Q}$, $\mathbb{R}$, and $\mathbb{C}$ satisfy properties 1–9, it is natural to give a name to those sets that have these properties in common.

However, before using these examples to motivate the definitions of fields and commutative rings, observe that if $a, b \in \mathbb{Z}$, then $a+b, a \cdot b \in \mathbb{Z}$. Similarly, if $c, d \in \mathbb{Q}$ and $e, f \in \mathbb{R}$, then $c+d, c \cdot d \in \mathbb{Q}$ and $e+f, e \cdot f \in \mathbb{R}$. Thus, according to Definition 2.12, addition and multiplication are both binary operations on the sets $\mathbb{Z}$, $\mathbb{Q}$, and $\mathbb{R}$. Therefore, yet another important property satisfied by $\mathbb{Z}$, $\mathbb{Q}$, and $\mathbb{R}$ is that they are *closed* under addition and multiplication. We now have the following.

Definition 5.2. *A set F where addition and multiplication are binary operations satisfying properties 1–9 is called a field.*

Note that we did not include $\mathbb{Z}$ with $\mathbb{Q}$, $\mathbb{R}$, and $\mathbb{C}$ when mentioning some of the sets that satisfy the nine properties satisfied by fields. To see the reason for this, let us consider various elements of $\mathbb{Z}$, such as $2, 15, -3$. The multiplicative inverses of these elements are, respectively, $\frac{1}{2}, \frac{1}{15}, -\frac{1}{3}$. However, none of these three multiplicative inverses is in $\mathbb{Z}$. In fact, the only integers whose multiplicative inverses are also in $\mathbb{Z}$ are 1 and -1. In light of this, $\mathbb{Z}$ does not satisfy property 7. In particular, for a set S to satisfy property 7, it is not enough for the nonzero elements of S to have multiplicative inverses in a larger set. Those multiplicative inverses *must* belong to S. There are many other sets that, like the integers, satisfy properties 1–6, 8, and 9 but may not satisfy property 7. We now give a special name to those sets that satisfy these properties.

Definition 5.3. *A set R where addition and multiplication are binary operations satisfying properties 1–6, 8, and 9 is called a commutative ring.*

Note that if R is a commutative ring, then R may satisfy, but is not required to satisfy, property 7. Some commutative rings, like $\mathbb{Q}$ and $\mathbb{R}$, satisfy property 7, whereas others, like $\mathbb{Z}$, do not. In particular, **every field is a commutative ring, but not every commutative ring is a field**.

For some additional examples of commutative rings that are not fields, let $\mathbb{Z}[x]$, $\mathbb{Q}[x]$, and $\mathbb{R}[x]$ denote, respectively, all polynomials with coefficients in $\mathbb{Z}$, $\mathbb{Q}$, and $\mathbb{R}$. Observe that when multiplying nonzero polynomials in $\mathbb{Z}[x]$, $\mathbb{Q}[x]$, and $\mathbb{R}[x]$, degrees never go down. In particular, if we multiply the polynomial $p(x) = x$ by other nonzero polynomials in $\mathbb{Z}[x]$, $\mathbb{Q}[x]$, and $\mathbb{R}[x]$, the result is always a polynomial of degree at least one. In $\mathbb{Z}[x]$, $\mathbb{Q}[x]$, and $\mathbb{R}[x]$, the constant polynomial $g(x) = 1$ is the identity element of multiplication. However, in $\mathbb{Z}[x]$, $\mathbb{Q}[x]$, and $\mathbb{R}[x]$, it is impossible to multiply x by another polynomial to obtain the polynomial 1. Thus, $p(x) = x$ does not have a multiplicative inverse in $\mathbb{Z}[x]$, $\mathbb{Q}[x]$, or $\mathbb{R}[x]$. In fact, this argument really tells us that any polynomial of degree at least one does not have a multiplicative inverse in $\mathbb{Z}[x]$, $\mathbb{Q}[x]$, or $\mathbb{R}[x]$. As a result, $\mathbb{Z}[x]$, $\mathbb{Q}[x]$, and $\mathbb{R}[x]$ are commutative rings that are not fields.

The previous examples can be generalized to allow us to produce even more examples of commutative rings that are not fields. If R is a commutative ring, then the set $R[x]$ of polynomials with coefficients in R is also a commutative ring. This tells us that every time we adjoin a new variable to a commutative ring, we obtain a new commutative ring.

In multivariable calculus, one studies the set $\mathbb{R}[x, y]$ of all polynomials in two variables with real coefficients. By our previous observation, $\mathbb{R}[x, y]$ is also a commutative ring. More generally, if R is any commutative ring, the set $R[x_1, x_2, x_3, \ldots, x_n]$ of polynomials in n variables, where n can be any natural number, is also a commutative ring. In fact, if we allow the number of variables to be infinite and consider the set $R[x_1, x_2, x_3, \ldots]$, we again obtain a commutative ring. However, in all of these examples, there are many polynomials that do not have multiplicative inverses, so all of these examples fail to be fields.

We can now turn our attention back to $\mathbb{C}$ and begin the work necessary to show that it is a field.

Lemma 5.4. $\mathbb{C}$ *is a commutative ring.*

Intuition.   As one becomes more adept at doing proofs, one needs to develop the skill of understanding not only what needs to be proved, but also what may be *assumed* in the proof. As we prove that $\mathbb{C}$ satisfies properties 1–6, 8, and 9, we will be assuming that $\mathbb{R}$ already satisfies these properties. Although the computations needed to show that $\mathbb{C}$ inherits these properties from $\mathbb{R}$ may occasionally become tedious, the thinking involved is always very straightforward. This is because the construction of $\mathbb{C}$ from $\mathbb{R}$ is much more straightforward and concrete than either the construction of $\mathbb{Q}$ from $\mathbb{Z}$ or the construction of $\mathbb{R}$ from $\mathbb{Q}$. Recall that every element of $\mathbb{Q}$ is not merely a pair of integers but an equivalence class of pairs of integers. Even more involved is the construction of $\mathbb{R}$ from $\mathbb{Q}$ as equivalence classes of Cauchy sequences of elements of $\mathbb{Q}$. Since both $\mathbb{Q}$ and $\mathbb{R}$ are defined in terms of equivalence classes, it took some work to justify that addition and multiplication in $\mathbb{Q}$ and $\mathbb{R}$ were well defined. On the other hand, every element of $\mathbb{C}$ corresponds to one and only one ordered pair of elements of $\mathbb{R}$. We do not need to use equivalence classes, and we know immediately that addition and multiplication are well defined. Thus, moving from an understanding of $\mathbb{R}$ to an understanding of $\mathbb{C}$ is, in many ways, much easier than making the analogous move from $\mathbb{Z}$ to $\mathbb{Q}$ or from $\mathbb{Q}$ to $\mathbb{R}$.

It will be fairly quick and easy to show that $\mathbb{C}$ inherits properties 1–4, 6, and 8. Showing that $\mathbb{C}$ inherits properties 5 and 9, although straightforward, will require a reasonable amount of calculating and bookkeeping. Before doing the actual proof, we will work through some examples to gain experience with the type of calculations needed to prove facts about the complex numbers. As you read along, you should do these calculations with paper and pencil. It is important to note that, in these examples, we will be illustrating that properties 1–6, 8, and 9 hold in these special cases but that these examples do *not* constitute a proof that they hold in *all* cases.

Let us consider the special case where

$$\alpha = 3 + 2i, \quad \beta = 5 - i, \quad \gamma = 7 + 8i.$$

In this case, $\alpha + \beta = 8 + i$, so $(\alpha + \beta) + \gamma = 15 + 9i$. Furthermore, $\beta + \gamma = 12 + 7i$, so $(\alpha + \beta) + \gamma = 15 + 9i$. Thus, $(\alpha + \beta) + \gamma = (\alpha + \beta) + \gamma$. Also, $\beta + \alpha = 8 + i$, so $\alpha + \beta = \beta + \alpha$. These examples illustrate (but do not prove) the associativity and commutativity of addition in $\mathbb{C}$. Furthermore, $0 + 0i$ has the property that

$$\alpha + (0 + 0i) = (3 + 2i) + (0 + 0i) = (3 + 0) + (2 + 0)i = 3 + 2i = \alpha$$

and

$$(0 + 0i) + \alpha = (0 + 0i) + (3 + 2i) = (0 + 3) + (0 + 2)i = 3 + 2i = \alpha.$$

This illustrates that $0 + 0i$ has the property that belongs to the additive identity of $\mathbb{C}$. If we consider the complex number $-3 - 2i$, it has the property that

$$\alpha + (-3 - 2i) = (3 + 2i) + (-3 - 2i) = 0 + 0i = (-3 - 2i) + (3 + 2i) = (-3 - 2i) + \alpha.$$

Thus $-3 - 2i$ is the additive inverse of α.

For the properties concerning multiplication, we have $\alpha \cdot \beta = 17 + 7i$, thus $(\alpha \cdot \beta) \cdot \gamma = 63 + 185i$. Furthermore, $\beta \cdot \gamma = 43 + 33i$, hence $\alpha \cdot (\beta \cdot \gamma) = 63 + 185i$. Thus $(\alpha \cdot \beta) \cdot \gamma = \alpha \cdot (\beta \cdot \gamma)$. Also observe that $\beta \cdot \alpha = 17 + 7i$, hence $\alpha \cdot \beta = \beta \cdot \alpha$. These examples illustrate (but do not prove) the associativity and commutativity of multiplication in $\mathbb{C}$. Note that the complex number $1 + 0i$ has the property that

$$\alpha \cdot (1 + 0i) = (3 + 2i) \cdot (1 + 0i) = 3 + 2i = \alpha = 3 + 2i = (1 + 0i) \cdot (3 + 2i) = (1 + 0i) \cdot \alpha.$$

This illustrates that $1 + 0i$ has the property belonging to the multiplicative identity of $\mathbb{C}$.

To illustrate one distributive law, we have

$$\alpha \cdot (\beta + \gamma) = (3 + 2i) \cdot ((5 - i) + (7 + 8i)) = (3 + 2i) \cdot (12 + 7i) = 22 + 45i$$

and

$$\alpha \cdot \beta + \alpha \cdot \gamma = (3 + 2i) \cdot (5 - i) + (3 + 2i) \cdot (7 + 8i) = (17 + 7i) + (5 + 38i) = 22 + 45i.$$

Therefore, $\alpha \cdot (\beta + \gamma) = \alpha \cdot \beta + \alpha \cdot \gamma$.

To illustrate the other distributive law, we have

$$(\alpha + \beta) \cdot \gamma = ((3 + 2i) + (5 - i)) \cdot (7 + 8i) = (8 + i) \cdot (7 + 8i) = 48 + 71i$$

and

$$\alpha \cdot \gamma + \beta \cdot \gamma = (3 + 2i) \cdot (7 + 8i) + (5 - i) \cdot (7 + 8i) = (5 + 38i) + (43 + 33i) = 48 + 71i.$$

Therefore $(\alpha + \beta) \cdot \gamma = \alpha \cdot \gamma + \beta \cdot \gamma$.

When reading the proof of Lemma 5.4, note how frequently we use the fact that $\mathbb{R}$ is a commutative ring. Although the proof is somewhat long and tedious, all the steps are straightforward.

Proof of Lemma 5.4. Let $\alpha, \beta, \gamma \in \mathbb{C}$, then we can write

$$\alpha = a + bi, \quad \beta = c + di, \quad \gamma = e + fi,$$

where $a, b, c, d, e, f \in \mathbb{R}$. To check that addition is associative, we have

$$(\alpha + \beta) + \gamma = ((a + bi) + (c + di)) + (e + fi) = ((a + c) + (b + d)i) + (e + fi) =$$

(1)
$$((a + c) + e) + ((b + d) + f)i$$

and

$$\alpha + (\beta + \gamma) = (a + bi) + ((c + di) + (e + fi)) = (a + bi) + ((c + e) + (d + f)i) =$$

(2)
$$(a + (c + e)) + (b + (d + f))i.$$

However, since addition in $\mathbb{R}$ is associative,

$$(a + c) + e = a + (c + e) \quad \text{and} \quad (b + d) + f = b + (d + f).$$

Referring back to (1) and (2), we see that

$$(\alpha + \beta) + \gamma = \alpha + (\beta + \gamma).$$

As a result, the associativity of addition in $\mathbb{R}$ tells us that addition in $\mathbb{C}$ is also associative.

To show that $\mathbb{C}$ has an additive identity, consider the complex number $0 + 0i$. We have

$$\alpha + (0 + 0i) = (a + bi) + (0 + 0i) = (a + 0) + (b + 0)i = a + bi = \alpha$$

and

$$(0 + 0i) + \alpha = (0 + 0i) + (a + bi) = (0 + a) + (0 + b)i = a + bi = \alpha.$$

Thus, $0 + 0i$ is the additive identity of $\mathbb{C}$. Technically speaking, $0 + 0i$ is the appropriate way of writing the additive identity of $\mathbb{C}$. However, if no confusion arises, we use the shorthand 0 for the additive identity element in a commutative ring.

Next, given $\alpha = a + bi$, consider the complex number $-a + (-b)i$. We now have

$$\alpha + (-a + (-b)i) = (a + bi) + (-a + (-b)i) = (a - a) + (b - b)i = 0 + 0i$$

and

$$(-a + (-b)i) + \alpha = (-a + (-b)i) + (a + bi) = (-a + a) + (-b + b)i = 0 + 0i.$$

Thus, $-a + (-b)i$ is the additive inverse of $\alpha = a + bi$. We usually use the shorthand $-a - bi$ for $-a + (-b)i$ and also let $-\alpha$ denote the additive inverse of α.

To show that addition in $\mathbb{C}$ is commutative, observe that

$$\alpha + \beta = (a + bi) + (c + di) = (a + c) + (b + d)i$$

and

$$\beta + \alpha = (c + di) + (a + bi) = (c + a) + (d + b)i.$$

Since addition in $\mathbb{R}$ is commutative, $a + c = c + a$ and $b + d = d + b$, which tells us that $\alpha + \beta = \beta + \alpha$. Thus, the commutativity of addition in $\mathbb{R}$ tells us that addition in $\mathbb{C}$ is also commutative.

Now, consider the complex number $1 + 0i$. We now have

$$\alpha \cdot (1 + 0i) = (a + bi) \cdot (1 + 0i) = (a \cdot 1 - b \cdot 0) + (a \cdot 0 + b \cdot 1)i = a + bi = \alpha$$

and

$$(1 + 0i) \cdot \alpha = (1 + 0i) \cdot (a + bi) = (1 \cdot a - 0 \cdot b) + (1 \cdot b + 0 \cdot a)i = a + bi = \alpha.$$

Therefore, $1 + 0i$ is the multiplicative identity. Once again, if no confusion arises, we will usually denote the multiplicative identity of a commutative ring as 1.

At this point, we have shown that $\mathbb{C}$ satisfies properties 1–4 and 6. To show that multiplication in $\mathbb{C}$ is associative, we observe that

$$(\alpha \cdot \beta) \cdot \gamma = ((a + bi) \cdot (c + di)) \cdot (e + fi) = ((ac - bd) + (ad + bc)i) \cdot (e + fi) =$$

(3)
$$((ac - bd)e - (ad + bc)f) + ((ac - bd)f + (ad + bc)e)i$$

and

$$\alpha \cdot (\beta \cdot \gamma) = (a + bi) \cdot ((c + di) \cdot (e + fi)) = (a + bi) \cdot ((ce - df) + (cf + de)i) =$$

(4)
$$(a(ce - df) - b(cf + de)) + (a(cf + de) + b(ce - df))i.$$

However, since addition and multiplication in $\mathbb{R}$ are both associative and addition in $\mathbb{R}$ is commutative, we see that both $(ac - bd)e - (ad + bc)f$ and $a(ce - df) - b(cf + de)$ are equal to $ace - bde - adf - bcf$. Similarly, we also see that both $(ac - bd)f + (ad + bc)e$ and $a(cf + de) + b(ce - df)$ are equal to $acf - bdf + ade + bce$. Plugging these facts into (3) and (4), shows us that

$$\alpha \cdot (\beta \cdot \gamma) = \alpha \cdot (\beta \cdot \gamma).$$

Thus, several properties of $\mathbb{R}$ combine to show that multiplication in $\mathbb{C}$ is associative.

To see that multiplication in $\mathbb{C}$ is commutative, observe that

$$\alpha \cdot \beta = (a + bi) \cdot (c + di) = (ac - bd) + (ad + bc)i$$

and

$$\beta \cdot \alpha = (c+di) \cdot (a+bi) = (ca-db)+(cb+da)i.$$

Since both addition and multiplication are commutative in $\mathbb{R}$, we see that $ac-bd = ca-db$ and $ad+bc = cb+da$. Plugging these facts into the preceding equations immediately shows us that $\alpha \cdot \beta = \beta \cdot \alpha$. Once again, $\mathbb{C}$ inherits properties from $\mathbb{R}$, and we see that multiplication in $\mathbb{C}$ is commutative.

Finally, we need to show that $\mathbb{C}$ satisfies the distributive laws. For one distributive law, observe that

$$\alpha \cdot (\beta+\gamma) = (a+bi) \cdot ((c+di)+(e+fi)) = (a+bi) \cdot ((c+e)+(d+f)i) =$$
$$(a(c+e)-b(d+f))+(a(d+f)+b(c+e))i$$

and

$$\alpha \cdot \beta + \alpha \cdot \gamma = (a+bi) \cdot (c+di)+(a+bi) \cdot (e+fi) =$$
$$((ac-bd)+(ad+bc)i)+((ae-bf)+(af+be)i) =$$
$$(ac-bd+ae-bf)+(ad+bc+af+be)i.$$

Using several properties of $\mathbb{R}$, including the distributive laws, it follows that $a(c+e) - b(d+f) = ac-bd+ae-bf$ and $a(d+f)+b(c+e) = ad+bc+af+be$. Plugging these facts into the preceding equations shows us that $\alpha \cdot (\beta+\gamma) = \alpha \cdot \beta + \alpha \cdot \gamma$. Thus, $\mathbb{C}$ satisfies one of the two distributive laws.

To show that $\mathbb{C}$ satisfies the second distributive law, we can save a good deal of work if we apply the first distributive law and the commutativity of multiplication. Applying these laws results in

$$(\alpha+\beta) \cdot \gamma = \gamma \cdot (\alpha+\beta) = \gamma \cdot \alpha + \gamma \cdot \beta = \alpha \cdot \gamma + \beta \cdot \gamma,$$

as desired.   □

In the preceding proof, we saw that by using the commutativity of multiplication, each distributive law easily follows from the other. Therefore, as long as multiplication is commutative, it is indeed redundant to list both distributive laws. However, it is very common in mathematics to study sets with addition and multiplication that satisfy all the properties of a commutative ring, except that multiplication may not be commutative. Such objects are known either as **rings** or **noncommutative rings**.

If you have taken a class in linear algebra, then you are familiar with matrix multiplication. If you go back and look at the properties satisfied by the addition and multiplication of the set of 2×2 matrices, you will see that they are an example of a noncommutative ring. In such sets, where multiplication is not commutative, then having one distributive law hold does not automatically imply that the other distributive law holds. For this reason, we usually mention both distributive laws even though it is only necessary to mention one of them for commutative rings.

In light of Lemma 5.4, all that remains to complete the proof that $\mathbb{C}$ is a field is to show that every nonzero element of $\mathbb{C}$ has a multiplicative inverse. As in the proof of Lemma 5.4, we will make strong use of properties of $\mathbb{R}$.

Theorem 5.5. $\mathbb{C}$ *is a field.*

Proof. Having proven in Lemma 5.4 that $\mathbb{C}$ is a commutative ring, it only remains to show that every nonzero element of $\mathbb{C}$ has a multiplicative inverse in $\mathbb{C}$. Since $0+0i$ is the additive identity element of $\mathbb{C}$, if $\alpha = a + bi$ is a nonzero element of $\mathbb{C}$, then $\alpha \neq 0 + 0i$, so at least one of a or b is nonzero. Thus, $a^2 + b^2 \neq 0$.

Since $a^2 + b^2$ has a multiplicative inverse in $\mathbb{R}$, we can consider the complex number

$$\beta = \frac{a}{a^2 + b^2} - \frac{b}{a^2 + b^2} i.$$

We now have

$$\alpha \cdot \beta = (a + bi) \left(\frac{a}{a^2 + b^2} - \frac{b}{a^2 + b^2} i \right) =$$

$$\left(\frac{a^2}{a^2 + b^2} + \frac{b^2}{a^2 + b^2} \right) + \left(\frac{-ab}{a^2 + b^2} + \frac{ab}{a^2 + b^2} \right) i = \frac{a^2 + b^2}{a^2 + b^2} + 0i = 1.$$

Since multiplication in $\mathbb{C}$ is commutative, we also have $\beta \cdot \alpha = 1$. Thus, β is the multiplicative inverse of α, so $\mathbb{C}$ is a field. $\qquad \square$

■ Examples

$$(1+i)^{-1} = \frac{1}{1^2 + 1^2} - \frac{1}{1^2 + 1^2} i = \frac{1}{2} - \frac{1}{2} i,$$

$$(3 - 4i)^{-1} = \frac{3}{3^2 + (-4)^2} - \frac{-4}{3^2 + (-4)^2} = \frac{3}{25} + \frac{4}{25} i,$$

$$\left(\frac{12}{193} + \frac{7}{193} i \right)^{-1} = \frac{\frac{12}{193}}{\left(\frac{12}{193}\right)^2 + \left(\frac{7}{193}\right)^2} - \frac{\frac{7}{193}}{\left(\frac{12}{193}\right)^2 + \left(\frac{7}{193}\right)^2} i = 12 - 7i,$$

$$(6i)^{-1} = \frac{0}{0^2 + 6^2} - \frac{6}{0^2 + 6^2} i = -\frac{1}{6} i, \qquad \left(\frac{4}{11} \right)^{-1} = \frac{11}{4}$$

Exercises for Sections 5.1 and 5.2

In these exercises, whenever your answer is a complex number, try to express it in the form $a + bi$, where $a, b \in \mathbb{R}$.

In exercises 1–20, let $\alpha = 2 + 3i$, $\beta = 5 - i$, $\gamma = -3 + 8i$, and $\delta = \sqrt{2} + \pi i$.

1. Compute $\alpha + \beta$.

2. Compute $6\alpha - 4\beta$.

3. Compute $\sqrt{3}\gamma - 11\delta$.

4. Compute $-12\gamma + \sqrt{7}\delta$.

5. Compute $-2\alpha + 5\beta + 7\gamma - 4\delta$.

6. Compute $3\alpha - 4\beta - 9\gamma + 12\delta$.

7. Compute $\alpha \cdot \beta$.

8. Use your answer from exercise 7 to compute $(\alpha \cdot \beta)^{-1}$.

9. Compute α^{-1} and β^{-1}.

10. Use your answer from exercise 9 to compute $\alpha^{-1} \cdot \beta^{-1}$. Compare your answer to your answer from exercise 8.

11. Compute $\gamma \cdot \delta$.

12. Use your answer from exercise 11 to compute $(\gamma \cdot \delta)^{-1}$.

13. Compute γ^{-1} and δ^{-1}.

14. Use your answer from exercise 13 to compute $\gamma^{-1} \cdot \delta^{-1}$. Compare your answer to your answer from exercise 12.

15. Compute $\alpha^{-1} + \beta^{-1}$ and compute $(\alpha + \beta)^{-1}$ and compare your answers. Can you draw any conclusions?

16. Compute $\gamma^{-1} + \delta^{-1}$ and compute $(\gamma + \delta)^{-1}$ and compare your answers. Can you draw any conclusions?

17. Compute $(\alpha \cdot \beta) \cdot \gamma$ and $\alpha \cdot (\beta \cdot \gamma)$ and compare your answers.

18. Compute $(\beta \cdot \gamma) \cdot \delta$ and $\beta \cdot (\gamma \cdot \delta)$ and compare your answers.

19. Compute $\alpha \cdot (\beta + \gamma)$ and $\alpha \cdot \beta + \alpha \cdot \gamma$ and compare your answers.

20. Compute $(\alpha + \beta) \cdot \gamma$ and $\alpha \cdot \gamma + \beta \cdot \gamma$ and compare your answers.

21. Solve for x: $(3 - 4i)x = 7 + 2i$.

22. Solve for y: $(2 + 5i)y = 3 - 15i$.

23. Solve for t: $(6 - i)t + (4 - 5i) = (-8 + 3i)t + (-2 + i)$.

24. Solve for v: $(5 - 2i)v + (7 + 11i) = (-6 - 81i)v + (23 - 6i)$.

25. Solve for x: $x^2 - 6x + 15 = 0$.

26. Solve for y: $y^2 + 12y + 50 = 0$.

27. Find the roots of the polynomial $x^2 - 5ix - 4$ by completing the square.

28. Find the roots of the polynomial $x^2 + 7ix + 18$ by completing the square.

29. Solve for z: $z^2 + (2 + 2i)z + 2i = 0$. You might want to use the fact that $(1 + i)^2 = 2i$.

30. Solve for w: $w^2 + (6 - 4i)w + (13 - 12i) = 0$. You might want to use the fact that $(3 - 2i)^2 = 13 - 12i$.

31. Solve for the ordered pair (x, y) in the following system of linear equations: $(3 + i)x + 4y = 7 + 8i$, $5x - (2 + i)y = 6$.

32. Solve for the ordered pair (A, B) in the following system of linear equations: $5A + (3 - 4i)B = 17$, $(9 - 4i)A + (8i)B = 1 - 2i$.

33. Check that $1 + \sqrt{2}i$, $-1 + \sqrt{2}i$, $1 - \sqrt{2}i$, $-1 - \sqrt{2}i$, are all roots of the polynomial $x^4 + 2x^2 + 9$ by plugging all four of them into the polynomial.

34. Use the substitution $t = x^2$ and then the quadratic formula to find the roots of the polynomial $x^4 + 2x^2 + 9$. At first, your four solutions may not be in the form $a + bi$, where $a, b \in \mathbb{R}$. However, check that your four roots are indeed the same four roots as in exercise 33.

35. Check that $\sqrt{3} + i$, $\sqrt{3} - i$, $-\sqrt{3} + i$, $-\sqrt{3} - i$, are all roots of the polynomial $x^4 - 4x^2 + 16$ by plugging all four of them into the polynomial.

36. Use the substitution $t = x^2$ and then the quadratic formula to find the roots of the polynomial $x^4 - 4x^2 + 16$. At first, your four solutions may not be in the form $a + bi$, where $a, b \in \mathbb{R}$. However, check that your four roots are indeed the same four roots as in exercise 35.

37. Check that $4 - 8i$, $4 + 8i$, $2 + \sqrt{3}i$, $2 - \sqrt{3}i$ are all roots of the polynomial $x^4 - 12x^3 + 119x^2 - 376x + 560$ by plugging all four of them into the polynomial.

38. Check that $3 + 5i$, $3 - 5i$, $7 - i$, $7 + i$ are all roots of the polynomial $x^4 - 20x^3 + 168x^2 - 776x + 1700$ by plugging all four of them into the polynomial.

39. If x, y are nonzero elements of a field F, show that $x^{-1}y^{-1}$ is the multiplicative inverse of xy.

40. Show that the nonzero elements of a field are closed under multiplication.

41. If R is a commutative ring, let $U(R)$ denote the elements in R which have a multiplicative inverse in R. Use the ideas in exercise 39 to show that $U(R)$ is closed under multiplication.

42. If $\mathbb{Q}[x]$ denotes the commutative ring consisting of all polynomials with rational coefficients, describe those elements in $\mathbb{Q}[x]$ that have a multiplicative inverse in $\mathbb{Q}[x]$.

43. If $\mathbb{Z}[x]$ denotes the commutative ring consisting of all polynomials with integer coefficients, list all the elements in $\mathbb{Z}[x]$ that have a multiplicative inverse in $\mathbb{Z}[x]$.

44. In Chapter 4, using equivalence relations and equivalence classes, we saw how the rational numbers can be constructed from the integers. Using similar ideas, we will now construct the set $\mathbb{R}(x)$ of rational functions with real coefficients from the set $\mathbb{R}[x]$ of polynomials with real coefficients.

    Let $S = \{(a(x), b(x)) \mid a(x), b(x) \in \mathbb{R}[x], b(x) \neq 0\}$. Define $\sim$ on S as

    $$(a(x), b(x)) \sim (c(x), d(x)) \text{ precisely when } a(x)d(x) = b(x)c(x).$$

    (a) Show that $\sim$ is an equivalence relation.

    (b) If we define the addition of equivalence classes as

    $$[(a(x), b(x))] + [(c(x), d(x))] = [(a(x)d(x) + b(x)c(x), b(x)d(x))],$$

    show that addition is well defined.

    (c) If we define the multiplication of equivalence classes as

    $$[(a(x), b(x))] \cdot [(c(x), d(x))] = [(a(x)c(x), b(x)d(x))],$$

    show that multiplication is well defined.

    (d) Show that every equivalence class, other than the equivalence class which contains $(0, 1)$, has a multiplicative inverse.

45. Let $\mathbb{R}[[x]] = \{a_0 + a_1x + a_2x^2 + \cdots + a_nx^n + \cdots \mid \text{each } a_i \in \mathbb{R}\}$ denote the set of power series with real coefficients. In calculus, you are concerned with where a power

series converges. However, even if one is not concerned with questions of convergence, $\mathbb{R}[[x]]$ is a commutative ring. If

$$\left(1 + x + \frac{x^2}{2} + \frac{x^3}{6} + \frac{x^4}{24} + \frac{x^5}{120} + \cdots\right)^2 =$$

$$a_0 + a_1 x + a_2 x^2 + a_3 x^3 + a_4 x^4 + \text{ higher-degree terms,}$$

where each $a_i \in \mathbb{R}$, compute a_0, a_1, a_2, a_3, a_4. Do you think $\mathbb{R}[[x]]$ is a field?

46. Let $\mathbb{R}[[x, x^{-1}]] = \{a_{-m} x^{-m} + a_{-m+1} x^{-m+1} + \cdots + a_{-1} x^{-1} + a_0 + a_1 x + a_2 x^2 + \cdots + a_n x^n + \cdots \mid \text{each } a_i \in \mathbb{R}\}$ denote the set of Laurent series with real coefficients. Laurent series are identical to power series except that a Laurent series is allowed to have a finite number of terms where the exponents are negative integers. The set $\mathbb{R}[[x, x^{-1}]]$ is also a commutative ring. If

$$\left(x^{-1} - 1 + \frac{x}{2} - \frac{x^2}{6} + \frac{x^3}{24} - \frac{x^4}{120} + \cdots\right)^2 =$$

$$b_{-2} x^{-2} + b_{-1} x^{-1} + b_0 + b_1 x + b_2 x^2 + b_3 x^3 + \text{ higher-degree terms,}$$

where each $b_i \in \mathbb{R}$, compute $b_{-2}, b_{-1}, b_0, b_1, b_2, b_3$. Do you think $\mathbb{R}[[x, x^{-1}]]$ is a field?

For exercises 47–54, please read the following:

As we remarked earlier, the addition of complex numbers is straightforward and is virtually identical to the addition of polynomials with real coefficients of degree at most 1. However, when we multiply complex numbers, we replace i^2 by -1 and this causes some people to question the validity and relevance of complex numbers. Since we can square an element of the complex numbers and obtain -1 as an answer, many students initially view the multiplication of complex numbers as an arbitrary piece of algebraic trickery that lacks any connection to the real world. However, the multiplication of complex numbers is not only a valid algebraic procedure but also has real-world applications, as it can be used to describe the opposition to the flow of electricity in electrical circuits. As electricity flows along an electrical circuit various objects, such as lightbulbs and toasters, can oppose the flow of the electrical current. The total opposition in an electrical circuit is called the **impedance** and is measured in ohms (which are denoted by the symbol Ω). The impedance provided by various objects along an electrical circuit can be represented by complex numbers. Two common types of electrical circuits are **series circuits** and **parallel circuits**.

If we let ／＼／＼／＼ denote an object along an electrical circuit that opposes the flow of the current, then two examples of series circuits are

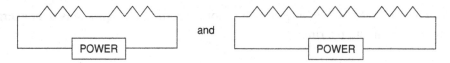

Observe that in a series circuit, all the objects that oppose the flow of electricity carry the same current. On the other hand, in a parallel circuit, the objects that oppose the flow of electricity have the same power source but no longer all carry the same current. Two examples of parallel circuits are

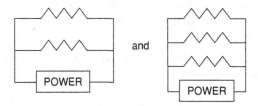

In addition, an electrical circuit can consist of a combination of series and parallel connections such as

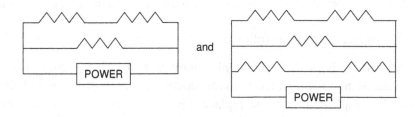

To compute the total impedance of a series circuit is straightforward and does not require the multiplication of complex numbers. In particular, if Z_1 and Z_2 represent the impedance of two objects in the series circuit

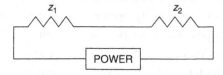

then the total impedance of the circuit, denoted as Z_T, is given by $Z_T = Z_1 + Z_2$. However, to find the total impedance of a parallel circuit does require the multiplication of complex numbers. Given the parallel circuit

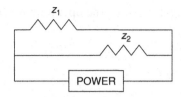

the total impedance of the circuit is given by $Z_T = \frac{Z_1 \cdot Z_2}{Z_1 + Z_2}$.

In exercises 47–52, let $Z_1 = 3 + 2i$ Ω, $Z_2 - 5 - 3i$ Ω, and $Z_3 = 2 + 7i$ Ω.

47. Given 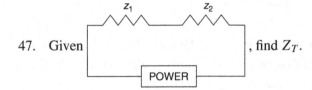 , find Z_T.

48. Given 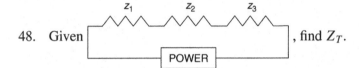 , find Z_T.

49. Given 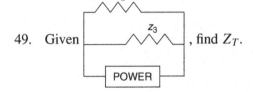 , find Z_T.

50. Given 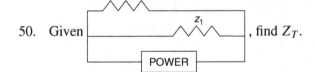 , find Z_T.

51. Given 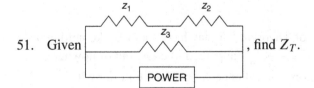 , find Z_T.

52. Given 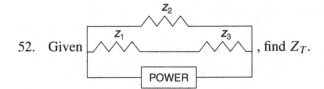 , find Z_T.

For exercises 53–54, if $\alpha, \beta \in \mathbb{C}$, let $\circ$ denote the operation $\alpha \circ \beta = \frac{\alpha \cdot \beta}{\alpha + \beta}$.

53.   In the circuit

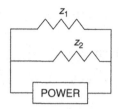

we could find Z_T by computing either $Z_1 \circ Z_2$ or $Z_2 \circ Z_1$. Observe that we would obtain different answers if $\circ$ was not commutative. In light of this, prove algebraically that the operation $\circ$ is commutative. (You may assume that $Z_1 + Z_2 \neq 0$.)

54.   In the circuit

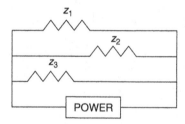

we could find Z_T by computing either $(Z_1 \circ Z_2) \circ Z_3$ or $Z_1 \circ (Z_2 \circ Z_3)$. Observe that we would obtain different answers if $\circ$ was not associative. In light of this, prove algebraically that the operation $\circ$ is associative. (You may assume that $Z_1 + Z_2$, $Z_2 + Z_3$, $Z_1 \cdot Z_2 + Z_1 \cdot Z_3 + Z_2 \cdot Z_3$ are all nonzero.)

5.3 Complex Conjugation

Coming up with the element β in the proof of Theorem 5.5 may have seemed like pulling a rabbit out of a hat. The complex number $\frac{a}{a^2+b^2} - \frac{b}{a^2+b^2}i$ appeared to come out of nowhere. However, $\frac{a}{a^2+b^2} - \frac{b}{a^2+b^2}i$ will seem much more natural once we become familiar with complex conjugation. Not only is complex conjugation extremely useful for studying $\mathbb{C}$, but generalizations of it will have applications throughout abstract algebra.

Definition 5.6. *Let $* : \mathbb{C} \to \mathbb{C}$ be the function defined as $(a + bi)^* = a - bi$, for all $a, b \in \mathbb{R}$. We call $*$ complex conjugation and if $\alpha \in \mathbb{C}$, we call α^* the complex conjugate of α.*

■ Examples

$(2+3i)^* = 2-3i$, $(2-3i)^* = 2+3i$, $(\sqrt{7}+\pi i)^* = \sqrt{7}-\pi i$, $(7)^* = 7$, $(8i)^* = -8i$, $(\sqrt{11})^* = \sqrt{11}$, $(\sqrt{11}i)^* = -\sqrt{11}i$.

■

Simply stated, $*$ fixes the real part of α and negates the complex part of α. We can now record some of the many properties of $*$.

Lemma 5.7. *Let* $* : \mathbb{C} \to \mathbb{C}$ *denote complex conjugation. Then*

(a) $*$ *is a bijection.*

(b) *If* $\alpha, \beta \in \mathbb{C}$, *then* $(\alpha + \beta)^* = \alpha^* + \beta^*$.

(c) *If* $\alpha, \beta \in \mathbb{C}$, *then* $(\alpha \cdot \beta)^* = \alpha^* \cdot \beta^*$.

Proof. In order to show that $*$ is a bijection, we will first show that it is surjective and will then show that it is injective. Let $\alpha = a + bi \in \mathbb{C}$ to show that $*$ is surjective; we need to find some element of $\mathbb{C}$ that, when plugged into $*$, gives α as the answer. However,

$$(a-bi)^* = a - (-bi) = a + bi = \alpha,$$

thus, $a - bi$ is the desired element.

To show that $*$ is injective, we must show that if two elements of $\mathbb{C}$ give the same answer when plugged into $*$, then those elements must have been equal. To this end, let $\alpha = a + bi$, $\beta = c + di \in \mathbb{C}$ such that $\alpha^* = \beta^*$. Since $\alpha^* = (a+bi)^* = a - bi$ and $\beta^* = (c+di)^* = c - di$, the fact that $\alpha^* = \beta^*$ tells us that $a = c$ and $-b = -d$. However, this certainly implies that $b = d$, which immediately tells us that $\alpha = \beta$, as desired. Thus, $*$ is a bijection.

To see that $*$ satisfies property (b), if $\alpha = a + bi$ and $\beta = c + di$, we have

$$(\alpha + \beta)^* = ((a+bi) + (c+di))^* = ((a+c) + (b+d)i)^* = (a+c) - (b+d)i =$$
$$(a-bi) + (c-di) = (a+bi)^* + (c+di)^* = \alpha^* + \beta^*.$$

The computation that shows that $*$ satisfies property (c) is only slightly more involved. On the one hand, we have

$$(\alpha \cdot \beta)^* = ((a+bi) \cdot (c+di))^* = ((ac-bd) + (ad+bc)i)^* = (ac-bd) - (ad+bc)i.$$

On the other hand,

$$\alpha^* \cdot \beta^* = (a+bi)^* \cdot (c+di)^* = (a-bi) \cdot (c-di) =$$
$$(ac - (-b)(-d)) + (a(-d) + (-b)c)i = (ac-bd) - (ad+bc)i.$$

Comparing the two sets of equations above, we see that $(\alpha \cdot \beta)^* = \alpha^* \cdot \beta^*$. □

Although ∗ satisfies many other important properties, we have singled out properties (a), (b), and (c) in Lemma 5.7. The reason for doing this is that functions that satisfy these three properties are the building blocks of group theory and Galois theory. An understanding of the interaction between these types of functions and the roots of polynomials will eventually lead us to the proof of the insolvability of the quintic. We now give a name to those functions, like complex conjugation, that satisfy the three conditions in Lemma 5.7.

Definition 5.8. *Let $\sigma : R \to R$ be a function defined on a commutative ring R. We say that σ is an automorphism of R if*

(a) *σ is a bijection,*

(b) *$\sigma(x + y) = \sigma(x) + \sigma(y)$, for all $x, y \in R$, and*

(c) *$\sigma(x \cdot y) = \sigma(x) \cdot \sigma(y)$, for all $x, y \in R$.*

Mathematical Induction will allow us to extend properties (b) and (c) from Lemma 5.7 and Definition 5.8 into an even more useful form.

Lemma 5.9. *Suppose $\sigma : R \to R$ is an automorphism of a commutative ring R. If $x_1, x_2, \ldots, x_n \in R$ then*

$$\sigma(x_1 + x_2 + \cdots + x_n) = \sigma(x_1) + \sigma(x_2) + \cdots + \sigma(x_n)$$

and

$$\sigma(x_1 \cdot x_2 \cdots x_n) = \sigma(x_1) \cdot \sigma(x_2) \cdots \sigma(x_n).$$

Proof. Let T be those $n \in \mathbb{N}$ such that, for all $x_1, x_2, \ldots, x_n \in R$,

$$\sigma(x_1 + x_2 + \cdots + x_n) = \sigma(x_1) + \sigma(x_2) + \cdots + \sigma(x_n)$$

and

$$\sigma(x_1 \cdot x_2 \cdots x_n) = \sigma(x_1) \cdot \sigma(x_2) \cdots \sigma(x_n).$$

It will suffice to show that $T = \mathbb{N}$, and we will do this by Mathematical Induction.

If we examine what it means to say that $1 \in T$; it simply means that $\sigma(x_1) = \sigma(x_1)$, which is certainly the case. Therefore, we may now assume that k is some natural number that belongs to T, and we need to show that $k + 1 \in T$. To this end, let $x_1, x_2, \ldots, x_k, x_{k+1} \in R$ and consider

$$\sigma(x_1 + x_2 + \cdots + x_k + x_{k+1}) \quad \text{and} \quad \sigma(x_1 \cdot x_2 \cdots x_k \cdot x_{k+1}).$$

To simplify the use of properties (b) and (c) in Definition 5.8, we will let $y = x_1 + x_2 + \cdots + x_k$ and $z = x_1 \cdot x_2 \cdots x_k$. Since $k \in T$, we have

$$\sigma(y) = \sigma(x_1 + x_2 + \cdots + x_k) = \sigma(x_1) + \sigma(x_2) + \cdots + \sigma(x_k)$$

and

$$\sigma(z) = \sigma(x_1 \cdot x_2 \cdots x_k) = \sigma(x_1) \cdot \sigma(x_2) \cdots \sigma(x_k).$$

Therefore, property (b) and the associativity of addition now imply that

$$\sigma(x_1 + x_2 + \cdots + x_k + x_{k+1}) = \sigma((x_1 + x_2 + \cdots + x_k) + x_{k+1}) =$$
$$\sigma(y + x_{k+1}) = \sigma(y) + \sigma(x_{k+1}) =$$
$$(\sigma(x_1) + \sigma(x_2) + \cdots + \sigma(x_k)) + \sigma(x_{k+1}) = \sigma(x_1) + \sigma(x_2) + \cdots + \sigma(x_k) + \sigma(x_{k+1}).$$

Similarly, property (c) and the associativity of multiplication now imply that

$$\sigma(x_1 \cdot x_2 \cdots x_k \cdot x_{k+1}) = \sigma((x_1 \cdot x_2 \cdots x_k) \cdot x_{k+1}) = \sigma(z \cdot x_{k+1}) = \sigma(z) \cdot \sigma(x_{k+1}) =$$
$$(\sigma(x_1) \cdot \sigma(x_2) \cdots \sigma(x_k)) \cdot \sigma(x_{k+1}) = \sigma(x_1) \cdot \sigma(x_2) \cdots \sigma(x_k) \cdot \sigma(x_{k+1}).$$

Thus, $k + 1 \in T$, thereby concluding the proof. $\qquad\square$

Here are some additional easily checked properties of $*$.

Lemma 5.10. *Let* $* : \mathbb{C} \to \mathbb{C}$ *be complex conjugation. If* $\alpha \in \mathbb{C}$, *then*

(a) $\alpha^* = \alpha$ *if and only if* $\alpha \in \mathbb{R}$;

(b) $\alpha \cdot \alpha^*, \alpha^* \cdot \alpha, \alpha + \alpha^*, \alpha^* + \alpha \in \mathbb{R}$;

(c) $(\alpha^n)^* = (\alpha^*)^n$, *for all* $n \in \mathbb{N}$;

(d) $(\alpha^*)^* = \alpha$.

Proof. Let $\alpha = a + bi$; then $\alpha^* = a - bi$. For part (a), if $\alpha \in \mathbb{R}$, then $b = 0$. Hence, α and α^* are both equal to a. On the other hand, if $\alpha^* = \alpha$, then $-bi = bi$, so $b = 0$. Thus, $\alpha \in \mathbb{R}$.

For part (b), we have

$$\alpha \cdot \alpha^* = \alpha^* \cdot \alpha = (a^2 + b^2) + 0i = a^2 + b^2 \in \mathbb{R}$$

and

$$\alpha + \alpha^* = \alpha^* + \alpha = 2a + 0i = 2a \in \mathbb{R}.$$

Thus, $\alpha \cdot \alpha^*, \alpha^* \cdot \alpha, \alpha + \alpha^*, \alpha^* + \alpha \in \mathbb{R}$.

For part (c), since $*$ is an automorphism, if we let

$$\alpha = x_1 = x_2 = \cdots = x_n$$

and $\sigma = *$ in the last part of Lemma 5.9, we see that $(\alpha^n)^* = (\alpha^*)^n$, for all $n \in \mathbb{N}$.

Finally, for part (d),

$$(\alpha^*)^* = (a - bi)^* = a - (-bi) = \alpha,$$

as desired. $\square$

In the proof of Lemma 5.10(b), we saw that if $\alpha = a + bi$, then $\alpha \cdot \alpha^* = a^2 + b^2$. Since $a^2 + b^2 \in \mathbb{R}$, if $\alpha \neq 0$, we can consider the complex number

$$\beta = \frac{1}{a^2 + b^2} \alpha^* = \frac{a}{a^2 + b^2} - \frac{b}{a^2 + b^2} i.$$

It now follows that

$$\alpha \cdot \beta = \alpha \left(\frac{1}{a^2 + b^2} \alpha^* \right) = \left(\frac{1}{a^2 + b^2} \right) (\alpha \alpha^*) = \left(\frac{1}{a^2 + b^2} \right) (a^2 + b^2) = 1.$$

Since multiplication in $\mathbb{R}$ is commutative, $\beta \cdot \alpha = 1$, thus β is the multiplicative inverse of α, and we now see where β came from in the proof of Theorem 5.5.

If you have dealt with complex numbers in the past, you have probably heard the saying that complex roots occur in complex pairs. For example, if $3 + 2i$ is a root of a polynomial with real coefficients, then so is $3 - 2i$. More formally, this really says that if $\alpha \in \mathbb{C}$ is a root of a polynomial $p(x) \in \mathbb{R}[x]$, then so is α^*. We can now prove this using facts about complex conjugation.

Proposition 5.11. *If $\alpha \in \mathbb{C}$ is a root of a polynomial $p(x)$ with real coefficients, then α^* is also a root of $p(x)$.*

Proof. Since $p(x) \in \mathbb{R}[x]$, we know that

$$p(x) = a_n x^n + a_{n-1} x^{n-1} + \cdots + a_1 x + a_0,$$

where each a_i is a real number. Plugging α into the $p(x)$, we obtain

(5) $$0 = a_n \alpha^n + a_{n-1} \alpha^{n-1} + \cdots + a_1 \alpha + a_0.$$

The idea behind this proof is to show that applying $*$ to $p(\alpha)$ results in $p(\alpha^*)$. Since $p(\alpha)^* = 0^* = 0$, this will suffice. Therefore, we begin by applying $*$ to both sides of (5) to obtain

$$0^* = (a_n \alpha^n + a_{n-1} \alpha^{n-1} + \cdots + a_1 \alpha + a_0)^*.$$

Lemma 5.10(a) tells us that $0^* = 0$. Combining this with the first part of Lemma 5.9, we see that

$$0 = (a_n\alpha^n)^* + (a_{n-1}\alpha^{n-1})^* + \cdots + (a_1\alpha)^* + a_0^*.$$

If we next apply property (b) of Lemma 5.7 to each term of the form $(a_i\alpha^i)^*$, we now have

$$0 = a_n^*(\alpha^n)^* + a_{n-1}^*(\alpha^{n-1})^* + \cdots + a_1^*\alpha^* + a_0^*.$$

Since each $a_i \in \mathbb{R}$, Lemma 5.10(a) tells us that $a_i^* = a_i$, and so

$$0 = a_n(\alpha^n)^* + a_{n-1}(\alpha^{n-1})^* + \cdots + a_1\alpha^* + a_0.$$

Finally, Lemma 5.10(c) tells us that for each i, $(\alpha^i)^* = (\alpha^*)^i$. Therefore, we now have

$$0 = a_n(\alpha^*)^n + a_{n-1}(\alpha^*)^{n-1} + \cdots + a_1\alpha^* + a_0.$$

However, the right-hand side of the previous equation is precisely what we get when we plug α^* into $p(x)$. Thus, $p(\alpha^*) = 0$, as desired. $\qquad\square$

■ Examples

1. Suppose $9 - 2i$ is the root of some polynomial $g(x)$ with real coefficients. Proposition 5.11 asserts that $g(x)$ must also have $(9 - 2i)^* = 9 + 2i$ as a root. Therefore, $g(x)$ must have degree at least 2. Observe that

    $$(x - (9 - 2i))(x - (9 + 2i)) = x^2 - 18x + 85$$

    has real coefficients and has both $9 - 2i$ and $9 + 2i$ as roots. Chapter 12 will show that $g(x)$ must be a multiple of $x^2 - 18x + 85$.

2. Suppose $4 - 8i$ and $2 + \sqrt{3}i$ are roots of a polynomial $p(x)$ with real coefficients. Proposition 5.11 asserts that

    $$(4 - 8i)^* = 4 + 8i \quad \text{and} \quad (2 + \sqrt{3}i)^* = 2 - \sqrt{3}i$$

    are also roots of $p(x)$. The polynomial

    $$(x - (4 - 8i))(x - (4 + 8i))(x - (2 + \sqrt{3}i))(x - (2 - \sqrt{3}i)) =$$
    $$x^4 - 12x^3 + 119x^2 - 376x + 560$$

    has real coefficients and has $4 - 8i, 4 + 8i, 2 + \sqrt{3}i$, and $2 - \sqrt{3}i$ as roots. It will follow from our work in Chapter 12 that $p(x)$ must be a multiple of $x^4 - 12x^3 + 119x^2 - 376x + 560$.

3. On the other hand, consider the polynomial $f(x) = x^2 - 5ix - 4$. You should check that i and $4i$ are roots of $f(x)$. However, you should also check that

$$i^* = -i \quad \text{and} \quad (4i)^* = -4i$$

are *not* roots of $f(x)$. Ask yourself: Does this contradict Proposition 5.11? The answer is *no* because $f(x)$ has a coefficient, $-5i$, that does not belong to $\mathbb{R}$, so Proposition 5.11 does not apply.

∎

When we go back and examine the proof of Proposition 5.11, it appears that the ideas and techniques used should apply not only to complex conjugation but also to automorphisms of any commutative ring. In an attempt to generalize Proposition 5.11 to other commutative rings, we proceed with

Definition 5.12. *If* $\sigma : R \to R$ *is an automorphism of a commutative ring* R*, define the set* R^σ *as* $R^\sigma = \{r \in R \mid \sigma(r) = r\}$.

In algebra or calculus, if we are given a function like $f(x) = 3 - x$, we can think of $\frac{3}{2}$ as a fixed point of f, since $f\left(\frac{3}{2}\right) = \frac{3}{2}$. In our more abstract setting, R^σ consists of those elements of R that are fixed points of σ. Note that Lemma 5.10(a) tells us that the set of fixed points of $*$ in $\mathbb{C}$ is equal to $\mathbb{R}$. We can now prove our more general version of Proposition 5.11.

Corollary 5.13. *Let* $\sigma : R \to R$ *be an automorphism of a commutative ring* R*. If* $\alpha \in R$ *is a root of a polynomial* $p(x)$ *with coefficients in* R^σ*, then* $\sigma(\alpha)$ *is also a root of* $p(x)$.

Intuition.   Certainly, Corollary 5.13 is stated in more general terms than Proposition 5.11. However, does Corollary 5.13 actually provide us with information about roots of polynomials that we could not obtain from Proposition 5.11? To begin to answer this, let us suppose that $5 - 3\sqrt{2}$ is the root of some $f(x) \in \mathbb{Q}[x]$. Proposition 5.11 no longer applies but can we say anything about additional roots of $f(x)$? Similarly, suppose $4 - 2\sqrt{3} + 5\sqrt{7}$ is the root of some $g(x) \in \mathbb{Q}[x]$. Can we say something about additional roots of $g(x)$?

It will turn out that Corollary 5.13 will soon enable us to assert that $5 + 3\sqrt{2}$ is also a root of $f(x)$. In addition, it will enable us to conclude that $4 + 2\sqrt{3} + 5\sqrt{7}, 4 - 2\sqrt{3} - 5\sqrt{7}$, $4 + 2\sqrt{3} - 5\sqrt{7}$ are also roots of $g(x)$. Thus, Corollary 5.13 can indeed by applied in many situations where Proposition 5.11 does not apply.

The proof of Corollary 5.13 will be virtually identical to the proof of Proposition 5.11. Only one new idea is needed for this proof. Other than that, all we will need to do is to go back through the proof of Proposition 5.11 and replace $*$ by σ, $\mathbb{C}$ by R, and $\mathbb{R}$ by R^σ.

Proof. In the proof of Proposition 5.11 we used the fact that $0^* = 0$. In order to modify the proof of Proposition 5.11 to our more general situation, the only new fact we will need to prove about commutative rings is that $\sigma(0) = 0$. To show this, we will use the following four facts that hold in R, since R is a commutative ring:

(i) R has an additive identity that is denoted as 0.

(ii) $\sigma(0)$ has an additive inverse in R that is denoted as $-\sigma(0)$.

(iii) σ satisfies property (b) of Definition 5.8.

(iv) Addition in R is associative.

Therefore, we now have

$$\sigma(0) = \sigma(0) + 0 = \sigma(0) + (\sigma(0) - \sigma(0)) = (\sigma(0) + \sigma(0)) - \sigma(0) =$$
$$\sigma(0 + 0) - \sigma(0) = \sigma(0) - \sigma(0) = 0,$$

as desired. You should be careful to check that you understand why each and every equality holds in the preceding computation. From this point on, the proof of this corollary is simply a rewriting of the proof of Proposition 5.11 with σ replacing $*$, R replacing $\mathbb{C}$, and R^σ replacing $\mathbb{R}$. Therefore, we leave it to you to check all the details. $\qquad\square$

In order to apply Corollary 5.13 to situations such as where $f(x) \in \mathbb{Q}[x]$ has $5 - 3\sqrt{2}$ as a root, we need an easy way to construct commutative rings and automorphisms. Proposition 5.15 will do this for us, but first we need

Lemma 5.14. *Let R be a commutative ring.*

(a) *R has only one additive identity element and only one multiplicative identity element.*

(b) *Each element of R has only one additive inverse.*

(c) *If $x \in R$, then $x \cdot 0 = 0$ and $0 \cdot x = 0$.*

(d) *If $x \in R$, then the product $(-1) \cdot x$ is equal to the additive inverse of x.*

Intuition. Some parts of this lemma, at first glance, might strike you as obvious or pedantic. In part (a), it may seem completely obvious that there can be only one 0 and only one 1. Similarly, in part (c), it might seem clear that multiplication by 0 always results in 0. Although it is not hard to prove these facts, they do require proof and the proofs must exploit the properties of commutative rings from Definition 5.3.

Remember that in a commutative ring, 0 and 1 are merely symbols used to represent the additive and multiplicative identities. However, the symbols 0 and 1 are not necessarily the numbers zero and one that belong to $\mathbb{Z}$, $\mathbb{Q}$, and $\mathbb{R}$. Therefore, it does require proof that a commutative ring can have only a single additive identity and a single multiplicative identity. Similarly, it requires proof that an element only has a single additive inverse and that if it has a multiplicative inverse, then it cannot have more than one multiplicative inverse.

The way that we show that a commutative ring has only one additive identity is that we prove that any two elements that satisfy the additive identity property must be equal. Similarly, to show that an element x of R has only one additive inverse, we prove that any two additive inverses of x must be equal. The proofs of these facts will be rather formal and are not always the most exciting aspects of algebra. However, it is necessary to prove these facts.

Proof. For part (a), suppose that e and f are both the additive identity of R. We need to show that $e = f$. Since e and f are both the additive identity, we can look at the sum $e + f$ in two different ways. Since e is the additive identity element, we have $e + f = f$. However, since f is also the additive identity, we also have $e + f = e$. Comparing these two sets of equations, we immediately see that $e = f$. The proof of the analogous fact for the multiplicative identity is virtually the same. Observe that if g, h are both the multiplicative identity of R, then $g \cdot h = h$ and $g \cdot h = g$. Thus, $g = h$.

For part (b), let $x \in R$ and suppose y, z are both additive inverses of x. We need to show that $y = z$. To do this, we will examine the expression $y + x + z$ two different ways using the associativity of addition. On the one hand, we have

$$(y + x) + z = 0 + z = z$$

and, on the other hand,

$$y + (x + z) = y + 0 = y.$$

Thus, $y = z$, as desired.

The ideas used to prove part (c) should remind you of the ideas used in the proof of Corollary 5.13 where we showed that $\sigma(0) = 0$. You should compare the following argument with the one used in that proof. If $x \in R$, we have

$$x \cdot 0 = x \cdot 0 + 0 = x \cdot 0 + (x \cdot 0 - x \cdot 0) = (x \cdot 0 + x \cdot 0) - x \cdot 0$$
$$= x \cdot (0 + 0) - x \cdot 0 = x \cdot 0 - x \cdot 0 = 0.$$

You should make sure that you understand why each and every equality in the preceding calculation follows from the properties in Definition 5.3. Since multiplication in R is commutative, it also follows that $0 \cdot x = 0$. It is worth noting that we could also directly prove $0 \cdot x = 0$ without using the commutativity of multiplication.

For part (d), to show that $(-1) \cdot x$ is the additive inverse of x, we will need to show that $x + (-1) \cdot x = 0$. That will suffice to complete the proof because the commutativity of addition would then tell us that $(-1) \cdot x + x = 0$. To this end, we have

$$x + (-1) \cdot x = (1) \cdot x + (-1) \cdot x = (1 - 1) \cdot x = 0 \cdot x = 0.$$

Remember that you should check that all of the equalities in the preceding calculation do indeed follow from the earlier parts of this proof and the properties in Definition 5.3. □

Technically, proving that a set is a commutative ring requires checking that eight different properties all hold. Since one of our goals is to come up with an easy way to produce examples of commutative rings, it would be nice to develop a shortcut that, in many circumstances, will greatly decrease the number of properties that need to be checked. Our next proposition does exactly that.

Proposition 5.15. *Let S be a subset of a commutative ring R with the following properties:*

(a) *For all $x, y \in S$, $x + y$ and $x \cdot y$ both belong to S.*

(b) *The additive and multiplicative identities 0 and 1 of R and -1, the additive inverse of 1, all belong to S.*

Then S is a commutative ring.

Proof. Remember that before we can begin checking if a set satisfies the eight properties of commutative rings, we first must establish that the set is closed with respect to both addition and multiplication. However, in this case, things are already taken care of as (a) states that S is closed with respect to both addition and multiplication.

Note that the associative laws of addition and multiplication, the commutative laws of addition and multiplication, and the distributive law hold for all elements of R. Since S is a subset of R, these five properties automatically also hold for all elements of S. As a result, we now only need to check on three of the eight properties of commutative rings. However, (b) stated that S contains both the additive and multiplicative identity elements. Therefore, all that remains is to show that the additive inverse of every element of S is also an element of S. However, (b) also states that -1 belongs to S. Therefore, if $x \in S$, then, by (a), the product $(-1) \cdot x$ also belongs to S. However, Lemma 5.14(d) asserts that $(-1) \cdot x$ is the additive inverse of x. Hence, S contains the additive inverse of each of its elements, and S is indeed a commutative ring. □

5.4 Automorphisms and Roots of Polynomials

We can now use Proposition 5.15 to construct examples that illustrate the interplay between automorphisms and roots of polynomials. These examples will illustrate that Corollary 5.13 does have many more applications than Proposition 5.11.

■ Example

Let $\mathbb{Q}(\sqrt{2}) = \{a + b\sqrt{2} \mid a, b \in \mathbb{Q}\}$. Some typical elements of $\mathbb{Q}(\sqrt{2})$ are $3 - 5\sqrt{2}$, $4\sqrt{2}$, $-\frac{7}{3} + 18\sqrt{2}$, and 50. We would like to see if we can apply Proposition 5.15 to assert that $\mathbb{Q}(\sqrt{2})$ is a commutative ring. The first observation to make is that since $\mathbb{Q}(\sqrt{2})$ is a subset of $\mathbb{R}$, we can use the associative, commutative, and distributive laws when adding and multiplying elements of $\mathbb{Q}(\sqrt{2})$. Therefore, if $a + b\sqrt{2}, c + d\sqrt{2} \in \mathbb{Q}(\sqrt{2})$, they must add and multiply as follows:

$$(a + b\sqrt{2}) + (c + d\sqrt{2}) = (a + c) + (b + d)\sqrt{2},$$

and

$$(a + b\sqrt{2}) \cdot (c + d\sqrt{2}) = ac + (ad + bc)\sqrt{2} + bd(\sqrt{2})^2 = (ac + 2bd) + (ad + bc)\sqrt{2}.$$

Looking at the preceding equations, we see that since $\mathbb{Q}$ is closed under addition and multiplication, so is $\mathbb{Q}(\sqrt{2})$. Therefore, part (a) from Proposition 5.15 is satisfied. Furthermore, since $\mathbb{Q}$ is a subset of $\mathbb{Q}(\sqrt{2})$, we see that $\mathbb{Q}(\sqrt{2})$ also satisfies part (b) from Proposition 5.15. Thus, $\mathbb{Q}(\sqrt{2})$ is a commutative ring. In fact, it is not difficult to go one more step and show that $\mathbb{Q}(\sqrt{2})$ is a field, but we will not need that additional information at this point.

Now suppose that σ is an automorphism of $\mathbb{Q}(\sqrt{2})$ that fixes all the elements of $\mathbb{Q}$. Therefore, properties (b) and (c) of Definition 5.8 imply that

$$\sigma(a + b\sqrt{2}) = \sigma(a) + \sigma(b\sqrt{2}) = \sigma(a) + \sigma(b)\sigma(\sqrt{2}) = a + b\sigma(\sqrt{2}),$$

for all $a, b \in \mathbb{Q}$. In light of the preceding equation, if we know what $\sigma(\sqrt{2})$ is equal to, then we will know the value of σ for each element of $\mathbb{Q}(\sqrt{2})$.

In order to compute $\sigma(\sqrt{2})$, we first observe that $\sqrt{2}$ is a root of the polynomial $p(x) = x^2 - 2$. The coefficients of $p(x)$ belong to $\mathbb{Q}$ and are therefore fixed by σ. As a result, we can apply Corollary 5.13 to conclude that $\sigma(\sqrt{2})$ is also a root of $p(x)$. Thus, either

$$\sigma(\sqrt{2}) = \sqrt{2} \quad \text{or} \quad \sigma(\sqrt{2}) = -\sqrt{2}.$$

Next, consider the functions $\sigma_1, \sigma_2 : \mathbb{Q}(\sqrt{2}) \to \mathbb{Q}(\sqrt{2})$ defined as

$$\sigma_1(a + b\sqrt{2}) = a + b\sqrt{2} \quad \text{and} \quad \sigma_2(a + b\sqrt{2}) = a - b\sqrt{2},$$

for all $a, b \in \mathbb{Q}$. Although we have not yet shown that the functions σ_1, σ_2 are automorphisms, the preceding argument indicates that no other function could possibly be an automorphism of $\mathbb{Q}(\sqrt{2})$ that fixes all elements of $\mathbb{Q}$. Thus, there are at most two automorphisms of $\mathbb{Q}(\sqrt{2})$ that fix all elements of $\mathbb{Q}$, and you should take a moment to check that σ_1, σ_2 are indeed automorphisms. This means that you need to check that σ_1, σ_2 satisfy all the conditions in Definition 5.8. Since σ_1 is the identity map on $\mathbb{Q}(\sqrt{2})$, there is very little to prove in this case. However, to show that σ_2 satisfies the conditions of Definition 5.8, you should first look at the proof of Lemma 5.7 as the ideas and techniques are quite similar.

Once we know that σ_2 is an automorphism of $\mathbb{Q}(\sqrt{2})$ that fixes all elements of $\mathbb{Q}$, we can use Corollary 5.13. For example, suppose $5 - 8\sqrt{2}$ is a root of some $p(x) \in \mathbb{Q}[x]$. Then, Corollary 5.13 immediately asserts that $\sigma_2(5 - 8\sqrt{2}) = 5 + 8\sqrt{2}$ must also be a root of $p(x)$. Using the quadratic formula, you can check that $x^2 - 10x - 103$ is an example of a polynomial that has both $5 - 8\sqrt{2}$ and $5 + 8\sqrt{2}$ as roots. In Chapter 12, we will show that any $p(x) \in \mathbb{Q}[x]$ that has $5 - 8\sqrt{2}$ as a root must be a multiple of $x^2 - 10x - 103$.

Similarly, if $f(x) \in \mathbb{Q}[x]$ has $-1/2 + 5\sqrt{2}$ as a root, then $f(x)$ must also have $\sigma_2(-1/2 + 5\sqrt{2}) = -1/2 - 5\sqrt{2}$ as a root. You can check that $4x^2 + 4x - 199$ has both $-1/2 + 5\sqrt{2}$ and $-1/2 - 5\sqrt{2}$ as roots, and in Chapter 12, you will see that $f(x)$ must be a multiple of $4x^2 + 4x - 199$.

Before leaving this example, consider the polynomial $g(x) = x^2 + 2\sqrt{2}x - 6$. You can check that $\sqrt{2}$ and $-3\sqrt{2}$ are roots of $g(x)$. However, neither $\sigma_2(\sqrt{2}) = -\sqrt{2}$ nor $\sigma_2(-3\sqrt{2}) = 3\sqrt{2}$ are roots of $g(x)$. Observe that this does not contradict Corollary 5.13, as $g(x)$ has a coefficient that is not fixed by σ_2.

We should certainly note that in this example we used Corollary 5.13 in two different ways. In one direction, we looked at the roots of $x^2 - 2$ and showed that there are only two automorphisms of $\mathbb{Q}(\sqrt{2})$ that fix every element of $\mathbb{Q}$. In the other direction, we used the automorphism σ_2 to find additional roots of some polynomials once we already knew one root. Thus, we used roots of polynomials to find automorphisms and used automorphisms to find roots of polynomials. It is this interplay between automorphisms and roots of polynomials that will be an important and recurring theme throughout this course and much of abstract algebra.

The computations in the next example will be more complicated than those in the previous example. However, the ideas and techniques used will be the same.

■ Example

Let

$$\mathbb{Q}(i, \sqrt{3}) = \{a + bi + c\sqrt{3} + di\sqrt{3} \mid a, b, c, d \in \mathbb{Q}\}.$$

Some typical elements of $\mathbb{Q}(i, \sqrt{3})$ are $2 - 3i + 4\sqrt{3} - 7i\sqrt{3}$, $\frac{3}{5}i + 9i\sqrt{3}$, 34, and $12i - 17\sqrt{3}$. Since $\mathbb{Q}(i, \sqrt{3})$ is a subset of $\mathbb{C}$, the addition and multiplication of elements in $\mathbb{Q}(i, \sqrt{3})$ must satisfy the associative, commutative, and distributive laws. Therefore, elements of $\mathbb{Q}(i, \sqrt{3})$ must add and multiply as follows:

$$(a_1 + b_1 i + c_1\sqrt{3} + d_1 i\sqrt{3}) + (a_2 + b_2 i + c_2\sqrt{3} + d_2 i\sqrt{3}) =$$
$$(a_1 + a_2) + (b_1 + b_2)i + (c_1 + c_2)\sqrt{3} + (d_1 + d_2)i\sqrt{3}$$

and

$$(a_1 + b_1 i + c_1\sqrt{3} + d_1 i\sqrt{3}) \cdot (a_2 + b_2 i + c_2\sqrt{3} + d_2 i\sqrt{3}) =$$
$$(a_1 a_2 - b_1 b_2 + 3c_1 c_2 - 3d_1 d_2) + (a_1 b_2 + b_1 a_2 + 3c_1 d_2 + 3d_1 c_2)i +$$
$$(a_1 c_2 - b_1 d_2 + c_1 a_2 - d_1 b_2)\sqrt{3} + (a_1 d_2 + b_1 c_2 + c_1 b_2 + d_1 a_2)i\sqrt{3},$$

for all $a_1, a_2, b_1, b_2, c_1, c_2, d_1, d_2 \in \mathbb{Q}$. It would be good practice for you to use the associative, commutative, and distributive laws to derive the preceding formulas.

Since $\mathbb{Q}$ is closed under addition and multiplication, the preceding formulas indicate that $\mathbb{Q}(i, \sqrt{3})$ is also closed under addition and multiplication. Thus, $\mathbb{Q}(i, \sqrt{3})$ satisfies part (a) from Proposition 5.15. However, since $\mathbb{Q}$ is a subset of $\mathbb{Q}(i, \sqrt{3})$, part (b) from Proposition 5.15 is also satisfied, and we see that $\mathbb{Q}(i, \sqrt{3})$ is a commutative ring. Now suppose σ is an automorphism of $\mathbb{Q}(i, \sqrt{3})$ that fixes every element of $\mathbb{Q}$. Then Lemma 5.9 asserts that

$$\sigma(a + bi + c\sqrt{3} + di\sqrt{3}) = \sigma(a) + \sigma(b)\sigma(i) + \sigma(c)\sigma(\sqrt{3}) + \sigma(d)\sigma(i)\sigma(\sqrt{3}) =$$
$$a + b\sigma(i) + c\sigma(\sqrt{3}) + d\sigma(i)\sigma(\sqrt{3}),$$

for all $a, b, c, d \in \mathbb{Q}$. The preceding equation tells us that once we know the values of $\sigma(i)$ and $\sigma(\sqrt{3})$, then we will know the value of σ for each element of $\mathbb{Q}(i, \sqrt{3})$.

Since i is a root of $f(x) = x^2 - 1$ and $\sqrt{3}$ is a root of $g(x) = x^2 - 3$, Corollary 5.13 asserts that $\sigma(i)$ must be a root of $f(x)$ and $\sigma(\sqrt{3})$ must be a root of $g(x)$. Therefore,

$$\sigma(i) = i \text{ or } -i \quad \text{and} \quad \sigma(\sqrt{3}) = \sqrt{3} \text{ or } -\sqrt{3}.$$

In light of this, there are *at most* four automorphisms of $\mathbb{Q}(i, \sqrt{3})$ that fix the elements of $\mathbb{Q}$. We will denote these four candidates for being automorphisms as $\sigma_1, \sigma_2, \sigma_3, \sigma_4$, and they have the properties that

$$\sigma_1(i) = i \text{ and } \sigma_1(\sqrt{3}) = \sqrt{3}; \quad \sigma_2(i) = -i \text{ and } \sigma_2(\sqrt{3}) = \sqrt{3};$$
$$\sigma_3(i) = i \text{ and } \sigma_3(\sqrt{3}) = -\sqrt{3}; \quad \sigma_4(i) = -i \text{ and } \sigma_4(\sqrt{3}) = -\sqrt{3}.$$

An expanded way of writing this is

$$\sigma_1(a + bi + c\sqrt{3} + di\sqrt{3}) = a + bi + c\sqrt{3} + di\sqrt{3},$$

$$\sigma_2(a + bi + c\sqrt{3} + di\sqrt{3}) = a - bi + c\sqrt{3} - di\sqrt{3},$$

$$\sigma_3(a + bi + c\sqrt{3} + di\sqrt{3}) = a + bi - c\sqrt{3} - di\sqrt{3},$$

$$\sigma_4(a + bi + c\sqrt{3} + di\sqrt{3}) = a - bi - c\sqrt{3} + di\sqrt{3},$$

for all $a, b, c, d \in \mathbb{Q}$.

It is clear that σ_1 is an automorphism of $\mathbb{Q}(i, \sqrt{3})$. However, at this point, in order to assert that $\sigma_2, \sigma_3, \sigma_4$ are automorphisms of $\mathbb{Q}(i, \sqrt{3})$, we need to check that they satisfy all three conditions of Definition 5.8. The first two conditions of Definition 5.8 are not difficult to check. However, part (c) of Definition 5.8 is extremely long and tedious to verify. The verification is straightforward but is still long and tedious.

However, good news awaits in Chapter 15. Indeed, Theorem 15.17 often tells us how many automorphisms a field has. In fact, Theorem 15.17 will tell us that $\mathbb{Q}(i, \sqrt{3})$ has exactly four automorphisms that fix every element of $\mathbb{Q}$. Similarly, if we apply Theorem 15.17 to our previous example, it tells us that $\mathbb{Q}(\sqrt{2})$ has exactly two automorphisms that fix every element of $\mathbb{Q}$.

As we have seen, Corollary 5.13 can supply us with a list of candidates of the automorphisms of a field. When we combine Corollary 5.13 and Theorem 15.17, we will often know all the automorphisms of a field without going through a long series of computations. At this point in this course, we should technically do all the computations needed to verify that $\sigma_2, \sigma_3, \sigma_4$ are all automorphisms of $\mathbb{Q}(i, \sqrt{3})$. However, when we get to Chapter 15, Theorem 15.17 will make life much easier for us. Thus, Theorem 15.17 is an example of a powerful piece of mathematical machinery that will help us avoid pages of computations.

Once we have reached the point that we believe that $\sigma_1, \sigma_2, \sigma_3, \sigma_4$ are all automorphisms of $\mathbb{Q}(i, \sqrt{3})$ that fix the elements of $\mathbb{Q}$, we are in a position to apply Corollary 5.13. Suppose $2i - 6\sqrt{3}$ is a root of some $p(x) \in \mathbb{Q}[x]$; then Corollary 5.13 tells us that

$$\sigma_1(2i - 6\sqrt{3}) = 2i - 6\sqrt{3}, \quad \sigma_2(2i - 6\sqrt{3}) = -2i - 6\sqrt{3},$$

$$\sigma_3(2i - 6\sqrt{3}) = 2i + 6\sqrt{3}, \quad \sigma_4(2i - 6\sqrt{3}) = -2i + 6\sqrt{3}$$

are all roots of $p(x)$. Thus, $p(x)$ must have at least four distinct roots. It turns out that the polynomial $x^4 - 208x^2 + 12,544$ is an example of a polynomial in $\mathbb{Q}[x]$ that has these four roots, and our work in Chapter 12 will imply that $p(x)$ must be a multiple of $x^4 - 208x^2 + 12,544$.

Now suppose $7+i\sqrt{3}$ is a root of some $q(x) \in \mathbb{Q}[x]$. Then Corollary 5.13 asserts that

$$\sigma_1(7+i\sqrt{3}) = 7+i\sqrt{3}, \quad \sigma_2(7+i\sqrt{3}) = 7-i\sqrt{3},$$

$$\sigma_3(7+i\sqrt{3}) = 7-i\sqrt{3}, \quad \sigma_4(7+i\sqrt{3}) = 7+i\sqrt{3}$$

are all roots of $q(x)$. Therefore, in this case, Corollary 5.13 only provides us with one additional root of $q(x)$. You can use the quadratic formula to check that $x^2 - 14x + 52$ has both $7+i\sqrt{3}$ and $7-i\sqrt{3}$ as roots. In Chapter 12, we will show that $q(x)$ must be a multiple of $x^2 - 14x + 52$.

■

In both of our previous examples, we exploited the dual nature of Corollary 5.13. In one direction, we used facts about the roots of polynomials to help us find automorphisms. In the other direction, we used automorphisms to help us find additional roots of some polynomials. Before moving on, we should look at one more example.

■ Example

Let $S = \{a + b2^{\frac{1}{3}} \mid a, b \in \mathbb{Q}\}$ and $T = \{a + b2^{\frac{1}{3}} + c2^{\frac{2}{3}} \mid a, b, c \in \mathbb{Q}\}$. In the exercises at the end of this chapter, you will be asked to show that $2^{\frac{2}{3}} \notin S$. In light of this fact, we see that S is not closed under multiplication as $2^{\frac{1}{3}} \in S$, yet $2^{\frac{1}{3}} \cdot 2^{\frac{1}{3}} = 2^{\frac{2}{3}} \notin S$. Thus, S is not a commutative ring.

If α, β are any elements of T, then they must be of the form

$$\alpha = a + b2^{\frac{1}{3}} + c2^{\frac{2}{3}}, \quad \beta = d + e2^{\frac{1}{3}} + f2^{\frac{2}{3}},$$

where $a, b, c, d, e, f \in \mathbb{Q}$. Since T is a subset of $\mathbb{R}$, when adding or multiplying elements of T, the associative, commutative, and distributive laws hold. Therefore, you should be able use these laws to derive formulas for $\alpha + \beta$ and $\alpha \cdot \beta$. These formulas will show that T is closed under both addition and multiplication. By combining this with the fact that T contains $\mathbb{Q}$, we can apply Proposition 5.15 to assert that T is a commutative ring.

We would now like to find all automorphisms of T that fix every element of $\mathbb{Q}$. If σ is such an automorphism and if $\alpha = a + b2^{\frac{1}{3}} + c2^{\frac{2}{3}}$, then Lemma 5.9 asserts that

$$\sigma\left(a + b2^{\frac{1}{3}} + c2^{\frac{2}{3}}\right) = \sigma\left(a + b2^{\frac{1}{3}} + c2^{\frac{1}{3}}2^{\frac{1}{3}}\right) = \sigma(a) + \sigma(b)\sigma\left(2^{\frac{1}{3}}\right) +$$

$$\sigma(c)\sigma\left(2^{\frac{1}{3}}\right)\sigma\left(2^{\frac{1}{3}}\right) = a + b\sigma\left(2^{\frac{1}{3}}\right) + c\sigma\left(2^{\frac{1}{3}}\right)^2.$$

Therefore, if we can find the value of $\sigma\left(2^{\frac{1}{3}}\right)$, then we will know the value of σ for every element of T. Since $2^{\frac{1}{3}}$ is a root of $x^3 - 2$, Corollary 5.13 tells us that $\sigma\left(2^{\frac{1}{3}}\right)$ must also be a root of $x^3 - 2$. However, $2^{\frac{1}{3}}$ is the only root of $x^3 - 2$ that is a real number. Since $T \subseteq \mathbb{R}$, it follows that $2^{\frac{1}{3}}$ is the *only* root of $x^3 - 2$ that belongs to T. However, $\sigma(2^{\frac{1}{3}})$ is a root of $x^3 - 2$ that belongs to T, so $\sigma(2^{\frac{1}{3}}) = 2^{\frac{1}{3}}$. Thus, if $\alpha \in T$, we have $\sigma(\alpha) = \alpha$. Therefore, σ must be the identity map, and it is the only automorphism of T that fixes every element of $\mathbb{Q}$.

∎

Exercises for Sections 5.3 and 5.4

In exercises 1–10, let $\alpha = 2 - i$, $\beta = 3 - 2i$, and $\gamma = -1 + i$.

1. Compute α^*, β^*, and $\alpha^* \cdot \beta^*$.

2. Compute $\alpha \cdot \beta$ and $(\alpha \cdot \beta)^*$ and compare your answers to those in exercise 1.

3. Compute $\alpha^* + \beta^* + \gamma^*$.

4. Compute $(\alpha + \beta + \gamma)^*$ and compare your answers to those in exercise 3.

5. Compute γ^*, $(\gamma^*)^2$, and $(\gamma^*)^4$.

6. Compute γ^2, γ^4, $(\gamma^2)^*$, $(\gamma^4)^*$ and compare your answers to those in exercise 5.

7. Compute α^{-1} and $(\alpha^{-1})^*$.

8. Compute α^* and $(\alpha^*)^{-1}$ and compare your answers to those in exercise 7.

9. Compute $\alpha^* \cdot \beta^* \cdot \gamma^*$.

10. Compute $(\alpha \cdot \beta \cdot \gamma)^*$ and compare your answer to that in exercise 9.

For exercises 11–47, please read the following:

A polynomial is called **monic** if the coefficient of the term with the largest exponent is one. Observe that every polynomial (other than the polynomial that is always zero) can be multiplied by exactly one nonzero constant to produce a monic polynomial. For example, $x^2 - 5ix - 4$ is monic, whereas $4x^2 + 4x - 199$ is not monic but can be multiplied by $\frac{1}{4}$ to produce a monic polynomial.

11. Suppose $7 - 2i$ is a root of some $f(x) \in \mathbb{R}[x]$.
    (a) Find another root of $f(x)$.

    (b) Find a monic polynomial of degree 1 in $\mathbb{C}[x]$ that has $7 - 2i$ as a root.

    (c) Find a monic polynomial of degree 2 in $\mathbb{R}[x]$ that has $7 - 2i$ as a root.

12. Suppose $5+3i$ is a root of some $g(x) \in \mathbb{R}[x]$.
     (a) Find another root of $g(x)$.

     (b) Find a monic polynomial of degree 1 in $\mathbb{C}[x]$ that has $5+3i$ as a root.

     (c) Find a monic polynomial of degree 2 in $\mathbb{R}[x]$ that has $5+3i$ as a root.

13. Suppose $\sqrt{3}+8i$ is a root of some $h(x) \in \mathbb{R}[x]$.
     (a) Find another root of $h(x)$.

     (b) Find a monic polynomial of degree 1 in $\mathbb{C}[x]$ that has $\sqrt{3}+8i$ as a root.

     (c) Find a monic polynomial of degree 2 in $\mathbb{R}[x]$ that has $\sqrt{3}+8i$ as a root.

14. Suppose $-4+\sqrt{5}i$ is a root of some $k(x) \in \mathbb{R}[x]$.
     (a) Find another root of $k(x)$.

     (b) Find a monic polynomial of degree 1 in $\mathbb{C}[x]$ that has $-4+\sqrt{5}i$ as a root.

     (c) Find a monic polynomial of degree 2 in $\mathbb{R}[x]$ that has $-4+\sqrt{5}i$ as a root.

15. Suppose $\frac{9}{2}-5i$ and $-7+6i$ are roots of some $f(x) \in \mathbb{R}[x]$.
     (a) Find two more roots of $f(x)$.

     (b) Find a monic polynomial of degree 2 in $\mathbb{C}[x]$ that has $\frac{9}{2}-5i$ and $-7+6i$ as roots.

     (c) Find a monic polynomial of degree 4 in $\mathbb{R}[x]$ that has $\frac{9}{2}-5i$ and $-7+6i$ as roots.

16. Suppose $-\sqrt{2}+i$ and $3-4i$ are roots of some $g(x) \in \mathbb{R}[x]$.
     (a) Find two more roots of $g(x)$.

     (b) Find a monic polynomial of degree 2 in $\mathbb{C}[x]$ that has $-\sqrt{2}+i$ and $3-4i$ as roots.

     (c) Find a monic polynomial of degree 4 in $\mathbb{R}[x]$ that has $-\sqrt{2}+i$ and $3-4i$ as roots.

17. Suppose $\pi+7i$ and $3+\sqrt{6}i$ are roots of some $h(x) \in \mathbb{R}[x]$.
     (a) Find two more roots of $h(x)$.

     (b) Find a monic polynomial of degree 2 in $\mathbb{C}[x]$ that has $\pi+7i$ and $3+\sqrt{6}i$ as roots.

     (c) Find a monic polynomial of degree 4 in $\mathbb{R}[x]$ that has $\pi+7i$ and $3+\sqrt{6}i$ as roots.

18. Suppose $5i$ and $\sqrt{2}-8i$ are roots of some $k(x) \in \mathbb{R}[x]$.
     (a) Find two more roots of $k(x)$.

     (b) Find a monic polynomial of degree 2 in $\mathbb{C}[x]$ that has $5i$ and $\sqrt{2}-8i$ as roots.

     (c) Find a monic polynomial of degree 4 in $\mathbb{R}[x]$ which has $5i$ and $\sqrt{2}-8i$ as roots.

19. Let $\mathbb{Q}(i) = \{a + bi \mid a, b \in \mathbb{Q}\}$.
    (a) If $a, b, c, d \in \mathbb{Q}$, find A, B such that
    $$(a + bi) + (c + di) = A + Bi.$$
    In your answers, A and B should be expressions in a, b, c, d.

    (b) If $a, b, c, d \in \mathbb{Q}$, find C, D such that
    $$(a + bi) \cdot (c + di) = C + Di.$$
    In your answers, C and D should be expressions in a, b, c, d.

    (c) Use parts (a) and (b) along with Proposition 5.15 to show that $\mathbb{Q}(i)$ is a commutative ring.

    (d) Show that $\mathbb{Q}(i)$ is a field.

20. Let $\mathbb{Q}(\sqrt{7}) = \{a + b\sqrt{7} \mid a, b \in \mathbb{Q}\}$.
    (a) If $a, b, c, d \in \mathbb{Q}$, find A, B such that
    $$(a + b\sqrt{7}) + (c + d\sqrt{7}) = A + B\sqrt{7}.$$
    In your answers, A and B should be expressions in a, b, c, d.

    (b) If $a, b, c, d \in \mathbb{Q}$, find C, D such that
    $$(a + b\sqrt{7}) \cdot (c + d\sqrt{7}) = C + D\sqrt{7}.$$
    In your answers, C and D should be expressions in a, b, c, d.

    (c) Use parts (a) and (b) along with Proposition 5.15 to show that $\mathbb{Q}(\sqrt{7})$ is a commutative ring.

    (d) Show that $\mathbb{Q}(\sqrt{7})$ is a field. (Hint: When you are looking for multiplicative inverses, think about the product $(a + b\sqrt{7}) \cdot (a - b\sqrt{7})$.)

21. Let $\mathbb{Q}(\sqrt{11}) = \{a + b\sqrt{11} \mid a, b \in \mathbb{Q}\}$.
    (a) If $a, b, c, d \in \mathbb{Q}$, find A, B such that
    $$(a + b\sqrt{11}) + (c + d\sqrt{11}) = A + B\sqrt{11}.$$
    In your answers, A and B should be expressions in a, b, c, d.

    (b) If $a, b, c, d \in \mathbb{Q}$, find C, D such that
    $$(a + b\sqrt{11}) \cdot (c + d\sqrt{11}) = C + D\sqrt{11}.$$
    In your answers, C and D should be expressions in a, b, c, d.

(c) Use parts (a) and (b) along with Proposition 5.15 to show that $\mathbb{Q}(\sqrt{11})$ is a commutative ring.

(d) Show that $\mathbb{Q}(\sqrt{11})$ is a field. (Hint: When you are looking for multiplicative inverses, think about the product $(a+b\sqrt{11}) \cdot (a-b\sqrt{11})$.)

22. Let $\mathbb{Q}(i\sqrt{19}) = \{a+bi\sqrt{19} \mid a, b \in \mathbb{Q}\}$.
    (a) If $a, b, c, d \in \mathbb{Q}$, find A, B such that

    $$(a+bi\sqrt{19}) + (c+di\sqrt{19}) = A + Bi\sqrt{19}.$$

    In your answers, A and B should be expressions in a, b, c, d.

    (b) If $a, b, c, d \in \mathbb{Q}$, find C, D such that

    $$(a+bi\sqrt{19}) \cdot (c+di\sqrt{19}) = C + Di\sqrt{19}.$$

    In your answers, C and D should be expressions in a, b, c, d.

    (c) Use parts (a) and (b) along with Proposition 5.15 to show that $\mathbb{Q}(i\sqrt{19})$ is a commutative ring.

    (d) Show that $\mathbb{Q}(i\sqrt{19})$ is a field. (Hint: When you are looking for multiplicative inverses, think about the product $(a+bi\sqrt{19}) \cdot (a-bi\sqrt{19})$.)

23. Let $\mathbb{Q}(\sqrt{2}, \sqrt{3}) = \{a+b\sqrt{2}+c\sqrt{3}+d\sqrt{6} \mid a, b, c, d \in \mathbb{Q}\}$.
    (a) If $a, b, c, d, e, f, g, h \in \mathbb{Q}$, find A, B, C, D such that

    $$(a+b\sqrt{2}+c\sqrt{3}+d\sqrt{6}) + (e+f\sqrt{2}+g\sqrt{3}+h\sqrt{6}) = A + B\sqrt{2}+C\sqrt{3}+D\sqrt{6}.$$

    In your answers A, B, C, D should be expressions in a, b, c, d, e, f, g, h.

    (b) If $a, b, c, d, e, f, g, h \in \mathbb{Q}$, find E, F, G, H such that

    $$(a+b\sqrt{2}+c\sqrt{3}+d\sqrt{6}) \cdot (e+f\sqrt{2}+g\sqrt{3}+h\sqrt{6}) = E + F\sqrt{2}+G\sqrt{3}+H\sqrt{6}.$$

    In your answers E, F, G, H should be expressions in a, b, c, d, e, f, g, h.

    (c) Use parts (a) and (b) along with Proposition 5.15 to show that $\mathbb{Q}(\sqrt{2}, \sqrt{3})$ is a commutative ring.

24. Let $\mathbb{Q}(\sqrt{7}, i) = \{a+b\sqrt{7}+ci+di\sqrt{7} \mid a, b, c, d \in \mathbb{Q}\}$.
    (a) If $a, b, c, d, e, f, g, h \in \mathbb{Q}$, find A, B, C, D such that

    $$(a+b\sqrt{7}+ci+di\sqrt{7}) + (e+f\sqrt{7}+gi+hi\sqrt{7}) = A + B\sqrt{7}+Ci+Di\sqrt{7}.$$

    In your answers A, B, C, D should be expressions in a, b, c, d, e, f, g, h.

(b) If $a, b, c, d, e, f, g, h \in \mathbb{Q}$, find E, F, G, H such that

$$(a + b\sqrt{7} + ci + di\sqrt{7}) \cdot (e + f\sqrt{7} + gi + hi\sqrt{7}) = E + F\sqrt{7} + Gi + Hi\sqrt{7}.$$

In your answers E, F, G, H should be expressions in a, b, c, d, e, f, g, h.

(c) Use parts (a) and (b) along with Proposition 5.15 to show that $\mathbb{Q}(\sqrt{7}, i)$ is a commutative ring.

25. Let σ be an automorphism of the field $\mathbb{Q}(i)$ that fixes every element of $\mathbb{Q}$. Show that either $\sigma(i) = i$ or $\sigma(i) = -i$.

26. Use your work from exercise 19 to show that both possibilities for σ in exercise 25 result in automorphisms of $\mathbb{Q}(i)$.

27. Let τ be an automorphism of the field $\mathbb{Q}(\sqrt{7})$ that fixes every element of $\mathbb{Q}$. Show that either $\tau(\sqrt{7}) = \sqrt{7}$ or $\tau(\sqrt{7}) = -\sqrt{7}$.

28. Use your work from exercise 20 to show that both possibilities for τ in exercise 27 result in automorphisms of $\mathbb{Q}(\sqrt{7})$.

29. Let ν be an automorphism of the field $\mathbb{Q}(\sqrt{11})$ that fixes every element of $\mathbb{Q}$. Show that either $\nu(\sqrt{11}) = \sqrt{11}$ or $\nu(\sqrt{11}) = -\sqrt{11}$.

30. Use your work from exercise 21 to show that both possibilities for ν in exercise 29 result in automorphisms of $\mathbb{Q}(\sqrt{11})$.

31. Let σ be an automorphism of the field $\mathbb{Q}(i\sqrt{19})$ that fixes every element of $\mathbb{Q}$. Show that either $\sigma(-i\sqrt{19}) = i\sqrt{19}$ or $\sigma(i\sqrt{19}) = -i\sqrt{19}$.

32. Use your work from exercise 22 to show that both possibilities for σ in exercise 31 result in automorphisms of $\mathbb{Q}(i\sqrt{19})$.

33. Show that every automorphism of the field $\mathbb{Q}(\sqrt{2}, \sqrt{3})$ from exercise 23 that fixes every element of $\mathbb{Q}$ is completely determined by where it sends $\sqrt{2}$ and $\sqrt{3}$. Use this fact to show that $\mathbb{Q}(\sqrt{2}, \sqrt{3})$ has at most four automorphisms that fix every element of $\mathbb{Q}$.

34. Let σ be the automorphism of $\mathbb{Q}(\sqrt{2}, \sqrt{3})$ such that $\sigma(a + b\sqrt{2} + c\sqrt{3} + d\sqrt{6}) = a - b\sqrt{2} + c\sqrt{3} - d\sqrt{6}$, where $a, b, c, d \in \mathbb{Q}$. Describe, in terms of a, b, c, d, all elements of $\mathbb{Q}(\sqrt{2}, \sqrt{3})$ that are fixed by σ.

35. Let τ be the automorphism of $\mathbb{Q}(\sqrt{2}, \sqrt{3})$ such that $\tau(a + b\sqrt{2} + c\sqrt{3} + d\sqrt{6}) = a + b\sqrt{2} - c\sqrt{3} - d\sqrt{6}$, where $a, b, c, d \in \mathbb{Q}$. Describe, in terms of a, b, c, d, all elements of $\mathbb{Q}(\sqrt{2}, \sqrt{3})$ that are fixed by τ.

36. Let v be the automorphism of $\mathbb{Q}(\sqrt{2}, \sqrt{3})$ such that $v(a+b\sqrt{2}+c\sqrt{3}+d\sqrt{6}) = a-b\sqrt{2}-c\sqrt{3}+d\sqrt{6}$, where $a, b, c, d \in \mathbb{Q}$. Describe, in terms of a, b, c, d, all elements of $\mathbb{Q}(\sqrt{2}, \sqrt{3})$ that are fixed by v.

37. Show that every automorphism of the field $\mathbb{Q}(\sqrt{7}, i)$ from exercise 24 that fixes every element of $\mathbb{Q}$ is completely determined by where it sends $\sqrt{7}$ and i. Use this fact to show that $\mathbb{Q}(\sqrt{7}, i)$ has at most four automorphisms that fix every element of $\mathbb{Q}$.

38. Let σ be the automorphism of $\mathbb{Q}(\sqrt{7}, i)$ such that $\sigma(a+b\sqrt{7}+ci+di\sqrt{7}) = a-b\sqrt{7}+ci-di\sqrt{7}$, where $a, b, c, d \in \mathbb{Q}$. Describe, in terms of a, b, c, d, all elements of $\mathbb{Q}(\sqrt{7}, \sqrt{i})$ that are fixed by σ.

39. Let τ be the automorphism of $\mathbb{Q}(\sqrt{7}, i)$ such that $\tau(a+b\sqrt{7}+ci+di\sqrt{7}) = a+b\sqrt{7} -ci-di\sqrt{7}$, where $a, b, c, d \in \mathbb{Q}$. Describe, in terms of a, b, c, d, all elements of $\mathbb{Q}(\sqrt{7}, \sqrt{i})$ that are fixed by τ.

40. Let v be the automorphism of $\mathbb{Q}(\sqrt{7}, i)$ such that $v(a+b\sqrt{7}+ci+di\sqrt{7}) = a-b\sqrt{7}- ci+di\sqrt{7}$, where $a, b, c, d \in \mathbb{Q}$. Describe, in terms of a, b, c, d, all elements of $\mathbb{Q}(\sqrt{7}, \sqrt{i})$ that are fixed by v.

41. Show that $2^{\frac{2}{3}}$ cannot be written in the form $a+b2^{\frac{1}{3}}$, where $a, b \in \mathbb{Q}$. Observe that this fact proves that the set S in the last example in Section 5.4 is not a commutative ring.

42. Let $\mathbb{Q}(2^{\frac{1}{3}}) = \{a+b2^{\frac{1}{3}}+c2^{\frac{2}{3}} \mid a, b, c \in \mathbb{Q}\}$.
    (a) If $a, b, c, d, e, f \in \mathbb{Q}$, find A, B, C such that

    $$\left(a+b2^{\frac{1}{3}}+c2^{\frac{2}{3}}\right)+\left(d+e2^{\frac{1}{3}}+f2^{\frac{2}{3}}\right) = A+B2^{\frac{1}{3}}+C2^{\frac{2}{3}}.$$

    In your final answer, A, B, C should be expressions in a, b, c, d, e, f.

    (b) If $a, b, c, d, e, f \in \mathbb{Q}$, find D, E, F such that

    $$\left(a+b2^{\frac{1}{3}}+c2^{\frac{2}{3}}\right) \cdot \left(d+e2^{\frac{1}{3}}+f2^{\frac{2}{3}}\right) = D+E2^{\frac{1}{3}}+F2^{\frac{2}{3}}.$$

    In your final answer, D, E, F should be expressions in a, b, c, d, e, f.

    (c) Use parts (a) and (b) along with Proposition 5.15 to show that $\mathbb{Q}(2^{\frac{1}{3}})$, that is the same as the set T in the last example in Section 5.4, is a commutative ring.

43. Let $\mathbb{Q}(7^{\frac{1}{3}}) = \{a+b7^{\frac{1}{3}}+c7^{\frac{2}{3}} \mid a, b, c \in \mathbb{Q}\}$.
    (a) If $a, b, c, d, e, f \in \mathbb{Q}$, find A, B, C such that

    $$\left(a+b7^{\frac{1}{3}}+c7^{\frac{2}{3}}\right)+\left(d+e7^{\frac{1}{3}}+f7^{\frac{2}{3}}\right) = A+B7^{\frac{1}{3}}+C7^{\frac{2}{3}}.$$

    In your final answer, A, B, C should be expressions in a, b, c, d, e, f.

(b) If $a, b, c, d, e, f \in \mathbb{Q}$, find D, E, F such that

$$\left(a + b7^{\frac{1}{3}} + c7^{\frac{2}{3}}\right) \cdot \left(d + e7^{\frac{1}{3}} + f7^{\frac{2}{3}}\right) = D + E7^{\frac{1}{3}} + F7^{\frac{2}{3}}.$$

In your final answer, D, E, F should be expressions in a, b, c, d, e, f.

(c) Use parts (a) and (b) along with Proposition 5.15 to show that $\mathbb{Q}(7^{\frac{1}{3}})$ is a commutative ring.

44. Let $\mathbb{Q}(7^{\frac{1}{3}})$ be the commutative ring from exercise 43. Show that the identity map is the only automorphism of $\mathbb{Q}(7^{\frac{1}{3}})$ which fixes every element of $\mathbb{Q}$.

45. Let $\alpha = 3 - 2\sqrt{7} \in \mathbb{Q}(\sqrt{7})$. You might want to refer to exercises 27 and 28 while doing this exercise.
    (a) If $f(x) \in \mathbb{Q}[x]$ has α as a root, what else can you say about the roots of $f(x)$?

    (b) Find a monic polynomial of degree 2 in $\mathbb{Q}[x]$ that has α as a root.

46. Let $\beta = \frac{3}{2} + 4\sqrt{11} \in \mathbb{Q}(\sqrt{11})$. You might want to refer to exercises 29 and 30 while doing this exercise.
    (a) If $g(x) \in \mathbb{Q}[x]$ has β as a root, what else can you say about the roots of $g(x)$?

    (b) Find a monic polynomial of degree 2 in $\mathbb{Q}[x]$ that has β as a root.

47. Let $\gamma = 4 + 5i\sqrt{19} \in \mathbb{Q}(i\sqrt{19})$. You might want to refer to exercises 31 and 32 while doing this exercise.
    (a) If $h(x) \in \mathbb{Q}[x]$ has γ as a root, what else can you say about the roots of $h(x)$?

    (b) Find a monic polynomial of degree 2 in $\mathbb{Q}[x]$ that has γ as a root.

In exercises 48–51, you will come across the sets $\mathbb{Q}(\sqrt{2})[x]$, $\mathbb{Q}(\sqrt{3})[x]$, $\mathbb{Q}(\sqrt{6})[x]$, $\mathbb{Q}(\sqrt{7})[x]$, $\mathbb{Q}(i)[x]$, $\mathbb{Q}(i\sqrt{7})[x]$. In each case, it represents the polynomials over a field F, where F consists of all terms of the form $\{a + bC \mid a, b \in \mathbb{Q}\}$, where C is either $\sqrt{2}$, $\sqrt{3}$, $\sqrt{6}$, $\sqrt{7}$, i, or $i\sqrt{7}$.

48. Let $\alpha = 5\sqrt{2} - 7\sqrt{3} \in \mathbb{Q}(\sqrt{2}, \sqrt{3})$. You might want to refer to exercise 33 while doing this problem.
    (a) If $f(x) \in \mathbb{Q}[x]$ has α as a root, what else can you say about the roots of $f(x)$?

    (b) Find a monic polynomial of degree 4 in $\mathbb{Q}[x]$ that has α as a root.

    (c) If $g(x) \in \mathbb{Q}(\sqrt{2})[x]$ has α as a root, what else can you say about the roots of $g(x)$?

    (d) Find a monic polynomial of degree 2 in $\mathbb{Q}(\sqrt{2})[x]$ that has α as a root.

    (e) If $h(x) \in \mathbb{Q}(\sqrt{3})[x]$ has α as a root, what else can you say about the roots of $h(x)$?

(f)   Find a monic polynomial of degree 2 in $\mathbb{Q}(\sqrt{3})[x]$ that has α as a root.

(g)   If $k(x) \in \mathbb{Q}(\sqrt{6})[x]$ has α as a root, what else can you say about the roots of $k(x)$?

(h)   Find a monic polynomial of degree 2 in $\mathbb{Q}(\sqrt{6})[x]$ that has α as a root.

49.   Let $\beta = 3 - 4\sqrt{2} + 2\sqrt{6} \in \mathbb{Q}(\sqrt{2}, \sqrt{3})$. You might want to refer to exercise 33 while doing this problem.

(a)   If $f(x) \in \mathbb{Q}[x]$ has β as a root, what else can you say about the roots of $f(x)$?

(b)   Find a monic polynomial of degree 4 in $\mathbb{Q}[x]$ that has β as a root.

(c)   If $g(x) \in \mathbb{Q}(\sqrt{2})[x]$ has β as a root, what else can you say about the roots of $g(x)$?

(d)   Find a monic polynomial of degree 2 in $\mathbb{Q}(\sqrt{2})[x]$ that has β as a root.

(e)   If $h(x) \in \mathbb{Q}(\sqrt{3})[x]$ has β as a root, what else can you say about the roots of $h(x)$?

(f)   Find a monic polynomial of degree 2 in $\mathbb{Q}(\sqrt{3})[x]$ that has β as a root.

(g)   If $k(x) \in \mathbb{Q}(\sqrt{6})[x]$ has β as a root, what else can you say about the roots of $k(x)$?

(h)   Find a monic polynomial of degree 2 in $\mathbb{Q}(\sqrt{6})[x]$ that has β as a root.

50.   Let $\gamma = 2\sqrt{7} - 5i \in \mathbb{Q}(\sqrt{7}, i)$. You might want to refer to exercise 37 while doing this problem.

(a)   If $f(x) \in \mathbb{Q}[x]$ has γ as a root, what else can you say about the roots of $f(x)$?

(b)   Find a monic polynomial of degree 4 in $\mathbb{Q}[x]$ that has γ as a root.

(c)   If $g(x) \in \mathbb{Q}(\sqrt{7})[x]$ has γ as a root, what else can you say about the roots of $g(x)$?

(d)   Find a monic polynomial of degree 2 in $\mathbb{Q}(i)[x]$ that has γ as a root.

(e)   If $h(x) \in \mathbb{Q}(i)[x]$ has γ as a root, what else can you say about the roots of $h(x)$?

(f)   Find a monic polynomial of degree 2 in $\mathbb{Q}(i)[x]$ that has γ as a root.

(g)   If $k(x) \in \mathbb{Q}(i\sqrt{7})[x]$ has γ as a root, what else can you say about the roots of $k(x)$?

(h)   Find a monic polynomial of degree 2 in $\mathbb{Q}(i\sqrt{7})[x]$ that has γ as a root.

51.   Let $\delta = 1 + 3\sqrt{7} - 5i\sqrt{7} \in \mathbb{Q}(\sqrt{7}, i)$. You might want to refer to exercise 37 while doing this problem.

(a)   If $f(x) \in \mathbb{Q}[x]$ has δ as a root, what else can you say about the roots of $f(x)$?

(b)   Find a monic polynomial of degree 4 in $\mathbb{Q}[x]$ that has δ as a root.

(c)   If $g(x) \in \mathbb{Q}(\sqrt{7})[x]$ has δ as a root, what else can you say about the roots of $g(x)$?

(d)   Find a monic polynomial of degree 2 in $\mathbb{Q}(\sqrt{7})[x]$ that has δ as a root.

(e) If $h(x) \in \mathbb{Q}(i)[x]$ has δ as a root, what else can you say about the roots of $h(x)$?

(f) Find a monic polynomial of degree 2 in $\mathbb{Q}(i)[x]$ that has δ as a root.

(g) If $k(x) \in \mathbb{Q}(i\sqrt{7})[x]$ has δ as a root, what else can you say about the roots of $k(x)$?

(h) Find a monic polynomial of degree 2 in $\mathbb{Q}(i\sqrt{7})[x]$ that has δ as a root.

52. Show that if R is a noncommutative ring, then $0 \cdot x = 0$, for all $x \in R$.

53. Let x, y, z be elements of a commutative ring such that both y and z are multiplicative inverses of x. Prove that $y = z$.

54. Let x, y be elements of a commutative ring.
    (a) Show that $x \cdot y$ and $(-x) \cdot (-y)$ are both additive inverses of $(-x) \cdot y$.

    (b) Use (a) to prove that $x \cdot y = (-x) \cdot (-y)$.

    (c) Does (b) help you understand why the product of two negative numbers is positive?

5.5 Groups of Automorphisms of Commutative Rings

In the examples at the end of the previous section, we examined the automorphisms of various commutative rings. In these examples, we used roots of polynomials to determine automorphisms and used automorphisms to find roots of polynomials. We will now begin to look at these automorphisms from a slightly different perspective.

The automorphisms $\sigma_1, \sigma_2, \sigma_3, \sigma_4$ of $\mathbb{Q}(i, \sqrt{3})$ all belong to $Bij(\mathbb{Q}(i, \sqrt{3}))$, which is the set of bijections from $\mathbb{Q}(i, \sqrt{3})$ to $\mathbb{Q}(i, \sqrt{3})$. Recall that Theorem 2.16 told us that if S is any set and if $\circ$ represents the composition of functions, then $\circ$ is a binary operation on $Bij(S)$ that is associative, has an identity element, and each element has an inverse in $Bij(S)$.

If we let $G = \{\sigma_1, \sigma_2, \sigma_3, \sigma_4\}$ then, when we compose the functions in G, we obtain the following table:

$\circ$	σ_1	σ_2	σ_3	σ_4
σ_1	σ_1	σ_2	σ_3	σ_4
σ_2	σ_2	σ_1	σ_4	σ_3
σ_3	σ_3	σ_4	σ_1	σ_2
σ_4	σ_4	σ_3	σ_2	σ_1

For example, the table tells us that

$$\sigma_2 \circ \sigma_3 = \sigma_4, \quad \sigma_3 \circ \sigma_4 = \sigma_2, \quad \sigma_1 \circ \sigma_2 = \sigma_2, \quad \sigma_3 \circ \sigma_3 = \sigma_1.$$

We will not verify all 16 entries on the table, but we will check that $\sigma_2 \circ \sigma_3 = \sigma_4$. If $a, b, c, d \in \mathbb{Q}$, we have

$$(\sigma_2 \circ \sigma_3)(a + bi + c\sqrt{3} + di\sqrt{3}) = \sigma_2(\sigma_3(a + bi + c\sqrt{3} + di\sqrt{3})) =$$

$$\sigma_2(a + bi - c\sqrt{3} - di\sqrt{3}) = a - bi - c\sqrt{3} + di\sqrt{3} =$$

$$\sigma_4(a + bi + c\sqrt{3} + di\sqrt{3}).$$

The functions $\sigma_2 \circ \sigma_3$ and σ_4 agree on all elements of $\mathbb{Q}(i, \sqrt{3})$, so they are indeed the same function.

At this point, it is beginning to look like $G = \{\sigma_1, \sigma_2, \sigma_3, \sigma_4\}$ has an algebraic structure that satisfies the same properties as $Bij(S)$. Our table indicates that when $g, h \in G$, we have $g \circ h \in G$. Next, since G is a subset of $Bij(\mathbb{Q}(i, \sqrt{3}))$, $\circ$ is certainly associative. It is easy to see that σ_1 is the identity element of G under $\circ$. Finally, since

$$\sigma_1 \circ \sigma_1 = \sigma_1, \quad \sigma_2 \circ \sigma_2 = \sigma_1, \quad \sigma_3 \circ \sigma_3 = \sigma_1, \quad \sigma_4 \circ \sigma_4 = \sigma_1,$$

we see that every element of G has an inverse in G under $\circ$.

As we remarked toward the beginning of Section 5.2, when several mathematical objects have a collection of properties in common, we often give the objects with these properties a special name. These lead us to

Definition 5.16. *A set G with a binary operation $\circ$ is called a group if $\circ$ satisfies the following:*

1. *Associative Law: For every $x, y, z \in G$, $(x \circ y) \circ z = x \circ (y \circ z)$.*

2. *Identity: There is an element in G, usually denoted as e, such that $x \circ e = x = e \circ x$, for every $x \in G$.*

3. *Inverses: For every $x \in G$ there is an element in G, usually denoted as x^{-1}, such that $x \circ x^{-1} = e = x^{-1} \circ x$.*

Although we did not use the term *group* at the time, we have already been introduced to three large classes of groups in Sections 2.4 and 5.2. In Section 2.4 we introduced $Bij(S)$ and, with composition of functions as the binary operation, it represents our first large class of groups. Next, in Section 5.2, we were introduced to commutative rings. When you look back at the properties satisfied by commutative rings, we obtain our second large class of groups, as you can see that every commutative ring is a group under addition. In Section 5.2, we were also introduced to fields. One of the exercises after Section 5.2 has you show that the product of any two nonzero elements of a field is also nonzero. From that fact, we obtain our third large class of groups, as it is now easy to see that the nonzero elements of a field form a group under multiplication.

In light of our work with the automorphisms of $\mathbb{Q}(i, \sqrt{3})$ that fix the elements of $\mathbb{Q}$, it appears that automorphisms of commutative rings will give us another large class of groups. We record this as

Definition 5.17. *If L and K are commutative rings with $K \subseteq L$, let $Gal(L/K)$ denote the set of all automorphisms of L that fix every element of K. We call $Gal(L/K)$ the Galois group of L over K.*

Although we refer to $Gal(L/K)$ as the Galois group, we haven't yet proved that the set $Gal(L/K)$ is actually a group under composition of functions. However, we have already proven some related facts. For example, we saw in Theorem 2.16 that $Bij(S)$ is always a group. In addition, our table from earlier in this section shows that $Gal(\mathbb{Q}(i, \sqrt{3})/\mathbb{Q})$ is a group. We will now show that, for any commutative rings $K \subseteq L$, the set $Gal(L/K)$ is always a group under composition of functions.

Theorem 5.18. *If L and K are commutative rings with $K \subseteq L$, then $Gal(L/K)$ satisfies the following properties:*

(a) *if $f, g \in Gal(L/K)$, then $f \circ g \in Gal(L/K)$;*

(b) *if $f, g, h \in Gal(L/K)$, then $(f \circ g) \circ h = f \circ (g \circ h)$;*

(c) *there is an element $e \in Gal(L/K)$ such that $f \circ e = f = e \circ f$, for all $f \in Gal(L/K)$;*

(d) *for every $f \in Gal(L/K)$, there is an element in $Gal(L/K)$, denoted as f^{-1}, such that $f \circ f^{-1} = e = f^{-1} \circ f$.*

Thus, $Gal(L/K)$ is a group under the composition of functions.

Proof. Suppose $f, g \in Gal(L/K)$; since both f and g are bijections, so is the composition $f \circ g$. Therefore, to show that $f \circ g \in Gal(L/K)$, we first need to show that $f \circ g$ satisfies properties (b) and (c) of Definition 5.8. This will tell us that $f \circ g$ is an automorphism. Then we will need to show that $f \circ g$ fixes every element of K. If $x, y \in L$, using the fact that both f and g satisfy property (b) of Definition 5.8, we have

$$(f \circ g)(x + y) = f(g(x + y)) = f(g(x) + g(y)) =$$

$$f(g(x)) + f(g(y)) = (f \circ g)(x) + (f \circ g)(y).$$

Thus, $f \circ g$ satisfies property (b) of Definition 5.8. Similarly, since f and g both satisfy property (c) of Definition 5.8, we also have

$$(f \circ g)(x \cdot y) = f(g(x \cdot y)) =$$

$$f(g(x) \cdot g(y)) = f(g(x)) \cdot f(g(y)) = (f \circ g)(x) \cdot (f \circ g)(y).$$

As a result, $f \circ g$ satisfies property (c) of Definition 5.8 and is therefore an automorphism of L. Next, if $x \in K$, since both f and g fix x, we have

$$(f \circ g)(x) = f(g(x)) = f(x) = x.$$

Hence, $f \circ g$ also fixes every $x \in K$, so $f \circ g \in Gal(L/K)$, thereby concluding the proof of part (a).

Since the composition of functions is associative, we know that part (b) holds. For part (c), suppose we let $e : L \to L$ be the function that fixes every element of L. Then e is certainly a bijection that also fixes every element of K. If $x, y \in L$, then

$$e(x + y) = x + y = e(x) + e(y)$$

and

$$e(x \cdot y) = x \cdot y = e(x) \cdot e(y).$$

Therefore, e also satisfies properties (b) and (c) of Definition 5.8, and we now know that $e \in Gal(L/K)$.

Observe that if $f \in Gal(L/K)$ and if $x \in L$, then

$$(f \circ e)(x) = f(e(x)) = f(x) \quad \text{and} \quad (e \circ f)(x) = e(f(x)) = f(x).$$

Thus,

$$f \circ e = f = e \circ f.$$

As a result, e is indeed the identity element of $Gal(L/K)$, thereby proving part (c).

Finally, let $f \in Gal(L/K)$; since $f \in Bij(L)$ and $Bij(L)$ is a group, we know that f has an inverse f^{-1} in $Bij(L)$. Therefore, to conclude this proof, it suffices to show that $f^{-1} \in Gal(L/K)$. To this end, if $x, y \in L$, let $u = f^{-1}(x)$ and $v = f^{-1}(y)$. Since $f(u) = x$ and $f(v) = y$, then the fact that f is an automorphism gives us

$$f^{-1}(x + y) = f^{-1}(f(u) + f(v)) = f^{-1}(f(u + v)) = (f^{-1} \circ f)(u + v) =$$

$$e(u + v) = u + v = f^{-1}(x) + f^{-1}(y)$$

and

$$f^{-1}(x \cdot y) = f^{-1}(f(u) \cdot f(v)) = f^{-1}(f(u \cdot v)) = (f^{-1} \circ f)(u \cdot v) =$$

$$e(u \cdot v) = u \cdot v = f^{-1}(x) \cdot f^{-1}(y).$$

Thus, f^{-1} satisfies all the conditions of Definition 5.8 and is an automorphism of L. In addition, if $x \in K$, then $f(x) = x$, and it immediately follows that $f^{-1}(x) = x$. Thus, f^{-1} also fixes every $x \in K$, and we have shown that $f^{-1} \in Gal(L/K)$, thereby proving part (d) and concluding the proof. □

We can now revisit two examples from the previous section and restate our work in the language of Galois groups.

■ Examples

1. If $\mathbb{Q}(\sqrt{2}) = \{a + b\sqrt{2} \mid a, b \in \mathbb{Q}\}$, let e denote the identity map on $\mathbb{Q}(\sqrt{2})$ and let σ denote the automorphism of $\mathbb{Q}(\sqrt{2})$ that sends $\sqrt{2}$ to $-\sqrt{2}$ and fixes every element of $\mathbb{Q}$. Then

$$Gal(\mathbb{Q}(\sqrt{2})/\mathbb{Q}) = \{e, \sigma\}.$$

2. If $T = \{a + b2^{\frac{1}{3}} + c2^{\frac{2}{3}} \mid a, b, c \in \mathbb{Q}\}$ and if e denotes the identity map on T, then

$$Gal(T/\mathbb{Q}) = \{e\}.$$

When we look at commutative rings under addition, the groups we obtain are commutative. Similarly, when we look at the nonzero elements of a field under multiplication, the groups we obtain are also commutative. But in general, groups need not be commutative. In the exercises after this section and in Chapter 15, we will see examples of fields $K \subseteq L$ such that $Gal(L/K)$ is not commutative. However, the easiest examples of groups that are not commutative are groups of the form $Bij(S)$, provided S has more than two elements.

Proposition 5.19. *If S is a set with more than two elements, then the group $Bij(S)$ of bijections of S is not commutative.*

Proof. Let a, b, c be distinct element of S. Then let $\sigma_1, \sigma_2 : S \to S$ be the bijections of S defined as

$$\sigma_1(a) = b, \quad \sigma_1(b) = a, \quad \sigma_1(c) = c, \quad \sigma_1(x) = x \text{ for every other } x \in S$$

and

$$\sigma_2(a) = a, \quad \sigma_2(b) = c, \quad \sigma_2(c) = b, \quad \sigma_2(x) = x \text{ for every other } x \in S.$$

Then

$$(\sigma_1 \circ \sigma_2)(a) = \sigma_1(\sigma_2(a)) = \sigma_1(a) = b,$$

whereas

$$(\sigma_2 \circ \sigma_1)(a) = \sigma_2(\sigma_1(a)) = \sigma_2(b) = c.$$

Thus,

$$(\sigma_1 \circ \sigma_2)(a) \neq (\sigma_2 \circ \sigma_1)(a).$$

Therefore, the functions $\sigma_1 \circ \sigma_2$ and $\sigma_2 \circ \sigma_1$ are different, so G is not commutative. $\square$

This chapter began with a specific goal: to introduce and study the complex numbers $\mathbb{C}$. Yet, we also ended up introducing the three most important objects in abstract algebra: fields, commutative rings, and groups. It may seem odd to introduce all three abstract objects at such an early stage in this course. However, there is method to this madness. The algebraic structure of groups and the interaction between groups and fields is an essential part of abstract algebra and is at the heart of the proof of the insolvability of the quintic. By gaining experience now with concrete examples of groups of automorphisms and their interaction with fields and roots of polynomials, you will be better equipped to handle some of the more theoretical aspects of abstract algebra needed to prove fundamental results like the insolvability of the quintic.

Exercises for Section 5.5

In exercises 1 and 2, we compute Galois groups. Both exercises have 12 parts, but each part is relatively short. Many parts of these exercises will use computations from earlier parts.

1. Let $\omega = -\frac{1}{2} + \frac{\sqrt{3}}{2}i$ and then let $\mathbb{Q}(\omega) = \{a + b\omega \mid a, b \in \mathbb{Q}\}$. In this exercise, we will examine $\mathbb{Q}(\omega)$ and eventually determine $Gal(\mathbb{Q}(\omega)/\mathbb{Q})$.

    (a) If $a, b, c, d \in \mathbb{Q}$ such that $a + b\omega = c + d\omega$, show that $a = b$ and $c = d$.

    (b) Compute ω^2, ω^3.

    (c) Show that both ω and ω^2 are roots of the polynomial $x^2 + x + 1$.

    (d) If $a, b, c, d \in \mathbb{Q}$, find A, B such that

    $$(a + b\omega) + (c + d\omega) = A + B\omega.$$

    A and B should be expressions in a, b, c, d.

(e)   If $a, b, c, d \in \mathbb{Q}$, find C, D such that

$$(a + b\omega) \cdot (c + d\omega) = C + D\omega.$$

C and D should be expressions in a, b, c, d.

(f)   Use Proposition 5.15 to show that $\mathbb{Q}(\omega)$ is a commutative ring.

(g)   If $a, b \in \mathbb{Q}$ such that $a + b\omega \neq 0$, show that $(a + b\omega)((a - b) - b\omega)$ is a nonzero element of $\mathbb{Q}$.

(h)   If $a, b \in \mathbb{Q}$ such that $a + b\omega \neq 0$, compute $(a + b\omega)^{-1}$, and then show that $\mathbb{Q}(\omega)$ is a field.

(i)   If $\sigma \in Gal(\mathbb{Q}(\omega)/\mathbb{Q})$, show that $\sigma(\omega) = \omega$ or $\sigma(\omega) = \omega^2$, and then show that $Gal(\mathbb{Q}(\omega)/\mathbb{Q})$ has at most two elements.

(j)   Let τ be defined as $\tau(a + b\omega) = a + b\omega^2$, for all $a, b \in \mathbb{Q}$. Show that $\tau(a + b\omega)$ is also equal to $(a - b) - b\omega$. Also show that τ^2 is the identity map on $\mathbb{Q}(\omega)$, and use this fact to show that τ is a bijection of the set $\mathbb{Q}(\omega)$.

(k)   Using the formulas you found in parts (d) and (e) of this problem, show that the function τ from part (j) is an automorphism of $\mathbb{Q}(\omega)$ that fixes every element of $\mathbb{Q}$.

(l)   Conclude that $Gal(\mathbb{Q}(\omega)/\mathbb{Q}) = \{e, \tau\}$, where e is the identity map and τ is the function from part (j).

2.   Let $\gamma = \frac{1}{2} + \frac{\sqrt{3}}{2}i$, and then let $\mathbb{Q}(\gamma) = \{a + b\gamma \mid a, b \in \mathbb{Q}\}$. In this exercise, we will examine $\mathbb{Q}(\gamma)$ and eventually determine $Gal(\mathbb{Q}(\gamma)/\mathbb{Q})$.

(a)   If $a, b, c, d \in \mathbb{Q}$ such that $a + b\gamma = c + d\gamma$, show that $a = b$ and $c = d$.

(b)   Compute γ^2, γ^3.

(c)   Show that both γ and $-\gamma^2$ are roots of the polynomial $x^2 - x + 1$.

(d)   If $a, b, c, d \in \mathbb{Q}$, find A, B such that

$$(a + b\gamma) + (c + d\gamma) = A + B\gamma.$$

A and B should be expressions in a, b, c, d.

(e)   If $a, b, c, d \in \mathbb{Q}$, find C, D such that

$$(a + b\gamma) \cdot (c + d\gamma) = C + D\gamma.$$

C and D should be expressions in a, b, c, d.

(f)   Use Proposition 5.15 to show that $\mathbb{Q}(\gamma)$ is a commutative ring.

(g) If $a, b \in \mathbb{Q}$ such that $a + b\gamma \neq 0$, show that $(a + b\gamma)((a + b) - b\gamma)$ is a nonzero element of $\mathbb{Q}$.

(h) If $a, b \in \mathbb{Q}$ such that $a + b\gamma \neq 0$, compute $(a + b\gamma)^{-1}$, and then show that $\mathbb{Q}(\gamma)$ is a field.

(i) If $\sigma \in Gal(\mathbb{Q}(\gamma)/\mathbb{Q})$, show that $\sigma(\gamma) = \gamma$ or $\sigma(\gamma) = -\gamma^2$, and then show that $Gal(\mathbb{Q}(\gamma)/\mathbb{Q})$ has at most two elements.

(j) Let τ be defined as $\tau(a + b\gamma) = a - b\gamma^2$, for all $a, b \in \mathbb{Q}$. Show that $\tau(a + b\gamma)$ is also equal to $(a + b) - b\gamma$. Also show that τ^2 is the identity map on $\mathbb{Q}(\gamma)$, and use this fact to show that τ is a bijection of the set $\mathbb{Q}(\gamma)$.

(k) Using the formulas you found in parts (d) and (e) of this problem, show that the function τ from part (j) is an automorphism of $\mathbb{Q}(\gamma)$ that fixes every element of $\mathbb{Q}$.

(l) Conclude that $Gal(\mathbb{Q}(\gamma)/\mathbb{Q}) = \{e, \tau\}$, where e is the identity map and τ is the function from part (j).

In exercises 3–26, we let $\mathbb{Q}(\sqrt{2}, \sqrt{3}, \sqrt{5})$ denote the field consisting of all elements of the form

$$\{a + b\sqrt{2} + c\sqrt{3} + d\sqrt{5} + e\sqrt{6} + f\sqrt{10} + g\sqrt{15} + h\sqrt{30} \mid a, b, c, d, e, f, g, h \in \mathbb{Q}\}.$$

If we let $G = Gal(\mathbb{Q}(\sqrt{2}, \sqrt{3}, \sqrt{5})/\mathbb{Q})$, then G consists of eight automorphisms, all of which are completely determined by their behavior on $\sqrt{2}, \sqrt{3}$, and $\sqrt{5}$. We can represent each element of G in the form $\sigma_{(i, j, k)}$, where $0 \leq i, j, k \leq 1$, and

$$\sigma_{(i, j, k)}\left(\sqrt{2}\right) = (-1)^i \sqrt{2}, \quad \sigma_{(i, j, k)}\left(\sqrt{3}\right) = (-1)^j \sqrt{3}, \quad \sigma_{(i, j, k)}\left(\sqrt{5}\right) = (-1)^k \sqrt{5}.$$

For example, using this notation, $\sigma_{(1, 0, 1)}(\sqrt{2}) = -\sqrt{2}$, $\sigma_{(1, 0, 1)}(\sqrt{3}) = \sqrt{3}$, and $\sigma_{(1, 0, 1)}(\sqrt{5}) = -\sqrt{5}$. You may assume different choices of a, b, c, d, e, f, g, h always produce different elements of $\mathbb{Q}(\sqrt{2}, \sqrt{3}, \sqrt{5})$.

3. For which of the eight triples of the form (i, j, k) is $\sigma_{(i, j, k)}$ the identity map?

4. Show that, for each of the eight triples of the form (i, j, k), the automorphism $(\sigma_{(i, j, k)})^2$ is the identity map.

5. In G, compute $\sigma_{(0, 1, 1)} \circ \sigma_{(1, 1, 0)}$.

6. In G, compute $\sigma_{(1, 1, 1)} \circ \sigma_{(0, 1, 0)}$.

7. In G, compute $\sigma_{(0, 0, 1)} \circ \sigma_{(0, 1, 0)}$.

8. In G, compute $\sigma_{(1,0,1)} \circ \sigma_{(1,0,0)}$.

9. In G, compute $\sigma_{(0,1,0)} \circ \sigma_{(1,0,1)}$.

10. In G, compute $\sigma_{(1,1,0)} \circ \sigma_{(0,1,1)}$.

11. In G, compute $\sigma_{(0,1,1)} \circ \sigma_{(0,1,0)}$.

12. In G, compute $\sigma_{(1,0,0)} \circ \sigma_{(1,1,0)}$.

13. In terms of a, b, c, d, e, f, g, h, describe all elements of $\mathbb{Q}(\sqrt{2}, \sqrt{3}, \sqrt{5})$ fixed by the automorphism $\sigma_{(1,0,0)}$.

14. In terms of a, b, c, d, e, f, g, h, describe all elements of $\mathbb{Q}(\sqrt{2}, \sqrt{3}, \sqrt{5})$ fixed by the automorphism $\sigma_{(0,1,0)}$.

15. In terms of a, b, c, d, e, f, g, h, describe all elements of $\mathbb{Q}(\sqrt{2}, \sqrt{3}, \sqrt{5})$ fixed by the automorphism $\sigma_{(0,0,1)}$.

16. In terms of a, b, c, d, e, f, g, h, describe all elements of $\mathbb{Q}(\sqrt{2}, \sqrt{3}, \sqrt{5})$ fixed by the automorphism $\sigma_{(1,1,0)}$.

17. In terms of a, b, c, d, e, f, g, h, describe all elements of $\mathbb{Q}(\sqrt{2}, \sqrt{3}, \sqrt{5})$ fixed by the automorphism $\sigma_{(1,0,1)}$.

18. In terms of a, b, c, d, e, f, g, h, describe all elements of $\mathbb{Q}(\sqrt{2}, \sqrt{3}, \sqrt{5})$ fixed by the automorphism $\sigma_{(0,1,1)}$.

19. In terms of a, b, c, d, e, f, g, h, describe all elements of $\mathbb{Q}(\sqrt{2}, \sqrt{3}, \sqrt{5})$ fixed by the automorphism $\sigma_{(1,1,1)}$.

20. In terms of a, b, c, d, e, f, g, h, describe all elements of $\mathbb{Q}(\sqrt{2}, \sqrt{3}, \sqrt{5})$ fixed by both $\sigma_{(1,0,0)}$. and $\sigma_{(0,1,0)}$.

21. In terms of a, b, c, d, e, f, g, h, describe all elements of $\mathbb{Q}(\sqrt{2}, \sqrt{3}, \sqrt{5})$ fixed by both $\sigma_{(1,0,0)}$. and $\sigma_{(0,0,1)}$.

22. In terms of a, b, c, d, e, f, g, h, describe all elements of $\mathbb{Q}(\sqrt{2}, \sqrt{3}, \sqrt{5})$ fixed by both $\sigma_{(0,1,0)}$. and $\sigma_{(0,0,1)}$.

23. In terms of a, b, c, d, e, f, g, h, describe all elements of $\mathbb{Q}(\sqrt{2}, \sqrt{3}, \sqrt{5})$ fixed by both $\sigma_{(1,1,0)}$. and $\sigma_{(0,0,1)}$.

24. In terms of a, b, c, d, e, f, g, h, describe all elements of $\mathbb{Q}(\sqrt{2}, \sqrt{3}, \sqrt{5})$ fixed by both $\sigma_{(0,1,0)}$. and $\sigma_{(1,0,1)}$.

25. In terms of a, b, c, d, e, f, g, h, describe all elements of $\mathbb{Q}(\sqrt{2}, \sqrt{3}, \sqrt{5})$ fixed by both $\sigma_{(0,1,1)}$. and $\sigma_{(1,0,0)}$.

26. In terms of a, b, c, d, e, f, g, h, describe all elements of $\mathbb{Q}(\sqrt{2}, \sqrt{3}, \sqrt{5})$ fixed by both $\sigma_{(0,1,1)}$. and $\sigma_{(1,0,1)}$.

In exercises 27–45, we let $\omega = -\frac{1}{2} + \frac{\sqrt{3}}{2}i$ and then let $\mathbb{Q}(2^{\frac{1}{3}}, \omega)$ denote the field consisting of all elements of the form

$$\{a + b2^{\frac{1}{3}} + c2^{\frac{2}{3}} + d\omega + e2^{\frac{1}{3}}\omega + f2^{\frac{2}{3}}\omega \mid a, b, c, d, e, f \in \mathbb{Q}\}.$$

If we let $G = Gal(\mathbb{Q}(2^{\frac{1}{3}}, \omega)/\mathbb{Q})$, then G consists of six automorphisms, all of which are completely determined by their behavior on $2^{\frac{1}{3}}$ and ω. In these exercises, we will see that G is not commutative, so Galois groups need not be commutative. We can represent each element of G in the form $\sigma_{(i,j)}$, where $0 \le i \le 2$, $1 \le j \le 2$, and

$$\sigma_{(i,j)}\left(2^{\frac{1}{3}}\right) = 2^{\frac{1}{3}}\omega^i, \quad \sigma_{(i,j)}(\omega) = \omega^j.$$

For example, using this notation, $\sigma_{(1,2)}(2^{\frac{1}{3}}) = 2^{\frac{1}{3}}\omega$ and $\sigma_{(1,2)}(\omega) = \omega^2$. Recall that since $\omega^3 = 1$ and $\omega^2 + \omega + 1 = 0$, we can use the terms ω^2 and $-1 - \omega$ interchangeably. You may assume different choices of a, b, c, d, e, f always produce different elements of $\mathbb{Q}(2^{\frac{1}{3}}, \omega)$.

27. For which of the six pairs of the form (i, j) is $\sigma_{(i,j)}$ the identity map?

28. In G, compute $\sigma_{(0,2)} \circ \sigma_{(0,2)}$.

29. In G, compute $\sigma_{(1,2)} \circ \sigma_{(1,2)}$.

30. In G, compute $\sigma_{(2,2)} \circ \sigma_{(2,2)}$.

31. In G, compute $\sigma_{(1,1)} \circ \sigma_{(1,1)}$ and $\sigma_{(1,1)} \circ \sigma_{(1,1)} \circ \sigma_{(1,1)}$.

32. In G, compute $\sigma_{(2,1)} \circ \sigma_{(2,1)}$ and $\sigma_{(2,1)} \circ \sigma_{(2,1)} \circ \sigma_{(2,1)}$.

33. In G, compute $\sigma_{(0,2)} \circ \sigma_{(1,2)}$ and $\sigma_{(1,2)} \circ \sigma_{(0,2)}$.

34. In G, compute $\sigma_{(0,2)} \circ \sigma_{(2,2)}$ and $\sigma_{(2,2)} \circ \sigma_{(0,2)}$.

35. In G, compute $\sigma_{(2,2)} \circ \sigma_{(1,2)}$ and $\sigma_{(1,2)} \circ \sigma_{(2,2)}$.

36. In G, compute $\sigma_{(0,2)} \circ \sigma_{(1,1)}$ and $\sigma_{(1,1)} \circ \sigma_{(0,2)}$.

37. In G, compute $\sigma_{(1,2)} \circ \sigma_{(1,1)}$ and $\sigma_{(1,1)} \circ \sigma_{(1,2)}$.

38. In G, compute $\sigma_{(2,2)} \circ \sigma_{(1,1)}$ and $\sigma_{(1,1)} \circ \sigma_{(2,2)}$.

39. In G, compute $\sigma_{(0,2)} \circ \sigma_{(2,1)}$ and $\sigma_{(2,1)} \circ \sigma_{(0,2)}$.

40. In G, compute $\sigma_{(1,2)} \circ \sigma_{(2,1)}$ and $\sigma_{(2,1)} \circ \sigma_{(1,2)}$.

41. In G, compute $\sigma_{(2,2)} \circ \sigma_{(2,1)}$ and $\sigma_{(2,1)} \circ \sigma_{(2,2)}$.

42. In terms of a, b, c, d, e, f, describe all elements of $\mathbb{Q}(2^{\frac{1}{3}}, \omega)$ fixed by the automorphism $\sigma_{(0,2)}$.

43. In terms of a, b, c, d, e, f, describe all elements of $\mathbb{Q}(2^{\frac{1}{3}}, \omega)$ fixed by the automorphism $\sigma_{(1,2)}$.

44. In terms of a, b, c, d, e, f, describe all elements of $\mathbb{Q}(2^{\frac{1}{3}}, \omega)$ fixed by the automorphism $\sigma_{(2,2)}$.

45. The same elements of $\mathbb{Q}(2^{\frac{1}{3}}, \omega)$ are fixed by the automorphisms $\sigma_{(1,1)}$ and $\sigma_{(2,1)}$. Describe these elements in terms of a, b, c, d, e, f.

For exercises 46–53, let $S = \{a, b, c, d\}$ and let σ, τ be the elements of $Bij(S)$ defined as

$$\sigma(a) = b, \quad \sigma(b) = c, \quad \sigma(c) = d, \quad \sigma(d) = a,$$
$$\tau(a) = d, \quad \tau(b) = c, \quad \tau(c) = b, \quad \tau(d) = a.$$

In each of the exercises below, compute the values of the given functions at $a, b, c,$ and d.

46. $\sigma \circ \sigma$

47. $\sigma \circ \sigma \circ \sigma$

48. $\sigma \circ \sigma \circ \sigma \circ \sigma$

49. $\tau \circ \tau$

50. $\sigma \circ \tau$

51. $\tau \circ \sigma$

52. $\sigma \circ \tau \circ \sigma$

53. $\tau \circ \sigma \circ \tau$

CHAPTER 6

The Fundamental Theorem of Algebra

In this chapter, we take a more geometric and concrete view of the complex numbers. We first take a geometric view of the real numbers and then see how this approach extends, in a natural way, to the complex numbers. This will provide us with the background we need to prove the Fundamental Theorem of Algebra.

In Chapter 4, we indicated that it is often very useful in mathematics to describe objects in terms of equivalence relations and equivalence classes. This philosophy will assist us in examining both the real and complex numbers from a more geometric perspective.

6.1 Representing Real Numbers and Complex Numbers Geometrically

We are accustomed to viewing a real number as a point on a number line. However, we can also view real numbers as arrows lying on a number line. For example, the number 2 can be viewed as the arrow beginning at 0 and ending at 2.

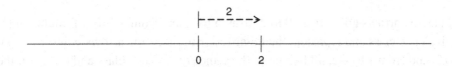

Similarly, we can view -3 as the arrow beginning at 0 and ending at -3.

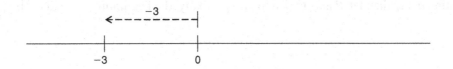

However, suppose we also consider the arrow which begins at 5 and ends at 7. Or we could examine the arrow that begins at −3 and ends at −1. Both of these arrows could also be used to represent the number 2.

Similarly, the arrow beginning at 5 and ending at 2 also represents the number −3, as does the arrow beginning at 1 and ending at −2.

We can see that there are an infinite number of arrows that represent each real number. In fact, given a real number a and a point P on the number line, there is exactly one arrow beginning at P that represents a. Since many different arrows can represent the same number, hopefully you can see that this is a perfect situation to apply equivalence relations and equivalence classes. In this context, we say that two arrows are equivalent if they have the same length and are going in the same direction. There are only two possible directions in which an arrow can go. Positive real numbers are represented by arrows that go to the right and negative real numbers are represented by arrows that go to the left. Zero is somewhat special, and it can be represented by an arrow consisting of only a single point. It is easy to see that we have described an equivalence relation on the set of arrows.

For example, the arrows going from 0 to 2, from 5 to 7, and from −3 to −1, all belong to the same equivalence class and represent the number 2. Similarly, the arrows going from 0 to −3, from 5 to 2, and from 1 to −2, all belong to the same equivalence class and represent the number −3.

We can now view the addition of real numbers geometrically. Suppose $a, b \in \mathbb{R}$; we will describe and illustrate how to find an arrow representing the sum $a + b$. First, choose any arrow representing a and then let P and Q denote, respectively, the beginning and end of the arrow.

Next, look at the arrow that represents b and starts at Q. Then let R denote the endpoint of this arrow.

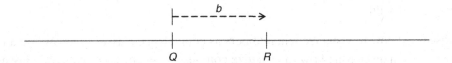

The arrow that begins at P and ends at R now represents $a + b$.

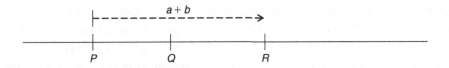

For example, suppose we wish to add the numbers 2 and -3. First, we choose any arrow that represents 2. There are many choices we could make, but suppose we pick the arrow that begins at 4 and ends at 6.

At the point 6, place an arrow that represents the number -3. This arrow will end at the point 3. Therefore, the sum $2 + (-3)$ can be represented by an arrow that begins at 4 and ends at 3.

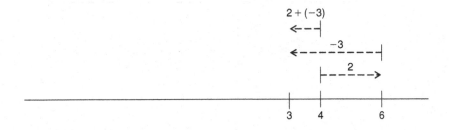

Given a real number a, we can let $|a|$ denote the length of the arrows that represent a. Observe that $|a|$ is the same as the ordinary absolute value of a. We now state, without proof, two familiar facts about the absolute values of real numbers that will be generalized to complex numbers later in this chapter.

Lemma 6.1. *If* $a, b \in \mathbb{R}$, *then*

(a) $\quad |a \cdot b| = |a| \cdot |b|$,

(b) $\quad |a + b| \le |a| + |b|$.

Having discussed how to visualize real numbers as equivalence classes of arrows on a number line, we now begin discussing how to visualize complex numbers as equivalence classes of arrows in the plane. Fortunately, examining arrows in the plane is, in some ways, easier and more natural than examining arrows on a line.

Given two arrows in the plane, we consider them to be equivalent if they have the same length and are going in the same direction. For example, the arrow beginning at the point $P_1 = (1, 2)$ and ending at the point $Q_1 = (4, 6)$ is equivalent to the arrow beginning at $P_2 = (-1, 0)$ and ending at the point $Q_2 = (2, 4)$. Both of these arrows have length 5. Furthermore, these arrows are going in the same direction as they are both going upward and both lie on lines whose slope is $\frac{4}{3}$.

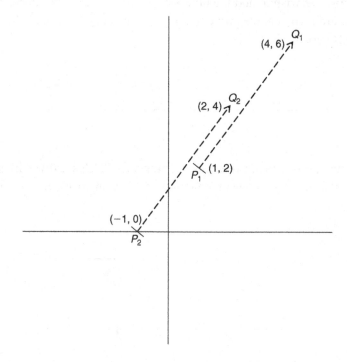

Equivalence classes of arrows can be added in the plane in virtually the same manner as they were on the line. Suppose $[\eta]$ and $[v]$ are two equivalence classes of arrows in the plane. Let η_1 be any arrow in the plane belonging to the equivalence class $[\eta]$. Then η_1 begins at some point P and ends at some point Q. There is some arrow, which we will call v_1, that belongs to the class $[v]$ and begins at the point Q. We will let R denote the endpoint of v_1 and now let ψ denote the arrow that begins at P and ends at R. The equivalence class $[\psi]$ that contains ψ is now the sum $[\eta] + [\nu]$.

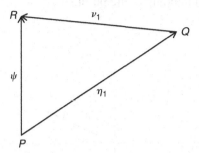

Since we are actually adding equivalence classes of arrows and not just arrows, we need to check that addition is well defined. To illustrate this, observe that we could have chosen a different element $\eta_2 \in [\eta]$. Then η_2 begins at some point P_1 and ends at Q_1. Now let ν_2 be the element of $[\nu]$ that begins at Q_1. We now let R_1 denote the endpoint of ν_2 and let ψ_1 denote the arrow that begins at P_1 and ends at R_1. Therefore, the equivalence class $[\psi_1]$ that contains ψ_1 is also the sum $[\eta] + [\nu]$. As a result, to show that addition is well defined, we need to show that the equivalence classes $[\psi]$ and $[\psi_1]$ are the same. This will be the case precisely if the arrows ψ and ψ_1 have the same length and go in the same direction. To see that this is the case, we refer to the following diagram.

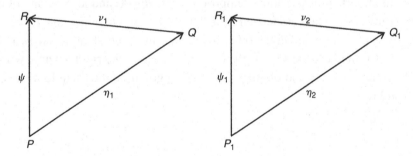

Since η_1 and η_2 are equivalent, line segments PQ and P_1Q_1 must have the same length. Similarly, since ν_1 and ν_2 are equivalent, line segments QR and Q_1R_1 also have the same length. Furthermore, since η_1 and η_2 go in the same direction, as do ν_1 and ν_2, it follows that angles PQR and $P_1Q_1R_1$ must be congruent. Using side-angle-side (SAS), we see that triangles PQR and $P_1Q_1R_1$ are congruent. As a result, line segments PR and P_1R_1 must have the same length and are parallel. Hence, the arrows ψ and ψ_1 have the same length and go in the same direction. Therefore, $[\psi]$ and $[\psi_1]$ are indeed the same equivalence class. A more formal argument would also handle the special case where P, Q, and R lie on the same line. The preceding discussion, although somewhat informal, hopefully convinces you that the addition of equivalence classes of arrows is well defined.

We can now associate to every complex number an equivalence class of arrows in the plane. If $\alpha = a + bi \in \mathbb{C}$, we associate to it the class consisting of all arrows equivalent to the arrow beginning at the point $(0, 0)$ and ending at the point (a, b).

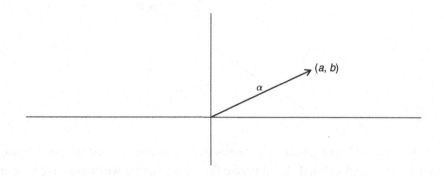

So far, we have shown how to add equivalence classes of arrows in the plane and also how to associate to every complex number an equivalence class of arrows in the plane. The next step in viewing addition in $\mathbb{C}$ geometrically is to show that the addition of two complex numbers corresponds to the addition of the equivalence classes that represent these numbers. In other words, we need to prove that if we add arrows that represent the complex numbers α and β, then we obtain an arrow that represents the complex number $\alpha + \beta$. To do this, we begin with a simple but important observation. Suppose $\beta = c + di \in \mathbb{C}$ and let v be any of the arrows that are in the equivalence class associated to β. Then v begins at some point $P = (x_1, y_1)$ and ends at some point $Q = (x_2, y_2)$. On the other hand, we can also look at the arrow representing β that begins at $(0, 0)$ and ends at (c, d). We can then consider the right triangle whose vertices are P, Q, and (x_2, y_1) and observe that it is congruent to the triangle with vertices $(0, 0)$, (c, d), and $(c, 0)$.

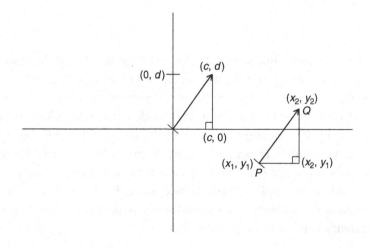

In light of this, it is easy to see that

$$c = x_2 - x_1 \quad \text{and} \quad d = y_2 - y_1.$$

As a result,

(1) $$x_2 = x_1 + c \quad \text{and} \quad y_2 = y_1 + d.$$

If we let $\alpha = a + bi$, $\beta = c + di \in \mathbb{C}$, first consider the arrow beginning at $(0, 0)$ and ending at (a, b). At the point (a, b), place an arrow that is associated to $\beta = c + di$. The equations in (1) now tell us that since this arrow begins at $(x_1, y_1) = (a, b)$, it must end at the point $(x_2, y_2) = (a + c, b + d)$.

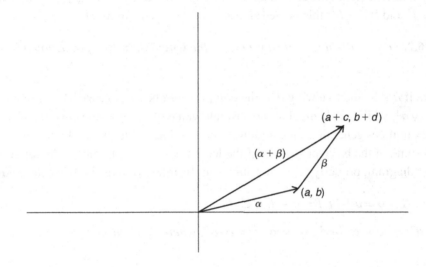

Thus, we have added an arrow that represents α to an arrow that represents β and obtained an arrow that begins at $(0, 0)$ and ends at $(a + c, b + d)$. However, the arrow that begins at $(0, 0)$ and ends at $(a + c, b + d)$ represents the complex number $\alpha + \beta = (a + c) + (b + d)i$. Therefore, when we add arrows that represent α and β, we obtain an arrow that represents $\alpha + \beta$. Thus, the addition of complex numbers does indeed correspond to the addition of equivalence classes of arrows in the plane.

When adding the complex numbers α and β geometrically in the preceding diagram, we saw that the arrow representing $\alpha + \beta$ is the third side of a triangle where the other two sides are arrows representing α and β. Recall that in a triangle, the length of one side of a triangle can never exceed the sum of the lengths of the other two sides. This is a geometric fact that we should certainly express algebraically. However, we first need to define the length of a complex number. If $\alpha = a + bi \in \mathbb{C}$, then one of the arrows representing α begins at $(0, 0)$ and ends at (a, b). Therefore, this arrow can be thought of as the hypotenuse of a right triangle where the lengths of the other two sides are $|a|$ and $|b|$.

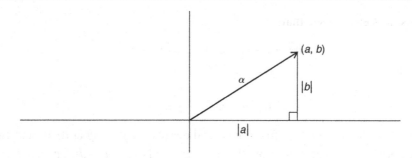

The Pythagorean Theorem tells us that the length of this arrow must be $\sqrt{|a|^2 + |b|^2}$. However, since $|a|^2 = a^2$ and $|b|^2 = b^2$, this is the same as $\sqrt{a^2 + b^2}$. This motivates

Definition 6.2. *If $\alpha = a + bi \in \mathbb{C}$, then we define the length or norm of α, denoted as $|\alpha|$, to be $\sqrt{a^2 + b^2}$.*

Observe that if $a \in \mathbb{R}$, then viewing a as the complex number $a + 0i$, we have $|a| = \sqrt{a^2 + 0^2} = \sqrt{a^2}$, which is identical to the absolute value of a. Thus, the concept of the length of a complex number generalizes the notion of the absolute value of a real number. We can now record some of the basic properties of the length of complex numbers. As suggested by an earlier diagram, property (c) is, not surprisingly, referred to as the **triangle inequality**.

Lemma 6.3. *Let $\alpha = a + bi$, $\beta = c + di \in \mathbb{C}$; then*

(a) $|\alpha| = |\alpha^*| = \sqrt{\alpha \cdot \alpha^*}$ *and* $|\alpha|^2 = |\alpha^*|^2 = \alpha \cdot \alpha^*$, *where α^* is the complex conjugate of α,*

(b) $|\alpha \cdot \beta| = |\alpha| \cdot |\beta|$,

(c) $|\alpha + \beta| \le |\alpha| + |\beta|$,

(d) *if $\alpha_1, \alpha_2, \ldots, \alpha_n \in \mathbb{C}$, then $|\alpha_1 + \alpha_2 + \cdots + \alpha_n| \le |\alpha_1| + |\alpha_2| + \cdots + |\alpha_n|$.*

Proof. For part (a), observe that

$$\alpha \cdot \alpha^* = (a + bi)(a - bi) = a^2 + b^2.$$

Thus $\alpha \cdot \alpha^* = |\alpha|^2$ and $\sqrt{\alpha \cdot \alpha^*} = \sqrt{a^2 + b^2} = |\alpha|$. Since $\alpha^* = a - bi$ and $a^2 + b^2 = a^2 + (-b)^2$, it immediately also follows that $|\alpha^*| = \sqrt{\alpha \cdot \alpha^*}$ and $|\alpha^*|^2 = \alpha \cdot \alpha^*$.

For part (b), recall, from Lemma 5.7(c), that $(\alpha \cdot \beta)^* = \alpha^* \cdot \beta^*$. Also recall, from Lemma 5.10(b), that $\alpha \cdot \alpha^*$ and $\beta \cdot \beta^*$ are nonnegative real numbers. Combining these facts with part (a), we see that

$$|\alpha \cdot \beta| = \sqrt{(\alpha \cdot \beta)(\alpha \cdot \beta)^*} = \sqrt{\alpha \cdot \beta \cdot \alpha^* \cdot \beta^*} = \sqrt{(\alpha \cdot \alpha^*)(\beta \cdot \beta^*)} =$$

$$\sqrt{\alpha \cdot \alpha^*} \cdot \sqrt{\beta \cdot \beta^*} = |\alpha| \cdot |\beta|,$$

thereby proving part (b).

From a geometric perspective, part (c), the triangle inequality, is intuitively quite clear. However, an algebraic proof is somewhat technical and not very intuitive. To prove part (c), we begin by observing that if $\gamma = e + fi \in \mathbb{C}$ then, by Lemma 5.10(b), $\gamma + \gamma^* \in \mathbb{R}$. Therefore,

$$\gamma + \gamma^* = (e + fi) + (e - fi) = 2e \leq 2|e| = 2\sqrt{e^2} \leq 2\sqrt{e^2 + f^2} = 2|\gamma|,$$

and so we can conclude that

(2) $$\gamma + \gamma^* \leq 2|\gamma|.$$

If we let $\gamma = \alpha \cdot \beta^*$ in (2) and use part (b) along with the fact that $|\beta| = |\beta^*|$, we obtain

(3) $$\alpha \cdot \beta^* + \alpha^* \cdot \beta = (\alpha \cdot \beta^*) + (\alpha \cdot \beta^*)^* \leq 2|\alpha \cdot \beta^*| = 2|\alpha||\beta^*| = 2|\alpha||\beta|.$$

Using part (a) along with (3), we now have

$$(|\alpha + \beta|)^2 = (\alpha + \beta)(\alpha + \beta)^* = (\alpha + \beta)(\alpha^* + \beta^*) = \alpha \cdot \alpha^* + \beta \cdot \beta^* + \alpha \cdot \beta^* + \alpha^* \cdot \beta =$$

$$|\alpha|^2 + |\beta|^2 + \alpha \cdot \beta^* + \alpha^* \cdot \beta \leq |\alpha|^2 + |\beta|^2 + 2|\alpha||\beta| = (|\alpha| + |\beta|)^2.$$

Taking square roots of the left-hand and right-hand terms in the preceding inequality yields $|\alpha + \beta| \leq |\alpha| + |\beta|$, as desired.

Part (d) is a generalization of part (c) and can be proved with Mathematical Induction. Let T be those positive integers k such that whenever $\alpha_1, \ldots, \alpha_k \in \mathbb{C}$, we have

$$|\alpha_1 + \cdots \alpha_k| \leq |\alpha_2| + \cdots + |\alpha_k|.$$

To see that $1 \in T$, we merely observe that if $\alpha_1 \in \mathbb{C}$, then $|\alpha_1| \leq |\alpha_1|$. Using Mathematical Induction, in order to show that $T = \mathbb{N}$, it now suffices to show that whenever T contains a positive integer k, then T also contains $k + 1$. To this end, let $\alpha_1, \ldots, \alpha_k, \alpha_{k+1} \in \mathbb{C}$. For convenience, let $\delta = \alpha_1 + \cdots + \alpha_k$. Now, using the part (c) and the fact that $k \in T$, we have

$$|\alpha_1 + \cdots + \alpha_k + \alpha_{k+1}| = |(\alpha_1 + \cdots + \alpha_k) + \alpha_{k+1}| = |\delta + \alpha_{k+1}| \leq |\delta| + |\alpha_{k+1}| \leq$$

$$(|\alpha_1| + \cdots + |\alpha_k|) + |\alpha_{k+1}| = |\alpha_1| + \cdots |\alpha_k| + |\alpha_{k+1}|.$$

Thus, $k + 1 \in T$, thereby concluding the proof. $\square$

In mathematics, we are often pleasantly surprised that ideas and techniques used to solve one problem can frequently be used to solve completely different problems. As an example, let us consider the problem of determining which integers can be written as a sum of squares of two integers. On the surface, this problem seems to deal only with the integers and doesn't appear to be related in any way to complex numbers. Let us begin by observing that

$$17 = 1^2 + 4^2 \quad \text{and} \quad 20 = 2^2 + 4^2.$$

Suppose we wish to determine if 340 can be written as a sum of two squares. At first glance, this problem seems to have nothing to do with our study of the complex numbers. However, if we let $\alpha = 1 + 4i$ and $\beta = 2 + 4i$, then we have

$$17 = \alpha \cdot \alpha^* \quad \text{and} \quad 20 = \beta \cdot \beta^*.$$

Performing a computation similar to one in the proof of Lemma 6.3(b), we have

$$340 = 17 \cdot 20 = (\alpha \cdot \alpha^*)(\beta \cdot \beta^*) = (\alpha \cdot \beta)(\alpha \cdot \beta)^*.$$

However, it is easy to compute that $\alpha \cdot \beta = -14 + 12i$. Therefore,

$$340 = (\alpha \cdot \beta)(\alpha \cdot \beta)^* = (-14 + 12i)(-14 - 12i) = (-14)^2 + 12^2 = 14^2 + 12^2.$$

Thus, 340 is also a sum of squares of two integers.

The preceding computation illustrates the fact that if two integers can each be expressed as a sum of squares of two integers, then their product can also be expressed this way. We now use ideas similar to those used in the proof of Lemma 6.3(b) to prove this fact.

Proposition 6.4.   *Suppose n, m are integers that can be written as a sum of squares of two integers. Then their product $n \cdot m$ can also be written as sum of squares of two integers. In particular, if $n = a^2 + b^2$ and $m = c^2 + d^2$, then $n \cdot m = (ac - bd)^2 + (ad + bc)^2$.*

Proof. Suppose $n = a^2 + b^2$ and $m = c^2 + d^2$, then

$$n \cdot m = (a^2 + b^2)(c^2 + d^2) = (a + bi)(a + bi)^*(c + di)(c + di)^* =$$
$$(a + bi)(c + di)(a + bi)^*(c + di)^* = ((a + bi)(c + di))((a + bi)(c + di))^* =$$
$$((ac - bd) + (ad + bc)i)((ac - bd) + (ad + bc)i)^* = (ac - bd)^2 + (ad + bc)^2.$$

$\square$

A complete description of those integers that can be expressed as a sum of squares of two integers can be found in a course in number theory, and Proposition 6.4 is an important tool used to find that description.

■ Examples

Since $58 = 3^2 + 7^2$ and $98 = 7^2 + 7^2$, we now have

$$58 \cdot 98 = (3^2 + 7^2)(7^2 + 7^2) = (3 + 7i)(3 - 7i)(7 + 7i)(7 - 7i) =$$
$$((3 + 7i)(7 + 7i))((3 - 7i)(7 - 7i)) = (-28 + 70i)(-28 - 70i) =$$
$$(-28)^2 + 70^2 = 28^2 + 70^2.$$

Similarly, $26 = 1^2 + 5^2$ and $74 = 5^2 + 7^2$, so

$$26 \cdot 74 = (1^2 + 5^2)(5^2 + 7^2) = (1 + 5i)(1 - 5i)(5 + 7i)(5 - 7i) =$$
$$((1 + 5i)(5 + 7i))((1 - 5i)(5 - 7i)) = (-30 + 32i)(-30 - 32i) =$$
$$(-30)^2 + 32^2 = 30^2 + 32^2.$$

■

Adding complex numbers is quite straightforward. It is no different from adding polynomials with real coefficients of degree at most 1. Thus, if there is anything you might find confusing or mysterious about the complex numbers, it would be multiplication. After all, the existence of two complex numbers, i and $-i$, whose square is -1 causes some people to be uncomfortable with the complex numbers. However, in the next section, we will see that when we view complex numbers geometrically, multiplication becomes very natural.

6.2 Rectangular and Polar Form

Since any complex number α can be written in the form $\alpha = a + bi$, we can describe α using the two real numbers a and b. We call $a + bi$ the **rectangular form** of α because, loosely speaking, the arrows that represent α correspond to one of the diagonals in a rectangle with sides a and b.

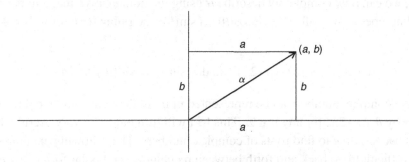

However, there is an alternative way to describe complex numbers using two real numbers. Recall, we consider two arrows in the plane to be equivalent if they have the same length and go in the same direction. Therefore, if we use one real number to describe its length and a second to describe its direction, then we have a new way of describing complex numbers with two real numbers that may be more natural than rectangular form. To be more precise, if $\alpha = a + bi$, then the length of α, $|\alpha|$, is equal to $\sqrt{a^2 + b^2}$. Next, in an attempt to describe the direction of the arrows representing α, let θ be the angle made by the positive x-axis and the arrow representing α that begins at $(0, 0)$ and ends at (a, b).

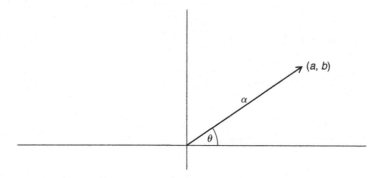

Observe that

$$\cos(\theta) = \frac{a}{\sqrt{a^2 + b^2}} = \frac{a}{|\alpha|} \quad \text{and} \quad \sin(\theta) = \frac{b}{\sqrt{a^2 + b^2}} = \frac{b}{|\alpha|}.$$

Therefore, we now have

$$\alpha = a + bi = \sqrt{a^2 + b^2} \left(\frac{a}{\sqrt{a^2 + b^2}} + \frac{b}{\sqrt{a^2 + b^2}} i \right) =$$

$$|\alpha| \left(\frac{a}{|\alpha|} + \frac{b}{|\alpha|} i \right) = |\alpha|(\cos(\theta) + \sin(\theta) \cdot i).$$

Since $\sin(\theta) \cdot i = i \sin(\theta)$, we can write $\alpha = |\alpha|(\cos(\theta) + i \sin(\theta))$.

In particular, we can now completely describe α using the nonnegative real number $|\alpha|$ along with the real number θ. We call $\alpha = |\alpha|(\cos(\theta) + i \sin(\theta))$ the **polar form** of α. If $n \in \mathbb{Z}$, observe that

$$\cos(\theta) = \cos(\theta + n \cdot 2\pi) \quad \text{and} \quad \sin(\theta) = \sin(\theta + n \cdot 2\pi).$$

Therefore, any complex number can be represented in more than one way in polar form as we can replace θ by $\theta + n \cdot 2\pi$, for any $n \in \mathbb{Z}$. This fact will turn out to be very useful when, in Theorem 6.8, we see how to find roots of complex numbers. The following examples indicate that it is not difficult to go back and forth between rectangular and polar form. However, we should be aware that if α is in rectangular form, then we may not be able to compute the value of θ by hand. Conversely, if α is in polar form, then we also may not be able to compute the values of $\cos(\theta)$ and $\sin(\theta)$ by hand.

■ Examples

When converting $\alpha \in \mathbb{C}$ between rectangular and polar forms, you should first draw the arrow representing α that begins at $(0, 0)$. Viewing α in this way will make the conversion more understandable. For two of the examples that follow, we include such a diagram. You should draw the appropriate diagram when you work through the other examples.

Converting from rectangular to polar form: When converting to polar form, it is almost always best to first compute and factor out the length of the complex number. This will be reflected in all of our examples.

$$-17i = 17(0 - i) = 17 \left(\cos\left(\frac{3}{2}\pi \right) + i \sin\left(\frac{3}{2}\pi \right) \right),$$

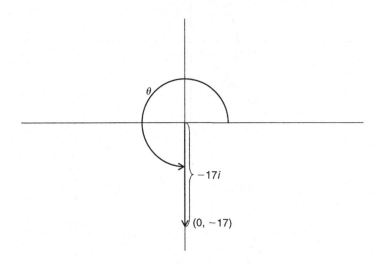

$$15 - 15\sqrt{3}i = 30 \left(\frac{1}{2} - i \cdot \frac{\sqrt{3}}{2} \right) = 30 \left(\cos\left(\frac{5}{3}\pi \right) + i \sin\left(\frac{5}{3}\pi \right) \right),$$

$$4\sqrt{2} - 4\sqrt{2}i = 8 \left(\frac{\sqrt{2}}{2} - i \cdot \frac{\sqrt{2}}{2} \right) = 8 \left(\cos\left(\frac{7}{4}\pi \right) + i \sin\left(\frac{7}{4}\pi \right) \right),$$

$$3 + 4i = 5 \left(\frac{3}{5} + i \cdot \frac{4}{5} \right) = 5 \left(\cos(\theta) + i \sin(\theta) \right), \text{ where } \theta = \arctan\left(\frac{4}{3} \right),$$

$$6 - 2i = \sqrt{40} \left(\frac{6}{\sqrt{40}} - i \cdot \frac{2}{\sqrt{40}} \right) = \sqrt{40} \left(\cos(\phi) + i \sin(\phi) \right),$$

where $\phi = 2\pi - \arctan\left(\frac{1}{3} \right)$.

The last two conversions cannot be simplified further, as we cannot, by hand, compute $\arctan\left(\frac{4}{3}\right)$ or $\arctan\left(\frac{1}{3}\right)$.

Converting from polar to rectangular form: When converting to rectangular form, if possible, you should first try to compute the values of the sine and cosine.

$$7\left(\cos\left(\frac{5}{4}\pi\right) + i\sin\left(\frac{5}{4}\pi\right)\right) = 7\left(\frac{-\sqrt{2}}{2} + i\cdot\frac{-\sqrt{2}}{2}\right) = -\frac{7\sqrt{2}}{2} - \frac{7\sqrt{2}}{2}i,$$

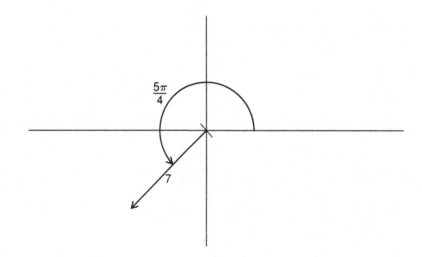

$$12\left(\cos\left(\frac{\pi}{2}\right) + i\sin\left(\frac{\pi}{2}\right)\right) = 12(0 + i) = 12i,$$

$$58\left(\cos\left(\frac{11}{6}\pi\right) + i\sin\left(\frac{11}{6}\pi\right)\right) = 58\left(\frac{\sqrt{3}}{2} + i\cdot\frac{-1}{2}\right) = 29\sqrt{3} - 29i,$$

$$10\left(\cos\left(\frac{\pi}{13}\right) + i\sin\left(\frac{\pi}{13}\right)\right) = 10\cos\left(\frac{\pi}{13}\right) + 10\sin\left(\frac{\pi}{13}\right)i.$$

The last conversion also cannot be simplified further as we cannot by hand compute $\cos\left(\frac{\pi}{13}\right)$ or $\sin\left(\frac{\pi}{13}\right)$. This situation is very similar to one we face when we solve various algebra problems. Many algebra problems have terms in their answer that we cannot compute by hand, like $\sqrt{7}$ or $35^{\frac{2}{3}}$. Similarly, other terms that we cannot compute by hand, like $\arctan\left(\frac{4}{3}\right)$ or $\cos\left(\frac{\pi}{13}\right)$, may well appear when converting between polar and rectangular form.

Exercises for Sections 6.1 and 6.2

1. Please refer to the following diagram.

    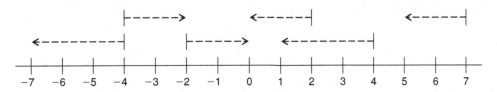

    (a) How many different equivalence classes of arrows are there?

    (b) What real numbers are represented by these arrows?

2. Please refer to the following diagram.

    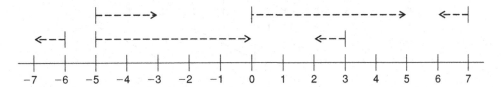

    (a) How many different equivalence classes of arrows are there?

    (b) What real numbers are represented by these arrows?

3. Please refer to the following diagram.

    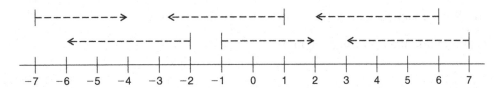

    (a) How many different equivalence classes of arrows are there?

    (b) What real numbers are represented by these arrows?

4. Please refer to the following diagram.

    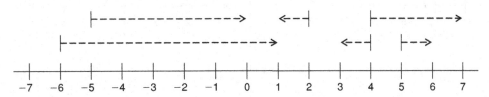

    (a)   How many different equivalence classes of arrows are there?

    (b)   What real numbers are represented by these arrows?

5.   Please refer to the following diagram.

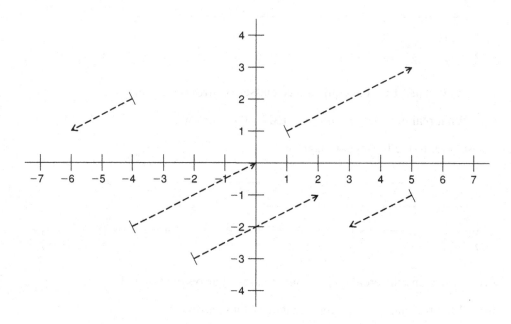

    (a)   How many different equivalence classes of arrows are there?

    (b)   What complex numbers are represented by these arrows?

6.   Please refer to the following diagram.

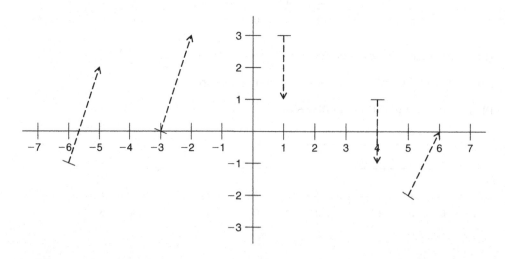

(a) How many different equivalence classes of arrows are there?

(b) What complex numbers are represented by these arrows?

7. Please refer to the following diagram.

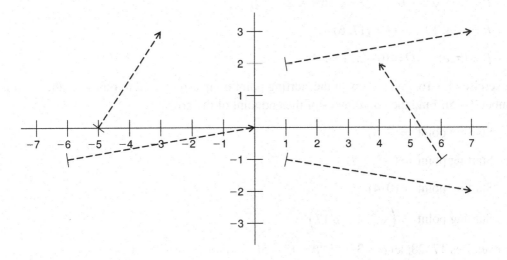

(a) How many different equivalence classes of arrows are there?

(b) What complex numbers are represented by these arrows?

8. Please refer to the following diagram.

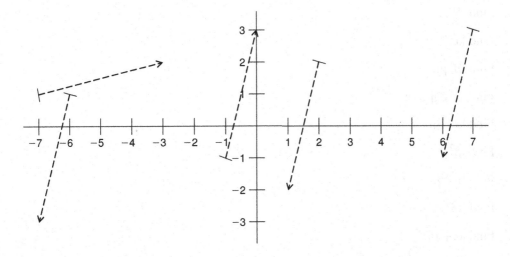

(a) How many different equivalence classes of arrows are there?

(b) What complex numbers are represented by these arrows?

In exercises 9–12, find the complex number represented by the arrow beginning at the point P and ending at the point Q.

9. $P = (3, 11), \quad Q = (-8, 14)$

10. $P = (1 + \sqrt{2}, -6), \quad Q = (-6 + \sqrt{2}, -19)$

11. $P = (-4, 12), \quad Q = (17, 6)$

12. $P = (\pi, e), \quad Q = (6 - \pi, 13 + 4e)$

In exercises 13–16, you are given the starting point of an arrow used to represent the complex number $7 - 5i$. Find the coordinates of the endpoint of the arrow.

13. Starting point $= (3, 8)$

14. Starting point $= (-3, -8)$

15. Starting point $= (0, 4)$

16. Starting point $= \left(\sqrt{2}, 4 - \sqrt{17} \right)$

For exercises 17–28, let $\alpha = 3 + 4i, \ \beta = 12 - 5i, \ \gamma = 1 + 2i, \ \delta = 5 + 8i$.

17. Find $|\alpha|$.

18. Find $|\beta|$.

19. Find $|\gamma|$.

20. Find $|\delta|$.

21. Find $|\alpha\delta|$.

22. Find $|\beta^*\gamma|$.

23. Find $|\alpha^*\beta\delta|$.

24. Find $|\alpha^{-1}|$.

25. Find $|\beta^2|$.

26. Find $|\gamma^{-2}|$.

27. Find $|(\delta^*)^4|$.

28. Find $|\alpha + \beta|$.

29. Find eight complex numbers $\alpha = a + bi$ such that $|\alpha| = \sqrt{5}$ and $a, b \in \mathbb{Z}$.

30. Find eight complex numbers $\beta = c + di$ such that $|\beta| = \sqrt{13}$ and $c, d \in \mathbb{Z}$.

31. Find twelve complex numbers $\alpha = a + bi$ such that $|\alpha| = 5$ and $a, b \in \mathbb{Z}$.

32. Find twelve complex numbers $\beta = c + di$ such that $|\beta| = 13$ and $c, d \in \mathbb{Z}$.

33. (a) If $a \in \mathbb{Z}$, what are the only remainders that can occur when a^2 is divided by 4?

    (b) If $a, b \in \mathbb{Z}$, what are the only remainders that can occur when $a^2 + b^2$ is divided by 4?

    (c) Use part (b) to show that there do not exist integers a, b such that $458931 = a^2 + b^2$.

    (d) Generalize part (c) to describe an infinite collection of positive integers such that none of them can be written as a sum of squares of two integers.

For exercises 34–39, you may wish to use the following:

$$2 = 1^2 + 1^2, \ 5 = 1^2 + 2^2, \ 13 = 2^2 + 3^2, \ 17 = 1^2 + 4^2, \ 29 = 2^2 + 5^2.$$

34. Use the ideas in Proposition 6.4 to find integers A, B such that $58 = 2 \cdot 29 = A^2 + B^2$.

35. Use the ideas in Proposition 6.4 to find integers C, D such that $85 = 5 \cdot 17 = C^2 + D^2$.

36. Use the ideas in Proposition 6.4 to find integers E, F such that $221 = 13 \cdot 17 = E^2 + F^2$.

37. Use the ideas in Proposition 6.4 to find integers G, H such that $493 = 17 \cdot 29 = G^2 + H^2$.

38. Use the ideas in Proposition 6.4 to find integers J, K such that $2210 = 2 \cdot 5 \cdot 13 \cdot 17 = J^2 + K^2$.

39. Use the ideas in Proposition 6.4 to find integers L, M such that $64090 = 2 \cdot 5 \cdot 13 \cdot 17 \cdot 29 = L^2 + M^2$.

40. (a) Give an example of two positive integers a, b such that neither a nor b nor ab can be written as a sum of squares of two integers.

    (b) Give an example of two positive integers a, b such that neither a nor b can be written as a sum of squares of two integers, but ab can be written as a sum of squares of two nonzero integers.

In exercises 41–44, write all four of the complex numbers in polar form.

41. $73, \ -73, \ 73i, \ -73i$.

42. $1 + \sqrt{3}i, \ -1 - \sqrt{3}i, \ -1 + \sqrt{3}i, \ 1 - \sqrt{3}i$.

43. $5\sqrt{3} + 5i, \ -5\sqrt{3} - 5i, \ -5\sqrt{3} + 5i, \ 5\sqrt{3} - 5i$.

44. $11 + 11i, \ -11 - 11i, \ -11 + 11i, \ 11 - 11i$.

For exercises 45–48, let $\theta = \arctan\left(\frac{3}{4}\right)$ and write all four of the complex numbers in polar form. You should leave your answer in terms of θ. Recall that if α is a positive real number, then $\arctan(\alpha) + \arctan\left(\frac{1}{\alpha}\right) = \frac{\pi}{2}$.

45. $4+3i, \quad -4-3i, \quad 4-3i, \quad -4+3i$.

46. $7+\frac{21}{4}i, \quad -7-\frac{21}{4}i, \quad 7-\frac{21}{4}i, \quad -7+\frac{21}{4}i$.

47. $3+4i, \quad -3-4i, \quad 3-4i, \quad -3+4i$.

48. $\frac{3}{5}+\frac{4}{5}i, \quad -\frac{3}{5}-\frac{4}{5}i, \quad \frac{3}{5}-\frac{4}{5}i, \quad -\frac{3}{5}+\frac{4}{5}i$.

In exercises 49–52, write all four of the complex numbers in rectangular form.

49. $\text{cis}\left(\frac{11\pi}{6}\right), \quad 4\text{cis}\left(\frac{11\pi}{6}\right), \quad 15\text{cis}\left(\frac{11\pi}{6}\right), \quad 2\sqrt{3}\text{cis}\left(\frac{11\pi}{6}\right)$.

50. $\text{cis}\left(\frac{3\pi}{4}\right), \quad 3\text{cis}\left(\frac{3\pi}{4}\right), \quad 80\text{cis}\left(\frac{3\pi}{4}\right), \quad 7\sqrt{2}\text{cis}\left(\frac{3\pi}{4}\right)$.

51. $\text{cis}\left(\frac{3\pi}{2}\right), \quad 8\text{cis}\left(\frac{3\pi}{2}\right), \quad 3\sqrt{5}\text{cis}\left(\frac{3\pi}{2}\right), \quad 17^{\frac{2}{3}}\text{cis}\left(\frac{3\pi}{2}\right)$.

52. $\text{cis}\left(\frac{4\pi}{3}\right), \quad 26\text{cis}\left(\frac{4\pi}{3}\right), \quad 11\sqrt{3}\text{cis}\left(\frac{4\pi}{3}\right), \quad 120^{\frac{3}{5}}\text{cis}\left(\frac{4\pi}{3}\right)$.

For exercises 53–56, let $\theta = \arctan\left(\frac{7}{5}\right)$ and write all four of the complex numbers in rectangular form. Once again, recall that if α is a positive real number, then $\arctan(\alpha) + \arctan\left(\frac{1}{\alpha}\right) = \frac{\pi}{2}$.

53. $\text{cis}(\theta), \quad \sqrt{74}\text{cis}(\theta), \quad 6\sqrt{74}\text{cis}(\theta), \quad 60\text{cis}(\theta)$.

54. $\text{cis}(\pi - \theta), \quad \text{cis}(\pi + \theta), \quad \text{cis}(2\pi - \theta), \quad \sqrt{148}\text{cis}(\pi - \theta)$.

55. $\text{cis}\left(\frac{\pi}{2} - \theta\right), \quad \text{cis}\left(\frac{\pi}{2} + \theta\right), \quad 2\sqrt{74}\text{cis}\left(\frac{\pi}{2} - \theta\right), \quad 9\sqrt{74}\text{cis}\left(\frac{\pi}{2} - \theta\right)$.

56. $\text{cis}\left(\frac{3\pi}{2} - \theta\right), \quad \text{cis}\left(\frac{3\pi}{2} + \theta\right), \quad 5\sqrt{74}\text{cis}\left(\frac{3\pi}{2} - \theta\right), \quad 23\sqrt{74}\text{cis}\left(\frac{3\pi}{2} - \theta\right)$.

6.3 Demoivre's Theorem and Roots of Complex Numbers

At this point, it is natural for you to wonder if we have gained anything by looking at complex numbers in polar form. Certainly, adding complex numbers is much easier when they are in rectangular form. However, as we will soon see, multiplying complex numbers is easier when they are in polar form. To see this, we first need to recall two basic facts from trigonometry about the cosine and sine of the sum of angles, which we state without proof.

Lemma 6.5

$$\cos(\theta + \phi) = \cos(\theta)\cos(\phi) - \sin(\theta)\sin(\phi)$$

and

$$\sin(\theta + \phi) = \cos(\theta)\sin(\phi) + \sin(\theta)\cos(\phi).$$

For convenience, we also introduce the following shorthand.

Definition 6.6. *cis(θ) is a shorthand for the complex number* $\cos(\theta) + i\sin(\theta)$. *Therefore, the polar form of any* $\alpha \in \mathbb{C}$ *can be written as* $|\alpha|cis(\theta)$.

■ Examples

$$3cis\left(\frac{\pi}{3}\right) = 3\left(\cos\left(\frac{\pi}{3}\right) + i\sin\left(\frac{\pi}{3}\right)\right) = \frac{3}{2} + \frac{3\sqrt{3}}{2}i,$$

$$12cis\left(\frac{7}{4}\pi\right) = 12\left(\cos\left(\frac{7}{4}\pi\right) + i\sin\left(\frac{7}{4}\pi\right)\right) = 6\sqrt{2} - 6\sqrt{2}i.$$

■

We can now prove the following.

Theorem 6.7—DeMoivre's Theorem. *If* $\alpha = |\alpha|cis(\theta)$ *and* $\beta = |\beta|cis(\phi)$ *are two complex numbers written in polar form, then*

$$\alpha \cdot \beta = |\alpha||\beta|cis(\theta + \phi).$$

Furthermore, if $n \in \mathbb{Z}$ *then*

$$\alpha^n = |\alpha|^n cis(n \cdot \theta).$$

Intuition. This theorem indicates how simple and natural multiplication in $\mathbb{C}$ is when we view complex numbers in polar form. The theorem says that when we multiply complex numbers in polar form, we simply multiply their lengths and add their angles. The proof is a direct and straightforward application of the addition formulas for the cosine and sine that we stated in Lemma 6.5. Also, observe how easy it is to find inverses in polar form. If $\alpha = |\alpha|cis(\theta) \neq 0$, then letting $n = -1$ in DeMoivre's Theorem immediately tells us that $\alpha^{-1} = |\alpha|^{-1}cis(-\theta)$.

Proof. If $\alpha = |\alpha|cis(\theta)$ and $\beta = |\beta|cis(\phi)$, then

(4) $$\alpha \cdot \beta = (|\alpha|cis(\theta))(|\beta|cis(\phi)) = |\alpha||\beta|(cis(\theta) \cdot cis(\phi))$$

However,

$$cis(\theta) \cdot cis(\phi) = (\cos(\theta) + i\sin(\theta))(\cos(\phi) + i\sin(\phi)) =$$
$$(\cos(\theta)\cos(\phi) - \sin(\theta)\sin(\phi)) + i(\cos(\theta)\sin(\phi) + \sin(\theta)\cos(\phi)).$$

Applying Lemma 6.5 to the preceding equation reveals to us that

$$cis(\theta) \cdot cis(\phi) = \cos(\theta + \phi) + i\sin(\theta + \phi) = cis(\theta + \phi).$$

Plugging this result into (4) results in

$$\alpha \cdot \beta = |\alpha||\beta|\text{cis}(\theta + \phi),$$

as desired.

To begin the proof of the second part of this theorem, let T be those natural numbers such that $\alpha^n = |\alpha|^n\text{cis}(n \cdot \theta)$. We would like to show that $T = \mathbb{N}$, and we will proceed by Mathematical Induction. To see that $1 \in T$, we merely observe that $\alpha^1 = |\alpha|^1\text{cis}(1 \cdot \theta)$. Therefore, we now need to show that if k is some natural number belonging to T, then $k+1$ also belongs to T.

Since $k \in T$, we have $\alpha^k = |\alpha|^k\text{cis}(k \cdot \theta)$. Applying the first part of this theorem along with the fact that $k \in T$, we now have

$$\alpha^{k+1} = \alpha \cdot \alpha^k = \left(|\alpha|\text{cis}\,(\theta)\right)\left(|\alpha|^k\text{cis}\,(k \cdot \theta)\right) =$$
$$\left(|\alpha| \cdot |\alpha|^k\right)\left(\text{cis}\,(\theta + k \cdot \theta)\right) = |\alpha|^{k+1}\text{cis}\,((k+1) \cdot \theta).$$

Thus, $k+1$ does indeed belong to T.

To conclude the proof of the second part of this theorem, we now need to verify that $\alpha^n = |\alpha|^n\text{cis}(n \cdot \theta)$ when $n = 0$ or n is a negative integer. Our result certainly holds when $n = 0$ because, in this case, both α^n and $|\alpha|^n\text{cis}(n \cdot \theta)$ are equal to 1. Finally, if n is a negative integer, then $-n$ is a positive integer and our previous argument tells us that $\alpha^{-n} = |\alpha|^{-n}\text{cis}(-n \cdot \theta)$. We now have

$$\alpha^{-n} \cdot \left(|\alpha|^n\text{cis}\,(n \cdot \theta)\right) = \left(|\alpha|^{-n}\text{cis}\,(-n \cdot \theta)\right) \cdot \left(|\alpha|^n\text{cis}\,(n \cdot \theta)\right)$$
$$= (|\alpha|^{-n}|\alpha|^n)\left(\text{cis}\,((-n+n)\,\theta)\right) = 1.$$

Observe that the previous equation tells us that $1 = \alpha^{-n} \cdot (|\alpha|^n\text{cis}(n \cdot \theta))$. Multiplying this equation by α^n results in $\alpha^n = |\alpha|^n\text{cis}(n \cdot \theta)$, as desired.   □

■ Examples

$$11\text{cis}\left(\frac{\pi}{7}\right) \cdot 6\text{cis}\left(\frac{\pi}{5}\right) = 66\text{cis}\left(\frac{12}{35}\pi\right),$$

$$5\text{cis}\left(\frac{\pi}{3}\right) \cdot 4\text{cis}\left(\frac{5}{6}\pi\right) = 20\text{cis}\left(\frac{7}{6}\pi\right) = 20\left(-\frac{\sqrt{3}}{2} - i\frac{1}{2}\right) = -10\sqrt{3} - 10i,$$

$$\left(\frac{1}{\sqrt{2}} + i\frac{1}{\sqrt{2}}\right)^2 = \left(\text{cis}\left(\frac{\pi}{4}\right)\right)^2 = \text{cis}\left(\frac{\pi}{2}\right) = i,$$

$$\left(-\frac{1}{\sqrt{2}} - i\frac{1}{\sqrt{2}}\right)^2 = \left(\text{cis}\left(\frac{5}{4}\pi\right)\right)^2 = \text{cis}\left(\frac{5}{2}\pi\right) = i,$$

$$(1-i)^{20} = \left(\sqrt{2} \left(\mathrm{cis} \left(\frac{7}{4} \right) \pi \right) \right)^{20} = \left(\sqrt{2} \right)^{20} \mathrm{cis}(35\pi) = 2^{10} \mathrm{cis}(\pi) = -2^{10} = -1024,$$

$$\left(4\mathrm{cis} \left(\frac{\pi}{9} \right) \right)^{3} = 4^{3}\mathrm{cis} \left(\frac{\pi}{3} \right) = 64 \left(\frac{1}{2} + i\frac{\sqrt{3}}{2} \right) = 32 + 32\sqrt{3}i.$$

■

These examples indicate that when raising a complex number to a positive integer power, it is often to our advantage to first write the number in polar form. In the examples, we saw that $\frac{1}{\sqrt{2}} + i\frac{1}{\sqrt{2}}$ and $-\frac{1}{\sqrt{2}} - i\frac{1}{\sqrt{2}}$ are both square roots of i. We also saw that $1 - i$ is a 20th root of -1024. In light of these examples, we will take another look at DeMoivre's Theorem with the goal of finding roots of complex numbers.

DeMoivre's Theorem says that raising a complex number to the nth power entails raising the length to the nth power and multiplying the angle by n. Conversely, it looks like in order to find an nth root of a complex number, we would need to take an nth root of the length and then divide the angle by n. Fortunately, the Intermediate Value Theorem (Theorem 4.4) guarantees that every positive real number has a positive nth root. Therefore, it appears that in $\mathbb{C}$ there are no obstacles to finding nth roots. This then raises the question as to how many nth roots a number can have in $\mathbb{C}$? Given $\alpha \in \mathbb{C}$, the nth roots of α are precisely the roots of the polynomial $x^n - \alpha$. In Chapter 12, we will show that any polynomial of degree n can have at most n roots in a field. Thus, the most nth roots an element of a field can have in that field is n. For example, in a field a number may have no square roots and can have as many as two square roots. In particular, -1 has no square roots in $\mathbb{Q}$ and $\mathbb{R}$, but it has two square roots in $\mathbb{C}$. Shortly we will see that whereas 1 has two fourth roots in $\mathbb{Q}$ and $\mathbb{R}$, it has four fourth roots in $\mathbb{C}$. In fact, as we will now see, the number of nth roots every nonzero complex number has in $\mathbb{C}$ is indeed n.

Theorem 6.8. *Let $\alpha = |\alpha|\mathrm{cis}(\theta)$ be a nonzero complex number written in polar form and let $n \in \mathbb{N}$. If we let $|\alpha|^{\frac{1}{n}}$ denote the positive real nth root of $|\alpha|$, then*

$$\alpha_0 = |\alpha|^{\frac{1}{n}} \mathrm{cis} \left(\frac{\theta}{n} \right), \quad \alpha_1 = |\alpha|^{\frac{1}{n}} \mathrm{cis} \left(\frac{\theta + 1 \cdot 2\pi}{n} \right), \ldots,$$

$$\alpha_j = |\alpha|^{\frac{1}{n}} \mathrm{cis} \left(\frac{\theta + j \cdot 2\pi}{n} \right), \ldots, \quad \alpha_{n-1} = |\alpha|^{\frac{1}{n}} \mathrm{cis} \left(\frac{\theta + (n-1) \cdot 2\pi}{n} \right)$$

are n different nth roots of α in $\mathbb{C}$.

Proof. We must first show that if $0 \leq j \leq n-1$, then each α_j mentioned above is indeed an nth root of α. To show this, we can apply the second part of DeMoivre's Theorem to see that

$$\alpha_j{}^n = \left(|\alpha|^{\frac{1}{n}} \operatorname{cis}\left(\frac{\theta + j \cdot 2\pi}{n}\right) \right)^n =$$

$$\left(|\alpha|^{\frac{1}{n}} \right)^n \operatorname{cis}\left(n \cdot \left(\frac{\theta + j \cdot 2\pi}{n}\right) \right) = |\alpha| \operatorname{cis}(\theta + j \cdot 2\pi) = |\alpha| \operatorname{cis}(\theta) = \alpha.$$

To see that $\alpha_0, \alpha_1, \ldots, \alpha_{n-1}$ are all different, we review some unit circle trigonometry. If ϕ is an angle in radian measure lying in the interval $[0, 2\pi)$, then the point $(\cos(\phi), \sin(\phi))$ lies on the unit circle $x^2 + y^2 = 1$.

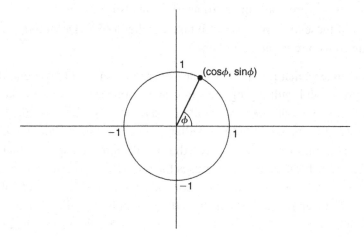

Furthermore, if ϕ_1, ϕ_2 are two angles in the interval $[0, 2\pi)$ that are not equal, then $(\cos(\phi_1), \sin(\phi_1))$ and $(\cos(\phi_2), \sin(\phi_2))$ are different points on the unit circle. Thus, the complex numbers $\operatorname{cis}(\phi_1)$ and $\operatorname{cis}(\phi_2)$ would be different.

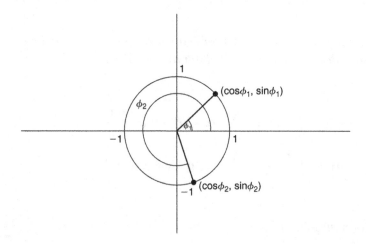

In light of this, since

$$0, \frac{1 \cdot 2\pi}{n}, \frac{2 \cdot 2\pi}{n}, \ldots, \frac{j \cdot 2\pi}{n}, \ldots, \frac{(n-1) \cdot 2\pi}{n}$$

are n different angles that belong to the interval $[0, 2\pi)$, we can conclude that the complex numbers

$$(5) \qquad \mathrm{cis}(0), \mathrm{cis}\left(\frac{1 \cdot 2\pi}{n}\right), \mathrm{cis}\left(\frac{2 \cdot 2\pi}{n}\right), \ldots, \mathrm{cis}\left(\frac{j \cdot 2\pi}{n}\right), \ldots, \mathrm{cis}\left(\frac{(n-1) \cdot 2\pi}{n}\right)$$

are all different.

Next, observe that if γ is a nonzero element of $\mathbb{C}$ and if $\beta_1, \beta_2 \in \mathbb{C}$ are not equal, then $\gamma \cdot \beta_1$ and $\gamma \cdot \beta_2$ are also not equal. This is easy to check, for if $\gamma \cdot \beta_1 = \gamma \cdot \beta_2$, then $\gamma^{-1} \cdot \gamma \cdot \beta_1 = \gamma^{-1} \cdot \gamma \cdot \beta_2$, which would immediately imply that $\beta_1 = \beta_2$. Note that this is not so much a fact about $\mathbb{C}$ but is a fact that holds in all fields.

If we multiply the n different complex numbers in (5) by the nonzero complex number $|\alpha|^{\frac{1}{n}} \mathrm{cis}(\frac{\theta}{n})$, we obtain $\alpha_0, \alpha_1, \ldots, \alpha_{n-1}$. However, the argument above tells us that they are all different, thereby concluding the proof. $\qquad\square$

■ Examples

Finding the 4 fourth roots of 1. In polar form, $1 = 1\,\mathrm{cis}(0)$, so Theorem 6.8 tells us that the 4 fourth roots of 1 are

$$1^{\frac{1}{4}} \mathrm{cis}\left(\frac{0}{4}\right) = 1 \cdot 1 = 1,$$

$$1^{\frac{1}{4}} \mathrm{cis}\left(\frac{1 \cdot 2\pi}{4}\right) = 1 \cdot \mathrm{cis}\left(\frac{\pi}{2}\right) = i,$$

$$1^{\frac{1}{4}} \mathrm{cis}\left(\frac{2 \cdot 2\pi}{4}\right) = 1 \cdot \mathrm{cis}(\pi) = -1,$$

$$1^{\frac{1}{4}} \mathrm{cis}\left(\frac{3 \cdot 2\pi}{4}\right) = 1 \cdot \mathrm{cis}\left(\frac{3\pi}{2}\right) = -i.$$

Thus, exactly 2 of the 4 fourth roots of 1 lie in $\mathbb{R}$.

Finding the 6 sixth roots of 64. In polar form, $64 = 64\mathrm{cis}(0)$, so Theorem 6.8 tells us that the 6 sixth roots of 64 are

$$64^{\frac{1}{6}}\mathrm{cis}\left(\frac{0}{6}\right) = 2 \cdot 1 = 2,$$

$$64^{\frac{1}{4}}\mathrm{cis}\left(\frac{1 \cdot 2\pi}{6}\right) = 2 \cdot \mathrm{cis}\left(\frac{\pi}{3}\right) = 2\left(\frac{1}{2} + i\frac{\sqrt{3}}{2}\right) = 1 + i\sqrt{3},$$

$$64^{\frac{1}{4}}\mathrm{cis}\left(\frac{2 \cdot 2\pi}{6}\right) = 2 \cdot \mathrm{cis}\left(\frac{2\pi}{3}\right) = 2\left(-\frac{1}{2} + i\frac{\sqrt{3}}{2}\right) = -1 + i\sqrt{3},$$

$$64^{\frac{1}{4}}\mathrm{cis}\left(\frac{3 \cdot 2\pi}{6}\right) = 2 \cdot \mathrm{cis}(\pi) = 2(-1) = -2,$$

$$64^{\frac{1}{4}}\mathrm{cis}\left(\frac{4 \cdot 2\pi}{6}\right) = 2 \cdot \mathrm{cis}\left(\frac{4\pi}{3}\right) = 2\left(-\frac{1}{2} - i\frac{\sqrt{3}}{2}\right) = -1 - i\sqrt{3},$$

$$64^{\frac{1}{4}}\mathrm{cis}\left(\frac{5 \cdot 2\pi}{6}\right) = 2 \cdot \mathrm{cis}\left(\frac{5\pi}{3}\right) = 2\left(\frac{1}{2} - i\frac{\sqrt{3}}{2}\right) = 1 + i\sqrt{3}.$$

For other problems involving nth roots, we may not be able to simplify the answer as much as we did in the previous two examples, as we may not be able to compute the exact value of either $|\alpha|^{\frac{1}{n}}$ or $\mathrm{cis}\left(\frac{\theta + j \cdot 2\pi}{n}\right)$.

Finding the 3 cube roots of $10i$. In polar form, $10i = 10\mathrm{cis}(\frac{\pi}{2})$, so Theorem 6.8 tells us that the 3 cube roots of $10i$ are

$$10^{\frac{1}{3}}\mathrm{cis}\left(\frac{\frac{\pi}{2}}{3}\right) = 10^{\frac{1}{3}}\mathrm{cis}\left(\frac{\pi}{6}\right) = 10^{\frac{1}{3}}\left(\frac{\sqrt{3}}{2} + i\frac{1}{2}\right) = \frac{10^{\frac{1}{3}} \cdot \sqrt{3}}{2} + i\frac{10^{\frac{1}{3}}}{2},$$

$$10^{\frac{1}{3}}\mathrm{cis}\left(\frac{\frac{\pi}{2} + 1 \cdot 2\pi}{3}\right) = 10^{\frac{1}{3}}\mathrm{cis}\left(\frac{5\pi}{6}\right) = 10^{\frac{1}{3}}\left(-\frac{\sqrt{3}}{2} + i\frac{1}{2}\right) = -\frac{10^{\frac{1}{3}} \cdot \sqrt{3}}{2} + i\frac{10^{\frac{1}{3}}}{2},$$

$$10^{\frac{1}{3}}\mathrm{cis}\left(\frac{\frac{\pi}{2} + 2 \cdot 2\pi}{3}\right) = 10^{\frac{1}{3}}\mathrm{cis}\left(\frac{3\pi}{2}\right) = 10^{\frac{1}{3}}(-i) = -i10^{\frac{1}{3}}.$$

Finding the 7 seventh roots of 1. In polar form, $1 = 1\mathrm{cis}(0)$, so Theorem 6.8 tells us that the 7 seventh roots of 1 are

$$\mathrm{cis}(0) = 1, \quad \mathrm{cis}\left(\frac{1 \cdot 2\pi}{7}\right), \quad \mathrm{cis}\left(\frac{2 \cdot 2\pi}{7}\right), \quad \mathrm{cis}\left(\frac{3 \cdot 2\pi}{7}\right), \quad \mathrm{cis}\left(\frac{4 \cdot 2\pi}{7}\right),$$

$$\mathrm{cis}\left(\frac{5 \cdot 2\pi}{7}\right), \quad \mathrm{cis}\left(\frac{6 \cdot 2\pi}{7}\right).$$

However, this is as far as we can simplify our answer as we cannot compute, by hand, the values of $\cos\left(\frac{j \cdot 2\pi}{7}\right)$ or $\sin\left(\frac{j \cdot 2\pi}{7}\right)$, where $1 \leq j \leq 6$.

6.4 A Proof of the Fundamental Theorem of Algebra

Let us now focus our attention on the Fundamental Theorem of Algebra. In Chapter 12, we will see that there are several equivalent ways to state this famous theorem. However, the version we will prove in this chapter asserts that every polynomial of degree at least 1 with coefficients in $\mathbb{C}$ must have a root in $\mathbb{C}$.

Before proving a theorem, one usually needs to come across enough examples and other types of evidence to develop an intuitive feeling that the theorem has a chance to be true. Therefore, before presenting a proof of the Fundamental Theorem of Algebra, let us look at some of the evidence that provides us with an intuitive feeling that the Fundamental Theorem of Algebra might be true. First, in Corollary 4.9, we showed that every polynomial of odd degree with coefficients in $\mathbb{R}$ must have a root in $\mathbb{R}$. Then, in Theorem 6.8, we showed that any polynomial of the form $x^n - \alpha$, where $\alpha \in \mathbb{C}$, has a root in $\mathbb{C}$. Thus, Corollary 4.9 and Theorem 6.8 indicate that there are two large classes of polynomials that must have roots in $\mathbb{C}$. Certainly, this is not a proof of the Fundamental Theorem of Algebra, but it is sufficient evidence to indicate that we should look for a proof.

Traditionally, when dealing with functions of a real variable, we usually denote the variable as x. Recall that every $\alpha \in \mathbb{C}$ can be written as $\alpha = a + bi$, with $a, b \in \mathbb{R}$. Therefore, if a variable ranges through the complex numbers, we can think of representing it as $x + yi$, where x, y are variables ranging through the real numbers. Thus, we traditionally write a complex variable as z and represent it as $z = x + yi$, where x and y are real variables.

A typical polynomial with coefficients in $\mathbb{C}$ would be

$$p(z) = 3z^2 + 5iz + (6 - 2i).$$

If we replace z by $x + yi$, we obtain

$$p(z) = p(x + yi) = 3(x + yi)^2 + 5i(x + yi) + (6 - 2i) =$$
$$((3x^2 - 3y^2) + 6xy \cdot i) + (-5y + 5x \cdot i) + (6 - 2i) =$$
$$(3x^2 - 3y^2 - 5y + 6) + (6xy + 5x - 2)i.$$

As in this example, if $p(z) = p(x + yi)$ is any polynomial with complex coefficients, we can always collect the real and complex parts of $p(z)$ in terms of x and y. Therefore, we can write

$$p(z) = p(x + yi) = f(x, y) + g(x, y)i.$$

In addition, it immediately follows from Lemma 6.3(a) that

$$|p(z)|^2 = (f(x, y) + g(x, y)i)(f(x, y) + g(x, y)i)^* =$$
$$(f(x, y) + g(x, y)i)(f(x, y) - g(x, y)i) = f(x, y)^2 + g(x, y)^2.$$

Furthermore, note that the only operations performed in finding $f(x, y)$ and $g(x, y)$ are addition, subtraction, and multiplication. Therefore, $f(x, y)$ and $g(x, y)$ must be polynomials with coefficients in $\mathbb{R}$. This observation, which will be used later in this chapter, can now be recorded as

Lemma 6.9. *Let $p(z)$ be a polynomial with coefficients in $\mathbb{C}$. If we let $z = x + yi$, then*

$$p(z) = p(x + yi) = f(x, y) + g(x, y)i,$$

where $f(x, y), g(x, y) \in \mathbb{R}[x, y]$, the set of polynomials with real coefficients in the variables x and y. Furthermore, $|p(z)|^2 = f(x, y)^2 + g(x, y)^2$.

When we proved Corollary 4.9, we informally discussed the fact that if $f(x) \in \mathbb{R}[x]$ has degree at least 1, then $|f(x)| \to +\infty$ as $|x| \to +\infty$. One of the tools we will need to prove the Fundamental Theorem of Algebra is the analogous result for polynomials with coefficients in $\mathbb{C}$. When we state this fact in the next lemma, it will look rather technical, but you should convince yourself that it is a formal way of saying that $|p(z)| \to +\infty$ as $|z| \to +\infty$.

Lemma 6.10. *Let $p(z)$ be a polynomial with degree at least 1 with coefficients in $\mathbb{C}$. For every real number $M > 0$, there exists a real number $R > 0$ such that $|p(z)| > M$, whenever $|z| > R$.*

Intuition. Once again, let us consider the polynomial $p(z) = 3z^2 + 5iz + (6 - 2i)$. Then

$$p(z) = z^2 \left(3 + \frac{5i}{z} + \frac{6 - 2i}{z^2} \right).$$

When $|z|$ is large $\frac{5i}{z}$ and $\frac{6-2i}{z^2}$ are close to 0. Therefore, the term $3 + \frac{5i}{z} + \frac{6-2i}{z^2}$ can be made as close to 3 as we wish by letting $|z|$ be large.

As a result, when $|z|$ is large, $p(z)$ is approximately the same as the polynomial $3z^2$. Since $|3z^2| \to +\infty$ as $|z| \to +\infty$, it follows that $|p(z)|$ also goes to $+\infty$ as $|z| \to +\infty$.

Proof. We can write

$$p(z) = \alpha_n z^n + \alpha_{n-1} z^{n-1} + \cdots + \alpha_1 z + \alpha_0,$$

where the $\alpha_i \in \mathbb{C}$, $n \in \mathbb{N}$, and $\alpha_n \neq 0$. Therefore,

$$p(z) = \alpha_n z^n \left(1 + \frac{\frac{\alpha_{n-1}}{\alpha_n}}{z} + \cdots + \frac{\frac{\alpha_1}{\alpha_n}}{z^{n-1}} + \frac{\frac{\alpha_0}{\alpha_n}}{z^n} \right).$$

Next, let T be the maximum value of $\left|\frac{\alpha_j}{\alpha_n}\right|$, where j ranges from 0 to $n-1$. If $z \in \mathbb{C}$ such that $|z| > 2nT$ and $|z| > 1$, we have

(6)
$$\left|\frac{\frac{\alpha_j}{\alpha_n}}{z^{n-j}}\right| = \left|\frac{\alpha_j}{\alpha_n}\right| \cdot \frac{1}{|z|} \cdot \frac{1}{|z|^{n-j-1}} \le \left|\frac{\alpha_j}{\alpha_n}\right| \cdot \frac{1}{|z|} < T \cdot \frac{1}{2nT} = \frac{1}{2n}.$$

Furthermore, if we let

$$\beta = \frac{\frac{\alpha_{n-1}}{\alpha_n}}{z} + \cdots + \frac{\frac{\alpha_1}{\alpha_n}}{z^{n-1}} + \frac{\frac{\alpha_0}{\alpha_n}}{z^n}$$

then

$$p(z) = \alpha_n z^n (1 + \beta)$$

and Lemma 6.3(d) and (6) tell us that

$$|\beta| = \left|\frac{\frac{\alpha_{n-1}}{\alpha_n}}{z} + \cdots + \frac{\frac{\alpha_1}{\alpha_n}}{z^{n-1}} + \frac{\frac{\alpha_0}{\alpha_n}}{z^n}\right| \le \left|\frac{\frac{\alpha_{n-1}}{\alpha_n}}{z}\right| + \cdots + \left|\frac{\frac{\alpha_1}{\alpha_n}}{z^{n-1}}\right| + \left|\frac{\frac{\alpha_0}{\alpha_n}}{z^n}\right| <$$

$$\frac{1}{2n} + \cdots + \frac{1}{2n} + \frac{1}{2n} = n\left(\frac{1}{2n}\right) = \frac{1}{2}.$$

Since $|\beta| < \frac{1}{2}$, when we write $\beta = a + bi$, we have

$$|a| = \sqrt{a^2} \le \sqrt{a^2 + b^2} = |\beta| < \frac{1}{2}.$$

Therefore, a is a real number that lies between $\frac{1}{2}$ and $-\frac{1}{2}$. As a result, $1 + a > \frac{1}{2}$ and it follows that

$$|1 + \beta| = |1 + (a + bi)| = |(1 + a) + bi| = \sqrt{(1+a)^2 + b^2} \ge$$

$$\sqrt{(1+a)^2} = |1 + a| > \frac{1}{2}.$$

Remembering that $|z| > 2nT$ and $|z| > 1$, we now have

(7)
$$|p(z)| = |\alpha_n z^n (1 + \beta)| = |\alpha_n||z|^n|1 + \beta| \ge |\alpha_n||z||1 + \beta| > \frac{|\alpha_n|}{2}|z|.$$

Finally, suppose we are given a real number $M > 0$. Then let R be a real number such that $R > 2nT$, $R > 1$, and $R > \frac{2M}{|\alpha_n|}$.

It now follows from (7), that if $|z| > R$, we have

$$|p(z)| > \frac{|\alpha_n|}{2}|z| > \frac{|\alpha_n|}{2} \cdot \frac{2M}{|\alpha_n|} = M.$$

Thus, $|p(z)| > M$ whenever $|z| > R$, as desired. $\qquad\square$

In Chapters 1 and 4, we remarked that the real numbers are not a purely algebraic object. Since the complex numbers are constructed directly from the real numbers, they too are not a purely algebraic object. Therefore, any proof of the Fundamental Theorem of Algebra must use some results related to calculus. The proof that is the most algebraic requires not only the Intermediate Value Theorem but also more group theory and Galois theory than we will need to prove the insolvability of the quintic. On the other hand, we will soon present a relatively short and easy proof that applies some results on continuous functions of one and two real variables. Since the continuity of function of two variables is a somewhat subtle concept, and since we only need to apply these results to polynomials, we will only state them for polynomials.

Proposition 6.11—The Extreme Value Theorem

(a) *Let $f(x)$ be a polynomial with real coefficients. Then $f(x)$ has a minimum and maximum value on the interval $a \leq x \leq b$.*

(b) *Let $g(x, y)$ be a polynomial in two variables with real coefficients. Then $g(x, y)$ has a minimum and maximum value on the interval $x^2 + y^2 \leq R$.*

We will not be providing a proof of Proposition 6.11, as it takes us rather far afield. However, hopefully you are familiar and comfortable with the Extreme Value Theorem from your courses in calculus. Proving the Fundamental Theorem of Algebra means that it will be available for us to use at any point throughout the remainder of this book. Indeed, the Fundamental Theorem of Algebra will allow us to reduce very abstract situations to the more concrete world of complex numbers, and this will greatly simplify the development of the Galois theory needed to prove the insolvability of the quintic.

Theorem 6.12—The Fundamental Theorem of Algebra. *If $p(z)$ is a polynomial of degree at least 1 with coefficients in $\mathbb{C}$, then $p(z)$ has a root in $\mathbb{C}$.*

Intuition. We will proceed with a proof by contradiction, and, as we mentioned in Chapter 2, such proofs are often lacking in intuition. Fortunately, for this particular proof by contradiction, it is still fairly easy to present an overview of the main ideas.

Using Lemmas 6.9 and 6.10 along with the Extreme Value Theorem, we first show that there exists some $z_0 \in \mathbb{C}$ that minimizes $|p(z)|$. If $p(z)$ has no root in $\mathbb{C}$, then $|p(z_0)|$ is some positive real number M. We then use $p(z)$ to construct a polynomial $q(z)$ such that the minimum value of $|q(z)|$ is also M, but it occurs at 0. Next, $q(z)$ is used to construct a polynomial $r(z)$ with the properties that the minimum value of $|r(z)|$ is 1, this minimum occurs at 0, and the constant term of $r(z)$ is 1.

Finally, $r(z)$ is used to construct a polynomial $H(z)$ that satisfies the same three conditions as $r(z)$ but also has the property that

$$H(z) = 1 - z^k + z^{k+1} s(z),$$

where $s(z)$ is yet another polynomial with coefficients in $\mathbb{C}$. However, we can then use the Extreme Value Theorem to show that there exists a real number a such that $|H(a)| < 1$. Since the minimum value of $|H(z)|$ is 1, this contradiction shows that the polynomial $p(z)$ must have had a root in $\mathbb{C}$.

Proof. By way of contradiction, suppose $p(z)$ has no root in $\mathbb{C}$. We will use $p(z)$ to construct a polynomial $H(z)$ that has two incompatible properties.

Let $N = |p(0)|$; by Lemma 6.10 there is a positive real number R such that $|p(z)| > N$, whenever $|z| > R$. Letting $z = x + yi$, Lemma 6.9 says that $|p(z)|^2$ is a polynomial in x and y with coefficients in $\mathbb{R}$. Therefore, the Extreme Value Theorem asserts that $|p(z)|^2$ has a minimum value on the interval $x^2 + y^2 \le R$. Since $|z|^2 = x^2 + y^2$, this is equivalent to saying that $|p(z)|^2$ has a minimum value on the interval $|z| \le R$. If we let z_0 be the point in the interval $|z| \le R$ that minimizes $|p(z)|^2$, then it is easy to see that z_0 is also the point in the interval $|z| \le R$ that minimizes $|p(z)|$. However, when $|z| > R$, we know that

$$|p(z)| > N = |p(0)| \ge |p(z_0)|.$$

Thus, $|p(z_0)|$ is the minimum value of $|p(z)|$ throughout all of $\mathbb{C}$, and we let $M = |p(z_0)|$. Observe that since $p(z_0) \ne 0$, we know that $M \ne 0$.

Next, let $q(z) = p(z + z_0)$. Certainly $q(z)$ is also a polynomial with coefficients in $\mathbb{C}$ that has the same degree as $p(z)$. Furthermore, every value of $q(z)$ is also a value of $p(z)$, therefore the minimum value of $|q(z)|$ is also M but it now occurs at 0 as

$$|q(0)| = |p(z_0)| = M.$$

If we let α_0 be the constant term of $q(z)$, then

$$\alpha_0 = q(0) \ne 0.$$

Therefore, we can now consider the polynomial $r(z) = \frac{q(z)}{\alpha_0}$. We can see that $r(z)$ is yet another polynomial with coefficients in $\mathbb{C}$ that has the same degree as $p(z)$. Furthermore, for any $z \in \mathbb{C}$, $|r(z)| = \frac{|q(z)|}{|\alpha_0|}$. Therefore, the minimum value of $|r(z)|$ also occurs at 0 and this minimum value is

$$|r(0)| = \frac{|q(0)|}{|\alpha_0|} = \frac{|\alpha_0|}{|\alpha_0|} = 1.$$

In addition, note that 1 is also the constant term of $r(z)$.

Since $r(z)$ has degree at least 1, we can let k be the smallest positive integer such that the coefficient of z^k in $r(z)$ is nonzero. Therefore, we can write

$$r(z) = 1 + \beta z^k + \text{terms of higher degree}.$$

Since $\beta \neq 0$, we know that $\beta^{-1} \in \mathbb{C}$, and Theorem 6.8 tells us that there is some $\gamma \in \mathbb{C}$ such that $\gamma^k = -\beta^{-1}$. We can now let $H(z) = r(\gamma z)$ and observe that

$$H(z) = r(\gamma z) = 1 + \beta(\gamma z)^k + \text{terms of higher degree} =$$

$$1 + (\beta)(-\beta^{-1})z^k + \text{terms of higher degree} = 1 - z^k + \text{terms of higher degree}.$$

Therefore, we can write

$$H(z) = 1 - z^k + z^{k+1}s(z),$$

where $s(z)$ is a polynomial with coefficients in $\mathbb{C}$. Every value of $H(z)$ is also a value of $r(z)$ and since

$$H(0) = r(\gamma \cdot 0) = r(0) = 1,$$

it follows that the minimum value of $|H(z)|$ is also 1.

Let us now turn to the polynomial $s(z)$. By Lemma 6.10, $|s(z)|^2$ is a polynomial with real coefficients in two real variables. Therefore the Extreme Value Theorem tells us that $|s(z)|^2$ has a maximum value on the interval $x^2 + y^2 \leq 1$. This immediately implies that $|s(z)|$ also has a maximum value on the interval $x^2 + y^2 \leq 1$, and we let T denote the maximum value of $|s(z)|$ on the interval $x^2 + y^2 \leq 1$. Finally, let a be a real number in the open interval $(0, 1)$ such that $aT < 1$. In this case, we can see that $1 - a^k$, $1 - aT$, and a^k are all positive real numbers. In particular, $a^k(1 - aT) > 0$, and this implies that $1 - a^k(1 - aT) < 1$.

At first, it may not be clear why the previous inequalities are useful. However, if we combine these inequalities with the formula $H(z) = 1 - z^k + z^{k+1}s(z)$ and the triangle inequality, we see that

$$|H(a)| = |1 - a^k + a^{k+1}s(a)| \leq |1 - a^k| + |a^{k+1}s(a)| = 1 - a^k + |a|^{k+1}|s(a)| \leq$$
$$1 - a^k + a^{k+1}T = 1 - a^k(1 - aT) < 1.$$

As a result, despite the fact that the minimum value of $|H(z)|$ is 1, we have shown that $|H(a)| < 1$. This is a contradiction, so the polynomial $p(z)$ must have a root in $\mathbb{C}$, thereby proving the theorem. $\qquad\square$

The Fundamental Theorem of Algebra is a beautiful and important theorem. However, we should keep in mind what the Fundamental Theorem of Algebra says and what it doesn't say. It says that any polynomial of degree at least 1 with coefficients in $\mathbb{C}$ must have a root in $\mathbb{C}$. But it is important to note that although it tells us of the existence of a root, it does not provide us with an algorithm or formula for finding a root. This is quite different from the situation with the Intermediate Value Theorem in Chapter 4. Recall that the Intermediate Value Theorem could not only be used to prove the existence of a root of certain polynomials but could also be used to compute the root to any desired degree of accuracy. In no way does this detract from the importance or the beauty of the Fundamental Theorem of Algebra. However, remember that it is always important to understand exactly what a theorem says and what it does not say.

Much of this chapter dealt with viewing the complex numbers as equivalence classes of arrows in the plane. Hopefully this approach made the complex numbers more "real" and less confusing to you. In abstract algebra, we try to understand abstract objects. In trying to better understand these objects, it is quite common and very natural to try to view these objects as concretely as possible. In fact, a very active field of algebraic research today is called **representation theory**, and it deals with trying to view and understand algebraic objects in terms of their relationships to more concrete objects.

Toward the beginning of Chapter 5, we remarked that it is much simpler to construct the complex numbers from the real numbers than it is to construct the real numbers from the rational numbers. At this point, we have enough experience with these two constructions to reflect back on them. Recall that given the rational numbers, we can construct the real numbers as equivalence classes of Cauchy sequences of rational numbers. The addition or multiplication of two real numbers therefore involves adding or multiplying all the terms from two Cauchy sequences. It takes an understanding of convergence simply to justify that addition and multiplication of equivalence classes of Cauchy sequences are well defined.

On the other hand, every complex number can easily be written in terms of two real numbers using either rectangular or polar form. Addition and multiplication in $\mathbb{C}$ are easily seen to be well defined. Furthermore, although checking that $\mathbb{C}$ satisfies the nine properties of a field can occasionally be somewhat tedious, the work involved is fairly straightforward, and it is easy to see that $\mathbb{C}$ inherits these nine properties directly from $\mathbb{R}$. Therefore, advancing from an understanding of the basic algebraic properties of $\mathbb{R}$ to a similar understanding of $\mathbb{C}$ is a step forward, but it is not a difficult or intimidating leap forward.

In this chapter, we chose to represent complex numbers as equivalence classes of arrows in the plane. However, there are other ways to represent $\mathbb{C}$ in terms of $\mathbb{R}$. We will now briefly mention two of these other approaches of representing $\mathbb{C}$. Both of these approaches are more algebraic and less geometric than using arrows in the plane, and both use ideas and techniques that we have not yet fully discussed. For these reasons, it was preferable to represent $\mathbb{C}$ using arrows in the plane.

In the first of these additional approaches, we let $\mathbb{R}[x]$ denote the set of polynomials with coefficients in $\mathbb{R}$. We then define $\sim$ on the set $\mathbb{R}[x]$ as $f(x) \sim g(x)$ precisely if $f(x) - g(x)$ is a multiple of $x^2 + 1$. It is easy to check that $\sim$ is indeed an equivalence relation. Observe that the equivalence class $[x^2]$ is the same as the equivalence class $[-1]$, as $x^2 - (-1)$ is certainly divisible by $x^2 + 1$. It turns out that we can associate every $a + bi \in \mathbb{C}$ with the equivalence class $[a + bx]$. In this representation, the equivalence class $[x]$ has the property that its square is $[-1]$.

For those of you familiar with matrix multiplication, the second additional approach deals with 2×2 matrices as we can associate every $a + bi \in \mathbb{C}$ with the matrix $\begin{pmatrix} a & -b \\ b & a \end{pmatrix}$. In this representation, the matrix $\begin{pmatrix} -1 & 0 \\ 0 & -1 \end{pmatrix}$ is the additive inverse of the multiplicative identity and the matrices $\begin{pmatrix} 0 & -1 \\ 1 & 0 \end{pmatrix}$ and $\begin{pmatrix} 0 & 1 \\ -1 & 0 \end{pmatrix}$ both have the property that their square is $\begin{pmatrix} -1 & 0 \\ 0 & -1 \end{pmatrix}$.

Exercises for Sections 6.3 and 6.4

For exercises 1–8, let $\alpha = \text{cis}\left(\frac{\pi}{7}\right)$, $\beta = 9\text{cis}\left(\frac{\pi}{10}\right)$, $\gamma = 7\text{cis}\left(\frac{\pi}{5}\right)$, $\delta = 10^{\frac{1}{4}}\text{cis}\left(\frac{7\pi}{5}\right)$. Unless your answer simplifies into the form a or the form bi, where $a, b \in \mathbb{R}$, you should leave your answer in polar form.

1. Compute each of the following: α^2, α^7, α^{14}.

2. Compute each of the following: α^{-1}, α^*, α^{-21}.

3. Compute each of the following: $\alpha\beta$, $(\alpha\beta)^2$, $(\alpha\beta)^{70}$.

4. Compute each of the following: $(\alpha\beta)^{-1}$, $(\alpha\beta)^*$, $((\alpha\beta)^*)^{-1}$.

5. Compute each of the following: γ^2, γ^5, γ^{10}.

6. Compute each of the following: $\beta\gamma$, $(\beta\gamma)^{-1}$, $(\beta\gamma)^*$.

7. Compute each of the following: δ^2, δ^5, δ^{-1}.

8. Compute each of the following: $\gamma\delta$, $(\gamma\delta)^2$, $\gamma^3\delta$.

In exercises 9–16, you should convert the complex number into polar form before doing any other computations. Express your answers in both polar and rectangular form.

9. $(1+i)^{20}$

10. $\left(\frac{1}{\sqrt{2}} - \frac{1}{\sqrt{2}}i\right)^{100}$

11. $(3-3i)^{-10}$

12. $\left(\frac{1}{2} - \frac{\sqrt{3}}{2}i\right)^{60}$

13. $(2\sqrt{3}+2i)^{12}$

14. $(-1-\sqrt{3}i)^{300}$

15. $(-5+5\sqrt{3}i)^9$

16. $\left(-\frac{\sqrt{3}}{7} - \frac{1}{7}i\right)^{10}$

In exercises 17–26, leave your answers in polar and rectangular form.

17. (a) Find the 4 fourth roots of -1.

    (b) Find the 4 fourth roots of -81.

18. (a) Find the 4 fourth roots of -25.

    (b) Find the 4 fourth roots of -7.

19. (a) Find the 6 sixth roots of -1.

    (b) Find the 6 sixth roots of -64.

20. (a) Find the 6 sixth roots of -8.

    (b) Find the 6 sixth roots of -99.

21. (a) Find the 12 twelfth roots of 1.

    (b) Find the 12 twelfth roots of 5^{12}.

22. (a) Find the 12 twelfth roots of 64.

    (b) Find the 12 twelfth roots of 75.

23. (a) Find the two square roots of $-\frac{1}{2} - \frac{\sqrt{3}}{2}i$.

    (b) Find the two square roots of $-18 - 18\sqrt{3}i$.

24. (a) Find the two square roots of $-5 - 5\sqrt{3}i$.

    (b) Find the two square roots of $-10\sqrt{3} - 30i$.

25. (a) Find the three cube roots of i.

    (b) Find the three cube roots of $1000i$.

26. (a) Find the three cube roots of $125i$.

    (b) Find the three cube roots of $10i$.

In exercises 27–34, leave your answers in polar form.

27. (a) Find the 5 fifth roots of $-i$.

    (b) Find the 5 fifth roots of $-44i$.

28. (a) Find the 5 fifth roots of $-243i$.

    (b) Find the 5 fifth roots of $-70i$.

29. (a) Find the 3 cube roots of $\frac{1}{\sqrt{2}} + \frac{1}{\sqrt{2}}i$.

    (b) Find the 3 cube roots of $500 + 500i$.

30. (a) Find the 3 cube roots of $53 + 53i$.

    (b) Find the 3 cube roots of $\frac{e}{4} + \frac{e}{4}i$.

31. (a) Find the 7 seventh roots of $\text{cis}\left(\frac{3\pi}{11}\right)$.

    (b) Find the 7 seventh roots of $20\text{cis}\left(\frac{3\pi}{11}\right)$.

32. (a) Find the 7 seventh roots of $6^7\text{cis}\left(\frac{3\pi}{11}\right)$.

    (b) Find the 7 seventh roots of $2^{\frac{1}{3}}\text{cis}\left(\frac{3\pi}{11}\right)$.

33. (a) Find the 5 fifth roots of $\text{cis}\left(\frac{2\pi}{9}\right)$.

    (b) Find the 5 fifth roots of $11\text{cis}\left(\frac{2\pi}{9}\right)$.

34. (a) Find the 5 fifth roots of $32\text{cis}\left(\frac{2\pi}{9}\right)$.

    (b) Find the 5 fifth roots of $100\text{cis}\left(\frac{2\pi}{9}\right)$.

35. (a) Find $\alpha_1, \alpha_2, \alpha_3, \alpha_4 \in \mathbb{C}$ such that $x^4 + 1 = (x - \alpha_1)(x - \alpha_2)(x - \alpha_3)(x - \alpha_4)$.

    (b) Use part (a) and Proposition 5.11 to write $x^4 + 1$ as a product of two quadratic polynomials with real coefficients.

36. (a) Find $\alpha_1, \alpha_2, \alpha_3, \alpha_4 \in \mathbb{C}$ such that $x^4 + 9 = (x - \alpha_1)(x - \alpha_2)(x - \alpha_3)(x - \alpha_4)$.

    (b) Use part (a) and Proposition 5.11 to write $x^4 + 9$ as a product of two quadratic polynomials with real coefficients.

37. Use your answer to part (b) of exercise 35 to write $x^8 - 1$ as a product of polynomials with real coefficients that are either quadratic with no real roots or are linear.

38. Use your answer to part (b) of exercise 36 to write $x^8 - 81$ as a product of polynomials with real coefficients that are quadratic with no real roots or are linear.

39. Find the 6 roots in $\mathbb{C}$ of $x^6 - 28x^3 + 27$.

40. Find the 6 roots in $\mathbb{C}$ of $x^6 - 133x^3 + 1000$.

41. Find the 8 roots in $\mathbb{C}$ of $x^8 - 13x^4 + 36$.

42. Find the 8 roots in $\mathbb{C}$ of $x^8 - 41x^4 + 400$.

43. Find the 10 roots in $\mathbb{C}$ of $x^{10} - 33x^5 + 32$.

44. Find the 10 roots in $\mathbb{C}$ of $x^{10} - 5x^5 + 6$.

For exercises 45–51, please first read the following:

DeMoivre's Theorem tells us that $(\cos(n\theta) + i\sin(n\theta)) = (\cos(\theta) + i\sin(\theta))^n$, for all $n \in \mathbb{N}$. On the other hand, we can also compute $(\cos(\theta) + i\sin(\theta))^n$ a second way by expanding it out using either the distributive law or the binomial theorem. If we compare the real and complex parts of our two expressions for $(\cos(\theta) + i\sin(\theta))^n$, we obtain formulas for $\cos(n\theta)$ and $\sin(n\theta)$ in terms of $\cos(\theta)$ and $\sin(\theta)$. In the next seven exercises, we derive and examine some of these formulas. To do this, we will often need to use the identity $\sin^2(\theta) + \cos^2(\theta) = 1$.

45. Expand the expression $(\cos(\theta) + i\sin(\theta))^3$ to find polynomials $f(x), g(x)$ such that $\cos(3\theta) = f(\cos(\theta))$ and $\sin(3\theta) = g(\sin(\theta))$.

46. Expand the expression $(\cos(\theta) + i\sin(\theta))^4$ to find polynomials $f(x), h(x)$ such that $\cos(4\theta) = f(\cos(\theta))$ and $\sin(4\theta) = \cos(\theta) \cdot h(\sin(\theta))$.

47. Expand the expression $(\cos(\theta) + i\sin(\theta))^5$ to find polynomials $f(x), g(x)$ such that $\cos(5\theta) = f(\cos(\theta))$ and $\sin(5\theta) = g(\sin(\theta))$.

48. Show that if $n \in \mathbb{N}$, then there exists a polynomial $f(x)$ such that $\cos(n\theta) = f(\cos(\theta))$.

49. Show that if $n \in \mathbb{N}$ is odd, then there exists a polynomial $g(x)$ such that $\sin(n\theta) = g(\sin(\theta))$. In addition, show that if n is even then there exists a polynomial $h(x)$ such that $\sin(n\theta) = \cos(\theta) \cdot h(\sin(\theta))$.

In light of exercises 48 and 49, we know that if n is an odd integer then there exist polynomials $f(x), g(x)$ such that $\cos(n\theta) = f(\cos(\theta))$ and $\sin(n\theta) = g(\sin(\theta))$. In the next two exercises, we examine the relationship between $f(x)$ and $g(x)$. We will need to apply the fact that $\sin(\theta) = \cos\left(\frac{\pi}{2} - \theta\right)$.

50. Suppose $n \in \mathbb{N}$ such that $n = 4k + 1$, for some $k \in \mathbb{Z}$ and let $f(x), g(x)$ be the polynomials in the formulas $\cos(n\theta) = f(\cos(\theta))$ and $\sin(n\theta) = g(\sin(\theta))$. Prove that $f(x) = g(x)$. (Hint: Replace θ by $\frac{\pi}{2} - \theta$ in the formula $\cos(n\theta) = f(\cos(\theta))$.)

51. Suppose $n \in \mathbb{N}$ such that $n = 4k + 3$, for some $k \in \mathbb{Z}$ and let $f(x), g(x)$ be the polynomials in the formulas $\cos(n\theta) = f(\cos(\theta))$ and $\sin(n\theta) = g(\sin(\theta))$. Prove that $f(x) = -g(x)$. (Hint: Replace θ by $\frac{\pi}{2} - \theta$ in the formula $\cos(n\theta) = f(\cos(\theta))$.)

In exercises 52–57, replace z by $x + yi$ in the polynomial $p(z)$ and then find polynomials $f(x, y), g(x, y)$ with real coefficients such that $p(z) = f(x, y) + g(x, y)i$.

52. $p(z) = (2 - 3i)z + (4 + 11i)$

53. $p(z) = (4 - i)z + (6 + 2i)$

54. $p(z) = 7z^2 + (1+i)z - 3$

55. $p(z) = (5+3i)z^2 + 8z + (-11+52i)$

56. $p(z) = z^3 + z^2 + z + 1$

57. $p(z) = (5-6i)z^3 - 4z^2 + 7iz + (31-8i)$

58. Let $\mathbb{R}[x]$ denote the set of polynomials with coefficients in $\mathbb{R}$. Then define the relation $\sim$ on the set $\mathbb{R}[x]$ as $f(x) \sim g(x)$ precisely if $f(x) - g(x)$ is a multiple of $x^2 + 1$.

    (a) Show that $\sim$ is an equivalence relation.

    (b) If we define the addition of equivalence classes as $[f(x)] + [g(x)] = [f(x) + g(x)]$, show that addition is well defined.

    (c) If we define the multiplication of equivalence classes as $[f(x)] \cdot [g(x)] = [f(x) \cdot g(x)]$, show that multiplication is well defined.

    (d) Show that $[0]$ is the additive identity and $[1]$ is the multiplicative identity.

    (e) Show that every equivalence class contains exactly one polynomial of the form $a + bx$, where $a, b \in \mathbb{R}$.

    (f) Show that the equivalence class $[x]$ has the property that its square is equal to $[-1]$.

    (g) If either a or b is nonzero, show that $\left[\frac{a}{a^2+b^2} - \frac{b}{a^2+b^2}x \right]$ is the multiplicative inverse of $[a+bx]$.

    (h) How similar does the set of equivalence classes seem to be to the set of complex numbers? Which equivalence class (or equivalence classes) seems to correspond to the complex number i?

CHAPTER 7

The Integers Modulo n

Before beginning this course, the number systems you were probably the most comfortable and familiar with were $\mathbb{Z}$ and $\mathbb{Q}$. Similarly, the sets of polynomials you were probably the most comfortable and familiar with were $\mathbb{Z}[x]$ and $\mathbb{Q}[x]$. Therefore, in Chapter 9, we will examine $\mathbb{Z}[x]$ and $\mathbb{Q}[x]$. This will lay the foundation for a more general and detailed study in Chapter 12 of polynomials with coefficients in any field. But first, we use our experience with $\mathbb{Z}$ and equivalence classes to introduce a new collection of commutative rings and fields known as the integers modulo n and denoted as $\mathbb{Z}_n$.

There are many reasons for introducing commutative rings of the form $\mathbb{Z}_n$ at this point in time:

1. They provide us with many concrete examples of fields, commutative rings, and groups.

2. They are useful tools for proving results about the existence of roots and the factoring of polynomials in $\mathbb{Z}[x]$ and $\mathbb{Q}[x]$.

3. An understanding of the construction of $\mathbb{Z}_n$ from $\mathbb{Z}$ will be very useful in understanding the solution, in Chapter 17, of an important problem that deals with the existence of roots of polynomials with coefficients in more general fields.

7.1 Definitions and Basic Properties

When we define $\mathbb{Z}_n$, we will not be defining one commutative ring but will actually be defining an infinite collection of rings. In particular, the commutative rings $\mathbb{Z}_2, \mathbb{Z}_3, \mathbb{Z}_4, \ldots, \mathbb{Z}_{100}, \ldots$ will all be different. To define $\mathbb{Z}_n$, we revisit an equivalence relation that appeared in Chapter 4. Once again, we will not be looking at a single equivalence relation but actually an infinite number of equivalence relations. If $n > 1$ is an integer, we let $\sim_n$ denote the equivalence relation where, for all $a, b \in \mathbb{Z}$, we say that $a \sim_n b$ precisely when n divides $a - b$. Note that for every n, we obtain a different equivalence relation.

■ Examples

> $1 \sim_4 5$, $1 \not\sim_6 5$, $2 \sim_4 -14$, $2 \sim_8 -14$, $2 \not\sim_5 -14$, $2 \not\sim_3 -14$, $-5 \sim_{11} 17$, $-5 \not\sim_7 17$,
> $-5 \sim_2 17$, $-5 \not\sim_4 17$.

■

Definition 7.1. *If $n > 1$ is an integer, we let $\mathbb{Z}_n$ denote the equivalence classes of $\mathbb{Z}$ corresponding to the equivalence relation $\sim_n$. For each $a \in \mathbb{Z}$, we let $[a]_n$ denote the equivalence class containing a. $\mathbb{Z}_n$ consists of n equivalence classes and, for convenience, we often refer to them using the names $[0]_n, [1]_n, [2]_n, \ldots, [n-2]_n, [n-1]_n$.*

Definition 7.1 mentions the fact that $\mathbb{Z}_n$ consists of exactly n equivalence classes. To see this, observe that if $a, b \in \mathbb{Z}$, then $[a]_n = [b]_n$ precisely if n divides $a - b$. However, if $0 \leq a, b \leq n - 1$, then the only way n can divide $a - b$ is for a and b to be equal. In light of this, the equivalence classes $[0]_n, [1]_n, [2]_n, \ldots, [n-2]_n, [n-1]_n$ must all be different. Therefore, $\mathbb{Z}_n$ consists of at least n different equivalence classes. On the other hand, the division algorithm asserts that if $a \in \mathbb{Z}$, then there exist $q, r \in \mathbb{Z}$ such that

$$a = q \cdot n + r \quad \text{and} \quad 0 \leq r \leq n - 1.$$

This tells us that n divides $a - r$, so $[a]_n = [r]_n$. But since $0 \leq r \leq n - 1$, this means that $[a]_n$ is equal to one of $[0]_n, [1]_n, [2]_n, \ldots, [n-2]_n, [n-1]_n$. As a result, $\mathbb{Z}_n$ does indeed consist of exactly n equivalence classes.

Even though $\mathbb{Z}_n$ consists of exactly n equivalence classes, it is important to be aware that each class has an infinite number of names. Traditionally, the equivalence classes obtained from $\sim_n$ are called **congruence classes**, and if a, b belong to the same class, we say that a and b are **congruent modulo** n, and we write this as

$$a \equiv b \pmod{n}.$$

■ Examples

> $\mathbb{Z}_4$ consists of four equivalence classes and, for convenience, we usually denote them as $[0]_4, [1]_4, [2]_4, [3]_4$. However, each class has an infinite number of elements and any element in a class can be used as the name of the class. For example,
>
> $$[0]_4 = [4]_4 = [-4]_4 = [8]_4 = [-8]_4 = [12]_4 = [-12]_4 = [16]_4 = [-16]_4 = \cdots$$
>
> $$[1]_4 = [5]_4 = [-3]_4 = [9]_4 = [-7]_4 = [13]_4 = [-11]_4 = [17]_4 = [-15]_4 = \cdots$$
>
> $$[2]_4 = [6]_4 = [-2]_4 = [10]_4 = [-6]_4 = [14]_4 = [-10]_4 = [18]_4 = [-14]_4 = \cdots$$
>
> $$[3]_4 = [7]_4 = [-1]_4 = [11]_4 = [-5]_4 = [15]_4 = [-9]_4 = [19]_4 = [-13]_4 = \cdots.$$

Similarly, $\mathbb{Z}_5$ consists of five equivalence classes, and, for convenience, we usually denote them as $[0]_5, [1]_5, [2]_5, [3]_5, [4]_5$. Each class has an infinite number of elements, and therefore each class has an infinite number of names. For example,

$$[0]_5 = [5]_5 = [-5]_5 = [10]_5 = [-10]_5 = [15]_5 = [-15]_5 = [20]_5 = [-20]_5 = \cdots$$

$$[1]_5 = [6]_5 = [-4]_5 = [11]_5 = [-9]_5 = [16]_5 = [-14]_5 = [21]_5 = [-19]_5 = \cdots$$

$$[2]_5 = [7]_5 = [-3]_5 = [12]_5 = [-8]_5 = [17]_5 = [-13]_5 = [22]_5 = [-18]_5 = \cdots$$

$$[3]_5 = [8]_5 = [-2]_5 = [13]_5 = [-7]_5 = [18]_5 = [-12]_5 = [23]_5 = [-17]_5 = \cdots$$

$$[4]_5 = [9]_5 = [-1]_5 = [14]_5 = [-6]_5 = [19]_5 = [-11]_5 = [24]_5 = [-16]_5 \cdots .$$

■ Examples

Using the $a \equiv b \pmod{n}$ notation, we have $17 \equiv -13 \pmod 6$, $17 \equiv -13 \pmod{15}$, $17 \not\equiv -13 \pmod{20}$, $17 \not\equiv -13 \pmod 8$.

At this point, for each $n > 1$, we know that $\mathbb{Z}_n$ is a set consisting of n elements. However, in order to say that $\mathbb{Z}_n$ is a commutative ring, we need to define addition and multiplication and then check that they satisfy the eight properties required of commutative rings. In mathematics, the simplest way to do something is often the right way to do it. That is the case when it comes to defining addition and multiplication in $\mathbb{Z}_n$. Given congruence classes $[a]_n$ and $[b]_n$, the simplest way to define addition and multiplication would be to add and multiply the names of the classes. As it turns out, that is exactly what we do.

Theorem 7.2. *For every integer $n > 1$, we define addition and multiplication in $\mathbb{Z}_n$ as*

$$[a]_n + [b]_n = [a+b]_n$$

and

$$[a]_n \cdot [b]_n = [a \cdot b]_n,$$

for all $a, b \in \mathbb{Z}$. With this addition and multiplication, $\mathbb{Z}_n$ is a commutative ring.

■ Examples

$$[2]_5 + [4]_5 = [2+4]_5 = [6]_5 \quad \text{and} \quad [6]_5 = [1]_5, \quad \text{thus} \quad [2]_5 + [4]_5 = [1]_5,$$

$$[2]_5 \cdot [4]_5 = [2 \cdot 4]_5 = [8]_5 \quad \text{and} \quad [8]_5 = [3]_5, \quad \text{thus} \quad [2]_5 \cdot [4]_5 = [3]_5,$$

$[8]_{10} + [5]_{10} = [8+5]_{10} = [13]_{10}$ and $[13]_{10} = [3]_{10}$, thus $[8]_{10} + [5]_{10} = [3]_{10}$,

$[8]_{10} \cdot [5]_{10} = [8 \cdot 5]_{10} = [40]_{10}$ and $[40]_{10} = [0]_{10}$, thus $[8]_{10} \cdot [5]_{10} = [0]_{10}$,

$$[7]_{41} + [6]_{41} = [7+6]_{41} = [13]_{41},$$

$[7]_{41} \cdot [6]_{41} = [7 \cdot 6]_{41} = [42]_{41}$ and $[42]_{41} = [1]_{41}$, thus $[7]_{41} \cdot [6]_{41} = [1]_{41}$.

■

Shortly, we will prove Theorem 7.2. In that proof we will show that $\mathbb{Z}_n$ has identity elements for both addition and multiplication, and we will also find the additive inverse for each equivalence class. Before reading the proof of Theorem 7.2, you should review the previous examples and then try to anticipate which elements of $\mathbb{Z}_n$ are the identity elements and also what the additive inverse of each class will be.

In several earlier chapters, we dealt with adding and multiplying equivalence classes. In those cases, it was important to show that addition and multiplication were well defined. When dealing with $\mathbb{Z}_n$, each equivalence class has an infinite number of different names. We will need to show that the choice of a name for an equivalence class does not affect the answer in addition and multiplication problems. The good news is that once we have succeeded in proving that addition and multiplication in $\mathbb{Z}_n$ are well defined, it will be very easy to verify that $\mathbb{Z}_n$ is a commutative ring.

■ Examples

In $\mathbb{Z}_5$, $[2]_5 = [17]_5$ and $[3]_5 = [-7]_5$. Observe that

$$[2]_5 + [3]_5 = [5]_5 = [0]_5 \quad \text{and} \quad [17]_5 + [-7]_5 = [10]_5 = [0]_5,$$

$$[2]_5 \cdot [3]_5 = [6]_5 = [1]_5 \quad \text{and} \quad [17]_5 \cdot [-7]_5 = [-119]_5 = [1]_5.$$

As a result,

$$[2]_5 + [3]_5 = [17]_5 + [-7]_5 \quad \text{and} \quad [2]_5 \cdot [3]_5 = [17]_5 \cdot [-7]_5.$$

Thus, in this example, changing the names of the equivalence classes did not change the answer when adding and multiplying.

For another example, in $\mathbb{Z}_8$ we have $[5]_8 = [-11]_8$ and $[7]_8 = [39]_8$. Note that

$$[5]_8 + [7]_8 = [12]_8 = [4]_8 \quad \text{and} \quad [-11]_8 + [39]_8 = [28]_8 = [4]_8,$$

$$[5]_8 \cdot [7]_8 = [35]_8 = [3]_8 \quad \text{and} \quad [-11]_8 \cdot [39]_8 = [-429]_8 = [3]_8.$$

As a result,

$$[5]_8 + [7]_8 = [-11]_8 + [39]_8 \quad \text{and} \quad [5]_8 \cdot [7]_8 = [-11]_8 \cdot [39]_8.$$

Once again, changing the names of the equivalence classes did not change the answer when adding and multiplying. ∎

Proof of Theorem 7.2. As just indicated, the main part of this proof will be showing that addition and multiplication in $\mathbb{Z}_n$ are well defined. To this end, suppose $[a]_n = [b]_n$ and $[c]_n = [d]_n$; we then need to show that $[a]_n + [c]_n = [b]_n + [d]_n$ and $[a]_n \cdot [c]_n = [b]_n \cdot [d]_n$.

By the definition of addition and multiplication in $\mathbb{Z}_n$, we have

$$[a]_n + [c]_n = [a+c]_n, \quad [b]_n + [d]_n = [b+d]_n,$$
$$[a]_n \cdot [c]_n = [a \cdot c]_n \quad \text{and} \quad [b]_n \cdot [d]_n = [b \cdot d]_n.$$

As a result, in order to verify that

$$[a]_n + [c]_n = [b]_n + [d]_n \quad \text{and} \quad [a]_n \cdot [c]_n = [b]_n \cdot [d]_n,$$

we must show that

$$[a+c]_n = [b+d]_n \quad \text{and} \quad [a \cdot c]_n = [b \cdot d]_n.$$

However, using the definition of the relation $\sim_n$, this means that we need to show that n divides $(a+c) - (b+d)$ and n also divides $a \cdot c - b \cdot d$.

Since $[a]_n = [b]_n$ and $[c]_n = [d]_n$, we know that n divides both $a - b$ and $c - d$. Thus,

$$(1) \qquad\qquad a - b = n \cdot m_1 \quad \text{and} \quad c - d = n \cdot m_2,$$

for some $m_1, m_2 \in \mathbb{Z}$. As a result,

$$(a+c) - (b+d) = (a-b) + (c-d) = n \cdot m_1 + n \cdot m_2 = n(m_1 + m_2).$$

Therefore, n divides $(a+c) - (b+d)$ and addition in $\mathbb{Z}_n$ is indeed well defined.

To see that multiplication is well defined, we first refer back to (1) and rewrite it as

$$a = b + n \cdot m_1 \quad \text{and} \quad c = d + n \cdot m_2.$$

Multiplying the preceding equations, we have

$$a \cdot c = (b + n \cdot m_1)(d + n \cdot m_2) = b \cdot d + n(b \cdot m_2 + d \cdot m_1 + n \cdot m_1 \cdot m_2).$$

As a result,

$$a \cdot c - b \cdot d = n(b \cdot m_2 + d \cdot m_1 + n \cdot m_1 \cdot m_2),$$

which immediately tells us that n divides $a \cdot c - b \cdot d$. Thus, multiplication in $\mathbb{Z}_n$ is also well defined.

Having shown that addition and multiplication in $\mathbb{Z}_n$ are well defined, showing that $\mathbb{Z}_n$ is a commutative ring will now be quite routine. Among the properties that we need to show are satisfied in $\mathbb{Z}_n$ are two associative laws, two commutative laws, and both parts of the distributive law. Fortunately, it is not hard to see that all of these properties are inherited by $\mathbb{Z}_n$ directly from $\mathbb{Z}$. For example, given classes $[a]_n$, $[b]_n$, and $[c]_n$, we have

$$([a]_n + [b]_n) + [c]_n = ([a+b]_n) + [c]_n = [(a+b)+c]_n = [a+(b+c)]_n =$$

(2)
$$[a]_n + [b+c]_n = [a]_n + ([b]_n + [c]_n).$$

Observe that in (2), the equality $[(a+b)+c]_n = [a+(b+c)]_n$ follows directly from the associativity of addition in $\mathbb{Z}$. All the other equalities in (2) follow from the definition of addition in $\mathbb{Z}_n$. Thus addition in $\mathbb{Z}_n$ is associative.

Similarly, using one part of the distributive law in $\mathbb{Z}$, we have

$$([a]_n + [b]_n)[c]_n = [a+b]_n \cdot [c]_n = [(a+b)c]_n = [a \cdot c + b \cdot c]_n = [a \cdot c]_n + [b \cdot c]_n =$$

$$[a]_n \cdot [c]_n + [b]_n \cdot [c]_n.$$

Note that the equality $[(a+b)c]_n = [a \cdot c + b \cdot c]_n$ follows from one part of the distributive law in $\mathbb{Z}$ and the other equalities above follow from the definition of addition and multiplication in $\mathbb{Z}_n$. Thus, $\mathbb{Z}_n$ satisfies one part of the distributive law.

It is no harder to show that $\mathbb{Z}_n$ also inherits the associativity of multiplication, both commutative laws, and the other part of the distributive law from $\mathbb{Z}$. In light of this, we will omit the details.

Also, observe that

$$[a]_n + [0]_n = [a+0]_n = [a]_n = [0+a]_n = [0]_n + [a]_n$$

and

$$[a]_n \cdot [1]_n = [a \cdot 1]_n = [a]_n = [1 \cdot a]_n = [1]_n \cdot [a]_n.$$

Thus, $[0]_n$ and $[1]_n$ are, respectively, the additive and multiplicative identity elements of $\mathbb{Z}_n$. Finally,

$$[a]_n + [-a]_n = [a-a]_n = [0]_n = [-a+a]_n = [-a]_n + [a]_n,$$

so $[-a]_n$ is the additive inverse $[a]_n$. As a result, $\mathbb{Z}_n$ does indeed satisfy all the properties of a commutative ring. $\qquad\square$

7.2 Zero Divisors and Invertible Elements

It is interesting to notice how different $\mathbb{Z}_2, \mathbb{Z}_3, \mathbb{Z}_4, \mathbb{Z}_5, \ldots$ are from our earlier examples of commutative rings like $\mathbb{Z}, \mathbb{Q}, \mathbb{R}, \mathbb{C}, \mathbb{Z}[x]$, and $\mathbb{Q}[x]$. The most glaring difference is that each $\mathbb{Z}_n$ is a finite set, whereas all of our previous examples were infinite. In addition, observe that in $\mathbb{Z}_{100}$,

$$[20]_{100}^2 = [20]_{100} \cdot [20]_{100} = [400]_{100} = [0]_{100}.$$

Thus, in $\mathbb{Z}_{100}$, we can multiply a nonzero element by itself and obtain 0. Certainly, nothing like this happened in our earlier examples of commutative rings. To better understand some of the differences between $\mathbb{Z}_n$ and our earlier examples, we need to introduce some terminology.

Many of the terms and concepts we are about to introduce do not require that multiplication be commutative. Recall that if a set satisfies all the properties of a commutative ring with the possible exception that multiplication may not be commutative, we simply refer to it as a ring. Thus, multiplication in a ring may be commutative, but it is not required to be commutative. In particular, **every commutative ring is a ring, but not every ring is a commutative ring**. If you are not familiar with the addition and multiplication of square matrices, then you may have never seen an example of a ring that is not a commutative ring. To remedy this, we present the following.

■ Example

Let R be the ring consisting of all polynomials in the variables x and y with coefficients in $\mathbb{R}$ with the special condition that $yx = 0$. Therefore, in R, multiplication is identical to the multiplication of polynomials in two variable calculus with the one huge difference. Namely, whenever we see a term of the form yx, instead of letting yx equal xy, we replace yx by 0. For example, in R, if we use the distributive law and the fact that $yx = 0$, we have the following:

$$(3 + 2x - 7y + 4x^2y - 8xy^4)(5x + y) =$$
$$15x + 10x^2 - 35yx + 20x^2yx - 40xy^4x + 3y + 2xy - 7y^2 + 4x^2y^2 - 8xy^5 =$$
$$15x + 10x^2 + 3y + 2xy - 7y^2 + 4x^2y^2 - 8xy^5.$$

Also note that in R, the order of elements in a multiplication problem can affect whether or not the product is 0 as $x \cdot y \neq 0$ and $y \cdot x = 0$. Furthermore, note that the element xy has the interesting property that it is nonzero, yet its square is 0 as

$$(xy)^2 = (xy)(xy) = x(yx)y = x \cdot 0 \cdot y = 0.$$

■

Definition 7.3. *Let R be a ring.*

(a) *A nonzero element $r \in R$ is called a zero divisor if there is a nonzero element $s \in R$ such that either $rs = 0$ or $sr = 0$.*

(b) *An element $r \in R$ is said to be invertible if there exists some $s \in R$ such that $rs = 1 = sr$. In this case, we write $s = r^{-1}$.*

■ Examples

In the previous example, the fact that $yx = 0$ tells us that both x and y are zero divisors. Furthermore, since

$$(1 + xy)(1 - xy) = 1 + xy - xy - xyxy = 1$$

and

$$(1 - xy)(1 + xy) = 1 - xy + xy - xyxy = 1,$$

we can see that both $1 + xy$ and $1 - xy$ are invertible in R as they are each the multiplicative inverse of the other.

■

Before this chapter, almost all of our examples of rings did not have zero divisors. However, depending on the value of n, $\mathbb{Z}_n$ may have many zero divisors.

■ Examples

In $\mathbb{Z}_{10}$, we have

$$[2]_{10} \cdot [5]_{10} = [4]_{10} \cdot [5]_{10} = [6]_{10} \cdot [5]_{10} = [8]_{10} \cdot [5]_{10} = [0]_{10}.$$

Thus, in $\mathbb{Z}_{10}$, $[2]_{10}, [4]_{10}, [5]_{10}, [6]_{10}, [8]_{10}$ are all zero divisors. Similarly, in $\mathbb{Z}_8$, we have

$$[2]_8 \cdot [4]_8 = [6] \cdot [4]_8 = [0]_8.$$

Therefore, in $\mathbb{Z}_8$, $[2]_8, [4]_8, [6]_8$ are all zero divisors.

■

Next, we observe that an element of a ring cannot be both invertible and a zero divisor.

Lemma 7.4. *If R is a ring and $r \in R$, then r cannot be both invertible and a zero divisor.*

Proof. Lemma 5.14(c) asserts that multiplying by 0 in a commutative ring always gives 0 as the answer. However, as we remarked at the time, the proof of that fact did not require that multiplication be commutative. Now suppose $r \in R$ such that r has a multiplicative inverse r^{-1}. We need to show that in this case, r cannot also be a zero divisor.

To this end, if $t \in R$ such that $tr = 0$, then Lemma 5.14(c) along with the associativity of multiplication, tells us that

$$0 = 0 \cdot r^{-1} = (tr)r^1 = t(r \cdot r^{-1}) = t \cdot 1 = t.$$

On the other hand, if $rt = 0$, an almost identical argument shows that

$$0 = r^{-1} \cdot 0 = r^{-1}(rt) = (r^{-1} \cdot r)t = 1 \cdot t = t.$$

As a result, if either $tr = 0$ or $rt = 0$, then $t = 0$. Since there does not exist a nonzero element t such that either $tr = 0$ or $rt = 0$, r is not a zero divisor. Hence, no element of R can simultaneously be invertible and a zero divisor. $\qquad\square$

Whereas Lemma 7.4 indicates that an element of a ring cannot simultaneously be invertible and a zero divisor, we should be aware that many elements of a ring are neither invertible nor a zero divisor. For example, in $\mathbb{Z}$, there are no zero divisors and the only invertible elements are 1 and -1. In $\mathbb{Q}[x]$, once again there are no zero divisors, and the only invertible elements are the nonzero polynomials of degree 0. In particular, every polynomial in $\mathbb{Q}[x]$ of degree at least 1 is neither invertible nor a zero divisor. However, as we examine the following examples, we will see that the situation is quite different in $\mathbb{Z}_n$.

■ Examples

In an earlier example, we saw that $[2]_{10}, [4]_{10}, [5]_{10}, [6]_{10}, [8]_{10}$ are all zero divisors in $\mathbb{Z}_{10}$. However, we also have

$$[1]_{10} \cdot [1]_{10} = [3]_{10} \cdot [7]_{10} = [9]_{10} \cdot [9]_{10} = [1]_{10}.$$

Since multiplication in $\mathbb{Z}_{10}$ is commutative, we see that $[1]_{10}, [3]_{10}, [7]_{10}, [9]_{10}$ are all invertible in $\mathbb{Z}_{10}$. Thus every nonzero element in $\mathbb{Z}_{10}$ is either invertible or a zero divisor. Returning to $\mathbb{Z}_8$, earlier we saw that $[2]_8, [4]_8, [6]_8$ are all zero divisors. On the other hand, we have

$$[1]_8 \cdot [1]_8 = [3]_8 \cdot [3]_8 = [5]_8 \cdot [5]_8 = [7]_8 \cdot [7]_8 = [1]_8.$$

Therefore, $[1]_8, [3]_8, [5]_8, [7]_8$ are all invertible in $\mathbb{Z}_8$. As a result, it is also the case in $\mathbb{Z}_8$ that every nonzero element is either invertible or a zero divisor.

Based on the preceding examples, it is reasonable to wonder if every nonzero element of $\mathbb{Z}_n$ must be either invertible or a zero divisor. In fact, not only is this true for $\mathbb{Z}_n$ but it is also true for all rings that are finite. Interestingly enough, it is no harder to prove this fact for all finite rings than it is to prove it solely for rings of the form $\mathbb{Z}_n$.

Theorem 7.5. *If a ring R is a finite set, then every nonzero element of R is either invertible or a zero divisor. Furthermore, if $a \in R$ is invertible, then there is a positive integer m such that $a^m = 1$.*

Proof. Suppose $a \in R$ such that there exists a positive integer m such that $a^m = 1$. Since $m - 1 \geq 0$, it follows that $a^{m-1} \in R$, and we immediately see that

$$a \cdot a^{m-1} = a^m = 1 = a^m = a^{m-1} \cdot a.$$

Thus, a^{m-1} is the multiplicative inverse of a, and so a is indeed invertible.

As a result, in order to prove our result, it suffices to show that if $a \in R$ is not a zero divisor and if R is finite, then there exists a positive integer m such that $a^m = 1$. We begin by considering the sequence

$$a^1, a^2, a^3, a^4, a^5, a^6, a^7, a^8, a^9, \ldots.$$

Since R is a finite set and every element of the sequence belongs to R, there must be elements of R that appear more than once in the sequence. As a result, we can let k be the smallest positive integer such that there exists a larger integer j such that $a^j = a^k$. In particular, a^k is the first term of the sequence that reappears later in the sequence.

Since $a^j = a^k$ and $j > k \geq 1$, we now have

$$(3) \qquad\qquad a(a^{j-1} - a^{k-1}) = a^j - a^k = 0.$$

There are now two possibilities, either $a^{j-1} - a^{k-1} \neq 0$ or $a^{j-1} - a^{k-1} = 0$. If the first possibility occurs, then (3) tells us that we have multiplied a by the nonzero element $a^{j-1} - a^{k-1}$ and obtained 0 as an answer. However, this contradicts our assumption that a is not a zero divisor.

As a result, it must be the case that $a^{j-1} - a^{k-1} = 0$, so

$$a^{j-1} = a^{k-1}.$$

However, recall that k had the property that it was the smallest positive integer such that a^k reappears later in the preceding sequence. If $k - 1$ was a positive integer, then the fact that $a^{j-1} = a^{k-1}$ would tell us that a^{k-1} is a term in the sequence that appears earlier than a^k, yet reappears later in the sequence. This would contradict the fact that a^k is the first element of the sequence that eventually reappears. In light of this, it must be the case that $k-1$ is not a positive integer. Since k is a positive integer and $k-1$ is not a positive integer, it must be the case that $k = 1$. If we now return to (3) and use the fact that $k = 1$, we now have

$$0 = a^j - a^k = a^j - a^1 = a \cdot a^{j-1} - a \cdot 1 = a(a^{j-1} - 1).$$

Since a is not a zero divisor, this tells us that $a^{j-1} - 1 = 0$, so $a^{j-1} = 1$. However, since $j > k$, we know that $j - 1 \geq 1$. Therefore, if we let $m = j - 1$, we now have

$$a^m = 1,$$

for some $m \geq 1$, thereby proving our result. $\square$

Specializing Theorem 7.5 to rings of the form $\mathbb{Z}_n$, we now have

Corollary 7.6. *In $\mathbb{Z}_n$, every nonzero element is either invertible or is a zero divisor.*

■ Examples

In $\mathbb{Z}_3$, $[1]_3$ and $[2]_3$ are the only invertible elements and

$$[2]_3^2 = [1]_3, \quad \text{thus} \quad [2]_3^{-1} = [2]_3^1 = [2]_3.$$

In $\mathbb{Z}_4$, $[1]_4$ and $[3]_4$ are the only invertible elements and

$$[3]_4^2 = [1]_4, \quad \text{thus} \quad [3]_4^{-1} = [3]_4^1 = [3]_4.$$

In $\mathbb{Z}_5$, $[1]_5$, $[2]_5$, $[3]_5$, $[4]_5$ are the only invertible elements and

$$[2]_5^4 = [1]_5, \quad \text{thus} \quad [2]_5^{-1} = [2]_5^3 = [2^3]_5 = [3]_5,$$

$$[3]_5^4 = [1]_5, \quad \text{thus} \quad [3]_5^{-1} = [3]_5^3 = [3^3]_5 = [2]_5,$$

$$[4]_5^2 = [1]_5, \quad \text{thus} \quad [4]_5^{-1} = [4]_5^1 = [4]_5.$$

In $\mathbb{Z}_6$, $[1]_6$ and $[5]_6$ are the only invertible elements and

$$[5]_6^2 = [1]_6, \quad \text{thus} \quad [5]_6^{-1} = [5]_6^1 = [5]_6.$$

■

As we know, every ring is a group under addition. However, as the next proposition indicates, the invertible elements of a ring form a group under multiplication. Therefore, not only do the rings $\mathbb{Z}_n$ provide us with new examples of rings, but they also provide us with new examples of groups.

Proposition 7.7. *If R is a ring, then the invertible elements of R, denoted as U(R), are a group under multiplication.*

Proof. The first thing we need to check is that $U(R)$ is closed under multiplication. If $a, b \in U(R)$, then a, b, a^{-1}, b^{-1} are all elements of R. As a result, $ab, b^{-1}a^{-1} \in R$. Next, observe that

$$(ab)(b^{-1}a^{-1}) = a(b \cdot b^{-1})a^{-1} = a \cdot 1 \cdot a^{-1} = a \cdot a^{-1} = 1$$

and

$$(b^{-1}a^{-1})(ab) = b^{-1}(a^{-1} \cdot a)b = b^{-1} \cdot 1 \cdot b = b^{-1} \cdot b = 1.$$

The preceding equations reveal that $b^{-1}a^{-1}$ is the multiplicative inverse of ab. Thus, ab is an element of R, which has a multiplicative inverse in R. Hence, $ab \in U(R)$, so $U(R)$ is indeed closed under multiplication.

Since multiplication in R is associative and $U(R)$ is a subset of R, it is clear that multiplication in $U(R)$ is also associative. If we let 1 denote the multiplicative identity of R, then $1 \cdot a = a = a \cdot 1$, for all $a \in R$. Therefore, in order to conclude that 1 is the identity element of $U(R)$, we only need to show that $1 \in U(R)$. However, since 1 is its own multiplicative inverse, it is certainly the case that $1 \in U(R)$. Finally, if $a \in U(R)$, then the equation

$$a \cdot a^{-1} = 1 = a^{-1} \cdot a$$

indicates that not only is a^{-1} the inverse of a but also that a is the inverse of a^{-1}. Thus, a^{-1} is an element of R whose multiplicative inverse belongs to R. Hence, $a^{-1} \in U(R)$, and we have shown that every element of $U(R)$ has an inverse in $U(R)$. Therefore, $U(R)$ satisfies all the properties of a group. $\square$

Although the proof of Proposition 7.7 is very abstract, we can represent one of the main ideas of the proof very concretely. If a and b are both invertible, then, at first glance, one might guess that the inverse of ab is $a^{-1}b^{-1}$. However, as we saw in the preceding proof, $b^{-1}a^{-1}$ is the inverse of ab. Therefore, in order to invert ab, not only do we need to invert both a and b to obtain a^{-1} and b^{-1}, but then we also need to reverse the order in which they appear to obtain $b^{-1}a^{-1}$. Here are some easy and concrete examples to help you understand why we need to reverse the order when finding the inverse of a product.

■ Examples

Suppose you are undressing. Earlier in the day, you had put on your socks and then your shoes. To undress, you must undo, or invert, this act. But note that when undressing, you must first take off your shoes and then take off your socks. Thus, whereas when dressing you deal first with the socks and then with the shoes, when inverting this procedure, you reverse the order and deal first with the shoes and then with the socks.

Similarly, when going to work in the morning, you might put on your coat, leave the house, and enter the car. At the end of the day, you invert this procedure, and when you get home, you leave the car, enter the house, and take off your coat. Thus, in the morning you deal with, in order, coat-house-car. But in the evening, each individual action is inverted, and the order is reversed as you deal with, in order, car-house-coat.

∎

■ Examples

In the following examples, for various rings R, we examine the group of invertible elements $U(R)$.

1. $U(\mathbb{Z}) = \{1, -1\}$.

2. $U(\mathbb{Q})$ is the set of all nonzero elements of $\mathbb{Q}$. In fact, observe that if R is a commutative ring, then R is a field if and only if $U(R)$ is equal to the entire set of nonzero elements of R.

3. $U(\mathbb{Z}[x]) = \{1, -1\}$, as no polynomial of degree at least one has an inverse in $\mathbb{Z}[x]$.

4. $U(\mathbb{R}[x])$ is the set of nonzero real numbers, as no polynomial of degree at least one has an inverse in $\mathbb{R}[x]$.

5. $U(\mathbb{Z}_2) = \{[1]_1\}$.

6. $U(\mathbb{Z}_3) = \{[1]_3, [2]_3\}$.

7. $U(\mathbb{Z}_4) = \{[1]_4, [3]_4\}$.

8. $U(\mathbb{Z}_5) = \{[1]_5, [2]_5, [3]_5, [4]_5\}$.

9. $U(\mathbb{Z}_6) = \{[1]_6, [5]_6\}$.

10. $U(\mathbb{Z}_8) = \{[1]_8, [3]_8, [5]_8, [7]_8\}$.

11. $U(\mathbb{Z}_{10}) = \{[1]_{10}, [3]_{10}, [7]_{10}, [9]_{10}\}$.

∎

Examining groups of the form $U(\mathbb{Z}_n)$ provides us with many examples of finite groups. In light of this, two question regarding $U(\mathbb{Z}_n)$ and the invertible elements of $\mathbb{Z}_n$ that you may have are

1. For each $n > 1$, is there an easy way to determine which of the elements of $U(\mathbb{Z}_n)$ are invertible and which are zero divisors?

2. For each $n > 1$, is there an easy way to determine the size of the set $U(\mathbb{Z}_n)$?

Before answering the first of our two questions, let us look back at some of our previous examples. In $\mathbb{Z}_6$, the invertible elements are $[1]_6$, $[5]_6$, and the zero divisors are $[2]_6$, $[3]_6$, $[4]_6$. In $\mathbb{Z}_{10}$ the invertible elements are $[1]_{10}$, $[3]_{10}$, $[7]_{10}$, $[9]_{10}$, and the zero divisors are $[2]_{10}$, $[4]_{10}$, $[5]_{10}$, $[6]_{10}$, $[8]_{10}$. It is usually a good idea to look for a pattern before trying to prove a general result. When we look at $\mathbb{Z}_6$, we can observe that the classes $[a]$ that were invertible were precisely those where a was relatively prime to 6. Similarly, when looking at $\mathbb{Z}_{10}$, we can see that the classes $[a]$ that were invertible were precisely those where a was relatively prime to 10. These observations lead us to

Theorem 7.8. *In $\mathbb{Z}_n$, the class $[a]_n$ is invertible if and only if a and n are relatively prime.*

Proof. In one direction, if a and n are relatively prime, then the greatest common divisor of a and n is 1, and Theorem 3.8 tells us that there exist integers s, t such that

$$s \cdot a + t \cdot n = 1.$$

This equation is equivalent to the equation $s \cdot a - 1 = -t \cdot n$. Therefore, the difference between $s \cdot a$ and 1 is divisible by n, which tells us that $s \cdot a \sim_n 1$. As a result, $[s \cdot a]_n = [1]_n$. Using the definition of multiplication in $\mathbb{Z}_n$, we now know that

$$[s]_n \cdot [a]_n = [1]_n.$$

Since multiplication in $\mathbb{Z}_n$ is commutative, we also know that $[a]_n \cdot [s]_n = [1]_n$. Therefore, $[s]_n$ is the multiplicative inverse of $[a]_n$ and so, $[a]_n$ is invertible.

In the other direction, if $[a]_n$ is invertible, then there exists some $u \in \mathbb{Z}$ such that

$$[u]_n \cdot [a]_n = [1]_n = [a]_n \cdot [u]_n.$$

Since $[u]_n \cdot [a]_n = [u \cdot a]_n$, we know that $u \cdot a \sim_n 1$. As a result, n divides $u \cdot a - 1$, which immediately tells us that there exists some $v \in \mathbb{Z}$ such that $u \cdot a - 1 = v \cdot n$. But this equation is equivalent to the equation

$$u \cdot a + (-v) \cdot n = 1.$$

However, having written 1 as a multiple of a plus a multiple of n, Theorem 3.8 tells us that we have indeed shown that a and n are relatively prime. □

Recall that every field is a commutative ring, but not every commutative ring is a field. In particular, for those n such that $\mathbb{Z}_n$ has zero divisors, we know that $\mathbb{Z}_n$ is a commutative ring that is not a field. This then raises the question, for which values of n is $\mathbb{Z}_n$ a field? If you look back at the examples in this chapter, you may already have a good idea what the answer to this question is. In fact, the answer to this question follows directly from Theorem 7.8.

Corollary 7.9. *$\mathbb{Z}_n$ is a field if and only if n is a prime number.*

Proof. In $\mathbb{Z}_n$, the nonzero equivalence classes are

$$[1]_n, [2]_n, [3]_n, \ldots, [n-2]_n, [n-1]_n.$$

In one direction, if n is prime then every one of the integers $1, 2, 3, \ldots, n-2, n-1$ is relatively prime to n. Therefore, Theorem 7.8 asserts that every nonzero class in $\mathbb{Z}_n$ has a multiplicative inverse. Thus, $\mathbb{Z}_n$ is a field.

In the other direction, if n is not prime, then there exists a positive integer a such that a divides n and $1 < a < n$. As a result, $[a]_n$ is a nonzero equivalence class in $\mathbb{Z}_n$ where a and n are not relatively prime. Thus, Theorem 7.8 tells us that $[a]_n$ is not invertible in $\mathbb{Z}_n$. However, since $\mathbb{Z}_n$ now contains a nonzero element that is not invertible, $\mathbb{Z}_n$ cannot be a field. $\qquad\square$

Exercises for Sections 7.1 and 7.2

1. In $\mathbb{Z}_8$, compute $[3]_8 + [7]_8$, $[3]_8 \cdot [7]_8$, $[4]_8 + [6]_8$, and $[4]_8 \cdot [6]_8$.

2. In $\mathbb{Z}_8$, compute $[5]_8 + [4]_8$, $[5]_8 \cdot [4]_8$, $[6]_8 + [2]_8$, and $[6]_8 \cdot [2]_8$.

3. In $\mathbb{Z}_{21}$, compute $[14]_{21} + [12]_{21}$, $[14]_{21} \cdot [12]_{21}$, $[18]_{21} + [13]_{21}$, and $[18]_{21} \cdot [13]_{21}$.

4. In $\mathbb{Z}_{21}$, compute $[9]_{21} + [18]_{21}$, $[9]_{21} \cdot [18]_{21}$, $[6]_{21} + [14]_{21}$, and $[6]_{21} \cdot [14]_{21}$.

5. In $\mathbb{Z}_6$, compute the square of each nonzero element.

6. In $\mathbb{Z}_8$, compute the square of each nonzero element.

7. In $\mathbb{Z}_5$, compute the cube of each nonzero element.

8. In $\mathbb{Z}_5$, compute the fourth power of each nonzero element.

9. In $\mathbb{Z}_{10}$, list all the invertible elements and find the inverse of each element.

10. In $\mathbb{Z}_{14}$, list all the invertible elements and find the inverse of each element.

11. In $\mathbb{Z}_{15}$, list all the invertible elements and find the inverse of each element.

12. In $\mathbb{Z}_{20}$, list all the invertible elements and find the inverse of each element.

13. If today is Monday, what day of the week will it be in 100 days?

14. If today is Tuesday, what day of the week will it be in 1000 days?

15. If today is Wednesday, what day of the week will it be in 1,000,000 days?

16. A child recites the alphabet over and over again. If she stops after reciting 187 letters, what was the last letter she said?

17. Let $n \geq 2$ be an integer and let $\sigma : \mathbb{Z} \to \mathbb{Z}_n$ be the function defined as $\sigma(a) = [a]_n$, for all $a \in \mathbb{Z}$.

    (a) Show that $\sigma(a+b) = \sigma(a) + \sigma(b)$, for all $a, b \in \mathbb{Z}$.

    (b) Show that $\sigma(a \cdot b) = \sigma(a) \cdot \sigma(b)$, for all $a, b \in \mathbb{Z}$.

    (c) Show that $\sigma(a_1 + a_2 + \cdots a_m) = \sigma(a_1) + \sigma(a_2) + \cdots \sigma(a_m)$, for all $m \in \mathbb{N}$ and $a_i \in \mathbb{Z}$.

    (d) Show that $\sigma(a_1 \cdot a_2 \cdots a_m) = \sigma(a_1) \cdot \sigma(a_2) \cdots \sigma(a_m)$, for all $m \in \mathbb{N}$ and $a_i \in \mathbb{Z}$.

18. Let $f(x) \in \mathbb{Z}[x]$ and let $n \geq 2$ be an integer. Use exercise 17 to prove that if $a, b \in \mathbb{Z}$ such that $[a]_n = [b]_n$, then $[f(a)]_n = [f(b)]_n$.

For exercise 19, please read the following:

Mathematicians have long searched for systematic ways of producing examples of prime numbers. This search leads naturally to the following question: Does there exists a polynomial $f(x)$ with integer coefficients of degree at least 1 such that $f(n)$ is prime, for *every* $n \in \mathbb{N}$? If such a polynomial existed, then the infinite sequence

$$f(1), \ f(2), \ f(3), \ f(4), \ f(5), \ldots.$$

would consist entirely of prime numbers. However, in the next exercise, we will show that no such polynomial can exist.

Even though we will prove that there does not exist a nonconstant polynomial that produces only values which are prime, there are some interesting results worth mentioning that deal with prime values of polynomials. Consider the polynomial

$$g(x) = x^2 - x + 41.$$

It will follow from our next exercise that there is some natural number n such that $g(n)$ is not prime. In fact, it is not hard to find such a natural number. Observe that if we let $x = 41$, then $g(41) = 41^2 - 41 + 41 = 41^2$ is not prime. However, the interesting thing to note about $g(x)$ is that

$$g(1) = 41, \ g(2) = 43, \ g(3) = 47, \ldots, \ g(39) = 1,523, \ g(40) = 1601$$

are all prime. Thus, whereas no nonconstant polynomial produces only prime values, we have exhibited a quadratic whose first 40 values are all prime. In light of this we can ask, is there a polynomial of small degree whose first 100 values are all prime? Or is there a polynomial of small degree whose first 1,000,000,000 values are all prime?

In 2004, B. Green and T. Tao proved a remarkable result in answering these questions. They proved that for any natural number N, there exists a linear polynomial $h(x)$ such that its first N values

$$h(1), h(2), h(3), h(4), \ldots, h(N-1), h(N)$$

are all prime.

19. Let $f(x) \in \mathbb{Z}[x]$ have degree at least 1 and suppose $a \in \mathbb{Z}$ such that $f(a) = p$, where p is a prime number.

    (a) Use exercise 18 to show that if $b \in \mathbb{Z}$ such that $[b]_p = [a]_p$, then $f(b)$ is a multiple of p.

    (b) Use part (a) to show that there exists some positive integer $b \in [a]_p$ such that $f(b)$ is not a prime number.

    (c) Conclude that there cannot exist a polynomial in $\mathbb{Z}[x]$ of degree at least 1 with the property that $f(n)$ is prime, for every $n \in \mathbb{N}$.

20. Let $f(x) \in \mathbb{Z}[x]$ and let $n \geq 2$ be an integer such than $[f(a)]_n \neq [0]_n$, for $a = 0, 1, 2 \ldots, n-2, n-1$. Prove that $f(x)$ has no roots in the integers. (Hint: Look at exercise 18.)

21. Use $n = 2$ in exercise 20 to show that $4x^5 - 17x^4 + 3x^3 - 8x^2 + 16x - 9$ has no roots in $\mathbb{Z}$.

22. Use $n = 2$ in exercise 20 to show that $9x^8 - 5x^6 + 14x^3 - 36x + 27$ has no roots in $\mathbb{Z}$.

23. Use $n = 3$ in exercise 20 to show that $10x^5 - 7x^4 - 4x^3 + 11x^2 - 21x + 85$ has no roots in $\mathbb{Z}$.

24. Use $n = 5$ in exercise 20 to show that $6x^8 - 3x^5 + 4x^4 + 8x - 72$ has no roots in $\mathbb{Z}$.

25. Use $n = 4$ in exercise 20 to show that $3x^4 - 9x^3 + 8x^2 + 17x + 63$ has no roots in $\mathbb{Z}$.

26. In the previous exercise, you used $n = 4$ and exercise 20 to show that $3x^4 - 9x^3 + 8x^2 + 17x + 63$ has no roots in $\mathbb{Z}$. Show that if you had used $n = 2$ and exercise 20 to examine $3x^4 - 9x^3 + 8x^2 + 17x + 63$, then you do not obtain enough information to determine whether $3x^4 - 9x^3 + 8x^2 + 17x + 63$ has any roots in $\mathbb{Z}$.

27. Show that using $n = 3$ and $n = 5$ in exercise 20 also do not give you enough information to determine whether $3x^4 - 9x^3 + 8x^2 + 17x + 63$ has any roots in $\mathbb{Z}$.

28. Find an integer $n \geq 2$ such that you can use exercise 20 to show that $4x^7 - 5x^6 + 11x^5 - 6x^4 + 9x^3 - 16x^2 - 23x + 105$ has no roots in $\mathbb{Z}$.

29. Find an integer $n \geq 2$ such that you can use exercise 20 to show that $2x^4 - 7x^3 + 3x^2 + 11x + 70$ has no roots in $\mathbb{Z}$.

30. Let $f(x) \in \mathbb{Z}[x]$ and let $n \geq 2$ be an integer such that $[f(a)]_n = [0]_n$, for $a = 0, 1, 2 \ldots, n-2, n-1$. Prove that for every $m \in \mathbb{Z}$, $f(m)$ is a multiple of n. (Hint: Look at exercise 18.)

31. Use exercise 30 to show that $7m^{11} - 5m^8 + 6m^5 + 4m^3 + 22m - 360$ is divisible by 2, for every $m \in \mathbb{Z}$.

32. Use exercise 30 to show that $8m^5 - 4m^4 - 17m^3 + 10m^2 - 45m + 90$ is divisible by 3, for every $m \in \mathbb{Z}$.

33. Use exercise 30 to show that $6m^{12} + 3m^{10} - 4m^9 + 7m^6 - 16m^4 + 9m - 100$ is divisible by 5, for every $m \in \mathbb{Z}$.

34. Use exercise 30 to show that $7m^5 - 4m^4 + 10m^3 - 23m^2 - 14m + 42$ is divisible by both 2 and 3 and is therefore divisible by 6, for every $m \in \mathbb{Z}$.

35. If a is an odd integer, show that $[a^2]_8 = [1]_8$.

36. For any $m \in \mathbb{N}$, examine $[11^m - 1]_{10}$ in $\mathbb{Z}_{10}$, and then prove that $11^m - 1$ is divisible by 10.

37. For any $m \in \mathbb{N}$ examine $[4^m + 2]_2$ in $\mathbb{Z}_2$ and $[4^m + 2]_3$ in $\mathbb{Z}_3$. Then prove that $4^m + 2$ is divisible by 6.

The goal of exercises 38–53 is to develop **divisibility tests**. By this we mean that given a positive integer N, we would like to develop shortcuts to determine if N is divisible by various integers. We will begin with relatively easy tests to determine if an integer is divisible by 2, 5, or powers of 10. Then we will develop tests to check divisibility by 3 and 9. Finally, we will develop more sophisticated tests for divisibility by 7 and 11.

38. Show that a positive integer is divisible by 2 if and only if its one's digit is divisible by 2.

39. Show that a positive integer is divisible by 5 if and only if its one's digit is divisible by 5.

40. (a) Show that a positive integer is divisible by 10 if and only if the one's digit is 0.

    (b) Generalize (a) and show that if $m \in \mathbb{N}$, then a positive integer is divisible by 10^m if and only if its rightmost m digits are all 0.

Given an integer n, let $S(n)$ denote the sum of its digits. If $S(n) \geq 10$, we can take the sum of the digits of $S(n)$. We can continue this process until we get an answer that is less than 10, and we denote this as $\tilde{S}(n)$. For example, if $n = 6, 574, 280, 357$, then $S(n) = 47$ and $\tilde{S}(n) = 2$. In order to do the next exercise, it will be useful to recall that if the base 10 expansion of the positive integer n is $a_m a_{m-1} \cdots a_1 a_0$, where $0 \leq a_i \leq 9$, then $S(n) = a_m + a_{m-1} + \cdots + a_1 + a_0$ and $n = 10^m a_m + 10^{m-1} a_{m-1} + \cdots + 10 a_1 + a_0$.

41. (a) If $n \in \mathbb{N}$, show that $[S(n)]_9 = [n]_9$.

    (b) If $n \in \mathbb{N}$, show that $[\tilde{S}(n)]_9 = [n]_9$.

    (c) Use (b) to prove that $n \in \mathbb{N}$ is divisible by 3 if and only if $\tilde{S}(n)$ is divisible by 3.

    (d) Use (b) to prove that $n \in \mathbb{N}$ is divisible by 9 if and only if $\tilde{S}(n)$ is divisible by 9.

42. For each value of n, find $S(n)$ and $\tilde{S}(n)$. Then determine if n is divisible by 3 or 9.
    (a) $n = 4,284$

    (b) $n = 51,782$

    (c) $n = 483,609$

43. For each value of n, find $S(n)$ and $\tilde{S}(n)$. Then determine if n is divisible by 3 or 9.
    (a) $n = 743,928,054$

    (b) $n = 687,012,867$

    (c) $n = 519,065,432$

44. (a) Show that if m is an odd positive integer, then $10^m + 1$ is divisible by 11.

    (b) Show that if m is an even positive integer, then $10^m - 1$ is divisible by 11.

    (c) Conclude that $10^m - (-1)^m$ is divisible by 11, for all $m \in \mathbb{N}$.

    If the base 10 expansion of the positive integer n is $a_m a_{m-1} \cdots a_1 a_0$, where $0 \le a_i \le 9$, then we define $B(n) = a_0 - a_1 + a_2 - a_3 + \cdots + (-1)^m a_m$ and can then define the alternating sum of the digits $A(n)$ as $A(n) = |B(n)|$. If $A(n) \ge 10$, we can also take the alternating sum of the digits of $A(n)$. We can continue this process until we get an answer that is less than 10, and we denote this as $\tilde{A}(n)$.

45. For part (a) of this exercise, you may want to refer to exercise 44.
    (a) Let $n \in \mathbb{N}$; show that $[B(n)]_{11} = [n]_{11}$.

    (b) Let $n \in \mathbb{N}$; show that $[A(n)]_{11} = [0]_{11}$ if and only if $[n]_{11} = [0]_{11}$.

    (c) Let $n \in \mathbb{N}$; show that $[\tilde{A}(n)]_{11} = [0]_{11}$ if and only if $[n]_{11} = [0]_{11}$.

    (d) Let $n \in \mathbb{N}$; show that n is divisible by 11 if and only if $\tilde{A}(n)$ is divisible by 11.

46. For each value of n, find $B(n)$, $A(n)$, and $\tilde{A}(n)$. Then determine if n is divisible by 11.
    (a) $n = 586$

    (b) $n = 486,329$

    (c) $n = 695,794$

47. For each value of n, find $B(n)$, $A(n)$, and $\tilde{A}(n)$. Then determine if n is divisible by 11.
    (a) $n = 845,009,314,764$

    (b) $n = 714,283,749,586$

    (c) $n = 283,716,295,417$

    If the base 10 expansion of the positive integer n is $a_m a_{m-1} \cdots a_1 a_0$, where $0 \le a_i \le 9$, then we define

    $$V(n) = a_0 + 3a_1 + 2a_2 - a_3 - 3a_4 - 2a_5 + a_6 + 3a_7 + 2a_8 - a_9 - 3a_{10} - 2a_{11} + \cdots$$

and can then define the twisted sum $T(n)$ as $T(n) = |V(n)|$. If $T(n) \geq 10$, we can take the twisted sum of the digits of $T(n)$. We can continue this process until we get an answer that is less than 10, and we denote this as $\tilde{T}(n)$. In order to easily compute $V(n)$ and $T(n)$, you should keep in mind the sequence

$$1, 3, 2, -1, -3, -2, 1, 3, 2, -1, -3, -2, 1, 3, 2, -1, -3, -2, \ldots$$

as these are the numbers that appear before the digits of n when computing $V(n)$ and $T(n)$.

48. Prove the following:
    (a) If $m \in \mathbb{N}$ and $[m]_6 = [0]_6$, then $[10^m]_7 = [1]_7$

    (b) If $m \in \mathbb{N}$ and $[m]_6 = [1]_6$, then $[10^m]_7 = [3]_7$

    (c) If $m \in \mathbb{N}$ and $[m]_6 = [2]_6$, then $[10^m]_7 = [2]_7$

    (d) If $m \in \mathbb{N}$ and $[m]_6 = [3]_6$, then $[10^m]_7 = [6]_7$

    (e) If $m \in \mathbb{N}$ and $[m]_6 = [4]_6$, then $[10^m]_7 = [4]_7$

    (f) If $m \in \mathbb{N}$ and $[m]_6 = [5]_6$, then $[10^m]_7 = [5]_7$

49. For part (a) of this exercise, you may want to refer to exercise 48.
    (a) Let $n \in \mathbb{N}$; show that $[V(n)]_7 = [n]_7$.

    (b) Let $n \in \mathbb{N}$; show that $[T(n)]_7 = [0]_7$ if and only if $[n]_7 = [0]_7$.

    (c) Let $n \in \mathbb{N}$; show that $[\tilde{T}(n)]_7 = [0]_7$ if and only if $[n]_7 = [0]_7$.

    (d) Let $n \in \mathbb{N}$; show that n is divisible by 7 if and only if $\tilde{T}(n)$ is divisible by 7.

50. For each value of n, find $V(n)$, $T(n)$, and $\tilde{T}(n)$. Then determine if n is divisible by 7.
    (a) $n = 955$

    (b) $n = 672$

    (c) $n = 864$

51. For each value of n, find $V(n)$, $T(n)$, and $\tilde{T}(n)$. Then determine if n is divisible by 7.
    (a) $n = 562, 980, 542$

    (b) $n = 398, 244, 728$

    (c) $n = 809, 824, 778$

52. For each value of n, use the divisibility tests developed in these exercises to determine which of the primes 2, 3, 5, 7, 11 divide n.
    (a) $n = 10, 857$

    (b) $n = 13, 706$

    (c) $n = 545, 445$

53. For each value of n, use the divisibility tests developed in these exercises to determine which of the primes $2, 3, 5, 7, 11$ divide n.

   (a) $n = 246, 642$

   (b) $n = 99, 015$

   (c) $n = 846, 279$

54. If G is a finite group and e denotes the identity element of G, show that for each $g \in G$, there is an integer $n \geq 1$ such that $g^n = e$. (Hint: Think about the proof of Theorem 7.5.)

55. If G is a finite group and e denotes the identity element of G, show that there is an integer $N \geq 1$ such that $g^N = e$, for all $g \in G$ (Note: In exercise 54, the integer n depended upon the element $g \in G$. However, in this exercise, the integer N works for *all* $g \in G$.)

56. Suppose R is a commutative ring with only a finite number of elements and no zero divisors. Show that R is a field.

57. Let R be a ring and suppose $r \in R$ such that $r^2 = 0$. Show that $1 + r$ has a multiplicative inverse in R.

58. Let R be a ring and suppose $r \in R$ and $n \in \mathbb{N}$ such that $r^n = 0$. Show that $1 + r$ has a multiplicative inverse in R.

59. Let G be a group such that $g^2 = e$, for all $g \in G$, where e denotes the identity element of G. Prove that G is commutative. (Note: You must prove that $ab = ba$, for *all* $a, b \in G$.)

60. Let R be a ring such that $r^2 = r$, for all $r \in R$. Prove that R is commutative. (Note: You must prove that $ab = ba$, for *all* $a, b \in R$.)

61. Let R be a ring such that $r^3 = r$, for all $r \in U(R)$. Prove that the group $U(R)$ is commutative.

62. In this exercise, you will provide the details we omitted in the proof of Theorem 7.2 that $\mathbb{Z}_n$ is a commutative ring.

   (a) Show that multiplication in $\mathbb{Z}_n$ is associative.

   (b) Show that addition in $\mathbb{Z}_n$ is commutative.

   (c) Show that multiplication in $\mathbb{Z}_n$ is commutative.

   (d) Show that $\mathbb{Z}_n$ also satisfies the other part of the distributive law.

63. In $\mathbb{Z}_5$, find all equivalence classes whose square is $[1]_5$.

64. In $\mathbb{Z}_7$, find all equivalence classes whose square is $[1]_7$.

65. In $\mathbb{Z}_{11}$, find all equivalence classes whose square is $[1]_{11}$.

66. In $\mathbb{Z}_{13}$, find all equivalence classes whose square is $[1]_{13}$.

67. If $p > 2$ is a prime number, show that $[1]_p$ and $[p-1]_p$ are only two equivalences in $\mathbb{Z}_p$ whose square is equal to $[1]_p$.

68. If p is a prime number, show that $[(p-1)!]_p = [p-1]_p$ in $\mathbb{Z}_p$. (Hint: You may want to refer to exercise 67.)

69. Let $g(x) \in \mathbb{Z}[x]$ have degree at least 2, and let p be a prime number such that
    (i) the leading coefficient of $g(x)$ is not divisible by p,

    (ii) every other coefficient of $g(x)$ is divisible by p,

    (iii) the constant term of $g(x)$ is not divisible by p^2.

    (a) Show that if $a \in \mathbb{Z}$ such that $[a]_p \neq [0]_p$, then $[g(a)]_p \neq [0]_p$.

    (b) Show that if $b \in \mathbb{Z}$ such that $[b]_p = [0]_p$, then $[g(b)]_{p^2} \neq [0]_{p^2}$.

    (c) Conclude that $g(x)$ has no roots in $\mathbb{Z}$.

70. Use exercise 69 with the prime $p = 3$ to show that $5x^6 + 9x^4 - 30x^2 + 15x - 33$ has no roots in $\mathbb{Z}$.

71. Use exercise 69 with the prime $p = 2$ to show that $9x^7 - 42x^6 + 30x^5 - 22x^3 + 16x + 70$ has no roots in $\mathbb{Z}$.

72. Use exercise 69 to show that $3x^4 + 10x^3 - 40x^2 + 25x - 20$ has no roots in $\mathbb{Z}$.

73. Use exercise 69 to show that $2x^7 - 6x^5 + 90x^4 - 21x^3 + 36x^2 - 57$ has no roots in $\mathbb{Z}$.

74. (a) Show that the criterion in exercise 69 does not apply to either $x^2 - 5x + 6$ or $x^2 - 5x + 7$.

    (b) Show that $x^2 - 5x + 6$ has roots in $\mathbb{Z}$ and $x^2 - 5x + 7$ has no roots in $\mathbb{Z}$.

    (c) Conclude that if exercise 69 does not apply, then you need more information to determine if a polynomial has roots in $\mathbb{Z}$.

7.3 The Euler ϕ Function

In Chapter 3, we remarked that prime numbers are the building blocks of the integers and various problems can be solved by understanding prime numbers. One of the problems that can be solved by exploiting properties of prime numbers is our question regarding the number of invertible elements in $\mathbb{Z}_n$. In light of Theorem 7.8, this problem requires us to determine how many of the integers from 1 to $n-1$ are relatively prime to n. To solve this, it will be useful to introduce some new terminology.

Definition 7.10. *If $n > 1$ is an integer, we define $\phi(n)$ to be the number of positive integers less than n that are relatively prime to n. The function ϕ is usually referred to as the Euler ϕ function.*

The first step in finding a formula for ϕ is to observe that if p is a prime number, then $\phi(p) = p - 1$. The reason for this is that all of the $p - 1$ positive integers less than p are relatively prime to p. The next step in finding a general formula for ϕ will be to consider $\phi(p^m)$, where p is prime and $m \in \mathbb{N}$.

Lemma 7.11. *If p is a prime number and $m \in \mathbb{N}$, then $\phi(p^m) = p^{m-1}(p-1)$.*

Intuition. Suppose we want to compute $\phi(125) = \phi(5^3)$. The only way an integer can fail to be relatively prime to 125 is if it is a multiple of 5. The only positive integers less than 125 that are multiples of 5 are

$$5 = 1 \cdot 5, \ 10 = 2 \cdot 5, \ 15 = 3 \cdot 5, \ 20 = 4 \cdot 5, \ldots, \ 115 = 23 \cdot 5, \ 120 = 24 \cdot 5.$$

Therefore, of the $124 = 5^3 - 1$ positive integers less than 125, we see that $24 = 5^2 - 1$ of them fail to be relative prime to 5^3. That means that the number of those that are relatively prime to 125 is

$$124 - 24 = (5^3 - 1) - (5^2 - 1) = 5^3 - 5^2 = 5^2(5 - 1).$$

The proof for the general case is virtually identical.

Proof. There are $p^m - 1$ positive integers less than p^m. The only ones that are not relatively prime to p^m are those that are multiples of p. If we list the positive integers that are less than p^m and divisible by p, we have

$$1 \cdot p, \ 2 \cdot p, \ 3 \cdot p, \ 4 \cdot p, \ldots, \ (p^{m-1} - 2) \cdot p, \ (p^{m-1} - 1) \cdot p.$$

Therefore, $p^{m-1} - 1$ of the positive integers less than p^m are divisible by p, meaning that the remaining

$$(p^m - 1) - (p^{m-1} - 1) = p^m - p^{m-1} = p^{m-1}(p - 1)$$

are relatively prime to p^m. Thus, $\phi(p^m) = p^{m-1}(p-1)$, as desired. $\square$

■ Examples

$\phi(256) = \phi(2^8) = 2^7(2 - 1) = 2^7 = 128$. Thus, the group $U(\mathbb{Z}_{256})$ has 128 elements.
$\phi(3125) = \phi(5^5) = 5^4(5 - 1) = 625 \cdot 4 = 2500$. Therefore, $U(\mathbb{Z}_{3125})$ is a group with 2500 elements.

Having found a formula for the special case where $n = p^m$, we need to handle the case where n cannot be written as p^m. One of the tools we will use to accomplish this is

Lemma 7.12. *If $a, s, t \in \mathbb{N}$, then a is relatively prime to $s \cdot t$ if and only if it is relatively prime to both s and t.*

Proof. In this proof, we will make repeated use of the fact that integers n, m are relatively prime if and only if there exist integers A, B such that $A \cdot n + B \cdot m = 1$. In one direction, suppose a is relatively prime to both s and t. Then there exist integers A, B, C, D such that

$$(4) \qquad\qquad A \cdot a + B \cdot s = 1 \quad \text{and} \quad C \cdot a + D \cdot t = 1.$$

Multiplying the second equation in (4) by B gives us $BC \cdot a + BD \cdot t = B$. This equation enables us to substitute for B in the first equation in (4) to obtain

$$A \cdot a + (BC \cdot a + BD \cdot t) \cdot s = 1.$$

We can rewrite this equation as

$$(A + BCs) \cdot a + (BD) \cdot (s \cdot t) = 1.$$

Having succeeded in writing 1 as a multiple of a plus a multiple of $s \cdot t$, we see that a is relatively prime to $s \cdot t$.

In the other direction, suppose a and $s \cdot t$ are relatively prime. Therefore, we know there exist integers A, B such that

$$A \cdot a + B \cdot (s \cdot t) = 1.$$

We can rewrite this equation as

$$A \cdot a + Bt \cdot s = 1 = A \cdot a + Bs \cdot t.$$

Having written 1 as a multiple of a plus a multiple of s, we see that a is relatively prime to s. The identical reasoning tells us that a is also relatively prime to t. $\qquad\square$

Suppose we can factor n as $n = s \cdot t$, and suppose we already know the values of $\phi(s)$ and $\phi(t)$. At first glance, it is not clear if this helps us in computing $\phi(n)$. If we look back at some of the earlier examples, we have

$$\phi(6) = 2 = 1 \cdot 2 = \phi(2) \cdot \phi(3) \quad \text{and} \quad \phi(10) = 4 = 1 \cdot 4 = \phi(2) \cdot \phi(5),$$

yet we also have

$$\phi(4) = 2 \neq 1 \cdot 1 = \phi(2) \cdot \phi(2) \quad \text{and} \quad \phi(25) = 20 \neq 4 \cdot 4 = \phi(5) \cdot \phi(5).$$

Table 7.13: *If $n = s \cdot t$, where $s, t > 1$, then we can list all the positive integers from 1 to n in a table consisting of s rows and t columns. As we move from left to right in any row, the numbers increase by 1. Then as we move down any of our columns, the numbers increase by t. Therefore, if a is the top number in a given column, then the s entries in the column would be $a, t+a, 2t+a, \ldots; (s-2)t+a; (s-1)t+a$.*

1	.	.	a	.	.	t
$t+1$	.	.	$t+a$	.	.	$2t$
.	.	.	.	.	.	.
.	.	.	.	.	.	.
.	.	.	.	.	.	.
$(s-2)t+1$	.	.	$(s-2)t+a$	.	.	$(s-1)t$
$(s-1)t+1$	.	.	$(s-1)t+a$	.	.	st

Therefore, it is sometimes but not always the case that $\phi(n) = \phi(s) \cdot \phi(t)$. However, it turns out that when s and t are relatively prime, $\phi(n)$ is indeed equal to $\phi(s) \cdot \phi(t)$. Proving this fact will require visualizing n as the product s times t, which we do in the table above.

Observe that computing $\phi(n)$ now reduces to counting the number of entries in our table that are relatively prime to n. In light of Lemma 7.12, we need to determine how many of the entries are relatively prime to both s and t. To do so, we need to analyze each column in our table.

Lemma 7.14. *If a is relatively prime to t then every entry in the column consisting of $a, t+a, 2t+a, \ldots, (s-2)t+a, (s-1)t+a$ is relatively prime to t. On the other hand, if a and t are not relatively prime, then none of the entries of this column are relatively prime to t.*

Proof. Suppose $A, B \in \mathbb{Z}$ such that

$$A \cdot a + B \cdot t = c.$$

Then, if $i \in \mathbb{Z}$, we can rewrite the previous equation as

$$A \cdot (i \cdot t + a) + (B - A \cdot i) \cdot t = c.$$

Therefore, any number that can be written as a multiple of a plus a multiple of t can also be written as a multiple of $i \cdot t + a$ plus a multiple of t.

In the other direction, if $C, D, i \in \mathbb{Z}$ such that

$$C \cdot (i \cdot t + a) + D \cdot t = d,$$

then we can rewrite this as

$$C \cdot a + (D + C \cdot i) \cdot t = d.$$

Therefore, any number that can be written as a multiple of $i \cdot t + a$ plus a multiple of t can also be written as a multiple of a plus a multiple of t.

These observations combine to tell us that regardless of our choice of $i \in \mathbb{Z}$, the smallest positive integer that can be written as a multiple of $i \cdot t + a$ plus a multiple of t remains the same. Theorem 3.8 now asserts that for every $i \in \mathbb{Z}$, the greatest common divisor of $i \cdot t + a$ and t is the same. As a result, if a and t are relatively prime, then t is relatively prime to every number in the column in Table 7.13 with a on top. Similarly, if a and t are not relatively prime, then t is not relatively prime to any of the numbers in the column with a on top.   □

As we move along the first row of the table, only $\phi(t)$ of the entries are relatively prime to t. Therefore, Lemma 7.14 tells us that in our search for entries that are relatively prime to both s and t, we can limit our search to the $\phi(t)$ columns where the top entry is relatively prime to t. Next, we analyze which entries in these $\phi(t)$ columns are relatively prime to s. Recall that if s and t are not relatively prime, then it need not be the case that $\phi(s \cdot t) = \phi(s) \cdot \phi(t)$. In the next lemma, you will see where the assumption that s and t are relatively prime comes into play.

Lemma 7.15. *If s and t are relatively prime, then the equivalence classes*

$$[a]_s, [t+a]_s, [2t+a]_s, \ldots, [(s-2)t+a]_s, [(s-1)t+a]_s$$

are the s different classes belonging to $\mathbb{Z}_s$.

Proof. It suffices to show that the s equivalence classes $[a]_s, [t+a]_s, [2t+a]_s, \ldots,$ $[(s-2)t+a]_s, [(s-1)t+a]_s$ are all different elements of $\mathbb{Z}_s$. To this end, let i, j be integers such that $0 \le i, j \le s-1$. If

$$[i \cdot t + a]_s = [j \cdot t + a]_s,$$

then s divides $(i \cdot t + a) - (j \cdot t + a) = t \cdot (i - j)$. Since $s \mid (t \cdot (i - j))$ and $gcd(s, t) = 1$, we can apply Lemma 3.9 to assert that s divides $i - j$. However, since $0 \le i, j \le s-1$, then the only way s can divide $i - j$ is for i and j to be equal. This tells us that as i ranges through the numbers from 0 to $s-1$, the equivalence classes of the form $[i \cdot t + a]_s$ are all different. But this is precisely what we needed to prove.   □

■ Examples

Let us examine what Lemma 7.15 says if we let $a = 3$, $s = 7$, and $t = 10$. It asserts that the classes

$$[3]_7, \ [13]_7, \ [23]_7, \ [33]_7, \ [43]_7, \ [53]_7, \ [63]_7$$

are the seven elements of the set $\mathbb{Z}_7$. To see this, observe that

$$[63]_7 = [0]_7, \ [43]_7 = [1]_7, \ [23]_7 = [2]_7, \ [3]_7 = [3]_7,$$
$$[53]_7 = [4]_7, \ [33]_7 = [5]_7, \ [13]_7 = [6]_7.$$

Note that whereas it is important that s be relatively prime to t, a can have factors in common with either s or t. In particular, if we instead let $a = 5$ in the preceding example, we see that

$$[5]_7, \ [15]_7, \ [25]_7, \ [35]_7, \ [45]_7, \ [55]_7, \ [65]_7$$

are also the seven elements of the set $\mathbb{Z}_7$. In this case,

$$[35]_7 = [0]_7, \ [15]_7 = [1]_7, \ [65]_7 = [2]_7, \ [45]_7 = [3]_7,$$
$$[25]_7 = [4]_7, \ [5]_7 = [5]_7, \ [55]_7 = [6]_7.$$

■

We can now combine Lemmas 7.14 and 7.15 to obtain

Lemma 7.16. *If $n = s \cdot t$, where $s, t > 1$ and $gcd(s, t) = 1$, then $\phi(n) = \phi(s) \cdot \phi(t)$.*

Proof. Recall that we need to determine the number of entries in Table 7.13 that are relatively prime to both s and t. Lemma 7.14 told us that only $\phi(t)$ of our t columns contain numbers relatively prime to t and every number in these $\phi(t)$ columns is relatively prime to t. On the other hand, Lemma 7.15 tells us that regardless of which column we look at, the s entries in the column correspond to the s different equivalent classes in $\mathbb{Z}_s$. However, in $\mathbb{Z}_s$ there are exactly $\phi(s)$ invertible elements, and Theorem 7.8 tells us that an element $[a]_s \in \mathbb{Z}_s$ is invertible precisely if a and s are relatively prime. Therefore, every column contains exactly $\phi(s)$ entries relatively prime to s. As a result, in each of the $\phi(t)$ columns consisting entirely of entries relatively prime to t, there are $\phi(s)$ entries that are also relatively prime to s. Thus, the number of entries in our table relatively prime to both s and t must be equal to $\phi(s) \cdot \phi(t)$. Hence, $\phi(n) = \phi(s) \cdot \phi(t)$. $\square$

In Lemma 7.11, we determined the formula for $\phi(n)$ in the special case where $n = p^m$. Combining that result with Lemma 7.16, we can now determine the general formula for $\phi(n)$.

Theorem 7.17. *If* $n = p_1{}^{a_1} \cdot p_2{}^{a_2} \cdots p_m{}^{a_m}$ *is the prime factorization of* n, *where the primes* p_i *are all different and each* $a_i \geq 1$, *then*

$$\phi(n) = p_1{}^{a_1-1}(p_1 - 1) \cdot p_2{}^{a_2-1}(p_2 - 1) \cdots p_m{}^{a_m-1}(p_m - 1).$$

Proof. Let $T = \{k \in \mathbb{N} \mid \text{the formula for } \phi \text{ holds for all } n \in \mathbb{N} \text{ with } k \text{ different prime divisors}\}$; it suffices to show that $T = \mathbb{N}$, and we will proceed by Mathematical Induction. To show that $1 \in T$, we need to verify that the formula for ϕ holds for all $n \in \mathbb{N}$ with only one prime divisor. But this means that n is of the form $p_1{}^{a_1}$. However, in this special case, Lemma 7.11 tells us that the formula for ϕ is correct. Thus, $1 \in T$.

Next, suppose T contains some positive integer k; we need to show that T also contains $k + 1$. Thus, we must show that the formula for ϕ holds for all positive integers with exactly $k + 1$ different prime divisors. To this end, suppose

$$n = p_1{}^{a_1} \cdot p_2{}^{a_2} \cdots p_k{}^{a_k} \cdot p_{k+1}{}^{a_{k+1}},$$

where all the primes p_i are different and each $a_i \geq 1$. If we let

$$s = p_1{}^{a_1} \cdot p_2{}^{a_2} \cdots p_k{}^{a_k} \quad \text{and} \quad t = p_{k+1}{}^{a_{k+1}},$$

then s and t are relatively prime as p_{k+1} is the only prime divisor of t, but p_{k+1} is not a divisor of s. Therefore, we can apply Lemma 7.16 to obtain the fact that $\phi(n) = \phi(s \cdot t) = \phi(s) \cdot \phi(t)$. However, since both 1 and k belong to T, we know that the formula for ϕ holds for s and t. Thus,

$$\phi(s) = p_1{}^{a_1-1}(p_1 - 1) \cdot p_2{}^{a_2-1}(p_2 - 1) \cdots p_k{}^{a_k-1}(p_k - 1)$$

and

$$\phi(t) = p_{k+1}{}^{a_{k+1}-1}(p_{k+1} - 1).$$

As a result, we now have

$$\phi(n) = \phi(s) \cdot \phi(t) =$$
$$(p_1{}^{a_1-1}(p_1 - 1) \cdot p_2{}^{a_2-1}(p_2 - 1) \cdots p_k{}^{a_k-1}(p_k - 1))(p_{k+1}{}^{a_{k+1}-1}(p_{k+1} - 1)) =$$
$$p_1{}^{a_1-1}(p_1 - 1) \cdot p_2{}^{a_2-1}(p_2 - 1) \cdots p_k{}^{a_k-1}(p_k - 1)p_{k+1}{}^{a_{k+1}-1}(p_{k+1} - 1).$$

Thus, $k + 1 \in T$, thereby proving the result. $\square$

■ Examples

$$\phi(100) = \phi(2^2 \cdot 5^2) = \phi(2^2) \cdot \phi(5^2) = 2^1(2-1) \cdot 5^1(5-1) = 40,$$

$$\phi(360) = \phi(2^3 \cdot 3^2 \cdot 5^1) = \phi(2^3) \cdot \phi(3^2) \cdot \phi(5^1) = 2^2(2-1) \cdot 3^1(3-1) \cdot 5^0(5-1) = 96,$$

$$\phi(847) = \phi(7^1 \cdot 11^2) = \phi(7^1) \cdot \phi(11^2) = 7^0(7-1) \cdot 11^1(11-1) = 660.$$

■

Theorem 7.5 told us that if $[a]_n \in U(\mathbb{Z}_n)$, then there exists a positive integer m such that $[a]_n^m = [1]_n$. We can now show that the exponent $\phi(n)$ works for every $[a]_n \in U(\mathbb{Z}_n)$.

Proposition 7.18. *In the group $U(\mathbb{Z}_n)$, $[a]_n^{\phi(n)} = [1]_n$, for every $[a]_n \in U(\mathbb{Z}_n)$.*

Proof. We begin with a very common argument which we already used back in the proof of Theorem 6.8. If g, h, k are elements of a group G such that $h \neq k$, then we claim that $gh \neq gk$. To see this, observe that if $gh = gk$, then $g^{-1} \cdot gh = g^{-1} \cdot gk$, which would lead to the contradiction $h = k$.

The preceding fact can be applied to $U(\mathbb{Z}_n)$ to assert that if $[a]_n \in U(\mathbb{Z}_n)$ and if $[b_1]_n, [b_2]_n, \ldots, [b_{\phi(n)}]_n$ are the $\phi(n)$ different elements of $U(\mathbb{Z}_n)$, then the elements

$$[a]_n \cdot [b_1]_n, \ [a]_n \cdot [b_2]_n, \ldots, \ [a]_n \cdot [b_{\phi(n)}]_n$$

are also the $\phi(n)$ different elements of $U(\mathbb{Z}_n)$. Since multiplication in $U(\mathbb{Z}_n)$ is commutative, the product of the $\phi(n)$ element of $U(\mathbb{Z}_n)$ gives us the same answer regardless of the order of the elements. In light of this, we now have

$$[b_1]_n \cdot [b_2]_n \cdots [b_{\phi(n)}]_n = ([a]_n \cdot [b_1]_n) \cdot ([a]_n \cdot [b_2]_n) \cdots ([a]_n \cdot [b_{\phi(n)}]_n).$$

If we let $[c]_n = [b_1]_n \cdot [b_2]_n \cdots [b_{\phi(n)}]_n$, then by applying the associativity and commutativity of multiplication in $\mathbb{Z}_n$, we have

$$[c]_n = [b_1]_n \cdot [b_2]_n \cdots [b_{\phi(n)}]_n = ([a]_n \cdot [b_1]_n) \cdot ([a]_n \cdot [b_2]_n) \ldots ([a]_n \cdot [b_{\phi(n)}]_n) =$$

$$[a]_n^{\phi(n)} \cdot ([b_1]_n \cdot [b_2]_n \cdots [b_{\phi(n)}]_n) = [a]_n^{\phi(n)} \cdot [c]_n.$$

If we multiply the preceding equation by $[c]_n^{-1}$, we immediately see that $[a]_n^{\phi(n)} = [1]_n$. □

■ Examples

The group $U(\mathbb{Z}_{12})$ has $\phi(12) = 4$ elements, and they are $[1]_{12}, [5]_{12}, [7]_{12}, [11]_{12}$. Proposition 7.18 now tells us that

$$[1]_{12}^4 = [1^4]_{12} = [1]_{12}, \quad [5]_{12}^4 = [5^4]_{12} = [625]_{12} = [1]_{12},$$

$$[7]_{12}^4 = [7^4]_{12} = [2401]_{12} = [1]_{12}, \quad [11]_{12}^4 = [11^4]_{12} = [14641]_{12} = [1]_{12}.$$

Similarly, the group $U(\mathbb{Z}_9)$ has $\phi(9) = 6$ elements, and they are $[1]_9, [2]_9, [4]_9, [5]_9, [7]_9, [8]_9$. Proposition 7.18 now tells us that

$$[1]_9^6 = [1^6]_9 = [1]_9, \quad [2]_9^6 = [2^6]_9 = [64]_9 = [1]_9,$$

$$[4]_9^6 = [4^6]_9 = [4096]_9 = [1]_9, \quad [5]_9^6 = [5^6]_9 = [15625]_9 = [1]_9,$$

$$[7]_9^6 = [7^6]_9 = [117649]_9 = [1]_9, \quad [8]_9^6 = [8^6]_9 = [262144]_9 = [1]_9.$$

■

7.4 Polynomials with Coefficients in $\mathbb{Z}_n$

In Chapter 5, we remarked that when R is a commutative ring, then so is $R[x]$, the set of polynomials with coefficients in R. As a result, by looking at $\mathbb{Z}_n[x]$, for various values of n, we obtain many additional examples of commutative rings. Before continuing, we need to discuss some of the notation and terminology associated with $\mathbb{Z}_n[x]$ in order to make sure that we fully understand what elements of $\mathbb{Z}_n[x]$ look like. For example, some typical elements of $\mathbb{Z}_5[x]$ are

$$[3]_5 x^2 + [2]_5 x + [4]_5, \quad [4]_5 x^6 + [3]_5 x^2, \quad [1]_5 x^{14} + [2]_5 x^7 + [4]_5 x^3 + [2]_5.$$

It is extremely important to understand that although the coefficients of the polynomials in $\mathbb{Z}_5[x]$ are elements of $\mathbb{Z}_5$, the exponents are ordinary integers. Therefore,

$$[8]_5 x^7 = [3]_5 x^7,$$

since $[8]_5 = [3]_5$. On the other hand,

$$[4]_5 x^8 \neq [4]_5 x^3,$$

since $x^8 \neq x^3$. Ordinarily, if a term in a polynomial is not the constant term and its coefficient is 1, then we need not write the 1. For example, in $\mathbb{Q}[x]$, we usually write

$$3x^2 + 1x + 7 \quad \text{as} \quad 3x^2 + x + 7.$$

Similarly, in $\mathbb{Z}_5[x]$, we can write

$$[3]_5 x^2 + [1]_5 x + [2]_5 \quad \text{as} \quad [3]_5 x^2 + x + [2]_5.$$

Furthermore, if the coefficient of a term in a polynomial is 0, then we usually skip writing the entire term. For example, in $\mathbb{R}[x]$, we can write

$$6x^3 - 17x^2 + 0x + 4 \quad \text{as} \quad 6x^3 - 17x^2 + 4.$$

Similarly, in $\mathbb{Z}_5[x]$, we can write

$$[2]_5 x^3 + [0]_5 x^2 + [3]_5 x + [0]_5 \quad \text{as} \quad [2]_5 x^3 + [3]_5 x.$$

If every coefficient of a polynomial is the additive identity, then we simply write the polynomial as 0.

Given an element $a \in \mathbb{Z}_n$ and a polynomial $p(x) \in \mathbb{Z}_n[x]$, we can easily check if a is a root of $p(x)$. For example, if we consider $p(x) = x^2 + [2]_5 x + [1]_5 \in \mathbb{Z}_5[x]$, then $[4]_5$ is a root because

$$p([4]_5) = [4]_5^2 + [2]_5 \cdot [4]_5 + [1]_5 = [16]_5 + [8]_5 + [1]_5 = [1]_5 + [3]_5 + [1]_5 = [5]_5 = [0]_5.$$

However, $[3]_5$ is not a root as

$$p([3]_5) = [3]_5^2 + [2]_5 \cdot [3]_5 + [1]_5 = [9]_5 + [6]_5 + [1]_5 = [4]_5 + [1]_5 + [1]_5 =$$
$$[6]_5 = [1]_5 \neq [0]_5.$$

As we consider examples where $\mathbb{Z}_n$ is not a field, some new and surprising things occur. In $\mathbb{Z}_8[x]$, consider the polynomial $g(x) = x^2 + [7]_8$. You can check that $[1]_8, [3]_8, [5]_8, [7]_8$ are all roots of $g(x)$. Thus, $g(x)$ is a polynomial of degree 2 with four different roots. This is a phenomenon that does not occur when we look at polynomials with coefficients in $\mathbb{Z}, \mathbb{Q}, \mathbb{R}$, or $\mathbb{C}$. In fact, as we will see in Chapter 12, the number of roots that a polynomial can have in a field cannot exceed the degree of the polynomial. But as this example shows, if the coefficients belong to a commutative ring that has zero divisors, the situation can be very different.

Next, in $\mathbb{Z}_8[x]$, consider the product of the polynomials $h(x) = [2]_8 x^2 + [6]_8 x$ and $j(x) = [4]_8 x$. We now have

$$h(x) \cdot j(x) = ([2]_8 x^2 + [6]_8 x)([4]_8 x) = [8]_8 x^3 + [24]_8 x^2 = [0]_8 x^3 + [0]_8 x^2 = 0.$$

In this example, we multiplied a polynomial of degree 2 with a polynomial of degree 1 and received 0 as the answer. In the past, when we multiplied polynomials, we added their degrees. This will certainly still be the case if our coefficients belong to a field. But, once again, if we are dealing with commutative rings with zero divisors, interesting new things can occur.

Since $\mathbb{Z}_p$ is field, for every prime number p, many of the algebraic ideas and techniques used to find roots of polynomials or solve equations in $\mathbb{Z}_p[x]$ will be familiar to you. For example, consider the equation $3x + 4 = 3$ in $\mathbb{Q}[x]$. To solve this, we first subtract 4 from both sides to obtain $3x = -1$ and then divide by 3 to conclude that $x = -\frac{1}{3}$. If we restate the steps taken in terms of addition and multiplication, we can say that we first added -4 to both sides and then multiplied by $\frac{1}{3}$.

Now let us examine a similar situation in $\mathbb{Z}_5[x]$ and consider the equation

$$[3]_5 x + [4]_5 = [3]_5.$$

Similar to the preceding situation, we first add $[-4]_5$ to both sides to obtain

$$[3]_5 x = [3]_5 + [-4]_5 = [-1]_5 = [4]_5.$$

In $\mathbb{Z}_5$, the multiplicative inverse of $[3]_5$ is $[2]_5$. Therefore, in this case we multiply the previous equation by $[2]_5$ to obtain

$$x = [2]_5 \cdot [4]_5 = [8]_5 = [3]_5.$$

As always, it is a good idea to check one's answer and if we plug $x = [3]_5$ into our equation, we have

$$[3]_5 \cdot [3]_5 + [4]_5 = [9]_5 + [4]_5 = [13]_5 = [3]_5,$$

as desired.

When looking for the roots of a polynomial, the first thing we often do is to try to factor the polynomial. For example, if we are looking for the roots in $\mathbb{R}$ of $x^2 - 7x + 10$, we can factor it as

(5) $x^2 - 7x + 10 = (x - 2)(x - 5).$

Replacing x by either 2 or 5 in (5) makes one of the factors of the right-hand side equal to 0. Since multiplying by 0 in a commutative ring always yields a product of 0, we can see that both 2 and 5 are roots of $x^2 - 7x + 10$. The same reasoning tells us that whenever we factor a polynomial into polynomials of smaller degree, any root of one of the factors must be a root of the original polynomial.

On the other hand, if in (5) we replace x with a real number other than 2 or 5, then both $x - 2$ and $x - 5$ will be nonzero real numbers. Since $\mathbb{R}$ has no zero divisors, the value of $x^2 - 7x + 10 = (x - 2)(x - 5)$ will also be a nonzero real number. Thus, the fact that $\mathbb{R}$ has no zero divisors is the reason that 2 and 5 are the *only* roots of $x^2 - 7x + 10$ in $\mathbb{R}$. Since $\mathbb{Z}, \mathbb{Q}, \mathbb{R}$, and $\mathbb{C}$ all contain no zero divisors, the same reasoning tells us whenever we can factor a polynomial in $\mathbb{Z}[x]$, $\mathbb{Q}[x]$, $\mathbb{R}[x]$, or $\mathbb{C}[x]$ into linear factors, we can then easily read off *all* of its roots.

Next, suppose we examine the polynomial $x^2 + [1]_5$ in $\mathbb{Z}_5[x]$. Observe that in $\mathbb{Z}_5[x]$, this polynomial does indeed factor, and we have

$$x^2 + [1]_5 = (x + [2]_5)(x + [3]_5).$$

Since $\mathbb{Z}_5$ also has no zero divisors, in order for an element of $\mathbb{Z}_5$ to be a root of $x^2 + [1]_5$, it must make either $x + [2]_5$ or $x + [3]_5$ equal to zero. Thus, the only roots in $\mathbb{Z}_5$ of $x^2 + [1]_5$ are $[3]_5$ and $[2]_5$.

Earlier we saw that $[1]_8$, $[3]_8$, $[5]_8$, and $[7]_8$ are all roots of the polynomial $x^2 + [7]_8$ in $\mathbb{Z}_8$. Note that if we factor $x^2 + [7]_8$ as $(x + [1]_8)(x + [7]_8)$ in $\mathbb{Z}_8[x]$, then it is clear that $[7]_8$ and $[1]_8$ are roots. However, these are not the only roots in $\mathbb{Z}_8$. Observe that there are elements in $\mathbb{Z}_8$ that fail to make either $(x + [1]_8)$ or $(x + [7]_8)$ equal to zero yet are still roots of $x^2 + [7]_8$. In particular, if we let $x = [3]_8$, then $(x + [1]_8)(x + [7]_8)$ becomes

$$([3]_8 + [1]_8)([3]_8 + [7]_8) = ([4]_8)([10]_8) = [4]_8 \cdot [2]_8 = [8]_8 = [0]_8.$$

Similarly, if we let $x = [5]_8$, then $(x + [1]_8)(x + [7]_8)$ becomes

$$([5]_8 + [1]_8)([5]_8 + [7]_8) = ([6]_8)([12]_8) = [6]_8 \cdot [4]_8 = [24]_8 = [0]_8.$$

Another surprising aspect of this example is that not only can we factor $x^2 + [7]_8$ as $(x + [1]_8)(x + [7]_8)$ in $\mathbb{Z}_8[x]$, but we can also factor it as $(x + [3]_8)(x + [5]_8)$. Thus, in the presence of zero divisors, there may be more than one way to factor a polynomial into linear factors. Hopefully, this example helps to illustrate why the number of roots of a polynomial can exceed its degree when working in a commutative ring with zero divisors.

In your earlier algebra courses, when you were asked to find all the roots of a polynomial, the first thing you probably did was to try to factor the polynomial into linear factors. When doing this, you weren't concerned with concepts and terms like *commutative rings* and *zero divisors*. But we can now see that using the linear factors to determine all the roots of a polynomial required that the coefficients of our polynomial belonged to a commutative ring without zero divisors.

In this chapter we observed that since $\mathbb{Z}_8$ has zero divisors, working with polynomials in $\mathbb{Z}_8[x]$ was very different from our previous experiences with $\mathbb{Z}[x]$, $\mathbb{Q}[x]$, $\mathbb{R}[x]$, and $\mathbb{C}[x]$. However, $\mathbb{Z}_p$ has no zero divisors whenever p is prime. Therefore, working with polynomials in $\mathbb{Z}_p[x]$ will be quite similar to our experiences with $\mathbb{Z}[x]$, $\mathbb{Q}[x]$, $\mathbb{R}[x]$, and $\mathbb{C}[x]$. In fact, in the next chapter, we will exploit properties of $\mathbb{Z}_p[x]$ to better understand the roots and factoring of polynomials in $\mathbb{Z}[x]$ and $\mathbb{Q}[x]$.

Exercises for Sections 7.3 and 7.4

1. Compute $\phi(3), \phi(9), \phi(27), \phi(81)$, and $\phi(243)$.

2. Compute $\phi(7), \phi(49), \phi(343), \phi(2401)$, and $\phi(16807)$.

3. Compute $\phi(6), \phi(18), \phi(54), \phi(162)$, and $\phi(486)$.

4. Compute $\phi(14), \phi(98), \phi(686), \phi(4802)$, and $\phi(33614)$.

5. Compute $\phi(4900)$.

6. Compute $\phi(64,000,000)$.

7. Compute $\phi(7!)$. Recall that if $n \in \mathbb{N}$, then $n! = 1 \cdot 2 \cdots (n-1) \cdot n$.

8. Compute $\phi(8!)$.

9. Compute $\phi(2^5 \cdot 3^4 \cdot 5^3 \cdot 7^2 \cdot 11^1)$.

10. Compute $\phi(5^3 \cdot 7^4 \cdot 13^2 \cdot 17^4)$.

11. Prove that if $n > 2$ is an integer, then $\phi(n)$ is even.

12. Prove that if $n \geq 3$ is an odd integer, then $\phi(2n) = \phi(n)$.

13. Prove that if $n \geq 2$ is an even integer, then $\phi(2n) = 2\phi(n)$.

14. Find all integers $n \geq 2$ such that $\phi(n) = 2$.

15. Find all integers $n \geq 2$ such that $\phi(n) = 4$.

16. Find all positive integers n such that $\phi(n)$ is a prime number.

17. Let $n \geq 2$ be an integer and let p be a prime number.
    (a) Prove that if p does not divide n, then $\phi(pn) = (p-1)\phi(n)$.

    (b) Prove that if p divides n, then $\phi(pn) = p\phi(n)$.

18. If $n \in \mathbb{N}$, then $\omega \in \mathbb{C}$ is called a **primitive** nth root of 1 if every other nth root of 1 in $\mathbb{C}$ can be written as ω raised to some positive integer power. For example, -1 is the only primitive square root of 1 in $\mathbb{C}$, whereas i and $-i$ are the only primitive 4th roots of 1 in $\mathbb{C}$. Show that if $n \geq 2$, then there are exactly $\phi(n)$ primitive nth roots of 1 in $\mathbb{C}$.

19. Compute the values of $p(x) = x^2 + [3]_4 x + [2]_4$ in $\mathbb{Z}_4$ as x takes on the values $[0]_4, [1]_4, [2]_4, [3]_4$.

20. Compute the values of $q(x) = x^2 + [5]_6 x + [2]_6$ in $\mathbb{Z}_6$ as x takes on the values $[0]_6, [1]_6, [2]_6, [3]_6, [4]_6, [5]_6$.

In exercises 21–24, find the solutions in $\mathbb{Z}_6$ of the given equations. Observe that since $\mathbb{Z}_6$ has only six elements, trial-and-error is often a reasonable way to proceed.

21. $[2]_6 x + [5]_6 = [3]_6$

22. $[2]_6 x + [5]_6 = [4]_6$

23. $x^2 + x = [0]_6$

24. $x^2 + x + [5]_6 = [0]_6$

In exercises 25–28, you will be doing computation in the ring $\mathbb{Z}_7[x]$.

25. $([4]_7 x + [2]_7) \cdot ([3]_7 x^2 + [5]_7 x + [6]_7) =$

26. $([5]_7 x^3 + [2]_7 x + [1]_7) \cdot ([6]_7 x^2 + [4]_7 x) =$

27. $(x + [5]_7)^3 =$

28. $([2]_7 x + [3]_7) \cdot (x + [6]_7) \cdot ([5]_7 x) =$

In exercises 29–32, you will be doing computation in the ring $\mathbb{Z}_{10}[x]$.

29. $([3]_{10} x + [8]_{10}) \cdot ([7]_{10} x^2 + [5]_{10} x + [4]_{10}) =$

30. $([5]_{10} x^3 + [8]_{10} x + [3]_{10}) \cdot ([4]_{10} x^2 + [8]_{10} x) =$

31. $(x + [4]_{10})^3 =$

32. $([6]_{10} x + [7]_{10}) \cdot ([5]_{10} x + [2]_{10}) \cdot ([7]_{10} x + [4]_{10}) =$

In exercises 33–36, find a solution to the linear equation in field $\mathbb{Z}_{11}$.

33. $[2]_{11} x + [9]_{11} = [6]_{11}$

34. $[7]_{11} x + [10]_{11} = [4]_{11}$

35. $[6]_{11} ([3]_{11} x + [2]_{11}) = [1]_{11}$

36. $[8]_{11} ([3]_{11} x + [5]_{11}) = [3]_{11} ([9]_{11} x + [6]_{11})$

In exercises 37–40, determine if the quadratic polynomial can be written as a product of two linear polynomials in $\mathbb{Z}_5[x]$.

37. $x^2 + [2]_5 x + [2]_5$

38. $[4]_5 x^2 + [2]_5 x + [4]_5$

39. $x^2 + [3]_5 x + [3]_5$

40. $[3]_5 x^2 + [3]_5 x + [4]_5$

In exercises 41–44, determine if the quadratic polynomial can be written as a product of two linear polynomials in $\mathbb{Z}_7[x]$.

41. $x^2 + [5]_7$

42. $[3]_7 x^2 + [5]_7 x + [6]_7$

43. $x^2 + [4]_7 x + [5]_7$

44. $[2]_7 x^2 + [5]_7 x + [6]_7$

45. Let $\pi : \mathbb{Z}_2[x] \to \mathbb{Z}_2[x]$ be defined as $\pi(f(x)) = f(x)^2$, for all $f(x) \in \mathbb{Z}_2[x]$.
    (a) Show that $\pi(f(x) + g(x)) = \pi(f(x)) + \pi(g(x))$, for all $f(x), g(x) \in \mathbb{Z}_2[x]$.

    (b) Show that $\pi(f(x) \cdot g(x)) = \pi(f(x)) \cdot \pi(g(x))$, for all $f(x), g(x) \in \mathbb{Z}_2[x]$.

    (c) Show that $a^2 = a$, for all $a \in \mathbb{Z}_2$.

    (d) Use parts (a), (b), and (c) to show that $\pi(h(x)) = h(x^2)$, for all $h(x) \in \mathbb{Z}_2[x]$.

46. Let $\rho : \mathbb{Z}_3[x] \to \mathbb{Z}_3[x]$ be defined as $\rho(f(x)) = f(x)^3$, for all $f(x) \in \mathbb{Z}_3[x]$.
    (a) Show that $\rho(f(x) + g(x)) = \rho(f(x)) + \rho(g(x))$, for all $f(x), g(x) \in \mathbb{Z}_3[x]$.

    (b) Show that $\rho(f(x) \cdot g(x)) = \rho(f(x)) \cdot \rho(g(x))$, for all $f(x), g(x) \in \mathbb{Z}_3[x]$.

    (c) Show that $a^3 = a$, for all $a \in \mathbb{Z}_3$.

    (d) Use parts (a), (b), and (c) to show that $\rho(h(x)) = h(x^3)$, for all $h(x) \in \mathbb{Z}_3[x]$.

47. List all the monic polynomials of degree 2 in $\mathbb{Z}_2[x]$.

48. In $\mathbb{Z}_2$, find the roots of each of the polynomials in your answer to the previous exercise.

49. List all the monic polynomials of degree 2 in $\mathbb{Z}_3[x]$.

50. In $\mathbb{Z}_3$, find the roots of each of the polynomials in your answer to the previous exercise.

51. List all the monic polynomials of degree 3 in $\mathbb{Z}_2[x]$.

52. In $\mathbb{Z}_2$, find the roots of each of the polynomials in your answer to the previous exercise.

53. If p is a prime number and $a \in \mathbb{Z}$ is relatively prime to p, show that $[a]_p^{p-1} = [1]_p$.

54. Use exercise 53 to show that if p is a prime number, then all p elements of $\mathbb{Z}_p$ are roots of the polynomial $x^p + [p-1]_p x$.

55. Use exercise 54 to show that if p is a prime number, then the polynomial $x^p + [p-1]_p x + [1]_p$ has no roots in $\mathbb{Z}_p$.

56. Let G be an abelian group with n elements. If we let e denote the identity element of G, prove that $g^n = e$, for all $g \in G$. (Hint: Think about the proof of Proposition 7.18.)

57. Let F be a field with n elements and let 1 denote the multiplicative identity of F.

    (a) Show that $a^n = a$, for all $a \in F$. You might want to look back at exercise 56.

    (b) Use part (a) to conclude that none of the elements of F are roots of the polynomial $x^n - x + 1$.

In exercises 58–66, we will look at another way to find the roots of quadratics in $\mathbb{Z}_p$. Remember that when you first saw the quadratic formula, it was probably derived by first completing the square. We can now apply similar techniques to look for the roots of polynomials with coefficients in $\mathbb{Z}_p$, where $p > 2$ is a prime number. For example, suppose we want to find the roots of $[2]_7 x^2 + [3]_7 x + [5]_7$ in $\mathbb{Z}_7$. We begin with the equation $[2]_7 x^2 + [3]_7 x + [5]_7 = 0$ and, in order to make the polynomial monic, multiply both sides of the equation by the multiplicative inverse of $[2]_7$ to obtain $x^2 + [5]_7 x + [6]_7 = [0]_7$. Next, we add the multiplicative inverse of $[6]_7$ to both sides to obtain $x^2 + [5]_7 x = [1]_7$. Recall that at this point in your earlier algebra classes, you would take half of the coefficient of the x term and add its square to both sides to turn the left side of the equation into the square of a linear polynomial. The analogous thing for us to do is to multiply $[5]_7$ by the multiplicative inverse of $[2]_7$ to obtain $[6]_7$ and then to add the square of $[6]_7$ to both sides, resulting in $x^2 + [5]_7 x + [1]_7 = [2]_7$. Observe that the left side of the equation is now the square of $x + [6]_7$. Thus, we are now dealing with the equation $(x + [6]_7)^2 = [2]_7$. Next, note that $[2]_7$ is a square in $\mathbb{Z}_7$ as $[3]_7^2 = [4]_7^2 = [2]_7$. Therefore, our equation now becomes $(x + [6]_7)^2 = [3]_7^2 = [4]_7^2$. This reduces to solving in $\mathbb{Z}_7$ the linear equations $x + [6]_7 = [3]_7$ and $x + [6]_7 = [4]_7$, which leads to the solutions $x = [4]_7$ and $x = [5]_7$. Keep in mind that in $\mathbb{Z}_p$, where $p > 2$ is a prime, not every element is a square. Thus, as you attempt to complete the square, you will find that some quadratics do not have a solution in $\mathbb{Z}_p$.

58. In $\mathbb{Z}_7$, find the square of each equivalence class.

59. Find the roots in $\mathbb{Z}_7$ of $x^2 + x + [1]_7$ by completing the square and using exercise 58.

60. Find the roots in $\mathbb{Z}_7$ of $[3]_7 x^2 + [4]_7 x + [3]_7$ by completing the square and using exercise 58.

61. Find the roots in $\mathbb{Z}_7$ of $x^2 + [5]_5 x + [2]_7$ by completing the square and using exercise 58.

62. Find the roots in $\mathbb{Z}_7$ of $[5]_7 x^2 + [3]_7 x + [4]_7$ by completing the square and using exercise 58.

63. Explain why the technique of completing the square in $\mathbb{Z}_p$ requires that the prime p must be greater than 2.

64. In $\mathbb{Z}_{11}$, find the square of each equivalence class.

65. Find the roots in $\mathbb{Z}_{11}$ of $[4]_{11} x^2 + [6]_{11} x + [1]_{11}$ by completing the square and using exercise 64.

66. Find the roots in $\mathbb{Z}_{11}$ of $[4]_{11} x^2 + [6]_{11} x + [8]_{11}$ by completing the square and using exercise 64.

CHAPTER 8

Group Theory

In the first seven chapters of this book, we examined $\mathbb{Z}$, $\mathbb{Q}$, $\mathbb{R}$, $\mathbb{C}$, $\mathbb{Z}_n$, and other examples of commutative rings and fields. Along the way, we also looked at the bijections of various sets, such as the roots of polynomials.

Groups are algebraic objects that occur, often in more than one way, in each of the examples just mentioned. We have briefly touched upon groups in Chapters 2, 5, and 7. In light of the essential role played by groups, especially finite groups, in Galois' proof of the insolvability of the quintic, it is time for groups to take center stage.

In the first section of this chapter, you will see some definitions and examples that have appeared in earlier chapters. However, repetition can be a good thing. The structure of finite groups is more abstract and theoretical than the topics you examined earlier in this book. Therefore, seeing some definitions and examples for a second time should help you deal with this greater degree of abstraction. When you finish this chapter, not only will you have seen the crown jewel of an introductory course in group theory, Sylow's Theorem, but you will also be equipped with all the group theory needed to understand Galois' work on fifth-degree polynomials.

8.1 Definitions and Examples

We begin this section with the definition of a group, which you have already seen as Definition 5.19.

Definition 8.1. *Let G be a set with a binary operation* ∘. *We say that G is a group if* ∘ *satisfies the following properties:*

1. *Associative Law: For every $x, y, z \in G$, $(x \circ y) \circ z = x \circ (y \circ z)$.*

2. *Identity: There is an element in G, usually denoted as e, such that $x \circ e = x = e \circ x$, for every $x \in G$.*

3. *Inverses: For every $x \in G$ there is an element in G, usually denoted as x^{-1}, such that $x \circ x^{-1} = e = x^{-1} \circ x$.*

If G is a group, we let $|G|$ be the number of elements in the set G and call it the *order* of the group G. More generally, for subsets M of G, we let $|M|$ denote the number of elements in M. In light of the experience we gained in earlier chapters, we will begin looking at examples of groups by focusing on three large classes:

(I)   Commutative Rings and Fields under Addition

(II)   Invertible Elements in Commutative Rings under Multiplication

(III)   Bijections of Sets

I. Commutative Rings and Fields under Addition

Recall that if R is a commutative ring, then addition is a binary operation that satisfies the three conditions required to be a group. Multiplication is also a binary operation on a commutative ring, but as we will see a little later, commutative rings do not form a group under multiplication. However, since every commutative ring is a group under addition, all ten sets on the following list are groups under addition:

(a) $\mathbb{Z}$, the set of integers

(b) $\mathbb{Q}$, the set of rational numbers

(c) $\mathbb{R}$, the set of real numbers

(d) $\mathbb{C}$, the set of complex numbers

(e) $\mathbb{Z}_n$, the set of integers modulo n

(f) $\mathbb{Q}[x]$, the set of polynomials with rational coefficients

(g) $\mathbb{R}[x, y]$, the set of polynomials in two variables with real coefficients

(h) $\mathbb{Q}(i) = \{a + bi \mid a, b \in \mathbb{Q}\}$

(i) $\mathbb{Z}(i) = \{a + bi \mid a, b \in \mathbb{Z}\}$

(j) $\mathbb{Q}(i, \sqrt{2}) = \{a + bi + c\sqrt{2} + di\sqrt{2} \mid a, b, c, d \in \mathbb{Q}\}$

At this point, there is another important group under addition worth noting: If $n \in \mathbb{Z}$, let $n\mathbb{Z} = \{na \mid a \in \mathbb{Z}\}$. Although $n\mathbb{Z}$ is not a ring, as it does not contain the identity element under multiplication, it will turn out to be a very useful group in motivating various concepts.

II. Invertible Elements in Commutative Rings under Multiplication

If R is a commutative ring, then multiplication is also a binary operation. However, R is not a group under multiplication, as elements of R can fail to have a multiplicative inverse in R.

Indeed, since $0 \cdot r = 0$, for all $r \in R$, it is impossible to multiply 0 by an element of R to obtain 1 as the answer. Thus, the additive identity of R does not have a multiplicative inverse, so R is not a group under multiplication.

Therefore, in our search for groups, it makes sense to consider the nonzero elements of commutative rings. Indeed, if R is a field, then the nonzero elements of R are a group. However, for other commutative rings, the situation can be somewhat different. If we consider $\mathbb{Z}$, then 1 and -1 are the only integers that have a multiplicative inverse in $\mathbb{Z}$. Remember, it is not enough that the nonzero elements of $\mathbb{Z}$ have multiplicative inverses in $\mathbb{Q}$. For a set to be a group, the inverse of every element needs to belong to the set and not to a larger set. Yet another problem arises when we consider the ring $\mathbb{Z}_6$. Note that $[2]_6, [3]_6, [4]_6$ are zero divisors in $\mathbb{Z}_6$, so they are nonzero elements that do not have inverses in $\mathbb{Z}_6$. In light of the examples of $\mathbb{Z}$ and $\mathbb{Z}_6$, it now makes sense to consider the elements in a commutative ring that have multiplicative inverses in the ring. In fact, in Proposition 7.7, we showed that $U(R)$, the set of elements in R with a multiplicative inverse in R, is a group under multiplication. Therefore, the following sets are all groups under multiplication:

(a) $\{1, -1\}$, the set of invertible elements in $\mathbb{Z}$

(b) $\mathbb{Q}^\times$, the set of nonzero rational numbers

(c) $\mathbb{R}^\times$, the set of nonzero real numbers

(d) $\mathbb{C}^\times$, the set of nonzero complex numbers

(e) $\{[1]_5, [2]_5, [3]_5, [4]_5\}$, the set of nonzero elements of the field $\mathbb{Z}_5$

(f) $\{[1]_6, [5]_6\}$, the set of invertible elements in $\mathbb{Z}_6$

(g) $\{[1]_{10}, [3]_{10}, [7]_{10}, [9]_{10}\}$, the set of invertible elements in $\mathbb{Z}_{10}$

(h) $\mathbb{Q}^\times$, which is also the set of invertible elements in $\mathbb{Q}[x]$

(i) $\{1, -1, i, -i\}$, the set of invertible elements in $\mathbb{Z}(i)$

(j) $\{1, -1\}$, which is also the set of invertible elements in $\mathbb{Z}[x]$

III. Bijections of Sets

In Section 2.4, we showed that if S is a set and $\circ$ represents composition of functions, then $\circ$ is a binary operation on each of the following sets:

(a) $F(S)$, the set of functions from S to S

(b) $Inj(S)$, the set of injective functions from S to S

(c)   *Sur(S)*, the set of surjective functions from *S* to *S*

(d)   *Bij(S)*, the set of bijective functions from *S* to *S*

If we look back at Theorem 2.16, we can see that it verifies that *Bij(S)* is a group under ∘. We now take a look at the sets *F(S)*, *Inj(S)*, and *Sur(S)*. Theorem 2.11 asserted that the composition of functions is associative. Furthermore, *F(S)*, *Inj(S)*, and *Sur(S)* all contain the identity map *e*. Therefore the only thing that could prevent *F(S)*, *Inj(S)*, *Sur(S)* from being groups under ∘ is if at least one element failed to have an inverse.

If we consider the very special case where *S* has only one element, then *F(S)*, *Inj(S)*, *Sur(S)*, and *Bij(S)* are all the same and consist only of the identity map *e*. Thus, in this case and only in this case, *F(S)*, *Inj(S)*, *Sur(S)*, and *Bij(S)* are all groups. Now let us consider the case where *S* contains more than one element. If $t \in S$, consider the function defined as $f(s) = t$, for all $s \in S$. Since *S* has more than one element, the function *f* takes on the value of *t* more than once. Thus, *f* does not have an inverse f^{-1}, as $f^{-1}(t)$ would not be defined. Thus, as long as *S* has more than one element, *F(S)* is not a group.

We now turn our attention to *Inj(S)* and *Sur(S)* and consider the important case where *S* is finite. If *S* has *n* elements and $f \in Inj(S)$, then the range of *f* consists of *n* different elements of *S*. Thus, the range of *f* is all of *S*, so *f* is also surjective.

On the other hand, if *S* has *n* elements and $g \in Sur(S)$, then *g* cannot repeat any values. To see this, observe that if *g* gave us the same value for two different elements of *S*, then the range of *g* would have at most $n-1$ elements of *S*. Hence, $g \in Inj(S)$, and we have shown, when *S* is finite, that $Inj(S) = Sur(S) = Bij(S)$. Thus, in this case, *Inj(S)*, *Sur(S)*, and *Bij(S)* are all groups.

When *S* is infinite, there are elements of *Inj(S)* that are not surjective and elements of *Sur(S)* that are not injective. However, for an element of *F(S)* to have an inverse, it must be both injective and surjective. Thus, when *S* is infinite, *F(S)*, *Inj(S)*, and *Sur(S)* are not groups. To better illustrate this, consider the following.

■ Example

Let $S = \mathbb{N}$, the set of natural numbers, and consider the function defined as

$$f(x) = 2x,$$

for all $x \in \mathbb{N}$. It is easy to see that $f \in Inj(\mathbb{N})$ but $f \notin Sur(\mathbb{N})$. Observe that 1 is not in the range of *f*, so f^{-1} does not exist as $f^{-1}(1)$ is not defined. Since *f* does not have an inverse, $Inj(\mathbb{N})$ is not a group.

However, before leaving this example, it is worth considering the function $g : \mathbb{N} \to \mathbb{N}$ defined as

$$g(x) = x, \text{ when } x \text{ is odd} \quad \text{and} \quad g(x) = \frac{x}{2}, \text{ when } x \text{ is even.}$$

We can see that $g \in Sur(\mathbb{N})$, but $g \notin Inj(\mathbb{N})$, as $g(1) = 1 = g(2)$. Since g repeats values, g^{-1} does not exist, as $g^{-1}(1)$ is not defined. Thus, $Sur(\mathbb{N})$ is not a group. On the other hand, it is interesting to observe that if $x \in \mathbb{N}$, then

$$(g \circ f)(x) = g(f(x)) = g(2x) = x = e(x),$$

where e is the identity.

As a result, $g \circ f = e$, which makes it look somewhat like f and g invert each other. However,

$$(f \circ g)(1) = f(g(1)) = f(1) = 2 \neq 1 = e(1).$$

Thus, $f \circ g \neq e$. This indicates that when dealing with binary operations, to show that an element has an inverse, you do indeed need to check that things work on both sides.

■

We will now look more closely at the group $Bij(S)$ for some small sets S. When S is finite, we will stop using the notation $Bij(S)$ and will adopt some terminology that is more common in abstract algebra.

Definition 8.2. *If $n \in \mathbb{N}$ and $A = \{1, 2, \ldots, n\}$, let S_n denote the group of bijections from A to itself. We refer to S_n as the symmetric group or call it the symmetric group of degree n.*

■ Examples

1. If $n = 1$ and $A = \{1\}$, then S_1 consists solely of the identity map e, where $e(1) = 1$ and $e \circ e = e$.

2. If $n = 2$ and $A = \{1, 2\}$, then S_2 consists of two elements. We will call these elements e, f, and they are defined as

$$e(1) = 1, \quad e(2) = 2,$$

$$f(1) = 2, \quad f(2) = 1.$$

Observe that $e \circ e = e$, $e \circ f = f$, $f \circ e = f$, and $f \circ f = e$. Note that each element of S_2 is its own inverse.

3. If $n = 3$ and $A = \{1, 2, 3\}$, then S_3 has six elements, which we will call e, f, g, h, j, k, and they are defined as

$$e(1) = 1, \quad e(2) = 2, \quad e(3) = 3,$$

$$f(1) = 1, \quad f(2) = 3, \quad f(3) = 2,$$

$$g(1) = 3, \quad g(2) = 2, \quad g(3) = 1,$$

$$h(1) = 2, \quad h(2) = 1, \quad h(3) = 3,$$

$$j(1) = 2, \quad j(2) = 3, \quad j(3) = 1,$$

$$k(1) = 3, \quad k(2) = 1, \quad k(3) = 2.$$

We can represent S_3 and $\circ$ in the following table:

$\circ$	e	f	g	h	j	k
e	e	f	g	h	j	k
f	f	e	j	k	g	h
g	g	k	e	j	h	f
h	f	j	k	e	f	g
j	j	h	f	g	k	e
k	k	g	h	f	e	j

For example, to find $g \circ h$, we look at the term on the table in the same row as the g under $\circ$ and the same column as the h to the right of $\circ$ to obtain $g \circ h = j$. Similarly, to find $h \circ g$, we look for the term on the table in the same row as the h under $\circ$ and the same column as the g to the right of $\circ$ to obtain $h \circ g = k$.

∎

In all the examples of groups that come from commutative rings and fields, the order of the terms does not matter when applying the binary operation. Therefore, the fact that $g \circ h \neq h \circ g$ in S_3 is worth taking note of. This leads to

Definition 8.3. *If G is a group such that $x \circ y = y \circ x$, for all $x, y \in G$, we say that G is abelian. When a group is not abelian, we call it nonabelian.*

Therefore, S_1 and S_2 are abelian and S_3 is nonabelian. In fact, as we will now see, S_n is nonabelian whenever $n \geq 3$. If this looks familiar, this is merely a reappearance of Proposition 5.20 using somewhat different notation.

Proposition 8.4. *S_n is nonabelian, for all $n \geq 3$.*

Proof. Let $x, y \in S_n$ be defined as

$$x(1) = 2, \quad x(2) = 1, \quad x(3) = 3, \quad x(m) = m, \text{ for any } m > 3,$$

$$y(1) = 1, \quad y(2) = 3, \quad y(3) = 2, \quad y(m) = m, \text{ for any } m > 3.$$

Therefore,

$$(x \circ y)(1) = x(y(1)) = x(1) = 2,$$

whereas

$$(y \circ x)(1) = y(x(1)) = y(2) = 3.$$

Hence, $x \circ y \neq y \circ x$ and so, S_n is nonabelian. $\square$

As you probably noticed, the definition of a group states that the identity element is usually written as e, and the inverse of the element x is usually written as x^{-1}. You have also noticed that when the context clearly dictates otherwise, we are willing to abandon this terminology. For example, when looking at the group $\mathbb{Q}$ under addition, we denote the identity as 0 and the inverse of a and $-a$. In fact, it would be silly and needlessly confusing to do otherwise.

However, as a convenience, we will typically stick with multiplicative notation when dealing with groups. In fact, we will often leave out the symbol $\circ$ and write ab in place of $a \circ b$. Similarly, we will use a^2 as a shorthand for $a \circ a$ and a^3 as a shorthand for $a \circ a \circ a$. More generally, if $n \in \mathbb{N}$, then a^n will be the shorthand for repeatedly applying $\circ$ to n copies of a, and a^{-n} will be the shorthand for applying $\circ$ to n copies of this a^{-1}. You should convince yourself that a^{-n} is indeed that inverse of a^n. By adopting this shorthand, the familiar rules of exponents, such as $a^m a^n = a^{m+n}$ and $(a^m)^n = a^{mn}$, for all $m, n \in \mathbb{Z}$, hold when doing computations in a group. In particular, you should convince yourself that if x belongs to a group G, then $(x^{-1})^{-1} = x$. This means that not only is x^{-1} the inverse of x, but x is also the inverse of x^{-1}.

Given a commutative ring R, Proposition 5.15 provided us with an easy way to find new commutative rings contained in R. We would like to do a similar thing for groups. In particular, given a group G, we would like an easy way to produce new groups that are contained in G.

Definition 8.5. *Let G be a group with binary operation $\circ$. If H is a subset of G that is also a group under $\circ$, then we call H a subgroup of G.*

If H is a subset of G, then in order to show that H is a subgroup of G, we need to show that

 (i) H is closed under $\circ$

 (ii) H contains an identity element

(iii) every element of H has an inverse in H

Observe that since G is a group, we already know that $\circ$ is associative, so we only need to check that H satisfies (i), (ii), and (iii). However, it would be nice if there was an easier way to show that H is a subgroup of G and that is the point of

Proposition 8.6.   *Suppose G is a group under $\circ$ and let H be a nonempty subset of G.*

(a)   *If $x \circ y^{-1} \in H$, for all $x, y \in H$, then H is a subgroup of G.*

(b)   *If H is finite and $x \circ y \in H$, for all $x, y \in H$, then H is a subgroup of G.*

(c)   *If $H = K_1 \cap K_2$, where K_1 and K_2 are subgroups of G, then H is a subgroup of G.*

Proof. For part (a), we will first show that H contains the identity element of G. Since H is nonempty, let $x \in H$. Therefore,

$$x \circ x^{-1} \in H.$$

But since $x \circ x^{-1} = e$, H does contain the identity element of G.

Next, we need to show that H contains the inverse of each of its elements. However, if $x \in H$, then the fact that both e and x belong to H tells us that

$$e \circ x^{-1} \in H.$$

Since $e \circ x^{-1} = x^{-1}$, H does contain the inverse of each of its elements.

For the final part of (a), we need to show that H is closed under $\circ$. If $x, y \in H$, let $z = y^{-1}$. We know that $z^{-1} = y$, and it follows from the previous argument that $z \in H$. Therefore,

$$x \circ z^{-1} \in H.$$

But since $x \circ z^{-1} = x \circ y$, we see that H is also closed under $\circ$.

For part (b), suppose that H had the property that it contained the inverse of each of its elements. Then, if $x, y \in H$, it would follow that $y^{-1} \in H$, and we could conclude that $x \circ y^{-1} \in H$. This tells us that H satisfies the condition in part (a) of this result, so H would indeed be a subgroup of G. Thus, it suffices to show that whenever $y \in H$, we have $y^{-1} \in H$.

Consider the list

$$y, y^2, y^3, \ldots.$$

Observe that every element of this list belongs to the finite set H. Since the list is infinite, some element on this list must occur more than once. In fact, some element on this list must occur an infinite number of times. Therefore, there is some $m \geq 1$ such that y^m occurs an infinite number of times on this list. However, this tells us that there is some n such that

$n > m + 1$ and $y^m = y^n$. If we let $t = n - m - 1$, then t is a positive integer, and we can rewrite $y^m = y^n$ as

$$y^m = y^m \cdot y \cdot y^t.$$

Observe that since $t > 0$, y^t is on the preceding list, so $y^t \in H$. We will now exploit the properties of the group G to show that y^t is actually the inverse of y. The following argument might strike you as longer than necessary. However, at this stage, it is important to include every step and to see that the only properties we are allowed to use are those dealing with associativity, identities, and inverses.

Since $y^m \in H \subseteq G$, y^m has an inverse in G, which we can denote as z. Using multiplicative notation and multiplicative terminology, we can multiply both sides of the previous equation on the left by z to obtain.

$$z \cdot y^m = z \cdot (y^m \cdot y \cdot y^t).$$

Applying the associative law, we obtain

$$z \cdot y^m = (z \cdot y^m) \cdot (y \cdot y^t).$$

Next, using properties of inverses and identities, the previous equation becomes

$$e = e(y \cdot y^t)$$

and then

$$e = y \cdot y^t.$$

Since $y^{-1} \in G$, we can multiply this equation on the left by y^{-1} to obtain

$$y^{-1} \cdot e = y^{-1} \cdot (y \cdot y^t).$$

Associativity and the properties of identities and inverses now tell us first that

$$y^{-1} = (y^{-1} \cdot y) \cdot y^t,$$

then

$$y^{-1} = e \cdot y^t,$$

and finally

$$y^{-1} = y^t.$$

Therefore, y^t is the inverse of y, so H does contain y^{-1}, as desired.

For part (c), suppose $x, y \in H = K_1 \cap K_2$; in light of part (a), it suffices to show that $xy^{-1} \in K_1 \cap K_2$. Since K_1 and K_2 are both subgroups and x, y both belong to K_1 and K_2, it is immediate that y^{-1} belongs to both K_1 and K_2. Hence, xy^{-1} belongs to both K_1 and K_1, so $xy^{-1} \in H$, as desired. $\square$

Proposition 8.6 opens the door to countless examples of groups that arise as subgroups of groups we have already examined in this section. In particular, we are immediately led to

Definition 8.7. *If G is a group and $g \in G$, let $< g >= \{g^n \mid n \in \mathbb{Z}\}$. We call $< g >$ the cyclic subgroup generated by g. If there exists some $a \in G$ such that $G =< a >$, then we say that G is a cyclic group.*

It is very important to check that the set $< g >$ satisfies the conditions in Proposition 8.6(a) and is therefore a subgroup. Indeed, if $g \in G$ and $x, y \in< g >$, then there exists $n, m \in \mathbb{Z}$ such that $x = g^n$ and $y = g^m$. We are using multiplicative notation and now have that $y^{-1} = g^{-m}$ and

$$xy^{-1} = g^n g^{-m} = g^{n-m}.$$

Since $n - m \in \mathbb{Z}$, it follows that $xy^{-1} \in < g >$, so $< g >$ is a subgroup of G.

On the other hand, if G is a group and we are using additive notation, then $< g >$ refers to the set $\{ng \mid n \in \mathbb{Z}\}$. This should be much clearer after we look at several examples.

■ Examples

1. $\mathbb{Z}_{10}$ is a group with 10 elements in which we use additive notation. As we compute the cyclic subgroup generated by each of the 10 elements, note that different elements can generate the same cyclic subgroup. You should check that

$$< [0]_{10} >= \{[0]_{10}\},$$

$$< [1]_{10} > = < [3]_{10} > = < [7]_{10} > = < [9]_{10} >= \mathbb{Z}_{10},$$

$$< [2]_{10} > = < [4]_{10} > = < [6]_{10} > = < [8]_{10} > =$$

$$\{[0]_{10}, [2]_{10}, [4]_{10}, [6]_{10}, [8]_{10}, \},$$

$$< [5]_{10} > = \{[0]_{10}, [5]_{10}, \}.$$

Since there exist elements $a \in \mathbb{Z}_{10}$ such that $< a >= \mathbb{Z}_{10}$, we see that $\mathbb{Z}_{10}$ is cyclic.

2. Generalizing the previous example, we look at the groups $\mathbb{Z}_n$, for $n \geq 2$. Since $< [1]_n >= \mathbb{Z}_n$, all groups of the form $\mathbb{Z}_n$ are cyclic.

3. $\mathbb{Z}$ is also a group under addition and is a cyclic group as

$$< 1 > = < -1 > = \mathbb{Z}.$$

For any $m \in \mathbb{Z}$, $< m >$ consists of all multiples of m. For example, $< 2 >$ is the set of all even integers. Also note that if $a \in \mathbb{Z}$ and $< a >= \mathbb{Z}$, then a must be either 1 or -1.

4. $\mathbb{Q}$ is also a group under addition, but it is not cyclic. If $r \in \mathbb{Q}$, we can write $r = \frac{a}{b}$, where $a \in \mathbb{Z}$ and $b \in \mathbb{N}$. Note that $\frac{1}{2b} \notin < r >$, as there does not exist any $n \in \mathbb{Z}$ such that $\frac{1}{2b} = n\left(\frac{a}{b}\right)$. Since there are elements of $\mathbb{Q}$ not in $< r >$, it follows that $\mathbb{Q}$ is not cyclic. For example, $< \frac{1}{3} >= \{\frac{n}{3} \mid n \in \mathbb{Z}\}$ and the subgroup $< \frac{1}{3} >$ does not contain $\frac{1}{6}$.

5. $\mathbb{C}^\times$ is a group under multiplication. Here are examples of some of its cyclic subgroups:

$$< 1 > = \{1\},$$

$$< -1 > = \{1, -1\},$$

$$< i > = < -i > = \{1, -1, i, -i\},$$

$$< 2 >= \left\{1, \ 2, \ \frac{1}{2}, \ 4, \ \frac{1}{4}, \ 8, \ \frac{1}{8}, \ 16, \ \frac{1}{16}, \ldots\right\},$$

$$< -2 >= \left\{1, \ -2, \ -\frac{1}{2}, \ 4, \ \frac{1}{4}, \ -8, \ -\frac{1}{8}, \ 16, \ \frac{1}{16}, \ldots\right\}.$$

Also, if $n \in \mathbb{N}$, then $< \text{cis}\left(\frac{2\pi}{n}\right) >$ consists of the n nth roots of 1 in the complex numbers.

The group $\mathbb{C}^\times$ is not cyclic. There are various ways to see this. One way is to note that if $\alpha \in \mathbb{C}^\times$ had the property that $< \alpha >$ contained -1, then some power of α would be equal to -1. However, it follows from DeMoivre's Theorem that the length of α must be 1. But now DeMoivre's Theorem asserts that every element of $< \alpha >$ also has length 1. As a result, we can see that if $-1 \in < \alpha >$, then $2 \notin < \alpha >$. Therefore, no cyclic subgroup of $\mathbb{C}^\times$ can equal all of $\mathbb{C}^\times$. Hence, $\mathbb{C}^\times$ is not cyclic.

6. $\mathbb{Z}_5^\times$ is a group under multiplication. Since it only contains four elements, it is easy to compute all its cyclic subgroups.

$$< [1]_5 > = \{[1]_5\},$$

$$< [2]_5 > = < [3]_5 > = \mathbb{Z}_5^\times,$$

$$< [4]_5 > = \{[1]_5, [4]_5\}.$$

Observe that $\mathbb{Z}_5^\times$ is cyclic and is generated by both $[2]_5$ and $[3]_5$.

7. Earlier in this section, we looked at S_3 and saw how each of the bijections e, f, g, h, j, k behaved on the set $\{1, 2, 3\}$. If you look back at the table for S_3 under

composition of functions, it is not hard to compute its cyclic subgroups.

$$< e > = \{e\},$$

$$< f > = \{e, f\},$$

$$< g > = \{e, g\},$$

$$< h > = \{e, h\},$$

$$< j > = < k > = \{e, j, k\}.$$

∎

When computing cyclic subgroups $< g >$ of a group G, one often comes across a positive integer m such that $g^m = e$, where e is the identity element. When this happens, the Well Ordering Principle tells us that there is a smallest positive integer n such that $g^n = e$. In this situation, let us consider the elements

$$e, g, g^2, \ldots, g^{n-1}.$$

If $m \in \mathbb{Z}$, then the division algorithm asserts there exist $q, r \in \mathbb{Z}$, with $0 \le r \le n - 1$ such that $m = qn + r$. Since the ordinary rules of exponents hold, we now have

$$g^m = g^{qn+r} = g^{qn} g^r = (g^n)^q g^r = e^q g^r = e g^r = g^r.$$

Therefore, every element of $< g >$ appears on this list. Next, we claim that no element of $< g >$ occurs more than once on this list. Observe that the only way the list could contain a repetition is if there exist integers s, t such that $0 \le s < t < n$ and $g^t = g^s$. But this implies that $g^{t-s} = e$. Since $t - s$ is a positive integer, this contradicts the minimality of n. Thus, all n elements of $< g >$ on the list are different. Formalizing these observations, we have

Definition 8.8. *Suppose G is a group and $g \in G$. If there exists a positive integer m such that $g^m = e$, let n be the smallest one with this property. In this case, we say that g has order n and denote this as $o(g) = n$. Furthermore, $< g > = \{e, g, g^2, \ldots, g^{n-1}\}$.*

As is often the case, symmetric groups provide excellent examples of new concepts.

∎ Example

Let $\{e, f, g, h, j, k\}$ be the six elements of S_3 that we described earlier in the section. Having computed all the cyclic subgroups of S_3 earlier in the section, we now have $o(e) = 1$, $o(f) = 2$, $o(g) = 2$, $o(h) = 2$, $o(j) = 3$, and $o(k) = 3$.

∎

Having calculated all the cyclic subgroups of S_3, we have seen that S_3 is not cyclic. However, the next result makes it quite clear, without calculating cyclic subgroups, that S_3 is not cyclic.

Proposition 8.9. *All cyclic groups are abelian.*

Proof. Suppose G is a group and $G = \langle g \rangle$, for some $g \in G$. We need to show that if $x, y \in G$, then $xy = yx$. We know that there exist integers n, m such that $x = g^n$ and $y = g^m$. Since the ordinary rules of exponents apply, we have

$$xy = g^n g^m = g^{n+m} = g^{m+n} = g^m g^n = yx,$$

as desired. $\qquad\qquad\qquad\qquad\qquad\qquad\qquad\qquad\qquad\qquad\qquad\qquad\qquad\qquad$ $\square$

Although all cyclic groups are abelian, there are certainly abelian groups that are not cyclic.

■ Example

Let $G = U(\mathbb{Z}_8)$; then $G = \{[1]_8, [3]_8, [5]_8, [7]_8\}$. Clearly, G is abelian. However, you should check that

$$o([1]_8) = 1, o([3]_8) = 2, o([5]_8) = 2, \quad \text{and} \quad o([7]_8) = 2.$$

Therefore, none of the cyclic subgroups of G contain all four elements of G. Hence, $U(\mathbb{Z}_8)$ is abelian but not cyclic.

■

We will now look at various other subgroups of groups of the form $Bij(S)$ and S_n.

■ Examples

In the next series of examples, we will be looking at some subgroups of $Bij(R)$, for various commutative rings R.

1. Let R be a commutative ring field and if $m \in U(R), b \in R$, let

$$T_{m,b} : R \to R$$

be defined as $T_{m,b}(r) = mr + b$, for all $r \in R$. It is not hard to see that the linear function $T_{m,b}$ is a bijection. Next, let

$$G = \{T_{m,b} \mid m \in U(R), b \in R\}.$$

We claim that G satisfies the conditions of Proposition 8.6(a) and is therefore a group. If $T_{m,b} \in G$, then $\frac{1}{m} \in U(R)$ and $-\frac{b}{m} \in R$, so we have

$$\left(T_{m,b} \circ T_{\frac{1}{m}, -\frac{b}{m}}\right)(r) = T_{m,b}\left(T_{\frac{1}{m}, -\frac{b}{m}}(r)\right) = T_{m,b}\left(\frac{r}{m} - \frac{b}{m}\right) =$$

$$m\left(\frac{r}{m} - \frac{b}{m}\right) + b = r - b + b = r$$

and

$$\left(T_{\frac{1}{m}, -\frac{b}{m}} \circ T_{m,b}\right)(r) = T_{\frac{1}{m}, -\frac{b}{m}}(T_{m,b}(r)) = T_{\frac{1}{m}, -\frac{b}{m}}(mr + b) =$$

$$\left(\frac{1}{m}\right)(mr + b) - \frac{b}{m} = r + \frac{b}{m} - \frac{b}{m} = r,$$

for all $r \in R$. Since $T_{\frac{1}{m}, -\frac{b}{m}}$ belongs to G and it is the inverse of $T_{m,b}$ in $Bij(R)$, G contains the inverse of each of its elements.

Now suppose $x, y \in G$; then there exist $n, m, a, b \in R$, with $n, m \in U(R)$ such that $x = T_{n,a}$ and $y = T_{m,b}$. Since $y^{-1} = T_{\frac{1}{m}, -\frac{b}{m}}$, we now have

$$(xy^{-1})(r) = \left(T_{n,a} \circ T_{\frac{1}{m}, -\frac{b}{m}}\right)(r) = T_{n,a}\left(T_{\frac{1}{m}, -\frac{b}{m}}(r)\right) = T_{n,a}\left(\frac{r}{m} - \frac{b}{m}\right) =$$

$$n\left(\frac{r}{m} - \frac{b}{m}\right) + a = \left(\frac{n}{m}\right)r + \left(a - \frac{nb}{m}\right),$$

for all $r \in R$. As a result, $xy^{-1} = T_{\frac{n}{m}, a - \frac{nb}{m}}$. Observe that $\frac{n}{m} = nm^{-1} \in U(R)$ and $a - \frac{nb}{m} = a - nbm^{-1} \in R$, thus $xy^{-1} \in G$. Thus, Proposition 8.6(a) tells us that G is indeed a group.

In this example, if R is a ring where 1 is not its own additive inverse—in other words, $1 \neq -1$—then the groups obtained in this way are not abelian. A direct computation shows that

$$(T_{-1,0} \circ T_{1,1})(0) = T_{-1,0}(T_{1,1}(0)) = T_{-1,0}(1) = -1,$$

whereas

$$(T_{1,1} \circ T_{-1,0})(0) = T_{1,1}(T_{-1,0}(0)) = T_{1,1}(0) = 1.$$

Since $1 \neq -1$, we see that $T_{-1,0} \circ T_{1,1} \neq T_{1,1} \circ T_{-1,0}$. Thus, G is not abelian.

2. We continue to examine the preceding example. If $T_{m,b}, T_{n,a} \in G$ give the same value when you plug in 0, it follows that $b = a$. If these functions also give the same value

when you plug in 1, then $m = n$. As a result, each different ordered pair $(m, b) \in U(R) \times R$ provides us with a different function.

Next, if $n > 2$, we will consider the special case where $R = \mathbb{Z}_n$; thus, $G = \{T_{m,b} \mid m \in U(\mathbb{Z}_n), b \in \mathbb{Z}_n\}$. Since $U(\mathbb{Z}_n)$ has $\phi(n)$ elements and $\mathbb{Z}_n$ has n elements, G is a nonabelian group with $\phi(n)n$ elements. For example, if $n = 3$, then G is nonabelian with 6 elements, whereas if $n = 4$, then G is nonabelian with 8 elements.

3. In the previous paragraph, we examined $G = \{T_{m,b} \mid m \in U(\mathbb{Z}_n), b \in \mathbb{Z}_n\}$. Every element of G is a bijection of $\mathbb{Z}_n$, so we can think of G as a subgroup of S_n. As we will see later in this chapter, every finite group can be viewed as a subgroup of S_t, for some appropriate value of $t \in \mathbb{N}$. That is one of the reasons that symmetric groups are such an important part of finite group theory.

    Let us now look at the following subset of G,

    $$H_n = \{T_{m,b} \mid m = [1]_n \text{ or } [-1]_n, b \in \mathbb{Z}_n\}.$$

    The set H_n is finite; in fact, it has exactly $2n$ elements. In order to show that H_n is a group, by Proposition 8.6(b), it suffices to show that H_n is closed under composition of function.

    Based on the experience you obtained working with elements of the form $T_{m,b} \in Bij(R)$, you should have little trouble verifying that

    $$T_{n,a} \circ T_{m,b} = T_{nm, nb+a}.$$

    Therefore, if $m, n \in \{[1]_n, [-1]_n\}$ and $a, b \in \mathbb{Z}_n$, it is easy to see that $nm \in \{[1]_n, [-1]_n\}$ and $nb + a \in \mathbb{Z}_n$. Thus, $T_{nm, nb+a} \in H_n$ and H_n is indeed a group. In fact, based on the work we have done with groups of this type, we can see that H is a nonabelian group with $2n$ elements.

∎

The preceding constructions shows us a way, for every $n \geq 3$, to construct a nonabelian group with $2n$ elements that can be viewed as a subgroup of S_n. We will now examine how these nonabelian subgroups of S_n also arise in a geometric context.

Suppose a jigsaw is used to cut an isosceles triangle, which is not equilateral, out of a piece of wood. We would like to analyze all the different ways in which that triangle can be removed from the wood, moved around in three dimensions, and then placed back into the piece of wood. Note that although the triangle must end up in its original location in the piece of wood, it need not be placed in its original position. To help keep track of any changes we make in the position of the triangle, we will label each vertex and corner of the triangle. The label for each vertex will be outside the triangle, whereas the label for each corner will be inside the triangle. Observe that the side of the triangle connecting vertex #2 to vertex #3 is shorter than the other two sides.

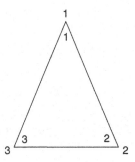

When you place the triangle back into the wood, it is certainly possible that you placed it back in its original position. Notice that each corner of the triangle ends up at a vertex with the same number, so we can represent this motion by the element $e \in S_3$ where $e(1) = 1$, $e(2) = 2$, $e(3) = 3$.

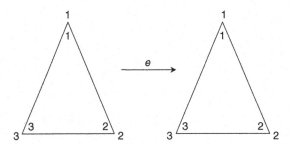

However, there is another way you could position the triangle. Obviously, corner #1 must end up at vertex #1; otherwise, the triangle will not fit back into its original location. Now, think about the line beginning at vertex #1 and heading down the middle of the triangle. We can flip the triangle around this line, thereby switching the part of the triangle that faces up with the part that faces down. The result of this is

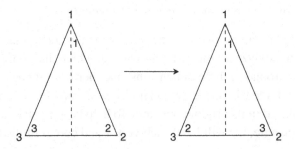

If we check where each corner of the triangle moves, we see that corner #1 remains at vertex #1, whereas corner #2 ends up at vertex #3 and corner #3 ends up at vertex #2. Therefore, we can represent this motion by the element $f \in S_3$, where $f(1) = 1$, $f(2) = 3$, $f(3) = 2$. As a result, the ways in which we can move this triangle is described by the two elements $\{e, f\}$, which is a subgroup S_3. Observe that the collection of ways in which we can move and reposition the triangle is not merely a set. It is a group because we can always follow one motion by another motion, and following one motion by another is really the same as composing functions in S_3.

For our next example, suppose an equilateral triangle is cut out of a piece of wood. Once again, we label the vertices and each corner of the triangle.

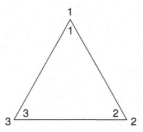

As before, after each movement of the triangle, we can record where each corner of the triangle ends up. In this way, we again see that the motions of this triangle are a subgroup of S_3. The question, at this point, is how many elements of S_3 are obtained as motions of this triangle?

Clearly, if we return the triangle to its original position, this motion corresponds to $e \in S_3$, where $e(1) = 1$, $e(2) = 2$, $e(3) = 3$. Next, consider the point at the center of the triangle. When we rotate the triangle 120° clockwise around this point, corner #1 ends up at vertex #2, corner #2 ends up at vertex #3, and corner #3 ends up at vertex #1. Therefore, this motion corresponds to $j \in S_3$, where $j(1) = 2$, $j(2) = 3$, $j(3) = 1$.

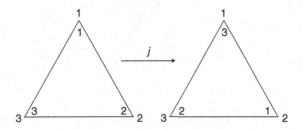

If we instead rotate the triangle 240° clockwise around this point, we obtain the motion corresponding to $k \in S_3$, where $k(1) = 3$, $k(2) = 1$, $k(3) = 2$.

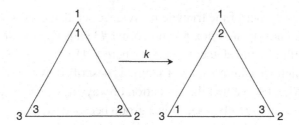

Next, consider the line starting at vertex #1. If we flip the triangle around this line, thereby switching the parts of the triangle that face up and down, we obtain the motion corresponding to $f \in S_3$, where $f(1) = 1$, $f(2) = 3$, $f(3) = 2$.

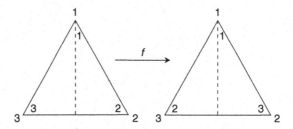

At this point, we have obtained four of the six elements of S_3. We can obtain the final two motions by combining or composing some of the previous motions. For example, to obtain the motion corresponding to $h \in S_3$, where $h(1) = 2$, $h(2) = 1$, $h(3) = 3$, we first apply the motion corresponding to f, then the motion corresponding to j.

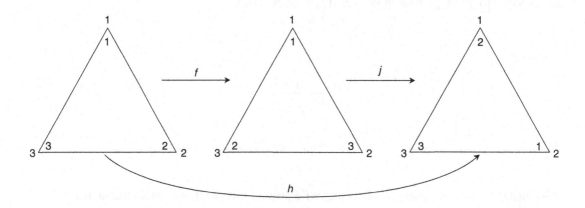

Finally, to obtain $g \in S_3$, where $g(1) = 3$, $g(2) = 2$, $g(3) = 1$, first apply the motion corresponding to f and then the one corresponding to k. As a result, we can see that the motions of an equilateral triangle is the group S_3.

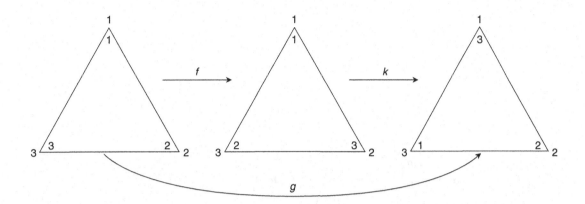

A regular n-gon is an n-sided object in the plane where all the angles and all the sides are equal. Thus, an equilateral triangle is a regular 3-gon, and a square is a regular 4-gon. We would now like to extend our analysis to describe the motions of all regular n-gons, for $n \geq 3$.

Given a regular n-gon, we begin by labeling each vertex and each corner of the n-gon in a clockwise fashion. By examining where a motion takes each corner, every motion of the n-gon corresponds to an element of S_n. Since any motion can be followed by another motion, the set of motions of an n-gon is closed under composition, so by Proposition 8.6(b), it is a subgroup of S_n. Our goal is to determine which elements of S_n belong to this subgroup. The types of motions we are discussing are commonly referred to as *rigid motions* or *symmetries*.

Definition 8.10. *If $n \geq 3$, the nth dihedral group is denoted as D_n and consists of all rigid motions or symmetries of a regular n-gon.*

In the next six diagrams, we will be looking at the rigid motions of a regular 12-gon. As we try to understand the situation for any $n \geq 3$, the number 12 should be large enough that these diagrams successfully illustrate the general situation.

Given any rigid motion, we first examine what happens to corner #1. It can end up at any of n different vertices. Observe that after applying this motion, corner #1 must still lie between corners #2 and #n. Therefore, as we can see following, there are at most $2n$ rigid motions of a regular n-gon.

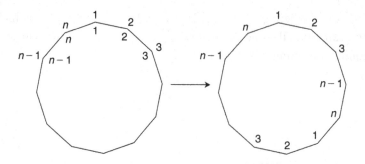

or

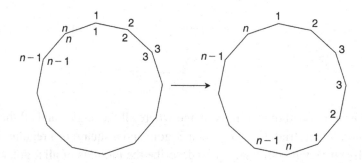

Now we will see that all $2n$ possibilities are indeed rigid motions of a regular n-gon. Consider the point in the center and look at the line starting at vertex #1 that passes through the center. Now, let x denote flipping the n-gon around this line, thus we have switched which part of the n-gon faces up and which faces down. Note that after applying x, corner #1 remains at vertex #1, but when we proceed clockwise around the n-gon, corner #n comes *after* corner #1 and not before.

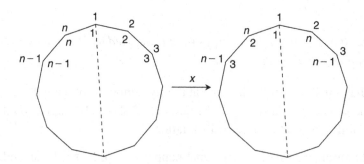

Next, let y denote rotating the entire n-gon $\frac{360°}{n}$ in a clockwise direction around the point in the center. Observe that this moves corner #1 to vertex #2 and keeps corner #2 after corner #1 when moving in a clockwise direction.

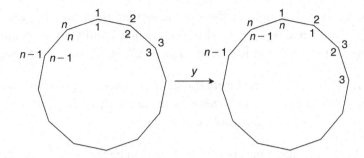

Since x and y belong to the group of motions of a regular n-gon, so does any series of repeated applications of x and y. Observe that if we apply y to the n-gon i times, with $0 \leq i \leq n - 1$, then we obtain the motion that moves corner #1 to vertex #$(i + 1)$ and corner #2 comes immediately after the corner labeled #1.

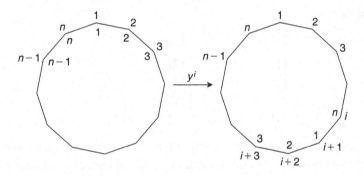

On the other hand, if we first apply x and then follow it by applying y^i times, where $0 \leq i \leq n - 1$, then this motion moves corner #1 to vertex #$(i + 1)$ while corner #2 now comes before corner #1.

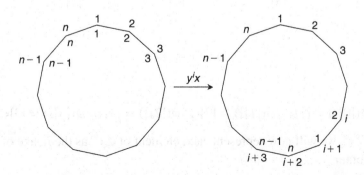

As a result, there are indeed $2n$ rigid motions of the regular n-gon and they are all of the form $y^i x^j$, where $0 \le i \le n-1$ and $0 \le j \le 1$. Also observe that x and y do not commute as motion xy is not the same as motion yx but is instead equal to $y^{n-1}x$. To summarize, we have shown

Proposition 8.11. *For every $n \ge 3$, the nth dihedral group is a nonabelian group with $2n$ elements. Every element of D_n can be represented uniquely in the form $y^i x^j$, where $0 \le i \le n-1, 0 \le j \le 1$, $y^n = x^2 = e$, and $xy = y^{n-1}x$.*

We now address the issue what it means for two groups to be the "same". To address this issue, we begin by looking at the tables for the group $\mathbb{Z}_4$ and $\mathbb{Z}_5{}^\times$. The table for $\mathbb{Z}_4$ is

$\circ$	$[0]_4$	$[1]_4$	$[2]_4$	$[3]_4$
$[0]_4$	$[0]_4$	$[1]_4$	$[2]_4$	$[3]_4$
$[1]_4$	$[1]_4$	$[2]_4$	$[3]_4$	$[0]_4$
$[2]_4$	$[2]_4$	$[3]_4$	$[0]_4$	$[1]_4$
$[3]_4$	$[3]_4$	$[0]_4$	$[1]_4$	$[2]_4$

and the table for $\mathbb{Z}_5{}^\times$ is

$\circ$	$[1]_5$	$[2]_5$	$[3]_5$	$[4]_5$
$[1]_5$	$[1]_5$	$[2]_5$	$[3]_5$	$[4]_5$
$[2]_5$	$[2]_5$	$[4]_5$	$[1]_5$	$[3]_5$
$[3]_5$	$[3]_5$	$[1]_5$	$[4]_5$	$[2]_5$
$[4]_5$	$[4]_5$	$[3]_5$	$[2]_5$	$[1]_5$

Admittedly, the next step will appear to be unmotivated, but it is quite necessary. In the table for $\mathbb{Z}_5{}^\times$, we will swap the positions of $[3]_5$ and $[4]_5$ in both the column under $\circ$ and the row to the right of $\circ$. This gives us a slightly different looking table for $\mathbb{Z}_5{}^\times$

$\circ$	$[1]_5$	$[2]_5$	$[4]_5$	$[3]_5$
$[1]_5$	$[1]_5$	$[2]_5$	$[4]_5$	$[3]_5$
$[2]_5$	$[2]_5$	$[4]_5$	$[3]_5$	$[1]_5$
$[4]_5$	$[4]_5$	$[3]_5$	$[1]_5$	$[2]_5$
$[3]_5$	$[3]_5$	$[1]_5$	$[2]_5$	$[4]_5$

Next, let

$$\phi : \mathbb{Z}_4 \to \mathbb{Z}_5{}^\times$$

be defined as

$$\phi([0]_4) = [1]_5, \quad \phi([1]_4) = [2]_5, \quad \phi([2]_4) = [4]_5, \quad \phi([3]_4) = [3]_5.$$

In the table for $\mathbb{Z}_5{}^\times$ we will now represent each element of $\mathbb{Z}_5{}^\times$ as the image of an element of $\mathbb{Z}_4$ under ϕ to obtain

$\circ$	$\phi([0]_4)$	$\phi([1]_4)$	$\phi([2]_4)$	$\phi([3]_4)$
$\phi([0]_4)$	$\phi([0]_4)$	$\phi([1]_4)$	$\phi([2]_4)$	$\phi([3]_4)$
$\phi([1]_4)$	$\phi([1]_4)$	$\phi([2]_4)$	$\phi([3]_4)$	$\phi([0]_4)$
$\phi([2]_4)$	$\phi([2]_4)$	$\phi([3]_4)$	$\phi([0]_4)$	$\phi([1]_4)$
$\phi([3]_4)$	$\phi([3]_4)$	$\phi([0]_4)$	$\phi([1]_4)$	$\phi([2]_4)$

Observe that the tables for $\mathbb{Z}_4$ and $\mathbb{Z}_5{}^\times$ are identical except that every g in the table for $\mathbb{Z}_4$ is replaced by $\phi(g)$ in the table for $\mathbb{Z}_5{}^\times$. Not only does ϕ provide us with a one-to-one correspondence between the elements of $\mathbb{Z}_4$ and $\mathbb{Z}_5{}^\times$, it also makes the tables for these groups look essentially the same.

Observe that if $g, h \in \mathbb{Z}_4$ then the product in $\mathbb{Z}_5{}^\times$ of $\phi(g)$ and $\phi(h)$ is $\phi(g)\phi(h)$. Even though $\mathbb{Z}_4$ is a group under addition, we will use multiplication notation. Therefore, when we look at the table for $\mathbb{Z}_5{}^\times$, the element corresponding to the product of $\phi(g)$ and $\phi(h)$ is $\phi(gh)$. Thus,

$$\phi(gh) = \phi(g)\phi(h)$$

and this equation is the key!

What this means is that $\mathbb{Z}_5{}^\times$ is essentially a carbon copy of $\mathbb{Z}_4$, except that the elements in $\mathbb{Z}_5{}^\times$ have slightly different names. In $\mathbb{Z}_4$, we may call elements g and h, whereas in $\mathbb{Z}_5{}^\times$, we may call them $\phi(g)$ and $\phi(h)$. But the equation $\phi(gh) = \phi(g)\phi(h)$ tells us that g and h add in $\mathbb{Z}_4$ the same exact way that $\phi(g)$ and $\phi(h)$ multiply in $\mathbb{Z}_5{}^\times$. Thus, not only do $\mathbb{Z}_4$ and $\mathbb{Z}_5{}^\times$ have the same number of elements, but each $g \in \mathbb{Z}_4$ behaves the same way in $\mathbb{Z}_4$ as $\phi(g)$ does in $\mathbb{Z}_5{}^\times$. We generalize this as

Definition 8.12. *Let G_1, G_2 be groups; G_1 and G_2 are said to be* isomorphic *if there exists a bijection $\phi : G_1 \to G_2$ such that*

$$\phi(gh) = \phi(g)\phi(h),$$

for all $g, h \in G_1$.

If ϕ is as in Definition 8.12, we say that ϕ is an *isomorphism* of groups. If two groups are isomorphic, then they are essentially the same object, except that the names of the elements in the two groups might be somewhat different. In particular, the table for G_2 under $\circ$ is identical to the table for G_1 except that each g in the table for G_1 is replaced by $\phi(g)$.

Not all groups of the same order are isomorphic. The groups $\mathbb{Z}_4$, $\mathbb{Z}_5{}^\times$, and $U(\mathbb{Z}_8)$ all have order 4, and we have already shown that $\mathbb{Z}_4$ and $\mathbb{Z}_5{}^\times$ are isomorphic. However, these groups are not isomorphic to $U(\mathbb{Z}_8)$. To see this, suppose we are in the situation where $\phi : G_1 \to G_2$ is an isomorphism of finite groups and G_1 is cyclic. If $g \in G_1$ such that $G_1 = < g >$, then we would expect that $\phi(g)$ plays the same role in G_2 that g plays in G_1. Therefore, not only should

G_2 be cyclic, but it should also be the case that $G_2 = < \phi(g) >$. Indeed, if $y \in G_2$, then there exists $x \in G_1$ such that $y = \phi(x)$. Since G_1 is cyclic, there exists $n \in \mathbb{N}$ such that $x = g^n$. Using arguments similar to the ones in Chapter 5 in our discussion of complex conjugation and automorphisms, you should convince yourself that $\phi(g^n) = (\phi(g))^n$. Therefore, we now have

$$y = \phi(x) = \phi(g^n) = (\phi(g))^n.$$

Thus, every $y \in G_2$ does indeed belong to $< \phi(g) >$, so G_2 is cyclic.

Since $\mathbb{Z}_4$ and $\mathbb{Z}_5{}^\times$ are cyclic, any group to which they are isomorphic must also be cyclic. However, $U(\mathbb{Z}_8)$ is not cyclic, so it is not isomorphic to $\mathbb{Z}_4$ and $\mathbb{Z}_5{}^\times$.

You should also convince yourself that if two groups are isomorphic and one of them is abelian, then so is the other one. Therefore, even though the groups $\mathbb{Z}_6$ and S_3 both have six elements, they are not isomorphic as $\mathbb{Z}_6$ is abelian and S_3 is not.

Earlier in this section, we remarked that the groups $H_n = \{T_{m,b} \mid m = [1]_n \text{ or } [-1]_n, b \in \mathbb{Z}_n\}$ also arise in a geometric context and are called D_n. In the language of isomorphisms, we are asserting that for $n \geq 3$, H_n is isomorphic to D_n. You will be asked to verify this in the exercises.

Suppose we define $\approx$ on the set of groups by saying $G_1 \approx G_2$ whenever G_1 and G_2 are isomorphic. In the exercises, you will be asked to show that $\approx$ is an equivalence relation on the set of groups. Among the things that researchers in the theory of finite groups try to do is to describe the structure of all finite groups. Having the concept of isomorphism at our disposal, we can be more formal about what this actually means. For every $n \geq 1$, mathematicians would like to determine the number of equivalence classes of groups of order n. Then, they would like to concretely describe one group in each equivalence class. Later in this chapter we shall see that, for certain values of n, this is fairly easy to do. But for some values of n, especially those with many different prime factors, this becomes a very, very difficult problem.

Exercises for Section 8.1

In exercises 1–26, determine if the set S is a group under $\circ$. Although $\circ$ often represents a binary operation, in these examples we are not assuming that S is closed under $\circ$. If S is not a group, briefly explain why it is not a group.

1. S is the complex numbers of length 1 and $\circ$ is multiplication.

2. S is the complex numbers of length 1 and $\circ$ is addition.

3. S is the complex numbers of length 1 and $\circ$ is division.

4. S is the complex numbers whose length is a positive integer and ∘ is multiplication.

5. S is the complex numbers whose length is a positive rational number and ∘ is multiplication.

6. S is the complex numbers whose length is a positive rational number and ∘ is addition.

7. $S = \{a + bi \mid a \in \mathbb{Q}, b \in \mathbb{R}\}$ and ∘ is addition.

8. $S = \{a + bi \mid a \in \mathbb{Q}, a \neq 0, b \in \mathbb{R}\}$ and ∘ is multiplication.

9. $S = \mathbb{Z}$ and ∘ is subtraction.

10. S is the set of nonzero rational numbers and ∘ is division.

11. S is the set of positive rational numbers and ∘ is addition.

12. S is the set of positive rational numbers and ∘ is multiplication.

13. $S = \{a + bx + cx^2 \mid a, b, c \in \mathbb{Z}\}$ and ∘ is addition.

14. $S = \{ax + bx^2 \mid a, b \in \mathbb{Q}\}$ and ∘ is addition.

15. $S = \{f(x) \in \mathbb{R}[x] \mid f(0) = 0\}$ and ∘ is addition.

16. $S = \{f(x) \in \mathbb{R}[x] \mid f(0) \neq 0\}$ and ∘ is addition.

17. S is the set of rational numbers of the form $\frac{a}{b}$ with $a, b \in \mathbb{Z}$ and b odd where ∘ is addition.

18. S is the set of rational numbers of the form $\frac{a}{b}$ with $a, b \in \mathbb{N}$ and b odd where ∘ is addition.

19. S is the set of rational numbers of the form $\frac{a}{b}$ with $a, b \in \mathbb{N}$ and b odd where ∘ is multiplication.

20. S is the set of rational numbers of the form $\frac{a}{b}$ with a, b positive odd integers and ∘ is multiplication.

21. S consists of those elements in S_4 which send 1 to 1 and ∘ is composition of functions.

22. S consists of those elements in S_4 which do not send 1 to 1 and ∘ is composition of functions.

23. S consists of those elements in S_3 of order 1 or 2 and ∘ is composition of functions.

24. If G is an abelian group, let S be the elements in G of order 1 or 2 and ∘ will be the binary operation in G.

25. S consists of those elements in S_3 of order 1 or 3 and ∘ is composition of functions.

26. If G is an abelian group, let S be the elements in G of order 1 or 3 and ∘ will be the binary operation in G.

Exercises 27–32 will be based on the following multiplication table for D_3, the third dihedral group. Recall that every element of D_3 can be written uniquely in the form $y^i x^j$, where $0 \le i \le 2, 0 \le j \le 1$, and $y^3 = x^2 = e$. In constructing the table, remember that the term xy can and should be replaced by $y^2 x$.

○	e	x	y	yx	y^2	$y^2 x$
e	e	x	y	yx	y^2	$y^2 x$
x	x	e	$y^2 x$	y^2	yx	y
y	y	yx	y^2	$y^2 x$	e	x
yx	yx	y	x	e	$y^2 x$	y^2
y^2	y^2	$y^2 x$	e	x	y	yx
$y^2 x$	$y^2 x$	y^2	yx	y	x	e

27. Write down the elements in the cyclic subgroup generated by x and determine the order of x.

28. Write down the elements in the cyclic subgroup generated by y and determine the order of y.

29. Write down the elements in the cyclic subgroup generated by yx and determine the order of yx.

30. Write down the elements in the cyclic subgroup generated by y^2 and determine the order of y^2.

31. Write down the elements in the cyclic subgroup generated by $y^2 x$ and determine the order of $y^2 x$.

32. Show that the only subgroup of D_3 which contains both $y^2 x$ and y is D_3.

Exercises 33–43 will be based on the following partially completed multiplication table for D_4, the fourth dihedral group. Recall that every element of D_4 can be written uniquely in the form $y^i x^j$, where $0 \le i \le 3, 0 \le j \le 1$, and $y^4 = x^2 = e$. In constructing the table, remember that the term xy can and should be replaced by $y^3 x$.

○	e	x	y	yx	y^2	$y^2 x$	y^3	$y^3 x$
e	e	x	y	yx	y^2	$y^2 x$	y^3	$y^3 x$
x	x							
y	y							
yx	yx							
y^2	y^2							
$y^2 x$	y^2							
y^3	y^3							
$y^3 x$	$y^3 x$							

33. Complete the two columns on the table that have x on top and y on top.

34. Complete the two columns on the table that have yx on top and y^2 on top.

35. Complete the two columns on the table that have y^2x on top and y^3 on top.

36. Complete the column on the table that has y^3x on top and observe that you are placing in each row the only element of D_4 that was missing.

37. Write down the elements in the cyclic subgroup generated by x and determine the order of x.

38. Write down the elements in the cyclic subgroup generated by y and determine the order of y.

39. Write down the elements in the cyclic subgroup generated by yx and determine the order of yx.

40. Write down the elements in the cyclic subgroup generated by y^2 and determine the order of y^2.

41. Write down the elements in the cyclic subgroup generated by y^2x and determine the order of y^2x.

42. Write down the elements in the cyclic subgroup generated by y^3 and determine the order of y^3.

43. Write down the elements in the cyclic subgroup generated by y^3x and determine the order of y^3x.

In exercises 44–52, let $f, g, h \in S_4$ be described in terms of their behavior on the set $\{1, 2, 3, 4\}$ as follows:

$$f(1) = 1, \quad f(2) = 4, \quad f(3) = 3, \quad f(4) = 2,$$
$$g(1) = 2, \quad g(2) = 3, \quad g(3) = 1, \quad g(4) = 4,$$
$$h(1) = 4, \quad h(2) = 3, \quad h(3) = 2, \quad h(4) = 1.$$

44. Compute the order of f.

45. Compute the order of g.

46. Compute the order of h.

47. Express $f \circ g$ in terms of its behavior on $\{1, 2, 3, 4\}$ and then find the order of $f \circ g$.

48. Express $g \circ f$ in terms of its behavior on $\{1, 2, 3, 4\}$ and then find the order of $g \circ f$.

49. Express $f \circ h$ in terms of its behavior on $\{1, 2, 3, 4\}$ and then find the order of $f \circ h$.

50. Express $h \circ f$ in terms of its behavior on $\{1, 2, 3, 4\}$ and then find the order of $h \circ f$.

51. Express $g \circ h$ in terms of its behavior on $\{1, 2, 3, 4\}$ and then find the order of $g \circ h$.

52. Express $h \circ g$ in terms of its behavior on $\{1, 2, 3, 4\}$ and then find the order of $h \circ g$.

53. Let G be a group with subgroups H, K. Suppose $h \in H$ and $k \in K$ such that $h \notin K$ and $k \notin H$.
    (a) Prove that $H \cup K$ does not contain hk.
    (b) Prove that $H \cup K$ is not a subgroup of G.

54. Use exercise 53 to show that if G is a group, then G is not the union of two subgroups that are not all of G.

55. Show that the group $U(\mathbb{Z}_8)$ is the union of three subgroups, none of which is all of $U(\mathbb{Z}_8)$.

56. If $H_1, H_2, \ldots, H_m$ are subgroups of a group G, show that the intersection $H_1 \cap H_2 \cap \cdots \cap H_m$ is also a subgroup of G.

57. If G is a group and $a \in G$, let $\pi : G \to G$ be the function defined as $\pi(g) = ag$, for all $g \in G$.
    (a) Show that π is a bijection.
    (b) Show that if π is an isomorphism, then a is the identity element of G.

58. If G is a group and $b \in G$, let $\rho : G \to G$ be the function defined as $\rho(g) = gb$, for all $g \in G$.
    (a) Show that ρ is a bijection.
    (b) Show that if ρ is an isomorphism, then b is the identity element of G.

59. If G is a group and $a, b \in G$, let $\tau : G \to G$ be the function defined as $\tau(g) = agb$, for all $g \in G$.
    (a) Show that τ is a bijection.
    (b) Show that if τ is an isomorphism, then $a = b^{-1}$.
    (c) Show that if $a = b^{-1}$, then τ is an isomorphism.

60. Explain how exercises 57 and 58 are special cases of exercise 59.

61. Let $\phi : \mathbb{R}^+ \to \mathbb{R}^+$ be defined as $\phi(x) = x^2$, for all $x \in \mathbb{R}^+$, where $\mathbb{R}^+$ is the set of positive real numbers.
    (a) Explain why $\mathbb{R}^+$ is a group under multiplication.
    (b) Is ϕ an isomorphism of groups? Explain your answer.

62. Let $\rho : \mathbb{R}^\times \to \mathbb{R}^\times$ be defined as $\rho(x) = x^2$, for all $x \in \mathbb{R}^\times$, where $\mathbb{R}^\times$ is the set of nonzero real numbers. Is ρ an isomorphism of groups? Explain your answer.

63. Let $\tau : \mathbb{R}^\times \to \mathbb{R}^\times$ be defined as $\tau(x) = x^3$, for all $x \in \mathbb{R}^\times$, where $\mathbb{R}^\times$ is the set of nonzero real numbers. Is τ an isomorphism of groups? Explain your answer.

64. Let $v : \mathbb{Q}^+ \to \mathbb{Q}^+$ be defined as $v(x) = x^2$, for all $x \in \mathbb{Q}^+$, where $\mathbb{Q}^+$ is the set of positive real numbers.
    (a) Explain why $\mathbb{Q}^+$ is a group under multiplication.

    (b) Is v an isomorphism of groups? Explain your answer.

65. Let G be a finite abelian group with no elements of order 2. Show that the function $\phi : G \to G$ defined as $\phi(g) = g^2$, for all $g \in G$, is an isomorphism.

66. Suppose G is a finite abelian group and p a prime number such that G has no elements of order p. Show that the function $\tau : G \to G$ defined as $\tau(g) = g^p$, for all $g \in G$, is an isomorphism.

67. Let $\phi : G_1 \to G_2$ be an isomorphism of groups. Show that G_1 is abelian if and only if G_2 is abelian.

68. Let $\phi : G_1 \to G_2$ be an isomorphism of groups.
    (a) If e_1 is the identity element of G_1 and e_2 is the identity element of G_2, show that $\phi(e_1) = e_2$.

    (b) If $n \in \mathbb{N}$ and $g \in G_1$, show that g has order n in G_1 if and only if $\phi(g)$ has order n in G_2.

69. Define $\approx$ on the set of groups as $G_1 \approx G_2$ precisely when G_1 and G_2 are isomorphic. Show that $\approx$ is an equivalence relation.

70. In this exercise, you will verify that for $n \geq 3$, the group
    $H_n = \{T_{m,b} \mid m = [1]_n \text{ or } [-1]_n, b \in \mathbb{Z}_n\}$ is isomorphic to the nth dihedral group D_n.
    (a) In D_n, show that if $m \in \mathbb{N}$ then $xy^m x = y^{-m}$. When doing this, you may want to look at part (c) of exercise 59.

    (b) In D_n, show that if $0 \leq i, k \leq n - 1$, and $0 \leq j, l \leq 1$, then $(y^i x^j)(y^k x^l) = y^{i+(-1)^j k} x^{j+l}$. It might help to consider the cases $j = 0$ and $j = 1$ separately.

    (c) In H_n, show that if $0 \leq i, k \leq n - 1$, and $0 \leq j, l \leq 1$, then

    $$T_{[(-1)^j]_n, [i]_n} \circ T_{[(-1)^l]_n, [k]_n} = T_{[(-1)^{j+l}]_n, [i+(-1)^j k]_n}.$$

    (d) Show that the function $\phi : D_n \to H_n$ defined as

    $$\phi(y^i x^j) = T_{[(-1)^j]_n, [i]_n},$$

    whenever $0 \leq i \leq n - 1$ and $0 \leq j \leq 1$, is an isomorphism.

8.2 Theorems of Lagrange and Sylow

One way to try to understand the structure of a finite group G is to look at its subgroups. Cyclic subgroups were introduced in the previous section because they arise in a very natural way. We now look at another collection of subgroups that arise very naturally.

Definition 8.13. *If G is a group and $g \in G$, let $C_G(g) = \{h \in G \mid gh = hg\}$, and we call this set the centralizer of g in G.*

Observe that $C_G(g)$ consists of the elements of G that commute with g. In particular, if G is abelian, then $C_G(g) = G$, for all $g \in G$. To look at a nonabelian example, we turn our attention to S_3. When no ambiguity arises, we will often write $C(g)$ instead of $C_G(g)$ to denote the centralizer.

■ Example

In Section 8.1, we looked at the table for S_3. If you take a look back at this table, you will see that

$$C(e) = S_3, \quad C(f) = \{e, f\}, \quad C(g) = \{e, g\},$$

$$C(h) = \{e, h\}, \quad C(j) = \{e, j, k\}, \quad C(k) = \{e, j, k\}.$$

■

Notice that the centralizer of each element of S_3 is a subgroup of S_3. We will now prove that this is merely a special case of the fact that in any group the centralizer of an element is always a subgroup. When reading the proof, keep in mind that the only properties we are allowed to use when doing computations with the elements of a group are the associative law and properties of the identity and inverses. When reading the proof, at various points, you should convince yourself that one line follows from the next using these properties and no others.

Proposition 8.14. *If G is a group and $g \in G$, then $C(g)$ is a subgroup of G.*

Proof. Let $x, y \in C(g)$; by Proposition 8.6(a) we need to show that $xy^{-1} \in C(g)$. We know that $gx = xg$ and $gy = yg$. Multiplying this last equation on the right and left by y^{-1}, we obtain

$$y^{-1}(gy)y^{-1} = y^{-1}(yg)y^{-1}.$$

Using the associative laws and properties of the identity and inverses, this simplifies to

$$y^{-1}g = gy^{-1}.$$

The previous equations combined with the repeated use of associativity tells us that

$$(xy^{-1})g = x(y^{-1}g) = x(gy^{-1}) = (xg)y^{-1} = (gx)y^{-1} = g(xy^{-1}).$$

Thus, $xy^{-1} \in C(g)$, as desired. $\qquad\square$

Observe that, for any group G and $g \in G$, it will always be the case that

$$<g> \subseteq C_G(g) \subseteq G.$$

In our preceding examples, $C_G(g)$ has always been equal to either $<g>$ or G. However, we will see some examples where this is no longer the case.

■ Example

Let D_4 be the fourth dihedral group. Recall it consists of $y^i x^j$, where $0 \le i \le 3, 0 \le j \le 1$, $x^2 = y^4 = e$, and $xy = y^3 x$. The table for this eight element group is

$\circ$	e	x	y	yx	y^2	$y^2 x$	y^3	$y^3 x$
e	e	x	y	yx	y^2	$y^2 x$	y^3	$y^3 x$
x	x	e	$y^3 x$	y^3	$y^2 x$	y^2	yx	y
y	y	yx	y^2	$y^2 x$	y^3	$y^3 x$	e	x
yx	yx	y	x	e	$y^3 x$	y^3	$y^2 x$	y^2
y^2	y^2	$y^2 x$	y^3	$y^3 x$	e	x	y	yx
$y^2 x$	$y^2 x$	y^2	yx	y	x	e	$y^3 x$	y^3
y^3	y^3	$y^3 x$	e	x	y	yx	y^2	$y^2 x$
$y^3 x$	$y^3 x$	y^3	$y^2 x$	y^2	yx	y	x	e

You should now do the computations needed to confirm that

$$C(e) = C(y^2) = D_4, \quad C(x) = C(y^2 x) = \{e, x, y^2, y^2 x\},$$

$$C(y) = C(y^3) = \{e, y, y^2, y^3\}, \quad C(yx) = C(y^3 x) = \{e, yx, y^3, y^3 x\}.$$

■

Centralizers tell us which elements of a group commute with a particular element of the group. In examining the structure of a group, we are often interested in those elements that commute with every element of the group.

Definition 8.15. *If G is a group, let $Z(G) = \{g \in G \mid gh = hg, \text{for all } h \in G\}$. We call $Z(G)$ the center of G.*

Observe that G is abelian if and only if $G = Z(G)$. In some sense, the larger $Z(G)$ is, the closer G is to being abelian. It should come as little surprise that $Z(G)$ is also a subgroup of G.

Proposition 8.16. *If G is a group, then Z(G) is a subgroup of G.*

Proof. This proof will be very similar to the proof of Proposition 8.14. Let $x, y \in Z(G)$; by Proposition 8.6(a) we need to show that $xy^{-1} \in Z(G)$. Therefore, if $h \in G$, we need to show that xy^{-1} commutes with h.

We know that $hx = xh$ and $hy = yh$. Multiplying this last equation on the right and left by y^{-1}, we obtain

$$y^{-1}(hy)y^{-1} = y^{-1}(yh)y^{-1}.$$

Using the associative laws and properties of the identity and inverses, this simplifies to

$$y^{-1}h = hy^{-1}.$$

The previous equations combined with the repeated use of associativity tell us that

$$(xy^{-1})h = x(y^{-1}h) = x(hy^{-1}) = (xh)y^{-1} = (hx)y^{-1} = h(xy^{-1}).$$

Thus, xy^{-1} commutes with every $h \in G$ and $xy^{-1} \in Z(G)$, as desired. $\qquad \square$

Another way to look at $Z(G)$ is as the intersection of every $C(g)$ as g ranges through the elements of G. Using this observation and the tables for S_3 and D_4, it is not hard to see that $Z(S_3) = \{e\}$ and $Z(D_4) = \{e, y^2\}$.

Let us now reflect on some of our examples. Keep in mind that every group G with more than one element always has at least two different subgroups, as $< e >$ and G are always subgroups.

■ Examples

1. S_3 has order 6, and we have already seen that S_3 has subgroups of order 1, 2, 3, and 6.

2. D_4 has order 8. When we look at the subgroups $< e >$ and $Z(D_4) = \{e, y^2\}$ along with the preceding list of centralizers, we can see that D_4 has subgroups of order 1, 2, 4, and 8.

■

In these examples, the order of every subgroup we found divides the order of the group. Furthermore, for every divisor of the order of the group, we can find a subgroup of that order. Although two examples may not appear to be an enormous amount of evidence, it does seem quite reasonable to ask the following two questions about the subgroups of a finite group G:

Questions:

1. If H is a subgroup of G, must $|H|$ divide $|G|$?

2. If $n \in \mathbb{N}$ and n divides $|G|$, must G have a subgroup of order n?

We will now begin to try to answer the first question. To answer this and other questions, we need

Lemma 8.17. *Let G be a group where $M \subseteq G$ and $a \in G$. Next, let $Ma = \{xa \mid x \in M\}$ and let $aM = \{ax \mid x \in M\}$.*

(a) *If M is finite, then the sets M, Ma, aM all have the same number of elements.*

(b) *If H is a subgroup and we define $\sim_r$ on the set G as $a \sim b$ whenever $ab^{-1} \in H$, then $\sim_r$ is an equivalence relation.*

(c) *If H is a subgroup and we define $\sim_l$ on the set G as $a \sim b$ whenever $a^{-1}b \in H$, then $\sim_l$ is an equivalence relation.*

At this point, you probably do not have a great deal of experience proving facts about groups. Therefore, at the risk of being repetitive, we again point out that the only properties we are allowed to use are those dealing with associativity, the identity, and inverses. In particular, we cannot assume that multiplication is commutative.

Proof. For part (a), we will first show that M and Ma have the same number of elements. It will be enough to find a bijection from M to Ma. Since Ma is defined as $\{ma \mid m \in M\}$, it seems natural to consider the function

$$\pi : M \rightarrow Ma,$$

where

$$\pi(m) = ma,$$

for all $m \in M$.

It is easy to see that π is surjective. Indeed, if $y \in Ma$ then, by the definition of Ma, $y = xa$, for some $x \in M$. Thus,

$$y = xa = \pi(x)$$

and y belongs to the range of π. Hence, π is surjective.

To prove that π is injective, suppose $g, h \in M$ such that $\pi(g) = \pi(h)$; we need to show that $g = h$. Observe that since $\pi(g) = \pi(h)$, we have

$$ga = ha.$$

Multiplying this equation on the right by a^{-1} yields

$$(ga)a^{-1} = (ha)a^{-1},$$

then

$$g(aa^{-1}) = h(aa^{-1}),$$

followed by

$$ge = he,$$

and finally

$$g = h.$$

Since π is both injective and surjective, it is a bijection, so M and Ma have the same number of elements. Not surprisingly, to show that M and aM have the same number of elements, we consider the function

$$\rho : M \to aM,$$

where

$$\rho(m) = am,$$

for all $m \in M$. It will be good practice for you to work through the details and verify that ρ is also a bijection.

Before proving part (b), you should observe that this is a generalization of the equivalence relation used in Chapter 7, where we defined $\mathbb{Z}_n$ using the group $\mathbb{Z}$ and the subgroup $n\mathbb{Z}$. First, to show that $\sim_r$ is reflexive, we need to show that $a \sim_r a$, for all $a \in G$. If $a \in G$, we have $aa^{-1} = e$. Since $e \in H$, we see that $a \sim_r a$ and so, $\sim_r$ is reflexive.

Next, suppose $a, b \in G$ such that $a \sim_r b$. To show that $\sim_r$ is symmetric, we need to show that $b \sim_r a$, so we may assume that $ab^{-1} \in H$, and we must verify that $ba^{-1} \in H$. Observe that

$$(ba^{-1})(ab^{-1}) = b(a^{-1}a)b^{-1} = beb^{-1} = e$$

and

$$(ab^{-1})(ba^{-1}) = a(b^{-1}b)a^{-1} = aea^{-1} = e.$$

These two equations tell us that ba^{-1} is the inverse of ab^{-1}. However, since H is a group, it contains the inverse of ab^{-1}. Thus, $ba^{-1} \in H$ and $\sim_r$ is symmetric.

For the final piece of part (b), suppose $a, b, c \in G$ such that $a \sim_r b$ and $b \sim_r c$. To show that $\sim_r$ is transitive, we need to show that $a \sim_r c$. Therefore, we may assume that $ab^{-1}, bc^{-1} \in H$, and we need to show that $ac^{-1} \in H$. Since H is a group, it contains the product of ab^{-1} and bc^{-1}. Thus,

$$(ab^{-1})(bc^{-1}) = a(b^{-1}b)c^{-1} = aec^{-1} = ac^{-1},$$

so $ac^{-1} \in H$, as desired.

The proof of part (c) is very similar to and uses the same ideas as the proof of part (b). Therefore, we will leave the proof to you. Providing all the details of the proof should reinforce your understanding of the proof of part (b). □

In the proof of Lemma 8.17(b), we showed that the inverse of ab^{-1} is ba^{-1}. Observe that this argument is similar to the one in the proof of Proposition 7.7 where we showed that the inverse of ab is $b^{-1}a^{-1}$. We can now state the first significant theorem about the structure of finite groups. Observe that it answers the first question we posed in this section in the affirmative.

Theorem 8.18—Lagrange's Theorem. *If G is a finite group and H is a subgroup of G, then $|H|$ divides $|G|$.*

Before proving Lagrange's Theorem, it is important to understand what it says and what it does not say. Note that it does not assert the existence of subgroups of G. It does quite the opposite, as it places limits on the orders of subgroups that can exist.

For example, since $|S_3| = 6$, Lagrange's Theorem tells us that *every* subgroup of S_3 must have order $1, 2, 3$, or 6. But it does not guarantee that subgroups of these orders exist. In fact, what it really tells us is that S_3 does not have any subgroups of order 4 or 5.

Lagrange's Theorem can also simplify the computations involved in finding various subgroups, such as centralizers. Let us consider the fifth dihedral group D_5. Every element of D_5 can be written in the form $y^i x^j$, where $0 \le i \le 4$, $0 \le j \le 1$, $y^5 = x^2 = e$, and $xy = y^4 x$. Since $|D_5| = 10$, Lagrange's Theorem tells us that the only possible subgroups of D_5 have order $1, 2, 5$, or 10.

Suppose we wanted to compute $C(y)$. Since $C(y)$ certainly contains $< y >$ and $o(y) = 5$, we know that $|C(y)| \ge 5$. Since y does not commute with x, we also know that $|C(y)| \ne 10$. Therefore, Lagrange's Theorem tells us that $|C(y)| = 5$. As a result, without having to do much computing, we see that

$$C(y) = < y > = \{e, y, y^2, y^3, y^4\}.$$

The proof of Lagrange's Theorem will once again indicate how useful equivalence relations and equivalence classes are in higher mathematics.

Proof of Lagrange's Theorem. Let $\sim_r$ be the equivalence relation on the set G defined in Lemma 8.17(b). Since every equivalence class is a subset of G, and G only has a finite number of subsets, there is only a finite number of equivalence classes, and we will call them $A_1, A_2, \ldots, A_m$. Recall that an equivalence class may have many names, but our list of equivalence classes will include each class only once.

Proposition 4.13 tells us that every element of G belongs to exactly one equivalence class, so

$$|G| = |A_1| + |A_2| + \cdots + |A_m|.$$

At this point, the question is how many elements belong to each class?

For every $b \in G$, let $[b]$ denote the equivalence class containing b. If $a \in [b]$, then the definition of $\sim_r$ tells us that there exists $h \in H$ such that $ab^{-1} = h$. This implies that $a = hb$, so $a \in Hb$. As a result, $[b] \subset Hb$.

On the other hand, if $a \in Hb$, then there exists $h \in H$ such that $a = hb$. This immediately tells us that $ab^{-1} = h \in H$, so $a \sim_r b$ and $a \in [b]$. Thus, $Hb \subseteq [b]$.

We can now see that for every $b \in G$, $[b] = Hb$. Lemma 8.17(a) told us that every set of the form Hb has the same number of elements as H. Therefore, every equivalence class has the same number of elements as H. The equation $|G| = |A_1| + |A_2| + \cdots + |A_m|$ now becomes

$$|G| = |H| + |H| + \cdots + |H| = m|H|.$$

Since $|G| = m|H|$, $|H|$ does indeed divide $|G|$.   □

The proof of Lagrange's Theorem used $\sim_r$ and sets of the form Hb from Lemma 8.17(b), but we could have equally well used $\sim_l$ and sets of the form bH from Lemma 8.17(c). Since sets of the form Hb and bH are so important, we have

Definition 8.19. *If H is a subgroup of a group G and $b \in H$, we call sets of the form Hb right cosets of H and sets of the form bH left cosets of H.*

When G is finite with subgroup H and $|G| = m|H|$, we call m the *index* of H in G. Thus, m tells us both the number of right cosets and left cosets corresponding to H. Let us now take a look at some examples of right and left cosets.

■ Examples

1. Let $G = \mathbb{Z}_4$; observe that in this example we will be using additive and not multiplicative notation.

1. If $H = \{[0]_4\}$, then $|H| = 1$ and $m = 4$. Therefore, H has four right cosets and four left cosets, each with only one element.

$$H + [0]_4 = \{[0]_4\} = [0]_4 + H,$$

$$H + [1]_4 = \{[1]_4\} = [1]_4 + H,$$

$$H + [2]_4 = \{[2]_4\} = [2]_4 + H,$$

$$H + [3]_4 = \{[3]_4\} = [3]_4 + H.$$

2. If $H = \{[0]_4, [2]_4\}$, then $|H| = 2$ and $m = 2$. Therefore, H has two right cosets and two left cosets, each with two elements. Observe that each right coset and each left coset now have two different names, as they are equivalence classes with two elements.

$$H + [0]_4 = H + [2]_4 = \{[0]_4, [2]_4\} = [2]_4 + H = [0]_4 + H,$$

$$H + [1]_4 = H + [3]_4 = \{[1]_4, [3]_4\} = [3]_4 + H = [1]_4 + H.$$

3. If H is all of $\mathbb{Z}_4$, then $|H| = 4$ and $m = 1$. Therefore, H has only one right coset and only one left coset, and they consist of all four elements of $\mathbb{Z}_4$. Therefore, each right coset and each left coset have four different names.

$$H + [0]_4 = H + [1]_4 = H + [2]_4 = H + [3]_4 = \{[0]_4, [1]_4, [2]_4, [3]_4\} =$$

$$[3]_4 + H = [2]_4 + H = [1]_4 + H = [0]_4 + H.$$

II. Let $G = S_3$; in this example things get much more interesting. Throughout this example and this section, we will use the notation for the six elements in S_3 developed in the previous section.

1. If $H = \{e\}$, then $|H| = 1$ and $m = 6$. Therefore, there are six right and six left cosets, each with one element.

$$He = \{e\} = eH, \quad Hf = \{f\} = fH, \quad Hg = \{g\} = gH,$$

$$Hh = \{h\} = hH, \quad Hj = \{j\} = jH, \quad Hk = \{k\} = kH.$$

2. If $H = \{e, f\}$, then $|H| = 2$ and $m = 3$. Now there are three right and three left cosets, each with two elements. But now when we look at the elements in the right and left cosets, we notice that something new is happening.

$$He = Hf = \{e, f\},$$

$$Hg = Hj = \{g, j\},$$

$$Hh = Hk = \{h, k\},$$

whereas

$$eH = fH = \{e, f\},$$

$$gH = kH = \{g, k\},$$

$$hH = jH = \{h, j\}.$$

Note that $Hg \neq gH$, $Hj \neq jH$, $Hh \neq hH$, and $Hk \neq kH$. For the first time, we see that for some $b \in G$, the right and left cosets Hb and bH need not be the same subset of G.

3. If $H = \{e, g\}$, then $|H| = 2$ and $m = 3$. There are again three right and three left cosets, each with two elements.

$$He = Hg = \{e, g\},$$

$$Hf = Hk = \{f, k\},$$

$$Hh = Hj = \{h, j\},$$

whereas

$$eH = gH = \{e, g\},$$

$$fH = jH = \{f, j\},$$

$$hH = kH = \{h, k\}.$$

Observe that $Hf \neq fH$, $Hk \neq kH$, $Hh \neq hH$, and $Hj \neq jH$.

4. If $H = \{e, h\}$, then $|H| = 2$ and $m = 3$. Again we have three right and three left cosets, each with two elements.

$$He = Hh = \{e, h\},$$

$$Hf = Hj = \{f, j\},$$

$$Hg = Hk = \{g, k\},$$

whereas

$$eH = hH = \{e, h\},$$

$$gH = jH = \{g, j\},$$

$$fH = kH = \{f, k\}.$$

In this example, $Hf \neq fH$, $Hj \neq jH$, $Hg \neq gH$, and $Hk \neq kH$.

5. If $H = \{e, j, k\}$, then $|H| = 3$ and $m = 2$. Therefore, there will be two right cosets and two left cosets, each with three elements.

$$He = Hj = Hk = \{e, j, k\},$$

$$Hf = Hg = Hh = \{f, g, h\},$$

and

$$eH = jH = kH - \{e, j, k\},$$

$$fH = gH = hH = \{f, g, h\}.$$

Note that we are back in the situation where, for every $b \in G$, $Hb = bH$.

6. If $H = S_3$, then $|H| = 6$ and $m = 1$. Thus, every right coset and every left coset is equal to all of S_3.

∎

We have seen that there exist subgroups H of groups G where $Hg \neq gH$, for some $g \in G$. This raises the question of whether it is important for a subgroup H to have the property that $Hg = gH$, for all $g \in G$?

To answer this question, we will look back at the construction of $\mathbb{Z}_n$ from $\mathbb{Z}$. We began with $\mathbb{Z}$ and some $n \geq 2$ and then defined $\sim_n$ on $\mathbb{Z}$ as $a \sim_n b$ whenever $a - b$ was a multiple of n. Rephrasing this in the language of groups, $\mathbb{Z}$ is a group under addition with subgroup $H = n\mathbb{Z}$. Then, since we are dealing with groups under addition, $\sim_n$ is the same as the equivalence relation in Lemma 8.17(b) in which $a \sim_r b$ whenever $a - b \in H$. Thus, the elements of $\mathbb{Z}_n$ are the left and right cosets corresponding to the subgroup $n\mathbb{Z}$. In Chapter 7, we showed that these cosets formed a group under addition. The key step was to first show that the addition of cosets was, in this situation, well defined. More generally, given a subgroup of a group, we can ask whether the left or right cosets form a group. In order to do this, if we are using multiplicative notation, we first need to see whether or not the multiplication of cosets is well defined. As we will soon see, the property that $Hg = gH$, for all $g \in G$, is exactly what is needed to make this happen.

Definition 8.20. *Let G be a group with subgroup H. We say that H is normal if $Hg = gH$, for all $g \in G$.*

Needless to say, if a mathematical property is called "normal," then it is considered a very good property to have. If G is abelian, then it is easy to see that all subgroups are normal. This was certainly the case when we constructed $\mathbb{Z}_n$ using the group $\mathbb{Z}$ and subgroup $n\mathbb{Z}$. However, for nonabelian groups, it can be the case that some subgroups are normal and others are not.

If we look back at our work with S_3, we can see that none of the three subgroups $\{e, f\}$, $\{e, g\}$, and $\{e, h\}$ are normal. On the other hand, the three subgroups $\{e\}$, $\{e, j, k\}$, and S_3 are all normal.

It is easy to see that for any group G, the subgroups G and $\{e\}$ are always normal. It is also not hard to see that any subgroup of G that is contained in $Z(G)$ must also be normal. Let us now begin to investigate the relationship between subgroups being normal and the multiplication of cosets.

■ Example

Let $G = S_3$ and let $H = \{e, f\}$. If we wanted to multiply right cosets, the most natural thing to do would be to define

$$Ha \cdot Hb = H(ab),$$

for all $a, b \in G$. Therefore, the way we multiply cosets is to multiply their names. The problem is that cosets can have many different names, and we need to see if changing the name of the cosets changes the answer when we multiply.

In this situation, we have $Hg = Hg$ and $Hg = Hj$; therefore, if the multiplication of right cosets is well defined, then the products $Hg \cdot Hg$ and $Hg \cdot Hj$ would need to be the same. However,

$$Hg \cdot Hg = H(gg) = He,$$

whereas

$$Hg \cdot Hj = H(gj) = Hh.$$

Since $He \neq Hh$, changing the names of the cosets did indeed change the answer when multiplying. Hence, in this case, the multiplication of cosets is not well defined.

■

In the previous example, H was not a normal subgroup of G. We will now show that being normal is exactly what is needed for the multiplication of cosets to be well defined.

Theorem 8.21. *Let G be a group with subgroup H and define the multiplication of right cosets as $Hg \cdot Hh = H(gh)$, for all $g, h \in G$. Then multiplication of right cosets is well defined if and only if H is normal.*

Proof. In one direction, suppose that H is normal. In order to show that multiplication of right cosets is well defined, we need to show that if $a, b, c, d \in G$ such that $Ha = Hb$ and $Hc = Hd$, then $Ha \cdot Hc = Hb \cdot Hd$. Since

$$Ha \cdot Hc = H(ac) \quad \text{and} \quad Hb \cdot Hd = H(bd),$$

we need to show that $H(ac) = H(bd)$.

Since a, b belong to the same right coset, $ab^{-1} \in H$. Similarly, since c, d belong to the same right coset, $cd^{-1} \in H$. Therefore,

$$(1) \qquad\qquad ab^{-1} = h \quad \text{and} \quad cd^{-1} = k,$$

where $h, k \in H$. Since the inverse of bd is $d^{-1}b^{-1}$, in order to show that $H(ac) = H(bd)$, we need to show that

$$(2) \qquad\qquad (ac)(bd)^{-1} = acd^{-1}b^{-1} \in H.$$

In light of equation (1), $a = hb$ and $c = kd$, so equation (2) becomes

$$(3) \qquad\qquad (ac)(bd)^{-1} = acd^{-1}b^{-1} = (hb)(kd)d^{-1}b^{-1} = hbkb^{-1}.$$

Clearly $bk \in bH$. However, since H is normal, $bH = Hb$ and so, $bk \in Hb$. As a result, there exist $l \in H$ such that $bk = lb$. Equation (3) now becomes

$$(ac)(bd)^{-1} = h(bk)b^{-1} = h(lb)b^{-1} = (hl)(bb^{-1}) = hl \in H.$$

Thus, $H(ac) = H(bd)$ and the multiplication of right cosets is well defined.

In the other direction, we will assume that the multiplication of right cosets is well defined, and we will show that H is normal. Therefore, if $a \in G$, we need to show that $Ha = aH$. It will suffice to show that if $h \in H$, then $ha \in aH$ and $ah \in Ha$, for these combine to tell us that $Ha \subseteq aH$ and $aH \subseteq Ha$.

Since $h \in H$, we know that $He = Hh$. Using the fact that coset multiplication is well defined, we have

$$Ha \cdot Hh = Ha \cdot He.$$

Since $Ha \cdot Hh = H(ah)$ and $Ha \cdot He = H(ae) = Ha$, we know that $H(ah) = Ha$. As a result, $(ah)a^{-1} \in H$ and so, there exists $k \in H$ such that $aha^{-1} = k \in H$. This immediately implies that $ah = ka \in Ha$.

Using an argument similar to the preceding one,

$$Ha^{-1} \cdot Hh = Ha^{-1} \cdot He.$$

However, $Ha^{-1} \cdot Hh = H(a^{-1}h)$ and $Ha^{-1} \cdot He = H(a^{-1}e) = Ha^{-1}$. This tells us that the right cosets $H(a^{-1}h)$ and Ha^{-1} are the same, so

$$a^{-1}h(a^{-1})^{-1} = a^{-1}ha \in H.$$

As a result, there exists $k \in H$ such that $a^{-1}ha = k$. This immediately implies that $ha = ak$, so $ha \in aH$, thereby concluding the proof. $\qquad\square$

We can now generalize the construction of $\mathbb{Z}_n$ from $\mathbb{Z}$ and $n\mathbb{Z}$ that we saw in Chapter 7.

Corollary 8.22. *Let G be a group with normal subgroup H. The set of right cosets is a group where multiplication is defined as $Ha \cdot Hb = H(ab)$, for all $a, b \in G$.*

Proof. Since G is closed under multiplication, it is easy to see that the set of right cosets is closed under multiplication. Furthermore, Theorem 8.21 asserts that the multiplication of cosets is well defined. As a result, it remains to verify that the multiplication of cosets is associative, that it has an identity element, and that each coset has an inverse. We will soon see that the set of right cosets inherits the desired properties from G in a very straightforward way.

Since G is associative, if $a, b, c \in G$, then $(ab)c = a(bc)$, which implies that

$$(Ha \cdot Hb) \cdot Hc = H(ab) \cdot Hc = H((ab)c) = H(a(bc)) = Ha \cdot H(bc) = Ha \cdot (Hb \cdot Hc).$$

Therefore, the multiplication of cosets is associative.

When we looked at $\mathbb{Z}_n$, the equivalence class $[0]_n$ was the identity element. In our more general situation, we would expect the coset containing the identity element of G to be the identity element when multiplying cosets. Indeed, if $a \in G$, since $ae = a = ea$, we have

$$Ha \cdot He = H(ae) = Ha = H(ea) = He \cdot Ha.$$

Thus, He, which is another name for H, is the identity element of coset multiplication.

Finally, we need to show that every coset has an inverse. Once again, we look at $\mathbb{Z}_n$ for some direction. The inverse of the class containing $a \in \mathbb{Z}$ was the class containing $-a$. Moving to the more general case and switching to multiplicative notation, we would then expect that the inverse of the coset containing $a \in G$ be the coset containing $a^{-1} \in G$. Checking that this is indeed the case, since $a \cdot a^{-1} = e = a^{-1} \cdot a$, we have

$$Ha \cdot Ha^{-1} = H(a \cdot a^{-1}) = He = H(a^{-1}a) = Ha^{-1} \cdot Ha.$$

Hence, Ha^{-1} is the inverse of Ha, and we have succeeded in showing that the set of right cosets is a group. $\qquad\square$

Groups of the type described in Corollary 8.22 are an important part of group theory, and we will return to them in greater detail in the next section. But now that we know what normal subgroups are, we can ask, how do we find normal subgroups? Certainly, if G is abelian, all subgroups are normal. Furthermore, for any group G, the subgroups $\{e\}$ and G can easily be seen to be normal. However, to find other normal subgroups, we look at functions similar to isomorphisms.

Definition 8.23. *If G_1, G_2 are groups, then a function $\phi : G_1 \to G_2$ is called a homomorphism of groups if $\phi(gh) = \phi(g)\phi(h)$, for all $g, h \in G$.*

Observe that homomorphisms satisfy the same defining equation as isomorphisms. In fact, every isomorphism is a homomorphism. However, there are homomorphisms that are not isomorphisms, as a homomorphism can fail to be injective, surjective, or both. We can now look at the connection between homomorphisms and normal subgroups.

Theorem 8.24. *Let $\phi : G_1 \to G_2$ be a homomorphism of groups.*

(a) *If e_1, e_2 are, respectively, the identity elements of G_1 and G_2, then $\phi(e_1) = e_2$.*

(b) *If $g \in G_1$, then $\phi(g^{-1}) = \phi(g)^{-1}$.*

(c) *If we let $Ker(\phi) = \{g \in G_1 \mid \phi(g) = e_2\}$ and $Im(\phi) = \{\phi(g) \mid g \in G_1\}$, then $Ker(\phi)$ is a subgroup of G_1 and $Im(\phi)$ is a subgroup of G_2.*

(d) *$Ker(\phi)$ is a normal subgroup of G_1.*

Proof. For part (a), we know that $e_1 = e_1 \cdot e_1$ and applying ϕ to this equation yields

$$\phi(e_1) = \phi(e_1 \cdot e_1) = \phi(e_1) \cdot \phi(e_1).$$

Multiplying the terms at the left and right of the previous equation by $\phi(e_1)^{-1}$ on the left gives us

$$\phi(e_1)^{-1} \cdot \phi(e_1) = \phi(e_1)^{-1}\phi(e_1) \cdot \phi(e_1),$$

which, using associativity, yields

$$e_2 = e_2 \cdot \phi(e_1),$$

and finally

$$e_2 = \phi(e_1).$$

For part (b), if $g \in G_1$, then $e_1 = g \cdot g^{-1}$ and applying ϕ and using (a) gives us

$$e_2 = \phi(e_1) = \phi(g \cdot g^{-1}) = \phi(g) \cdot \phi(g^{-1}).$$

Multiplying the terms at the far left and right of the previous equation by $\phi(g)^{-1}$ on the left tells us that

$$\phi(g)^{-1} \cdot e_2 = \phi(g)^{-1} \cdot \phi(g) \cdot \phi(g^{-1}),$$

which, using associativity, simplifies to

$$\phi(g)^{-1} = e_2 \cdot \phi(g^{-1}),$$

and then

$$\phi(g)^{-1} = \phi(g^{-1}).$$

For the first half of part (c), to show that $Ker(\phi)$ is a subgroup of G_1, we again use Proposition 8.6(a). Therefore, if $x, y \in Ker(\phi)$, we need to show that $xy^{-1} \in Ker(\phi)$. In light of (a) and (b), we know that

$$\phi(y^{-1}) = \phi(y)^{-1} = e_2^{-1} = e_2.$$

Therefore,

$$\phi(x \cdot y^{-1}) = \phi(x) \cdot \phi(y^{-1}) = e_2 \cdot e_2 = e_2.$$

Hence, $xy^{-1} \in Ker(\phi)$, as desired.

Next, to show that $Im(\phi)$ is a subgroup, we once again use Proposition 8.6(a). Therefore, if $x, y \in Im(\phi)$, we must show that $xy^{-1} \in Im(\phi)$. We know that there exist $a, b \in G_1$ such that $x = \phi(a)$ and $y = \phi(b)$. Using (b), we now have

$$x \cdot y^{-1} = \phi(a) \cdot \phi(b)^{-1} = \phi(a) \cdot \phi(b^{-1}) = \phi(ab^{-1}).$$

Thus, $x \cdot y^{-1} \in Im(\phi)$ as required.

For part (d), we need to show that if $g \in G_1$, then $gKer(\phi) = Ker(\phi)g$. We will do this by showing that for every $h \in Ker(\phi)$, we have $gh \in Ker(\phi)g$ and $hg \in gKer(\phi)$. In order to do this, we look at the elements $ghg^{-1}, g^{-1}hg$ and observe that

$$\phi(ghg^{-1}) = \phi(g) \cdot \phi(h) \cdot \phi(g^{-1}) = \phi(g) \cdot e_2 \cdot \phi(g^{-1}) =$$

$$\phi(g) \cdot \phi(g^{-1}) = \phi(gg^{-1}) = \phi(e_1) = e_2$$

and

$$\phi(g^{-1}hg) = \phi(g^{-1}) \cdot \phi(h) \cdot \phi(g) = \phi(g^{-1}) \cdot e_2 \cdot \phi(g) =$$

$$\phi(g^{-1}) \cdot \phi(g) = \phi(g^{-1}g) = \phi(e_1) = e_2.$$

The preceding equations tell us that both ghg^{-1} and $g^{-1}hg$ belong to $Ker(\phi)$. This enables us to express gh, hg as

$$gh = \left(ghg^{-1}\right) g \in Ker(\phi)g \quad \text{and} \quad hg = g\left(g^{-1}hg\right) \in gKer(\phi).$$

As a result, it is indeed the case that $gh \in Ker(\phi)g$ and $hg \in gKer(\phi)$. $\qquad\qquad\square$

Homomorphisms have uses far beyond being used to produce normal subgroups. For example, suppose $\phi : G_1 \to G_2$ is a homomorphism of groups and suppose $a, b \in G_1$ commute when multiplied. Since $a \cdot b = b \cdot a$, we have

$$\phi(a) \cdot \phi(b) = \phi(a \cdot b) = \phi(b \cdot a) = \phi(b) \cdot \phi(a).$$

Therefore, we have shown that since a and b commute in G_1, then $\phi(a)$ and $\phi(b)$ must commute in G_2. You should convince yourself that this tells us that if G_1 is abelian and ϕ is surjective, then G_2 must also be abelian. More generally, a homomorphism $\phi : G_1 \to G_2$ provides us with a link between algebraic properties of G_1 and those of G_2.

For another example, suppose $\phi : G_1 \to G_2$ is a homomorphism of groups that is also injective. Also suppose that G_2 has the property that there is a fixed positive integer m such that $h^m = h$, for all $h \in G_2$. We claim that G_1 must also have this property. First observe that if $g \in G_1$ then $\phi(g) \in G_2$, so $\phi(g)^m = \phi(g)$. However, since ϕ is a homomorphism, the identical argument used in the proof of Lemma 5.9 shows us that $\phi(g^m) = \phi(g)^m$. As a result, $\phi(g^m) = \phi(g)$. But since ϕ is injective, this tells us that $g^m = g$. Thus, G_1 does inherit this property from G_2.

■ Examples

Homomorphisms of Groups
1. If $n \geq 2$, then $\mathbb{Z}$ and $\mathbb{Z}_n$ are both groups under addition. Let

$$\phi : \mathbb{Z} \to \mathbb{Z}_n$$

be defined as

$$\phi(a) = [a]_n,$$

for all $a \in \mathbb{Z}$. Observe that

$$\phi(a+b) = [a+b]_n = [a]_n + [b]_n = \phi(a) + \phi(b),$$

for all $a, b \in \mathbb{Z}$. Thus, ϕ is a homomorphism. In this example, $Ker(\phi) = n\mathbb{Z}$, which is a normal subgroup of $\mathbb{Z}$, and $Im(\phi) = \mathbb{Z}_n$, which is certainly a subgroup of $\mathbb{Z}_n$.

2. $\mathbb{R}^\times$ is a group under multiplication. Let

$$\phi : \mathbb{R}^\times \to \mathbb{R}^\times$$

be defined as $\phi(x) = x^2$, for all $x \in \mathbb{R}^\times$. Since $\mathbb{R}^\times$ is abelian, we have

$$\phi(xy) = (xy)^2 = xyxy = x^2 y^2 = \phi(x)\phi(y).$$

Thus, ϕ is a homomorphism. In this case, $Ker(\phi) = \{1, -1\}$, which is a normal subgroup of $\mathbb{R}^\times$. Furthermore, $Im(\phi)$ is the set of positive real numbers, which is also a subgroup of $\mathbb{R}^\times$.

3. Let G be any abelian group and let $n \in \mathbb{Z}$. Define

$$\phi : G \to G$$

as $\phi(g) = g^n$, for all $g \in G$. Since G is abelian, we have

$$\phi(gh) = (gh)^n = g^n h^n = \phi(g)\phi(h),$$

so ϕ is a homomorphism. Since $Ker(\phi) = \{g \in G \mid g^n = e\}$ and $Im(\phi) = \{g^n \mid g \in G\}$, Theorem 8.24(c) tells us that the sets $\{g \in G \mid g^n = e\}$ and $\{g^n \mid g \in G\}$ are always subgroups of an abelian group.

4. $\mathbb{Z}$ is a group under addition and $U(\mathbb{Q})$ is a group under multiplication. Let

$$\phi : \mathbb{Z} \to U(\mathbb{Q})$$

be defined as $\phi(n) = (-1)^n$, for all $n \in \mathbb{Z}$. Then ϕ is a homomorphism as

$$\phi(n+m) = (-1)^{n+m} = (-1)^n(-1)^m = \phi(n) \cdot \phi(m),$$

for all $n, m \in \mathbb{Z}$. Observe that $\phi(n) = 1$ if n is even and $\phi(n) = -1$ if n is odd. It is easy to see that $Ker(\phi) = 2\mathbb{Z}$ and $Im(\phi) = \{1, -1\}$.

5. $\mathbb{Z}$ is a group under addition and $U(\mathbb{C})$ is a group under multiplication. Let

$$\phi : \mathbb{Z} \to U(\mathbb{C})$$

be defined as $\phi(n) = i^n$, for all $n \in \mathbb{Z}$. Then ϕ is a homomorphism as

$$\phi(n+m) = i^{n+m} = i^n \cdot i^m = \phi(n) \cdot \phi(m),$$

for all $n, m \in \mathbb{Z}$. Observe that $\phi(n) = 1$ if $n \equiv 0 \pmod 4$, $\phi(n) = i$ if $n \equiv 1 \pmod 4$, $\phi(n) = -1$ if $n \equiv 2 \pmod 4$, and $\phi(n) = -i$ if $n \equiv 3 \pmod 4$. We now have $Ker(\phi) = 4\mathbb{Z}$ and $Im(\phi) = \{1, -1, i, -1\}$.

6. $\mathbb{R}^\times$ is a group under multiplication. Let

$$\phi : \mathbb{R}^\times \to \mathbb{R}^\times$$

be defined as $\phi(x) = |x|$, for all $x \in \mathbb{R}^\times$, where $|x|$ denotes the absolute value of x. Then ϕ is a homomorphism as

$$\phi(x \cdot y) = |x \cdot y| = |x| \cdot |y| = \phi(x) \cdot \phi(y),$$

for all $x, y \in \mathbb{R}^\times$. As in our second example, $Ker(\phi) - \{1, -1\}$ and $Im(\phi)$ is the set of positive real numbers.

7. $\mathbb{R}^+$ is a group under multiplication and $\mathbb{R}$ is a group under addition. Let

$$\phi : \mathbb{R}^+ \to \mathbb{R}$$

be defined as $\phi(x) = \ln(x)$, for all $x \in \mathbb{R}^+$, where $\ln(x)$ denotes the logarithm of x using the base e. Using a basic property of log functions, it follows that ϕ is a homomorphism as

$$\phi(x \cdot y) = \ln(x \cdot y) = \ln(x) + \ln(y) = \phi(x) + \phi(y),$$

for all $x, y \in \mathbb{R}^+$. In this example, $Ker(\phi) = \{1\}$ and $Im(\phi) = \mathbb{R}$.

8. $\mathbb{Z}$ is a group under addition. Let

$$\phi : \mathbb{Z} \to \mathbb{Z}$$

be defined as $\phi(n) = 10n$, for all $n \in \mathbb{Z}$. Then ϕ is a homomorphism as

$$\phi(n + m) = 10(n + m) = 10n + 10m = \phi(n) + \phi(m),$$

for all $n, m \in \mathbb{Z}$. In this example, $Ker(\phi) = \{0\}$ and $Im(\phi) = 10\mathbb{Z}$.

We now turn our attention to the second question we posed earlier in this section. Namely, if G is a finite group and $n \in \mathbb{N}$ divides $|G|$, must G have a subgroup with n elements? For the examples we looked at earlier, such as S_3 and D_4, the answer was yes. In fact, although it is not obvious at first glance, the answer turns out to be yes for all abelian groups. However, in general, the answer turns out to be no. The smallest example that shows that the answer is no is a group with 12 elements that does not contain a subgroup with 6 elements. We will take a detailed look at this example in Section 8.4 when we examine symmetric groups.

What does turn out to be true is that if G is a finite group and p^a divides $|G|$, where p is a prime, then G must contain a subgroup with p^a elements. This result is known as Sylow's

Theorem, and at the beginning of this chapter, we described it as the crown jewel of an introductory course in group theory. It has countless applications for group theorists in their attempts to understand the structure of finite groups. Even though this is a fundamental result that provides deep insights into finite group theory, it does not require advanced techniques to prove. In fact, we will be able to prove it after three relatively short lemmas.

As we attempt to find a subgroup of G with p^a elements, we will need to derive some facts about the number of subsets of G with p^a elements. Recall that if a set G has n elements and $1 \le t \le n$, then the number of subsets of G with exactly t elements is

$$\binom{n}{t} = \frac{n!}{(n-t)! \cdot t!}.$$

To motivate our next lemma, let us look at some examples,

$$24 = 2^1 \cdot 12 \quad \text{and} \quad \binom{24}{2^1} = 276,$$

$$24 = 2^2 \cdot 6 \quad \text{and} \quad \binom{24}{2^2} = 10626,$$

$$24 = 2^3 \cdot 3 \quad \text{and} \quad \binom{24}{2^3} = 735471.$$

Admittedly, it is hard to find a pattern in these examples. In an attempt to find a pattern, we note that in all three cases, we can write 24 as $24 = 2^a m$. Observe that in all three cases, the largest power of 2 that divides $\binom{24}{2^a}$ is the same as the largest power of 2 that divides m. For example, when we write $24 = 2^2 \cdot 6$, then the largest power of 2 that divides 6 and the largest power of 2 that divides 10626 is 2^1. All of this is merely a special case of

Lemma 8.25. *Suppose* $n = p^a m$, *where* p *is a prime,* $a \ge 0$, *and* $m \in \mathbb{N}$. *If* p^b *is the largest power of* p *that divides* m, *then* p^b *is also the largest power of* p *that divides* $\binom{n}{p^a}$.

Proof. Since $\binom{n}{p^a} = \frac{p^a m!}{p^a! \cdot (p^a m - p^a)!}$, we have

$$\binom{n}{p^a} = \frac{(p^a m)(p^a m - 1)(p^a m - 2) \cdots (p^a m - (p^a - 1))}{(p^a)(p^a - 1)(p^a - 2) \cdots (1)} =$$

$$m \left(\frac{(p^a m - 1)(p^a m - 2) \cdots (p^a m - (p^a - 1))}{(p^a - 1)(p^a - 2) \cdots (1)} \right).$$

Therefore, it suffices to show that when we look at the fraction

$$\left(\frac{(p^a m - 1)(p^a m - 2) \cdots (p^a m - (p^a - 1))}{(p^a - 1)(p^a - 2) \cdots (1)} \right),$$

the power of p that occurs in the prime factorizations of both the numerator and the denominator is the same.

At first, this looks like it might involve some difficult computations. However, things will not be too difficult because we will be able to do this one term at a time. More precisely, for any $1 \leq j \leq p^a - 1$, we will show p^t divides $p^a - j$ if and only if it divides $p^a m - j$. This will suffice as the power of p that divides the numerator is the product of all the powers of p that divide terms of the form $p^a m - j$, and the power of p that divides the denominator is the product of all the powers of p that divide terms of the form $p^a - j$.

In one direction, suppose p^t divides $p^a - j$. Since $p^t \leq p^a - j < p^a$, it is clear that $t \leq a$. Therefore, p^t divides $p^a - j$, p^a, and $p^a m$, so p^t divides $j = (p^a) - (p^a - j)$. But this immediately implies that p^t also divides $p^a m - j$.

In the other direction, suppose p^t divides $p^a m - j$. Observe that if $t > a$, then p^a divides both $p^a m - j$ and p^a, which would imply that p^a implies $j = p^a - (p^a m - j)$. But this contradicts the fact that $1 \leq j \leq p^a - 1$. Thus, $t \leq a$, which implies that p^t divides p^a, $p^a m$, and $p^a m - j$. As a result, p^t divides $j = p^a m - (p^a m - j)$. Since p^t divides both p^a and j, p^t divides $p^a - j$, as desired.

In light of the preceding argument, p^b divides $\binom{n}{p^a}$ if and only if p^b divides m. $\qquad\square$

The second lemma we need to prove Sylow's Theorem is

Lemma 8.26. *Let G be a group with subset M.*

(a) *If $a, b \in G$, then $(Ma)b = M(ab)$.*

(b) *If $n \in \mathbb{N}$ and if we define $\sim$ on the set of subsets of G with n elements as $M \sim N$ whenever there exists $a \in G$ such that $M = Na$, then $\sim$ is an equivalence relation.*

Proof. For part (a), we know that $M(ab) = \{m(ab) \mid m \in M\}$. On the other hand, $Ma = \{ma \mid m \in M\}$ and $(Ma)b = \{nb \mid n \in Ma\}$, which combine to imply that $(Ma)b = \{(ma)b \mid m \in M\}$. The associative law tells us that $m(ab) = (ma)b$, so the sets $M(ab)$ and $(Ma)b$ consist of the same elements of G. Thus, $(Ma)b = M(ab)$.

For part (b), we need to show that $\sim$ is reflexive, symmetric, and transitive. Observe that if $N \subseteq G$, with $|N| = n$, then $N = Ne$. Therefore, $N \sim N$ and $\sim$ is reflexive. Next, to show that $\sim$ is symmetric, suppose we are given $M, N \subseteq G$ such that $|M| = |N| = n$ and $M \sim N$; we need to show that $N \sim M$. By the definition of $\sim$, $M = Na$, for some $a \in G$. In light of part (a), when we multiply this equality of sets on the right by a^{-1}, we obtain

$$Ma^{-1} = (Na)a^{-1} = N(aa^{-1}) = Ne = N.$$

Therefore, $N = Ma^{-1}$, which tells us that $N \sim M$, and so $\sim$ is symmetric.

Finally, to show that $\sim$ is transitive, suppose we are given $M, N, P \subseteq G$ such that $|M| = |N| = |P| = n$, $M \sim N$, and $N \sim P$; we need to show that $M \sim P$. We know there exist $a, b \in G$ such that $M = Na$ and $N = Pb$, and we need to find some $c \in G$ such that $M = Pc$. If we multiply the equality $N = Pb$ on the right by a and apply (a), we obtain

$$Na = (Pb)a = P(ba).$$

Combining this with the fact that $M = Na$, if we let $c = ba$, we can see that $M = P(ba) = Pc$. Hence, $M \sim P$. Thus, $\sim$ is also transitive and is therefore an equivalence relation.   □

The final piece of the puzzle we need before proving Sylow's Theorem is

Lemma 8.27.   *Let M be a subset of a group G and let $H = \{g \in G \mid Mg = M\}$.*

(a)   *H is a subgroup of G.*

(b)   *If $a, b \in G$ then $Ha = Hb$ if and only if $Ma = Mb$.*

(c)   *If M is finite, then $|M| \geq |H|$.*

Proof. For part (a), Proposition 8.6(a) asserts that to show H is a subgroup, it suffices to show that if $x, y \in H$, then $xy^{-1} \in H$. This means that we must verify that $M(xy^{-1}) = M$. Using Lemma 8.26(a), we have

$$M = Me = M(yy^{-1}) = (My)y^{-1} = My^{-1}.$$

The preceding equation, along with an additional application of Lemma 8.26(a), tells us that

$$M(xy^{-1}) = (Mx)y^{-1} = My^{-1} = M,$$

so $xy^{-1} \in H$.

For part (b), suppose $a, b \in G$ such that $Ha = Hb$. As we saw in the proof of Lagrange's Theorem, a and b belong to the same equivalence class under the equivalence relation $\sim_r$ that was defined in Lemma 8.17(b). Therefore, $ab^{-1} = h$, for some $h \in H$. This implies that $a = hb$, which, using Lemma 8.26(a), implies that

$$Ma = M(hb) = (Mh)b = Mb.$$

In the other direction, suppose $a, b \in G$ such that $Ma = Mb$. Multiplying this on the right by b^{-1} and using Lemma 8.26(a), we obtain

$$M(ab^{-1}) = (Ma)b^{-1} = (Mb)b^{-1} = M(bb^{-1}) = Me = M.$$

But this means that $ab^{-1} \in H$, which implies that $a \sim_r b$ under the equivalence relation just mentioned. Since Ha is the equivalence class of a and Hb is the equivalence class of b, we see that $Ha = Hb$, as desired.

For part (c), if $m \in M$, consider the set mH. In light of Lemma 8.17(a), $|mH| = |H|$. On the other hand, by the definition of H, every element of mH belongs to M. Since mH is a subset of M, we have $|M| > |mH| = |H|$, as desired. $\qquad\square$

We can now prove the main result of this section.

Theorem 8.28—Sylow's Theorem. *Let G be a finite group and suppose p^a divides $|G|$, where p is a prime. Then G contains a subgroup with p^a elements.*

Proof. We know that G has many subsets with exactly p^a elements; in fact, if we let $n = |G|$, then we know that G has $\binom{n}{p^a}$ of them. The question is, how do we produce a subgroup with p^a elements? Once again, equivalence relations and equivalence classes play a key role.

Since p^a divides n, we can write $n = p^a m$, for some $m \in \mathbb{N}$. Furthermore, we can let p^b denote the largest power of p that divides m, keeping in mind that it is possible that $p^b = 1$. As shown in Lemma 8.25, p^b is also the largest power of p that divides $\binom{n}{p^a}$.

Let $\mathbb{T}$ denote all subsets of G with p^a elements and, in light of Lemma 8.26(b), we can let $\sim$ be the equivalence relation on $\mathbb{T}$ defined as $M \sim N$, whenever there exists $a \in G$ such that $M = Na$. Since $\mathbb{T}$ is finite, there are only a finite number of equivalence classes, and we can let $A_1, A_2, \ldots A_t$ denote these classes. Remember that each class may have many different names, but each element of $\mathbb{T}$ belongs to only one equivalence class. Therefore,

$$|\mathbb{T}| = |A_1| + |A_2| + \cdots + |A_t|.$$

Since $|\mathbb{T}| = \binom{n}{p^a}$, we know that p^b divides $|\mathbb{T}|$, but p^{b+1} does not divide $|\mathbb{T}|$. In light of the previous equation, we can see that there must be at least one equivalence class where p^{b+1} does not divide the number of elements in it. Therefore, there exists some $M \in \mathbb{T}$ such that p^{b+1} does not divide the number of elements in the equivalence class containing M.

What does $[M]$, the equivalence class containing M, look like? By the definition of $\sim$, $N \in [M]$ whenever $N = Ma$, for some $a \in G$. Therefore,

$$[M] = \{Ma_1, Ma_2, \ldots, Ma_l\},$$

for various $a_i \in G$.

We can now ask, how many different sets belong to $[M]$? Remember that each element of $[M]$ can have many different names, but we are really asking the question, how many different sets of the form Mg are there as g ranges through the elements of G?

If we let $H = \{g \in G \mid Mg = M\}$, then Lemma 8.27(a) told us that H is a subgroup of G. Furthermore, Lemma 8.27(b) asserted that if $a, b \in G$, then $Ma = Mb$ if and only if $Ha = Hb$. This means that the number of sets of the form Mg, as g ranges through G, is the same as the number of sets of the form Hg, as g ranges through G. However, we already called this number the index of H in G and saw, since G is finite, that it is equal to $\frac{|G|}{|H|}$.

We can now determine the size of the subgroup H. Our previous work indicated that

$$|G| = l \cdot |H|,$$

where l is not only the index of H in G but is also the number of elements in $[M]$. Since $|G| = n = p^a m$ and p^b divides m, we know that p^{a+b} divides $|G|$. On the other hand, we have already seen that p^{b+1} does not divide l. Therefore, the only way that p^{a+b} can divide $l \cdot |H|$ is if p^a divides $|H|$.

Since p^a divides $|H|$ and H has at least one element, we now know that

$$|H| \geq p^a.$$

On the other hand, Lemma 8.27(c) indicated that $|H| \leq |M|$, and we know $|M| = p^a$, so

$$|H| \leq p^a.$$

Combining the preceding observations, we can see that H is a subgroup of G that contains exactly p^a elements, as desired.   □

We will conclude this section by briefly examining some of the things that Lagrange's Theorem and Sylow's Theorem tell us about finite groups. But first, we need some terminology.

If G_1, G_2 are groups, we define the *direct product*

$$G_1 \times G_2 = \{(a, b) \mid a \in G_1, b \in G_2\}.$$

It is not hard to check that $G_1 \times G_2$ is also a group. Observe that if e_1 and e_2 are, respectively, the identity elements of G_1 and G_2, then (e_1, e_2) is the identity element of $G_1 \times G_2$. Furthermore, if $a \in G_1$ and $b \in G_2$, then (a^{-1}, b^{-1}) is the inverse in $G_1 \times G_2$ of (a, b). More generally, if $G_1, G_2, \ldots, G_n$ are groups we can form the direct product

$$G_1 \times G_2 \times \cdots \times G_n = \{(a_1, a_2, \ldots, a_n) \mid a_i \in G_i, \quad \text{for } 1 \leq i \leq n\}.$$

Next, suppose G is cyclic and $|G| = n$, where $n \geq 1$. If $g \in G$ such that $< g > \; = \; G$, then

$$G = \{e, g, g^2, \ldots, g^{n-1}\}.$$

Observe that this tells us that any two finite cyclic groups with the same number of elements are isomorphic. Indeed, if H is another cyclic group with n elements and $H \; = \; < h >$, then the function

$$\phi : G \to H$$

defined as $\phi(g^i) = h^i$, for $0 \le i \le n - 1$, is an isomorphism. We can now use Lagrange's Theorem to prove

Corollary 8.29. *Every group with p elements, where p is a prime, is cyclic. Therefore, up to isomorphism, there is only one group of order p.*

Proof. If G is a group with p elements, where p prime, then G contains at least one element that is not the identity element e. Therefore, if $g \in G$ such that $g \ne e$, then $| < g > | \ge 2$. However, by Lagrange's Theorem, $| < g > |$ must divide the prime number p. Hence, $| < g > | = p$, so $< g >$ is all of G, so G is cyclic.

Since all finite cyclic groups of the same order are isomorphic and all groups of order p are cyclic, it follows that, up to isomorphism, there is only one group of order p. ☐

Earlier in this section, we remarked that there exists a group with 12 elements that does not contain a subgroup with 6 elements. However, we can now use Sylow's Theorem to see that any group with fewer than 12 elements has the property that if n divides $|G|$, then G has a subgroup with n elements.

Corollary 8.30. *Let G be a group such that $|G| \le 11$. If $n \in \mathbb{N}$ such that n divides $|G|$, then G contains a subgroup of order n.*

Proof. If $|G| = p^m$ and n divides $|G|$, where p is prime, then $n = p^a$, where $0 \le a \le m$. This certainly occurs when $|G| = 1, 2, 3, 4, 5, 7, 8, 9, 11$, and, in these cases, Sylow's Theorem asserts that G has a subgroup of order n. We should point out that in these cases we really only need to use Sylow's Theorem when $|G| = 4, 8, 9$ as the cases where $|G| = 1, 2, 3, 5, 7, 11$ follow immediately from the fact that $\{e\}$ and G are always subgroups of G.

In remains to consider the cases where $|G| = 6, 10$. In both of these cases, $|G| = pq$, where p and q are different primes. Since $\{e\}$ and G are subgroups, G certainly has subgroups of order 1 and pq. However, we can now use Sylow's Theorem to assert that G also has subgroups of order p and q, thereby covering the remaining $n \in \mathbb{N}$, which divide $|G|$. ☐

We should point out that since 13, 14, 15, 16, 17 are also of the form p^m or pq, where p is a prime and q is a different prime, Corollary 8.30 holds for groups of these orders. Corollary 8.30 indicates that groups of order at most 11 are, in some ways, not as complicated as some other groups. We will now present a table of all groups, up to isomorphism, of order $n \le 11$. Since being isomorphic is an equivalence relation, the table will provide an example of one group per equivalence class. Before presenting this table, we need one more example.

Suppose we let $G = \{1, -1, i, -i\}$ be the 4 fourth roots of 1 in $\mathbb{C}$. We will use G to construct a larger group. Let j have the property that $j^2 = -1$. Thus, j behaves like i. However, when we multiply i and j, the order matters as $ji = -ij$. In this situation, the set

$$Qu = \{1, -1, i, -i, j, -j, ij, -ij\}$$

is a nonabelian group with eight elements. We use the notation Qu as these eight elements are a subset of a very important noncommutative ring known as the Quaternions.

The proofs that our table contains all groups, up to isomorphism, of order at most 11 will appear in several exercises at the end of the next section. They follow from corollaries in this and the next section. In this table, for $n \geq 1$, we will let C_n denote a cyclic group of order n.

$\|G\| = 1$	C_1
$\|G\| = 2$	C_2
$\|G\| = 3$	C_3
$\|G\| = 4$	C_4 and $C_2 \times C_2$ (both abelian)
$\|G\| = 5$	C_5
$\|G\| = 6$	C_6 and S_3 (one abelian, one nonabelian)
$\|G\| = 7$	C_7
$\|G\| = 8$	$C_8, C_4 \times C_2, C_2 \times C_2 \times C_2, D_4$, and Qu (three abelian, two nonabelian)
$\|G\| = 9$	C_9 and $C_3 \times C_3$ (both abelian)
$\|G\| = 10$	C_{10} and D_5 (one abelian, one nonabelian)
$\|G\| = 11$	C_{11}

Exercises for Section 8.2

Exercises 1–13, as well as several others, will refer to the following multiplication table for the group Qu.

$\circ$	1	-1	i	$-i$	j	$-j$	ij	$-ij$
1	1	-1	i	$-i$	j	$-j$	ij	$-ij$
-1	-1	1	$-i$	i	$-j$	j	$-ij$	ij
i	i	$-i$	-1	1	ij	$-ij$	$-j$	j
$-i$	$-i$	i	1	-1	$-ij$	ij	j	$-j$
j	j	$-j$	$-ij$	ij	-1	1	i	$-i$
$-j$	$-j$	j	ij	$-ij$	1	-1	$-i$	i
ij	ij	$-ij$	j	$-j$	$-i$	i	-1	1
$-ij$	$-ij$	ij	$-j$	j	i	$-i$	1	-1

1. In Qu, write down the elements of the subgroups $< 1 >$ and $C(1)$.

2. In Qu, write down the elements of the subgroups $< -1 >$ and $C(-1)$.

3. In Qu, write down the elements of the subgroups $< i >$ and $C(i)$.

4. In Qu, write down the elements of the subgroups $< -i >$ and $C(-i)$.

5. In Qu, write down the elements of the subgroups $< j >$ and $C(j)$.

6. In Qu, write down the elements of the subgroups $< -j >$ and $C(-j)$.

7. In Qu, write down the elements of the subgroups $< ij >$ and $C(ij)$.

8. In Qu, write down the elements of the subgroups $< -ij >$ and $C(-ij)$.

9. In Qu, write down the elements in all the right cosets and all the left cosets of $< -1 >$.

10. In Qu, write down the elements in all the right cosets and all the left cosets of $< i >$.

11. In Qu, write down the elements in all the right cosets and all the left cosets of $< j >$.

12. In Qu, write down the elements in all the right cosets and all the left cosets of $< ij >$.

13. In Qu, determine the number of elements of order 1, 2, 4, and 8.

14. In D_4, determine the number of elements of order 1, 2, 4, and 8. You might want to look at exercises 37–43 from the previous section.

15. Based on the previous two exercises and exercise 68 from the previous section, show that Qu and D_4 are not isomorphic.

16. In the group C_8, determine the number of elements of order 1, 2, 4, and 8.

17. In the group $C_4 \times C_2$, determine the number of elements of order 1, 2, 4, and 8.

18. In the group $C_2 \times C_2 \times C_2$, determine the number of elements of order 1, 2, 4, and 8.

19. Use exercises 15–18 to show that there are at least five different groups, up to isomorphism, with eight elements.

20. Find all the elements in $Z(Qu)$.

21. In $n \geq 3$, show that $y^m x$ has order 2 in D_n.

22. If $n \geq 3$ is odd, determine the number of elements of order 2 in D_n.

23. If $n \geq 3$ is even, determine the number of elements of order 2 in D_n.

24. In D_4, write down the elements in all the right cosets and left cosets of the subgroup $\{e, x\}$. Is $\{e, x\}$ normal?

25. In D_4, write down the elements in all the right cosets and left cosets of the subgroup $\{e, yx\}$. Is $\{e, yx\}$ normal?

26. In D_4, write down the elements in all the right cosets and left cosets of the subgroup $\{e, y^2 x\}$. Is $\{e, y^2 x\}$ normal?

27. In D_4, write down the elements in all the right cosets and left cosets of the subgroup $\{e, y^3 x\}$. Is $\{e, y^3 x\}$ normal?

28. In D_4, write down the elements in all the right cosets and left cosets of the subgroup $\{e, y^2\}$. Is $\{e, y^2\}$ normal?

29. In D_4, write down the elements in all the right cosets and left cosets of the subgroup $\{e, y, y^2, y^3\}$. Is $\{e, y, y^2, y^3\}$ normal?

30. If $n \geq 3$, show that $Z(D_n) = C(x) \cap C(y)$.

31. If $n \geq 3$ is odd, use exercise 30 to show that $Z(D_n) = \{e\}$.

32. If $n \geq 3$ is even, use exercise 30 show that $Z(D_n) = \left\{e, y^{\frac{n}{2}}\right\}$.

33. Let $\tau \in S_n$, where $n \geq 3$.

    (a) If τ is not the identity map, show that there exist $i, j \in \{1, 2, \ldots, n\}$ such that $i \neq j$ and $\tau(i) = j$.

    (b) Suppose $k \in \{1, 2, \ldots n\}$ such that $k \neq i$ and $k \neq j$. If $\sigma \in S_n$ such that $\sigma(i) = i$ and $\sigma(j) = k$, show that $(\tau \circ \sigma)(i) \neq (\sigma \circ \tau)(i)$.

    (c) Use parts (a) and (b) to show that if $n \geq 3$, then $Z(S_n)$ consists only of the identity map.

34. For which values of n does $U(\mathbb{Z}_{10})$ have a subgroup of order n?

35. For which values of n does $U(\mathbb{Z}_{12})$ have a subgroup of order n?

36. For which values of n does $U(\mathbb{Z}_{14})$ have a subgroup of order n?

37. For which values of n does $U(\mathbb{Z}_{16})$ have a subgroup of order n?

38. For which values of n does $U(\mathbb{Z}_{18})$ have a subgroup of order n?

39. For which values of n does $U(\mathbb{Z}_{20})$ have a subgroup of order n?

40. If G is a group with p^m elements, where p is prime and $m \in \mathbb{N}$, for which values of n does G have a subgroup of order n?

41. If G is a group with n elements and $g \in G$, show that $g^n = e$, where e is the identity element.

42. Suppose H, K are subgroups of a group G where $|H| = 2$ and $|K| = 3$. Show that $|H \cap K| = 1$.

43. Suppose H, K are subgroups of a group G where $|H| = p$ and $|K| = q$, where p and q are different primes. Show that $|H \cap K| = 1$.

44. Suppose H, K are finite subgroups of a group G such that the orders of H and K are relatively prime. Show that $H \cap K$ contains only one element.

45. If G is a nonabelian finite group and $g \in G$, show that $|C(g)| > |Z(G)|$. You might want to consider separately the cases where $g \in Z(G)$ and $g \notin Z(G)$.

46. If G is a group of order pm, where p is a prime and $m \in \mathbb{N}$, show that $Z(G)$ cannot have exactly m elements. You might want to use exercise 45 and then separately consider the cases where G is abelian and where G is nonabelian.

47. Suppose G is a group of order pq, where p and q are primes which are not necessarily different. Show that if $Z(G)$ has more than one element, then G is abelian.

48. If G is a group where every element has order 1 or 2, show that G must be abelian.

49. Use exercise 48 to show that every group of order 4 must be abelian.

50. Let G be a group and let $\phi : G \to G$ be defined as $\phi(g) = g^{-1}$, for all $g \in G$. Show that if ϕ is a homomorphism, then G is abelian.

51. Let G be a group and let $\rho : G \to G$ be defined as $\rho(g) = g^2$, for all $g \in G$.
    (a) Show that if G is abelian, then ρ is a homomorphism.

    (b) Show that if ρ is a homomorphism, then G is abelian.

52. Let $G = H_1 \times H_2$, where H_1 and H_2 are groups. Show that G is abelian if and only if both H_1 and H_2 are abelian.

53. Let $G = H_1 \times H_2$, where H_1 and H_2 are groups. Show that if G is cyclic then both H_1 and H_2 are cyclic.

54. Suppose H_1 and H_2 are finite cyclic groups. Show that $H_1 \times H_2$ is cyclic if and only if $|H_1|$ and $|H_2|$ are relatively prime.

In exercises 55–62, ϕ is a homomorphism of groups. Describe, as simply as possible, the elements of $Ker(\phi)$ and $Im(\phi)$.

55. $\phi : \mathbb{C}^\times \to \mathbb{C}^\times$, where $\phi(\alpha) = |\alpha|$, for all $\alpha \in \mathbb{C}^\times$. Note that $\mathbb{C}^\times$ is a group under multiplication.

56. $\phi : \mathbb{C}^\times \to \mathbb{C}^\times$, where $\phi(\alpha) = \alpha^4$, for all $\alpha \in \mathbb{C}^\times$. Note that $\mathbb{C}^\times$ is a group under multiplication.

57. $\phi : \mathbb{R}^\times \to \mathbb{R}^\times$, where $\phi(\alpha) = \alpha^4$, for all $\alpha \in \mathbb{R}^\times$. Note that $\mathbb{R}^\times$ is a group under multiplication.

58. $\phi : \mathbb{R}[x] \to \mathbb{R}$, where $\phi(f(x)) = f(0)$, for all $f(x) \in \mathbb{R}[x]$. Note that $\mathbb{R}[x]$ and $\mathbb{R}$ are groups under addition.

59. $\phi : \mathbb{Z}[x] \to \mathbb{Q}$, where $\phi(g(x)) = g(0)$, for all $g(x) \in \mathbb{Z}[x]$. Note that $\mathbb{Z}[x]$ and $\mathbb{Q}$ are groups under addition.

60. $\phi : D_3 \to \mathbb{R}^\times$, where $\phi(y^i x^j) = (-1)^j$, for $0 \le i \le 2$ and $0 \le j \le 1$. Note that $\mathbb{R}^\times$ is a group under multiplication.

61. $\phi : \mathbb{Z}_{20} \to \mathbb{Z}_4$, where $\phi([i]_{20}) = [i]_4$, for all $i \in \mathbb{Z}$. Note that $\mathbb{Z}_{20}$ and $\mathbb{Z}_4$ are groups under addition. Before examining $Ker(\phi)$ and $Im(\phi)$, you should first check that the function ϕ is well defined.

62. $\phi : \mathbb{Z} \to \mathbb{C}^\times$, where $\phi(n) = \text{cis}(\frac{n \cdot \pi}{6})$, for all $n \in \mathbb{Z}$. Note that $\mathbb{Z}$ is a group under addition and $\mathbb{C}^\times$ is a group under multiplication.

63. Show that any subgroup of a cyclic group is cyclic.

8.3 Solvable Groups

In the previous two sections, we were introduced to some of the basic concepts of finite group theory. In the next two sections, we focus on the group-theoretic topics needed for Galois' proof of the insolvability of the quintic.

Remember that for a normal subgroup the right cosets and left cosets corresponding to the subgroup are equal. Furthermore, Lagrange's Theorem told us how many cosets there are, and Theorem 8.21 showed that these cosets form a group in their own right. This leads us to

Definition 8.31. *If G is a group with normal subgroup N; we let G/N denote the set of cosets corresponding to N. Then G/N is a group and we call it a factor group or quotient group. In addition, if G is finite, then $|G/N| = \frac{|G|}{|N|}$.*

Recall that for any group G, both G and $\{e\}$ are normal subgroups. Therefore, we can always form the factor groups G/G and $G/\{e\}$. Observe that G/G is merely a group with one element, and this element is a coset containing every element of G. At the other extreme, every element of $G/\{e\}$ is a coset containing only one element of G. Furthermore, the function

$$\phi : G \to G/\{e\}$$

defined as $\phi(g) = \{e\}g$ is an isomorphism, so $G/\{e\}$ is isomorphic to G.

Factor groups are often difficult for students to deal with as they present us with a new level of abstraction. Groups are sets whose elements satisfy various nice properties under a binary operation. Often, the elements of a group are very concrete objects like integers, rational numbers, or functions. However, when dealing with factor groups, the elements of our groups are cosets, so they are sets in their own right. When dealing with new concepts that are quite abstract, it is always helpful to have concrete examples to look at. Fortunately, in trying to understand factor groups, we can always reexamine the construction of $\mathbb{Z}_n$ from $\mathbb{Z}$ to help reinforce these new ideas.

Let us now look at three other examples of factor groups.

■ Examples

1. Let $G = S_3$, and, using the notation for S_3 of the previous two sections, let N be the three element subgroup $\{e, j, k\}$. As we saw in Section 8.2, N is a normal subgroup, and there are two cosets, each with three names: $Ne = Nj = Nk$ and $Nf = Ng = Nh$. Since N is normal, there is no difference between a right coset and a left coset, and, more importantly, the name of the coset does not affect the answer when multiplying. If we use the names Ne and Nf for our two cosets, then we obtain the following table for the factor group S_3/N:

$\circ$	Ne	Nf
Ne	Ne	Nf
Nf	Nf	Ne

2. Now let $G = C_6 = \{e, g, g^2, g^3, g^4, g^5\}$ be the cyclic group with six elements and let $H = <g^2> = \{e, g^2, g^4\}$ be the cyclic subgroup generated by g^2. Since G is abelian, all of its subgroups are normal. As in the previous example, there are again two cosets, each with three names: $He = Hg^2 = Hg^4$ and $Hg = Hg^3 = Hg^5$. If we use the names He and Hg for the cosets, then the factor group C_6/H has table

$\circ$	He	Hg
He	He	Hg
Hg	Hg	He

3. Let $G = D_4$, the fourth dihedral group. Recall that every element of D_4 is of the form $y^i x^j$, where $0 \leq i \leq 3, 0 \leq j \leq 1, y^4 = x^2 = e$, and $xy = y^3 x$. The center of every group is always a normal subgroup and, in this case, $Z(D_4) = \{e, y^2\}$. Since D_4 has order 8 and $Z(D_4)$ has two elements, the factor group $D_4/Z(D_4)$ consists of four cosets each with two names:

$$Z(D_4)e = Z(D_4)y^2, \quad Z(D_4)x = Z(D_4)y^2 x,$$

$$Z(D_4)y = Z(D_4)y^3, \quad \text{and} \quad Z(D_4)yx = Z(D_4)y^3 x.$$

Using the names $Z(D_4)e$, $Z(D_4)x$, $Z(D_4)y$, $Z(D_4)yx$ for the cosets, then the factor group $D_4/Z(D_4)$ has table

$\circ$	$Z(D_4)e$	$Z(D_4)x$	$Z(D_4)y$	$Z(D_4)yx$
$Z(D_4)e$	$Z(D_4)e$	$Z(D_4)x$	$Z(D_4)y$	$Z(D_4)yx$
$Z(D_4)x$	$Z(D_4)x$	$Z(D_4)e$	$Z(D_4)yx$	$Z(D_4)y$
$Z(D_4)y$	$Z(D_4)y$	$Z(D_4)yx$	$Z(D_4)e$	$Z(D_4)x$
$Z(D_4)yx$	$Z(D_4)yx$	$Z(D_4)y$	$Z(D_4)x$	$Z(D_4)e$

Let us now examine what the previous examples tell us. If G is a group with normal subgroup N, then it should come as little surprise that both the factor group G/N and the subgroup N inherit properties from G. For example, if G is abelian, then both G/N and N are abelian. This is certainly the case in our second example where $G = C_6$.

It is now natural to wonder if information about G/N and N can be used to obtain information about G. One piece of information about G that can be obtained by looking at G/N and N is its size. If G is a finite group, then the proof of Lagrange's Theorem tells us that

$$|G| = |G/N| \cdot |N|.$$

In fact, if we merely assumed that $|G/N|$ and $|N|$ were finite but did not assume that $|G|$ was finite, the proof of Lagrange's Theorem would still show that $|G| = |G/N| \cdot |N|$. Thus, the sizes of G/N and N do tell us the size of G.

However, the situation is quite different if we look at the property of being abelian. First we introduce some notation. If G_1 and G_2 are isomorphic, we write $G_1 \approx G_2$, but if G_1 and G_2 are not isomorphic, we write $G_1 \not\approx G_2$. In our first example,

$$S_3/N \approx C_2 \quad \text{and} \quad N \approx C_3.$$

Therefore, both S_3/N and N are abelian, yet S_3 is not abelian. In fact, when we compare the first two examples, we can see that

$$S_3/N \approx C_6/H \quad \text{and} \quad N \approx H.$$

Thus, the factor groups in the first two examples are isomorphic, and the normal subgroups in the first two examples are also isomorphic. However, we certainly know that $S_3 \not\approx C_6$.

Based on these examples, we see that a complete understanding of both the factor group G/N and the normal subgroup N still does not completely determine the structure of G. When we look at our third example, we have

$$D_4/Z(D_4) \approx C_2 \times C_2 \quad \text{and} \quad Z(D_4) \approx C_2.$$

Thus, both $D_4/Z(D_4)$ and $Z(D_4)$ are abelian, yet D_4 is not abelian. When you look at the table for $D_4/Z(D_4)$, you can see that the order of every element of the factor group is either one or two. Thus, $D_4/Z(D_4)$ is not cyclic. This explains why $D_4/Z(D_4) \not\approx C_4$. Since there are only two groups of degree 4, it immediately follows that $D_4/Z(D_4) \approx C_2 \times C_2$.

In all three of our examples, G contained a normal subgroup N such that both G/N and N were abelian. Although G need not be abelian, we can say that G is, at worst, only one step away from being abelian. Soon, we will generalize this idea and look at groups that may not be abelian but are only a finite number of steps away from being abelian. You might wonder if this is merely generalizing for the sake of generalizing or whether such a concept has applications. It turns out that this is exactly the group-theoretic concept needed for Galois' work on the insolvability of the quintic. This leads us to

Definition 8.32. *If G is a group, we say that G is solvable if there exists a chain of subgroups*

$$G = G_0 \supseteq G_1 \supseteq G_2 \cdots \supseteq G_{n-1} \supseteq G_n = \{e\}$$

such that, for $0 \leq i \leq n - i$, G_{i+1} is a normal subgroup of G_i and the factor group G_i/G_{i+1} is abelian.

As we will see in Chapters 15 and 17, to every polynomial $p(x)$ in $\mathbb{Q}[x]$, we can associate a field L such that $\mathbb{Q} \subseteq L \subseteq \mathbb{C}$. Galois showed, and we will show in Chapter 17, that if $p(x)$ is solvable in radicals then the Galois group $Gal(L/\mathbb{Q})$ is a solvable group. By producing fifth-degree polynomials $p(x)$ such that the Galois groups $Gal(L/\mathbb{Q})$ were not solvable, Galois showed that fifth-degree polynomials are not solvable in radicals.

In mathematics, it is often useful to have several different ways to look at the same concept. Therefore, we now present another way to look at solvable groups.

Proposition 8.33. *A group G is solvable if and only if there exists a chain of subgroups*

$$G = G_0 \supseteq G_1 \supseteq G_2 \cdots \supseteq G_{n-1} \supseteq G_n = \{e\}$$

such that whenever $a, b \in G_i$, we have $aba^{-1}b^{-1} \in G_{i+1}$, for $0 \leq i \leq n - 1$.

Proof. In one direction, suppose G is a group with a chain of subgroups

$$G = G_0 \supseteq G_1 \supseteq G_2 \cdots \supseteq G_{n-1} \supseteq G_n = \{e\}$$

that satisfy the conditions in Definition 8.32. We need to show that for all $0 \leq i \leq n - 1$, if $a, b \in G_i$ then $aba^{-1}b^{-1} \in G_{i+1}$.

By Definition 8.32, the factor group G_i/G_{i+1} is abelian. Therefore,

$$(G_{i+1}a)(G_{i+1}b) = (G_{i+1}b)(G_{i+1}a).$$

By the definition of coset multiplication, this means that $G_{i+1}ab = G_{i+1}ba$. Since ab and ba belong to the same coset corresponding to G_{i+1}, we know that $(ab)(ba)^{-1} \in G_{i+1}$. However, $(ba)^{-1} = a^{-1}b^{-1}$, so

$$aba^{-1}b^{-1} = (ab)(ba)^{-1} \in G_{i+1},$$

as desired.

In the other direction, suppose

$$G = G_0 \supseteq G_1 \supseteq G_2 \cdots \supseteq G_{n-1} \supseteq G_n = \{e\}$$

satisfies the conditions of this proposition; then we need to show that G_{i+1} is a normal subgroup of G_i and the factor group G_i/G_{i+1} is abelian, for $0 \le i \le n-1$.

In order to prove that G_{i+1} is a normal subgroup of G_i, we need to show that if $a \in G_i$, then $aG_{i+1} = G_{i+1}a$. Therefore, it suffices to show that if $g \in G_{i+1}$, then $ag \in G_{i+1}a$ and $ga \in aG_{i+1}$. Since both a and g belong to G_i, we know that $aga^{-1}g^{-1} \in G_{i+1}$, so $aga^{-1}g^{-1} = h$, for some $h \in G_{i+1}$. We now have

$$ag = (ag)e = (ag)(a^{-1}g^{-1}ga) = (aga^{-1}g^{-1})(ga) = h(ga) = (hg)a.$$

Since both h and g belong to G_{i+1}, we see that $hg \in G_{i+1}$ and

$$ag = (hg)a \in G_{i+1}a.$$

Similarly, since $g^{-1}, a^{-1} \in G_i$, it follows that

$$g^{-1}a^{-1}ga = g^{-1}a^{-1}(g^{-1})^{-1}(a^{-1})^{-1} \in G_{i+1}.$$

Therefore, $g^{-1}a^{-1}ga = k$, for some $k \in G_{i+1}$. Thus,

$$ga = e(ga) = (agg^{-1}a^{-1})(ga) = (ag)(g^{-1}a^{-1}ga) = (ag)k = a(gk).$$

Since both g and k belong to G_{i+1}, we know that $gk \in G_{i+1}$, so

$$ga = a(gk) \in aG_{i+1}.$$

To conclude the proof, we need to show that the factor group G_i/G_{i+1} is abelian. Therefore, if $a, b \in G_i$, we need to show that

$$(G_{i+1}a)(G_{i+1}b) = (G_{i+1}b)(G_{i+1}a).$$

We know that $aba^{-1}b^{-1} \in G_{i+1}$. However, $aba^{-1}b^{-1} = (ab)(ba)^{-1}$, which tells us that ab and ba belong to the same coset corresponding to G_{i+1}. As a result,

$$G_{i+1}ab = G_{i+1}ba,$$

which, by the definition of coset multiplication, immediately implies that $(G_{i+1}a)(G_{i+1}b) = (G_{i+1}b)(G_{i+1}a)$, as desired. $\square$

Having defined solvable groups and having briefly discussed their connection to Galois' work, it is time to try to determine which familiar groups are solvable and which are not. Certainly, abelian groups are solvable. We next turn our attention to dihedral groups. To examine them, we need

Lemma 8.34. *Let G be a finite group with a subgroup H such that $|G| = 2 \cdot |H|$. Then H is a normal subgroup of G.*

Proof. We need to show that for every $g \in G$, $Hg = gH$. The proof will be surprisingly easy if we look at things the right way, and in this situation, the easiest way to proceed is by separately considering the cases where $g \in H$ and $g \notin H$.

If $g \in H$, then g and e belong to the same right coset, so $Hg = He$. Similarly, if $g \in H$, then g and e also belong to the same left coset, so $gH = eH$. However, it is easy to see that the sets He and eH are equal to the set H, so

$$Hg = He = H = eH = gH.$$

Observe that this part of our proof did not require that $|G| = 2 \cdot |H|$.

Now suppose $g \notin H$; since the index of H in G is two, it follows that G is the union of the two disjoint cosets Hg and He. Therefore, Hg consists of the $|H|$ elements of G that do not belong to $He = H$. Using the same reasoning, gH also consists of the $|H|$ elements of G that do not belong to $eH = H$. Thus, the sets Hg and gH are the same, as needed. $\square$

It is now easy to prove

Corollary 8.35. *The dihedral group D_n is solvable, for all $n \geq 1$.*

Proof. The group D_n consists of the $2n$ terms of the form $y^i x^j$, where $0 \leq i \leq n-1, 0 \leq j \leq 1$, $y^n = x^2 = e$, and $xy = y^{n-1}x$. If we let H be the cyclic subgroup generated by y, then

$$H = <y> = \{e, y, y^2, \ldots, y^{n-1}\}.$$

Since $|H| = n$ and $|G| = 2n$, we can apply Lemma 8.34 to assert that H is a normal subgroup of G. Therefore, we now have the chain

$$G \supseteq H \supseteq \{e\},$$

where H is normal in G, $\{e\}$ is normal in H, and the factor groups

$$G/H \approx C_2, \quad H/\{e\} \approx H \approx C_n$$

are both abelian. As a result, D_n is solvable. □

Having shown that D_n is solvable for all $n \geq 2$, it is reasonable to investigate when symmetric groups are solvable. Since S_1 and S_2 are abelian, we know they are solvable. Earlier in this section, we saw that S_3 is solvable, and at the end of Section 8.4, we will show that S_4 is also solvable. However, this is as far as we can go with symmetric groups. In Section 8.4, we will show that S_n is *not* solvable for all $n \geq 5$. The fact that S_5 is not solvable will turn out to be a very important piece of the puzzle in proving the insolvability of the quintic.

The next result will be very useful in examining groups of order p^n. It is another illustration of the importance of equivalence relations and equivalence classes in abstract algebra.

Theorem 8.36—The Class Equation. *If G is a finite group, then G contains a subset $\{g_1, \ldots, g_m\}$ of $m \geq 0$ elements that do not belong to $Z(G)$, such that*

$$|G| = |Z(G)| + \frac{|G|}{|C_G(g_1)|} + \frac{|G|}{|C_G(g_2)|} + \cdots + \frac{|G|}{|C_G(g_m)|}.$$

For an abelian group G, the Class Equation does not provide any new information about G as in this case $m = 0$, and the equation merely says that $|G| = |Z(G)|$. However, for various nonabelian groups, the Class Equation is very useful in examining the size of $Z(G)$. At this point, it is probably unclear how we go about using the Class Equation. Therefore, we will provide several applications before proving Theorem 8.36.

Theorem 8.37. *Let G be a group with p^n elements, where p is prime and $n \geq 1$. Then $Z(G)$ has at least p elements.*

Proof. If G is abelian, there is nothing to prove as $|Z(G)| = |G| \geq p$. On the other hand, if G is nonabelian, we can look at the class equation

$$|G| = |Z(G)| + \frac{|G|}{|C_G(g_1)|} + \frac{|G|}{|C_G(g_2)|} + \cdots + \frac{|G|}{|C_G(g_m)|}$$

and note that $m \geq 1$.

Since $|G| = p^n$, Lagrange's Theorem tells us that if $1 \leq i \leq m$, then

$$\frac{|G|}{|C_G(g_i)|} = p^t,$$

for some $t \geq 0$. However, since $g_i \notin Z(G)$, we know that $C_G(g_i)$ is not all of G. Hence, $\frac{|G|}{|C_G(g_i)|} > 1$, so $\frac{|G|}{|C_G(g_i)|}$ is a multiple of p.

When we look at the class equation, $|G|$ is a multiple of p, and every term of the form $\frac{|G|}{|C_G(g_i)|}$ is also a multiple of p. Since we can rewrite the class equation as

$$|Z(G)| = |G| - \left(\frac{|G|}{|C_G(g_1)|} + \frac{|G|}{|C_G(g_2)|} + \cdots + \frac{|G|}{|C_G(g_m)|} \right),$$

we see that $|Z(G)|$ is also a multiple of p. Since $e \in Z(G)$, we know that $|Z(G)|$ is both at least one and also a multiple of p. This tells us that $|Z(G)| \geq p$, as desired. $\square$

Note that Theorem 8.37 says nothing about groups of order 6 but does guarantee that for groups with eight elements, the center must have at least two elements. This is perfectly consistent with our previous observations that $|S_3| = 6$, $|Z(S_3)| = 1$, $|D_4| = 8$, and $|Z(D_4)| = 2$. We can now use Theorem 8.37 to completely describe up to isomorphism groups of order p^2.

Corollary 8.38. *If G is a group of order p^2, where p is prime, then G is abelian. Furthermore, either $G \approx C_{p^2}$ or $G \approx C_p \times C_p$.*

Proof. By way of contradiction, suppose G is not abelian. There exists $b \in G$ such that $b \notin Z(G)$. We will now examine $C_G(b)$.

Certainly $C_G(b)$ contains $Z(G)$. By Theorem 8.37, $|Z(G)| \geq p$; therefore,

$$|C_G(b)| \geq |Z(G)| \geq p.$$

On the other hand, b belongs to $C_G(b)$ and does not belong to $Z(G)$, so the previous inequality now becomes

$$|C_G(b)| > |Z(G)| \geq p.$$

Lagrange's Theorem tells us $C_G(b)$ must divide p^2. However, we know that $|C_G(b)| > p$, so it must be the case that $|C_G(b)| = p^2 = |G|$. As a result, $C_G(b)$ is equal to all of G, which means that b commutes with every element of G. But this contradicts the fact that $b \notin Z(G)$, so G is abelian.

By Lagrange's Theorem, every element of G has order 1, p, or p^2. If G contains an element of order p^2, then G is cyclic, and we know that $G \approx C_{p^2}$. Therefore, it only remains to consider the case where every element of G, other than the identity, has order p.

Let $a \in G$ such that $o(a) = p$ and then let $b \in G$ such that $b \notin < a >$. It is clear that b also has order p, and we now consider the set

$$H = \{a^i b^j \mid 0 \leq i, j \leq p - 1\}.$$

Since $a^p = b^p = e$ and $ab = ba$, we can see that H is closed under multiplication, so by Proposition 8.6(b), H is a subgroup of G. By Lagrange's Theorem, $|H|$ must divide p^2. However, not only does H contain the p elements in $< a >$, but it also contains b. Therefore, H has more than p elements, so $|H| = p^2 = |G|$. Thus, every element of G can be written uniquely in the form $a^i b^j$, where $0 \leq i, j \leq p - 1$.

Next, if we let $g \in C_p$ such that $< g > = C_p$, we can define the function

$$\phi : G \rightarrow C_p \times C_p$$

as

$$\phi(a^i b^j) = (g^i, g^j),$$

for all $0 \leq i, j \leq p - 1$.

It is easy to see that ϕ is a bijection from G to $C_p \times C_p$. To show that ϕ is an isomorphism, let $h_1, h_2 \in G$; we need to show that $\phi(h_1 h_2) = \phi(h_1)\phi(h_2)$. There exist i, j, k, l, with $0 \leq i, j, k, l \leq p - 1$, such that $h_1 = a^i b^j$ and $h_2 = a^k b^l$. If we let n and m be, respectively, the remainders when $i + k$ and $j + l$ are divided by p, we now have

$$\phi(h_1 h_2) = \phi((a^i b^j)(a^k b^l)) = \phi(a^{i+k} b^{j+l}) = \phi(a^n b^m) = (g^n, g^m) =$$

$$(g^{i+k}, g^{j+l}) = (g^i, g^j)(g^k, g^l) = \phi(a^i b^j)\phi(a^k b^l) = \phi(h_1)\phi(h_2),$$

as required. $\qquad \square$

Having seen some applications of the Class Equation, it is now time for its proof.

Proof of Theorem 8.36—The Class Equation. As mentioned earlier, this proof will be another illustration of the importance of equivalence relations and equivalence classes in abstract. If $g, h \in G$, we define $\sim$ as $g \sim h$ when there exists $a \in G$ such that $h = a^{-1} ga$. We want to show that $\sim$ is an equivalence relation, so we need to show that it is reflexive, symmetric, and transitive.

If $g \in G$ and if e is the identity element of G, then

$$g = ege = e^{-1} ge.$$

As a result, for every $g \in G$, $g \sim g$, thus $\sim$ is reflexive.

To show that $\sim$ is symmetric, suppose $g, h \in G$ such that $g \sim h$. We need to show that $h \sim g$. Since $g \sim h$, there exists $a \in G$ such that $h = a^{-1}ga$. If we take this equation and solve for g, we obtain

$$g = aha^{-1} = (a^{-1})^{-1}ha^{-1}.$$

If we let $b = a^{-1}$, then the previous equation becomes $g = b^{-1}hb$, and since $b \in G$, we see that $h \sim g$. Thus, $\sim$ is also symmetric.

Next, to show that $\sim$ is transitive, suppose $g, h, k \in G$ such that $g \sim h$ and $h \sim k$. We need to show that $g \sim k$. We know that there exist $c, d \in G$ such that $h = c^{-1}gc$ and $k = d^{-1}hd$. Substituting the first equation into the second results in

$$k = d^{-1}hd = d^{-1}(c^{-1}gc)d = (d^{-1}c^{-1})g(cd) = (cd)^{-1}g(cd).$$

Since $cd \in G$, the preceding equation tells us that $g \sim k$, so $\sim$ is also transitive and is therefore an equivalence relation.

Having shown that $\sim$ is an equivalence relation, the next step is to start looking at the equivalence classes. Observe that if $g \in G$, then the equivalence class $[g]$ consists of the elements of G of the form $a^{-1}ga$, where a ranges through all $|G|$ elements of G. Certainly not all $|G|$ choices for a produce different elements of $[g]$. In fact, it is not hard to see that $[g]$ consists of the single element g if and only if $g \in Z(G)$. Next, let $A_1, A_2, \ldots, A_m$ be the equivalence classes that contain more than one element. The facts that every element of G belongs to exactly one equivalence class and the number of elements whose equivalence class contains only one element equals $|Z(G)|$ can now be expressed in the equation

(4) $$|G| = |Z(G)| + |A_1| + |A_2| + \cdots + |A_m|.$$

The question remains, if $g \notin Z(G)$, then how do we determine the size of $[g]$? If $a, b \in G$ such that $a^{-1}ga = b^{-1}gb$, then multiplying on the left by b and on the right by a^{-1} yields

$$ba^{-1}g = gba^{-1}.$$

As a result, ba^{-1} commutes with g, so there exists $h \in C(g)$ such that $ba^{-1} = h$. This tells us that

$$b = ha \in C(g)a.$$

Therefore, we now know that a and b belong to the same right coset corresponding to $C(g)$, so $C(g)a = C(g)b$.

On the other hand, if $C(g)a = C(g)b$, then $a = hb$, for some $h \in C(g)$. Therefore,

$$a^{-1}ga = (hb)^{-1}g(hb) = b^{-1}h^{-1}ghb = b^{-1}h^{-1}hgb = b^{-1}gb.$$

The preceding arguments combine to show that if $a, b \in G$, then $a^{-1}ga = b^{-1}gb$ if and only if $C(g)a = C(g)b$. As a result, the number of elements of the form $a^{-1}ga$ is the same as the number of right cosets corresponding to the subgroup $C(g)$. However, Lagrange's Theorem already told us that the number of right cosets corresponding to $C(g)$ is equal to $\frac{|G|}{|C(g)|}$.

For $1 \le i \le m$, choose some $g_i \in A_i$. Therefore $A_i = [g_i]$ and, using our previous observations, we have succeeded in showing that $|A_i| = \frac{|G|}{|C(g_i)|}$. Substituting this into (4) now gives us

$$|G| = |Z(G)| + \frac{|G|}{|C_G(g_1)|} + \frac{|G|}{|C_G(g_2)|} + \cdots + \frac{|G|}{|C_G(g_m)|},$$

thereby concluding the proof. □

As we saw earlier in this section, a group G need not be abelian even if a normal subgroup N and the factor group G/N are both abelian. If we think of solvable as being, at most, a finite number of steps away from being abelian, then it is reasonable to ask, if both N and G/N are solvable, must G also be solvable? After all, if N is n steps away from being abelian and G/N is m steps away, then perhaps G is, at most, $n + m$ steps away. In order to prove this, we need a technical result which has many applications in group theory.

Theorem 8.39—The Isomorphism Theorem. *Let $\phi : G_1 \to G_2$ be a homomorphism of groups.*

(a) *If we let N denote the kernel of ϕ, then the factor group G_1/N is isomorphic to $Im(\phi)$.*

(b) *If H_1 is a subgroup of G_1, then $\phi(H_1)$ is a subgroup of G_1.*

(c) *If H_2 is a subgroup of G_2, then $\phi^{-1}(H_2) = \{g \in G_1 \mid \phi(g) \in H_2\}$ is a subgroup of G_1 containing N.*

(d) *If H_2 is a normal subgroup of G_2, then $\phi^{-1}(H_2)$ is a normal subgroup of G_1.*

Proof. By Theorem 8.24, we know that N is a normal subgroup of G_1 and $Im(\phi)$ is a subgroup of G_2. In this proof, we will let e_1, e_2 denote, respectively, the identity elements of G_1 and G_2.

For part (a), we need to find a homomorphism

$$v : G/N \to Im(\phi)$$

which is also a bijection. When we multiply the cosets in G/N, we do so by multiplying their names. Therefore, it is reasonable to define v by using the names of the cosets. Since ϕ is defined on all elements of G, this suggests defining v as

$$v(Ng) = \phi(g),$$

for all $g \in G$.

We first need to show that v is well defined. This means that changing the name of a coset will not change the value of v. To this end, suppose $Ng = Nh$, for some $g, h \in G_1$, and we need to show that $v(Ng) = v(Nh)$. Since $Ng = Nh$, we know that $g \in Ng = Nh$, so $g = xh$, for some $x \in N$. As a result,

$$v(Ng) = \phi(g) = \phi(xh) = \phi(x)\phi(h) = e_2\phi(h) = \phi(h) = v(Nh),$$

as required.

Next, to show that v is surjective, we must show that if $y \in Im(\phi)$, then $y \in Im(v)$. There exist some $g \in G_1$ such that $y = \phi(g)$, so

$$y = \phi(g) = v(Ng) \in Im(v),$$

as required.

In order to prove that v is injective, we must show that if $g, h \in G_1$ such that $v(Ng) = v(Nh)$, then $Ng = Nh$. Since $v(Ng) = v(Nh)$, we know that $\phi(g) = \phi(h)$. Therefore,

$$\phi(gh^{-1}) = \phi(g)\phi(h^{-1}) = \phi(g)\phi(h)^{-1} = \phi(h)\phi(h)^{-1} = e_2.$$

As a result, $gh^{-1} \in N$, so the cosets Ng and Nh are the same. Thus, v is indeed a bijection.

To complete the proof of part (a), if $g, h \in G_1$, we can use the definition of coset multiplication and the fact that ϕ is a homomorphism to see

$$v((Ng)(Nh)) = v(N(gh)) = \phi(gh) = \phi(g)\phi(h) = v(Ng)v(Nh).$$

Thus, v is also a homomorphism, so v is an isomorphism. Therefore, G/N and $Im(\phi)$ are isomorphic.

For part (b), in light of Proposition 8.6(a), it suffices to show that if $x, y \in \phi(H_1)$, then $xy^{-1} \in \phi(H_1)$. We know that there exists $g, h \in H_1$ such that $x = \phi(g)$ and $y = \phi(h)$. Since ϕ is a homomorphism and $gh^{-1} \in H_1$, it follows that

$$xy^{-1} = \phi(g)\phi(h)^{-1} = \phi(g)\phi(h^{-1}) = \phi(gh^{-1}) \in \phi(H_1).$$

For part (c), we again use Proposition 8.6(a), and it suffices to show that if $g, h \in \phi^{-1}(H_2)$, then $gh^{-1} \in \phi^{-1}(H_2)$. By the definition of $\phi^{-1}(H_2)$, both $\phi(g)$ and $\phi(h)$ belong to H_2. Since H_2 is a subgroup of G_2, we know that $\phi(g)\phi(h)^{-1} \in H_2$. Using that ϕ is a homomorphism, it follows that

$$\phi(gh^{-1}) = \phi(g)\phi(h^{-1}) = \phi(g)\phi(h)^{-1} \in H_2.$$

The preceding equation indicates that $gh^{-1} \in \phi^{-1}(H_2)$, so $\phi^{-1}(H_2)$ is a subgroup. Furthermore,

$$\phi(N) = \{e\} \subseteq H_2,$$

so $\phi^{-1}(H_2)$ contains N, as required.

For part (d), we may now assume that the subgroup H_2 from part (c) is also normal. If $a \in G_1$ and $g \in \phi^{-1}(H_2)$, we must show that

$$ag \in \phi^{-1}(H_2)a \quad \text{and} \quad ga \in a\phi^{-1}(H_2).$$

The fact that H_2 is normal implies that

$$\phi(a)\phi(g) \in \phi(a)H_2 = H_2\phi(a) \quad \text{and} \quad \phi(g)\phi(a) \in H_2\phi(a) = \phi(a)H_2.$$

Therefore, there exist $x, y \in H_2$ such that

$$\phi(a)\phi(g) = x\phi(a) \quad \text{and} \quad \phi(g)\phi(a) = \phi(a)y.$$

Since ϕ is a homomorphism, the previous equations imply that

$$\phi(aga^{-1}) = \phi(a)\phi(g)\phi(a)^{-1} = x\phi(a)\phi(a)^{-1} = x \in H_2$$

and

$$\phi(a^{-1}ga) = \phi(a^{-1})\phi(g)\phi(a) = \phi(a)^{-1}\phi(a)y = y \in H_2.$$

Therefore, $aga^{-1}, a^{-1}ga \in \phi^{-1}(H_2)$. We now have

$$ag = (aga^{-1})a \in \phi^{-1}(H_2)a \quad \text{and} \quad ga = a(a^{-1}ga) \in a\phi^{-1}(H_2),$$

as desired.   □

We now have the tools needed to prove

Corollary 8.40. *If G is a group with normal subgroup N, then G is solvable if and only if both G/N and N are solvable.*

Proof. In one direction, suppose G is solvable. Then, by Proposition 8.33, there exists a chain of subgroups

$$G = G_0 \supseteq G_1 \supseteq G_2 \cdots \supseteq G_{n-1} \supseteq G_n = \{e\}$$

where $aba^{-1}b^{-1} \in G_{i+1}$, whenever $a, b \in G_i$ and $0 \le i \le n-1$.

Let $N_i = N \cap G_i$, for $0 \le i \le n-1$; since N is a subgroup, it follows that if $a, b \in N_i$ then

$$aba^{-1}b^{-1} \in N \cap G_{i+1} = N_{i+1}.$$

Therefore, the chain of subgroups of N,

$$N = N_0 \supseteq N_1 \supseteq N_2 \cdots \supseteq N_{n-1} \supseteq N_n = \{e\},$$

shows that N is solvable.

Next, we need to show that if G is solvable, then so is G/N. Let $H = G/N$ and consider the function

$$\phi : G \to H,$$

where $\phi(g) = Ng$. Observe that ϕ is a homomorphism as

$$\phi(gh) = N(gh) = (Ng)(Nh) = \phi(g)\phi(h),$$

for all $g, h \in G$. Therefore, we are in a position to apply Theorem 8.39.

By Theorem 8.39(b), $\phi(G_i)$ is a subgroup of H, for $0 \le i \le n-1$. Therefore, H has a chain of subgroups

$$H = \phi(G_0) \supseteq \phi(G_1) \supseteq \phi(G_2) \cdots \supseteq \phi(G_{n-1}) \supseteq \phi(G_n) = \{e_H\},$$

where e_H is the identity element of H. Furthermore, if $a, b \in \phi(G_i)$, for $0 \le i \le n-1$, then $a = \phi(x)$ and $b = \phi(y)$, where $x, y \in G_i$. Since ϕ is a homomorphism and $xyx^{-1}y^{-1} \in G_{i+1}$, we have

$$aba^{-1}b^{-1} = \phi(x)\phi(y)\phi(x)^{-1}\phi(y)^{-1} = \phi(x)\phi(y)\phi(x^{-1})\phi(y^{-1}) =$$

$$\phi(xyx^{-1}y^{-1}) \in \phi(G_{i+1}).$$

Therefore, our chain of subgroups for H satisfies the conditions of Proposition 8.33, so $G/N = H$ is solvable.

Note that the previous portion of our argument actually shows that the image under a homomorphism of a solvable group is also a solvable group.

In the other direction, let us assume that both N and G/N are solvable. Since N is normal, there exist a chain of subgroups

$$N = N_0 \supseteq N_1 \supseteq N_2 \cdots \supseteq N_{n-1} \supseteq N_n = \{e\}$$

such that $aba^{-1}b^{-1} \in N_{i+1}$, whenever $a, b \in N_i$ and $0 \le i \le n-1$. If we let $H = G/N$ and ϕ be as previously, then there also exists a chain of subgroups

$$H = H_0 \supseteq H_1 \supseteq H_2 \cdots \supseteq H_{m-1} \supseteq H_m = \{e_H\},$$

where e_H is the identity of H and $aba^{-1}b^{-1} \in H_{i+1}$, whenever $a, b \in H_i$ and $0 \le i \le m - 1$.

By Theorem 8.39(b), $\phi^{-1}(H_i)$ is a subgroup of G, for $0 \le i \le m - 1$ and G has the chain of subgroups

$$G = \phi^{-1}(H_0) \supseteq \phi^{-1}(H_1) \supseteq \phi^{-1}(H_2) \cdots \supseteq \phi^{-1}(H_{m-1}) \supseteq \phi^{-1}(H_m) = N.$$

Next, if $0 \le i \le m - 1$ and if we let $a, b \in \phi(H_i)^{-1}$, then we have $\phi(a) = x$ and $\phi(b) = y$, where $x, y \in H_i$. Since ϕ is a homomorphism and $xyx^{-1}y^{-1} \in H_{i+1}$, we have

$$\phi(aba^{-1}b^{-1}) = \phi(a)\phi(b)\phi(a^{-1})\phi(b^{-1}) = \phi(a)\phi(b)\phi(a)^{-1}\phi(b)^{-1} =$$

$$xyx^{-1}y^{-1} \in H_{i+1}.$$

Therefore, $aba^{-1}b^{-1} \in \phi^{-1}(H_{i+1})$. As a result, the chain of subgroups for G satisfies all the conditions of Proposition 8.33 *except* that it does not go all the way down to the subgroup $\{e\}$. However, if we place the chain of subgroups we obtained for N at the end of the chain of subgroups for G, we obtain

$$G = \phi^{-1}(H_0) \supseteq \phi^{-1}(H_1) \supseteq \phi^{-1}(H_2) \cdots \supseteq \phi^{-1}(H_{m-1}) \supseteq \phi^{-1}(H_m) = N =$$

$$N_0 \supseteq N_1 \supseteq N_2 \cdots \supseteq N_{n-1} \supseteq N_n = \{e\}.$$

Observe that this new chain satisfies all the properties required in Proposition 8.33, so G is indeed solvable. $\qquad \square$

We have now developed quite a few tools that, when combined with Lagrange's Theorem and Sylow's Theorem, allow us to prove that many large classes of finite groups are solvable. In particular, we can now use Lemma 8.34, Theorem 8.37, and Corollary 8.40 to show

Corollary 8.41. *All groups of order p^n and $2p^n$, where p is prime, are solvable.*

Proof. If G is a group of order p^n, where p is prime, we will show that G is solvable using a proof by contradiction. Observe that if the result does not hold, then the Well Ordering Principle asserts that there is a smallest integer $n \ge 0$ such that there exists a group G with p^n elements that is not solvable. Certainly groups with only one element are solvable, so we may assume that $n \ge 1$. As a result, Theorem 8.37 and Lagrange's Theorem guarantee that

$$|Z(G)| = p^m > 1.$$

Note that $Z(G)$ is both a normal subgroup of G and a solvable group. When we examine the factor group $G/Z(G)$, we have

$$|G/Z(G)| = \frac{|G|}{|Z(G)|} = p^{n-m} < p^n$$

As a result, the minimality of n implies that $G/Z(G)$ is solvable. Since $G/Z(G)$ and $Z(G)$ are both solvable, Corollary 8.40 tells us that G is indeed solvable, contradicting the assumption that G was not solvable. Thus, our proof by contradiction does indeed tells us that G is solvable.

We now consider the case where G is a group of order $2p^n$, where p is a prime. If $p = 2$, then $|G| = 2^{n+1}$, and the first part of this result tells us that G is solvable. On the other hand, if $p \neq 2$, then Sylow's Theorem tells us that G contains a subgroup H of order p^n. By the first part of this result, we know that H is solvable. Furthermore, since the index of H in G is 2, Lemma 8.34 tells us that H is a normal subgroup of G. We can now form the factor group G/H, and since this group only has two elements, it is certainly solvable. Therefore, we are in the situation where both H and G/H are solvable, so Corollary 8.40 asserts that G is solvable, as desired. □

To indicate the power and usefulness of Corollary 8.41, suppose you were interested in examining whether all groups of order 512 are solvable. It turns out that there are over 8,000,000 isomorphism classes of groups of order 512. Therefore, it would be virtually impossible to do deal with this problem on a case by case basis. However, since $512 = 2^9$, Corollary 8.41 immediately tells us that every group of order 512 is solvable. This indicates that although some results in algebra may, at first, appear to be very technical or abstract, they often provide quick and easy solutions to problems that we would not want to deal with on a case by case basis.

In order to prove that other classes of finite groups are solvable, we need two more tools. The first is

Lemma 8.42. *Let G be a group with finite subgroups H and K.*

(a) *If $|H|$ and $|K|$ are relatively prime, then $H \cap K = \{e\}$.*

(b) *If $H \cap K = \{e\}$, then $\{hk \mid h \in H, k \in K\}$ consists of $|H| \cdot |K|$ different elements of G.*

Proof. For part (a), Proposition 8.6(c) tells us that $H \cap K$ is a subgroup of both H and K. By Lagrange's Theorem, $|H \cap K|$ must divide both $|H|$ and $|K|$. Since $|H|$ and $|K|$ are relatively prime, it must now be the case that $|H \cap K| = 1$. Hence, $H \cap K = \{e\}$.

For part (b), when we look at $\{hk \mid h \in H, k \in K\}$, there are certainly $|H|$ choices for h and $|K|$ choices for k. Therefore, there are $|H| \cdot |K|$ possible ways to form products of this type. However, we need to make sure that all $|H| \cdot |K|$ products do indeed produce different elements of G. To this end, suppose $h_1, h_2 \in H$ and $k_1, k_2 \in K$ such that

$$h_1 k_1 = h_2 k_2.$$

By multiplying this equation on the left by $h_2{}^{-1}$ and on the right by $k_1{}^{-1}$, we obtain

$$h_2{}^{-1}h_1 = k_2k_1{}^{-1} \in H \cap K = \{e\}.$$

However, since both $h_2{}^{-1}h_1$ and $k_2k_1{}^{-1}$ are equal to e, we immediately obtain $h_1 = h_2$ and $k_1 = k_2$. Thus, different pairs of elements from H and K always produce different products in G. Thus, $\{hk \mid h \in H, k \in K\}$ does contain $|H| \cdot |K|$ elements.   □

Earlier in this section, Definition 8.32 and Proposition 8.33 gave us two equivalent ways to look at being solvable. As we commented at the time, it is frequently useful to have more than one way to look at a concept. Continuing with this theme, it also useful to have an additional way to look at being normal. The second part of our next lemma provides us with this.

Lemma 8.43. *If H is a subgroup of a group G and if $a \in G$, then $aHa^{-1} = \{aha^{-1} \mid h \in H\}$ is also a subgroup of G. Furthermore, H is normal if and only if $aHa^{-1} = H$, for all $a \in G$.*

Proof. If $a \in G$, Proposition 8.6(a) tells us that in order to show that aHa^{-1} is a subgroup, we must show that if $x, y \in aHa^{-1}$, then $xy^{-1} \in aHa^{-1}$. Since $x, y \in aHa^{-1}$, there exist $g, h \in H$ such that $x = aga^{-1}$ and $y = aha^{-1}$. Recall that to find the inverse of a product, we invert each term and reverse the order. Therefore,

$$y^{-1} = (a^{-1})^{-1}h^{-1}a^{-1} = ah^{-1}a^{-1}.$$

Since H is a subgroup, $gh^{-1} \in H$, which implies that

$$xy^{-1} = (aga^{-1})(ah^{-1}a^{-1}) = ag(a^{-1}a)h^{-1}a^{-1} =$$
$$ageh^{-1}a^{-1} = a(gh^{-1})a^{-1} \in aHa^{-1}.$$

Thus, aHa^{-1} is indeed a subgroup.

Next, if H is normal and $a \in G$, then $aH = Ha$. Multiplying this equality of sets on the right by a^{-1} results in

$$(aH)a^{-1} = (Ha)a^{-1}.$$

However, Lemma 8.26 tells us that $(Ha)a^{-1} = H(aa^{-1}) = He = H$, so $aHa^{-1} = H$.

In the other direction, suppose $a \in G$ and $aHa^{-1} = H$. Multiplying this equality of sets on the right by a gives us

$$(aHa^{-1})a = Ha.$$

However, Lemma 8.26(a) tells us that $(aHa^{-1})a = (aH)(a^{-1}a) = (aH)e = aH$. As a result, $aH = Ha$ and H is indeed normal.   □

We can now prove that another large class of groups is solvable.

Theorem 8.44. *If G is a group of order pq, where p and q are prime, then G is solvable.*

Proof. If $p = q$, then $|G| = p^2$ and both Corollary 8.38 and Corollary 8.41 tell us that G is solvable. Therefore, we may assume that $p \neq q$. At this point, we let p denote the larger of the two primes and q the smaller.

By Sylow's Theorem, G contains a subgroup H with p elements. Suppose for the moment that G also contains a subgroup K such that $|K| = p$ and $H \neq K$. Note that $H \cap K$ is a subgroup of H, so Lagrange's Theorem asserts that $|H \cap K|$ equals 1 or p. However, if $|H \cap K| = p$, then $H \cap K$ is a subgroup of both H and K with the same number of elements as H and K. This tells us that $H \cap K$ is equal to both H and K, contradicting the fact that $H \neq K$. Therefore, $|H \cap K| = 1$.

By Lemma 8.42(b), since $H \cap K = \{e\}$, G contains p^2 different elements of the form hk, where $h \in H$ and $k \in K$. However, since $p > q$, this leads to the contradiction

$$|G| \geq p^2 > pq = |G|.$$

Therefore, G only contains one subgroup with p elements.

Next, if $a \in G$, consider $aHa^{-1} = \{aga^{-1} \mid g \in H\}$; by Lemma 8.42(a), aHa^{-1} is a subgroup of G. Furthermore, Lemma 8.17(a) asserts that aHa^{-1} has the same number of elements as aH and aH has the same number of elements as H, so $|aHa^{-1}| = p$. However, since H is the only subgroup of G with p elements, it follows that $aHa^{-1} = H$ and Lemma 8.43 now tells us that H is a normal subgroup of G.

We now have the following chain of subgroups:

$$G \supseteq H \supseteq \{e\}.$$

Observe that H is normal in G and $\{e\}$ is normal in H. Since $|G/H| = q$ and $H/\{e\} = p$, both quotient groups have prime order and are therefore abelian. Therefore, the preceding chain satisfies the conditions of Definition 8.32, and G is indeed solvable, as required. $\qquad\square$

In light of Theorem 8.44, we immediately know that groups of order 15, 21, and 35 are solvable. In fact, if we are interested in groups of small order, Corollary 8.40 and Theorem 8.44 handle most cases. For example, if we wanted to show that all groups of order at most 23 are solvable, the combination of these two results would handle all cases except for groups of order 12 and 20. However, our next result will handle these two cases. The proof will illustrate that proofs in finite group theory often split into many cases and might require the careful counting of the number of elements in various subsets and subgroups.

Theorem 8.45. *If G is a group of order* $4p$, *where* p *is prime, then G is solvable.*

Proof. If $p = 2$, then $|G| = 2^3$ and G is solvable by Corollary 8.40. Therefore, we may assume that $p \neq 2$. Sylow's Theorem now asserts that G contains subgroups H and K such that $|H| = p$ and $|K| = 4$.

We now define the set $N = \{g \in G \mid gHg^{-1} = H\}$. If $g \in H$, then $gHg^{-1} \subseteq H$. However, using the argument used in the proof of Theorem 8.44, $|gHg^{-1}| = |H|$. But this implies that $gHg^{-1} = H$, so $g \in N$. As a result, the set N contains H.

We would like to show that N is a subgroup of G. Since N is a subset of the finite set G, Proposition 8.6(b) tells us that it suffices to show that N is closed under multiplication. To this end, if $g, h \in N$, we have

$$(gh)H(gh)^{-1} = (gh)H(h^{-1}g^{-1}) = g(hHh^{-1})g^{-1} = gHg^{-1} = H,$$

so $gh \in N$. Now that we know that N is a subgroup with $H \subseteq N \subseteq G$, Lagrange's Theorem tells us that the only possibilities for $|N|$ are $4p$, $2p$, and p.

If $|N| = 4p$, then N is equal to all of G. Therefore, $gHg^{-1} = H$, for all $g \in G$, and Lemma 8.43 tells us that H is a normal subgroup of G. This means that we have the chain of subgroups

$$G \supseteq H \supseteq \{e\},$$

where H is normal in G and $\{e\}$ is normal in H. Furthermore, $|G/H| = 4$ and $H/\{e\} = p$, so both factor groups are abelian. This chain now satisfies the conditions of Definition 8.32, so, in this case, G is solvable.

For the next case, let us suppose that $|N| = 2p$. Now we have the chain of subgroups

$$G \supseteq N \supseteq H \supseteq \{e\}.$$

Since the index of N in G is 2 and the index of H in N is 2, Lemma 8.34 asserts that N is a normal subgroup of G and H is a normal subgroup of N. Certainly $\{e\}$ is normal in H and the factor groups G/N, N/H, and $H/\{e\}$, respectively, have orders 2, 2, and p. Therefore, all of the factor groups are abelian, and the chain satisfies the conditions of Definition 8.32. Hence, in this case, G is solvable.

For the final case, we are in the situation where $|N| = p$, so we know that $N = H$. We now define the set

$$T = \{aha^{-1} \mid h \in H, h \neq e, a \in K\},$$

and we will be interested in determining the size of T. Since H has $p - 1$ elements other than the identity, and K has four elements, the set T is formed by computing $4(p-1)$ products.

However, we need check whether all $4(p-1) = 4p - 4$ of these products determine different elements of G. To this end, suppose $g, h \in H$ are not the identity and $a, b \in K$ such that

$$aga^{-1} = bhb^{-1}.$$

Then we have

$$b^{-1}aga^{-1}b = h,$$

which implies that

$$(b^{-1}a)g(b^{-1}a)^{-1} = h.$$

Proposition 8.6(c) tells us that $H \cap (b^{-1}a)H(b^{-1}a)^{-1}$ is a subgroup of H. Since $|H| = p$, Lagrange's Theorem implies that $|H \cap (b^{-1}a)H(b^{-1}a)^{-1}|$ is either 1 or p. However, we have seen that $H \cap (b^{-1}a)H(b^{-1}a)^{-1}$ contains both h and e. As a result, $|H \cap (b^{-1}a)H(b^{-1}a)^{-1}| > 1$, so $|H \cap (b^{-1}a)H(b^{-1}a)^{-1}| = p$. Since $H \cap (b^{-1}a)H(b^{-1}a)^{-1}$, $(b^{-1}a)H(b^{-1}a)^{-1}$, and H all contain p elements, it follows that $H = (b^{-1}a)H(b^{-1}a)^{-1}$. Thus, $b^{-1}a \in N$.

Since $N = H$ and $a, b \in K$, we see that

$$b^{-1}a \in N \cap K = H \cap K.$$

However, $|H|$ and $|K|$ are relatively prime, so Lemma 8.42(a) asserts that $H \cap K = \{e\}$. Thus, $b^{-1}a = e$, which implies that $a = b$.

The facts that $a = b$ and $aga^{-1} = bhb^{-1}$ combine to tell us that $g = h$. Therefore, different pairs of elements of K and nonidentity elements of H do indeed produce different elements of T. As a result, T consists of $4p - 4$ different elements of G. Observe that every element of T belongs to a group of order p and is not the identity. Therefore, T does not contain any elements of order 1, 2, or 4. Since every element of K has order 1, 2, or 4, K consists of the four elements of G that do not belong to T.

If $a \in G$, let us look at the subgroup aKa^{-1}. This group has order 4, so every one of its elements has order 1, 2, or 4. However, the previous argument indicates that all such elements must belong to K. Hence, $aKa^{-1} = K$, so K is normal. Therefore, we now have the chain of subgroups

$$G \supseteq K \supseteq \{e\}.$$

In our present situation, K is normal in G, $\{e\}$ is normal in K, and the factor groups G/K and $K/\{e\}$ have, respectively, orders p and 4. Therefore, the factor groups are abelian, and the chain satisfies the condition of Definition 8.32. As a result, in all possible cases, G is solvable, as desired. $\qquad \square$

We conclude this section by formalizing our observation that all groups of order at most 23 are solvable.

Corollary 8.46. *If G is a group and* $|G| \leq 23$, *then G is solvable.*

Proof. Corollary 8.41 covers groups of order p^n and $2p^n$, where p is prime. Therefore, we know that groups of order

$$1, 2, 3, 4, 5, 6, 7, 8, 9, 10, 11, 13, 14, 16, 17, 18, 19, 22, 23$$

are solvable. Since Theorem 8.44 covers groups of order pq, where p and q are prime, we also know that groups of order $15, 21$ are solvable. All that remains are groups of order 12 and 20 but, in light of Theorem 8.45, these are also solvable. □

Exercises for Section 8.3

Before doing exercises 1–22, please read the following:

In the proof of the class equation, given $g \in G$, we looked at all elements of the form $a^{-1}ga$, where $a \in G$. We call elements of this form **conjugates** of g. The proof of the class equation examines the relationship between the number of conjugates of g and the size of $C(g)$. Throughout these exercises, you will often need to take advantage of this relationship.

In exercises 1–6, we use the notation for elements of S_3 used throughout Sections 8.1–8.3.

1. In S_3, first determine the number of conjugates of f and then find all the conjugates of f.

2. In S_3, first determine the number of conjugates of g and then find all the conjugates of g.

3. In S_3, first determine the number of conjugates of h and then find all the conjugates of h.

4. In S_3, first determine the number of conjugates of j and then find all the conjugates of j.

5. In S_3, first determine the number of conjugates of k and then find all the conjugates of k.

6. Check that your answers to exercises 1–5 are consistent with the class equation.

7. In D_4, first determine the number of conjugates of y and then find all the conjugates of y.

8. In D_4, first determine the number of conjugates of x and then find all the conjugates of x.

9. In D_4, first determine the number of conjugates of yx and then find all the conjugates of yx.

10. In D_4, first determine the number of conjugates of y^2 and then find all the conjugates of y^2.

11. In D_4, first determine the number of conjugates of y^2x and then find all the conjugates of y^2x.

12. In D_4, first determine the number of conjugates of y^3 and then find all the conjugates of y^3.

13. In D_4, first determine the number of conjugates of y^3x and then find all the conjugates of y^3x.

14. Check that your answers to exercises 7–13 are consistent with the class equation.

15. In Qu, first determine the number of conjugates of -1 and then find all the conjugates of -1.

16. In Qu, first determine the number of conjugates of i and then find all the conjugates of i.

17. In Qu, first determine the number of conjugates of $-i$ and then find all the conjugates of $-i$.

18. In Qu, first determine the number of conjugates of j and then find all the conjugates of j.

19. In Qu, first determine the number of conjugates of $-j$ and then find all the conjugates of $-j$.

20. In Qu, first determine the number of conjugates of ij and then find all the conjugates of ij.

21. In Qu, first determine the number of conjugates of $-ij$ and then find all the conjugates of $-ij$.

22. Check that your answers to exercises 15–21 are consistent with the class equation.

23. In D_n, determine how many conjugates y has and then find them.

24. In D_n, if n is odd, determine how many conjugates x has and then find them.

25. In D_n, if n is even, determine how many conjugates x has and then find them.

26. Show that all groups of order 25, 26, 27, and 28 are solvable.

27. Show that all groups of order 29, 31, 32, and 33 are solvable.

28. Show that all groups of order 34, 35, 37, and 38 are solvable.

29. Show that all groups of order 39, 41, 43, and 44 are solvable.

30. Show that all groups of order 46, 47, 49, and 50 are solvable.

31. Show that all groups of order 51, 52, 53, and 54 are solvable.

32. Show that all groups of order 55, 57, 58, and 59 are solvable.

In exercises 33–36, we let D_n be the nth dihedral group, where $n \geq 4$ is an even integer. We also let $H = <y^2>$, the cyclic subgroup generated by y^2.

33. Find the order of H and the index of H in D_n.

34. Show that H is a normal subgroup of D_n.

35. If we let H, Hx, Hy, Hyx be the names of the four cosets corresponding to H, complete the multiplication table for the factor group D_n/H.

$\circ$	H	Hx	Hy	Hyx
H				
Hx				
Hy				
Hyx				

36. Based on your answer to exercise 35, is D_n/H isomorphic to C_4 or to $C_2 \times C_2$?

In exercises 37–40, we examine the group Qu and let $K = <-1>$, the cyclic subgroup generated by -1.

37. Find the order of K and the index of K in Qu.

38. Show that K is a normal subgroup of Qu.

39. If we let K, Ki, Kj, Kij be the names of the four cosets corresponding to K, complete the multiplication table for the factor group Qu/K.

$\circ$	K	Ki	Kj	Kij
K				
Ki				
Kj				
Kij				

40. Based on your answer to exercise 39, is Qu/K isomorphic to C_4 or to $C_2 \times C_2$?

For exercises 41–50, if G is a group and $c \in G$, we say that c is a **commutator** in G if $c = a \cdot b \cdot a^{-1} \cdot b^{-1}$, for some $a, b \in G$. We refer to the term $a \cdot b \cdot a^{-1} \cdot b^{-1}$ as the commutator of a and b.

41. If G is a group and $a, b \in G$, show that the commutator $a \cdot b \cdot a^{-1} \cdot b^{-1}$ is equal to the identity if and only if a and b commute.

42. In Qu, compute the commutator $i \cdot j \cdot i^{-1} \cdot j^{-1}$.

43. In Qu, compute the commutator $ij \cdot i \cdot (ij)^{-1} \cdot i^{-1}$.

44. In Qu, compute the commutator $j \cdot ij \cdot j^{-1} \cdot (ij)^{-1}$.

45. In S_3, compute the commutator $f \cdot g \cdot f^{-1} \cdot g^{-1}$.

46. In S_3, compute the commutator $g \cdot h \cdot g^{-1} \cdot h^{-1}$.

47. In S_3, compute the commutator $k \cdot f \cdot k^{-1} \cdot f^{-1}$.

48. In S_3, compute the commutator $h \cdot j \cdot h^{-1} \cdot j^{-1}$.

49. If $\phi : G_1 \to G_2$ is a homomorphism of groups, show that if $c \in G_1$ is a commutator then $\phi(c)$ is a commutator in G_2.

50. Show that the inverse of every commutator in a group is also a commutator.

51. Let G be a group with normal subgroups H, K. If $h \in H$ and $k \in K$, show that $h \cdot k \cdot h^{-1} \cdot k^{-1} \in H \cap K$.

52. Let G be a group with normal subgroups H, K. If $|H \cap K| = 1$, show that $hk = kh$, for all $h \in H$ and $k \in K$.

53. If G is a group, show that $Z(G)$ is a normal subgroup of G.

54. If G is a nonabelian group of order p^3, where p is a prime, show that $|Z(G)| = p$.

For exercises 55–58, if G is a group with subgroup H, then the set $N(H) = \{g \in G \mid gHg^{-1} = H\}$ is called the **normalizer** of H.

55. If G is a group with subgroup H, show that H is normal if and only if $G = N(H)$.

56. If G is a group with subgroup H, show that $N(H)$ is a subgroup of G that contains H.

57. If G is a group with subgroup H, show that $Z(G) \subseteq N(H)$.

58. Suppose G is a nonabelian group with a subgroup H such that $|N(H)| = p$, where p is a prime. Show that the $Z(G)$ contains only one element.

59. Let $\phi : G_1 \to G_2$ be a homomorphism of groups. Show that ϕ is injective if and only if $Ker(\phi) = \{e_1\}$, where e_1 is the identity element of G_1.

60. Let $\phi : G_1 \to G_2$ be a homomorphism of finite groups such that $|G_1|$ and $|G_2|$ are relatively prime. Show that $Ker(\phi) = G_1$.

61. If $\phi : G_1 \to G_2$ is a surjective homomorphism of groups and H is a normal subgroup of G_1, show that $\phi(H)$ is a normal subgroup of G_2.

At the end of Section 8.2, we presented a table that listed all groups, up to isomorphism, of order at most 11. We now have the tools needed to verify that our table is indeed complete for groups of order at most 11. Corollary 8.29 told us that any group of prime order must be cyclic. Combined with the fact that cyclic groups with the same number of elements must be isomorphic, we can see that the table is complete for groups of order 1, 2, 3, 5, 7, 11. Next,

Corollary 8.38 verifies the table is complete for groups of order 4 and 9. Therefore, it remains to examine groups of order 6, 8, and 10.

In exercises 62–67, we will assume that G is a group of order $2p$, where p is an odd prime. The classification of groups of order 6 and 10 will be a special case of the results in these exercises.

62. Show that there exist $x, y \in G$ such that
    (a) $o(x) = 2$,

    (b) $o(y) = p$,

    (c) $< y >$ is normal,

    (d) every element of G can be expressed uniquely in the form $y^i x^j$, where $0 \le i \le p - 1$ and $0 \le j \le 1$, and

    (e) $xyx = y^m$, where $1 \le m \le p - 1$.

63. If $m = 1$ in part (e) of exercise 62, show that $G =< yx >$, so $G \approx C_{2p}$.

64. Use part (e) of exercise 62 to show that $y = xy^m x$.

65. Use part (e) of exercise 62 to show that $xy^m x = y^{m^2}$.

66. Use exercises 64 and 65 to show that $m^2 - 1$ is a multiple of p.

67. Show that if $m \ne 1$, then $m = p - 1$ and $G \approx D_p$.

In exercises 68–76, we will assume that G is a group of order 8. These exercises will show that G must be isomorphic to one of the five groups on our table. In these exercises, we will let $t(G)$ denote the largest integer m such that G contains an element of order m.

68. If G is nonabelian, show that $t(G) = 4$.

69. If $t(G) = 8$, show that $G \approx C_8$.

70. If G is abelian and $t(G) = 4$, show that $G \approx C_4 \times C_2$.

71. If $t(G) = 2$, show that $G \approx C_2 \times C_2 \times C_2$.

72. If G is nonabelian, show that
    (a) there exists some $y \in G$ such that $o(y) = 4$,

    (b) $< y >$ is normal,

    (c) either (i) there exists $x \in G$, $x \notin < y >$ such that $o(x) = 2$, or (ii) $o(x) = 4$, for every $x \in G$, $x \notin < y >$.

73. If (i) occurs in part (c) of exercise 72, show that every element of G can be written uniquely in the form $y^i x^j$, where $0 \le i \le 3$ and $0 \le j \le 1$.

74. If x, y are as in exercise 73, show that $xyx = y^3$ and $G \approx D_4$.

75. If (ii) occurs in part (c) of exercise 72, let $x \in G$ such that $x \notin < y >$ and $o(x) = 4$.
    (a) Show that $x^2 = y^2$.

    (b) Show that every element of G can be expressed uniquely as $y^k x^l$, where $0 \le k \le 3$ and $0 \le l \le 1$.

    (c) Show that $xy = y^3 x$.

76. If x, y are as in exercise 75, let $\phi : G \to Qu$ be the function defined as $\phi(y^k x^l) = i^k j^l$, for $0 \le k \le 3$ and $0 \le l \le 1$. Show that ϕ is an isomorphism.

In exercises 77–80, G will be a group of order pq, where $p > q$ are primes and q does not divide $p - 1$. We will show that G must be cyclic. For example, we will have succeeded in classifying groups of order 15, 33, and 35. However, since q divides $p - 1$ when $p = 7, q = 3$ and $p = 11, q = 5$, these exercises do not fully classify groups of order 21 or 55.

77. Show that there exist $x, y \in G$ such that
    (a) $o(x) = q$,

    (b) $o(y) = p$,

    (c) $< y >$ is normal,

    (d) every element of G can be expressed uniquely as $y^i x^j$, where $0 \le i \le p - 1$ and $0 \le j \le q - 1$,

    (e) $xyx^{-1} = y^m$, where $1 \le m \le p - 1$.

78. If x, y, m are as in exercise 77, show that $y = y^{m^q}$.

79. If m is as in exercises 77 and 78, show that the order of $[m]_p$ in $U(\mathbb{Z}_p)$ is a divisor of both q and $p - 1$.

80. Use exercise 79 to show that $m = 1$ and conclude that $G = < yx >$ and $G \approx C_{pq}$.

8.4 Symmetric Groups

Symmetric groups are perhaps the most interesting and important class of groups in finite group theory. So far, virtually all of the finite groups we have looked at have been solvable. However, in this section, we will show that S_n is not solvable, for $n \ge 5$. In fact, the $n = 5$ case is a big part of the proof of the insolvability of the quintic. Not only can we think of S_n as being more complicated than other nonabelian groups like D_n, but we will also show that S_n contains an isomorphic copy of *every* group of order n. Thus, it quickly becomes apparent that symmetric groups represent an essential piece of finite group theory.

The good news when studying S_n is that there are some very convenient ways of representing its elements. We begin with

Definition 8.47. *If* $\{a_1, a_2, \ldots, a_t\}$ *is a subset of* $\{1, 2, \ldots, n\}$, *then we let*

$$(a_1 \; a_2 \; \ldots \; a_t)$$

be a shorthand for the element $f \in S_n$ *such that*

$$f(a_1) = a_2, \; f(a_2) = a_3, \ldots, f(a_{t-1}) = a_t, \; f(a_t) = a_1,$$

and $f(b) = b$, *for every other element of* $\{1, 2, \ldots, n\}$. *We refer to functions of this type as* *t-cycles.*

■ Example

The 3-cycle (1 4 3), when viewed as an element of S_4, is a shorthand for the function f such that

$$f(1) = 4, \quad f(4) = 3, \quad f(3) = 1, \quad \text{and} \quad f(2) = 2.$$

Note that we could also consider (1 4 3) as an element of S_5, in which it would also be the case that $f(5) = 5$. In fact, (1 4 3) can be considered as an element of S_n, for all $n \geq 4$, and it would act as the identity on any $m \geq 5$.

■

As we will see in the following examples, every element of S_n that can be represented by a t-cycle can actually be represented by t different t-cycles. This is because we can begin the t-cycle with any of the t terms that appears in the cycle.

■ Examples

$$(1 \; 4 \; 3) = (4 \; 3 \; 1) = (3 \; 1 \; 4),$$

$$(6 \; 7) = (7 \; 6),$$

$$(1 \; 2 \; 3 \; 4 \; 5) = (2 \; 3 \; 4 \; 5 \; 1) = (3 \; 4 \; 5 \; 1 \; 2) = (4 \; 5 \; 1 \; 2 \; 3) = (5 \; 1 \; 2 \; 3 \; 4).$$

■

When cycles, such as (4 1 7), (3 8 2), and (5 6), have no elements of $\{1, 2, \ldots, n\}$ in common, we say that they are *disjoint cycles*. It is often the case in algebra that simple objects serve as the building blocks of more complicated objects. For example, prime numbers are the building blocks of the natural numbers. In a similar manner, we can consider disjoint cycles to be the building blocks of S_n, as every $f \in S_n$ other than the identity can be written as a product of disjoint cycles. This is done by following what f does to each element of $\{1, 2, \ldots, n\}$.

■ Example

Let $f \in S_{10}$ be the function such that

$$f(1) = 8, \quad f(2) = 2, \quad f(3) = 9, \quad f(4) = 10, \quad f(5) = 1,$$

$$f(6) = 5, \quad f(7) = 4, \quad f(8) = 6, \quad f(9) = 3, \quad f(10) = 7.$$

For any $i \in \{1, 2, \ldots, n\}$ and $\sigma \in S_n$, we can examine the list

$$i \mapsto \sigma(i) \mapsto \sigma^2(i) \mapsto \cdots$$

until the first repetition occurs. Every element of $\{1, 2, \ldots, n\}$ will appear exactly once on a list of this type and it gives us a shorthand for representing elements of S_n. In this example, it allows us to abbreviate the action of f on $\{1, 2, \ldots, 10\}$ as

$$1 \mapsto 8 \mapsto 6 \mapsto 5 \mapsto 1, \quad 2 \mapsto 2,$$

$$3 \mapsto 9 \mapsto 3, \quad 4 \mapsto 10 \mapsto 7 \mapsto 4.$$

Having broken f into four pieces, at first glance it appears that f consists of four disjoint cycles. However, we need not include the piece that tells us that f sends 2 to 2. Therefore, we can express f as a product of three disjoint cycles as

$$f = (1\ 8\ 6\ 5)(3\ 9)(4\ 10\ 7).$$

■

Observe that disjoint cycles always commute. However, in general, elements of S_n do not commute.

■ Examples

1. Let $g = (3\ 7\ 1\ 2\ 5)$ and $h = (4\ 3\ 2\ 1\ 6)$; recall that when we compose functions, we begin with the function on the far right. Therefore, as we examine how the function gh acts on the set $\{1, 2, 3, 4, 5, 6, 7\}$, we see that

$$(gh)(1) = 6, \quad (gh)(2) = 2, \quad (gh)(3) = 5, \quad (gh)(4) = 7,$$

$$(gh)(5) = 3, \quad (gh)(6) = 4, \quad (gh)(7) = 1.$$

Therefore, we can abbreviate the action of gh as

$$1 \mapsto 6 \mapsto 4 \mapsto 7 \mapsto 1,$$

$$2 \mapsto 2, \quad \text{and} \quad 3 \mapsto 5 \mapsto 3.$$

As a result, the product $(3\ 7\ 1\ 2\ 5)(4\ 3\ 2\ 1\ 6)$ can be now rewritten using disjoint cycles as

$$(3\ 7\ 1\ 2\ 5)(4\ 3\ 2\ 1\ 6) = (1\ 6\ 4\ 7)(3\ 5).$$

2. Consider the product

$$w = (1\ 4\ 3)(2\ 4)(2\ 3)(1\ 4\ 3)(2\ 4\ 3)(1\ 2\ 4);$$

at first, it is not clear where w sends each element of $\{1, 2, 3, 4\}$. However, when we compute the composition of these six functions, we can see that

$$1 \mapsto 3 \mapsto 1, \quad 2 \mapsto 4 \mapsto 2.$$

Therefore,

$$w = (1\ 4\ 3)(2\ 4)(2\ 3)(1\ 4\ 3)(2\ 4\ 3)(1\ 2\ 4) = (1\ 3)(2\ 4).$$

■

Observe that when we rewrite a product of functions in S_n as a product of disjoint cycles, it is easy to see how the product acts on the elements of $\{1, 2, \ldots, n\}$. Another advantage of using disjoint cycles is that it makes it easier to compute the order of the functions in S_n. We record this as

Lemma 8.48

(a) *If an element g of a group G has order $n \geq 1$, then g^m is the identity if and only if m is a multiple of n.*

(b) *Every t-cycle in S_n has order t.*

(c) *If $f \in S_n$ is written as the product of disjoint cycles $f_1 f_2 \cdots f_m$, where each f_i is a t_i-cycle, then the order of f is equal to the least common multiple of $t_1, t_2, \ldots, t_m$.*

Proof. In one direction, if m is a multiple of n, then there exists $q \in \mathbb{Z}$ such that $m = qn$. Therefore,

$$g^m = g^{qn} = (g^n)^q = e^q = e,$$

as desired. In the other direction, suppose that $g^m = e$. The division algorithm tells us that there exist $q, r \in \mathbb{Z}$ such that

$$m = qn + r \quad \text{and} \quad 0 \leq r < n.$$

We now have

$$e = g^m = g^{qn+r} = g^{qn}g^r = (g^n)^q g^r = e^q g^r = eg^r = g^r.$$

Since $r < n$, the minimality of n tells us that $r = 0$. Hence, $m = qn$, so m is a multiple of n, as needed.

For part (b), if $g = (i_1 \ i_2 \ \cdots \ i_t)$ is a t-cycle belonging to S_n, then

$$g(i_1) = i_2 \neq i_1, \ g^2(i_1) = i_3 \neq i_1, \ldots, \ g^{t-1}(i_1) = i_t \neq i_1.$$

As a result, the order of g cannot be less than t. On the other hand, it is clear that g^t is the identity map, so g has order t.

For part (c), let n be the least common multiple of $t_1, t_2, \ldots, t_m$. In light of part (b), n is a multiple of the order of every f_i. Therefore, part (a) tells us that $f_i^n = e$, for $1 \leq i \leq m$. Since the f_i's are disjoint cycles, they all commute. As a result, we now have

$$f^n = (f_1 f_2 \cdots f_m)^n = f_1^n f_2^n \cdots f_m^n = e \cdot e \cdots e = e.$$

Since $f^n = e$, in order to show that $o(f) = n$, it suffices to show that if $1 \leq l < n$, then $f^l \neq e$. To this end, since the f_i commute

$$f^l = (f_1 f_2 \cdots f_m)^l = f_1^l f_2^l \cdots f_m^l.$$

Since l is less than n, we know that l is not a multiple of some t_i. In light of part (a), we now know that $f_i^l \neq e$. Therefore, there exists some $j \in \{1, 2, \ldots, n\}$ such that $f_i^l(j) = k \neq j$. Since $f_1, f_2, \ldots f_m$ are disjoint cycles and the cycle f_i must mention both j and k, it follows that none of the other cycles mention either j or k. As a result, every other cycle in f fixes both j and k. Therefore, when we look at

$$f^l(j) = (f_1 f_2 \cdots f_m)^l(j) = (f_1^l f_2^l \cdots f_m^l)(j),$$

we can see that any term to the right of f_i^l fixes j and any term to the left of f_i^l fixes k. Thus,

$$f^l(j) = f_i^l(j) = k \neq j,$$

so $f^l \neq e$, as needed. $\qquad\qquad\qquad\qquad\qquad\qquad\qquad\qquad\qquad\qquad\qquad\qquad\square$

To illustrate how to use Lemma 8.48, we revisit some of our earlier examples.

■ Examples

1. To compute the order of $(3\ 7\ 1\ 2\ 5)(4\ 3\ 2\ 1\ 6)$, Lemma 8.48 and the fact that

   $$(3\ 7\ 1\ 2\ 5)(4\ 3\ 2\ 1\ 6) = (1\ 6\ 4\ 7)(3\ 5)$$

tell us that $o((3\ 7\ 1\ 2\ 5)(4\ 3\ 2\ 1\ 6))$ is the least common multiple of $o((1\ 6\ 4\ 7)) = 4$ and $o((3\ 5)) = 2$. Hence, $o((3\ 7\ 1\ 2\ 5)(4\ 3\ 2\ 1\ 6)) = 4$.

2. If $w = (1\ 4\ 3)(2\ 4)(2\ 3)(1\ 4\ 3)(2\ 4\ 3)(1\ 2\ 4)$, then since

$$(1\ 4\ 3)(2\ 4)(2\ 3)(1\ 4\ 3)(2\ 4\ 3)(1\ 2\ 4) = (1\ 3)(2\ 4),$$

we have that $o(w)$ is the least common multiple of $o((1\ 3)) = 2$ and $o((2\ 4)) = 2$. Thus, $o(w) = 2$.

∎

Another name for a 2-cycle is a *transposition*. As we will now see, if we look at S_n from a somewhat different perspective, we can also view transpositions as the building blocks of S_n. Since we already have concrete ways of representing elements in S_n, proofs about the structure of S_n often follow from some very short and simple formulas. That is certainly the case with our next proof. It will follow easily from a simple formula that, unfortunately, seems to appear from out of nowhere. However, the formula is actually the result of a great deal of experimentation and trial and error. Since none of this experimentation appears in the proof, the argument is quite short but does not add to our intuition.

Lemma 8.49. *Let $n \geq 2$;*

(a) *every element of S_n is equal to a product of transpositions and*

(b) *if $f \in S_n$, then $f \circ (i\ j) \circ f^{-1} = (f(i)\ f(j))$, for any $1 \leq i, j \leq n$.*

Proof. For part (a), we have already observed that every element of S_n can be written as a product of disjoint cycles. Therefore, it suffices to show that every cycle can be written as a product of transpositions. To this end, consider the t-cycle $(i_1\ i_2\ \cdots\ i_j\ i_{j+1}\ \cdots\ i_{t-1}\ i_t)$; we claim that

$$(i_1\ i_2\ \cdots\ i_{t-1}\ i_t) = (i_1\ i_2)(i_2\ i_3)\cdots(i_j\ i_{j+1})\cdots(i_{t-1}\ i_t).$$

If we let g be the function obtained by composing the $t - 1$ transpositions from the previous equation, then it is clear that both $(i_1\ i_2\ \cdots\ i_{t-1}\ i_t)$ and g act as the identity on any element of $\{1, 2, \ldots, n\}$ not mentioned in $(i_1\ i_2\ \cdots\ i_{t-1}\ i_t)$. In order to show that $(i_1\ i_2\ \cdots\ i_{t-1}\ i_t)$ and g are the same element of S_n, there are two remaining cases. First, when we compute $g(i_t)$, as we go from right to left in composing the $t - 1$ transpositions, we can see that each transposition drops the subscript by 1 and

$$i_t \mapsto i_{t-1} \mapsto \cdots \mapsto i_2 \mapsto i_1.$$

Thus, $g(i_t) = i_1$, and we see that g and $(i_1\ i_2\ \cdots\ i_{t-1}\ i_t)$ do the same thing to i_t.

Next, if $1 \le j \le t-1$, then i_j does not appear in any transposition until $(i_j \ i_{j+1})$ and this transposition sends i_j to i_{j+1}. However, i_{j+1} does not appear in any of the transpositions to the left of $(i_j \ i_{j+1})$, so i_{j+1} remains fixed the rest of the way. As a result, $g(i_j) = i_{j+1}$, so g and $(i_1 \ i_2 \ \cdots \ i_{t-1} \ i_t)$ also do the same thing to all i_j, where $1 \le j \le t-1$. Thus, we have indeed shown that $(i_1 \ i_2 \ \cdots \ i_{t-1} \ i_t)$ and $(i_1 \ i_2)(i_2 \ i_3) \cdots (i_j \ i_{j+1}) \cdots (i_{t-1} \ i_t)$ are the same element of S_n.

For part (b), we must examine what the composition $f \circ (i \ j) \circ f^{-1}$ does to every element of $\{1, 2, \ldots, n\}$. First, if $1 \le k \le n$ and k is not equal to either $f(i)$ or $f(j)$, then $f^{-1}(k)$ is not equal to either i or j. Therefore, the transposition $(i \ j)$ fixes $f^{-1}(k)$ and the composition $f \circ (i \ j) \circ f^{-1}$ does the following to k:

$$k \mapsto f^{-1}(k) \mapsto f^{-1}(k) \mapsto k.$$

On the other hand, the composition $f \circ (i \ j) \circ f^{-1}$ does the following to $f(i)$:

$$f(i) \mapsto i \mapsto j \mapsto f(j).$$

Similarly, $f \circ (i \ j) \circ f^{-1}$ does the following to $f(j)$:

$$f(j) \mapsto j \mapsto i \mapsto f(i).$$

The preceding computations show that $f \circ (i \ j) \circ f^{-1}$ interchanges $f(i)$ and $f(j)$ while fixing every other element of $\{1, 2, \ldots, n\}$. However, this is exactly what the transposition $(f(i) \ f(j))$ does, so $f \circ (i \ j) \circ f^{-1} = (f(i) \ f(j))$, as desired. $\qquad\square$

To illustrate the formula we verified in part (a) of Lemma 8.49, we have

■ Example

$$(3 \ 2 \ 1 \ 4 \ 7 \ 5) = (3 \ 2)(2 \ 1)(1 \ 4)(4 \ 7)(7 \ 5) \text{ and}$$

$$(8 \ 6 \ 2 \ 4 \ 1 \ 9 \ 3 \ 7) = (8 \ 6)(6 \ 2)(2 \ 4)(4 \ 1)(1 \ 9)(9 \ 3)(3 \ 7).$$

These combine to tell us that

$$(3 \ 2 \ 1 \ 4 \ 7 \ 5)(8 \ 6 \ 2 \ 4 \ 1 \ 9 \ 3 \ 7) =$$

$$(3 \ 2)(2 \ 1)(1 \ 4)(4 \ 7)(7 \ 5)(8 \ 6)(6 \ 2)(2 \ 4)(4 \ 1)(1 \ 9)(9 \ 3)(3 \ 7).$$

■

When we showed that prime numbers were the building blocks of the natural numbers, we showed that the representation of each $n \ge 2$ as a product of primes was unique up to order. Similarly, when expressing an element of S_n as a product of disjoint cycles, this representation is also unique up to the order of the cycles and the fact that every t-cycle can be written using any of t terms from the cycle in the first position. However, there is no analogous statement regarding uniqueness when it comes to representing cycles as products of transpositions. For example, we could replace the formula

$$(i_1 \; i_2 \; \cdots \; i_{t-1} \; i_t) = (i_1 \; i_2)(i_2 \; i_3) \cdots (i_j \; i_{j+1}) \cdots (i_{t-1} \; i_t)$$

in the proof of Lemma 8.49(a) by the formula

$$(i_1 \; i_2 \; \cdots \; i_{t-1} \; i_t) = (i_{t-1} \; i_t)(i_{t-2} \; i_t) \cdots (i_j \; i_t) \cdots (i_1 \; i_t).$$

This formula is another example of a fact about S_n which, although straightforward to verify, takes a great deal of experimentation to find. If we use this new formula, we now have

$$(3 \; 2 \; 1 \; 4 \; 7 \; 5) = (7 \; 5)(4 \; 5)(1 \; 5)(2 \; 5)(3 \; 5)$$

and

$$(8 \; 6 \; 2 \; 4 \; 1 \; 9 \; 3 \; 7) = (3 \; 7)(9 \; 7)(1 \; 7)(4 \; 7)(2 \; 7)(6 \; 7)(8 \; 7).$$

Our next lemma is another example of how short formulas in S_n can be exploited to obtain a good deal of information about S_n.

Lemma 8.50. *If H is a subgroup of S_n that contains $(1 \; 2)$ and $(1 \; 2 \; 3 \; \cdots \; n)$, then $H = S_n$.*

Proof. Lemma 8.49(a) tells us that every element of S_n is a product of transpositions, so it suffices to show that H contains all transpositions.

Every transposition in S_n can be written as $(i \; j)$, where $1 \le i < j \le n$. We will first show that H contains all transposition of the form $(i \; i+1)$, where $1 \le i \le n-1$. When $i = 1$, there is nothing to prove, so we may assume that $i > 1$. If we let $g = (1 \; 2 \; 3 \; \cdots \; n)$ and let $f = g^{i-1}$, then we can see that $f \in H$ and

$$f(1) = i \quad \text{and} \quad f(2) = i+1.$$

Since H is a subgroup, we have $f \circ (1 \; 2) \circ f^{-1} \in H$ and Lemma 8.49(b) tells us that

$$f \circ (1 \; 2) \circ f^{-1} = (f(1) \; f(2)) = (i \; i+1).$$

The preceding computations tell us that H contains the transpositions

$$(1 \; 2), \; (2 \; 3), \; (3 \; 4), \ldots, \; (n-1 \; n).$$

We need to finish showing that H contains all transpositions of the form $(i\ j)$, where $1 \leq i < j \leq n$. Observe that we have already handled the case where $j = i+1$, so we may assume that $j > i+1$. If we let

$$h = (j-1\ j) \cdots (i+2\ i+3)(i+1\ i+2),$$

then $h \in H$, and it is not hard to see that

$$h(i) = i \quad \text{and} \quad h(i+1) = j.$$

As a result,

$$h \circ (i\ i+1) \circ h^{-1} \in H$$

and Lemma 8.49(b) asserts that

$$h \circ (i\ i+1) \circ h^{-1} = (h(i)\ h(i+1)) = (i\ j).$$

Thus, H does indeed contain all transpositions in S_n, so $H = S_n$, as desired. $\square$

The next result is one of two in this section that will play a key role in the proof of the insolvability of the quintic.

Theorem 8.51. *If H is a subgroup of S_p which contains both a transposition and an element of order p, where p is prime, then $H = S_p$.*

Proof. In light of Lemma 8.50, it suffices to show that H contains $(1\ 2)$ and $(1\ 2\ 3\ \cdots\ p)$. The transposition in H flips two elements of $\{1, 2, \ldots, p\}$ and leaves the other $p-2$ elements fixed. Without loss of generality, we may rename the two elements flipped by the transposition as 1 and 2. Therefore, we may now assume that H contains $(1\ 2)$. Note that we still reserve the right to rename, if necessary, the other $p-2$ elements of $\{1, 2, \ldots, p\}$, provided we do not use the names 1 or 2.

Let $f \in H$ have order p; by Lemma 8.48(b)(c), f must be a product of disjoint p-cycles. However, since each p-cycle mentions every element of $\{1, 2, \ldots, p\}$, it follows that f can be represented by a single p-cycle.

When we represent f as a p-cycle, we can choose the cycle that begins with the number 1. Therefore,

$$f = (1\ \cdots\ 2\ \cdots).$$

Next, let j represent the number of elements from $\{1, 2, \ldots, p\}$ that appear between 1 and 2 in f. For example, if

$$f = (1\ 6\ 7\ 4\ 2\ \cdots),$$

then $j = 3$. It is easy to see that $0 \le j \le p - 2$, and if we let $g = f^{j+1}$, we will examine some of the properties of g. First, g is also an element of both H and the cyclic subgroup generated by f. Since f has order p and $1 \le j + 1 \le p - 1$, we can see that $g = f^{j+1}$ is not the identity. Note that since every element of $< f >$ other than the identity has order p, it follows that g has order p. However, we already saw that every element of S_p of order p can be represented by a p-cycle.

The p-cycle representing f has j terms between 1 and 2, so

$$g(1) = f^{j+1}(1) = 2.$$

As a result, when we write g as a p-cycle, we can write

$$g = (1 \ 2 \ \cdots \ ).$$

The $p - 2$ terms in the cycle for g that are to the right of 2 all belong to the set $\{3, \ldots, p\}$. However, as we remarked earlier, we reserved the right to rename the elements of this set with other elements of the set. Therefore, we may assume that the terms in our p-cycle for g following 2 are, in order, $3, 4, \ldots, p$. As a result,

$$g = (1 \ 2 \ 3 \ \cdots \ p).$$

Now that we know that H contains both $(1 \ 2)$ and $(1 \ 2 \ 3 \ \cdots \ p)$, we can apply Lemma 8.50 to assert that $H = S_p$, as desired. $\square$

Our next lemma is yet another example of a significant fact about S_n that, once you have found the right formula, is quite easy to prove.

Lemma 8.52. *Let H, K be subgroups of S_n, where $n \ge 5$, such that*

(a) *H contains all 3-cycles and*

(b) *$ghg^{-1}h^{-1} \in K$, for all $g, h \in H$.*

Then K also contains all 3-cycles.

Proof. It suffices to show that, if $f \in S_n$ is a 3-cycle, then there exist $g, h \in H$ such that $f = ghg^{-1}h^{-1}$. To this end, since f is a 3-cycle, there exist three different elements in $\{1, 2, \ldots, n\}$, which we will call a_1, a_2, a_3, such that $f = (a_1 \ a_2 \ a_3)$. Since $n \ge 5$, there are at least two elements in $\{1, 2, \ldots, n\}$ that are not in the set $\{a_1, a_2, a_3\}$, and we will call them a_4 and a_5. If we let

$$g = (a_1 \ a_2 \ a_4) \quad \text{and} \quad h = (a_1 \ a_3 \ a_5),$$

then both g and h are 3-cycles and therefore belong to H. Since

$$(a_1 \ a_2 \ a_4)^{-1} = (a_1 \ a_4 \ a_2) \quad \text{and} \quad (a_1 \ a_3 \ a_5)^{-1} = (a_1 \ a_5 \ a_3),$$

a simple computation tells us that

$$ghg^{-1}h^{-1} = (a_1 \ a_2 \ a_4)(a_1 \ a_3 \ a_5)(a_1 \ a_2 \ a_4)^{-1}(a_1 \ a_3 \ a_5)^{-1} =$$

$$(a_1 \ a_2 \ a_4)(a_1 \ a_3 \ a_5)(a_1 \ a_4 \ a_2)(a_1 \ a_5 \ a_3) = (a_1 \ a_2 \ a_3) = f.$$

Thus, $f \in K$, so K does indeed contain all 3-cycles, as desired. $\qquad\qquad\square$

We can now prove the important fact, which we mentioned at the beginning of this section, that S_n is not solvable when $n \geq 5$.

Theorem 8.53.   *If $n \geq 5$, then S_n is not solvable.*

Proof. Suppose S_n contains a chain of subgroups

$$S_n = G_0 \supseteq G_1 \supseteq G_2 \cdots \supseteq G_{m-1} \supseteq G_m$$

such that whenever $g, h \in G_i$, we have $ghg^{-1}h^{-1} \in G_{i+1}$, for $0 \leq i \leq m - 1$.

Lemma 8.52 tells us that whenever G_i contains all 3-cycles, then so does G_{i+1}, for $0 \leq i \leq m - 1$. Since $G_0 = S_n$, it contains all 3-cycles, so every subgroup on this chain also contains all 3-cycles. If S_n were solvable, there would exist some chain of this type where $G_m = \{e\}$. But since G_m must contain all 3-cycles, $G_m \neq \{e\}$, so S_n is not solvable. $\qquad\square$

At the beginning of this section, we said that we would prove for any $n \geq 1$ that every group of order n is isomorphic to a subgroup of S_n. This reinforces the notion that symmetric groups are an extremely rich and interesting collection of groups. It might lead you to believe that symmetric groups are the only finite groups you need to study, as a complete understanding of the subgroups of symmetric groups would allow you to understand the structure of all finite groups. For example, suppose you wanted to understand all groups G such that $|G| = 11$. It will follow from the result we are about to prove, that G is isomorphic to a subgroup of S_{11}. However, S_{11} contains

$$11! = 39,916,800$$

elements. Trying to find and understand the structure of all subgroups of a group with $39,916,800$ elements is an incredibly difficult job. On the other hand, since 11 is prime, groups of order 11 are very easy to understand, as up to isomorphism there is only one. Thus, as a practical matter, there are usually easier and more direct ways to examine a group of order n than using the fact that an isomorphic copy of it is contained in S_n.

Theorem 8.54—Cayley's Theorem. *If G is a group of order n, then S_n contains a subgroup isomorphic to G.*

Proof. Since G is a set with n elements, we can think of S_n as the set of bijections from G to G. We begin by associating to every element of G an element of S_n. If $g \in G$, define the function

$$\pi_g : G \to G$$

as $\pi_g(a) = ga$, for all $a \in G$. Certainly π_g is a function from G to G and we need to show that it is a bijection.

In order to show that π_g is a injection, suppose $a, b \in G$ such that $\pi_g(a) = \pi_g(b)$; we must prove that $a = b$. However,

$$ga = \pi_g(a) = \pi_g(b) = gb$$

and multiplying the outer terms of this equation on the left by g^{-1} yields

$$g^{-1}(ga) = g^{-1}(gb),$$

from which it quickly follows that $a = b$. Thus, π_g is injective.

There are two ways to show that π_g is a surjection. First, since G is finite, any injection of G is also a surjection. However, we can also directly prove that π_g is surjective without using the fact that G is finite. Observe that if $a \in G$, then

$$\pi(g^{-1}a) = g(g^{-1}a) = (gg^{-1})a = ea = a.$$

Therefore, every $a \in G$ belongs to the image of π_g, so π_g is a surjection. As a result, for all $g \in G$, π_g is a bijection.

We can now define the function

$$\phi : G \to S_n,$$

as $\phi(g) = \pi_g$, for all $g \in G$. We first need to show that ϕ is a homomorphism. If $g, h \in G$, observe that

$$\phi(gh) = \pi_{gh} \quad \text{and} \quad \phi(g)\phi(h) = \pi_g \pi_h.$$

Both π_{gh} and $\pi_g \pi_h$ are functions from G to G. To check that they are the same function, if $a \in G$, we have

$$\pi_{gh}(a) = (gh)a = g(ha) = g(\pi_h(a)) = \pi_g(\pi_h(a)) = (\pi_g \pi_h)(a).$$

Thus, $\pi_{gh} = \pi_g \pi_h$, so $\phi(gh) = \phi(g)\phi(h)$, so ϕ is indeed a homomorphism.

Since ϕ is a homomorphism, Theorem 8.39 asserts that $Im(\phi)$ is a subgroup of S_n. We can think of ϕ as being a homomorphism from G to $Im(\phi)$. In this context, ϕ is certainly surjective, and in order to conclude that G is isomorphic to $Im(\phi)$, it suffices to show that ϕ is injective. To this end, suppose $g, h \in G$ such that $\phi(g) = \phi(h)$; we need to show that $g = h$. Since

$$\pi_g = \phi(g) = \phi(h) = \pi_h,$$

π_g and π_h are the same function from G to G. If we apply both functions to $e \in G$, we obtain

$$\pi_g(e) = ge = g \quad \text{and} \quad \pi_h(e) = he = h,$$

so $g = \pi_g(e) = \pi_h(e) = h$. Thus, ϕ is also injective.

Therefore, $\phi : G \to Im(\phi)$ is not only a homomorphism but also a bijection. As a result, G is isomorphic to $Im(\phi)$ and $Im(\phi)$ is a subgroup of S_n, as required. $\square$

In Section 8.2, we remarked that the converse of Lagrange's Theorem does not hold. In other words, there exists a group G of order n that does not contain any subgroup of order m, for some m that divides n. The smallest example is a group G of order 12, which is contained in S_4, that contains no subgroup with six elements. We now present this example and show that it has the desired properties.

Proposition 8.55. *S_4 contains a subgroup G of order* 12 *that does not contain a subgroup of order* 6.

Proof. Let

$$G = \{e, \ (1\,2\,3), \ (1\,3\,2), \ (1\,2\,4), \ (1\,4\,2), \ (1\,3\,4), \ (1\,4\,3), \ (2\,3\,4),$$
$$(2\,4\,3), \ (1\,2)(3\,4), \ (1\,3)(2\,4), \ (1\,4)(2\,3)\};$$

it is somewhat tedious but not hard to check that G is closed under multiplication. Therefore, by Proposition 8.6(b), G is indeed a group with 12 elements.

By way of contradiction, suppose G contained a subgroup H with six elements. By Sylow's Theorem, there must exist $g, h \in H$ such that $o(g) = 2$ and $o(h) = 3$. Lemma 8.42(b) now asserts that the set

$$\{g^i h^j \mid 0 \le i \le 1, 0 \le j \le 2\}$$

contains six different elements of H, so this set is equal to H.

Since $|o(g)|$ and $|o(h)|$ are relatively prime, $<g> \cap <h> = \{e\}$. Therefore, if $hg \in <g>$, we would obtain the contradiction

$$h \in <g> \cap <h> = \{e\}.$$

Similarly, if $hg \in <h>$, then we would obtain the contradiction

$$g \in <g> \cap <h> = \{e\}.$$

However, we know that $hg \in H$, so out of the six elements in H, the only remaining possibilities are that either $hg = gh$ or $hg = gh^2$.

If $hg = gh$, then H is abelian, and it is easy to see that $o(gh) = 6$. However, when we look at the list of elements of G, we can see that every element of G has order $1, 2, 3$, or 4. This is a contradiction, so we may now assume that $hg = gh^2$.

The only elements of order three in G are 3-cycles, so h is a 3-cycle. Observe that not only does G contain every 3-cycle in S_4, but it also contains every product of two disjoint transpositions in S_4. In light of this, by appropriately naming the four elements of the set that the functions in S_4 permute, we may assume that $h = (1\ 2\ 3)$. On the other hand, g has order 2, and G only contains three elements of order 2. As a result, g must be equal to one of $(1\ 2)(3\ 4)$, $(1\ 3)(2\ 4)$, or $(1\ 4)(2\ 3)$.

By multiplying the equation $hg = gh^2$ on the left by g, we obtain the equation $ghg = h^2$. We now examine the values of ghg that occur when $h = (1\ 2\ 3)$ and g ranges through the three possibilities $(1\ 2)(3\ 4)$, $(1\ 3)(2\ 4)$, and $(1\ 4)(2\ 3)$. If $g = (1\ 2)(3\ 4)$, we obtain

$$ghg = ((1\ 2)(3\ 4))(1\ 2\ 3)((1\ 2)(3\ 4)) = (1\ 4\ 2) \neq (1\ 3\ 2) = h^2;$$

if $g = (1\ 3)(2\ 4)$, we obtain

$$ghg = ((1\ 3)(2\ 4))(1\ 2\ 3)((1\ 3)(2\ 4)) = (1\ 3\ 4) \neq (1\ 3\ 2) = h^2;$$

and finally, if $g = (1\ 4)(2\ 3)$, we obtain

$$ghg = ((1\ 4)(2\ 3))(1\ 2\ 3)((1\ 4)(2\ 3)) = (2\ 4\ 3) \neq (1\ 3\ 2) = h^2.$$

Therefore, regardless of our choice of g, we obtain the contradiction $ghg \neq h^2$. As a result, G cannot contain a subgroup H with six elements. □

We conclude this chapter by verifying the remark we made in Section 8.3 that S_4 is solvable. If G is the subgroup of S_4 from Proposition 8.55, then $|S_4| = 24$ and $|G| = 12$. Since G has index 2 in S_4, Lemma 8.34 asserts that G is a normal subgroup of S_4. Observe that the quotient group S_4/G has order 2, so it is solvable. In addition, it follows from Theorem 8.45 that groups of order 12 are solvable, so G is solvable. As a result, both S_4/G and G are solvable, and Corollary 8.40 now asserts that S_4 is indeed solvable.

Exercises for Section 8.4

1. In S_n, show that the inverse of the t-cycle $(a_1 \, a_2 \, \cdots \, a_{t-1} \, a_t)$ is the t-cycle $(a_t \, a_{t-1} \, \cdots \, a_2 \, a_1)$.

2. If $n \geq 1$, what is the order of the group S_n?

Exercises 3–41 will deal with functions $f, g, h, k \in S_8$. The behavior of f, g, h, k on the set $\{1, 2, 3, 4, 5, 6, 7, 8\}$ is described by the following.

$$f(1) = 3, \quad f(2) = 5, \quad f(3) = 8, \quad f(4) = 6,$$
$$f(5) = 2, \quad f(6) = 1, \quad f(7) = 7, \quad f(8) = 4.$$

$$g(1) = 1, \quad g(2) = 6, \quad g(3) = 8, \quad g(4) = 5,$$
$$g(5) = 4, \quad g(6) = 7, \quad g(7) = 2, \quad g(8) = 3.$$

$$h(1) = 8, \quad h(2) = 1, \quad h(3) = 5, \quad h(4) = 6,$$
$$h(5) = 2, \quad h(6) = 4, \quad h(7) = 3, \quad h(8) = 7.$$

$$k(1) = 5, \quad k(2) = 3, \quad k(3) = 8, \quad k(4) = 1,$$
$$k(5) = 4, \quad k(6) = 6, \quad k(7) = 7, \quad k(8) = 2.$$

3. Write f as a product of disjoint cycles and find the order of f.

4. Write g as a product of disjoint cycles and find the order of g.

5. Write h as a product of disjoint cycles and find the order of h.

6. Write k as a product of disjoint cycles and find the order of k.

7. Write f^{-1} as a product of disjoint cycles.

8. Write g^{-1} as a product of disjoint cycles.

9. Write h^{-1} as a product of disjoint cycles.

10. Write k^{-1} as a product of disjoint cycles.

11. Write f^2 as a product of disjoint cycles and find the order of f^2.

12. Write g^2 as a product of disjoint cycles and find the order of g^2.

13. Write h^2 as a product of disjoint cycles and find the order of h^2.

14. Write k^2 as a product of disjoint cycles and find the order of k^2.

15. Write $f \circ g$ as a product of disjoint cycles and find the order of $f \circ g$.

16. Write $g \circ f$ as a product of disjoint cycles and find the order of $g \circ f$.

17. Write $f \circ h$ as a product of disjoint cycles and find the order of $f \circ h$.

18. Write $h \circ f$ as a product of disjoint cycles and find the order of $h \circ f$.

19. Write $f \circ k$ as a product of disjoint cycles and find the order of $f \circ k$.

20. Write $k \circ f$ as a product of disjoint cycles and find the order of $k \circ f$.

21. Write $g \circ h$ as a product of disjoint cycles and find the order of $g \circ h$.

22. Write $h \circ g$ as a product of disjoint cycles and find the order of $h \circ g$.

23. Write $g \circ k$ as a product of disjoint cycles and find the order of $g \circ k$.

24. Write $k \circ g$ as a product of disjoint cycles and find the order of $k \circ g$.

25. Write $h \circ k$ as a product of disjoint cycles and find the order of $h \circ k$.

26. Write $k \circ h$ as a product of disjoint cycles and find the order of $k \circ h$.

27. In the exercises above, you have seen many examples where $a, b \in S_8$ and $ab \neq ba$. In the cases you examined, how do $o(ab)$ and $o(ba)$ compare?

28. Let G be a group and suppose $a, b \in G$ such that $(ab)^n = e$, for some $n \geq 1$, where e is the identity element of G.

    (a) Show that $(ba)^{n-1}b = a^{-1}$.

    (b) Use part (a) to show that $(ba)^n = e$.

    (c) How do $o(ab)$ and $o(ba)$ compare?

    (d) Is the answer from part (c) consistent with your observations in exercise 27?

29. Write f two different ways as a product of transpositions.

30. Write g two different ways as a product of transpositions.

31. Write h two different ways as a product of transpositions.

32. Write k two different ways as a product of transpositions.

33. Write f^2 two different ways as a product of transpositions.

34. Write g^2 two different ways as a product of transpositions.

35. Write h^2 two different ways as a product of transpositions.

36. Write k^2 two different ways as a product of transpositions.

37. If $\sigma_1, \sigma_2, \ldots, \sigma_t$ are transpositions, show that the inverse of the composition $\sigma_1 \circ \sigma_2 \circ \cdots \circ \sigma_{t-1} \circ \sigma_t$ is $\sigma_t \circ \sigma_{t-1} \circ \cdots \circ \sigma_2 \circ \sigma_1$.

38. Use exercise 37 to write f^{-1} in two different ways as a product of transposition.

39. Use exercise 37 to write g^{-1} in two different ways as a product of transposition.

40. Use exercise 37 to write h^{-1} in two different ways as a product of transposition.

41. Use exercise 37 to write k^{-1} in two different ways as a product of transposition.

42. If $(a\ b\ c)$ is a 3-cycle in S_n and $\sigma \in S_n$, show that $\sigma \circ (a\ b\ c) \circ \sigma^{-1} = (\sigma(a)\ \sigma(b)\ \sigma(c))$.

43. Generalize exercise 42 and show that if $(a_1\ a_2\ \cdots\ a_t)$ is a t-cycle in S_n and $\sigma \in S_n$, then

$$\sigma \circ (a_1\ a_2\ \cdots\ a_t) \circ \sigma^{-1} = (\sigma(a_1)\ \sigma(a_2)\ \cdots\ \sigma(a_t)).$$

44. The groups S_4 and D_{12} are both nonabelian groups of order 24.
    (a) What is the largest order of an element in S_4?

    (b) What is the largest order of an element of D_{12}?

    (c) Are S_4 and D_{12} isomorphic? Explain your answer.

45. The group G in the proof of Proposition 8.55 and D_6 are both nonabelian groups of order 12. Use the fact that G does not contain any subgroup with six elements to show that G and D_6 are not isomorphic.

46. For which $n \in \mathbb{N}$ does S_4 contain a cyclic subgroup with n elements?

47. For which $n \in \mathbb{N}$ does S_5 contain a cyclic subgroup with n elements?

48. For which $n \in \mathbb{N}$ does S_6 contain a cyclic subgroup with n elements?

49. If σ_1, σ_2 are different transpositions in S_n which are not disjoint, show that $\sigma_1 \circ \sigma_2$ is a 3-cycle.

50. If i, j, k, l are distinct positive integers, write the composition $(i\ k\ j) \circ (i\ k\ l)$ as a product of disjoint cycles.

51. Use exercise 50 to show that if σ_1, σ_2 are disjoint transpositions, then there exist 3-cycles τ_1, τ_2 such that $\tau_1 \circ \tau_2 = \sigma_1 \circ \sigma_2$.

52. Use exercises 49 and 51 to show that any product of an even number of transpositions can be rewritten as a product of 3-cycles.

53. Find the smallest $n \in \mathbb{N}$ such that S_n contains an element of order 10.

54. Find the smallest $n \in \mathbb{N}$ such that S_n contains an element of order 15.

55. Find the smallest $n \in \mathbb{N}$ such that S_n contains an element of order 16.

56. Find the smallest $n \in \mathbb{N}$ such that S_n contains an element of order 30.

In exercises 57–69, you will be asked to prove various facts about the group G from the proof of Proposition 8.55. At first glance, these exercises may appear to require a good deal of computation. However, if you use Lagrange's Theorem, Sylow's Theorem, and the fact that G does not contain any subgroups with six elements, relatively little computation should be necessary. These exercises will refer to the **normalizer** of a subgroup, which was a concept introduced in exercises 55–58 in Section 8.3.

57. Without finding all the subgroups of G, for which $n \in \mathbb{N}$ does G contain a subgroup of order n?

58. If $g, h \in G$ such that $o(g) = 2$ and $o(h) = 3$, show that $gh \neq hg$.

59. If $g, h \in G$ such that $o(g) = 2$ and $o(h) = 3$, show that $gh \neq h^2 g$.

60. If $h \in G$ such that $o(h) = 3$, show that subgroups $< h >$, $C(h)$, and $N(< h >)$ are all the same.

61. By looking at the number of elements of order 2, show that G has exactly one subgroup with four elements.

For the remainder of these exercises, K will denote the unique subgroup of G with the four elements referred to in exercise 61.

62. If $g \in G$ such that $o(g) = 2$, show that $C(g) = K$ and $N(K) = G$.

63. If $g \in G$ has order 2, determine the number of conjugates of g and write down the elements of G that are conjugate to g. In light of exercises 61 and 62, you should not need to do any computations inside of G.

64. If $h \in G$ has order 3 and a, b are distinct elements of K, show that $a^{-1}ha \neq b^{-1}hb$.

65. If $h_1 \in G$ has order 3 and $h_2 \in G$ is conjugate to h_1, show that there exists some $a \in K$ such that $h_2 = a^{-1}h_1 a$.

66. If $h \in G$ has order 3, show that h and h^2 are not conjugate.

67. Find all conjugates of the 3-cycle (1 2 3). Exercise 65 can reduce the number of computations needed.

68. Find all conjugates of the 3-cycle (1 3 2). Exercise 65 can reduce the number of computations needed.

69. Look at your answers to exercises 63, 67, and 68, and check to see if they are consistent with the class equation.

CHAPTER 9

Polynomials over the Integers and Rationals

In your previous algebra courses, many problems and questions probably arose that dealt with the roots and factoring of polynomials in $\mathbb{Z}[x]$ and $\mathbb{Q}[x]$. We are now in a position to apply our work on $\mathbb{Z}$ and $\mathbb{Z}_n$ from Chapters 3 and 7 to deal with many of these same problems. Although the focus of this chapter will be $\mathbb{Z}[x]$ and $\mathbb{Q}[x]$, occasionally we will work in the more general context of fields and commutative rings. Hopefully, by continuing to look at concrete examples, you will become more and more comfortable working with fields and commutative rings.

9.1 Integral Domains and Homomorphisms of Rings

When asked to multiply polynomials consisting of several terms, students often forget that the key to multiplying polynomials is the distributive law. Names like "FOIL" are sometimes given in certain situations to help students remember a procedure. However, it is much more important to realize that the distributive law can always be used, regardless of how many terms appear in the polynomials that are being multiplied.

■ **Example**

$$(3x^2 + 2x + 8)(4x^5 - 6x^3 + 5x) =$$
$$3x^2(4x^5 - 6x^3 + 5x) + 2x(4x^5 - 6x^3 + 5x) + 8(4x^5 - 6x^3 + 5x) =$$
$$(12x^7 - 18x^5 + 15x^3) + (8x^6 - 12x^4 + 10x^2) + (32x^5 - 48x^3 + 40x) =$$
$$12x^7 + 8x^6 + 14x^5 - 12x^4 - 33x^3 + 10x^2 + 40x.$$

■

More generally, the distributive law tells us that when multiplying the polynomials

$$f(x) = a_n x^n + a_{n-1} x^{n-1} + \cdots + a_1 x + a_0$$

and

$$g(x) = b_m x^m + b_{m-1} x^{m-1} + \cdots + b_1 x + b_0,$$

the coefficient of x^t in $f(x) \cdot g(x)$ is $a_t \cdot b_0 + a_{t-1} \cdot b_1 + \cdots + a_1 \cdot b_{t-1} + a_0 \cdot b_t$.

Unusual things can happen if the coefficients of our polynomials belong to a ring with zero divisors. Recall that if $\mathbb{Z}_n$ has zero divisors, then the degrees of polynomials need not add when we multiply in $\mathbb{Z}_n[x]$. For example, in $\mathbb{Z}_{10}[x]$ we have

$$([4]_{10} x^7 + [3]_{10} x^3 + [8]_{10} x)([5]_{10} x^2) = [20]_{10} x^9 + [15]_{10} x^5 + [40]_{10} x^3 =$$

$$[0]_{10} x^9 + [5]_{10} x^5 + [0]_{10} x^3 = [5]_{10} x^5.$$

However, at the moment, we are primarily interested in understanding polynomials in $\mathbb{Z}[x]$ and $\mathbb{Q}[x]$. Therefore, in this chapter, we will usually look at commutative rings without zero divisors.

Definition 9.1. *A commutative ring R is called an integral domain if it has no zero divisors.*

■ Examples

Some examples of integral domains are $\mathbb{Z}$, $\mathbb{Q}$, $\mathbb{R}$, $\mathbb{C}$, $\mathbb{Z}[x]$, $\mathbb{R}[x]$, $\mathbb{C}[x]$, $\mathbb{R}[x, y]$. Also, when p is a prime number, $\mathbb{Z}_p$ and $\mathbb{Z}_p[x]$ are also integral domains. In particular, every field is an integral domain, and whenever R is an integral domain, then so is $R[x]$, the set of polynomials with coefficients in R.

■

Before listing some of the basic properties of multiplication in $R[x]$, for integral domains R, we need some terminology.

Definition 9.2. *Let $f(x) = a_n x^n + a_{n-1} x^{n-1} + \cdots + a_{m+1} x^{m+1} + a_m x^m$ be a polynomial where $a_n, a_m \neq 0$ and $n \geq m$. We call $a_n x^n$ the leading term of $f(x)$ and $a_m x^m$ the trailing term of $f(x)$. In this case, we also call a_n the leading coefficient of $f(x)$ and a_m the trailing coefficient of $f(x)$. Finally, we say that n is the degree of $f(x)$, and we write this as $deg(f(x)) = n$.*

■ Examples

If $f(x) = 23x^{14} - 6x^9 + \frac{11}{13}x^5 - 4x^3$ in $\mathbb{Q}[x]$, then the leading term is $23x^{14}$, the leading coefficient is 23, the trailing term is $-4x^3$, the trailing coefficient is -4, and $deg(f(x)) = 14$.

In $\mathbb{Z}_5[x]$, if $g(x) = [4]_5 x^8$, then $[4]_5 x^8$ is both the leading and trailing term and $deg(g(x)) = 8$.

■

We can now record some standard facts about the multiplication of polynomials with coefficients in an integral domain.

Lemma 9.3. *Let*

$$f(x) = a_n x^n + a_{n-1} x^{n-1} + \cdots + a_{s+1} x^{s+1} + a_s x^s$$

and

$$g(x) = b_m x^m + b_{m-1} x^{m-1} + \cdots + b_{t+1} x^{t+1} + b_t x^t$$

be polynomials with coefficients in an integral domain where $a_n, a_s, b_m, b_t \neq 0$, $n \geq s$, and $m \geq t$. Then

(a) *the leading term of $f(x) \cdot g(x)$ is $a_n b_m x^{n+m}$ and $deg(f(x) \cdot g(x)) = n + m$;*

(b) *the trailing term of $f(x) \cdot g(x)$ is $a_s b_t x^{s+t}$;*

(c) *$f(x) \cdot g(x)$ consists of only one term with a nonzero coefficient if and only if both $f(x)$ and $g(x)$ consist of only one term with a nonzero coefficient.*

Intuition. Although multiplying polynomials is a straightforward procedure, computing all the terms in the product of two polynomials can sometimes be a long, tedious job. However, it is always quick and easy to find both the leading and trailing terms of a product if the coefficients of our polynomials belong to an integral domain. In the product $f(x) \cdot g(x)$, the leading term is the product of the leading terms of $f(x)$ and $g(x)$ and the trailing term is the product of the trailing terms of $f(x)$ and $g(x)$. For example, consider

$$f(x) = 3x^{12} + 6x^{10} - 4x^5 + 2x^4 - 8x^3 \quad \text{and} \quad g(x) = 4x^6 - 11x^5 + 9x^4$$

in $\mathbb{Q}[x]$. We can immediately see that the leading term of $f(x) \cdot g(x)$ is $(3x^{12})(4x^6) = 12x^{18}$ and the trailing term is $(-8x^3)(9x^4) = -72x^7$.

Proof. If

$$f(x) = a_n x^n + a_{n-1} x^{n-1} + \cdots + a_{s+1} x^{s+1} + a_s x^s$$

and

$$g(x) = b_m x^m + b_{m-1} x^{m-1} + \cdots + b_{t+1} x^{t+1} + b_t x^t,$$

then the largest exponent that appears in $f(x)$ is n and the largest exponent that appears in $g(x)$ is m. Therefore, the largest exponent that could possibly appear in $f(x) \cdot g(x)$ is $n + m$. The only way to obtain a term of degree $n + m$ in $f(x) \cdot g(x)$ is by multiplying the $a_n x^n$ term from $f(x)$ and the $b_m x^m$ term from $g(x)$. Since a_n and b_m are nonzero elements of an integral domain, the product $(a_n x^n)(b_m x^m) = a_n b_m x^{n+m}$ is nonzero. Thus, the term $a_n b_m x^{n+m}$ does indeed appear in $f(x) \cdot g(x)$ with a nonzero coefficient, and it is therefore the leading term of $f(x) \cdot g(x)$. In addition, this also tells us that $deg(f(x) \cdot g(x)) = n + m$, thereby proving part (a).

The proof of part (b) is very similar. The smallest exponent that appears in $f(x)$ is s and the smallest exponent that appears in $g(x)$ is t. Therefore, the smallest exponent that could possibly appear in $f(x) \cdot g(x)$ is $s + t$. The only way to obtain a term of degree $s + t$ in $f(x) \cdot g(x)$ is by multiplying the $a_s x^s$ term from $f(x)$ and the $b_m x^m$ term from $g(x)$. Since a_s and b_t are nonzero elements of an integral domain, the product $(a_s x^s)(b_t x^t) = a_s b_t x^{s+t}$ is nonzero. Thus, the term $a_s b_t x^{s+t}$ does indeed appear in $f(x) \cdot g(x)$ with a nonzero coefficient, and it is therefore the trailing term of $f(x) \cdot g(x)$.

For part (c), one direction is quite easy. If both $f(x)$ and $g(x)$ consist of a single nonzero term, then $f(x) = a_n x^n$ and $g(x) = b_m x^m$. Thus, $f(x) \cdot g(x) = a_n b_m x^{n+m}$, so $f(x) \cdot g(x)$ certainly consists of a single nonzero term. In the other direction, if $f(x) \cdot g(x)$ consists of only one nonzero term, then the leading term $a_n b_m x^{n+m}$ and the trailing term $a_s b_t x^{s+t}$ must be the same term. This tells us that $n + m = s + t$. However, since $n \geq s$ and $m \geq t$, this can only happen if $n = s$ and $m = t$. As a result, $f(x) = a_n x^n$ and $g(x) = b_m x^m$, as desired. $\square$

Observe that part (a) of Lemma 9.3 also tells us that if $f(x)$ and $g(x)$ are both nonzero, then $f(x) \cdot g(x) \neq 0$. That is the reason why $R[x]$ is an integral domain whenever R is an integral domain.

■ Examples

In $\mathbb{Z}_{11}[x]$, if

$$f(x) = [6]_{11} x^7 + [10]_{11} x^4 + [2]_{11} x$$

and

$$g(x) = [8]_{11} x^{14} + [5]_{11} x^8 + [8]_{11} x^6 + [9]_{11} x^4,$$

then the leading term of $f(x) \cdot g(x)$ is $([6]_{11}x^7)([8]_{11}x^{14}) = [4]_{11}x^{21}$ and the trailing term is $([2]_{11}x)([9]_{11}x^4) = [7]_{11}x^5$.

In $\mathbb{Z}_{29}[x]$, if

$$f(x) = [25]_{29}x^8 + [4]_{29}x^7 + [15]_{29}x^5$$

and

$$g(x) = [6]_{29}x^{24} + [17]_{29}x^{19} + [20]_{29}x^{17} + [5]_{29}x^{10} + [21]_{29}x^9,$$

then the leading term of $f(x) \cdot g(x)$ is $([25]_{29}x^8)([6]_{29}x^{24}) = [5]_{29}x^{32}$ and the trailing term is $([15]_{29}x^5)([21]_{29}x^9) = [25]_{29}x^{14}$.

If we revisit an example from earlier in this section, we can see why our coefficients need to belong to an integral domain for the conclusions of Lemma 9.3 to hold. Recall that in $\mathbb{Z}_{10}[x]$ we had

$$([4]_{10}x^7 + [3]_{10}x^3 + [8]_{10}x)([5]_{10}x^2) = [20]_{10}x^9 + [15]_{10}x^5 + [40]_{10}x^3 =$$

$$[0]_{10}x^9 + [5]_{10}x^5 + [0]_{10}x^3 = [5]_{10}x^5.$$

Observe that in this example the conclusions in Lemma 9.3 do not hold. In particular, $[5]_{10}x^5$ has degree 5, whereas the polynomials being multiplied have degrees 7 and 2. Furthermore, $[5]_{10}x^5$ is the only term in the product with a nonzero coefficient, yet the polynomial $[4]_{10}x^7 + [3]_{10}x^3 + [8]_{10}x$ has three terms with nonzero coefficients.

■

Since we will soon use $\mathbb{Z}_n$ and $\mathbb{Z}_n[x]$ to obtain information about polynomials in $\mathbb{Z}[x]$ and $\mathbb{Q}[x]$, it is reasonable to wonder how does one use $\mathbb{Z}_n$ to learn about $\mathbb{Z}$? To begin answering this question, recall that addition and multiplication in $\mathbb{Z}_n$ are defined as

(1) $$[a]_n + [b]_n = [a+b]_n \quad \text{and} \quad [a]_n \cdot [b]_n = [a \cdot b]_n,$$

for all $a, b \in \mathbb{Z}$. Next, if $n \geq 2$, consider the function

$$\sigma_n : \mathbb{Z} \to \mathbb{Z}_n$$

defined as

$$\sigma_n(a) = [a]_n,$$

for all $a \in \mathbb{Z}$. By (1), it follows that

$$\sigma_n(a+b) = \sigma_n(a) + \sigma_n(b) \quad \text{and} \quad \sigma_n(a \cdot b) = \sigma_n(a) \cdot \sigma_n(b),$$

for all $a, b \in \mathbb{Z}$. Observe that the equations satisfied by σ_n are identical to those satisfied by automorphisms, and this leads us to

Definition 9.4. *If R and S are rings, we call a function $\tau : R \to S$ a homomorphism if*

$$\tau(a+b) = \tau(a) + \tau(b) \ \ and \ \ \tau(a \cdot b) = \tau(a) \cdot \tau(b),$$

for all $a, b \in R$.

At various points in this chapter, such as in Definition 9.4, we refer to rings instead of commutative rings. When we do this, we are no longer assuming that multiplication is commutative. The reason for occasionally dealing with rings instead of commutative rings is because various ideas and concepts, such as homomorphisms, make perfect sense even when not assuming that multiplication is commutative.

In Definition 8.23, we defined homomorphisms of groups. It should come as little surprise that Definitions 8.23 and 9.4 are identical except that homomorphisms of rings need to deal with two binary operations instead of one. In particular, if $\tau : R \to S$ is a homomorphism of rings, and if we ignore multiplication, then τ is still a homomorphism of groups under addition.

As we saw in Chapter 8, group homomorphisms are strongly connected to normal subgroups and quotient groups. In a very similar manner, homomorphisms of rings are connected to objects known as ideals and quotient rings. To obtain information about the roots and factoring of polynomials in $\mathbb{Z}[x]$ and $\mathbb{Q}[x]$, we are primarily interested in homomorphisms from $\mathbb{Z}$ to $\mathbb{Z}_n$ and from $\mathbb{Z}[x]$ to $\mathbb{Z}_n[x]$. However, when we prove Kronecker's Theorem in Chapter 17, we will take a deeper and more general look at homomorphisms of rings and their connections with ideals and quotient rings.

If you look back at Lemma 5.9, you will see that it is stated in terms of automorphisms of commutative rings. However, having now been introduced to homomorphisms of rings, it is easy to see that the proof of Lemma 5.9 actually yields a more general result. Since this more general result will be needed throughout this chapter, we record it as

Lemma 9.5. *Suppose $\tau : R \to S$ is a homomorphism of rings. If $x_1, x_2, \ldots, x_n \in R$, then*

$$\tau(x_1 + x_2 + \cdots + x_n) = \tau(x_1) + \tau(x_2) + \cdots + \tau(x_n)$$

and

$$\tau(x_1 \cdot x_2 \cdots x_n) = \tau(x_1) \cdot \tau(x_2) \cdots \tau(x_n).$$

■ Examples: Homomorphisms of Rings

1. Our motivating example was, for every $n \geq 2$, to let

$$\sigma_n : \mathbb{Z} \to \mathbb{Z}_n$$

be defined as $\sigma(a) = [a]_n$, for all $a \in \mathbb{Z}$.

We can use σ_n to now define, for every $n \geq 2$,

$$\tau_n : \mathbb{Z}[x] \to \mathbb{Z}_n[x]$$

as

$$\tau_n \left(a_m x^m + a_{m-1} x^{m-1} + \cdots + a_1 x + a_0 \right) =$$

$$[a_m]_n x^m + [a_{m-1}]_n x^{m-1} + \cdots + [a_1]_n x + [a_0]_n,$$

for all $a_i \in \mathbb{Z}$.

Observe that if $f(x) = a_m x^m + \cdots + a_1 x + a_0$ and $g(x) = b_l x^l + \cdots + b_1 x + b_0$ belong to $\mathbb{Z}[x]$, then the coefficient of x^t in $f(x) + g(x)$ and $f(x) \cdot g(x)$ are, respectively, $a_t + b_t$ and $a_t b_0 + a_{t-1} b_1 + \cdots + a_1 b_{t-1} + a_0 b_t$. As a result, the coefficient of x^t in $\tau_n(f(x) + g(x))$ is $[a_t + b_t]_n$, whereas the coefficient of x^t in $\tau_n(f(x)) + \tau_n(g(x))$ is $[a_t]_n + [b_t]_n$. Since σ_n is a homomorphism, it follows that x^t has the same coefficient in both $\tau_n(f(x) + g(x))$ and $\tau_n(f(x)) + \tau_n(g(x))$, for every $t \geq 0$. Thus,

$$\tau_n(f(x) + g(x)) = \tau_n(f(x)) + \tau_n(g(x)),$$

for all $f(x), g(x) \in \mathbb{Z}[x]$.

Similarly, the coefficient of x^t in $\tau_n(f(x) \cdot g(x))$ is

$$[a_t b_0 + a_{t-1} b_1 + \cdots + a_1 b_{t-1} + a_0 b_t]_n,$$

whereas the coefficient of x^t in $\tau_n(f(x)) \cdot \tau_n(g(x))$ is

$$[a_t]_n [b_0]_n + [a_{t-1}]_n [b_1]_n + \cdots + [a_1]_n [b_{t-1}]_n + [a_0]_n [b_t]_n.$$

However, since σ_n is a homomorphism, Lemma 9.5 asserts that

$$[a_t b_0 + a_{t-1} b_1 + \cdots + a_1 b_{t-1} + a_0 b_t]_n =$$

$$[a_t b_0]_n + [a_{t-1} b_1]_n + \cdots + [a_1 b_{t-1}]_n + [a_0 b_t]_n =$$

$$[a_t]_n [b_0]_n + [a_{t-1}]_n [b_1]_n + \cdots + [a_1]_n [b_{t-1}]_n + [a_0]_n [b_t]_n.$$

Therefore, the coefficient of x^t is the same in both $\tau_n(f(x)\cdot g(x))$ and $\tau_n(f(x))\cdot\tau_n(g(x))$, for every $t\geq 0$, so

$$\tau_n(f(x)\cdot g(x)) = \tau_n(f(x))\cdot\tau_n(g(x)),$$

for all $f(x), g(x) \in \mathbb{Z}[x]$. As a result, τ_n preserves both addition and multiplication and is indeed a homomorphism of rings.

To illustrate this, let $f(x) = 6x^3 + 2x^2 - 7$ and $g(x) = 4x + 6$. If $n = 5$, we now have

$$\tau_5(f(x)) = x^3 + [2]_5x^2 + [3]_5 \quad \text{and} \quad \tau_5(g(x)) = [4]_5x + [1]_5.$$

Therefore,

$$\tau_5(f(x) + g(x)) = \tau_5(6x^3 + 2x^2 + 4x - 1) = x^3 + [2]_5x^2 + [4]_5x + [4]_5$$

and

$$\tau_5(f(x)) + \tau_5(g(x)) = (x^3 + [2]_5x^2 + [3]_5) + ([4]_5x + [1]_5) =$$
$$x^3 + [2]_5x^2 + [4]_5x + [4]_5.$$

Similarly,

$$\tau_5(f(x)\cdot g(x)) = \tau_5(24x^4 + 44x^3 + 12x^2 - 28x - 42) =$$
$$[4]_5x^4 + [4]_5x^3 + [2]_5x^2 + [2]_5x + [3]_5$$

and

$$\tau_5(f(x))\cdot\tau_5(g(x)) = (x^3 + [2]_5x^2 + [3]_5)\cdot([4]_5x + [1]_5) =$$
$$[4]_5x^4 + [4]_5x^3 + [2]_5x^2 + [2]_5x + [3]_5.$$

2. We can now generalize our previous example. Suppose $\sigma: R \to S$ is a homomorphism of rings. Next, define

$$\tau: R[x] \to S[x]$$

as

$$\tau(a_mx^m + a_{m-1}x^{m-1} + \cdots + a_1x + a_0) =$$
$$\sigma(a_m)x^m + \sigma(a_{m-1})x^{m-1} + \cdots + \sigma(a_1)x + \sigma(a_0),$$

for all $a_i \in R$. The same argument that shows that each τ_n in the previous example is a homomorphism of $\mathbb{Z}[x]$ also shows that τ is a homomorphism of $R[x]$.

3. $\mathbb{R}[x]$ is a ring and we can let

$$\tau : \mathbb{R}[x] \to \mathbb{R}[x]$$

be defined as $\tau(f(x)) = f(x^2)$, for all $f(x) \in \mathbb{R}[x]$. Then τ is a homomorphism as

$$\tau(f(x) + g(x)) = f(x^2) + g(x^2) = \tau(f(x)) + \tau(g(x))$$

and

$$\tau(f(x) \cdot g(x)) = f(x^2) \cdot g(x^2) = \tau(f(x)) \cdot \tau(g(x))$$

for all $f(x), g(x) \in \mathbb{R}[x]$.

For an example of how τ behaves, let $f(x) = 7x - 5$ and $g(x) = 3x + 23$. Then

$$\tau(f(x) + g(x)) = \tau(10x + 18) = 10(x^2) + 18 = 10x^2 + 18$$

and

$$\tau(f(x)) + \tau(g(x)) = \tau(7x - 5) + \tau(3x + 23) =$$
$$(7(x^2) - 5) + (3(x^2) + 23) = 10x^2 + 18.$$

Furthermore,

$$\tau(f(x) \cdot g(x)) = \tau(21x^2 + 146x - 115) =$$
$$21(x^2)^2 + 146(x^2) - 115 = 21x^4 + 146x^2 - 115$$

and

$$\tau(f(x)) \cdot \tau(g(x)) = \tau(7x - 5) \cdot \tau(3x + 23) =$$
$$(7(x^2) - 5) \cdot (3(x^2) + 23) = 21x^4 + 146x^2 - 115.$$

4. $\mathbb{R}[x]$ and $\mathbb{R}$ are both rings, and we let

$$\tau : \mathbb{R}[x] \to \mathbb{R}$$

be defined as $\tau(f(x)) = f(2)$, for all $f(x) \in \mathbb{R}[x]$. Then τ is a homomorphism as

$$\tau(f(x) + g(x)) = f(2) + g(2) = \tau(f(x)) + \tau(g(x))$$

and

$$\tau(f(x) \cdot g(x)) = f(2) \cdot g(2) = \tau(f(x)) \cdot \tau(g(x));$$

for all $f(x), g(x) \in \mathbb{R}[x]$.

For an example of the behavior of τ, let $f(x) = 5x - 11$ and $g(x) = 6x^2 + 2x + 9$. Then

$$\tau(f(x) + g(x)) = \tau(6x^2 + 7x - 2) = 6(2)^2 + 7(2) - 2 = 36$$

and

$$\tau(f(x)) + \tau(g(x)) = (5(2) - 11) + (6(2)^2 + 2(2) + 9) = (-1) + (37) = 36.$$

Similarly,

$$\tau(f(x) \cdot g(x)) = \tau(30x^3 - 56x^2 + 23x - 99) = 30(2)^3 - 56(2)^2 + 23(2) - 99 = -37$$

and

$$\tau(f(x)) \cdot \tau(g(x)) = \tau(5x - 11) \cdot \tau(6x^2 + 2x + 9) =$$
$$(5(2) - 11) \cdot (6(2)^2 + 2(2) + 9) = (-1) \cdot (37) = -37.$$

5. Every automorphism of a commutative ring or field is an example of a homomorphism of rings. In Chapter 5, we were introduced to $Gal(L/K)$, the automorphisms of a commutative ring L that are the identity map on the commutative ring K. Therefore, all the elements of $Gal(L/K)$ are examples of homomorphisms of rings.

■

Exercises for Section 9.1

In exercises 1–3, find the degree, leading term, and trailing term of each product in $\mathbb{Q}[x]$.

1. $(7x^5 - 4x^3 - 11x^2 + 54x - 13) \cdot (9x^8 - 16x^7 + 4x^6 - 43x^3 - 22x + 6)$

2. $(5x^3 - 7x^2 + 8x - 3) \cdot (11x^7 + 24x^5 - 16x^4 + 61x^2 - 20x) \cdot (4x^5 - 10x^3 + 57x^2 - 19x + 7)$

3. $(2x^7 - 6x + 5)^4$

In exercises 4–6, find the degree, leading term, and trailing term of each product in $\mathbb{Z}_5[x]$.

4. $([3]_5 x^{12} + [2]_5 x^4 + [2]_5 x^2 + [4]_5) \cdot ([2]_5 x^7 + x^6 + [3]_5 x^4 + [4]_5 x^2)$

5. $([3]_5 x^5 + x^3 + [4]_5 x^2) \cdot ([4]_5 x^8 + [2]_5 x^5 + [3]_5 x^3 + [3]_5) \cdot ([2]_5 x^4 + [2]_5 x^3 + [3]_5 x^2)$

6. $([2]_5 x^7 + x^3 + [3]_5 x)^{13}$

In exercises 7 and 8, find the degree, leading term, and trailing term of each product in $\mathbb{Z}_8[x]$.

7. $([5]_8 x^7 + [4]_8 x^2) \cdot ([6]_8 x^3 + [3]_8 x^2 + [2]_8 x)$

8. $([2]_8 x + 1)^3$

In exercises 9 and 10, find the degree, leading term, and trailing term of each product in $\mathbb{Z}_{10}[x]$.

9. $([8]_{10}x^3 + [5]_{10}x + [2]_{10}) \cdot ([5]_{10}x^2 + [6]_{10})$

10. $([4]_{10}x + [7]_{10})^5$

In exercises 11–18, let $\sigma : \mathbb{Q}[x] \to \mathbb{Q}$ be the homomorphism of rings defined as $\sigma(p(x)) = p(5)$, for all $p(x) \in \mathbb{Q}[x]$. In addition, let

$$f(x) = x^2 + 1, \quad g(x) = 2x - 7, \quad h(x) = x^2 - 6x + 5.$$

11. Compute $\sigma(f(x))$, $\sigma(g(x))$, and $\sigma(h(x))$.

12. Compute $f(x) + g(x)$, $\sigma(f(x) + g(x))$, and $\sigma(f(x)) + \sigma(g(x))$. Then compare $\sigma(f(x) + g(x))$ and $\sigma(f(x)) + \sigma(g(x))$.

13. Compute $g(x) + h(x)$, $\sigma(g(x) + h(x))$, and $\sigma(g(x)) + \sigma(h(x))$. Then compare $\sigma(g(x) + h(x))$ and $\sigma(g(x)) + \sigma(h(x))$.

14. Compute $f(x) + h(x)$, $\sigma(f(x) + h(x))$, and $\sigma(f(x)) + \sigma(h(x))$. Then compare $\sigma(f(x) + h(x))$ and $\sigma(f(x)) + \sigma(h(x))$.

15. Compute $f(x) \cdot g(x)$, $\sigma(f(x) \cdot g(x))$, and $\sigma(f(x)) \cdot \sigma(g(x))$. Then compare $\sigma(f(x) \cdot g(x))$ and $\sigma(f(x)) \cdot \sigma(g(x))$.

16. Compute $g(x) \cdot h(x)$, $\sigma(g(x) \cdot h(x))$, and $\sigma(g(x)) \cdot \sigma(h(x))$. Then compare $\sigma(g(x) \cdot h(x))$ and $\sigma(g(x)) \cdot \sigma(h(x))$.

17. Compute $f(x) \cdot h(x)$, $\sigma(f(x) \cdot h(x))$, and $\sigma(f(x)) \cdot \sigma(h(x))$. Then compare $\sigma(f(x) \cdot h(x))$ and $\sigma(f(x)) \cdot \sigma(h(x))$.

18. Describe, as simply as possible, the elements of $Ker(\sigma)$ and $Im(\sigma)$.

In exercises 19–26, let $\tau : \mathbb{R}[x] \to \mathbb{C}$ be the homomorphism of rings defined as $\sigma(p(x)) = p(i)$, for all $p(x) \in \mathbb{R}[x]$. In addition, let

$$f(x) = 5x + 4, \quad g(x) = 2x^3 + 3x^2 + 2x + 3, \quad h(x) = 4x^2 - 6x + 7.$$

19. Compute $\tau(f(x))$, $\tau(g(x))$, and $\tau(h(x))$.

20. Compute $f(x) + g(x)$, $\tau(f(x) + g(x))$, and $\tau(f(x)) + \tau(g(x))$. Then compare $\tau(f(x) + g(x))$ and $\tau(f(x)) + \tau(g(x))$.

21. Compute $g(x) + h(x)$, $\tau(g(x) + h(x))$, and $\tau(g(x)) + \tau(h(x))$. Then compare $\tau(g(x) + h(x))$ and $\tau(g(x)) + \tau(h(x))$.

22. Compute $f(x) + h(x)$, $\tau(f(x) + h(x))$, and $\tau(f(x)) + \tau(h(x))$. Then compare $\tau(f(x) + h(x))$ and $\tau(f(x)) + \tau(h(x))$.

23. Compute $f(x) \cdot g(x)$, $\tau(f(x) \cdot g(x))$, and $\tau(f(x)) \cdot \tau(g(x))$. Then compare $\tau(f(x) \cdot g(x))$ and $\tau(f(x)) \cdot \tau(g(x))$.

24. Compute $g(x) \cdot h(x)$, $\tau(g(x) \cdot h(x))$, and $\tau(g(x)) \cdot \tau(h(x))$. Then compare $\tau(g(x) \cdot h(x))$ and $\tau(g(x)) \cdot \tau(h(x))$.

25. Compute $f(x) \cdot h(x)$, $\tau(f(x) \cdot h(x))$, and $\tau(f(x)) \cdot \tau(h(x))$. Then compare $\tau(f(x) \cdot h(x))$ and $\tau(f(x)) \cdot \tau(h(x))$.

26. Describe, as simply as possible, the elements of $Ker(\tau)$ and $Im(\tau)$.

In exercises 27–34, let $\sigma : \mathbb{Z}[x] \to \mathbb{C}$ be the homomorphism of rings defined as $\sigma(p(x)) = p(2+i)$, for all $p(x) \in \mathbb{Q}[x]$. In addition, let

$$f(x) = 6x - 11, \quad g(x) = 3x^2 - 2x + 9, \quad h(x) = x^3 - 3x^2 - x + 5.$$

27. Compute $\sigma(f(x))$, $\sigma(g(x))$, and $\sigma(h(x))$.

28. Compute $f(x) + g(x)$, $\sigma(f(x) + g(x))$, and $\sigma(f(x)) + \sigma(g(x))$. Then compare $\sigma(f(x) + g(x))$ and $\sigma(f(x)) + \sigma(g(x))$.

29. Compute $g(x) + h(x)$, $\sigma(g(x) + h(x))$, and $\sigma(g(x)) + \sigma(h(x))$. Then compare $\sigma(g(x) + h(x))$ and $\sigma(g(x)) + \sigma(h(x))$.

30. Compute $f(x) + h(x)$, $\sigma(f(x) + h(x))$, and $\sigma(f(x)) + \sigma(h(x))$. Then compare $\sigma(f(x) + h(x))$ and $\sigma(f(x)) + \sigma(h(x))$.

31. Compute $f(x) \cdot g(x)$, $\sigma(f(x) \cdot g(x))$, and $\sigma(f(x)) \cdot \sigma(g(x))$. Then compare $\sigma(f(x) \cdot g(x))$ and $\sigma(f(x)) \cdot \sigma(g(x))$.

32. Compute $g(x) \cdot h(x)$, $\sigma(g(x) \cdot h(x))$, and $\sigma(g(x)) \cdot \sigma(h(x))$. Then compare $\sigma(g(x) \cdot h(x))$ and $\sigma(g(x)) \cdot \sigma(h(x))$.

33. Compute $f(x) \cdot h(x)$, $\sigma(f(x) \cdot h(x))$, and $\sigma(f(x)) \cdot \sigma(h(x))$. Then compare $\sigma(f(x) \cdot h(x))$ and $\sigma(f(x)) \cdot \sigma(h(x))$.

34. Describe, as simply as possible, the elements of $Ker(\sigma)$ and $Im(\sigma)$.

35. Let $f : R \to S$ and $g : S \to T$ be homomorphisms of rings. Show that the composition $g \circ f : R \to T$ is also a homomorphism.

36. If F is a field and $F[x]$ is the set of polynomials with coefficients in F, let $\sigma : F[x] \to F[x]$ and $\tau : F[x] \to F[x]$ be defined as $\sigma(p(x)) = p(x+1)$ and $\tau(p(x)) = p(x-1)$, for all $p(x) \in F[x]$.
    (a) Show that both σ and τ are homomorphisms of rings.

    (b) Show that the compositions $\sigma \circ \tau$ and $\tau \circ \sigma$ are both the identity map on $F[x]$.

(c) Show that both σ and τ are automorphisms of $F[x]$.

(d) Show that $f(x) \in F[x]$ is reducible in $F[x]$ if and only if $f(x+1)$ is reducible in $F[x]$.

37. Let R be an integral domain with only a finite number of elements.
    (a) If $a \in R$ and $a \neq 0$, show that the function $\tau : R \to R$ defined as $\tau(r) = ar$, for all $r \in R$, is an injection.

    (b) Show that if $a \in R$ and $a \neq 0$, then a has a multiplicative inverse in R.

    (c) Prove that R must be a field.

The remaining exercises in this section will deal with various properties of rings. In these exercises, unless stated otherwise, we will not be assuming that multiplication is commutative. To avoid confusion, when dealing with rings R, S, we will let $0_R, 0_S$ denote, respectively, the additive identities of R and S. Similarly, we will let $1_R, 1_S$ denote, respectively, the multiplicative identities of R and S.

38. Let r be an element of an integral domain R such that $r^2 = r$. Show that either $r = 0_R$ or $r = 1_R$.

39. Let $\tau : R \to S$ be a homomorphism of rings. If S is an integral domain, show that either $\tau(1_R) = 0_S$ or $\tau(1_R) = 1_S$.

40. Let r be an element of an integral domain R such that $r^3 = r$. Show that either $r = 0_R$, $r = 1_R$, or $r = -1_R$, where -1_R is the additive inverse of 1_R.

41. Let $\tau : R \to S$ be a homomorphism of rings. Show that $\tau(0_R) = 0_S$.

42. Let $\tau : R \to S$ be a homomorphism of rings which is also a bijection. Show that R is an integral domain if and only if S is an integral domain.

43. If $\sigma : R \to S$ is a homomorphism of rings and $a \in Ker(\sigma)$, show that $ar, ra \in Ker(\sigma)$, for all $r \in R$.

44. Let $\tau : R \to S$ be a homomorphism of rings which is also an injection. Show that if S is a commutative ring then R is also a commutative ring.

45. Let $\sigma : R \to S$ be a homomorphism of rings which is also a surjection. Show that if R is a commutative ring then S is also a commutative ring.

46. Let $\sigma : \mathbb{R}[x] \to \mathbb{R}[x]$ be the homomorphism of rings where $\sigma(f(x)) = f(2x - 1)$, for all $f(x) \in \mathbb{R}[x]$.
    (a) Find $\sigma(x^2)$.

    (b) Find $\sigma(3x^2 - 4x + 1)$.

(c)   Find $g(x) \in \mathbb{R}[x]$ such that $\sigma(g(x)) = x$.

(d)   Explain why σ^{-1} exists and then find the formula for $\sigma^{-1}(f(x))$, for all $f(x) \in \mathbb{R}[x]$.

47.   Let $\sigma : \mathbb{R}[x] \to \mathbb{R}[x]$ be the homomorphism of rings where $\sigma(f(x)) = f(x^4)$, for all $f(x) \in \mathbb{R}[x]$.
   (a)   Is σ injective? Explain your answer.
   (b)   Is σ is surjective? Explain your answer.

48   Let $\tau : \mathbb{Z} \to \mathbb{Z}$ be a homomorphism of rings.
   (a)   Show that either $\tau(1) = 0$ or $\tau(1) = 1$.
   (b)   Conclude that either $\tau(n) = 0$, for all $n \in \mathbb{Z}$, or τ is the identity map.

For exercises 49–50, please read the following:

A subset I of a ring R is called an **ideal** if I is a subgroup of R under addition and whenever $a \in I$ and $r \in R$, we have $ar, ra \in I$. Observe that exercise 43 along with the fact that the kernel of every group homomorphism is a subgroup combine to show us the kernel of every ring homomorphism is an ideal.

49.   If $n \in \mathbb{N}$, let $n\mathbb{Z} = \{n \cdot t \mid t \in \mathbb{Z}\}$. Show that the set $n\mathbb{Z}$ is an ideal of $\mathbb{Z}$.

50.   Suppose I is an ideal of the ring $\mathbb{Z}$ such that I consists of more than the single element 0.
   (a)   Show that I must contain a positive integer.
   (b)   Show that I contains a smallest positive integer.
   (c)   If n is the smallest positive integer in I, show that any $a \in I$ must be a multiple of n. (Hint: Apply the division algorithm and divide a by n and then examine the remainder.)
   (d)   Conclude that I is equal to the set $n\mathbb{Z} = \{n \cdot t \mid t \in \mathbb{Z}\}$.
   (e)   Use this exercise and the previous exercise to describe all ideals of the ring $\mathbb{Z}$.

51.   Let I be an ideal of a ring R. Show that $I = R$ if and only if $1_R \in I$.

52.   Generalize exercise 49 and show that if R is a commutative ring and $a \in R$ then the set $aR = \{a \cdot t \mid t \in R\}$ is an ideal of R.

53.   Let R be a commutative ring. Show that R is a field if and only of the sets $\{0_R\}$ and R are the only ideals of R.

54.   Let $\sigma : \mathbb{Z}_p[x] \to \mathbb{Z}_p[x]$ be defined as $\sigma(f(x)) = f(x)^p$, for all $f(x) \in \mathbb{Z}_p[x]$.
   (a)   Show that σ is a homomorphism of rings.
   (b)   Contrast σ to the function $\tau : \mathbb{Z}_p[x] \to \mathbb{Z}_p[x]$ defined as $\tau(f(x)) = f(x^p)$, for all $f(x) \in \mathbb{Z}_p[x]$.

9.2 Rational Root Test and Irreducible Polynomials

We now have the tools needed to examine the roots and factoring of polynomials in $\mathbb{Z}[x]$ and $\mathbb{Q}[x]$. First, we will deal with the question of whether there is an algorithm for finding all the rational roots of a polynomial with rational coefficients. We will show that there is an algorithm such that for any $f(x) \in \mathbb{Q}[x]$, we can write down a short list of rational numbers that are **candidates** for being roots of $f(x)$. Usually, not all of our candidates will be roots of $f(x)$. In fact, sometimes none of our candidates will be roots of $f(x)$. However, the important property of our list of candidates will be that it is guaranteed to contain **all** the rational roots of $f(x)$. Therefore, if we simply plug each of our candidates into $f(x)$ and see which ones give 0 as an answer, we will have succeeded in finding all the rational roots of $f(x)$.

Before stating how we come up with our list of candidates, we need to make an observation that will simplify the problem. If $f(x) \in \mathbb{Q}[x]$, then obviously all the coefficients of $f(x)$ are rational numbers. Therefore, there exists a positive integer M such that the polynomial $M \cdot f(x)$ has only integer coefficients. Furthermore, $M \cdot f(x)$ has the same roots as $f(x)$. Thus, whenever we are looking for the roots of a polynomial in $\mathbb{Q}[x]$, we can always simplify the problem by looking at an appropriate polynomial in $\mathbb{Z}[x]$.

For example, if we want to find the roots of

$$f(x) = \frac{1}{6}x^3 - \frac{1}{3}x^2 - \frac{1}{3}x - \frac{1}{2},$$

we can multiply $f(x)$ by 6 to obtain the polynomial

$$F(x) = x^3 - 2x^3 - 2x - 3.$$

To find the roots of $f(x)$, it suffices to find the roots of $F(x)$. In light of this, we can state our result in terms of polynomials in $\mathbb{Z}[x]$.

Theorem 9.6—The Rational Root Test. *Let*

$$f(x) = a_n x^n + a_{n-1} x^{n-1} + \cdots + a_1 x + a_0 \in \mathbb{Z}[x],$$

where $a_n, a_0 \neq 0$. If a rational number is a root of $f(x)$ and is written in lowest terms as $\frac{r}{s}$, then $r \mid a_0$ and $s \mid a_n$.

Before proving Theorem 9.6, we will look at several examples.

■ Examples

1. Let us return to the polynomial $F(x) = x^3 - 2x^2 - 2x - 3$. The Rational Root Test asserts that if $\frac{r}{s}$ is a rational root of $F(x)$ in lowest terms, then $r \mid -3$ and $s \mid 1$. Therefore, the only possibilities for r are $\pm 1, \pm 3$, and the only possibilities for s are

± 1. As a result, the only possibilities for $\frac{r}{s}$ are $\pm 1, \pm 3$. Having produced a list of four candidates, we now plug each candidate into $F(x)$ and observe that $F(1) = -6$, $F(-1) = -4$, $F(3) = 0$, and $F(-3) = -42$. Therefore, 3 is the only rational root of $x^3 - 2x^2 - 2x - 3$.

2. Let $g(x) = \frac{3}{2}x^3 - \frac{7}{5}$; in order to apply the Rational Root Test, we multiply $g(x)$ by 10 and turn our attention to the polynomial

$$G(x) = 10g(x) = 15x^3 - 14.$$

The test now asserts that if $\frac{r}{s}$ is a rational root of $G(x)$ in lowest terms, then $r| - 14$ and $s|15$. Now the only possibilities for r are $\pm 1, \pm 2, \pm 7, \pm 14$ and the only possibilities for s are $\pm 1, \pm 3, \pm 5, \pm 15$. As a result, the only possibilities for $\frac{r}{s}$ are

$$\pm 1, \pm 2, \pm 7, \pm 14, \pm \frac{1}{3}, \pm \frac{2}{3}, \pm \frac{7}{3}, \pm \frac{14}{3},$$

$$\pm \frac{1}{5}, \pm \frac{2}{5}, \pm \frac{7}{5}, \pm \frac{14}{5}, \pm \frac{1}{15}, \pm \frac{2}{15}, \pm \frac{7}{15}, \pm \frac{14}{15}.$$

If $G(x)$ has any rational roots, they must be on our list of 32 candidates. But none of these candidates give 0 as an answer when plugged into $G(x)$. Therefore, we can conclude that $G(x)$ has no rational roots.

Note that the Intermediate Value Theorem tells us that $G(x)$ does indeed have a real root. In fact, it is not hard to see that $\left(\frac{14}{15}\right)^{\frac{1}{3}}$ is the real root of this polynomial. Therefore, the Rational Root Test also tells us that $\left(\frac{14}{15}\right)^{\frac{1}{3}}$ is irrational.

3. In light of the previous example, it is not surprising that the Rational Root Test can be used to show that many real numbers are irrational. For example, if we look at $x^2 - 2$, then the test says that the only possible rational roots of $x^2 - 2$ are $\pm 1, \pm 2$. Since none of these four candidates are roots of $x^2 - 2$, we know that $x^2 - 2$ has no rational roots. But since $\sqrt{2}$ is a root of $x^2 - 2$, we have yet another proof that $\sqrt{2}$ is irrational. Similarly, the Rational Root Test says that the only possible rational roots of $x^5 - 3$ are $\pm 1, \pm 3$. But since none of these four candidates are roots of $x^5 - 3$, and since $3^{\frac{1}{5}}$ is a root of $x^5 - 3$, we see that $3^{\frac{1}{5}}$ is also irrational.

4. Let $h(x) = 2x^4 - 7x^3 - 18x^2 - 20x - 11$. The Rational Root Test tells us that if $\frac{r}{s}$ is a rational root in lowest terms, then $r| - 11$ and $s|2$. Therefore, the only possible values of r are $\pm 1, \pm 11$, and the only possible values of s are $\pm 1, \pm 2$. As a result, our list of candidates for $\frac{r}{s}$ is $\pm 1, \pm 11, \pm \frac{1}{2}, \pm \frac{11}{2}$. If we plug each of these eight candidates into $h(x)$, we will find that -1 and $\frac{11}{2}$ are the only rational roots of $2x^4 - 7x^3 - 18x^2 - 20x - 11$.

5. Let $w(t) = t^5 + 3t^4 - 4t^3 - 12t^2$; the Rational Root Test discusses polynomials with nonzero constant term. However, if the constant term is not zero, we can factor out as many powers of t as necessary to obtain a new polynomial where the constant term is zero. For example, $w(t) = t^2(t^3 + 3t^2 - 4t - 12)$. Applying the test to $t^3 + 3t^2 - 4t - 12$, the only possible rational roots of $t^3 + 3t^2 - 4t - 12$ are $\pm 1, \pm 2, \pm 3, \pm 4, \pm 6, \pm 12$. If we check each of these 12 candidates, we will see that $-3, -2, 2$ are the only rational roots of $t^3 + 3t^2 - 4t - 12$. Since t^2 is also a factor of $w(t)$, we know that 0 is also a root of $w(t)$. Therefore, $-3, -2, 0, 2$ are all the rational roots of $w(t)$.

■

Proof of Theorem 9.6. Suppose $\frac{r}{s}$ is a rational root of $f(x)$, which is in lowest terms. Plugging $\frac{r}{s}$ into $f(x)$ yields

$$0 = a_n \left(\frac{r}{s}\right)^n + a_{n-1} \left(\frac{r}{s}\right)^{n-1} + \cdots + a_1 \left(\frac{r}{s}\right) + a_0.$$

Multiplying this equation by s^n gives us

(2) $$0 = a_n r^n + a_{n-1} r^{n-1} s + \cdots + a_1 r s^{n-1} + a_0 s^n.$$

At various points in the proof, it might be helpful to assume that $r > 0$ or that $s > 0$. Since $\frac{r}{s} = \frac{-r}{-s}$, this will not be a problem. Furthermore, if r or s is equal to 1, then it is certainly a divisor of every integer. Therefore, when attempting to show that $r|a_0$, we may assume that $r \geq 2$. Similarly, when attempting to show that $s|a_n$, we may assume that $s \geq 2$.

Since we may assume that $r \geq 2$, let σ_r be the homomorphism of rings from $\mathbb{Z}$ to $\mathbb{Z}_r$ where $\sigma_r(a) = [a]_r$, for all $a \in \mathbb{Z}$. Since r and s are relatively prime, Theorem 7.8 tells us that $[s]_r$ is invertible in $\mathbb{Z}_r$. Also remember that if $b \in \mathbb{Z}$, then $[b]_r = [0]_r$ if and only if b is a multiple of r. Now, if we plug both sides of (2) into the function σ_r and then use Lemma 9.5, we obtain

$$[0]_r = \sigma_r(0) = \sigma_r(a_n r^n + a_{n-1} r^{n-1} s + \cdots + a_1 r s^{n-1} + a_0 s^n) =$$
$$\sigma_r(a_n r^n) + \sigma_r(a_{n-1} r^{n-1} s) + \cdots + \sigma_r(a_1 r s^{n-1}) + \sigma_r(a_0 s^n) =$$
$$[0]_r + [0]_r + \cdots [0]_r + [a_0]_r \cdot [s]_r^n.$$

Thus,

$$[a_0]_r \cdot [s]_r^n = [0]_r$$

in $\mathbb{Z}_r$. However, since $[s]_r$ is invertible in $\mathbb{Z}_r$, Proposition 7.7 tells us that $[s]_r^n$ is also invertible in $\mathbb{Z}_r$. Multiplying our last equation by the inverse of $[s]_r^n$ now tells us that $[a_0]_r = [0]_r$. As a result, a_0 is a multiple of r, so $r|a_0$, as desired.

The proof that $s|a_n$ is very similar to the preceding argument. In this case, we may assume that $s \geq 2$, and we then let $\sigma_s : \mathbb{Z} \to \mathbb{Z}_s$ be the homomorphism where $\sigma_s(a) = [a]_s$, for all $a \in \mathbb{Z}$. Analogous to the computation in the preceding paragraph, if we apply σ_s to both sides of (2), we obtain

$$[a_n]_s \cdot [r]_s^n = [0]_s.$$

Since r and s are relatively prime, $[r]_s$ is invertible in $\mathbb{Z}_s$. Therefore, $[r]_s^n$ is also invertible in $\mathbb{Z}_s$, and if we multiply the previous equation by the inverse of $[r]_s^n$, we obtain $[a_n]_s = [0]_s$. Thus, a_n is a multiple of s and $s|a_n$. $\quad\square$

If we apply the Rational Root Test to a polynomial with leading coefficient 1 and nonzero constant term, the test says that any rational root must be of the form $\frac{r}{s}$, where s is a divisor of 1 and r divides the constant term. But this means that s must be ±1, which immediately tells us that $\frac{r}{s}$ is an integer that divides the constant term. We can use this observation to prove

Corollary 9.7. *If $f(x) \in \mathbb{Z}[x]$ has leading coefficient 1, then any nonzero rational root of $f(x)$ must be an integer which divides the trailing coefficient.*

Proof. By factoring out the largest power of x that is a common factor of all the terms of $f(x)$, we can write $f(x) = x^m g(x)$, where $m \geq 0$ and the constant term of $g(x)$ is nonzero. Therefore, the roots of $g(x)$ are precisely the nonzero roots of $f(x)$. Our previous observation says that any rational root of $g(x)$ must be an integer that divides the trailing coefficient of $g(x)$. However, $f(x)$ and $g(x)$ have the same leading and trailing coefficients. Thus, any nonzero rational root of $f(x)$ must be an integer that divides the trailing coefficient of $f(x)$. $\quad\square$

■ Examples

1. Let $f(x) = x^6 + 17x^5 + 53x^4 + 13x^3$; then Corollary 9.7 says that the only possible nonzero rational roots of $f(x)$ are divisors of 13. As a result, $\pm1, \pm13$ are the only four candidates for nonzero rational roots. However, $f(1) = 84$, $f(-1) = 24$, $f(13) = 12{,}681{,}084$, and $f(-13) = 0$. Thus, 0 and -13 are the only rational roots of $f(x)$.

2. Let $h(x) = x^5 + 9x^3 + 2$; Corollary 9.7 asserts that the only possible rational roots of $h(x)$ are divisors of 2. Therefore, $\pm1, \pm2$ are the only candidates. However, $h(1) = 12$, $h(-1) = -8$, $h(2) = 106$, and $h(-2) = -102$. As a result, $h(x)$ has no rational roots. On the other hand, the Intermediate Value Theorem tells us that $h(x)$ has a real root between -1 and 0. Furthermore, since the derivative of $h(x)$ is $5x^4 + 27x^2$, which is never negative, we know that $h(x)$ is always increasing. Thus,

$h(x)$ cannot have more than one real root. In summary, we know that $h(x)$ has exactly one real root, and this root is an irrational number that lies between -1 and 0.

■

We can now begin to address the question of which polynomials in $\mathbb{Z}[x]$ and $\mathbb{Q}[x]$ can be written as a product of polynomials of smaller degree. To do so, we need to introduce some terminology that will then be followed by a large number of examples.

Definition 9.8. *Let $f(x)$ be a polynomial of degree at least one with coefficients in an integral domain R. We say that $f(x)$ is reducible in $R[x]$ if there exist $g(x), h(x) \in R[x]$, both with smaller degree than $f(x)$, such that $f(x) = g(x) \cdot h(x)$. If $f(x) \in R[x]$ has degree at least one and is not reducible in $R[x]$, then we say that $f(x)$ is irreducible in $R[x]$.*

It is important to be aware that whether $f(x) \in R[x]$ is reducible or irreducible depends not only on $f(x)$ but also on R. This will be illustrated in the following examples. In these examples we will make frequent use of the fact, which we will prove in Chapter 12, that if F is a field and if $\alpha \in F$ is a root of $f(x) \in F[x]$, then there is some $g(x) \in F[x]$ such that $f(x) = (x - \alpha)g(x)$. Observe that if, in addition, $f(x)$ has degree at least 2, then both $x - \alpha$ and $g(x)$ have smaller degree than $f(x)$. As a result, we now know that if F is a field, then **any** $f(x) \in F[x]$ **of degree at least 2 with a root in F, must be reducible in** $F[x]$.

1. **Polynomials of Degree 1 with Coefficients in a Field**
   All polynomials of degree 1 are irreducible. To see this, if we could factor $f(x) = g(x) \cdot h(x)$, where $f(x)$ has degree 1, then

$$1 = deg(f(x)) = deg(g(x) \cdot h(x)) = deg(g(x)) + deg(h(x)).$$

   Therefore, either $g(x)$ or $h(x)$ would have degree 1. But this means it is impossible to write $f(x)$ as a product of two polynomials, both of which have smaller degree than $f(x)$. Thus, $f(x)$ is irreducible.

2. **Polynomials of Degree 2 with Coefficients in a Field**
   We already know that if $f(x) \in F[x]$ is of degree 2 and has a root in F, then $f(x)$ is reducible in $F[x]$. Conversely, if $f(x)$ is reducible in $F[x]$ there exist $g(x), h(x) \in F[x]$, both of which have smaller degree than $f(x)$, such that $f(x) = g(x) \cdot h(x)$. Since $deg(f(x)) = 2$, it follows that both $g(x)$ and $h(x)$ must have degree 1. As a result, $g(x) = ax + b$, where $a, b \in F$ and $a \neq 0$. However, since F is a field, this tells us that $-ba^{-1}$ belongs to F and is a root of $g(x)$. Therefore,

$$f(-ba^{-1}) = g(-ba^{-1}) \cdot h(-ba^{-1}) = 0 \cdot h(-ba^{-1}) = 0.$$

   Hence, $f(x)$ does have a root in F. As a result, **if $f(x) \in F[x]$ has degree 2, then $f(x)$ is irreducible in $F[x]$ if and only if it has no roots in F.**

Now let us look at several examples of polynomials of degree 2.

(a) $f(x) = x^2 - 7x - 18$; this polynomial is reducible in $\mathbb{Z}[x]$ as $f(x) = (x-9)(x+2)$. Since $\mathbb{Q}$, $\mathbb{R}$, and $\mathbb{C}$ are all larger than $\mathbb{Z}$, this polynomial is also reducible in $\mathbb{Q}[x]$, $\mathbb{R}[x]$, and $\mathbb{C}[x]$.

(b) $g(x) = x^2 - 2$; this polynomial is irreducible in $\mathbb{Q}[x]$ as $g(x)$ has no roots in $\mathbb{Q}$. Since $\mathbb{Q}$ is larger than $\mathbb{Z}$, $g(x)$ is also irreducible in $\mathbb{Z}[x]$. However, $g(x)$ does have roots in $\mathbb{R}$. Therefore, $g(x)$ is reducible in $\mathbb{R}$ and we have

$$g(x) = (x - \sqrt{2})(x + \sqrt{2}).$$

Since $\mathbb{C}$ is larger than $\mathbb{R}$, $g(x)$ is also reducible in $\mathbb{C}[x]$.
If we consider the field $\mathbb{Q}(\sqrt{2}) = \{a + b\sqrt{2} \mid a, b \in \mathbb{Q}\}$, then $g(x)$ also has roots in this field. Therefore, $g(x)$ is also reducible in $\mathbb{Q}(\sqrt{2})[x]$.

(c) $h(x) = x^2 + 1$; this polynomial is irreducible in $\mathbb{R}[x]$ as it has no real roots. Since $\mathbb{Z}$ and $\mathbb{Q}$ are smaller than $\mathbb{R}$, $h(x)$ is also irreducible in $\mathbb{Z}[x]$ and $\mathbb{Q}[x]$. However, $h(x)$ has roots in $\mathbb{C}$, so $h(x)$ is reducible in $\mathbb{C}[x]$ and we have

$$h(x) = (x - i)(x + i).$$

(d) $k(x) = 2x^2 - 7x + 1$; this polynomial is irreducible in $\mathbb{Q}[x]$, since, by the Rational Root Test, it has no rational roots. Observe that the only candidates for rational roots are $\pm 1, \pm \frac{1}{2}$, and none of these make $k(x)$ equal to zero. Since $\mathbb{Z}$ is smaller than $\mathbb{Q}$, $k(x)$ is also irreducible in $\mathbb{Z}[x]$.

However, $k(0) = 1$ and $k(1) = -4$, so the Intermediate Value Theorem tells us that $k(x)$ has a real root between 0 and 1. Therefore, $k(x)$ is reducible in $\mathbb{R}[x]$. Since $\mathbb{C}$ is larger than $\mathbb{R}$, we also know that $k(x)$ is reducible in $\mathbb{C}[x]$.

(e) $v(x) = x^2 + [2]_5 x + [2]_5$; this polynomial is reducible in $\mathbb{Z}_5[x]$, since it has roots in $\mathbb{Z}_5$. You should check that $[1]_5$ and $[2]_5$ are indeed roots of $v(x)$. Therefore, we now have

$$v(x) = x^2 + [2]_5 x + [2]_5 = (x - [1]_5)(x - [2]_5) = (x + [4]_5)(x + [3]_5).$$

(f) $w(x) = x^2 + x + [1]_2$; this polynomial is irreducible in $\mathbb{Z}_2[x]$, since it has no roots in $\mathbb{Z}_2$. To see this, observe that

$$w([0]_2) = [0]_2^2 + [0]_2 + [1]_2 = [1]_2 \text{ and } w([1]_2) = [1]_2^2 + [1]_2 + [1]_2 = [3]_2 = [1]_2.$$

3. **Polynomials with Coefficients in $\mathbb{R}$ or $\mathbb{C}$**

By the Fundamental Theorem of Algebra, every polynomial of degree at least 1 in $\mathbb{C}[x]$ has a root in $\mathbb{C}$. Since polynomials of degree at least 2 with a root in a field are reducible, it follows that **the only irreducible polynomials in $\mathbb{C}[x]$ are those of degree** 1.

Moving to $\mathbb{R}[x]$, we know that whereas there are no polynomials of degree 2 that are irreducible in $\mathbb{C}[x]$, there are certainly polynomials of degree 2 that are irreducible in $\mathbb{R}[x]$. In fact, the quadratic formula tells us that if

$$f(x) = ax^2 + bx + c \in \mathbb{R}[x],$$

then $f(x)$ has a root in $\mathbb{R}$ if and only if $\sqrt{b^2 - 4ac} \geq 0$. Therefore, $f(x)$ is reducible in $\mathbb{R}[x]$ if and only if $\sqrt{b^2 - 4ac} \geq 0$. As a result, it is easy to determine which polynomials of degree 2 are reducible in $\mathbb{R}[x]$.

Next, we turn our attention to polynomials of degree 3 or more. Recall that the Intermediate Value Theorem told us that any polynomial of odd degree in $\mathbb{R}[x]$ has a root in $\mathbb{R}$. Therefore, every polynomial in $\mathbb{R}[x]$ whose degree is odd and exceeds 1 is reducible in $\mathbb{R}[x]$.

As a result, we now need to examine polynomials in $\mathbb{R}[x]$ whose degree are even and exceed 2. Consider the polynomial

$$f(x) = x^4 + 4x^2 + 3.$$

Note that $f(x)$ has no roots in $\mathbb{R}$, but we can factor it in $\mathbb{R}[x]$ as

$$f(x) = x^4 + 4x^2 + 3 = (x^2 + 1)(x^2 + 3).$$

Observe that the roots of $f(x)$ in $\mathbb{C}$ are $\pm i, \pm i\sqrt{3}$. Therefore, there certainly exist polynomials of degree at least 4 with no roots in $\mathbb{R}[x]$ that are reducible in $\mathbb{R}[x]$.

More generally, let $f(x) \in \mathbb{R}[x]$ have degree at least 4. If $f(x)$ has a real root, then it is certainly reducible in $\mathbb{R}[x]$. So let us consider the case where $f(x)$ has no real roots. The Fundamental Theorem of Algebra tells us that there is some $\alpha \in \mathbb{C}$ such that α is a root of $f(x)$. Since $f(x)$ has no real roots, $\alpha \notin \mathbb{R}$. Therefore, α^*, the complex conjugate of α, is not equal to α. Next, consider the polynomial

$$g(x) = (x - \alpha)(x + \alpha^*) = x^2 - (\alpha + \alpha^*) + \alpha \cdot \alpha^*.$$

By Lemma 5.10(b), $\alpha + \alpha^*$ and $\alpha \cdot \alpha^*$ belong to $\mathbb{R}$, so $g(x) \in \mathbb{R}[x]$. In Chapter 12, we will show that we can write $f(x) = g(x) \cdot h(x)$, where $h(x) \in \mathbb{R}[x]$. Using this fact, observe that since $deg(f(x)) \geq 4$, then $h(x)$ must also have degree at least 2. Therefore, $f(x)$ has been written as a product of two polynomials in $\mathbb{R}[x]$ of smaller degree. Thus, $f(x)$ is reducible in $\mathbb{R}[x]$. As a result, **every polynomial of degree at least 3 in $\mathbb{R}[x]$ is reducible in $\mathbb{R}[x]$**.

4. **Polynomials of Degree 3 with Coefficients in a Field**

   The situation for polynomials of degree 3 is virtually the same as for those of degree 2. We know that if $f(x) \in F[x]$ is of degree 3 and has a root in F, then $f(x)$ is reducible in F. Conversely, if $f(x) = g(x) \cdot h(x)$, where $g(x), h(x)$ both have smaller degree than $f(x)$, then either $g(x)$ or $h(x)$ must have degree 1. Therefore, either $g(x)$ or $h(x)$ must have a root in F, which immediately tells us that $f(x)$ has a root in F. As a result, **if $f(x) \in F[x]$ has degree 3, then $f(x)$ is irreducible if and only if it has no roots in F.**

Given a polynomial in $\mathbb{Q}[x]$, we can use the Rational Root Test to determine if it has any rational roots. Therefore, in $\mathbb{Q}[x]$, we now a have straightforward procedure to determine when polynomials of degree 3 are reducible. We now consider several examples.

(a) $f(x) = 5x^3 + 3x^2 - 35x - 21$; by the Rational Root Test, the 16 candidates for rational roots of $f(x)$ are $\pm 1, \pm 3, \pm 7, \pm 21, \pm\frac{1}{5}, \pm\frac{3}{5}, \pm\frac{7}{5}, \pm\frac{21}{5}$. You can check that $-\frac{3}{5}$ is a root of $f(x)$, so $f(x)$ is reducible in $\mathbb{Q}[x]$. In fact,

$$f(x) = 5x^3 + 3x^2 - 35x - 21 = (5x + 3)(x^2 - 7)$$

in $\mathbb{Q}[x]$.

(b) $g(x) = 7x^3 - 8x^2 + 5x + 2$; by the Rational Root Test the eight candidates for rational roots of $g(x)$ are $\pm 1, \pm 2, \pm\frac{1}{7}, \pm\frac{2}{7}$. None of these candidates make $g(x)$ equal to zero, so $g(x)$ is irreducible in $\mathbb{Q}[x]$.

(c) $h(x) = x^3 + x + [1]_2$; this polynomial is irreducible in $\mathbb{Z}_2[x]$, since it has no roots in $\mathbb{Z}_2$. To see that neither $[0]_2$ nor $[1]_2$ are roots, observe that

$$h([0]_2) = [0]_2^3 + [0]_2 + [1]_2 = [1]_2 \quad \text{and} \quad h([1]_2) = [1]_2^3 + [1]_2 + [1]_2 = [3]_2 = [1]_2.$$

(d) $j(x) = [2]_7 x^3 + x + [1]_7$; this polynomial has the root $[4]_7 \in \mathbb{Z}_7$. Observe that

$$j([4]_7) = [2]_7 \cdot [4]_7^3 + [4]_7 + [1]_7 = [128]_7 + [4]_7 + [1]_7 =$$
$$[2]_7 + [4]_7 + [1]_7 = [7]_7 = [0]_7.$$

Therefore, $x - [4]_7 = x + [3]_7$ is a factor of $j(x)$, and we have

$$j(x) = [2]_7 x^3 + x + [1]_7 = (x + [3]_7)([2]_7 x^2 + x + [5]_7)$$

in $\mathbb{Z}_7[x]$.

5.   **Polynomials of Degree at Least 4 with Coefficients in a Field**
The situation for polynomials of degree 4 and the preceding is different from those of degree 2 or 3. Certainly, if $f(x) \in F[x]$ has a root in F, then $f(x)$ is reducible in $F[x]$. However, **a polynomial of degree 4 or more can fail to have a root in F yet still be reducible in $F[x]$.**

(a) $f(x) = x^4 + x^3 + 6x^2 + 5x + 1$; the Rational Root Test shows that this polynomial has no roots in $\mathbb{Q}$. The only candidates are ± 1 but $f(1) = 14$ and $f(-1) = 2$. On the other hand, $f(x)$ is reducible in $\mathbb{Q}[x]$ as

$$f(x) = x^4 + x^3 + 6x^2 + 5x + 1 = (x^2 + 5)(x^2 + x + 1).$$

Note that neither of the factors of $f(x)$ can have a root in $\mathbb{Q}$, since any root of a factor of $f(x)$ would automatically be a root of $f(x)$.

(b) $g(x) = x^4 + x^2 + [1]_2$; this polynomial has no roots in $\mathbb{Z}_2$ as

$$g([0]_2) = [0]_2^4 + [0]_2^2 + [1]_2 = [1]_2 \quad \text{and} \quad g([1]_2) = [1]_2^4 + [1]_2^2 + [1]_2 = [3]_2 = [1]_2.$$

However, $g(x)$ is reducible in $\mathbb{Z}_2[x]$ as

$$g(x) = x^4 + x^2 + [1]_2 = (x^2 + x + [1]_2)(x^2 + x + [1]_2) = (x^2 + x + [1]_2)^2.$$

Before continuing on, we should mention an easy but important point that is worth keeping in mind. Suppose E, F, L are fields with $E \subseteq F \subseteq L$. If $f(x) \in F[x]$ is reducible in $F[x]$, then $f(x)$ is certainly reducible in $L[x]$. Similarly, if $f(x) \in F[x]$ is irreducible in $F[x]$ and if $f(x) \in E[x]$, then $f(x)$ is also irreducible in $E[x]$.

As indicated in our examples, there is no difficulty in determining when polynomials in $\mathbb{R}[x]$ and $\mathbb{C}[x]$ are reducible. In addition, we have a straightforward procedure for determining when polynomials of degree at most 3 in $\mathbb{Q}[x]$ are reducible. However, in $\mathbb{Q}[x]$ the situation for polynomials of degree at least 4 can become quite difficult. Although the Rational Root Test provides us with an algorithm for finding rational roots, it cannot, by itself, be used to determine whether a polynomial of degree at least 4 is reducible.

Exercises for Section 9.2

In exercises 1–14, find all rational roots of the given polynomial.

1. $2x^3 + 5x^2 - x - 1$

2. $3x^3 - 19x^2 + 34x - 14$

3. $4x^6 - 11$

4. $x^7 - 5x^6 + 2x - 10$

5. $x^4 + 10x^3 + 35x^2 + 50x + 24$

6. $x^3 + x^2 - 7x - 15$

7. $x^4 + 7x^3 + 2x + 14$

8. $x^3 + 4x + 3$

9. $x^4 - 8x^2 - 9$

10. $6x^3 + x^2 - 18x + 8$

11. $3x^4 - 6x^3 + 5x - 10$

12. $6x^3 - 8x^2 + 5$

13. $x^3 + 2x^2 - 5x - 6$

14. $48x^3 - 74x^2 - 17x + 30$

15. Use the rational root test to prove that $7^{\frac{1}{3}}$ is irrational.

16. Use the rational root test to prove that $11^{\frac{1}{4}}$ is irrational.

17. Let $g(x) = x^4 - 16x^2 + 4$.
    (a) Verify that $\sqrt{3} + \sqrt{5}$ is a root of $g(x)$.

    (b) Use the rational root test to show that $g(x)$ has no rational roots.

    (c) Use parts (a) and (b) to show that $\sqrt{3} + \sqrt{5}$ is irrational.

    (d) Why do you now immediately know that $\sqrt{3} - \sqrt{5}$, $-\sqrt{3} + \sqrt{5}$, and $-\sqrt{3} - \sqrt{5}$ are also irrational?

18. Let $h(x) = x^4 - 14x^2 + 9$.
    (a) Verify that $\sqrt{2} + \sqrt{5}$ is a root of $h(x)$.

    (b) Use the rational root test to show that $h(x)$ has no rational roots.

    (c) Use parts (a) and (b) to show that $\sqrt{2} + \sqrt{5}$ is irrational.

    (d) Why do you now immediately know that $\sqrt{2} - \sqrt{5}$, $-\sqrt{2} + \sqrt{5}$, and $-\sqrt{2} - \sqrt{5}$ are also irrational?

19. (a) Find a monic polynomial in $\mathbb{Z}[x]$ of degree 4 which has $\sqrt{2} + \sqrt{3}$ as a root.

    (b) Use the rational root test to show that your answer to part (a) has no rational roots.

    (c) Use parts (a) and (b) to show that $\sqrt{2} + \sqrt{3}$ is irrational.

20. (a) Find a monic polynomial in $\mathbb{Z}[x]$ of degree 4 that has $2\sqrt{2} - \sqrt{3}$ as a root.

    (b) Use the rational root test to show that your answer to part (a) has no rational roots.

    (c) Use parts (a) and (b) to show that $2\sqrt{2} - \sqrt{3}$ is irrational.

21. List all monic irreducible polynomials of degree 2 in $\mathbb{Z}_2[x]$.

22. List all monic irreducible polynomials of degree 3 in $\mathbb{Z}_2[x]$.

23. List all monic irreducible polynomials of degree 4 in $\mathbb{Z}_2[x]$.

24. List all monic irreducible polynomials of degree 2 in $\mathbb{Z}_3[x]$.

25. List all monic irreducible polynomials of degree 3 in $\mathbb{Z}_3[x]$.

26. List all monic irreducible polynomials of degree 2 in $\mathbb{Z}_5[x]$.

27. Find all integers A such that $x^2 - Ax + 1$ is reducible in $\mathbb{Q}[x]$.

28. Find all integers A such that $x^2 - Ax - 1$ is reducible in $\mathbb{Q}[x]$.

29. Find all integers A such that $x^2 - Ax + 2$ is reducible in $\mathbb{Q}[x]$.

30. Find all integers A such that $x^2 - Ax - 2$ is reducible in $\mathbb{Q}[x]$.

31. If p is a prime number, find all integers A such that $x^2 - Ax + p$ is reducible in $\mathbb{Q}[x]$.

32. If p is a prime number, find all integers A such that $x^2 - Ax - p$ is reducible in $\mathbb{Q}[x]$.

33. Find all real numbers k such that $x^2 + kx + k$ is reducible in $\mathbb{R}[x]$.

34. Find all real numbers k such that $x^2 + kx - 5$ is reducible in $\mathbb{R}[x]$.

35. Find all real numbers k such that $x^2 + kx + 5$ is reducible in $\mathbb{R}[x]$.

36. Find all real numbers k such that $x^2 + kx + k^2$ is reducible in $\mathbb{R}[x]$.

37. Write $x^4 + 1$ as a product of two monic quadratic polynomials in $\mathbb{R}[x]$.

38. Write $x^4 + 2$ as a product of two monic quadratic polynomials in $\mathbb{R}[x]$.

39. Write $x^8 - 1$ as a product of monic irreducible polynomials in $\mathbb{R}[x]$.

40. Write $x^8 - 4$ as a product of monic irreducible polynomials in $\mathbb{R}[x]$.

In exercises 41–48, find all the roots in $\mathbb{C}$ of the given polynomial.

41. $2x^2 + 3x - 7$

42. $2x^2 + 3x + 7$

43. $2x^3 - 9x^2 - 2x - 15$

44. $3x^4 + 2x^3 - 5x^2 + 2x - 8$

45. $x^4 - 4x^2 - 21$

46. $x^8 - x^4 - 2$

47. $6x^3 - 5x^2 - 29x + 10$

48. $10x^3 - 2x^2 + 45x - 9$

In exercises 49–54, determine if the given polynomials are reducible in $\mathbb{Q}(\sqrt{2})[x]$, where $\mathbb{Q}(\sqrt{2})$ denote the field consisting of all real numbers of the form $\{a + b\sqrt{2} \mid a, b \in \mathbb{Q}\}$.

49. $x^4 - 8$

50. $x^4 + 1$

51. $x^2 + 5x + 8$

52. $x^3 + 5$

53. $x^2 - 6x - 41$

54. $x^2 - 3x + 1$

9.3 Gauss' Lemma and Eisenstein's Criterion

When a polynomial $f(x) \in R[x]$ is irreducible in $R[x]$, we often use the expression that $f(x)$ **is irreducible over** R. Similarly, if $f(x) \in R[x]$ is reducible in $R[x]$, we often say that $f(x)$ **is reducible over** R. In this chapter, we have seen many examples of polynomials that are irreducible over one field but are reducible over a larger field. For example, $x^2 - 11$ is irreducible over $\mathbb{Q}$ but is reducible over $\mathbb{R}$. Similarly, $x^2 + 5$ is irreducible over $\mathbb{R}$ but is reducible over $\mathbb{C}$. Since $\mathbb{Q}$ is larger than $\mathbb{Z}$, this raises the question as to what happens to the reducibility of polynomials when we move from $\mathbb{Z}[x]$ and $\mathbb{Q}[x]$? In particular, if a polynomial $f(x) \in \mathbb{Z}[x]$ is irreducible in $\mathbb{Z}[x]$, can it possibly be reducible in $\mathbb{Q}[x]$? This question is answered in

Theorem 9.9—Gauss' Lemma. *Let $f(x)$ be a polynomial with integer coefficients. If $f(x)$ is reducible in $\mathbb{Q}[x]$, then $f(x)$ is reducible in $\mathbb{Z}[x]$. More precisely, if $f(x) = g(x) \cdot h(x)$, where $g(x), h(x) \in \mathbb{Q}[x]$, then there exist $G(x), H(x) \in \mathbb{Z}[x]$ such that $f(x) = G(x) \cdot H(x)$, $deg(G(x)) = deg(g(x))$, and $deg(H(x)) = deg(h(x))$.*

Intuition. In the proof of Gauss' Lemma, we will need to show that if we can factor $f(x)$ as

$$f(x) = g(x) \cdot h(x)$$

in $\mathbb{Q}[x]$, then we can modify $g(x)$ and $h(x)$ to obtain polynomials $G(x)$ and $H(x)$ such that

(a) $f(x) = G(x) \cdot H(x)$;

(b) $deg(G(x)) = deg(g(x))$, $deg(H(x)) = deg(h(x))$; and

(c) $G(x), H(x) \in \mathbb{Z}[x]$.

To see how the procedure for modifying $g(x)$ and $h(x)$ works, consider the polynomial

$$f(x) = 2x^2 + 13x - 7 \in \mathbb{Z}[x].$$

Suppose $f(x)$ is factored in $\mathbb{Q}[x]$ as

$$f(x) = 2x^2 + 13x - 7 = \left(\frac{3}{2}x + \frac{21}{2}\right)\left(\frac{4}{3}x - \frac{2}{3}\right);$$

we need to appropriately modify $\frac{3}{2}x + \frac{21}{2}$ and $\frac{4}{3}x - \frac{2}{3}$ to obtain polynomials $G(x)$ and $H(x)$ with the three properties listed above.

First, since all the coefficients of $\frac{3}{2}x + \frac{21}{2}$ belong to $\mathbb{Q}$, we can multiply this polynomial by a positive integer such that all the coefficients will become integers. In particular, the number 2 will do this trick in this case. Therefore, if we multiply $\frac{3}{2}x + \frac{21}{2}$ by 2 and divide $\frac{4}{3}x - \frac{2}{3}$ by 2, we obtain

$$f(x) = 2x^2 + 13x - 7 = \left(2\left(\frac{3}{2}x + \frac{21}{2}\right)\right) \cdot \left(\frac{1}{2}\left(\frac{4}{3}x - \frac{2}{3}\right)\right) = (3x + 21)\left(\frac{2}{3}x - \frac{1}{3}\right).$$

Next, we let c be the greatest common divisor of all the coefficients of $3x + 21$. We then divide $3x + 21$ by c and multiply the second polynomial ($\frac{2}{3}x - \frac{1}{3}$) by c. In this case, $c = 3$ and we obtain

$$f(x) = 2x^2 + 13x - 7 = (3x + 21)\left(\left(\frac{2}{3}x - \frac{1}{3}\right)\right) =$$
$$\left(\frac{1}{3}(3x + 21)\right) \cdot \left(3\left(\frac{2}{3}x - \frac{1}{3}\right)\right) = (x + 7)(2x - 1).$$

Now observe that both $x + 7$ and $2x - 1$ belong to $\mathbb{Z}[x]$, so we have appropriately modified $\frac{3}{2}x + \frac{21}{2}$ and $\frac{4}{3}x - \frac{2}{3}$.

As we look back at this procedure, the steps taken guaranteed that $\frac{3}{2}x + \frac{21}{2}$ would be modified into a polynomial with integer coefficients such that the greatest common divisor of the coefficients was 1. But there was no guarantee that this procedure would modify $\frac{4}{3}x - \frac{2}{3}$ into a polynomial with integer coefficients. Certainly, in this particular case, $\frac{4}{3}x - \frac{2}{3}$ was modified into a polynomial with integer coefficients. However, the real work in proving Gauss' Lemma is showing that the modified version of the second polynomial always belongs to $\mathbb{Z}[x]$.

Proof. Suppose in $\mathbb{Q}[x]$ we can write

$$f(x) = g(x) \cdot h(x),$$

where both $g(x)$ and $h(x)$ have smaller degree than $f(x)$. Since every coefficient of $g(x)$ belongs to $\mathbb{Q}$, there is a positive integer m such that all the coefficients of $m \cdot g(x)$ are integers.

Therefore, we now have

$$f(x) = g(x) \cdot h(x) = (m \cdot g(x)) \left(\frac{1}{m} \cdot h(x) \right).$$

Next, let c be the greatest common divisor of all the coefficients in $m \cdot g(x)$. We now have

$$f(x) = g(x) \cdot h(x) = (m \cdot g(x)) \left(\frac{1}{m} \cdot h(x) \right) = \left(\frac{m}{c} \cdot g(x) \right) \left(\frac{c}{m} \cdot h(x) \right).$$

If we let $G(x) = \frac{m}{c} \cdot g(x)$ and $H(x) = \frac{c}{m} \cdot h(x)$, then we can easily see that

(a) $f(x) = G(x) \cdot H(x)$;

(b) $deg(G(x)) = deg(g(x))$, $deg(H(x)) = deg(h(x))$; and

(c) $G(x)$ has integer coefficients and the greatest common divisor of the coefficients of $G(x)$ is 1.

In light of this, it now suffices to show that $H(x)$ also has integer coefficients. Since $H(x)$ has rational coefficients, there is a smallest positive integer n such that $n \cdot H(x)$ has integer coefficients. Therefore, our goal is to show that $n = 1$. We know that every integer greater than 1 is divisible by some prime number. Thus, the only positive integer not divisible by a prime number is 1. As a result, the way that we will show that $n = 1$ is to show that n is not divisible by any prime number.

Therefore, by way of contradiction, let us assume that there is some prime number p that divides n. Let

$$\rho : \mathbb{Z}[x] \to \mathbb{Z}_p[x]$$

be the homomorphism, where

$$\rho(a_m x^m + \cdots + a_1 x + a_0) = [a_m]_p x^m + \cdots + [a_1]_p x + [a_0]_p,$$

for all $a_i \in \mathbb{Z}$. Since $f(x) = G(x) \cdot H(x)$, multiplying this equation by n results in

(4) $n \cdot f(x) = (G(x)) \cdot (n \cdot H(x))$.

Every coefficient of $n \cdot f(x)$ is divisible by p, so $\rho(n \cdot f(x)) = [0]_p$. Furthermore, since the greatest common divisor of the coefficients of $G(x)$ is 1, at least one of the coefficients of $G(x)$ is not divisible by p. Hence, $\rho(G(x)) \neq [0]_p$.

Since ρ is a homomorphism defined on elements of $\mathbb{Z}[x]$, when we apply ρ to (4) and use that both $G(x)$ and $n \cdot H(x)$ belong to $\mathbb{Z}[x]$, we obtain

$$[0]_p = \rho(n \cdot f(x)) = \rho((G(x)) \cdot (n \cdot H(x))) = \rho(G(x)) \cdot \rho(n \cdot H(x)).$$

However, $\mathbb{Z}_p[x]$ is an integral domain, so the previous equation and the fact that $\rho(G(x)) \neq [0]_p$ combine to tell us that $\rho(n \cdot H(x)) = [0]_p$. As a result, every coefficient of $n \cdot H(x)$ must be a multiple of p. This then tells us that every coefficient of $\frac{n}{p} \cdot H(x)$ must be an integer. Since p divides n, we now know that $\frac{n}{p}$ is a positive integer that, when multiplied by $H(x)$, gives us a polynomial with integer coefficients. But $\frac{n}{p}$ is less than n, and n was chosen to be the smallest positive integer that when multiplied by $H(x)$ gives us a polynomial with integer coefficients. This is a contradiction, so we can conclude that $n = 1$ and $H(x)$ does indeed have integer coefficients, thereby concluding the proof. $\qquad\square$

■ Examples

Consider the polynomial $f(x) = x^4 + 2x^2 - 1$; we will use the Rational Root Test along with Gauss' Lemma to show that $f(x)$ is irreducible over $\mathbb{Q}$. If $f(x)$ were reducible over $\mathbb{Q}$, one possibility is that it has a factor of degree 1. But having a factor of degree 1 in $\mathbb{Q}[x]$ is equivalent to having a rational root. However, the Rational Root Test tells us that the only possible rational roots of $f(x)$ are ± 1. Since $f(1) = f(-1) = 2$, we now know that $f(x)$ has no factors of degree 1 in $\mathbb{Q}[x]$.

Now, since $f(x)$ has degree 4 and has no factor of degree 1 in $\mathbb{Q}[x]$, the only way it could possibly be reducible over $\mathbb{Q}$ is to be the product of two polynomials in $\mathbb{Q}[x]$, both of degree 2. However, Gauss' Lemma now tells us that if $f(x)$ could be factored this way in $\mathbb{Q}[x]$, then $f(x)$ would also be the product of two polynomials of degree 2 with coefficients in $\mathbb{Z}$. Let us now suppose that $f(x) = g(x) \cdot h(x)$, where $g(x)$ and $h(x)$ are quadratic polynomials in $\mathbb{Z}[x]$. Since the trailing coefficient of $f(x)$ is -1 , the product of the trailing coefficients of $g(x)$ and $h(x)$ is -1. Therefore, one of $g(x)$ or $h(x)$ has a trailing coefficient of 1, and the other has a trailing coefficient of -1. Similarly, the product of the leading coefficients of $g(x)$ and $h(x)$ must be 1, so either both $g(x)$ and $h(x)$ have leading coefficients of 1, or they both have leading coefficients of -1. However, in the latter case, $f(x) = (-g(x)) \cdot (-h(x))$, and both $-g(x)$ and $-h(x)$ have leading coefficients of 1. As a result, if $f(x)$ is reducible in $\mathbb{Q}[x]$, then we can write

$$f(x) = x^4 + 2x^2 - 1 = (x^2 + ax + 1)(x^2 + bx - 1),$$

where $a, b \in \mathbb{Z}$. However,

$$(x^2 + ax + 1)(x^2 + bx - 1) = x^4 + (a+b)x^3 + (ab)x^2 + (-a+b)x - 1.$$

Therefore , if

$$x^4 + 2x^2 - 1 = x^4 + (a+b)x^3 + (ab)x^2 + (-a+b)x - 1,$$

by comparing the coefficients of each term, we can conclude that a and b must satisfy

$$a+b=0, \quad ab=2, \quad \text{and} \quad -a+b=0.$$

Since there are no integers that can simultaneously satisfy these three equations, we can conclude that $f(x)$ is indeed irreducible over $\mathbb{Q}$.

■

In the preceding argument, if we did not know Gauss' Lemma, then we could not have immediately reduced to the case where $g(x)$ and $h(x)$ had leading coefficients of 1 with trailing coefficients of 1 and -1. As a result, the computations involved in showing that $f(x)$ was irreducible over $\mathbb{Q}$ would have been lengthier and more involved. Before leaving this example, observe that the Intermediate Value Theorem tells us that $f(x)$ has a real root between -1 and 0 and a real root between 0 and 1. Also note that the derivative of $f(x)$ is equal to $4x^3 + 4x$. Thus, $f(x)$ is increasing whenever x is positive and decreasing whenever x is negative. Hence, $f(x)$ cannot have more than one negative real root or more than one positive real root. In light of this, we know that $f(x)$ has exactly two real roots, neither of which is a rational number.

When dealing with a polynomial of degree 4 or more in $\mathbb{Q}[x]$, it is often quite difficult to determine whether or not it is reducible over $\mathbb{Q}$. For example, consider the following nine very similar looking polynomials:

$$x^4 + 2x^2 + 6, \quad x^4 + 3x^2 + 6, \quad x^4 + 4x^2 + 6, \quad x^4 + 5x^2 + 6, \quad x^4 + 6x^2 + 6,$$

$$x^4 + 7x^2 + 6, \quad x^4 + 8x^2 + 6, \quad x^4 + 9x^2 + 6, \quad x^4 + 10x^2 + 6.$$

One easy observation is that none of these polynomials have any real roots as x^4 and x^2 are never negative. So, in particular, none of these polynomials have any rational roots. However, this does not guarantee that they are all irreducible over $\mathbb{Q}$. In particular, you may have noticed that

$$x^4 + 5x^2 + 6 = (x^2 + 2)(x^2 + 3) \quad \text{and} \quad x^4 + 7x^2 + 6 = (x^2 + 1)(x^2 + 6).$$

But what about the other seven polynomials from our collection?

The bad news is that there is no algorithm that can determine, for all polynomials $f(x) \in \mathbb{Q}[x]$, whether or not $f(x)$ is reducible over $\mathbb{Q}$. On the other hand, the good news is that there are several reducibility tests that, when combined with Gauss' Lemma, can handle many, many special cases. In fact, the next test can be used on the seven remaining polynomials from our collection.

Theorem 9.10—Eisenstein's Criterion. *Let*

$$f(x) = a_n x^n + a_{n-1} x^{n-1} + \cdots + a_1 x + a_0 \in \mathbb{Z}[x].$$

If there exists a prime number p with the following properties:

(a) *p does not divide a_n,*

(b) *p divides every coefficient of $f(x)$ other than a_n,*

(c) *p^2 does not divide a_0,*

then $f(x)$ is irreducible in $\mathbb{Q}[x]$.

Before proving Eisenstein's Criterion, let us see how useful it can be.

■ Examples

First, consider the five polynomials

$$x^4 + 2x^2 + 6, \quad x^4 + 4x^2 + 6, \quad x^4 + 6x^2 + 6, \quad x^4 + 8x^2 + 6, \quad x^4 + 10x^2 + 6.$$

For all of these polynomials, we can apply Eisenstein's Criterion with the prime $p = 2$. The leading coefficient of each polynomial is 1, so 2 does not divide the leading coefficient. Next, all the other coefficients are even, so 2 does divide the other coefficients. The constant term of each polynomial is 6, and $2^2 = 4$ does not divide 6. Therefore, $p = 2$ satisfies all three of the criterion's properties, and we can conclude that all five polynomials are irreducible over $\mathbb{Q}$. Observe that for the polynomial $x^4 + 6x^2 + 6$, we could also have applied Eisenstein's Criterion with the prime $p = 3$.

We still need to consider the polynomials $x^4 + 3x^2 + 6$ and $x^4 + 9x^2 + 6$. Note that in these cases we cannot apply Eisenstein's Criterion with $p = 2$ as 2 does not divide the coefficient of x^2 in either of these polynomials. However, for these two polynomials, the prime number $p = 3$ satisfies the three properties needed to apply Eisenstein's Criterion. Therefore, these two polynomials are also irreducible over $\mathbb{Q}$.

■

It is certainly the case that if Eisenstein's Criterion applies to a polynomial, then it must be irreducible over $\mathbb{Q}$. But it is very important to keep in mind that the converse does not hold. In particular, if you are unable to apply Eisenstein's Criterion to a polynomial, it does not mean that the polynomial is reducible over $\mathbb{Q}$. If you are unable to apply Eisenstein's Criterion, it simply means that you need to find some other test or technique to deal with that particular polynomial. For example, let us compare the two similar looking polynomials

$$x^4 + 2x^2 - 1 \quad \text{and} \quad x^4 + 2x^2 + 1.$$

Eisenstein's Criterion does not apply to either of these polynomials. Whereas we have already shown that $x^4 + 2x^2 - 1$ is irreducible over $\mathbb{Q}$, $x^4 + 2x^2 + 1$ is reducible as

$$x^4 + 2x^2 + 1 = (x^2 + 1)(x^2 + 1) = (x^2 + 1)^2.$$

Proof of Theorem 9.10. Let $f(x) = a_n x^n + a_{n-1} x^{n-1} + \cdots + a_1 x + a_0 \in \mathbb{Z}[x]$ and suppose that p is a prime number such that p satisfies the three properties listed in the statement of the criterion. By way of contradiction, let us also suppose that $f(x)$ is reducible over $\mathbb{Q}$. Therefore, by Gauss' Lemma, $f(x)$ is also reducible over $\mathbb{Z}$. As a result, there exist $g(x), h(x) \in \mathbb{Z}[x]$ such that

$$f(x) = g(x) \cdot h(x),$$

where $deg(g(x)), deg(h(x)) < n$.

Observe that if $b_s x^s$ and $c_t x^t$ are the leading terms of, respectively, $g(x)$ and $h(x)$, then

$$a_n = b_s \cdot c_t \quad \text{and} \quad 0 < s, t < n.$$

In particular, since $p \nmid a_n$, it follows that $p \nmid b_s$ and $p \nmid c_t$. Also note that if b_0 and c_0 are the constant terms of, respectively, $g(x)$ and $h(x)$, then $a_0 = b_0 \cdot c_0$.

As in the proof of Gauss' Lemma, let

$$\rho : \mathbb{Z}[x] \to \mathbb{Z}_p[x]$$

be the homomorphism where

$$\rho(a_m x^m + \cdots + a_1 x + a_0) = [a_m]_p x^m + \cdots + [a_1]_p x + [a_0]_p,$$

for all $a_i \in \mathbb{Z}$. Since p divides all the coefficients of $f(x)$ other than a_n, we have

$$\rho(f(x)) = [a_n]_p x^n.$$

Using the fact that ρ is a homomorphism, we have

$$\rho(f(x)) = \rho(g(x) \cdot h(x)) = \rho(g(x)) \cdot \rho(h(x)).$$

By Lemma 9.3(c), since $\rho(f(x))$ consists of only a single term with a nonzero coefficient, the same must be true of $\rho(g(x))$ and $\rho(h(x))$. Having already shown that $[b_s]_p \neq [0]_p$ and $[c_t]_p \neq [0]_p$, it now follows that

$$\rho(g(x)) = [b_s]_p x^s \quad \text{and} \quad \rho(h(x)) = [c_t]_p x^t.$$

Since $s > 0$ and the coefficient of x^s is the only nonzero coefficient in $\rho(g(x))$, it follows that $[b_0]_p = [0]_p$. Using the same reasoning, it also follows that $[c_0]_p = [0]_p$. As a result, $p | b_0$ and

$p|c_0$. But this immediately implies that $p^2|(b_0 \cdot c_0)$. Since $a_0 = b_0 \cdot c_0$, this results in the contradiction $p^2|a_0$, thereby proving the result. $\qquad\qquad\qquad\qquad\qquad\qquad\qquad\qquad\quad$ $\square$

Exercises for Section 9.3

In exercises 1–28, determine if the given polynomial is reducible in $\mathbb{Q}[x]$. Remember, if a polynomial in $\mathbb{Q}[x]$ has degree 2 or 3, then the rational root test might not be the most efficient test available but it can always be used to determine if the polynomial is reducible. For polynomials in $\mathbb{Q}[x]$ of degree 4 and above, you can try to apply the rational root test or Eisenstein's Criterion. However, both of these tests could be inconclusive. If so, it might be necessary to apply Gauss' Lemma and try to solve a series of equations. For example, if you are trying to determine if a monic polynomial of degree 4 in $\mathbb{Z}[x]$ factors into a product of two quadratics, you could try to find $a, b, c, d \in \mathbb{Z}$ such that your polynomial can be written as $(x^2 + ax + b)(x^2 + cx + d) = x^4 + (a+c)x^3 + (ac+b+d)x^2 + (ad+bc)x + bd$. At this point, you would need to find **integer solutions** to four equations in the four unknowns a, b, c, d.

1. $x^4 - 2$

2. $x^4 - 4$

3. $x^4 - 8$

4. $x^5 + 2x^3 + 2x^2 + 2$

5. $x^5 + x^3 + 2x^2 + 2$

6. $x^4 + x^3 + 4x^2 + 3x + 3$

7. $x^4 + x^3 + 9x^2 + 3x + 3$

8. $x^4 + 3x^3 + 9x^2 + 3x + 3$

9. $2x^4 - 13x^3 + 13x^2 + 7x - 6$

10. $x^4 - 9x^2 + 14$

11. $x^4 - 8x^2 + 14$

12. $x^4 - 7x^2 + 14$

13. $105x^3 - 21x + 429$

14. $x^3 - 24x + 5$

15. $x^3 - 25x + 5$

16. $x^4 + 18x^2 + 77$

17. $x^4 + 18x^2 + 78$

18. $x^4 + 4x^2 - x + 6$

19. $x^4 + 4x^2 - 2x + 6$

20. $x^4 + 4x^2 - x + 625$

21. $x^6 - 63$

22. $x^6 - 64$

23. $x^6 - 65$

24. $x^6 - 8x^3 - 65$

25. $x^6 + 20x^3 + 96$

26. $x^6 + 21x^3 + 96$

27. $x^4 + 4x^3 + 3x^2 + 2x + 1$

28. $x^4 + x^3 + 2x^2 + 3x + 4$

29. Find all integers A such that $x^4 - Ax^2 + 2$ is reducible in $\mathbb{Q}[x]$.

30. Find all integers A such that $x^4 - Ax^2 - 2$ is reducible in $\mathbb{Q}[x]$.

31. Find all integers A such that $x^4 - Ax^2 - 1$ is reducible in $\mathbb{Q}[x]$.

32. If p is a prime number, find all integers A such that $x^4 - Ax^2 + p$ is reducible in $\mathbb{Q}[x]$.

33. If p is a prime number, find all integers A such that $x^4 - Ax^2 - p$ is reducible in $\mathbb{Q}[x]$.

34. Find all integers A such that $x^4 - Ax^2 + 10$ is reducible in $\mathbb{Q}[x]$.

35. Find all integers A such that $x^4 - Ax^2 - 10$ is reducible in $\mathbb{Q}[x]$.

36. Find all integers A such that $x^4 - Ax^2 + 15$ is reducible in $\mathbb{Q}[x]$.

37. Find all integers A such that $x^4 - Ax^2 - 15$ is reducible in $\mathbb{Q}[x]$.

38. Find all integers A such that $x^4 - Ax^2 - 4$ is reducible in $\mathbb{Q}[x]$.

9.4 Reduction Modulo p

Let us now consider whether the polynomial $f(x) = 273x^3 - 491x^2 + 935$ is reducible over $\mathbb{Q}$. Eisenstein's Criterion does not apply to $f(x)$ as the only primes which divide 935 are 5, 11, 17 and none of them divide -491. Since the degree of $f(x)$ is only 3, we could apply the Rational Root Test as $f(x)$ will be reducible over $\mathbb{Q}$ if and only if it has a rational root.

For this polynomial, the Rational Root Test provides us with 128 candidates, as every rational number of the form $\pm\frac{5^a\cdot11^b\cdot17^c}{3^d\cdot7^e\cdot13^f}$ is a candidate, where a, b, c, d, e, f can take on any values from the set $\{0, 1\}$. However, as a practical matter, we might hope to find a test that is simpler to use in this case. In fact, the following test can make short work of this and many other polynomials.

Theorem 9.11. *Suppose* $f(x) = a_nx^n + \cdots + a_1x + a_0 \in \mathbb{Z}[x]$ *and also suppose that* p *is a prime number such that* p *does not divide* a_n. *Let* $\rho : \mathbb{Z}[x] \to \mathbb{Z}_p[x]$ *be the homomorphism defined as*

$$\rho(a_nx^n + \cdots + a_1x + a_0) = [a_n]_px^n + \cdots + [a_1]_px + [a_0]_p.$$

If $\rho(f(x))$ *is irreducible in* $\mathbb{Z}_p[x]$, *then* $f(x)$ *is irreducible in* $\mathbb{Q}[x]$.

Before proving Theorem 9.11, let us return to the polynomial $f(x) = 273x^3 - 491x^2 + 935$. If we consider the prime $p = 2$, observe that $[273]_2 = [1]_2$, $[-491]_2 = [1]_2$, and $[935]_2 = [1]_2$. Therefore,

$$\rho(f(x)) = x^3 + x^2 + [1]_2.$$

Since the degree of $x^3 + x^2 + [1]_2$ is 3, it will be reducible over $\mathbb{Z}_2$ if and only if it has a root in $\mathbb{Z}_2$. However, plugging both $[0]_2$ and $[1]_2$ into $x^3 + x^2 + [1]_2$ gives an answer of $[1]_2$. Thus, $x^3 + x^2 + [1]_2$ is irreducible over $\mathbb{Z}_2$, and Theorem 9.11 now asserts that $273x^3 - 491x^2 + 935$ is irreducible over $\mathbb{Q}$.

At this point, it is important to give you a warning similar to the one we discussed after Eisenstein's Criterion. Certainly, if you can apply Theorem 9.11 to a polynomial, then the polynomial is irreducible over $\mathbb{Q}$. However, the converse does not hold. If you try to apply Theorem 9.11 to a polynomial $f(x)$ and the polynomial $\rho(f(x))$ is reducible in $\mathbb{Z}_p[x]$, it does not tell you whether or not $f(x)$ is reducible in $\mathbb{Q}[x]$. It simply means that you will need to find another technique or test to determine whether or not $f(x)$ is reducible over $\mathbb{Q}$. For example, consider the polynomials $f(x) = x^2 + 2$ and $g(x) = x^2 + 3x + 2$. If we use the prime $p = 3$, then

$$\rho(f(x)) = x^2 + [2]_3 = \rho(g(x))$$

in $\mathbb{Z}_3[x]$.

Therefore, although the polynomials $f(x)$ and $g(x)$ are quite different in $\mathbb{Q}[x]$, they become indistinguishable in $\mathbb{Z}_3[x]$. In this case, $\rho(f(x))$ and $\rho(g(x))$ are reducible in $\mathbb{Z}_3[x]$ as

$$\rho(f(x)) = \rho(g(x)) = x^2 + [2]_3 = (x + [1]_3)(x + [2]_3).$$

However, the fact that $\rho(f(x))$ and $\rho(g(x))$ are reducible in $\mathbb{Z}_3[x]$ gives us no information about the reducibility of $f(x)$ and $g(x)$ over $\mathbb{Q}$. In fact, $f(x)$ is irreducible over $\mathbb{Q}$, as it has no rational roots, whereas $g(x)$ is reducible and factors as $g(x) = x^2 + 3x + 2 = (x+1)(x+2)$.

Proof of Theorem 9.11. Suppose $f(x) = a_n x^n + \cdots + a_1 x + a_0 \in \mathbb{Z}[x]$ and also suppose that p is a prime number such that p does not divide a_n and $\rho(f(x))$ is irreducible in $\mathbb{Z}_p[x]$. By way of contradiction, let us assume that $f(x)$ is reducible over $\mathbb{Q}$. Therefore, by Gauss' Lemma, $f(x)$ is also reducible over $\mathbb{Z}$. As a result, there exist $g(x), h(x) \in \mathbb{Z}[x]$ such that

$$f(x) = g(x) \cdot h(x) \text{ and } deg(g(x)), deg(h(x)) > 0.$$

As in the proof of Theorem 9.10, since p does not divide the leading coefficient of $f(x)$, it follows that p divides neither the leading coefficient of $g(x)$ nor the leading coefficient of $h(x)$. In light of this, $deg(\rho(g(x))) = \deg(g(x)) > 0$ and $deg(\rho(h(x))) = \deg(h(x)) > 0$. Since ρ is a homomorphism, we have

$$\rho(f(x)) = \rho(g(x) \cdot h(x)) = \rho(g(x)) \cdot \rho(h(x)).$$

However, this says that $\rho(f(x))$ has been written in $\mathbb{Z}_p[x]$ as the product of two polynomials of degree greater than 0. This contradicts the fact that $\rho(f(x))$ is irreducible in $\mathbb{Z}_p[x]$, thereby proving the result. $\qquad\square$

■ Examples

Let us consider whether the polynomial $f(x) = 36x^3 + 34x + 12$ is reducible over $\mathbb{Q}$. Observe that we cannot apply Eisenstein's Criterion. The reason is that the only prime p such that p divides the constant term and p^2 does not divide the constant term is 3. However, we cannot use the prime 3 for two reasons. First, it divides the leading coefficient, and second, it doesn't divide the coefficient of x. Since the degree of $f(x)$ is only 3, we could use the Rational Root Test as $f(x)$ will be reducible over $\mathbb{Q}$ if and only if it has a rational root. For this polynomial, it looks like the Rational Root Test might produce a large number of candidates. Therefore, we might try to find a simpler solution and only use the Rational Root Test as a last resort. If we try to apply Theorem 9.11, we cannot use the primes $p = 2$ or $p = 3$, since they are both divisors of the leading coefficient. Next, we look at what happens if we try the prime $p = 5$. In this case, $[36]_5 = [1]_5$, $[34]_5 = [4]_5$, and $[12]_5 = [2]_5$. As a result, we now have

$$\rho(f(x)) = x^3 + [4]_5 x + [2]_5.$$

Theorem 9.11 tells us that if $\rho(f(x))$ is irreducible in $\mathbb{Z}_5[x]$, then $f(x)$ is irreducible in $\mathbb{Q}[x]$. But be aware that if $\rho(f(x))$ turns out to be reducible over $\mathbb{Z}_5$, then we will have obtained no useful information, and we will need to go back to the drawing board.

Since the degree of $x^3 + [4]_5 x + [2]_5$ is only 3, it will be reducible over $\mathbb{Z}_5$ if and only if it has a root in $\mathbb{Z}_5$. However, you can check that the values when we plug $[0]_5, [1]_5, [2]_5,$ $[3]_5, [4]_5$ into $x^3 + [4]_5 x + [2]_5$ are, respectively, $[2]_5, [2]_5, [3]_5, [1]_5, [2]_5$. Therefore, $\rho(f(x))$ is irreducible over $\mathbb{Z}_5$, and Theorem 9.11 asserts that $f(x)$ is irreducible over $\mathbb{Q}$.

■

After looking at the examples in this chapter, it should be clear that whereas tests like the Rational Root Test and Eisenstein's Criterion are extremely useful for determining when many polynomials are irreducible in $\mathbb{Q}[x]$, there are also many other polynomials for which these tests are inconclusive and give us no information.

Similarly, Theorem 9.11 is useful for many polynomials and inconclusive for many others. This situation raises an interesting question: If $f(x) \in \mathbb{Z}[x]$ is irreducible, does there exist a prime p such that p does not divide the leading coefficient of $f(x)$ and $\rho(f(x))$ is irreducible in $\mathbb{Z}_p$? In other words, if $f(x) \in \mathbb{Z}[x]$ is irreducible, is there some p out there that will enable us to apply Theorem 9.11? As we will soon see, the answer is **no**. Therefore, there exist polynomials in $\mathbb{Z}[x]$ that are irreducible over $\mathbb{Q}$ that are also reducible in $\mathbb{Z}_p[x]$, for every prime p. These examples will once again indicate, in a very strong way, that if $\rho(f(x))$ is reducible in $\mathbb{Z}_p[x]$, we cannot draw any conclusions as to whether or not $f(x)$ is reducible over $\mathbb{Q}$. In order to produce these examples, we will need three lemmas.

Lemma 9.12. *If A is an integer then the polynomial $f(x) = x^4 - Ax^2 + 1$ is reducible over $\mathbb{Q}$ if and only if either $A + 2$ or $A - 2$ is a perfect square in $\mathbb{Z}$.*

Proof. If $f(x)$ has a factor of degree 1, then it has a rational root. However, the Rational Root Test asserts that the only possible rational roots of $f(x)$ are ± 1. Since $f(1) = f(-1) = 2 - A$, the only time $f(x)$ has a factor of degree 1 is when $2 - A = 0$, so $A = 2$. Observe that when $A = 2$, we have $A - 2 = 0 = 0^2$, so $A - 2$ is a perfect square in $\mathbb{Z}$.

In light of the previous paragraph, it now suffices to show that $f(x)$ factors into the product of two quadratic polynomials in $\mathbb{Q}[x]$ if and only if either $A + 2$ or $A - 2$ is a perfect square in $\mathbb{Z}$. By Gauss' Lemma, if $g(x), h(x)$ are quadratics in $\mathbb{Q}[x]$ such that $f(x) = g(x) \cdot h(x)$, then we may assume that $g(x), h(x) \in \mathbb{Z}[x]$. Since the constant term of $f(x)$ is 1, the product of the constant terms of $g(x)$ and $h(x)$ is 1. Similarly, the product of the leading coefficients of $g(x)$ and $h(x)$ must also be 1. It is not hard to see that this implies that $f(x)$ must factor as either

$$(x^2 + ax + 1)(x^2 + bx + 1) \quad \text{or} \quad (x^2 + ax - 1)(x^2 + bx - 1),$$

where $a, b \in \mathbb{Z}$. If $f(x) = (x^2 + ax + 1)(x^2 + bx + 1)$, then we have

$$x^4 - Ax^2 + 1 = (x^2 + ax + 1)(x^2 + bx + 1) = x^4 + (a+b)x^3 + (ab+2)x^2 + (a+b)x + 1.$$

Comparing the coefficients of the various terms, this tells us that a and b must satisfy the equations

$$a+b=0 \quad \text{and} \quad ab+2=-A.$$

But these equations imply that $b = -a$ and $a^2 = A+2$. Therefore, if $f(x)$ factors in this way, then there exist $a, b \in \mathbb{Z}$ satisfying these equations and $A+2$ is a perfect square in $\mathbb{Z}$. In the opposite direction, if $a \in \mathbb{Z}$ such that $A+2 = a^2$, then the preceding calculations indicate that

$$f(x) = x^4 - Ax^2 + 1 = x^4 - (a^2 - 2)x^2 + 1 = (x^2 + ax + 1)(x^2 - ax + 1)$$

and $f(x)$ is the product of two quadratics.

Similarly, if $f(x) = (x^2 + ax - 1)(x^2 + bx - 1)$, we now have

$$x^4 - Ax^2 + 1 = (x^2 + ax - 1)(x^2 + bx - 1) = x^4 + (a+b)x^3 + (ab - 2)x^2 - (a+b)x + 1.$$

If we again compare the coefficients of the various terms, we see that a and b must satisfy the equations

$$a+b=0 \quad \text{and} \quad ab-2=-A.$$

These equations imply that $b = -a$ and $a^2 = A - 2$. Thus, if $f(x)$ factors in this way, then there exist $a, b \in \mathbb{Z}$ satisfying these equations and $A - 2$ is a perfect square in $\mathbb{Z}$. Going in the other direction, if $a \in \mathbb{Z}$ such that $A - 2 = a^2$, then our previous calculations show that

$$f(x) = x^4 - Ax^2 + 1 = x^4 - (a^2 + 2)x^2 + 1 = (x^2 + ax - 1)(x^2 - ax - 1)$$

and $f(x)$ is the product of two quadratics. $\qquad \square$

■ Examples

In the next several examples, we will apply Lemma 9.12 to polynomials of the form $x^4 - Ax^2 + 1$, for various values of A.

1. If $A = 4$, then $x^4 - 4x^2 + 1$ is irreducible over $\mathbb{Q}$ as $A + 2 = 6$ and $A - 2 = 2$ are not perfect squares in $\mathbb{Z}$.

2. If $A = 3$, then $x^4 - 3x^2 + 1$ is reducible over $\mathbb{Q}$ as $A - 2 = 1 = 1^2$. Looking back at the proof of Lemma 9.11, $x^4 - 3x^2 + 1 = (x^2 + ax - 1)(x^2 + bx - 1)$, where $b = -a$ and $a^2 = A - 2$. Therefore, $a = 1$, $b = -1$ and

$$x^4 - 3x^2 + 1 = (x^2 + x - 1)(x^2 - x - 1).$$

3. If $A = 2$, then $x^4 - 2x^2 + 1$ is reducible over $\mathbb{Q}$ as $A - 2 = 0 = 0^2$. In this case, 1 and -1 are roots of $x^4 - 2x^2 + 1$ and we have

$$x^4 - 2x^2 + 1 = (x+1)(x+1)(x-1)(x-1) = (x+1)^2(x-1)^2.$$

4. If $A = 1$, then $x^4 - x^2 + 1$ is irreducible over $\mathbb{Q}$ as $A + 2 = 3$ and $A - 2 = -1$ are not perfect squares in $\mathbb{Z}$.

5. If $A = 0$, then $x^4 + 1$ is irreducible over $\mathbb{Q}$ as $A + 2 = 2$ and $A - 2 = -2$ are not perfect squares in $\mathbb{Z}$.

6. If $A = -1$, then $x^4 + x^2 + 1$ is reducible over $\mathbb{Q}$ as $A + 2 = 1 = 1^2$. Looking back at the proof of Lemma 9.12, $x^4 + x^2 + 1 = (x^2 + ax + 1)(x^2 + bx + 1)$, where $b = -a$ and $a^2 = A + 2$. Therefore, $a = 1$, $b = -1$ and

$$x^4 + x^2 + 1 = (x^2 + x + 1)(x^2 - x + 1).$$

7. If $A = -2$, then $x^4 + 2x^2 + 1$ is reducible over $\mathbb{Q}$ as $A + 2 = 0 = 0^2$. The proof of Lemma 9.12 tells us that $x^4 + 2x^2 + 1 = (x^2 + ax + 1)(x^2 + bx + 1)$, where $b = -a$ and $a^2 = A + 2$. Therefore, $a = 0$, $b = 0$ and

$$x^4 + 2x^2 + 1 = (x^2 + 1)(x^2 + 1) = (x^2 + 1)^2.$$

8. If $A < -2$, then $x^4 - Ax^2 + 1$ is always irreducible over $\mathbb{Q}$ as $A + 2$ and $A - 2$ are both negative and therefore cannot be perfect squares in $\mathbb{Z}$.

∎

Recall that we usually use the symbol 1 to denote the multiplicative identity element of a field. Going one step further, we often use the symbol 2 to represent the sum $1 + 1$. However, observe that in the field $\mathbb{Z}_2$, we have $[1]_2 + [1]_2 = [2]_2 = [0]_2$. Therefore, there exist fields where the symbol 2 is equal to the additive identity. Thus, depending on the field in which you are working, the element 2 might not have a multiplicative inverse.

Lemma 9.13. *Let F be a field where 2 has a multiplicative inverse. If $A \in F$ and if either $A + 2$, $A - 2$, or $A^2 - 4$ is the square of an element in F, then the polynomial $x^4 - Ax^2 + 1$ is reducible in $F[x]$.*

Proof. Let us first consider the case where $A + 2$ is the square of an element in F. In this case, we let $\sqrt{A + 2}$ denote an element of F whose square is $A + 2$ and observe that

$$(x^2 + (\sqrt{A+2})x + 1)(x^2 - (\sqrt{A+2})x + 1) =$$
$$x^4 + (\sqrt{A+2} - \sqrt{A+2})x^3 + (-(\sqrt{A+2})^2 + 2)x + (\sqrt{A+2} - \sqrt{A+2})x + 1 =$$
$$x^4 - Ax^2 + 1.$$

Therefore, in this case, we have factored $x^4 - Ax^2 + 1$ into the product of two quadratics in $F[x]$.

In the next case, suppose $A - 2$ is the square of an element in F and let $\sqrt{A-2}$ denote an element of F whose square is $A - 2$. We now have

$$(x^2 + (\sqrt{A-2})x - 1)(x^2 - (\sqrt{A-2})x - 1) =$$
$$x^4 + (\sqrt{A-2} - \sqrt{A-2})x^3 + (-(\sqrt{A-2})^2 - 2)x + (\sqrt{A-2} - \sqrt{A-2})x + 1 =$$
$$x^4 - Ax^2 + 1.$$

Once again, we have factored $x^4 - Ax^2 + 1$ into the product of two quadratics in $F[x]$.

For the final case, suppose that $A^2 - 4$ is the square of an element in F and let $\sqrt{A^2-4}$ denote an element of F whose square is $A^2 - 4$. This is the only case where we will need that 2 has a multiplicative inverse. In this case, we have

$$\left(x^2 + \frac{-A+\sqrt{A^2-4}}{2}\right)\left(x^2 + \frac{-A-\sqrt{A^2-4}}{2}\right) =$$
$$x^4 + \left(\frac{-A+\sqrt{A^2-4}}{2} + \frac{-A-\sqrt{A^2-4}}{2}\right)x^2 +$$
$$\left(\frac{-A+\sqrt{A^2-4}}{2}\right) \cdot \left(\frac{-A-\sqrt{A^2-4}}{2}\right) =$$
$$x^4 - Ax^2 + 1.$$

Thus, once again, we have factored $x^4 - Ax^2 + 1$ into the product of two quadratics in $F[x]$.

□

Lemmas 9.12 and 9.13 indicate that the reducibility of polynomials of the form $x^4 - Ax^2 + 1$ is related to various elements being squares of other elements. To apply these results, we will need to examine the squares of elements in $\mathbb{Z}_p$.

Lemma 9.14. *If $a, b \in Z$ and if p is a prime number, then at least one of $[a]_p$, $[b]_p$, $[a \cdot b]_p$ is the square of an element of $\mathbb{Z}_p$.*

Proof. If either $[a]_p$ or $[b]_p$ is equal to $[0]_p$, then there is nothing to prove. Therefore, without loss of generality, we may consider the case where $[a]_p$, $[b]_p$, and $[a \cdot b]_p$ are all elements of $U(\mathbb{Z}_p)$.

We begin by supposing that $c, d \in Z$ such that $1 \leq c, d \leq \frac{p-1}{2}$, and we claim that if $[c]_p^2 = [d]_p^2$, then $c = d$. To see this, observe that if $[c]_p^2 = [d]_p^2$, then $[c^2]_p = [d^2]_p$, which implies that

$[c^2 - d^2]_p = [0]_p$. Thus,

$$[0]_p = [c^2 - d^2]_p = [c - d]_p \cdot [c + d]_p.$$

Since $\mathbb{Z}_p$ is a field, this tells us that either

$$[c - d]_p = [0]_p \quad \text{or} \quad [c + d]_p = [0]_p.$$

Therefore, either p divides $c - d$ or p divides $c + d$.

Since $1 \leq c, d \leq \frac{p-1}{2}$, we see that both $c - d$ and $c + d$ must lie between p and $-p$. Therefore, the only way $c - d$ or $c + d$ could be divisible by p is to be equal to 0. But since c and d are both positive, it is impossible for $c + d$ to equal 0. Thus, the only remaining possibility is that $c - d = 0$, so $c = d$.

The preceding argument tells us that all $\frac{p-1}{2}$ elements of the list

$$[1]_p^2, \quad [2]_p^2, \quad \ldots, \quad \left[\frac{p-1}{2} \right]_p^2$$

are different. Now suppose that $[a]_p$ and $[b]_p$ are both elements of $U(\mathbb{Z}_p)$ that are not squares of elements of $U(\mathbb{Z}_p)$. We claim that the new list

$$[1]_p^2, [2]_p^2, \ldots, \left[\frac{p-1}{2} \right]_p^2, \; [a]_p \cdot [1]_p^2, [a]_p \cdot [2]_p^2, \ldots, [a]_p \cdot \left[\frac{p-1}{2} \right]_p^2$$

is a complete listing of the $p - 1$ elements of $U(\mathbb{Z}_p)$. Since $U(\mathbb{Z}_p)$ has exactly $p - 1$ elements, we only need to show that all $p - 1$ elements of our list are different. Earlier in this proof, we already showed that the first $\frac{p-1}{2}$ elements of the list are all different from one another. In addition, a now familiar argument, already used in the proofs of Theorem 6.8 and Proposition 7.18, tells us that since we are dealing with elements of a group, the last $\frac{p-1}{2}$ elements on our list are also all different from one another. Therefore, it only remains to show that none of the first $\frac{p-1}{2}$ elements of our list can be equal to any of the last $\frac{p-1}{2}$ elements of our list. By way of contradiction, suppose there exist $[u]_p, [v]_p \in U(\mathbb{Z}_p)$ such that

$$[a] \cdot [u]_p^2 = [v]_p^2.$$

If we let $[w]_p$ be the multiplicative inverse of $[u]_p$ then, using the previous equation, we have

$$[a]_p = [a]_p \cdot [1]_p^2 = [a]_p \cdot ([u]_p \cdot [w]_p)^2 = ([a]_p \cdot [u]_p^2) \cdot [w]_p^2 =$$
$$[v]_p^2 \cdot [w]_p^2 = ([v]_p \cdot [w]_p)^2 = [v \cdot w]_p^2.$$

But this contradicts the fact that $[a]_p$ is not the square of an element in $U(\mathbb{Z}_p)$. Thus, all $p - 1$ elements on our list are different from each other, and every element of $U(\mathbb{Z}_p)$ appears exactly once on the list.

Since we are assuming that $[b]_p$ is not the square of any element of $U(\mathbb{Z}_p)$, $[b]_p$ must appear on our list as one of the elements of the form $[a] \cdot [x]_p^2$, for some $[x]_p \in U(\mathbb{Z}_p)$. Using the fact that $[b]_p = [a] \cdot [x]_p^2$, we have

$$[a \cdot b]_p = [a]_p \cdot [b]_p = [a]_p \cdot ([a] \cdot [x]_p^2) = ([a]_p \cdot [x]_p)^2 = [a \cdot x]_p^2.$$

Therefore, in this case, $[a \cdot b]_p$ is indeed the square of an element of $U(\mathbb{Z}_p)$. Thus, it is always the case that at least one of $[a]_p$, $[b]_p$, $[a \cdot b]_p$ is the square of an element of $\mathbb{Z}_p$. $\square$

We can now use Lemmas 9.12, 9.13, and 9.14 to produce an infinite number of polynomials in $\mathbb{Z}[x]$ that are irreducible over $\mathbb{Q}$ but become reducible when we reduce down to $\mathbb{Z}_p$, for every prime p.

Theorem 9.15. *If A is an integer then, for every prime number p, the polynomial*

$$x^4 - [A]_p x^2 + [1]_p$$

is reducible over $\mathbb{Z}_p$. However, the polynomial

$$x^4 - Ax^2 + 1$$

is irreducible over $\mathbb{Q}$, except when either $A + 2$ or $A - 2$ is a perfect square in $\mathbb{Z}$.

Proof. Lemma 9.12 told us that if A is an integer then $x^4 - Ax^2 + 1$ is irreducible over $\mathbb{Q}$, except when either $A + 2$ or $A - 2$ is a perfect square in $\mathbb{Z}$. Therefore, to prove our result, we only need to consider the situation over $\mathbb{Z}_p$. As is often the case when studying $\mathbb{Z}_p$, we will need to examine the $p = 2$ case separately.

In $\mathbb{Z}_2$, $[0]_2^2 = [0]_2$ and $[1]_2^2 = [1]_2$. Therefore, for any integer A, $[A]_2^2 = [A]_2$. Combining this with the fact that $[A]_2 = -[A]_2$, we have

$$(x^2 + [A]_2 + [1]_2)^2 = (x^2 + [A]_2 + [1]_2)(x^2 + [A]_2 + [1]_2) =$$

$$x^4 + ([A]_2 + [A]_2)x^3 + ([A]_2^2 + [1]_2 + [1]_2)x^2 + ([A]_2 + [A]_2)x + [1]_2 =$$

$$x^4 + [A]_2 x^2 + [1]_2 = x^4 - [A]_2 x^2 + [1]_2.$$

Thus, $x^4 - [A]_2 x^2 + [1]_2$ is reducible over $\mathbb{Z}_2$.

Next, if A is an integer and p is a prime number other than 2, let $a = [A]_p + [2]_p$ and $b = [A]_p - [2]_p$. Therefore,

$$a \cdot b = ([A]_p + [2]_p) \cdot ([A]_p - [2]_p) = [A]_p^2 - [4]_p.$$

Applying Lemma 9.14, it follows that at least one of $[A]_p + [2]_p$, $[A]_p - [2]_p$, $[A]_p^2 - [4]_p$ is the square of an element of $\mathbb{Z}_p$. Furthermore, since $p \neq 2$, 2 has a multiplicative inverse in $\mathbb{Z}_p$. As a result, we can apply Lemma 9.13 to conclude that $x^4 - [A]_p x^2 + [1]_p$ is reducible over $\mathbb{Z}_p$. □

■ Examples

As we indicated in the examples that followed Lemma 9.12, if A is an integer that is less than -2, then $x^4 - Ax^2 + 1$ is irreducible over $\mathbb{Q}$. Thus, all polynomials of the form

$$x^4 + 3x^2 + 1, \quad x^4 + 4x^2 + 1, \quad x^4 + 5x^2 + 1, \quad x^4 + 6x^2 + 1,$$

$$x^4 + 7x^2 + 1, \quad x^4 + 8x^2 + 1, \ldots$$

have the property that they are irreducible over $\mathbb{Q}$, but Theorem 9.11 does not apply, as they become reducible whenever we look at them over $\mathbb{Z}_p$, for all primes p.

■

We conclude this chapter with an example of a polynomial that appears to be beyond the tools and techniques we have developed so far. However, we will show how to adapt our tools to handle this and some other polynomials. To this end, suppose $f(x) \in \mathbb{Q}[x]$ is reducible over $\mathbb{Q}$. Therefore, there exist $g(x), h(x) \in \mathbb{Q}[x]$ such that

$$f(x) = g(x) \cdot h(x), \quad \text{where} \quad deg(g(x)), deg(h(x)) < deg(f(x)).$$

Now, replace x by $t + 1$ and let $F(t) = f(t+1)$, $G(t) = g(t+1)$, and $H(t) = h(t+1)$. The fact that $f(x) = g(x) \cdot h(x)$ now tells us that $F(t) = G(t) \cdot H(t)$. Since $G(t)$ and $g(x)$ have the same degree, as do $H(t)$ and $h(x)$, we see that $F(t)$ is also reducible over $\mathbb{Q}$. Looking at this from another perspective, it tells us that if $F(t)$ is irreducible over $\mathbb{Q}$, then so is $f(x)$. We now apply this observation.

■ Examples

Let $f(x) = x^4 + x^3 + x^2 + x + 1$. At first glance, it is unclear as to whether $f(x)$ is reducible over $\mathbb{Q}$. However, if we let $x = t + 1$, we have

$$f(x) = x^4 + x^3 + x^2 + x + 1 = \frac{x^5 - 1}{x - 1} = \frac{(t+1)^5 - 1}{(t+1) - 1} =$$

$$\frac{(t^5 + 5t^4 + 10t^3 + 10t^2 + 5t + 1) - 1}{t} = \frac{t^5 + 5t^4 + 10t^3 + 10t^2 + 5t}{t} =$$

$$t^4 + 5t^3 + 10t^2 + 10t + 5 = F(t).$$

We can use Eisenstein's Criterion with the prime $p = 5$ to conclude that $F(t)$ is irreducible over $\mathbb{Q}$. Thus $f(x)$ is also irreducible over $\mathbb{Q}$. In the exercises, we will extend this argument to all polynomials of the form $x^{p-1} + x^{p-2} + \cdots + x + 1$, where p is a prime.

∎

Exercises for Section 9.4

In exercises 1–27, use any of the techniques in this section to determine if the given polynomial is reducible in $\mathbb{Q}[x]$. These exercises should indicate that two polynomials can appear to be very similar and have graphs that are virtually identical, but one can be reducible and the other irreducible.

1. $x^4 + 4x^2 - x + 6$

2. $x^4 + 4x^2 - x + 5$

3. $x^4 + 4x^2 + x + 6$

4. $x^4 + 4x^2 + x + 7$

5. $x^4 + 4x^2 + 2x + 6$

6. $3x^4 - 4x^3 - 21x + 28$

7. $3x^4 - 7x^3 - 21x + 28$

8. $3x^4 - 7x^3 - 12x + 28$

9. $3x^4 - 7x^3 - 14x + 28$

10. $3x^4 - 5x^3 - 9x + 15$

11. $3x^4 - 6x^3 - 9x + 15$

12. $3x^4 - 5x^3 - 10x + 15$

13. $x^4 + x^3 + 2x - 4$

14. $x^4 + x^3 + 2x - 3$

15. $x^4 + 2x^3 + 2x - 2$

16. $x^4 + x^3 - 2x - 4$

17. $x^4 + x^3 - 2x - 5$

18. $2x^4 - 3x^3 + 10x - 15$

19. $2x^4 - 3x^3 + 9x - 15$

20. $2x^4 - 5x^3 + 10x - 15$

21. $2x^4 - 5x^3 + 6x - 15$

22. $2x^4 - 6x^3 + 6x - 15$

23. $2x^4 - 5x^3 + 5x - 15$

24. $8x^3 + 4x^2 + 6x + 20$

25. $8x^3 - 10x^2 - 16x + 20$

26. $8x^3 - 6x^2 - 21x + 20$

27. $8x^3 - 5x^2 - 20x + 20$

28. How many monic irreducible polynomials of degree 1 are there in $\mathbb{Z}_p[x]$?

29. How many monic irreducible polynomials of degree 2 are there in $\mathbb{Z}_p[x]$? It might be easier if you first count the number of monic reducible polynomials of degree 2.

30. How many monic irreducible polynomials of degree 3 are there in $\mathbb{Z}_p[x]$? It might be easier if you first count the number of monic reducible polynomials of degree 3.

For exercises 31–32, please read the following:

A nonzero polynomial $f(x) \in \mathbb{Z}[x]$ is called **primitive** if the greatest common divisor of the coefficients of $f(x)$ is 1.

31. Suppose $f(x), g(x), h(x) \in \mathbb{Z}[x]$ such that $f(x) = g(x) \cdot h(x)$. Show that $f(x)$ is primitive if and only if both $g(x)$ and $h(x)$ are primitive. For various primes p, you might want to think about the ring homomorphisms we have examined that send $\mathbb{Z}[x]$ to $\mathbb{Z}_p[x]$.

32. Suppose $f(x), g(x) \in \mathbb{Z}[x]$ and $h(x) \in \mathbb{Q}[x]$ such that $f(x) = g(x) \cdot h(x)$. Show that if $g(x)$ is primitive, then $h(x) \in \mathbb{Z}[x]$.

33. Let p be a prime number and let $f(x) = x^{p-1} + x^{p-2} + \cdots + x^2 + x + 1$. In doing this exercise, you might want to refer to exercise 36 from Section 9.1.

    (a) Show that $f(x) = \frac{x^p - 1}{x - 1}$.

    (b) If we let $F(x) = f(x+1)$, show that $F(x) = x^{p-1} + \binom{p}{1}x^{p-2} + \binom{p}{2}x^{p-3} + \cdots + \binom{p}{p-3}x^2 + \binom{p}{p-2}x + p$.

    (c) Prove that $F(x)$ is irreducible in $\mathbb{Q}[x]$.

    (d) Conclude that $f(x)$ is irreducible in $\mathbb{Q}[x]$.

In exercises 34–42, express the given polynomial as a product of monic irreducible polynomials in $\mathbb{Q}[x]$, $\mathbb{R}[x]$, and $\mathbb{C}[x]$. If necessary, elements of $\mathbb{C}$ can be expressed in polar form.

34. $x^4 - 1$.

35. $x^4 + 1$.

36. $x^5 - 1$.

37. $x^5 + 1$.

38. $x^6 - 1$.

39. $x^6 + 1$.

40. $x^8 - 1$.

41. $x^{10} - 1$.

42. $x^{12} - 1$.

CHAPTER 10

Roots of Polynomials of Degree Less than 5

In Chapter 1, we mentioned that there exist formulas for the roots of all polynomials of degree at most 4. These formulas involve combining the polynomial's coefficients using only addition, subtraction, multiplication, division, and taking nth roots, for $n \geq 2$. In this rather brief chapter, we will derive the formulas for the roots of all polynomials of degree at most 4. More precisely, we will provide a series of steps and procedures that will lead us to the roots of these polynomials by manipulating the coefficients using only the operations just listed. We will not explicitly write out the final versions of these formulas as that would take many pages. But more important than writing out extremely long formulas, we provide algorithms that prove that one can indeed find the roots of any polynomial of degree at most 4 by doing nothing more than combining the polynomial's coefficients using addition, subtraction, multiplication, division, and taking nth roots, for $n \geq 2$.

Galois proved that analogous formulas cannot exist for polynomials of degree 5 and above. In Chapter 17, we will prove Galois' famous result on the insolvability of the quintic. Although we do not yet have enough mathematical machinery to prove this result, we are in a position to apply his result. Therefore, we will conclude this chapter with examples of fifth-degree polynomials whose roots cannot be found by adding, subtracting, multiplying, dividing, and taking nth roots of combinations of the coefficients.

10.1 Finding Roots of Polynomials of Small Degree

The strategy for finding formulas for the roots of a polynomial of small degree will be to make a series of substitutions, each of which will simplify the polynomial into a new polynomial with simpler coefficients. In some ways, this will be analogous to integration problems in calculus where you made several substitutions before finally obtaining a problem that could be easily handled. Here is an outline of our strategy:

Step I Divide the polynomial by its leading coefficient so that the new leading coefficient is equal to 1.

Step II   Make a substitution so that the new coefficient immediately following the leading coefficient is equal to 0.

Step III   Make a substitution so that the new trailing coefficient is equal to 1.

For polynomials of degree 1, performing Step I will be enough for us to easily find the root. In the degree 2 case, we will be able to find the roots after performing Steps I and II. At that point, we will have succeeded in deriving familiar formulas that you have undoubtedly come across in your earlier algebra courses. Not surprisingly, the situation for polynomials of degree 3 will be more difficult. Not only will we need to apply Steps I, II, and III, but we will also need to make a substitution that appears to come out of nowhere. The degree 4 case will involve even more computations, but in some ways, these computations will be more natural and motivated than those required to complete the degree 3 case.

Roots of Polynomials of Degree 1

Given

$$ax + b = 0$$

with $a \neq 0$, apply Step I and divide both sides of the equation by a to obtain

$$x + \frac{b}{a} = 0.$$

Then subtract $\frac{b}{a}$ from both sides to obtain the root

$$x = -\frac{b}{a}.$$

Roots of Polynomials of Degree 2

Given

$$ax^2 + bx + c = 0$$

with $a \neq 0$, apply Step I and divide both sides of the equation by a to obtain

$$x^2 + \frac{b}{a}x + \frac{c}{a} = 0.$$

Next, to apply Step II, make the substitution $x = y - \frac{b}{2a}$. Note that if we can find y, then we can certainly find x, thus it suffices to find y. Our previous equation now becomes

$$\left(y - \frac{b}{2a}\right)^2 + \frac{b}{a}\left(y - \frac{b}{2a}\right) + \frac{c}{a} = 0.$$

Expanding out the terms in this equation, we obtain

$$y^2 - \frac{b}{a}y + \frac{b^2}{4a^2} + \frac{b}{a}y - \frac{b^2}{2a^2} + \frac{c}{a} = 0,$$

which simplifies to

$$y^2 + \frac{4ac - b^2}{4a^2} = 0.$$

We then subtract $\frac{4ac-b^2}{4a^2}$ from both sides to obtain

$$y^2 = \frac{b^2 - 4ac}{4a^2}.$$

Taking square roots is an allowable operation and when we take the square root of both sides we obtain

$$y = \pm\frac{\sqrt{b^2 - 4ac}}{2a}.$$

Having successfully solved for y, we can now go back to the substitution $x = y - \frac{b}{2a}$ and see that

$$x = \frac{-b \pm \sqrt{b^2 - 4ac}}{2a}.$$

This is, of course, the familiar quadratic formula.

In your previous algebra courses, the techniques used to derive the quadratic formula might have been simpler than the procedure just outlined. The reason for this is that we are not necessarily looking for the simplest technique in the degree 2 case but are instead looking for a technique that can be easily generalized and extended to help us study polynomials of degree 3 and 4. The key step in the previous procedure was making the substitution $x = y - \frac{b}{2a}$. Often this step is referred to as "completing the square." In order to find the roots of polynomials of degree 3 and 4, we will need to generalize this substitution to those situations. In other words, in the degree 3 case, we will need to find an α such that substituting $y - \alpha$ for x will result in a polynomial in y with no degree 2 term. Let us examine what value of α will do the trick. If we begin with

$$ax^3 + bx^2 + cx + d$$

and let $x = y - \alpha$, then we obtain

$$a(y-\alpha)^3 + b(y-\alpha)^2 + c(y-\alpha) + d = ay^3 + (-3a\alpha + b)y^2 + \text{ terms of lower degree.}$$

Therefore, for the coefficient of y^2 to be 0, we need $-3a\alpha + b = 0$. As a result, α must be equal to $\frac{b}{3a}$. Thus, in order to apply Step II in the degree 3 case, we will make the substitution $x = y - \frac{b}{3a}$.

Roots of Polynomials of Degree 3

Given

$$ax^3 + bx^2 + cx + d = 0$$

with $a \neq 0$, apply Step I and divide both sides of the equation by a to obtain

$$x^3 + \frac{b}{a}x^2 + \frac{c}{a}x + \frac{d}{a} = 0.$$

To apply Step II, we now make the substitution $x = y - \frac{b}{3a}$. Once again, if we can find y, then we can certainly find x, so it suffices to find y. You should check that our previous equation now becomes

$$y^3 + \frac{-b^2 + 3ac}{3a^2}y + \frac{2b^3 - 9abc + 27a^2d}{27a^3} = 0.$$

To simplify matters, we will let $B = \frac{-b^2 + 3ac}{3a^2}$ and $C = \frac{2b^3 - 9abc + 27a^2d}{27a^3}$. Our previous equation now simplifies to

$$y^3 + By + C = 0.$$

If $C = 0$, then we can factor and obtain

$$0 = y^3 + By = y(y^2 + B).$$

Therefore, one of the roots is 0 and the others can be found by applying the quadratic formula to $y^2 + B = 0$. As a result, we will assume that $C \neq 0$ and, to apply Step III, we will make the substitution $y = C^{\frac{1}{3}}z$. Note that if we can solve for z, then we will know the value of y, so it suffices to solve for z. Our substitution turns the equation $y^3 + By + C = 0$ into

$$0 = (C^{\frac{1}{3}}z)^3 + B(C^{\frac{1}{3}}z) + C = Cz^3 + BC^{\frac{1}{3}}z + C.$$

Since $C \neq 0$, we can divide this equation by C to obtain

$$z^3 + \frac{B}{C^{\frac{2}{3}}}z + 1 = 0.$$

We can now further simplify this equation by letting $D = \frac{B}{C^{\frac{2}{3}}}$, thereby giving us the equation

$$z^3 + Dz + 1 = 0.$$

This is as far as Steps I, II, and III can take us. Therefore, in order to finish this problem, we will need another substitution or idea. There is a substitution that works at this point, but, unfortunately, it is rather unmotivated and appears to come out of nowhere. However, it does get the job done, and that is the most important thing. We let $z = v - \frac{D}{3v}$ and once again note that if we can find v, then we will know the value of z. Thus, it suffices to find v. Our previous equation now becomes

$$0 = \left(v - \frac{D}{3v}\right)^3 + D\left(v - \frac{D}{3v}\right) + 1 = \left(v^3 - Dv + \frac{D^2}{3v} - \frac{D^3}{27v^3}\right) + \left(Dv - \frac{D^2}{3v}\right) + 1,$$

which simplifies to

$$v^3 - \frac{D^3}{27v^3} + 1 = 0.$$

Multiplying this equation by v^3 and then letting $E = -\frac{D^3}{27}$ gives us

$$v^6 + v^3 + E = 0.$$

Finally, let $v = w^{\frac{1}{3}}$; if we can solve for w, then we can certainly find v. With this substitution, the previous equation now becomes

$$\left(w^{\frac{1}{3}}\right)^6 + \left(w^{\frac{1}{3}}\right)^3 + E = 0,$$

which immediately simplifies to

$$w^2 + w + E = 0.$$

At various points in this procedure, we have taken cube roots, which is an allowable operation. We can now use the quadratic formula to find w. Knowing w, we can then, in order, find v, z, y, and then x. The actual expression for x in terms of a, b, c, d is quite long and complicated. However, at this point, we are much more interested in the fact that such an expression actually exists. Thus, we have succeeded in showing that there is indeed a formula for the roots of $ax^3 + bx^2 + cx + d$, which only involves combining a, b, c, d using addition, subtraction, multiplication, division, taking square roots, and taking cube roots.

Roots of Polynomials of Degree 4

Given

$$ax^4 + bx^3 + cx^2 + dx + e = 0$$

with $a \neq 0$, apply Step I and divide the equation by a to obtain

$$x^4 + \frac{b}{a}x^3 + \frac{c}{a}x^2 + \frac{d}{a}x + \frac{e}{a} = 0.$$

In light of our discussion before the degree 3 case, it should now come as no surprise that to apply Step II, we make the substitution $x = y - \frac{b}{4a}$. As before, in order to find x, it suffices to find y. Our equation now becomes

$$\left(y - \frac{b}{4a}\right)^4 + \frac{b}{a}\left(y - \frac{b}{4a}\right)^3 + \frac{c}{a}\left(y - \frac{b}{4a}\right)^2 + \frac{d}{a}\left(y - \frac{b}{4a}\right) + \frac{e}{a} = 0.$$

If we expand out the terms in the previous equation, we obtain

$$y^4 + \left(-4 \cdot \frac{b}{4a} + \frac{b}{a}\right)y^3 + \text{ terms of lower degree } = 0.$$

Therefore, this equation can be written as

$$y^4 + By^2 + Cy + D = 0,$$

where B, C, D are obtained from a, b, c, d, e using only addition, subtraction, multiplication, and division. If $D = 0$, then we can factor and obtain

$$y(y^3 + By + C) = 0.$$

Thus, $y = 0$ is one of the roots, and the other three can be found by applying our work on cubic polynomials to $y^3 + By + C$. As a result, we may assume that $D \neq 0$ and can apply Step III, by letting $y = D^{\frac{1}{4}}z$, to obtain

$$0 = (D^{\frac{1}{4}}z)^4 + B(D^{\frac{1}{4}}z)^2 + C(D^{\frac{1}{4}}z) + D = Dz^4 + BD^{\frac{1}{2}}z^2 + CD^{\frac{1}{4}}z + D.$$

Dividing this equation by D gives us

$$z^4 + \frac{B}{D^{\frac{1}{2}}}z^2 + \frac{C}{D^{\frac{3}{4}}}z + 1 = 0.$$

We can simplify further by letting $E = \frac{B}{D^{\frac{1}{2}}}$ and $F = \frac{C}{D^{\frac{3}{4}}}$ to obtain

$$z^4 + Ez^2 + Fz + 1 = 0.$$

If $F = 0$, then the substitution $z = t^{\frac{1}{2}}$ turns our equation into

$$t^2 + Et + 1 = 0.$$

Since the roots of this polynomial can be found using the quadratic formula, we would be done in this case. Therefore, we may now assume that $F \neq 0$.

This is as far as we can go using only Steps I, II, and III. However, the idea we will use to finish this problem is much more natural than the one used at the comparable point in our study of the degree 3 case. The idea is to try to factor $z^4 + Ez^2 + Fz + 1$ into two quadratic polynomials as

$$z^4 + Ez^2 + Fz + 1 = (z^2 + \alpha z + \beta)(z^2 + \gamma z + \delta),$$

where $\alpha, \beta, \gamma, \delta$ can be obtained from E and F using the usual operations of addition, subtraction, multiplication, division, and taking nth roots, for various $n \geq 2$. If we are successful in factoring $z^4 + Ez^2 + Fz + 1$, then we can easily find all of its roots by using the quadratic formula to find the roots of both $z^2 + \alpha z + \beta$ and $z^2 + \gamma z + \delta$.

Observe that

$$(z^2 + \alpha z + \beta)(z^2 + \gamma z + \delta) = z^4 + (\alpha + \gamma)z^3 + (\beta + \delta + \alpha\gamma)z^2 + (\alpha\delta + \beta\gamma)z + \beta\delta.$$

By comparing the coefficients of like terms, the above product will be equal to $z^4 + Ez^2 + Fz + 1$ precisely if we can find $\alpha, \beta, \gamma, \delta$ such that

(1) $\qquad \alpha + \gamma = 0, \quad \beta + \delta + \alpha\gamma = E, \quad \alpha\delta + \beta\gamma = F, \quad \text{and} \quad \beta\delta = 1.$

It is clear that the first and fourth equations from (1) tell us that

$$\gamma = -\alpha \quad \text{and} \quad \delta = \beta^{-1}.$$

As a result, the second and third equations from (1) can now be simplified to

$$\beta + \beta^{-1} - \alpha^2 = E \quad \text{and} \quad \alpha(\beta^{-1} - \beta) = F.$$

Since we are in the case where $F \neq 0$, it now follows that $\alpha \neq 0$. Therefore, we are allowed to divide the second of the preceding two equations by α, and it is easy to see that it now suffices to find α, β such that

(2) $\qquad \beta + \beta^{-1} = E + \alpha^2 \quad \text{and} \quad \beta^{-1} - \beta = \dfrac{F}{\alpha}.$

If we add the second equation from (2) to the first and then also subtract the second equation from the first, we obtain

(3) $\qquad 2\beta^{-1} = (E + \alpha^2) + \dfrac{F}{\alpha} \quad \text{and} \quad 2\beta = (E + \alpha^2) - \dfrac{F}{\alpha}.$

If we multiply the two equations in (3), we see that

$$4 = (E + \alpha^2)^2 - \left(\dfrac{F}{\alpha}\right)^2.$$

Therefore, it now suffices to find an α satisfying this equation, since plugging that value of α into the second equation in (3) will immediately tell us the value of β.

You should check that if we expand the terms from our previous equation, multiply both sides by α^2, and then collect like terms, we obtain

$$\alpha^6 + 2E\alpha^4 + (E^2 - 4)\alpha^2 - F^2 = 0.$$

Finally, if we make the substitution $\alpha = \theta^{\frac{1}{2}}$, then the previous equation becomes

$$(\theta^{\frac{1}{2}})^6 + 2E(\theta^{\frac{1}{2}})^4 + (E^2 - 4)(\theta^{\frac{1}{2}})^2 - F^2 = 0.$$

But this immediately simplifies to

$$\theta^3 + 2E\theta^2 + (E^2 - 4)\theta - F^2 = 0.$$

However, this is now a cubic polynomial in the variable θ. Thus, using our work on cubic polynomials, we can find θ. Working backward, knowing θ, we can then find, in order, α, β, γ, δ, z, y, and then x. At every point along the way, the only operations we have used are addition, subtraction, multiplication, division, and taking nth roots for various $n \geq 2$. Thus, the roots of $ax^4 + bx^3 + cx^2 + dx + e$ are indeed obtained from a, b, c, d, e using only addition, subtraction, multiplication, division, and taking nth roots, for various $n \geq 2$. Admittedly, if we wrote out the final formula for the roots of $ax^4 + bx^3 + cx^2 + dx + e$, it would be incredibly long and complicated. In fact, it would probably take several pages. But the more important point is that we have shown, in a fairly straightforward way, that it is indeed possible to find the roots of all polynomials of degree at most 4 by combining the coefficients using only addition, subtraction, multiplication, division, and taking nth roots, for various $n \geq 2$.

10.2 A Brief Look at Some Consequences of Galois' Work

If we wanted to try to find a formula for the roots of polynomials of degree 5, it would be logical to once again begin by applying Steps I, II, and III.

Given

$$ax^5 + bx^4 + cx^3 + dx^2 + ex + f = 0$$

with $a \neq 0$, divide the equation by a to obtain

$$x^5 + \frac{b}{a}x^4 + \frac{c}{a}x^3 + \frac{d}{a}x^2 + \frac{e}{a}x + \frac{f}{a} = 0.$$

Next, make the substitution $x = y - \frac{b}{5a}$. After simplifying we obtain

$$y^5 + By^3 + Cy^2 + Dy + E = 0,$$

where B, C, D, E are obtained from a, b, c, d, e, f using only addition, subtraction, multiplication, and division. If $E = 0$, then we can factor and obtain

$$y(y^4 + By^2 + Cy + D) = 0.$$

Thus, $y = 0$ is one of the roots, and the other four can be found by applying our work on quartic polynomials to $y^4 + By^2 + Cy + D$. As a result, we may assume $E \neq 0$, and if we let $y = E^{\frac{1}{5}}z$ and then simplify, we obtain

$$z^5 + Fz^3 + Gz^2 + Hz + 1 = 0,$$

where F, G, H are obtained from a, b, c, d, e, f using only addition, subtraction, multiplication, division, and taking fifth roots.

Unfortunately, this is a far as Steps I, II, III can take us. Needless to say, many people worked long and hard for many years trying to factor or simplify the polynomial $z^5 + Fy^3 + Gy^2 + Hy + 1$ in an attempt to find its roots. But Galois showed that there is no formula for the roots of polynomials of degree 5 that combines the coefficients using only addition, subtraction, multiplication, division, and taking nth roots, for various n.

Certainly, there are many polynomials of degree 5 and above whose roots can be found by combining the coefficients using only addition, subtraction, multiplication, division, and taking nth roots. For example, using our study of the complex numbers from Chapter 6, we know how to find the five roots of the polynomial $x^5 - 1$. We can also find the roots of the polynomial $x^5 + 5x^3 + 5x^2 + 6x + 1$, since it factors as

$$x^5 + 5x^3 + 5x^2 + 6x + 1 = (x^2 + x + 1)(x^3 - x^2 + 5x + 1).$$

More generally, you can check that we can find the roots of any polynomial of the form $x^5 + Bx^3 + Bx^2 + (B+1)x + 1$, since it factors as

$$x^5 + Bx^3 + Bx^2 + (B+1)x + 1 = (x^2 + x + 1)(x^3 - x^2 + Bx + 1).$$

These examples in no way contradict Galois' work. Indeed, there are many polynomials of degree 5 whose roots can be found by combining the coefficients using the preceding operations listing. However, Galois's work shows that there exist some polynomials of degree 5 where it is impossible to find their roots by combining the coefficients in the preceding ways. More precisely, Galois proved that if a polynomial satisfies certain properties, then there does not exist a formula for its roots that combines its coefficients using only addition, subtraction, multiplication, division, and taking nth roots, for various n.

At this point, it is worth looking at an important special case of Galois's work, which we will prove in Chapter 17.

Theorem 10.1—A Special Case of Galois' Work. *Let $f(x) \in \mathbb{Z}[x]$ such that*

(i) $f(x)$ has degree 5,

(ii) $f(x)$ is irreducible in $\mathbb{Q}[x]$, and

(iii) $f(x)$ has exactly three real roots.

Then there does not exist a formula for the roots of $f(x)$ that only involves combinations of the coefficients using addition, subtraction, multiplication, division, and taking nth roots, for various $n \geq 2$.

By using Eisenstein's Criterion, the Intermediate Value Theorem, and Rolle's Theorem, it will be easy to find many degree 5 polynomials with these properties. Recall that the Intermediate Value Theorem asserts that if a polynomial changes sign three times, then it has at least three real roots. On the other hand, Rolle's Theorem asserts that if the derivative of a polynomial is equal to zero at only two points, then the original polynomial cannot have more than three real roots.

Let us now consider the polynomials

$$x^5 - 6x + 3, \quad x^5 - 4x + 2, \quad x^5 - 8x + 6.$$

We will show that all three satisfy the properties that allow us to apply Galois' work. First, observe that Eisenstein's Criterion implies that all three polynomials are irreducible in $\mathbb{Q}[x]$. Next, if you look at the values of all three polynomials at $x = -2, -1, 0, 1, 2$, then you can see that they all change sign three times, so the Intermediate Value Theorem tells us that they must all have at least three real roots. On the other hand, the derivatives of these three polynomials are all of the form $5x^4 - \alpha$, where α is a positive real number. It is easy to see that these derivatives are equal to 0 exactly twice, once when x is positive and once when x is negative. Therefore, Rolle's Theorem tells us that all three polynomials each have at most three real roots. Combining all of the preceding observations, we can see that Galois' work can indeed by applied to all three of these polynomials. Therefore, for all three of these polynomials, it is impossible to find its roots by combining the coefficients using only addition, subtraction, multiplication, division, and taking nth roots, for various n. Hopefully, you can see that it is not difficult to find many degree 5 polynomials to which Galois' work can be applied.

Exercises for Sections 10.1 and 10.2

In exercises 1–14, find all the roots in $\mathbb{C}$ of the given polynomial. In each case, you may first want to apply Steps I and II from this chapter to simplify the polynomial. You may leave your answers in the form $a\text{cis}(\theta) + b$, where $a, b, \theta \in \mathbb{R}$, $a > 0$, and $0 \leq \theta < 2\pi$.

1. $x^3 + 6x^2 + 12x + 1$

2. $x^3 - 9x^2 + 27x - 17$

3. $8x^3 - 12x^2 + 6x - 201$

4. $125x^3 + 225x^2 + 135x + 12$

5. $x^4 + 4x^3 + 6x^2 + 4x - 5$

6. $x^4 - 20x^3 + 150x^2 - 500x + 614$

7. $16x^4 - 96x^3 + 216x^2 - 216x + 60$

8. $81x^4 + 108x^3 + 54x^2 + 12x - 49$

9. $x^5 - 5x^4 + 10x^3 - 10x^2 + 5x + 16$

10. $x^5 + 10x^4 + 40x^3 + 80x^2 + 80x - 1$

11. $243x^5 - 405x^4 + 270x^3 - 90x^2 + 15x - 13$

12. $32x^5 + 240x^4 + 720x^3 + 1080x^2 + 810x + 143$

13. $x^6 + 6x^5 + 15x^4 + 20x^3 + 15x^2 + 6x - 1950$

14. $64x^6 - 192x^5 + 240x^4 - 160x^3 + 60x^2 - 12x - 1954$

In exercises 15–19, show that the polynomial satisfies the properties required to apply Theorem 10.1. As a result, we can then conclude that there is no formula for the roots of the polynomial that combines the coefficients using only addition, subtraction, multiplication, division, and taking nth roots, for various $n \geq 2$.

15. $3x^5 - 5x^3 - 30x + 5$

16. $3x^5 - 5x^3 - 90x - 15$

17. $x^5 - 10x^3 - 80x + 6$

18. $x^5 - 85x + 34$

19. $x^5 - 14x + 7$

20. In this exercise, we will generalize exercise 19.

   (a) Show that for any prime p, the polynomial $x^5 - 2px + p$ satisfies the properties required to apply Theorem 9.1.

   (b) Conclude that there are an infinite number of polynomials of degree 5 for which there is no formula for the roots of the polynomial that combines the coefficients using only addition, subtraction, multiplication, division, and taking nth roots, for various $n \geq 2$.

In exercises 21–26, factor the polynomial into a product of two quadratic polynomials in $\mathbb{R}[x]$. Then use this factorization to find all four roots in $\mathbb{C}$ of the original polynomial.

21. $x^4 + 10x^3 + 23x^2 + 10x + 1$

22. $x^4 - 32x^2 + 12x + 3$

23. $x^4 + 12x^3 - 6x^2 - 84x - 7$

24. $x^4 + 16x^3 + 64x^2 - 1$

25. $x^4 - 9x^2 - 36$

26. $x^4 + 2x^3 + 26x + 7$

In exercises 27–32, factor the polynomial into a product of a cubic and a quadratic polynomial in $\mathbb{R}[x]$.

27. $x^5 - 18x^3 - 44x^2 - 13x - 1$

28. $x^5 + x^3 + x^2 + 1$

29. $x^5 - x^4 + x^3 + 2x^2 - 2x + 2$

30. $x^5 - 2x^3 + x^2 - 15x - 5$

31. $x^5 - 3x^4 + 2x^3 - 5x^2 + 2$

32. $x^5 - 41x^3 + 5x^2 + 3$

CHAPTER 11

Rational Values of Trigonometric Functions

In this chapter, we apply the Rational Root Test along with Mathematical Induction to prove some surprising results about the values of trigonometric functions. Virtually every trigonometry student has used the Pythagorean Theorem to examine $30°-60°-90°$ and $45°-45°-90°$ triangles. In fact, most students are expected to derive, or at least memorize, the following table:

θ	$0°$	$30°$	$45°$	$60°$	$90°$
$\sin(\theta)$	0	$\frac{1}{2}$	$\frac{\sqrt{2}}{2}$	$\frac{\sqrt{3}}{2}$	1
$\cos(\theta)$	1	$\frac{\sqrt{3}}{2}$	$\frac{\sqrt{2}}{2}$	$\frac{1}{2}$	0
$\tan(\theta)$	0	$\frac{\sqrt{3}}{3}$	1	$\sqrt{3}$	undefined

Observe that with the obvious exception of $\tan(90°)$, every entry on the table is either a rational number or the square root of a rational number. Using formulas like the double-angle and triple-angle formulas, it is not hard to compute the values of the trigonometric functions at various other angles in the first quadrant. For example, $\cos(36°) = \frac{1+\sqrt{5}}{4}$ and $\cos(20°)$ is the unique positive root of the polynomial $8x^3 - 6x - 1$. Note that neither $\cos(36°)$ nor $\cos(20°)$ is rational. Furthermore, neither $\cos(36°)$ nor $\cos(20°)$ is the square root of a rational number.

At this point, it is natural to wonder which values of the sine, cosine, and tangent functions are rational. Or which values, at the very least, are the square root of a rational number. Since the sine and cosine functions are continuous everywhere, the Intermediate Value Theorem tells us that they both assume all values in the interval $[-1, 1]$. For example, if $\frac{2}{3}, \frac{\sqrt{5}}{4}$, and $\frac{e}{\pi}$ are your favorite numbers in the interval $[0, 1]$, then there exists angles $\theta_1, \theta_2, \theta_3$ in the first quadrant such that

$$\cos(\theta_1) = \sin(90° - \theta_1) = \frac{2}{3}, \quad \cos(\theta_2) = \sin(90° - \theta_2) = \frac{\sqrt{5}}{4},$$

$$\text{and} \quad \cos(\theta_3) = \sin(90° - \theta_3) = \frac{e}{\pi}.$$

Similarly, when plugging in all the angles between $-90°$ and $90°$, the tangent function takes on all real values. Thus, if 3, $\sqrt{11}$, and $e + \pi$ are your favorite positive numbers, then there exist angles $\theta_4, \theta_5, \theta_6$ in the first quadrant such that

$$\tan(\theta_4) = 3, \quad \tan(\theta_5) = \sqrt{11}, \quad \text{and} \quad \tan(\theta_6) = e + \pi.$$

In light of the preceding observations, it might seem curious that in trigonometry courses we don't use techniques like the Pythagorean Theorem along with the formulas for $\cos(n\theta)$, where $n \in \mathbb{N}$, to find the angle in the first quadrant such that the cosine is equal to $\frac{2}{3}$. Similarly, why don't we find the angles in the first quadrant such that the sine is equal to $\frac{\sqrt{5}}{4}$ or the tangent is equal to 3?

As we shall soon see, in all three of these cases, the degree measure of the appropriate angle is not a rational number. Observe that in a first course in trigonometry, the only angles for which we have the tools required to compute exact values of trigonometric functions are angles where the degree measure is a rational number. Therefore, there would be no way in a basic trigonometry course for us to compute the degree measure of the angles that makes the cosine equal to $\frac{2}{3}$, the sine equal to $\frac{\sqrt{5}}{4}$, or the tangent equal to 3.

11.1 Values of Trigonometric Functions

The main results of this chapter will assert that the table we saw earlier in this chapter provides us with essentially the only cases where the degree measure of an angle is rational and the value of a trigonometric function is rational or the square root of a rational number. This explains why the $30°-60°-90°$ and $45°-45°-90°$ triangles tend to be the only right triangles studied in trigonometry courses.

We now make the standard transition to radian measure. Recall that the radian measure of the five angles in the table presented earlier are $0, \frac{\pi}{6}, \frac{\pi}{4}, \frac{\pi}{3}, \frac{\pi}{2}$. In fact, it is easy to see that the degree measure of an angle is a rational number if and only if its radian measure is of the form $\alpha\pi$, where $\alpha \in \mathbb{Q}$. Using radian measure, we can now state the main result of this chapter.

Theorem 11.1. *If $\alpha \in \mathbb{Q}$ such that $\cos^2(\alpha\pi) \in \mathbb{Q}$, then $\cos(\alpha\pi)$ belongs to the set* $\{0, \pm\frac{1}{2}, \pm\frac{\sqrt{2}}{2}, \pm\frac{\sqrt{3}}{2}, \pm 1\}$.

Before proving Theorem 11.1, let us make several observations. First, Theorem 11.1 asserts that angles in the four quadrants with reference angles of $30°, 45°$, and $60°$ along with angles that are integer multiples of $90°$ are the *only* angles whose degree measure is a rational number and the value of the cosine is either rational or the square root of a rational number. Next, one would expect a result analogous to Theorem 11.1 to also hold for the sine. This is indeed the case. But there is a surprising lack of symmetry. Whereas we will first prove Theorem 11.1 and then use it to obtain a result about the sine, it does not appear that we can reverse the order and first prove a result about the sine. However, once we do prove Theorem 11.1, we will immediately verify

Corollary 11.2. *If* $\alpha \in \mathbb{Q}$ *such that* $\sin^2(\alpha\pi) \in \mathbb{Q}$, *then* $\sin(\alpha\pi)$ *belongs to the set* $\{0, \pm\frac{1}{2}, \pm\frac{\sqrt{2}}{2}, \pm\frac{\sqrt{3}}{2}, \pm 1\}$.

After proving Corollary 11.2, we will apply Theorem 11.1 to the tangent function to obtain

Corollary 11.3. *If* $\alpha \in \mathbb{Q}$ *such that* $\tan^2(\alpha\pi) \in \mathbb{Q}$, *then* $\tan(\alpha\pi)$ *belongs to the set* $\{0, \pm\frac{\sqrt{3}}{3}, \pm 1, \pm\sqrt{3}\}$.

At first glance, it is unclear how the Rational Root Test can be used to obtain information about values of trigonometric functions. Therefore, before presenting all the details behind the proof of Theorem 11.1, we proceed with

Intuition behind the proof of Theorem 11.1. To gain some insight into this problem, let us suppose that both α and $\cos(\alpha\pi)$ are rational numbers. We want to understand how the Rational Root Test helps us in determining the possible values of $\cos(\alpha\pi)$. Since $\alpha \in \mathbb{Q}$, when we consider the sequence

$$\alpha\pi, 2\alpha\pi, 2^2\alpha\pi, 2^3\alpha\pi, 2^4\alpha\pi, \ldots.$$

there will be terms that differ by an integer multiple of 2π. This tells us that there exist positive integers $n > m$ such that $\cos(2^m\alpha\pi) = \cos(2^n\alpha\pi)$.

Next, recall that the double-angle formula for the cosine asserts that

$$\cos(2\theta) = 2\cos^2(\theta) - 1.$$

Therefore, if we let $f(x) = 2x^2 - 1$ and also let $f^n(x)$ denote the composition of n copies of the function $f(x)$, then we have

$$\cos(2\theta) = f(\cos(\theta)),$$

$$\cos(4\theta) = f(\cos(2\theta)) = f^2(\cos(\theta)),$$

and, more generally,

$$\cos(2^t\theta) = f^t(\cos(\theta)),$$

for all $t \in \mathbb{N}$.

Therefore, if we let $c = \cos(2^m\theta)$ and also let $t = n - m$, then we have

$$c = \cos(2^m\alpha\pi) = \cos(2^n\alpha\pi) = f^t(\cos(2^m\alpha\pi)) = f^t(c).$$

The preceding equation indicates that c is a root of the polynomial $f'(x) - x$. On the other hand, since $\cos(\alpha\pi) \in \mathbb{Q}$, it also follows that $c = f^m(\cos(\alpha\pi)) \in \mathbb{Q}$.

Thus, c is a rational root of a polynomial with integer coefficients. We can now apply the Rational Root Test to greatly limit the possible values of c. Once this is done, we can then determine the possible values of $\cos(\alpha\pi)$. In the proof of Theorem 11.1, instead of using the formula for $\cos(2\theta)$, we will actually use a formula for $2\cos(2\theta)$. This minor change will allow us to deal with monic polynomials, and this will make our use of the Rational Root Test even more efficient.

Similar reasoning will allow us to show that if α and $\cos^2(\alpha\pi)$ are both rational, then $2\cos^2(\alpha\pi)$ must belong to the set $\{0, \frac{1}{2}, 1, \frac{3}{2}, 2\}$. The remainder of the proof of Theorem 11.1 will then consist of some simple computations. $\qquad\square$

We can now start down the path that will lead us to the proof of Theorem 11.1. Many of the steps along the way will make heavy use of Mathematical Induction.

Lemma 11.4. *Let b_0 be a rational number in the interval $[-2, 2]$ and use it to define a sequence by letting $b_{n+1} = b_n{}^2 - 2$, for all $n \geq 0$. Then*

(a) *b_n is a rational number in the interval $[-2, 2]$, for all $n \geq 0$;*

(b) *if $b_n = b_m$, where $n > m$, then b_m belongs to the set $\{\pm 1, \pm 2\}$;*

(c) *if there exists some $m \geq 0$ such that b_m belongs to the set $\{0, \pm 1, \pm 2\}$, then b_0 also belongs to the set $\{0, \pm 1, \pm 2\}$.*

Proof. Part (a) will follow using a fairly simple Mathematical Induction argument. First, the result clearly holds when $n = 0$. Next, suppose $k \geq 0$ such that b_k has the desired properties. We need to show that b_{k+1} also has the desired properties. Since $b_k \in \mathbb{Q}$, it follows that $b_{k+1} = b_k{}^2 - 2$ is also rational. Next, since $b_k \in [-2, 2]$, we see that $b_k{}^2 \in [0, 4]$, so $b_{k+1} = b_k{}^2 - 2 \in [-2, 2]$. Thus, b_{k+1} does inherit the desired properties from b_k, thereby concluding the proof of (a).

For part (b), let $f(x) = x^2 - 2$ and let $f^l(x)$ denote the composition of l copies of $f(x)$, for all $l \in \mathbb{N}$. Observe that $b_l = f^l(b_0)$, for all $l \geq 1$ and if $t = n - m$, then $b_n = f^t(b_m)$. We now claim that if $l \geq 2$, then $f^l(x)$ is a monic polynomial of degree 2^l with a constant term equal to 2. First, observe that

$$f^2(x) = f(f(x)) = f(x^2 - 2) = (x^2 - 2)^2 - 2 = (x^4 - 4x^2 + 4) - 2 = x^{2^2} - 4x^2 + 2,$$

thereby settling the $l = 2$ case.

Next, suppose that $k \geq 2$ such that $f^k(x)$ has the desired properties. Then

$$f^{k+1}(x) = f(f^k(x)) = f\left(x^{2^k} + \text{lower-degree terms} + 2\right) =$$

$$(x^{2^k} + \text{lower-degree terms} + 2)^2 - 2 = \left((x^{2^k})^2 + \text{lower-degree terms} + 4\right) - 2 =$$

$$x^{2^{k+1}} + \text{lower-degree terms} + 2.$$

Therefore, $f^{k+1}(x)$ inherits the desired properties from $f^k(x)$. Thus, for all $l \geq 2$, $f^l(x)$ is indeed monic of degree 2^l with a constant term of 2.

Now, let $c = b_m$ and also let $t = n - m$. Since $b_n = b_m$, we have

$$c = b_m = b_n = f^t(b_m) = f^t(c).$$

Therefore, c is a root of the polynomial $f^t(x) - x$.

From part (a), c is a rational number in the interval $[-2, 2]$. Using our preceding argument, if $t \geq 2$, then $f^t(x) - x$ is monic with constant term 2. On the other hand, if $t = 1$, then $f^t(x) - x$ is monic with constant term -2. Regardless of which case we are in, since c is a root of $f^t(x) - x$, we can apply the Rational Root Test to assert that c must be an integer that divides 2. Hence, $b_m = c \in \{\pm 1, \pm 2\}$, concluding the proof of (b).

For part (c), we will proceed using a proof by contradiction. If $b_0 \notin \{0, \pm 1, \pm 2\}$, we can let k be the largest integer such that $0 \leq k \leq m - 1$ and $b_k \notin \{0, \pm 1, \pm 2\}$. Observe that if $k < m - 1$, then our choice of k guarantees that $b_{k+1} \in \{0, \pm 1, \pm 2\}$. On the other hand, if $k = m - 1$, then $k + 1 = m$ and $b_{k+1} = b_m \in \{\pm 1, \pm 2\}$. Therefore, in both cases, $b_{k+1} \in \{0, \pm 1, \pm 2\}$. Since $b_k^2 - 2 = b_{k+1} \in \{-2, -1, 0, 1, 2\}$, we immediately see that

$$b_k^2 \in \{0, 1, 2, 3, 4\}.$$

Using the fact that $b_k \in \mathbb{Q}$, it now follows that

$$b_k \in \{0, \pm 1, \pm 2\},$$

which contradicts our choice of k. Thus, $b_0 \in \{0, \pm 1, \pm 2\}$, as desired. $\square$

The most important piece of information obtained from Lemma 11.4 is that if the sequence $b_0, b_1, b_2, b_3, \ldots$ *ever* repeats a term, then the initial term b_0 must belong to $\{0, \pm 1, \pm 2\}$. We now show how the sequence in Lemma 11.4 is related to the double-angle formula for the cosine.

Lemma 11.5. *If $\cos(\theta) \in \mathbb{Q}$, let $b_0 = 2\cos(\theta)$ in the sequence defined in Lemma 11.4. Then $b_n = 2\cos(2^n \theta)$, for all $n \geq 0$.*

Proof. As you probably expected, we will prove this using Mathematical Induction. The $n = 0$ case is clear, so let us now suppose that $k \geq 0$ has the property that $b_k = 2\cos(2^k\theta)$. We need to show that $b_{k+1} = 2\cos(2^{k+1}\theta)$.

If we apply the double-angle formula to the angle $2^k\theta$, we obtain

$$b_{k+1} = b_k{}^2 - 2 = \left(2\cos(2^k\theta)\right)^2 - 2 = 4\cos^2(2^k\theta) - 2 =$$
$$2(2\cos^2(2^k\theta) - 1) = 2\left(\cos(2 \cdot 2^k\theta)\right) = 2\cos(2^{k+1}\theta),$$

as desired. $\qquad\qquad\qquad\qquad\qquad\qquad\qquad\qquad\qquad\qquad\qquad\qquad\qquad\qquad\qquad\quad\square$

We now have all the pieces we need to prove Theorem 11.1.

Proof of Theorem 11.1. Suppose both α and $\cos^2(\alpha\pi)$ are rational. Let $\theta = 2\alpha\pi$; then the double-angle formula for the cosine tells us that

$$\cos(\theta) = \cos(2 \cdot \alpha\pi) = 2\cos^2(\alpha\pi) - 1 \in \mathbb{Q}.$$

Since $\cos(\theta) \in \mathbb{Q}$, we can let $b_0 = 2\cos(\theta)$ and use b_0 to generate the sequence described in Lemma 11.4.

We know that since $\alpha \in \mathbb{Q}$, we can write $\alpha = \frac{a}{b}$, where $a, b \in \mathbb{Z}$ and $b > 0$. Next, consider the sequence

$$a, 2a, 2^2a, 2^3a, 2^3a, \ldots.$$

When we divide integers by b, there are only b possible remainders. Therefore, there exist terms in the preceding sequence that yield the same remainder when divided by b. As a result, there exist integers $n > m \geq 0$ such that 2na and 2ma have the same remainder when divided by b. Therefore, we can write

$$2^na = 2^ma + sb,$$

where $s \in \mathbb{Z}$.

Doubling this equation results in

$$2^n(2a) = 2^m(2a) + 2sb.$$

Dividing by b yields

$$2^n\left(\frac{2a}{b}\right) = 2^m\left(\frac{2a}{b}\right) + 2s.$$

Then multiplying by π gives us

$$2^n\left(\frac{2a}{b}\pi\right) = 2^m\left(\frac{2a}{b}\pi\right) + s(2\pi).$$

Since $2^n\left(\frac{2a}{b}\pi\right)$ and $2^m\left(\frac{2a}{b}\pi\right)$ differ by an integer multiple of 2π, they give us the same value when plugged into the cosine. Combining this with the facts that $\theta = 2\alpha\pi$ and $\alpha = \frac{a}{b}$, we obtain

$$\cos(2^n\theta) = \cos(2^n \cdot 2\alpha\pi) = \cos\left(2^n \cdot \frac{2a}{b}\pi\right) =$$

$$\cos\left(2^m \cdot \frac{2a}{b}\pi\right) = \cos(2^m \cdot 2\alpha\pi) = \cos(2^m\theta).$$

It now follows from Lemma 11.5 and the previous equation that

$$b_n = 2\cos(2^n\theta) = 2\cos(2^m\theta) = b_m.$$

As a result, the sequence defined in Lemma 11.4 has repetitions and part (c) of the lemma tells us that $b_0 \in \{0, \pm1, \pm2\}$. The double-angle formula now tells us that

$$b_0 = 2\cos(\theta) = 2\cos(2 \cdot \alpha\pi) = 2(2\cos^2(\alpha\pi) - 1).$$

Thus,

$$2(2\cos^2(\alpha\pi) - 1) \in \{-2, -1, 0, 1, 2\}.$$

A series of simple calculations now yields

$$2\cos^2(\alpha\pi) - 1 \in \left\{-1, -\frac{1}{2}, 0, \frac{1}{2}, 1\right\},$$

$$2\cos^2(\alpha\pi) \in \left\{0, \frac{1}{2}, 1, \frac{3}{2}, 2\right\},$$

$$\cos^2(\alpha\pi) \in \left\{0, \frac{1}{4}, \frac{1}{2}, \frac{3}{4}, 1\right\},$$

and finally,

$$\cos(\alpha\pi) \in \left\{0, \pm\frac{1}{2}, \pm\frac{\sqrt{2}}{2}, \pm\frac{\sqrt{3}}{2}, \pm1\right\},$$

thereby concluding the proof. $\qquad\qquad\qquad\square$

It is now an easy task to use Theorem 11.1 to prove Corollary 11.2.

Proof of Corollary 11.2. Suppose both α and $\sin^2(\alpha\pi)$ are rational. In light of the identity $\sin^2(\theta) + \cos^2(\theta) = 1$, it follows that $\cos^2(\alpha\pi)$ is also rational. Theorem 11.1 now tells us that

$$\cos(\alpha\pi) \in \left\{0, \pm\frac{1}{2}, \pm\frac{\sqrt{2}}{2}, \pm\frac{\sqrt{3}}{2}, \pm1\right\}.$$

Thus,

$$\cos^2(\alpha\pi) \in \left\{0, \frac{1}{4}, \frac{1}{2}, \frac{3}{4}, 1\right\}$$

and the identity $\sin^2(\theta) + \cos^2(\theta) = 1$ implies that

$$\sin^2(\alpha\pi) \in \left\{0, \frac{1}{4}, \frac{1}{2}, \frac{3}{4}, 1\right\}.$$

It now immediately follows that

$$\sin(\alpha\pi) \in \left\{0, \pm\frac{1}{2}, \pm\frac{\sqrt{2}}{2}, \pm\frac{\sqrt{3}}{2}, \pm 1\right\}.$$

$\Box$

Using Theorem 11.1 to prove Corollary 11.3 is also a fairly straightforward task.

Proof of Corollary 11.3. Suppose both α and $\tan^2(\alpha\pi)$ are both rational. Using the identity $\sec^2(\theta) = \tan^2(\theta) + 1$, we see that $\sec^2(\alpha\pi)$ is also rational. Since the secant function is the reciprocal of the cosine function, we know that $\cos^2(\alpha\pi)$ is rational.

Observe that since $\tan^2(\alpha\pi)$ is rational, $\cos(\alpha\pi)$ cannot be equal to 0. Therefore, in our present situation, Theorem 11.1 tells us that

$$\cos(\alpha\pi) \in \left\{\pm\frac{1}{2}, \pm\frac{\sqrt{2}}{2}, \pm\frac{\sqrt{3}}{2}, \pm 1\right\}.$$

A short series of calculations tells us that

$$\cos^2(\alpha\pi) \in \left\{\frac{1}{4}, \frac{1}{2}, \frac{3}{4}, 1\right\},$$

$$\sec^2(\alpha\pi) \in \left\{4, 2, \frac{4}{3}, 1\right\},$$

$$\sec^2(\alpha\pi) - 1 \in \left\{3, 1, \frac{1}{3}, 0\right\}.$$

Using the fact that $\tan^2(\theta) = \sec^2(\theta) - 1$, we see that

$$\tan^2(\alpha\pi) \in \left\{3, 1, \frac{1}{3}, 0\right\},$$

which leads to

$$\tan(\alpha\pi) = \left\{0, \pm\frac{\sqrt{3}}{3}, \pm 1, \pm\sqrt{3}\right\},$$

as desired.

$\Box$

At this point, we need a little bit of terminology in order to make some additional observations about values of trigonometric functions. Real numbers that are the root of a nonconstant polynomial with integer coefficients are known as **algebraic**. Certainly every rational number is algebraic. To see this, observe that if $\alpha \in \mathbb{Q}$, then we can write $\alpha = \frac{a}{b}$, where $a \in \mathbb{Z}$ and $b \in \mathbb{N}$. Then α is a root of the polynomial $bx - a$.

There are also many irrational numbers that are algebraic. As we have seen, numbers such as $2^{\frac{1}{2}}, 5^{\frac{1}{3}}$, and $7^{\frac{3}{4}}$ are all irrational. Yet $2^{\frac{1}{2}}$ is a root of $x^2 - 2$, $5^{\frac{1}{3}}$ is a root of $x^3 - 5$, and $7^{\frac{3}{4}}$ is a root of $x^4 - 243$. Thus, $2^{\frac{1}{2}}, 5^{\frac{1}{3}}$, and $7^{\frac{3}{4}}$ are all algebraic. More complicated-looking numbers like $\sqrt{3} + \sqrt{5}$ and $\left(\frac{2-11^{\frac{6}{7}}}{19^{\frac{1}{2}}} \right)^{\frac{1}{3}}$ are also algebraic. By doing a series of computations, it is not too hard to find nonconstant polynomials that have these numbers as roots.

Real numbers that are not algebraic are known as **transcendental**. Although transcendental numbers abound, it is often quite difficult to determine whether a particular number is transcendental. It turns out that both e and π are transcendental. Whereas it takes a good deal of work to verify that e is transcendental, it takes even more to do the same for π.

In light of Theorem 11.1, we know that $\cos(1°)$, $\sin(40°)$, and $\tan(83°)$ are all irrational. However, we will soon show that all three are algebraic. More precisely, after returning to radian measure, we will see that if $\alpha \in \mathbb{Q}$ such that the value of one of the six trigonometric functions exists when plugging in $\alpha\pi$, then that value must be algebraic. We will begin by looking at the cosine and sine. The idea will be to use DeMoivre's Theorem to find a polynomial that has both $\cos(\alpha\pi)$ and $\sin(\alpha\pi)$ as roots.

Theorem 11.6. *If $\alpha \in \mathbb{Q}$, then both $\cos(\alpha\pi)$ and $\sin(\alpha\pi)$ are both algebraic.*

Proof. Since $\alpha \in \mathbb{Q}$, we can write $\alpha = \frac{a}{b}$, where $a \in \mathbb{Z}$ and $b \in \mathbb{N}$. DeMoivre's Theorem tells us that

$$(\cos(\theta) + i \sin(\theta))^{4b} = \cos(4b\theta) + i \sin(4b\theta).$$

When we use the binomial theorem to expand the left-hand side of the equation and then compare the real parts of both sides of the equation, we obtain

$$\cos^{4b}(\theta) - \binom{4b}{2} \cos^{4b-2}(\theta) \sin^2(\theta) + \binom{4b}{4} \cos^{4b-4}(\theta) \sin^4(\theta) - \cdots$$

$$- \binom{4b}{4b-2} \cos^2(\theta) \sin^{4b-2}(\theta) + \sin^{4b}(\theta) = \cos(4b\theta).$$

On the left-hand side of the equation, we can use the identity $\sin^2(\theta) + \cos^2(\theta) = 1$ to replace every term of the form $\sin^{2t}(\theta)$ by $(1 - \cos^2(\theta))^t$. This will result in the left-hand side becoming a sum of terms of the form $c\cos^{2t}(\theta)$, where $c \in \mathbb{Z}$ and $t \geq 0$. As a result, there exists a nonconstant polynomial $f(x) \in \mathbb{Z}[x]$, in which x only appears with even exponents, such that

$$f(\cos(\theta)) = \cos(4b\theta).$$

If we let $\theta = \alpha\pi$ in the previous equation, then we see that

$$f(\cos(\alpha\pi)) = \cos(4b \cdot \alpha\pi) = \cos\left(4b \cdot \frac{a}{b}\pi\right) = \cos(4a\pi) = \cos(2a \cdot 2\pi) = 1.$$

Therefore, $\cos(\alpha\pi)$ is a root of the polynomial $f(x) - 1$, so $\cos(\alpha\pi)$ is algebraic.

It now suffices to show that $\sin(\alpha\pi)$ is also a root of $f(x) - 1$. Since $\frac{\pi}{2} - \alpha\pi$ and $\alpha\pi$ add up to $\frac{\pi}{2}$, it follows that $\cos\left(\frac{\pi}{2} - \alpha\pi\right) = \sin(\alpha\pi)$. Therefore, if we let $\theta = \frac{\pi}{2} - \alpha\pi$ in the equation $f(\cos(\theta)) = \cos(4b\theta)$, we obtain

$$f(\sin(\alpha\pi)) = f\left(\cos\left(\frac{\pi}{2} - \alpha\pi\right)\right) = \cos\left(4b\left(\frac{\pi}{2} - \alpha\pi\right)\right) =$$
$$\cos\left(4b\left(\frac{\pi}{2} - \frac{a}{b}\pi\right)\right) = \cos((b - 2a)2\pi) = 1.$$

Thus, $f(\sin(\alpha\pi)) - 1 = 0$, so $\sin(\alpha\pi)$ is also a root of $f(x) - 1$.   ☐

In order to examine the other trigonometric functions, we need a lemma that essentially says that if β is a nonzero algebraic number, then β^{-1} is also algebraic.

Lemma 11.7. *Suppose $\beta, a_0, a_1, \ldots, a_{n-1}, a_n$ belong to a field such that β is a nonzero root of the polynomial $a_n x^n + a_{n-1}x^{n-1} + \cdots + a_1 x + a_0$. Then β^{-1} is a root of the polynomial $a_0 x^n + a_1 x^{n-1} + \cdots + a_{n-1}x + a_n$.*

Proof. Plugging β into the polynomial $a_n x^n + a_{n-1}x^{n-1} + \cdots + a_1 x + a_0$ results in

$$a_n \beta^n + a_{n-1}\beta^{n-1} + \cdots + a_1\beta + a_0 = 0.$$

Multiplying both sides of this equation by $(\beta^{-1})^n$ yields

$$a_n + a_{n-1}\beta^{-1} + \cdots + a_1 (\beta^{-1})^{n-1} + a_0 (\beta^{-1})^n = 0.$$

However, it is easy to see that the preceding equation says that β^{-1} is indeed a root of $a_0 x^n + a_1 x^{n-1} + \cdots + a_{n-1}x + a_n$.   ☐

We can now extend Theorem 11.6 to the other four trigonometric functions. When reading the statement of this result, remember that whereas the cosine and sine function are defined for all real numbers, there exist $\alpha \in \mathbb{Q}$ such that either $\sec(\alpha\pi)$, $\csc(\alpha\pi)$, $\tan(\alpha\pi)$, or $\cot(\alpha\pi)$ may fail to be defined.

Theorem 11.8. *If* $\cos(\alpha\pi)$, $\sin(\alpha\pi)$, $\tan(\alpha\pi)$, $\sec(\alpha\pi)$, $\csc(\alpha\pi)$, *or* $\cot(\alpha\pi)$ *is defined, where* $\alpha \in \mathbb{Q}$, *then it must be algebraic.*

Proof. Since $\alpha \in \mathbb{Q}$, Theorem 11.6 asserts that both $\cos(\alpha\pi)$ and $\sin(\alpha\pi)$ are algebraic. Recall that the secant and the cosecant are the reciprocals of the cosine and sine. Therefore, Lemma 11.7 tells us that if $\sec(\alpha\pi)$ or $\csc(\alpha\pi)$ is defined, then it must be algebraic.

Next, if $\cot(\alpha\pi)$ is defined, then either $\cot(\alpha\pi) = 0$ or $\cot(\alpha\pi) = \frac{1}{\tan(\alpha\pi)}$. In the first case, $\cot(\alpha\pi)$ is clearly algebraic. Therefore, in light of Lemma 11.7, to conclude the proof, it suffices to show that if $\tan(\alpha\pi)$ is defined, then it is algebraic. At this point, we need to refer back to the proof of Theorem 11.6 and look at the nonconstant polynomial $f(x) \in \mathbb{Z}[x]$ that had the property that $\cos(\alpha\pi)$ was a root of $f(x) - 1$. The idea behind this proof is to modify $f(x) - 1$ to find a nonconstant element of $\mathbb{Z}[x]$ that has $\tan(\alpha\pi)$ as a root.

Since $\tan(\alpha\pi)$ is defined, we know that $\cos(\alpha\pi) \neq 0$. Therefore, $\sec(\alpha\pi) = \frac{1}{\cos(\alpha\pi)}$ is also defined. Lemma 11.7 allows us to modify the polynomial $f(x) - 1$ to obtain a polynomial $g(x)$ which has $\sec(\alpha\pi)$ as a root. Recall that whenever x appeared in $f(x)$, it had an even exponent. It is easy to see that $f(x) - 1$ has this property and the proof of Lemma 11.7 indicates that this property is also inherited by $g(x)$. Since all the exponents of x in $g(x)$ are even, there exists a nonconstant $h(x) \in \mathbb{Z}[x]$ such that $h(x^2) = g(x)$. Therefore,

$$h(\sec^2(\alpha\pi)) = g(\sec(\alpha\pi)) = 0.$$

Now, let $M(x) = h(x^2 + 1)$. Clearly, $M(x)$ is a nonconstant polynomial with integer coefficient. Using the identity $\sec^2(\theta) = \tan^2(\theta) + 1$, we now have

$$M(\tan(\alpha\pi)) = h(\tan^2(\alpha\pi) + 1) = h(\sec^2(\alpha\pi)) = 0.$$

Thus, $\tan(\alpha\pi)$ is a root of $M(x)$, hence $\tan(\alpha\pi)$ is algebraic, thereby concluding the proof. $\square$

Exercises for Section 11.1

Many of the exercises in this chapter will use DeMoivre's Theorem. You will frequently be required to find polynomials in $\mathbb{Z}[x]$ that, for various values of θ, have either $\cos(\theta)$ or $\sin(\theta)$ as a root.

1. Derive the formula $\cos(3\theta) = 4\cos^3(\theta) - 3\cos(\theta)$.

2. Derive the formula $\sin(3\theta) = -4\sin^3(\theta) + 3\sin(\theta)$.

3. Use exercise 1 to show that $\cos(20°)$ is a root of the polynomial $8x^3 - 6x - 1$.

4. Show that $8x^3 - 6x - 1$ is irreducible in $\mathbb{Q}[x]$ and explain why it has three distinct roots in $\mathbb{C}$.

5. Use exercise 1 to show that $\cos(100°)$ and $\cos(140°)$ are the other two roots of $8x^3 - 6x - 1$.

6. Show that the polynomial $8x^3 - 6x + 1$ is also irreducible in $\mathbb{Q}[x]$ and has three distinct roots in $\mathbb{C}$.

7. Use exercise 1 to show that $\cos(40°)$, $\cos(80°)$, and $\cos(160°)$ are the three roots of $8x^3 - 6x + 1$.

8. Use exercise 2 to show that $\sin(10°)$, $\sin(50°)$, and $\sin(250°)$ are also the three roots of $8x^3 - 6x + 1$. Pair up these roots with the three roots you found in exercise 7.

For exercises 9–13, please read the following:

In Chapter 11, we indicated that there was a surprising lack of symmetry between the sine and cosine functions as the main results of Chapter 11 could be obtained by using the cosine but not the sine. However, a slightly weaker version of Theorem 11.1 can be obtained using the sine function. In particular, we can prove that the only rational values of the sine function that occur when plugging in angles of the form $\alpha\pi$, where $\alpha \in \mathbb{Q}$, must belong to the set $\{0, \pm\frac{1}{2}, \pm 1\}$. We will work through the details of this proof in the next five exercises. To begin, let b_0 be a rational number in the interval $[-2, 2]$ and then use it to define a sequence by letting $b_{n+1} = -b_n^3 + 3b_n$, for $n \geq 0$.

9. Show that, for every $n \geq 0$, b_n is a rational number in the interval $[-2, 2]$.

10. Show that if $n > m$ and $b_n = b_m$, then $b_m \in \{0, \pm 1, \pm 2\}$.

11. Show that if $b_m \in \{0, \pm 1, \pm 2\}$, for some $m \geq 0$, then $b_0 \in \{0, \pm 1, \pm 2\}$.

12. Show that if $b_0 = 2\sin(\theta)$, then $b_n = 2\sin(3^n\theta)$, for all $n \geq 0$.

13. Show that if both α and $\sin(\alpha\pi)$ belong to $\mathbb{Q}$, then $\sin(\alpha\pi) \in \{0, \pm\frac{1}{2}, \pm 1\}$.

For exercises 14–20, please read the following:

In this next set of exercises, we will compute the exact values of $\cos(36°)$ and $\cos(72°)$. In light of Theorem 11.1, neither of these values of the cosine is rational, nor is it the square root of a rational number. However, as we will soon see, they are both roots of quadratic polynomials in $\mathbb{Z}[x]$.

14. Derive the formula $2\cos(4\theta) = (2\cos(\theta))^4 - 4(2\cos(\theta))^2 + 2$.

15. Show that $2\cos(36°)$ is a root of the polynomial $x^4 - 4x^2 + x + 2$.

16. Use the Rational Root Test to find all rational roots of $x^4 - 4x^2 + x + 2$.

17. Find a monic, quadratic polynomial with integer coefficients which has $2\cos(36°)$ as a root.

18. Find the exact value of $\cos(36°)$.

19. Show that $2\cos(108°)$ is the other root of the polynomial in exercise 17.

20. Find the exact value of $\cos(72°)$.

For exercises 21–25, please read the following:

In this set of exercises, we will compute the exact values of $\cos(18°)$ and $\cos(54°)$. As we shall see, these values of the cosine satisfy polynomials in $\mathbb{Q}[x]$ of degree 4 but do not satisfy any nonconstant polynomial of degree less than 4.

21. Derive the formula $2\cos(5\theta) = (2\cos(\theta))^5 - 5(2\cos(\theta))^3 + 5(2\cos(\theta))$.

22. Show that $2\cos(18°)$ is a root of $x^4 - 5x^2 + 5$.

23. Show that the four roots of $x^4 - 5x^2 + 5$ are $2\cos(18°)$, $2\cos(54°)$, $2\cos(126°)$, $2\cos(162°)$.

24. Find the exact value of $\cos(18°)$ and the exact value of $\cos(54°)$.

25. Find a monic irreducible polynomial in $\mathbb{Q}[x]$ which has $\cos(18°)$ and $\cos(54°)$ as two of its roots.

For exercises 26–30, please read the following:

Next, we will compute the exact values of $\cos(22.5°)$ and $\cos(67.5°)$. These values of the cosine will also satisfy polynomials in $\mathbb{Q}[x]$ of degree 4 but satisfy no nonconstant polynomial of degree less than 4.

26. Use the formula in exercise 14 to find a monic polynomial of degree 4 in $\mathbb{Z}[x]$ that has $2\cos(22.5°)$ as a root.

27. Show that the polynomial in exercise 26 is irreducible over $\mathbb{Q}$.

28. Show the other three roots of the polynomial obtained in exercise 26 are $2\cos(67.5°)$, $2\cos(112.5°)$, and $2\cos(157.5°)$.

29. Find the exact values of $\cos(22.5°)$ and $\cos(67.5°)$.

30. Find a monic irreducible polynomial in $\mathbb{Q}[x]$ that has $\cos(22.5°)$ and $\cos(67.5°)$ as two of its roots.

31. If $n \in \mathbb{N}$ is even, show that there exists a polynomial of degree n in $\mathbb{Q}[x]$ that has $\cos\left(\frac{\pi}{2n}\right)$ as a root.

32. If $n \geq 3$ is odd, show that there exists a polynomial of degree $n - 1$ in $\mathbb{Q}[x]$ that has $\cos\left(\frac{\pi}{2n}\right)$ as a root.

33. If p is an odd prime, examine the polynomial obtained in exercise 32, and then show that there exists an irreducible polynomial of degree $p-1$ in $\mathbb{Q}[x]$ that has $\cos\left(\frac{\pi}{2p}\right)$ as a root.

34. Show that the polynomial in exercise 33 has $p-1$ distinct roots and these roots are all of the form $\cos\left(\left(\frac{2i-1}{2p}\right)\pi\right)$, where i takes on all integer values from 1 to p *except* for $i = \frac{p+1}{2}$.

35. Use exercise 34 to find the smallest degree of a nonconstant polynomial over $\mathbb{Q}$ that has $\cos\left(\frac{\pi}{14}\right)$ as a root, and then find all the roots of this polynomial.

36. Use exercise 34 to find the smallest degree of a nonconstant polynomial over $\mathbb{Q}$ that has $\cos\left(\frac{\pi}{22}\right)$ as a root, and then find all the roots of this polynomial.

CHAPTER 12

Polynomials over Arbitrary Fields

To this point, we have looked at various examples of commutative rings. In Chapter 3, we took an in-depth look at our first example, the integers. While studying $\mathbb{Z}$, we examined four concepts of particular importance: prime numbers, the division algorithm, the Euclidean Algorithm, and the existence and uniqueness of prime factorization. In many commutative rings, analogs of these concepts do not exist in any recognizable form. However, it may come as a pleasant surprise that for any field F, similar concepts exist in $F[x]$. In fact, when studying $F[x]$, many of the proofs and ideas will be almost identical to the ones you saw when we studied $\mathbb{Z}$. Since $F[x]$ is a more abstract object, as you work through the proofs in this chapter, it might be worthwhile to go back and review some of the proofs in Chapter 3.

There are many important facts about polynomials that are stated without proof in high school algebra courses. In this chapter, we will prove several of these, such as

(i)   if $f(x) \in F[x]$ and if $\alpha \in F$, then α is a root of $f(x)$ if and only if $x - \alpha$ is a factor of $f(x)$;

(ii) if $f(x) \in \mathbb{R}[x]$, then $f(x)$ can be written as a product of linear polynomials and irreducible quadratics in $\mathbb{R}[x]$;

(iii) if $f(x) \in F[x]$ has degree n, then $f(x)$ has at most n roots in F.

12.1 Similarities between Polynomials and Integers

We now start down a path that will reveal remarkable similarities between $\mathbb{Z}$ and $F[x]$ by looking at divisibility. Throughout this chapter, F will always be a field, and $F[x]$ will always be the set of polynomials with coefficients if F. In many of the proofs in this chapter, we will use notation similar to that used in Chapter 3. Hopefully, this should make the similarities between $\mathbb{Z}$ and $F[x]$ even more apparent.

Definition 12.1. *Given polynomials $a(x), b(x) \in F[x]$ (with $a(x) \neq 0$), we say that $a(x)$ divides $b(x)$, written $a(x) \mid b(x)$, if there exists a polynomial $m(x) \in F[x]$ such that $b(x) = a(x) \cdot m(x)$. In this case, we also say that $a(x)$ is a divisor of $b(x)$.*

If $a(x)$ is not a divisor of $b(x)$, we write $a(x) \nmid b(x)$.

■ Examples

We will now look at examples of divisibility in $\mathbb{Q}[x]$, $\mathbb{R}[x]$, $\mathbb{C}[x]$, and $\mathbb{Z}_5[x]$.

1. In $\mathbb{Q}[x]$,

$$(x^2+1) \mid (x^3-5x^2+x-5) \quad \text{as} \quad x^3-5x^2+x-5 = (x^2+1)(x-5)$$

and

$$(2x+7) \mid (6x^3+29x^2+26x-7) \quad \text{as} \quad 6x^3+29x^2+26x-7 = (2x+7)(3x^2+4x-1).$$

2. In $\mathbb{R}[x]$,

$$(x-\sqrt{2}) \mid (x^2-2) \quad \text{as} \quad x^2-2 = (x-\sqrt{2})(x+\sqrt{2})$$

and

$$(3x^2+5) \mid (21x^4+11x^2-40) \quad \text{as} \quad 21x^4+11x^2-40 = (3x^2+5)(7x^2-8).$$

3. In $\mathbb{C}[x]$,

$$(x+i) \mid (x^2+1) \quad \text{as} \quad x^2+1 = (x+i)(x-i),$$
$$(x+5) \mid (x^2-2x-35) \quad \text{as} \quad x^2-2x-35 = (x+5)(x-7),$$

and

$$(x+2+i) \mid (2x^3+9x^2+14x+5) \quad \text{as} \quad 2x^3+9x^2+14x+5 =$$
$$(x+2+i)(x+2-i)(2x+1) = (x+2+i)\left(2x^2+(5-2i)x+(2-i)\right).$$

4. In $\mathbb{Z}_5[x]$,

$$(x+[2]_5) \mid (x^2+[1]_5) \quad \text{as} \quad x^2+[1]_5 = (x+[2]_5)(x+[3]_5)$$

and

$$(x+[3]_5) \mid (x^5+[3]_5) \quad \text{as} \quad x^5+[3]_5 = (x+[3]_5)^5.$$

Note that if $b(x) = a(x) \cdot m(x) \in F[x]$ then, for any nonzero $\alpha \in F$, we have

$$b(x) = (\alpha a(x)) \cdot \left(\alpha^{-1}m(x)\right).$$

This tells us if $a(x)$ is a divisor of $b(x)$, then so is $\alpha a(x)$, for every nonzero $\alpha \in F$.

■

We can also list many polynomials $a(x), b(x)$ where $a(x)$ is not a divisor of $b(x)$. For example, in $\mathbb{Q}[x]$, it is not hard to see that

$$(x+1) \nmid (x+2), \quad (x^3+3) \nmid (x^2+5x+2), \quad \text{and} \quad x \nmid (x^2-1).$$

Let us now consider the following question: Is 5 a divisor of 2? Certainly one is tempted to immediately answer no. This is certainly the correct answer if we are working in $\mathbb{Z}$. On the other hand, suppose we are working in $\mathbb{Q}[x]$. In this situation, $2 = 5 \cdot \frac{2}{5}$, so 5 is a divisor of 2 in $\mathbb{Q}[x]$. One of the differences between working in $\mathbb{Z}$ and working in $\mathbb{Q}[x]$ is that 5 has a multiplicative inverse in $\mathbb{Q}[x]$ but not in $\mathbb{Z}$. Observe that in $\mathbb{Z}$ the only elements with multiplicative inverses are ± 1, whereas in $\mathbb{Q}[x]$ the elements with multiplicative inverses are precisely the nonzero elements of $\mathbb{Q}$.

The preceding example indicates, once again, that the answers to many questions in algebra often depend on the context. Previously we had observed that the question of whether a polynomial is irreducible also depends on the context. For example, $x^2 - 2$ is irreducible in $\mathbb{Q}[x]$ but is reducible in $\mathbb{R}[x]$. Since enlarging a field can certainly affect whether a polynomial is irreducible, it is natural to wonder if enlarging a field can affect whether one polynomial is a divisor of the other. To investigate this question, suppose $F \subseteq K$ are fields and suppose $a(x), b(x)$ belong to $F[x]$. In one direction, it is not hard to see that if $a(x)$ is a divisor of $b(x)$ in $F[x]$, then $a(x)$ is also a divisor of $b(x)$ in $K[x]$. To see this, observe that there exists some $m(x) \in F[x]$ such that $b(x) = a(x) \cdot m(x)$. But since $m(x)$ also belongs to $K[x]$, it is now clear that $a(x)$ is also a divisor of $b(x)$ in $K[x]$. The other direction is the harder one. If $a(x)$ is a divisor of $b(x)$ in $K[x]$, we need to determine if $a(x)$ is also a divisor of $b(x)$ in $F[x]$. The answer to this question will appear later in this chapter as one of the applications of the division algorithm.

Since every nonzero element of F has a multiplicative inverse in F, every nonzero element of F is a divisor of every polynomial in $F[x]$. To see this, observe that if $f(x) \in F[x]$ and if α is a nonzero element of F, then $f(x) = \alpha \cdot (\alpha^{-1} f(x))$. We should also note that, similar to the situation in $\mathbb{Z}$, every nonzero polynomial in $F[x]$ is a divisor of 0.

Whereas the building blocks of the integers are the prime numbers, the analogous role in $F[x]$ will be played by the irreducible polynomials. The next definition is essentially a restatement of Definition 9.8, and, for convenience, we include it here as

Definition 12.2. *Let $f(x) \in F[x]$ be a polynomial of degree at least 1. We say that $f(x)$ is reducible in $F[x]$ if there exist $g(x), h(x) \in F[x]$, both with smaller degree than $f(x)$, such that $f(x) = g(x) \cdot h(x)$. If $f(x) \in F[x]$ has degree at least 1 and is not reducible in $F[x]$, then we say that $f(x)$ is irreducible in $F[x]$.*

Recall that to prove the uniqueness of prime factorization in $\mathbb{Z}$, it was important that we not consider 1 to be a prime number. For virtually identical reasons, we do not consider polynomials to be irreducible in $F[x]$ unless their degree is at least 1. To see this, observe that in $\mathbb{Q}[x]$, if we considered the degree 0 polynomial $g(x) = 2$ to be irreducible, then polynomials could not be factored uniquely. For example, we could factor the polynomial $f(x) = x$ in many different ways as

$$x = 2 \cdot \frac{1}{2} \cdot x = 2 \cdot 2 \cdot \frac{1}{2} \cdot \frac{1}{2} \cdot x = 2 \cdot 2 \cdot 2 \cdot \frac{1}{2} \cdot \frac{1}{2} \cdot \frac{1}{2} \cdot x = 2 \cdot 2 \cdot 2 \cdot 2 \cdot \frac{1}{2} \cdot \frac{1}{2} \cdot \frac{1}{2} \cdot \frac{1}{2} \cdot x.$$

Therefore, in order to prove a unique factorization theorem in $F[x]$, we will not consider polynomials of degree 0 to be irreducible.

As pointed out in Chapter 9, all polynomials of degree 1 in $F[x]$ are irreducible. In light of this, we can factor $6x^2 - 6$ into a product of two irreducible polynomials in $\mathbb{Q}[x]$ in many different ways. For example, we have

$$6x^2 - 6 = (x+1)(6x-6) = (2x+2)(3x-3) = (3x+3)(2x-2) =$$

$$(6x+6)(x-1) = \left(\frac{1}{2}x + \frac{1}{2}\right)(12x - 12) = (60x + 60)\left(\frac{1}{10}x - \frac{1}{10}\right).$$

This example indicates that in order to prove a unique factorization theorem, we will need the following concept.

Definition 12.3. *We say that $f(x) \in F[x]$ is monic if its leading coefficient is* 1.

Returning to the polynomial $6x^2 - 6$, observe that the only monic polynomials of degree 1 in $\mathbb{Q}[x]$ which are divisors of $6x^2 - 6$ are $x + 1$ and $x - 1$. We now have

$$6x^2 - 6 = 6 \cdot (x+1) \cdot (x-1).$$

Note that, except for juggling the order of 6, $x + 1$, and $x - 1$, there appears to be no other way to factor $6x^2 - 6$ in $\mathbb{Q}[x]$ as the product of an element of $\mathbb{Q}$ and monic, irreducible polynomials in $\mathbb{Q}[x]$. This indicates that the only irreducible polynomials a unique factorization theorem should refer to are the monic ones. We can now state the main result of this chapter.

Theorem 12.4—Unique Factorization Theorem. *Every polynomial $f(x) \in F[x]$ of degree at least one can be written uniquely (up to order) as a product of an element of F and monic, irreducible polynomials in $F[x]$.*

As was the case for Theorem 3.3, this theorem has two parts. First, we need to show that every $f(x) \in F[x]$ of degree at least 1 *can* be written as a product of an element of F and monic, irreducible polynomials. That will be the easier task, and it will be accomplished by using a proof by contradiction along with the Well Ordering Principle. When a polynomial is written as a product of an element of F and monic, irreducible polynomials from $F[x]$, we say that it has been **completely factored**. Looking back at the proof of the second part of Theorem 3.3, it should come as no surprise that the proof of the uniqueness part of Theorem 12.4 will require some additional mathematical machinery.

■ Examples

We will examine the factorization of $f(x) = 7x^4 - 7x^2 - 140$ in $\mathbb{Q}[x]$, $\mathbb{R}[x]$, and $\mathbb{C}[x]$.

1. In $\mathbb{Q}[x]$, we begin by factoring out the leading coefficient to obtain

$$f(x) = 7(x^4 - x^2 - 20).$$

Next, we need to factor $x^4 - x^2 - 20$ into monic, irreducible polynomials in $\mathbb{Q}[x]$. The Rational Root Test shows that $x^4 - x^2 - 20$ has no rational roots, so $x^4 - x^2 - 20$ has no factor in $\mathbb{Q}[x]$ of degree 1. However, it is not hard to see that we can factor $x^4 - x^2 - 20$ into two quadratics in $\mathbb{Q}[x]$, giving us

$$f(x) = 7(x^2 - 5)(x^2 + 4).$$

Since $x^2 - 5$ and $x^2 + 4$ are both irreducible in $\mathbb{Q}[x]$, we have now completely factored $f(x)$ in $\mathbb{Q}[x]$.

2. In $\mathbb{R}[x]$, since $\mathbb{Q} \subseteq \mathbb{R}$, we go back to the factorization of $f(x)$ in $\mathbb{Q}[x]$ and examine whether the factors $x^2 - 5$ and $x^2 + 4$ remain irreducible in $\mathbb{R}[x]$. It easy to see that in $\mathbb{R}[x]$, $x^2 - 5$ has monic, irreducible factors $x - \sqrt{5}$ and $x + \sqrt{5}$. It is also easy to see that $x^2 + 4$ remains irreducible in $\mathbb{R}[x]$. Therefore, the complete factorization of $f(x)$ in $\mathbb{R}[x]$ is

$$f(x) = 7(x - \sqrt{5})(x + \sqrt{5})(x^2 + 4).$$

3. In $\mathbb{C}[x]$, since $\mathbb{R} \subseteq \mathbb{C}$, we go back to the factorization of $f(x)$ in $\mathbb{R}[x]$ and examine whether the factor $x^2 + 4$ remains irreducible in $\mathbb{R}[x]$. However, $\pm 2i$ are roots in $\mathbb{C}$ of $x^2 + 4$, so $x^2 + 4$ has monic, irreducible factors $x - 2i$ and $x + 2i$ in $\mathbb{C}[x]$. Therefore, the complete factorization of $f(x)$ in $\mathbb{C}[x]$ is

$$f(x) = 7(x - \sqrt{5})(x + \sqrt{5})(x - 2i)(x + 2i).$$

As this example indicates, by enlarging the field, the factorization of $f(x)$ can change. But this does not contradict the uniqueness portion of Theorem 12.4. Note that Theorem 12.4 asserts that the factorization of $f(x)$ is unique provided the field remains fixed. Thus, the only way to factor $7x^4 - 7x^2 - 140$ in $\mathbb{Q}[x]$ is as $7(x^2 - 5)(x^2 + 4)$, the only way to factor $7x^4 - 7x^2 - 140$ in $\mathbb{R}[x]$ is as $7(x - \sqrt{5})(x + \sqrt{5})(x^2 + 4)$, and the only way to factor $7x^4 - 7x^2 - 140$ in $\mathbb{C}[x]$ is as $7(x - \sqrt{5})(x + \sqrt{5})(x - 2i)(x + 2i)$.

4. For another example, consider the polynomial $g(x) = [2]_3 x^4 + [1]_3$ in $\mathbb{Z}_3[x]$. This polynomial can be factored as

$$g(x) = [2]_3(x + [1]_3)(x + [2]_3) \left(x^2 + [1]_3\right).$$

Since $x^2 + [1]_3$ is a quadratic with no roots in $\mathbb{Z}_3$, it is irreducible in $\mathbb{Z}_3[x]$. Therefore, $x + [1]_3$, $x + [2]_3$, and $x^2 + [1]_3$ are all monic and irreducible in $\mathbb{Z}_3[x]$. Thus, the preceding factorization is the complete factorization of $g(x) = [2]_3 x^4 + [1]_3$ in $\mathbb{Z}_3[x]$.

■

When completely factoring a polynomial, as a matter of convenience, we often do not factor out the leading term. Instead, we sometimes write our polynomial as a product of irreducible polynomials, some of which might not be monic. For example, if asked to completely factor $6x^2 + 7x - 5$ in $\mathbb{Q}[x]$, we are more likely to leave the answer as $6x^2 + 7x - 5 = (2x - 1)(3x + 5)$ as opposed to writing $6x^2 + 7x - 5 = 6\left(x - \frac{1}{2}\right)\left(x + \frac{5}{3}\right)$.

In our proof of the first part of Theorem 12.4, you should look for the similarities between this proof and the proof of the first part of Theorem 3.3.

Proof of the first part of Theorem 12.4. We proceed with a proof by contradiction and begin by supposing that there exists a polynomial $f(x) \in F[x]$ of degree at least 1 that cannot be completely factored in $F[x]$. The Well Ordering Principle now guarantees that there is a smallest positive integer m such that there exists a polynomial $g(x) \in F[x]$ of degree m that cannot be completely factored in $F[x]$.

If α is the leading coefficient of $g(x)$, we can let $h(x) = \alpha^{-1} g(x)$. Then $h(x)$ is a monic polynomial that belongs to $F[x]$ and

$$g(x) = \alpha h(x).$$

Let us now examine the nature of $h(x)$. One possibility is that $h(x)$ is irreducible, but, in this case, we have completely factored $g(x)$ in $F[x]$ as $g(x) = \alpha h(x)$. In this case, we see that, simultaneously, $g(x)$ can and cannot be completely factored in $F[x]$. This is certainly a contradiction, so the case of $h(x)$ being irreducible cannot occur.

The only remaining possibility is that $h(x)$ is not irreducible, so we can write $h(x)$ as a product of two polynomials in $F[x]$, both of which have smaller degree. Since $h(x)$ is monic, there exist monic polynomials $a(x), b(x) \in F[x]$ such that

$$h(x) = a(x) \cdot b(x), \quad \text{where} \quad 1 \le deg(a(x)), deg(b(x)) < m.$$

Since m is the smallest positive integer such that there exists a polynomial in $F[x]$ of this degree that cannot be completely factored in $F[x]$, we see that $a(x)$ and $b(x)$ can be completely factored in $F[x]$. Since $a(x)$ and $b(x)$ are both monic, there exist monic, irreducible polynomials

$$p_1(x), p_2(x), \ldots, p_k(x), q_1(x), q_2(x), \ldots, q_l(x) \in F[x]$$

such that

$$a(x) = p_1(x) \cdot p_2(x) \cdots p_k(x) \quad \text{and} \quad b(x) = q_1(x) \cdot q_2(x) \cdots q_l(x).$$

Note that the list of polynomials $p_1(x), p_2(x), \ldots, p_k(x), q_1(x), q_2(x), \ldots, q_l(x)$ is allowed to have the same polynomial occurring more than once.

Since $h(x) = a(x) \cdot b(x)$, we now have

$$g(x) = \alpha h(x) = \alpha \cdot a(x) \cdot b(x) =$$

$$\alpha \cdot (p_1(x) \cdot p_2(x) \cdots p_k(x)) \cdot (q_1(x) \cdot q_2(x) \cdots q_l(x)) =$$

$$\alpha \cdot p_1(x) \cdot p_2(x) \cdots p_k(x) \cdot q_1(x) \cdot q_2(x) \cdots q_l(x).$$

However, the preceding equation illustrates that $g(x)$ can be completely factored in $F[x]$. This is a contradiction, since $g(x)$, simultaneously, can and cannot be completely factored in $F[x]$. We have now shown that the case of $h(x)$ being irreducible as well as the case of $h(x)$ not being irreducible both lead to a contradiction. Therefore, we can now conclude that every $f(x) \in F[x]$ of degree at least 1 can be completely factored in $F[x]$. $\qquad\square$

If the field F is infinite, then the set

$$\{x - \alpha \mid \alpha \in F\}$$

is an infinite set of monic, irreducible polynomials in $F[x]$. Thus, in this case, there are certainly an infinite number of monic, irreducible polynomials in $F[x]$. But if the field F is finite, it is not immediately clear whether there are an infinite number of monic, irreducible polynomials in $F[x]$. Whereas Theorem 3.4 told us that there are an infinite number of prime numbers, a similar argument will show that $F[x]$ always contains an infinite number of monic, irreducible polynomials.

Corollary 12.5. *There are an infinite number of monic, irreducible polynomials in $F[x]$.*

Proof. We will proceed with a proof by contradiction and begin by supposing there are only a finite number of monic, irreducible polynomials in $F[x]$. Then there is a finite list

$$p_1(x), p_2(x), \ldots, p_m(x),$$

which consists of all the monic, irreducible polynomials in $F[x]$. Next, let

$$f(x) = p_1(x) \cdot p_2(x) \cdots p_m(x) + 1.$$

By Theorem 12.4, there is a monic, irreducible polynomial $q(x) \in F[x]$ which divides $f(x)$. But since $q(x)$ belongs to the list $p_1(x), p_2(x), \ldots, p_m(x)$, it follows that $q(x)$ also divides $p_1(x) \cdot p_2(x) \cdots p_m(x)$. As a result, $q(x)$ must divide

$$f(x) - p_1(x) \cdot p_2(x) \cdots p_m(x) = 1.$$

However, since the degree of $q(x)$ is at least one, it is impossible for $q(x)$ to divide 1, so we have reached a contradiction. □

As we will now see, the ideas behind the proofs of Theorem 3.4 and Corollary 12.5 have some further consequences.

■ Examples

In $\mathbb{Z}_2[x]$ the only polynomials of degree 1 are x and $x + [1]_2$. If we multiply the two of them and then add $[1]_2$, we obtain

$$f(x) = ((x) \cdot (x + [1]_2)) + [1]_2 = x^2 + x + [1]_2.$$

Observe that since $f(x)$ has degree 2, but no factor of degree 1, it follows that $f(x)$ must be irreducible in $\mathbb{Z}_2[x]$. You should now convince yourself that $f(x)$ is the *only* monic, irreducible polynomial in $\mathbb{Z}_2[x]$ of degree 2.

If we instead multiply two copies of x with $x + [1]_2$ or two copies of $x + [1]_2$ with x before adding $[1]_2$, we obtain

$$g(x) = \left(x^2 \cdot (x + [1]_2)\right) + [1]_2 = x^3 + x^2 + [1]_2$$

and

$$h(x) = \left(x \cdot (x + [1]_2)^2\right) + [1]_2 = x^3 + x + [1]_2.$$

Note that both $g(x)$ and $h(x)$ have degree 3, yet have no factors of degree 1. Thus, $g(x)$ and $h(x)$ are both irreducible in $\mathbb{Z}_2[x]$. Again, you should convince yourself that $g(x)$ and $h(x)$ are the *only* monic, irreducible polynomials in $\mathbb{Z}_2[x]$ of degree 3.

■

12.2 Division Algorithm

Let us recall that when the degree of a polynomial was defined in Definition 9.2, the definition only referred to polynomials with at least one nonzero coefficient. Therefore, we did not assign any degree to the polynomial $f(x) = 0$. Some books handle this problem by referring to the degree of the polynomial $f(x) = 0$ as $-\infty$. However, the approach that we will take is that the polynomial, all of whose coefficients are zero, is not assigned a degree.

Theorem 12.6—The Division Algorithm for $F[x]$. *If $a(x)$ is a nonzero element of $F[x]$ and if $f(x) \in F[x]$, then there exist unique polynomials $q(x)$ and $r(x)$ in $F[x]$ with the properties that*

$$f(x) = q(x) \cdot a(x) + r(x) \quad \text{and either} \quad deg(r(x)) < deg(a(x)) \quad \text{or} \quad r(x) = 0.$$

We call $q(x)$ the quotient and $r(x)$ the remainder.

Intuition. Let us consider the existence of $q(x)$ and $r(x)$ in the case where we divide $3x^4 + 8x^3 + x^2 + 7x - 5$ by $2x^2 + 1$ in $\mathbb{Q}[x]$. If you have ever performed long division with polynomials, the ideas and computation involved will look familiar. This example will illustrate why it is necessary for the coefficients of our polynomials to belong to a field and not just a commutative ring. First, we need to find a monomial $q_1(x)$ such that the leading term of $q_1(x) \cdot (2x^2 + 1)$ is equal to the leading term of $3x^4 + 8x^3 + x^2 + 7x - 5$. Certainly, $q_1(x) = \frac{3}{2}x^2$ does the trick as

$$q_1(x) \cdot \left(2x^2 + 1\right) = 3x^4 + \frac{3}{2}x^2.$$

Observe that, in this case, $q_1(x) \in \mathbb{Q}[x]$, but $q_1(x) \notin \mathbb{Z}[x]$. In fact, there is no polynomial in $\mathbb{Z}[x]$ that, when multiplied by $2x^2 + 1$, produces a leading term of $3x^4$. Since it is necessary to divide by 2 to find $q_1(x)$, we can see that our coefficients need to belong to a field. Next, let

$$g_1(x) = \left(3x^4 + 8x^3 + x^2 + 7x - 5\right) - q_1(x) \cdot \left(2x^2 + 1\right) =$$

$$\left(3x^4 + 8x^3 + x^2 + 7x - 5\right) - \left(3x^4 + \frac{3}{2}x^2\right) = 8x^3 - \frac{1}{2}x^2 + 7x - 5.$$

For the moment, let us suppose that there exist $q_*(x), r(x) \in \mathbb{Q}[x]$ such that

(1) $g_1(x) = q_*(x) \cdot \left(2x^2 + 1\right) + r(x) \quad \text{and either} \quad deg(r(x)) < 2 \quad \text{or} \quad r(x) = 0.$

Since

$$3x^4 + 8x^3 + x^2 + 7x - 5 = q_1(x) \cdot \left(2x^2 + 1\right) + g_1(x),$$

it would follow from (1) that

$$3x^4 + 8x^3 + x^2 + 7x - 5 = q_1(x) \cdot \left(2x^2 + 1\right) + g_1(x) =$$

$$q_1(x) \cdot \left(2x^2 + 1\right) + q_*(x) \cdot \left(2x^2 + 1\right) + r(x) = (q_1(x) + q_*(x)) \cdot \left(2x^2 + 1\right) + r(x).$$

Thus, in this case, $q(x) = q_1(x) + q_*(x)$ would be the quotient and $r(x)$ the remainder. Essentially, this argument illustrates that if all polynomials of degree less than 4 produced an appropriate quotient and remainder when divided by $2x^2 + 1$, then polynomials of degree 4 would also produce an appropriate quotient and remainder when divided by $2x^2 + 1$. This is precisely the type of reasoning used in Mathematical Induction proofs when we show that $k + 1 \in T$ whenever $\{1, 2, \ldots, k\} \subseteq T$. In light of this, it should come as no surprise that our formal proof of the existence of $q(x)$ and $r(x)$ will use Mathematical Induction.

We now return to the explicit computation of $q(x)$ and $r(x)$ in our example. Having already found $q_1(x)$ and $g_1(x)$, we now need to find a monomial $q_2(x)$ such that the leading term of $q_2(x) \cdot (2x^2 + 1)$ is the same as the leading term of $g_1(x) = 8x^3 - \frac{1}{2}x^2 + 7x - 5$. Therefore, $q_2(x) = 4x$ and we let

$$g_2(x) = g_1(x) - q_2(x) \cdot (2x^2 + 1) = \left(8x^3 - \frac{1}{2}x^2 + 7x - 5\right) - (8x^3 + 4x) = -\frac{1}{2}x^2 + 3x - 5.$$

Next, let $q_3(x)$ be a monomial such that the leading term of $q_3(x) \cdot (2x^2 + 1)$ is the same as the leading term of $g_2(x) = -\frac{1}{2}x^2 + 3x - 5$. Therefore, $q_3(x) = -\frac{1}{4}$ and we let

$$g_3(x) = g_2(x) - q_3(x) \cdot (2x^2 + 1) = \left(-\frac{1}{2}x^2 + 3x - 5\right) - \left(-\frac{1}{2}x^2 - \frac{1}{4}\right) = 3x - \frac{19}{4}.$$

Since $3x - \frac{19}{4}$ has smaller degree than $2x^2 + 1$, it will be our remainder, and it is just a matter of doing a little bookkeeping to find $q(x)$. Putting the pieces together, we obtain

$$3x^4 + 8x^3 + x^2 + 7x - 5 = q_1(x) \cdot (2x^2 + 1) + g_1(x) =$$

$$q_1(x) \cdot (2x^2 + 1) + (q_2(x) \cdot (2x^2 + 1) + g_2(x)) =$$

$$(q_1(x) + q_2(x)) \cdot (2x^2 + 1) + (q_3(x) \cdot (2x^2 + 1) + g_3(x)) =$$

$$(q_1(x) + q_2(x) + q_3(x)) \cdot (2x^2 + 1) + g_3(x) = \left(\frac{3}{2}x^2 + 4x - \frac{1}{4}\right) \cdot (2x^2 + 1) + \left(3x - \frac{19}{4}\right).$$

In our example, $deg(g_1(x)) > deg(g_2(x)) > deg(g_3(x))$ and $g_3(x)$ has smaller degree than $2x^2 + 1$. In all cases, unless some $g_i(x) = 0$, the degrees of the $g_i(x)$'s will always be decreasing. Therefore, these will always eventually be some positive integer k such that either $g_k(x) = 0$ or $deg(g_k(x))$ is smaller than the degree of the polynomial we are dividing by. The polynomial $g_k(x)$ is now the remainder, and we can then perform the same type of bookkeeping as previously to find $q(x)$. Note that $q(x)$ will always be the sum of the $q_i(x)$'s that we found along the way.

For another example, suppose we wish to divide $x^2 + 5x - 4$ by $3x - 1$. First we need to find monomial $q_1(x)$ such that the leading term of $q_1(x) \cdot (3x - 1)$ is the same as the leading term of $x^2 + 5x - 4$. Therefore, $q_1(x) = \frac{1}{3}x$ and we let

$$g_1(x) = (x^2 + 5x - 4) - q_1(x) \cdot (3x - 1) = (x^2 + 5x - 4) - \left(x^2 - \frac{1}{3}x\right) = \frac{16}{3}x - 4.$$

Next, $q_2(x)$ must be a monomial such that the leading term of $q_2(x) \cdot (3x - 1)$ is the same as the leading term of $g_1(x) = \frac{16}{3}x - 4$. Therefore, $q_2(x) = \frac{16}{9}$ and we let

$$g_2(x) = \left(\frac{16}{3}x - 4\right) - q_2(x) \cdot (3x - 1) = \left(\frac{16}{3}x - 4\right) - \left(\frac{16}{3}x - \frac{16}{9}\right) = -\frac{20}{9}.$$

Observe that when finding $q_1(x)$ and $q_2(x)$, it was necessary to use the multiplicative inverses of 3 and 9, once again pointing out the need to be working in a field. Since $-\frac{20}{9}$ has smaller degree than $3x + 1$, it is our remainder. As noted earlier, our quotient will be $q_1(x) + q_2(x) = \frac{1}{3}x + \frac{16}{9}$. This results in

$$x^2 + 5x - 4 = \left(\frac{1}{3}x + \frac{16}{9}\right) \cdot (3x - 1) - \frac{20}{9},$$

as desired.

Proof. If $f(x) = 0$, then $q(x) = 0$ and $r(x) = 0$ are easily seen to be the quotient and remainder. Next, we will use Mathematical Induction to prove that for every nonzero polynomial $f(x)$, there always exist a quotient $q(x)$ and remainder $r(x)$ with the desired properties. To this end, we will let $a(x) \neq 0$ be an element of $F[x]$ and will let T be the set of integers $n \geq 0$ such that every element of $F[x]$ of degree n yields a quotient and remainder with the desired properties when divided by $a(x)$. It suffices to show that T contains all integers greater than or equal to 0.

To show that $0 \in T$, suppose $f(x) \in F[x]$ has degree 0. Therefore, $f(x) = \alpha$, for some $0 \neq \alpha \in F$. If $deg(a(x)) > 0$, then $q(x) = 0$ and $r(x) = f(x)$ are our quotient and remainder as

$$f(x) = 0 \cdot a(x) + f(x) \quad \text{and} \quad deg(f(x)) < deg(a(x)).$$

On the other hand, if $deg(a(x)) = 0$, then $a(x) = \beta$ for some $0 \neq \beta \in F$. In this case $q(x) = \frac{\alpha}{\beta}$ and $r(x) = 0$ are our quotient and remainder as

$$f(x) = \frac{\alpha}{\beta} \cdot a(x) + 0 \quad \text{and} \quad r(x) = 0.$$

Thus, in both cases, $0 \in T$.

Using the Second Version of Mathematical Induction, it now suffices to show that whenever there is an integer $k \geq 0$ such that $\{0, 1, \ldots, k\} \subseteq T$, then T also contains $k + 1$. We now suppose that $f(x) \in F[x]$ has degree $k + 1$. There are two cases to consider as either $deg(a(x)) > deg(f(x))$ or $deg(a(x)) \leq deg(f(x))$. In the first case, $q(x) = 0$ and $r(x) = f(x)$ are our quotient and remainder as

$$f(x) = 0 \cdot a(x) + f(x) \quad \text{and} \quad deg(f(x)) < deg(a(x)).$$

In the second case, we can let αx^{k+1} be the leading term of $f(x)$ and βx^m be the leading term of $a(x)$. Since $k+1 \geq m$, we can let $q_1(x) = \frac{\alpha}{\beta} x^{k+1-m} \in F[x]$ and can also let

(2)
$$g_1(x) = f(x) - q_1(x) \cdot a(x).$$

The leading terms of $f(x)$ and $q_1(x) \cdot a(x)$ are both αx^{k+1}, so either $g_1(x)$ has degree less than $k+1$ or $g_1(x) = 0$. If $deg(g_1(x)) < k+1$, then $deg(g_1(x)) \in T$, and this implies that there exist $q_*(x), r(x) \in F[x]$ such that

$$g_1(x) = q_*(x) \cdot a(x) + r(x) \quad \text{and} \quad deg(r(x)) < deg(a(x)) \quad \text{or} \quad r(x) = 0.$$

Along with (2), the previous equation implies that

$$f(x) = q_1(x) \cdot a(x) + g_1(x) = q_1(x) \cdot a(x) + q_*(x) \cdot a(x) + r(x) =$$
$$(q_1(x) + q_*(x)) \cdot a(x) + r(x) \quad \text{and} \quad deg(r(x)) < deg(a(x)) \quad \text{or} \quad r(x) = 0.$$

Thus, $q(x) = q_1(x) + q_*(x)$ is the quotient and $r(x)$ the remainder.

Finally, if $g_1(x) = 0$, then $q_1(x)$ is the quotient and $r(x) = 0$ the remainder as

$$f(x) = q_1(x) \cdot a(x) + 0 \quad \text{and} \quad r(x) = 0.$$

To complete the proof, we need to show that the quotient $q(x)$ and remainder $r(x)$ are unique. Suppose, on two different occasions, we divide $f(x)$ by $a(x)$ and obtain

$$f(x) = q_1(x) \cdot a(x) + r_1(x) \quad \text{and} \quad f(x) = q_2(x) \cdot a(x) + r_2(x),$$

where $r_1(x)$ and $r_2(x)$ are both either 0 or have smaller degree than $a(x)$. To prove that the quotient and remainder are unique, we need to show that $q_1(x) = q_2(x)$ and $r_1(x) = r_2(x)$.

Since

$$q_1(x) \cdot a(x) + r_1(x) = f(x) = q_2(x) \cdot a(x) + r_2(x),$$

if we subtract both $q_2(x) \cdot a(x)$ and $r_1(x)$ from the previous equation, we obtain

$$q_1(x) \cdot a(x) - q_2(x) \cdot a(x) = r_2(x) - r_1(x).$$

Therefore,

(3)
$$(q_1(x) - q_2(x)) \cdot a(x) = r_2(x) - r_1(x).$$

If $q_2(x) - q_1(x) \neq 0$, then the fact that $a(x) \neq 0$ implies that the left-hand side of (3) has degree greater than or equal to the degree of $a(x)$. However, since $r_1(x)$ and $r_2(x)$ are either 0 or have smaller degree than $a(x)$, it follows that the right-hand side of (3) is 0 or has smaller degree than $a(x)$. This is a contradiction, so it must be the case that $q_2(x) - q_1(x) = 0$. Using (3), it immediately follows that $q_1(x) = q_2(x)$ and $r_1(x) = r_2(x)$. $\qquad\square$

Earlier we indicated that our coefficients need to belong to a field for the division algorithm to hold. Our next example should drive this point home.

■ Example—There Is No Division Algorithm in $\mathbb{Z}[x]$

In $\mathbb{Z}[x]$, suppose we wish to divide x^2 by $2x + 1$. Observe that if $f(x) \in \mathbb{Z}[x]$, then the leading coefficient of $f(x) \cdot (2x + 1)$ will always be even. In particular, the leading term of $f(x) \cdot (2x + 1)$ can never be x^2. As a result, the degree of $x^2 - f(x) \cdot (2x + 1)$ will always be at least 2. Now suppose that $q(x), r(x) \in \mathbb{Z}[x]$ such that

$$x^2 = q(x) \cdot (2x + 1) + r(x).$$

Then $r(x) = x^2 - q(x) \cdot (2x + 1)$ and the preceding argument says that the degree of $r(x)$ must be at least 2. As a result, it is impossible to find a quotient and remainder in $\mathbb{Z}[x]$ such that the remainder is equal to 0 or has degree less than 1. On the other hand, since there is a division algorithm in $\mathbb{Q}[x]$, you can check that when dividing x^2 by $2x + 1$ in $\mathbb{Q}[x]$, the quotient is $\frac{1}{2}x - \frac{1}{4}$ and the remainder is $\frac{1}{4}$.

■

The division algorithm has many interesting and useful applications. Some of these applications appear in the next collection of examples as well as in Corollaries 12.7, 12.8, and 12.10.

■ Examples

1. If we did not have the division algorithm at our disposal, then if we were given $a(x), b(x) \in F[x]$, with $a(x) \neq 0$, it would not be easy to determine if $a(x)$ was a divisor of $b(x)$. However, we can now apply the division algorithm to systematically determine whether $a(x)$ is a divisor of $b(x)$. Observe that if $a(x) \mid b(x)$, then $b(x) = m(x) \cdot a(x)$, for some $m(x) \in F[x]$. The uniqueness aspect of the division algorithm tells us that $m(x)$ is the *only* quotient and 0 is the *only* remainder that can occur when dividing $b(x)$ by $a(x)$. Therefore, to determine whether $a(x)$ is a divisor of $b(x)$, we simply perform long division of polynomials and check whether the remainder is 0. If you are uncomfortable with doing long division of polynomials, you should review the examples presented before our proof of the division algorithm.

   For example, let us consider whether $x^2 + 3x + 5$ is a divisor of $2x^3 + 7x^2 + 13x + 5$ in $\mathbb{Q}[x]$. You should check that long division results in

   $$2x^3 + 7x^2 + 13x + 5 = (2x + 1) \cdot (x^2 + 3x + 5).$$

Thus, $(x^2 + 3x + 5) \mid (2x^3 + 7x^2 + 13x + 5)$.

For another example, let us consider whether $[3]_5 x^2 + [2]_5 x + [1]_5$ is a divisor of $[2]_5 x^3 + [4]_5 x^2 + x$ in $\mathbb{Z}_5[x]$. If you perform long division, you will see that

$$[2]_5 x^3 + [4]_5 x^2 + x = ([4]_5 x + [2]_5) \cdot ([3]_5 x^2 + [2]_5 x + [1]_5) + ([3]_5 x + [3]_5).$$

Since the division algorithm yields a nonzero remainder, we can see that $([3]_5 x^2 + [2]_5 x + [1]_5) \nmid ([2]_5 x^3 + [4]_5 x^2 + x)$.

2. Another question we briefly discussed before proving the division algorithm was if $a(x)$ is not a divisor of $b(x)$ in $F[x]$, is it possible that $a(x)$ is a divisor of $b(x)$ in $K[x]$, where K is a field that contains F? Observe that the division algorithm asserts that there exist $q(x), r(x) \in F[x]$ such that

    (4) $b(x) = q(x) \cdot a(x) + r(x)$ where $deg(r(x)) < deg(a(x))$ or $r(x) = 0$.

    If $a(x) \mid b(x)$ in $K[x]$, then there exists some $m(x) \in K[x]$ such that $b(x) = m(x) \cdot a(x)$. Since $F \subseteq K$, the previous equation and (4) provide us with two ways to divide $b(x)$ by $a(x)$ in $K[x]$. However, the uniqueness aspect of the division algorithm asserts that each time we divide $b(x)$ by $a(x)$ in $K[x]$, we must obtain the same quotient and remainder. Thus, $m(x) = q(x) \in F[x]$ and $r(x) = 0$. Hence, $a(x) \mid b(x)$ in $F[x]$. As a result, we now see that $a(x)$ is a divisor of $b(x)$ using the larger field K only if it was already a divisor of $b(x)$ using the smaller field F.   ∎

The first corollary of the division algorithm will be useful, not only in this chapter but also in later examinations of fields and roots of polynomials.

Corollary 12.7. *Suppose $F \subseteq K$ are fields, $\alpha \in K$, and $g(x)$ is a polynomial of smallest possible degree in $F[x]$ that has α as a root. If $f(x) \in F[x]$ has α as a root, then $g(x)$ must be a divisor of $f(x)$ in $F[x]$.*

At first glance, Corollary 12.7 looks like a very abstract statement. However, it can easily be applied to some very concrete examples. In light of this, we will present some of these examples before proving the corollary. This will enable us to have a greater understanding and appreciation of this corollary before we work through its proof. Examples 2–5 will provide proofs to various statements made back in Chapters 5 and 9.

■ Examples

1. Since $i \notin \mathbb{R}$ and $i^2 + 1 = 0$, we see that $x^2 + 1$ is a polynomial of smallest possible degree in $\mathbb{R}[x]$ that has i as a root. Therefore, Corollary 12.7 now asserts that any

$f(x) \in \mathbb{R}[x]$ that has i as a root is a multiple of $x^2 + 1$. Combining this with the fact that any multiple of $x^2 + 1$ must have i as a root, we now know that $f(x) \in \mathbb{R}[x]$ **has i as a root if and only if $x^2 + 1$ is a divisor of** $f(x)$.

2. Our next example generalizes the previous one. Suppose $\alpha \in \mathbb{C}$ such that $\alpha \notin \mathbb{R}$; then there is no polynomial of degree 1 in $\mathbb{R}[x]$ with α as a root. However, α is certainly a root of

$$g(x) = (x - \alpha) \cdot (x - \alpha^*) = x^2 - (\alpha + \alpha^*)x + \alpha\alpha^*,$$

where α^* is the complex conjugate of α. By Lemma 5.10(b), $\alpha + \alpha^*, \alpha\alpha^* \in \mathbb{R}$. In light of this, $g(x)$ is a polynomial of smallest possible degree in $\mathbb{R}[x]$ that has α as a root. With the help of Corollary 12.7, we can now assert that $f(x) \in \mathbb{R}[x]$ has α as a root if and only if $f(x)$ is a multiple of $g(x)$.

Now suppose that $f(x)$ is any polynomial in $\mathbb{R}[x]$ of degree at least 3. The Fundamental Theorem of Algebra guarantees that $f(x)$ has some root $\alpha \in \mathbb{C}$. If $\alpha \notin \mathbb{R}$, then the polynomial $g(x)$ just constructed is a divisor of $f(x)$ of degree 2 in $\mathbb{R}[x]$. Therefore, in this case, $f(x)$ is reducible in $\mathbb{R}[x]$. On the other hand, if $\alpha \in \mathbb{R}$, then $x - \alpha$ is a polynomial of smallest possible degree in $\mathbb{R}[x]$ having α as a root, so Corollary 12.7 now asserts that $x - \alpha$ is a divisor of $f(x)$ in $\mathbb{R}[x]$. Once again, we can see that $f(x)$ is reducible in $\mathbb{R}[x]$.

In light of the preceding argument, we have now proven the statement, first made in Chapter 9, that **every polynomial of degree at least 3 in $\mathbb{R}[x]$ is reducible in $\mathbb{R}[x]$**.

3. The number $5 - 8\sqrt{2}$ is an element of the field $\mathbb{Q}(\sqrt{2})$ but it does not belong to $\mathbb{Q}$. Therefore, $5 - 8\sqrt{2}$ cannot be the root of any polynomial of degree 1 in $\mathbb{Q}[x]$. However, it is not hard to check that $5 - 8\sqrt{2}$ is a root of the polynomial $x^2 - 10x - 103 \in \mathbb{Q}[x]$. Thus, $x^2 - 10x - 103$ is a polynomial of smallest possible degree in $\mathbb{Q}[x]$ that has $5 - 8\sqrt{2}$ as a root. Therefore, Corollary 12.7 asserts that any $f(x) \in \mathbb{Q}[x]$ that has $5 - 8\sqrt{2}$ as a root must be a multiple of $x^2 - 10x - 103$.

4. The numbers $-\frac{1}{2} \pm 5\sqrt{2}$ are elements of the field $\mathbb{Q}(\sqrt{2})$ that do not belong to $\mathbb{Q}$. Therefore, neither of $-\frac{1}{2} \pm 5\sqrt{2}$ can be the root of any polynomial of degree 1 in $\mathbb{Q}[x]$. However, it is not hard to check that $-\frac{1}{2} \pm 5\sqrt{2}$ are the roots of the polynomial $4x^2 + 4x - 199 \in \mathbb{Q}[x]$. Thus, $4x^2 + 4x - 199$ is a polynomial of smallest possible degree in $\mathbb{Q}[x]$ that has either $-\frac{1}{2} \pm 5\sqrt{2}$ as a root. Therefore, Corollary 12.7 asserts that any $f(x) \in \mathbb{Q}[x]$ that has either $-\frac{1}{2} \pm 5\sqrt{2}$ as a root must be a multiple of $4x^2 + 4x - 199$.

5. The numbers $7 \pm i\sqrt{3}$ are elements of the field $\mathbb{C}$ that do not belong to $\mathbb{Q}$. Therefore, neither of $7 \pm i\sqrt{3}$ can be the root of any polynomial of degree 1 in $\mathbb{Q}[x]$. However, it is not hard to check that $7 \pm i\sqrt{3}$ are the roots of the polynomial

$x^2 - 14x + 52 \in \mathbb{Q}[x]$. Thus, $x^2 - 14x + 52$ is a polynomial of smallest possible degree in $\mathbb{Q}[x]$ that has either $7 \pm i\sqrt{3}$ as a root. Therefore, Corollary 12.7 asserts that any $f(x) \in \mathbb{Q}[x]$ that has either $7 \pm i\sqrt{3}$ as a root must be a multiple of $x^2 - 14x + 52$.

6. Let us consider the number $2^{\frac{1}{3}} \in \mathbb{R}$ and let $g(x)$ be a polynomial of smallest possible degree in $\mathbb{Q}[x]$ that has $2^{\frac{1}{3}}$ as a root. We know that $x^3 - 2$ has $2^{\frac{1}{3}}$ as a root, and using either Eisenstein's Criterion or the Rational Root Test, we know that $x^3 - 2$ is irreducible in $\mathbb{Q}[x]$. But Corollary 12.7 asserts that $g(x)$ is a divisor of $x^3 - 2$. The irreducibility of $x^3 - 2$ now implies that $g(x)$ and $x^3 - 2$ must have the same degree, so $x^3 - 2 = \alpha g(x)$, for some nonzero $\alpha \in F$. Thus, $x^3 - 2$ is a polynomial of smallest degree in $\mathbb{Q}[x]$ having $2^{\frac{1}{3}}$ as a root. As a result, if $f(x) \in \mathbb{Q}[x]$ has $2^{\frac{1}{3}}$ as a root, then $f(x)$ must be a multiple of $x^3 - 2$.

∎

Proof of Corollary 12.7. If $f(x) \in F[x]$ has α as a root, we can apply the division algorithm to obtain $q(x), r(x) \in F[x]$ such that

$$f(x) = q(x) \cdot g(x) + r(x) \quad \text{where} \quad deg(r(x)) < deg(g(x)) \quad \text{or} \quad r(x) = 0.$$

Therefore, $r(x) = f(x) - q(x) \cdot g(x)$ and plugging α into this equation yields

$$r(\alpha) = f(\alpha) - q(\alpha) \cdot g(\alpha) = 0 - q(\alpha) \cdot 0 = 0.$$

Thus, $r(x)$ also has α as a root. However, it is impossible for $r(x)$ to have smaller degree than $g(x)$, since $g(x)$ has the smallest possible degree among polynomials in $F[x]$ that have α as a root. Therefore, the only remaining possibility is that $r(x) = 0$. As a result, $f(x) = q(x) \cdot g(x)$, so $g(x)$ is a divisor of $f(x)$. $\square$

In an example that preceded the proof of Corollary 12.7, we showed that all polynomials of degree at least 3 in $\mathbb{R}[x]$ are reducible in $\mathbb{R}[x]$. Combining this with the existence portion of Theorem 12.4, it immediately follows that

Corollary 12.8. *Every polynomial in $\mathbb{R}[x]$ of degree at least 1 can be written as a product of linear polynomials and irreducible quadratic polynomials in $\mathbb{R}[x]$.*

It is important to realize that although every $f(x)$ in $\mathbb{R}[x]$ can be factored into a product of linear and irreducible quadratics in $\mathbb{R}[x]$, there is, in many cases, no algorithm for finding these factors. For example, in Section 10.2 we pointed out that the polynomials $x^5 - 6x + 3$, $x^5 - 4x + 2$, and $x^5 - 8x + 6$ all have exactly three real roots. Therefore, in $\mathbb{R}[x]$, each of these polynomials can be factored in $\mathbb{R}[x]$ into a product of three linear factors and one irreducible quadratic. However, as noted in Section 10.2, Galois' work on the insolvability of the quintic implies that there is no algorithm for finding these factors that only involves various combinations of the polynomial's coefficients using addition, subtraction, multiplication, division, and taking nth roots, for various $n \in \mathbb{N}$.

Exercises for Sections 12.1 and 12.2

1. Let $f(x) = x^4 - 7x^2 - 44$.
   (a) Factor $f(x)$ completely in $\mathbb{Q}[x]$.
   (b) Factor $f(x)$ completely in $\mathbb{R}[x]$.
   (c) Factor $f(x)$ completely in $\mathbb{C}[x]$.

2. Let $g(x) = x^4 + 46$.
   (a) Factor $g(x)$ completely in $\mathbb{Q}[x]$.
   (b) Factor $g(x)$ completely in $\mathbb{R}[x]$.
   (c) Factor $g(x)$ completely in $\mathbb{C}[x]$.

3. Let $h(x) = 2x^3 + 3x^2 - 9x - 5$.
   (a) Factor $h(x)$ completely in $\mathbb{Q}[x]$.
   (b) Factor $h(x)$ completely in $\mathbb{R}[x]$.
   (c) Factor $h(x)$ completely in $\mathbb{C}[x]$.

4. Let $j(x) = x^4 + 4x^3 + 4x^2 - 4$.
   (a) Factor $j(x)$ completely in $\mathbb{Q}[x]$.
   (b) Factor $j(x)$ completely in $\mathbb{R}[x]$.
   (c) Factor $j(x)$ completely in $\mathbb{C}[x]$.

5. Let $k(x) = x^4 - 15x + 14$.
   (a) Factor $k(x)$ completely in $\mathbb{Q}[x]$.
   (b) Factor $k(x)$ completely in $\mathbb{R}[x]$.
   (c) Factor $k(x)$ completely in $\mathbb{C}[x]$.

6. Let $l(x) = x^4 - 15x + 21$.
   (a) Factor $k(x)$ completely in $\mathbb{Q}[x]$.
   (b) How many irreducible factors are there when $l(x)$ is factored completely in $\mathbb{R}[x]$?
   (c) How many irreducible factors are there when $l(x)$ is factored completely in $\mathbb{C}[x]$?

7. Let $m(x) = 2x^3 + 3x^2 - 9x - 6$.
    (a) Factor $m(x)$ completely in $\mathbb{Q}[x]$.

    (b) How many irreducible factors are there when $m(x)$ is factored completely in $\mathbb{R}[x]$?

    (c) How many irreducible factors are there when $m(x)$ is factored completely in $\mathbb{C}[x]$?

8. Let $w(x) = 2x^3 + 3x^2 - 9x + 12$.
    (a) Factor $w(x)$ completely in $\mathbb{Q}[x]$.

    (b) How many irreducible factors are there when $w(x)$ is factored completely in $\mathbb{R}[x]$?

    (c) How many irreducible factors are there when $w(x)$ is factored completely in $\mathbb{C}[x]$?

For exercises 9–10, you might first want to refer to Theorem 9.15.

9. Let $a(x) = x^4 - 61x^2 + 1$.
    (a) Factor $a(x)$ completely in $\mathbb{Q}[x]$.

    (b) Factor $a(x)$ completely in $\mathbb{R}[x]$.

    (c) Factor $a(x)$ completely in $\mathbb{C}[x]$.

10. Let $b(x) = x^4 + 61x^2 + 1$.
    (a) Factor $b(x)$ completely in $\mathbb{Q}[x]$.

    (b) Factor $b(x)$ completely in $\mathbb{R}[x]$.

    (c) Factor $b(x)$ completely in $\mathbb{C}[x]$.

In exercises 11–16, completely factor the given polynomial in $\mathbb{Z}_3[x]$.

11. $x^2 + x + [1]_3$

12. $[2]_3 x^3 + [2]_3 x^2 + x + [1]_3$

13. $x^3 + x^2 + x + [1]_3$

14. $[2]_3 x^3 + x^2 + [2]_3 x + [1]_3$

15. $[2]_3 x^2 + x + [1]_3$

16. $x^3 + x + [1]_3$

In exercises 17–22, completely factor the given polynomial in $\mathbb{Z}_5[x]$.

17. $x^2 + x + [3]_5$

18. $x^2 + [4]_5 x + [1]_5$

19. $[3]_5 x^2 + [2]_5 x + [1]_5$

20. $x^3 + [3]_5 x^2 + [3]_5 x + [4]_5$

21. $[2]_5 x^3 + [4]_5 x^2 + [2]_5$

22. $[4]_5 x^3 + [2]_5 x^2 + x + [3]_5$

In exercises 23–28, completely factor the given polynomial in $\mathbb{Z}_7[x]$.

23. $x^2 + [3]_7$

24. $x^2 + [4]_7$

25. $x^2 + [2]_7 x + [5]_7$

26. $x^2 + [2]_7 x + [6]_7$

27. $x^3 + [2]_7 x^2 + [5]_7 x + [3]_7$

28. $x^3 + [6]_7 x^2 + [4]_7$

In exercises 29–32, find the quotient and remainder in $\mathbb{Z}_5[x]$ when $f(x)$ is divided by $a(x)$.

29. $f(x) = [4]_5 x^2 + x + [2]_5$ and $a(x) = x + [3]_5$

30. $f(x) = [4]_5 x^2 + x + [2]_5$ and $a(x) = x + [4]_5$

31. $f(x) = [2]_5 x^3 + [4]_5 x + [3]_5$ and $a(x) = x + [2]_5$

32. $f(x) = [2]_5 x^3 + [4]_5 x + [3]_5$ and $a(x) = [2]_5 x^2 + [1]_5$

In exercises 33–36, find the quotient and remainder in $\mathbb{Z}_{11}[x]$ when $f(x)$ is divided by $a(x)$.

33. $f(x) = [7]_{11} x^2 + [8]_{11} x + [10]_{11}$ and $a(x) = x + [5]_{11}$

34. $f(x) = [7]_{11} x^2 + [8]_{11} x + [10]_{11}$ and $a(x) = x + [7]_{11}$

35. $f(x) = [5]_{11} x^3 + [3]_{11} x^2 + [6]_{11}$ and $a(x) = x + [8]_{11}$

36. $f(x) = [5]_{11} x^3 + [3]_{11} x^2 + [6]_{11}$ and $a(x) = [4]_{11} x^2 + [8]_{11} x$

In exercises 37–39, p is a prime number. For exercises 38 and 39, it will probably be easier to first count the number of monic, reducible polynomials of the desired degree.

37. Determine the number of monic, irreducible polynomials of degree 1 in $\mathbb{Z}_p[x]$.

38. Determine the number of monic, irreducible polynomials of degree 2 in $\mathbb{Z}_p[x]$.

39. Determine the number of monic, irreducible polynomials of degree 3 in $\mathbb{Z}_p[x]$.

40. Let $f(x) = x^n + a_{n-1} x^{n-1} + \cdots + a_1 x + a_0 \in F[x]$, where F is a field, and suppose $\alpha_1, \alpha_2, \ldots, \alpha_n \in F$ are the n (not necessarily distinct) roots of $f(x)$.

(a) Express a_{n-1} in terms of the α_i.

(b) Express a_0 in terms of the α_i.

41. Let $F(x) = x^4 + 6x^3 - 7x - 3$.
    (a) In $\mathbb{C}$, what is the sum of the roots of $F(x)$?

    (b) In $\mathbb{C}$, what is the product of the roots of $F(x)$?

42. Let $G(x) = 3x^5 - 4x^3 + 11x - 19$.
    (a) In $\mathbb{C}$, what is the sum of the roots of $G(x)$?

    (b) In $\mathbb{C}$, what is the product of the roots of $G(x)$?

43. Let $H(x) = [6]_7 x^8 + [4]_7 x^5 + [2]_7 x^3 + [3]_7 x + [5]_7 \in \mathbb{Z}_7[x]$ and suppose F is a field which contains $\mathbb{Z}_7$ such that all the irreducible factors of $H(x)$ in F are of degree 1.
    (a) In F, what is the sum of the roots of $H(x)$?

    (b) In F, what is the product of the roots of $H(x)$?

44. Let $f(x) = (x^2 - 5)^3 (x^2 + 3)^2 \in \mathbb{Q}[x]$.
    (a) In $\mathbb{Q}[x]$, how many different monic polynomials are divisors of $f(x)$?

    (b) In $\mathbb{R}[x]$, how many different monic polynomials are divisors of $f(x)$?

    (c) In $\mathbb{C}[x]$, how many different monic polynomials are divisors of $f(x)$?

45. Let $g(x) = 24(x+3)^4 (x-2)^3 (x^2 + 2x + 3)^5 (x^3 - 7) \in \mathbb{Q}[x]$.
    (a) In $\mathbb{Q}[x]$, how many different monic polynomials are divisors of $g(x)$?

    (b) In $\mathbb{R}[x]$, how many different monic polynomials are divisors of $g(x)$?

    (c) In $\mathbb{C}[x]$, how many different monic polynomials are divisors of $g(x)$?

46. Let $h(x) = (4x - 5)^2 (2x^2 + 1)^5 (3x^2 - 7)^4 \in \mathbb{Q}[x]$.
    (a) In $\mathbb{Q}[x]$, how many different monic polynomials are divisors of $h(x)$?

    (b) In $\mathbb{R}[x]$, how many different monic polynomials are divisors of $h(x)$?

    (c) In $\mathbb{C}[x]$, how many different monic polynomials are divisors of $h(x)$?

47. In $\mathbb{Z}_2[x]$, how many different monic polynomials divide $x^5(x + [1]_2)^2 (x^2 + x + [1]_2)^3$?

48. Suppose a field F has n elements and $F = \{a_1, a_2, \ldots, a_n\}$. Show that the polynomial $w(x) = (x - a_1)(x - a_2) \cdots (x - a_n) + 1_F$ has no roots in F, where 1_F denotes the multiplicative identity in F.

49. Show that if a field F has n elements and $m \geq n$, then there exists some polynomial $g(x) \in F[x]$ of degree m such that $g(x)$ has no roots in F.

12.3 Irreducible and Minimum Polynomials

Let us now consider the situation where $F \subseteq K$ and $\alpha \in K$ is the root of some nonzero polynomial in $F[x]$. If $g(x)$ and $h(x)$ are both polynomials of smallest possible degree in $F[x]$ which have α as a root, then Corollary 12.7 asserts that $g(x) \mid h(x)$ and $h(x) \mid g(x)$. This immediately implies that $g(x)$ and $h(x)$ have the same degree, so there is some $\beta \in F$ such that $h(x) = \beta g(x)$. If γ is the leading coefficient of $g(x)$, then $m(x) = \gamma^{-1}g(x)$ is a monic polynomial in $F[x]$ that has the same degree as $g(x)$ and also has α as a root. Combining these facts, we have $h(x) = (\beta\gamma^{-1})m(x)$. This tells us that any polynomial in $F[x]$ of smallest possible degree that has α as a root must be equal to an element of F times $m(x)$. In particular, if $w(x) \in F[x]$ is also a monic polynomial of this same degree having α as a root, then $w(x)$ must be equal to $m(x)$. In light of this observation, $m(x)$ is the *unique* monic polynomial in $F[x]$ of this minimal degree that has α as a root. The polynomial $m(x)$ is often referred to as the **minimum polynomial for α over** F and minimum polynomials will be of great importance in our examination of field extensions in Chapter 15. If we are given some $\alpha \in K$, it is often quite easy to find monic polynomials in $F[x]$ that have α as a root. However, at this point, it is not clear how to determine which of these monic polynomials is the minimum polynomial. Recall that before the proof of Corollary 12.7, we observed that $x^3 - 2$ is a monic, irreducible polynomial in $\mathbb{Q}[x]$ that has $2^{\frac{1}{3}}$ as a root. We then used the irreducibility of $x^3 - 2$ to show that $x^3 - 2$ is the minimum polynomial for $2^{\frac{1}{3}}$ over $\mathbb{Q}$. Indeed, as the next lemma will confirm, irreducibility is a criterion that can be used to determine if a polynomial is the minimum polynomial.

Lemma 12.9.   *Let $F \subseteq K$ be fields and let $\alpha \in K$. If $m(x) \in F[x]$ is monic and has α as a root, then $m(x)$ is the minimum polynomial for α over F if and only if $m(x)$ is irreducible in $F[x]$.*

Proof. In one direction, suppose $m(x)$ is the minimum polynomial for α over F. By way of contradiction, suppose $m(x)$ is reducible in $F[x]$. Thus, $m(x) = a(x) \cdot b(x)$, where $a(x), b(x) \in F[x]$ and $deg(a(x)), deg(b(x)) < deg(m(x))$. Plugging in α, we see that

$$0 = m(\alpha) = a(\alpha) \cdot b(\alpha).$$

Since F has no zero divisors, this implies that $a(\alpha) = 0$ or $b(\alpha) = 0$. However, this says that either $a(x)$ or $b(x)$ has smaller degree than $m(x)$ and has α as a root. But this contradicts the fact that $m(x)$ has the smallest possible degree from among all polynomials in $F[x]$ which have α as a root. Thus, we can conclude that $m(x)$ is indeed irreducible in $F[x]$.

In the other direction, suppose $m(x) \in F[x]$ is monic, irreducible, and has α as a root. If $g(x)$ is the minimum polynomial for α over F, then Corollary 10.7 implies that $g(x) \mid m(x)$ in $F[x]$. However, since $m(x)$ is irreducible, it must be the case that $deg(g(x)) = deg(m(x))$, so $m(x) = \gamma g(x)$, for some $\gamma \in F$. But since $m(x)$ and $g(x)$ are both monic, this immediately implies that $m(x) = g(x)$. As a result, $m(x)$ is the minimum polynomial for α over F. $\qquad \square$

■ Examples

1. Consider $7^{\frac{1}{5}} \in \mathbb{R}$; certainly $x^5 - 7$ is a monic polynomial in $\mathbb{Q}[x]$ having $7^{\frac{1}{5}}$ as a root. By Eisenstein's Criterion, $x^5 - 7$ is irreducible over $\mathbb{Q}$. Therefore, $x^5 - 7$ is the minimum polynomial for $7^{\frac{1}{5}}$ over $\mathbb{Q}$.

2. Consider $i \in \mathbb{C}$; certainly $x^2 + 1$ is a monic polynomial in $\mathbb{R}[x]$ having i as a root. Since $x^2 + 1$ is irreducible over $\mathbb{R}$, it is the minimum polynomial for i over $\mathbb{R}$.

3. Given $\alpha = 5 + 7i \in \mathbb{C}$, let us try to find the minimum polynomial for $5 + 7i$ over $\mathbb{Q}$ and over $\mathbb{R}$. The first thing we need to do is to find polynomials in $\mathbb{Q}[x]$ and $\mathbb{R}[x]$ that have $5 + 7i$ as a root. Then we will check if the polynomials we found are irreducible. Observe that if $\alpha = 5 + 7i$, then $\alpha - 5 = 7i$ and squaring both sides yields

    $$\alpha^2 - 10\alpha + 25 = -49.$$

    Thus, $\alpha^2 - 10\alpha + 74 = 0$, and we see that $5 + 7i$ is a root of $x^2 - 10x + 74$. Since $x^2 - 10x + 74$ has no real roots, it is irreducible over both $\mathbb{Q}$ and $\mathbb{R}$. As a result, $x^2 - 10x + 74$ is the minimum polynomial for $5 + 7i$ over both $\mathbb{Q}$ and $\mathbb{R}$.

4. Consider $\alpha = \sqrt{2} + \sqrt{3} \in \mathbb{R}$; we would like to find the minimum polynomial for $\sqrt{2} + \sqrt{3}$ over $\mathbb{Q}$. To find a monic polynomial in $\mathbb{Q}[x]$ which has $\sqrt{2} + \sqrt{3}$ as a root will require some computations. If $\alpha = \sqrt{2} + \sqrt{3}$, then

    $$\alpha^2 = \left(\sqrt{2} + \sqrt{3}\right)^2 = 2 + 2\sqrt{6} + 3 = 5 + 2\sqrt{6}.$$

This implies that

$$\alpha^2 - 5 = 2\sqrt{6}$$

and squaring both sides yields

$$\alpha^4 - 10\alpha^2 + 25 = 24.$$

Thus, $\alpha^4 - 10\alpha^2 + 1 = 0$. As a result, $\sqrt{2} + \sqrt{3}$ is a root of $x^4 - 10x^2 + 1 \in \mathbb{Q}[x]$. However, at this point, it is not clear if $x^4 - 10x^2 + 1$ is the minimum polynomial for $\sqrt{2} + \sqrt{3}$ over $\mathbb{Q}$. In light of Lemma 12.9, it is sufficient to determine if $x^4 - 10x^2 + 1$ is irreducible over $\mathbb{Q}$. Note that neither the Rational Root Test nor Eisenstein's Criterion gives us any information at this point. On the other hand, $x^4 - 10x^2 + 1$ is one of the polynomials described by Lemma 9.12. If we look back at Lemma 9.12, we have $A = 10$, $A + 2 = 12$, and $A - 2 = 8$, so the theorem tells us that $x^4 - 10x^2 + 1$ is irreducible over $\mathbb{Q}$. Hence, $x^4 - 10x^2 + 1$ is the minimum polynomial for $\sqrt{2} + \sqrt{3}$ over $\mathbb{Q}$.

■

Now let us look at a related question, what is the minimum polynomial for $\sqrt{2}+\sqrt{3}$ over the field $\mathbb{Q}(\sqrt{2})$? It is not difficult to find a quadratic polynomial in $\mathbb{Q}(\sqrt{2})[x]$ which has $\sqrt{2}+\sqrt{3}$ as a root. To see this, observe that

$$\alpha - \sqrt{2} = \sqrt{3}$$

and squaring both sides gives us

$$\alpha^2 - 2\sqrt{2}\alpha + 2 = 3.$$

This implies that

$$\alpha^2 - 2\sqrt{2}\alpha - 1 = 0,$$

which tells us that $\sqrt{2}+\sqrt{3}$ is a root of $x^2 - 2\sqrt{2}x - 1 \in \mathbb{Q}(\sqrt{2})[x]$. To determine if $x^2 - 2\sqrt{2}x - 1$ is the minimum polynomial for $\sqrt{2}+\sqrt{3}$ over $\mathbb{Q}(\sqrt{2})$, we must determine if $x^2 - 2\sqrt{2}x - 1$ is irreducible in $\mathbb{Q}(\sqrt{2})[x]$.

You can easily check that $\sqrt{2}\pm\sqrt{3}$ are the roots of $x^2 - 2\sqrt{2}x - 1$. Therefore, the only way $x^2 - 2\sqrt{2}x - 1$ could be reducible over $\mathbb{Q}(\sqrt{2})$ would be for $\sqrt{2}\pm\sqrt{3}$ to belong to $\mathbb{Q}(\sqrt{2})$. In particular, this would imply that $\sqrt{3} \in \mathbb{Q}(\sqrt{2})$. There are several ways to determine whether $\sqrt{3}$ belongs to $\mathbb{Q}(\sqrt{2})$. Our approach will be to apply some of the tools involving automorphisms and roots of polynomials introduced in Chapter 5. Recall that the function σ defined as $\sigma(a+b\sqrt{2}) = a - b\sqrt{2}$, for all $a, b \in \mathbb{Q}$, is an automorphism of $\mathbb{Q}(\sqrt{2})$. If $\sqrt{3} \in \mathbb{Q}(\sqrt{2})$, then there exist $a, b \in \mathbb{Q}$ such that $\sqrt{3} = a + b\sqrt{2}$. Since $\sqrt{3}$ is a root of $x^2 - 3 \in \mathbb{Q}[x]$, Corollary 5.13 asserts that $\sigma(\sqrt{3})$ is also a root of $x^2 - 3$. Therefore, either $\sigma(\sqrt{3}) = \sqrt{3}$ or $\sigma(\sqrt{3}) = -\sqrt{3}$.

In the first case,

$$\sqrt{3} = a + b\sqrt{2} \quad \text{and} \quad \sqrt{3} = \sigma(\sqrt{3}) = \sigma(a+b\sqrt{2}) = a - b\sqrt{2}.$$

Adding these two equations yields

$$2\sqrt{3} = 2a,$$

which implies that $\sqrt{3} = a \in \mathbb{Q}$. But this is a contradiction as $\sqrt{3}$ is not rational. In the second case, we have

$$\sqrt{3} = a + b\sqrt{2} \quad \text{and} \quad -\sqrt{3} = \sigma(\sqrt{3}) = \sigma(a+b\sqrt{2}) = a - b\sqrt{2}.$$

Subtracting the second equation from the first yields

$$2\sqrt{3} = 2b\sqrt{2}.$$

Multiplying both sides of this equation by $\sqrt{2}$ and dividing by 2 results in

$$\sqrt{6} = 2b \in \mathbb{Q}.$$

But this is also a contradiction as $\sqrt{6}$ is not rational. In light of this, $x^2 - 2\sqrt{2}x - 1$ is indeed irreducible over $\mathbb{Q}(\sqrt{2})$, so $x^2 - 2\sqrt{2}x - 1$ is the minimum polynomial for $\sqrt{2} + \sqrt{3}$ over $\mathbb{Q}(\sqrt{2})$.

If F is a field and $\alpha \in F$ such that $x - \alpha$ is a divisor of some $f(x) \in F[x]$, then certainly $f(\alpha) = 0$. However, observe that $x - \alpha$ is the minimum polynomial for α over F. Therefore, Corollary 12.7 asserts that $x - \alpha$ is a divisor of every polynomial in $F[x]$ that has α as a root. We have now proven a familiar fact that we record as

Corollary 12.10. *Suppose F is a field, $\alpha \in F$, and $f(x) \in F[x]$. Then α is a root of $f(x)$ if and only if $x - \alpha$ is a divisor of $f(x)$.*

12.4 Euclidean Algorithm and Greatest Common Divisors

In order to prove the uniqueness part of Theorem 12.4, we return to the path we followed in Chapter 3 and begin by examining greatest common divisors in $F[x]$. In $\mathbb{Z}$, the greatest common divisor of nonzero integers a, b was the largest integer that was a divisor of both a and b. Therefore, one might suspect that, in $F[x]$, we would define the greatest common divisor of nonzero polynomials $a(x), b(x)$ to be the polynomial of largest degree that is a divisor of both $a(x)$ and $b(x)$. Let us now consider what this would mean if we looked at the polynomials $30x^4 - 30x^3$ and $70x^4 - 70x^2$ in $\mathbb{Q}[x]$. There are many polynomials of degree 3 in $\mathbb{Q}[x]$ that are divisors of both $30x^4 - 30x^3$ and $70x^4 - 70x^2$. For example, $x^3 - x^2$, $2x^3 - 2x^2$, $5x^3 - 5x^2$, and $10x^3 - 10x^2$ are all common divisors of $30x^4 - 30x^3$ and $70x^4 - 70x^2$. In fact, for every nonzero $\alpha \in \mathbb{Q}$, the polynomial $\alpha x^3 - \alpha x^2$ is a common divisor of $30x^4 - 30x^3$ and $70x^4 - 70x^2$. Therefore, there is certainly no unique polynomial of degree 3 in $\mathbb{Q}[x]$ that is a common divisor of $30x^4 - 30x^3$ and $70x^4 - 70x^2$. However, all of the degree 3 polynomials that are common divisors of $30x^4 - 30x^3$ and $70x^4 - 70x^2$ are of the form $\alpha \cdot (x^3 - x^2)$, where α is a nonzero element of $\mathbb{Q}$. Thus, $x^3 - x^2$ is the only monic polynomial of degree 3 that is a common divisor of $30x^4 - 30x^3$ and $70x^4 - 70x^2$ in $\mathbb{Q}[x]$. As a result, it appears that we should define the greatest common divisor of nonzero polynomials $a(x), b(x) \in F[x]$ to be the unique monic polynomial of largest degree that is a common divisor of $a(x)$ and $b(x)$. However, even now, we are not entirely finished with the problem of uniqueness.

In $\mathbb{Z}$, if $c \neq d$, then either $c > d$ or $c < d$. This means that if c and d are two different common divisors of a and b, then either c or d is a greater common divisor than the other. However, the situation is quite different in $F[x]$ as a polynomial can have many monic divisors of the same degree. For example, if we go back to $30x^4 - 30x^3$ and $70x^4 - 70x^2$, both x^2 and $x^2 - x$ are common divisors that are monic and have the same degree. Therefore, at this point, given

$a(x), b(x) \in F[x]$, it is not obvious that there will be a *unique* monic polynomial of largest degree that is a common divisor of $a(x)$ and $b(x)$. Fortunately, using our work in $\mathbb{Z}$ as a guide, there is another way to view greatest common divisors in $F[x]$. In $\mathbb{Z}$, it was very helpful to consider $gcd(a, b)$ as the smallest positive integer that could be written in the form $s \cdot a + t \cdot b$, where $s, t \in \mathbb{Z}$. This motivates the following.

Lemma 12.11. *If $a(x), b(x) \in F[x]$ then there is a unique monic polynomial $c(x)$ of smallest possible degree that can be written in the form $c(x) = s(x) \cdot a(x) + t(x) \cdot b(x)$, where $s(x), t(x) \in F[x]$.*

Proof. Let S be the set of nonnegative integers n with the property that there exists some $d(x) \in F[x]$ of degree n such that $d(x) = s(x) \cdot a(x) + t(x) \cdot b(x)$, for some $s(x), t(x) \in F[x]$. Since S is certainly nonempty, the Well Ordering Principle guarantees that S contains a smallest element $m \geq 0$. Therefore, there exist some $d(x) \in F[x]$ of degree m such that $d(x) = s(x) \cdot a(x) + t(x) \cdot b(x)$, where $s(x), t(x) \in F[x]$. If we let α be the leading coefficient of $d(x)$, then the previous equation implies that

$$\alpha^{-1} d(x) = \left(\alpha^{-1} s(x)\right) \cdot a(x) + \left(\alpha^{-1} t(x)\right) \cdot b(x).$$

Next, if we let $c(x) = \alpha^{-1} d(x)$, $s_1(x) = \alpha^{-1} s(x)$, and $t_1(x) = \alpha^{-1} t(x)$, then we now have

$$(5) \qquad c(x) = s_1(x) \cdot a(x) + t_1(x) \cdot b(x).$$

To complete the proof of this lemma, we need to show that $c(x)$ is the only monic polynomial of degree m in $F[x]$ that can be written as a multiple of $a(x)$ plus a multiple of $b(x)$. To this end, suppose

$$(6) \qquad e(x) = s_2(x) \cdot a(x) + t_2(x) \cdot b(x)$$

is a monic polynomial of degree m, where $s_2(x), t_2(x) \in F[x]$. Subtracting equation (6) from equation (5) results in

$$c(x) - e(x) = (s_1(x) - s_2(x)) \cdot a(x) + (t_1(x) - t_2(x)) \cdot b(x).$$

Since both $c(x)$ and $e(x)$ have leading term x^m, we see that either

$$deg(c(x) - e(x)) < m \quad \text{or} \quad c(x) - e(x) = 0.$$

In the first case, $c(x) - e(x)$ has degree less than m, yet it can be written as a multiple of $a(x)$ plus a multiple of $b(x)$. This contradicts the fact that m is the smallest integer in S. Therefore, the only remaining possibility is that $c(x) - e(x) = 0$. Thus, $c(x) = e(x)$ and $c(x)$ is indeed the unique monic polynomial of smallest degree that can be written as a multiple of $a(x)$ plus a multiple of $b(x)$. □

Lemma 12.11 tells us of the existence of the polynomial $c(x)$, but it doesn't provide us with an algorithm for finding it. In order to show that $c(x)$ has the additional property of being the unique monic polynomial of largest degree in $F[x]$ that is a common divisor of $a(x)$ and $b(x)$, we will first need to develop an algorithm for finding $c(x)$. This algorithm will enable us to better understand the properties possessed by $c(x)$. It should come as no surprise that the algorithm will be very similar to the Euclidean Algorithm we saw in Chapter 3, but this version applies to $F[x]$. But first, we need the analog of Lemma 3.7 for $F[x]$.

Lemma 12.12.   *Let $a(x), b(x), s(x), t(x)$, and $c(x)$ belong to $F[x]$ such that $c(x)$ is a divisor of $a(x)$ and $b(x)$. Then $c(x)$ is also a divisor of $s(x) \cdot a(x) + t(x) \cdot b(x)$.*

Proof. Since $c(x)$ is a divisor of both $a(x)$ and $b(x)$, there exist polynomials $u(x)$ and $v(x)$ such that $a(x) = u(x) \cdot c(x)$ and $b = v(x) \cdot c(x)$. We now have

$$s(x) \cdot a(x) + t(x) \cdot b(x) = s(x) \cdot (u(x) \cdot c(x)) + t(x) \cdot (v(x) \cdot c(x)) =$$

$$(s(x) \cdot u(x) + t(x) \cdot v(x)) \cdot c(x).$$

As a result, $c(x)$ is a divisor of $s(x) \cdot a(x) + t(x) \cdot b(x)$. □

The next result develops the Euclidean Algorithm for $F[x]$ and proves that the polynomial $c(x)$ from Lemma 12.11 is indeed the greatest common divisor of $a(x)$ and $b(x)$. The proof resembles the proof of Theorem 3.8.

Theorem 12.13—The Euclidean Algorithm and Greatest Common Divisors in F[x].   *If $a(x), b(x) \in F[x]$ are nonzero, let $c(x)$ be the monic polynomial of smallest possible degree in $F[x]$ that can be written as a multiple of $a(x)$ plus a multiple of $b(x)$. Then*

(a)   *$c(x)$ is a common divisor of $a(x)$ and $b(x)$,*

(b)   *$c(x)$ is a multiple of every other common divisor of $a(x)$ and $b(x)$ and therefore has the largest degree of any common divisor of $a(x)$ and $b(x)$, and*

(c)   *$c(x)$ is the only monic polynomial of its degree that is a common divisor of $a(x)$ and $b(x)$.*

We call $c(x)$ the greatest common divisor of $a(x)$ and $b(x)$ and denote it as $gcd(a(x), b(x))$.

Proof. The proof consists of developing an algorithm to find the polynomial discussed in Lemma 12.11. If at any point you have difficulty understanding this proof, you should go back and review the more concrete proof of Theorem 3.8. We begin by applying the division algorithm to $a(x)$ and $b(x)$ to obtain polynomials $q_1(x)$ and $r_1(x)$ such that

$$b(x) = q_1(x) \cdot a(x) + r_1(x) \quad \text{and} \quad deg(r_1(x)) < deg(a(x)) \quad \text{or} \quad r_1(x) = 0.$$

If $r_1(x) \neq 0$, divide $a(x)$ by $r_1(x)$ to obtain polynomials $q_2(x), r_2(x)$ such that

$$a(x) = q_2(x) \cdot r_1(x) + r_2(x) \quad \text{and} \quad deg(r_2(x)) < deg(r_1(x)) \quad \text{or} \quad r_2(x) = 0.$$

Next, if $r_2(x) \neq 0$, divide $r_1(x)$ by $r_2(x)$ to obtain polynomials $q_3(x), r_3(x)$ such that

$$r_1(x) = q_3(x) \cdot r_2(x) + r_3(x) \quad \text{and} \quad deg(r_3(x)) < deg(r_2(x)) \quad \text{or} \quad r_3(x) = 0.$$

Observe that every time we apply the division algorithm, we obtain a remainder that is either equal to 0 or has smaller degree than the remainder in the previous step. In particular, if none of our remainders are yet equal to 0, we have $deg(a(x)) > deg(r_1(x)) > deg(r_2(x)) > deg(r_3(x))$. Therefore, if we continue this process of dividing remainder $r_i(x)$ by the next remainder $r_{i+1}(x)$, we will eventually obtain a remainder of 0. Let us now suppose that n is the positive integer such that $r_n(x)$ is the *last* remainder that is *not* 0. This says that if we continue to apply this procedure, we will eventually obtain the equations

$$r_{n-2}(x) = q_n(x) \cdot r_{n-1}(x) + r_n(x) \quad \text{and} \quad r_{n-1}(x) = q_{n+1}(x) \cdot r_n(x) + 0,$$

where

$$deg(a(x)) > deg(r_1(x)) > deg(r_2(x)) > deg(r_3(x)) > \cdots >$$
$$deg(r_{n-2}(x)) > deg(r_{n-1}(x)) > deg(r_n(x)).$$

The last equation tells us that $r_n(x)$ is a divisor $r_{n-1}(x)$. Applying Lemma 12.12 to the next to last equation tells us that $r_n(x)$ is a divisor of $r_{n-2}(x)$. We can continue to move upward through our list of equations, and if we apply Lemma 12.12 at every step, we see that

$$r_n(x) \mid r_{n-1}(x), \quad r_n(x) \mid r_{n-2}(x), \quad r_n(x) \mid r_{n-3}(x), \ldots, \quad r_n(x) \mid r_2(x),$$
$$r_n(x) \mid r_1(x), \quad r_n(x) \mid a(x), \quad r_n(x) \mid b(x).$$

Therefore, $r_n(x)$ is a common divisor of $a(x)$ and $b(x)$. Observe that if α is the leading coefficient of $r_n(x)$, then the polynomial $c(x) = \alpha^{-1} r_n(x)$ is monic and is also a common divisor of $a(x)$ and $b(x)$. Therefore, $c(x)$ satisfies property (a).

The next to last equation shows that $r_n(x)$ can be written as a multiple of $r_{n-2}(x)$ plus a multiple of $r_{n-1}(x)$. Moving up to the next equation enables us to replace $r_{n-1}(x)$ by a multiple of $r_{n-3}(x)$ plus a multiple of $r_{n-2}(x)$, which shows that $r_n(x)$ can be written as a multiple of $r_{n-3}(x)$ plus a multiple of $r_{n-2}(x)$. Continuing in this way, we eventually see that

$r_n(x)$ can be written as a multiple of $a(x)$ plus a multiple of $r_1(x)$, and, finally, $r_n(x)$ can be written as a multiple of $b(x)$ plus a multiple of $a(x)$. Therefore, we can write

$$r_n(x) = s(x) \cdot a(x) + t(x) \cdot b(x),$$

where $s(x), t(x) \in F[x]$. But if we multiply this equation by α^{-1}, we see that

$$c(x) = \alpha^{-1} r_n(x) = (\alpha^{-1} s(x)) \cdot a(x) + (\alpha^{-1} t(x)) \cdot b(x).$$

Thus, $c(x)$ can also be written as a multiple of $b(x)$ plus a multiple of $a(x)$.

If $d(x)$ is any common divisor of $a(x)$ and $b(x)$, then Lemma 12.12 asserts that $d(x)$ is a divisor of any multiple of $a(x)$ plus a multiple of $b(x)$. Therefore, $d(x)$ is also a divisor of $c(x)$ and the degree of $c(x)$ must be greater than or equal to the degree of $d(x)$. In addition, if $d(x)$ is monic and has the same degree as $c(x)$, then $c(x) = d(x)$. Thus, $c(x)$ also satisfies properties (b) and (c).

Having succeeded in showing that $c(x)$ can be written as a multiple of $a(x)$ plus a multiple of $b(x)$, it now suffices to show that it has the smallest degree of any monic polynomial that can be written this way. This will show that $c(x)$ is indeed the polynomial discussed in Lemma 12.11. To this end, suppose $f(x)$ is a polynomial that can be also written as a multiple of $a(x)$ plus a multiple of $b(x)$. Since $c(x)$ is a common divisor of $a(x)$ and $b(x)$, Lemma 12.12 implies that $c(x)$ is also a divisor of $f(x)$. However, since $c(x)$ is a divisor of $f(x)$, we see that the degree of $c(x)$ is less than or equal to the degree of $f(x)$. Thus, $c(x)$ is indeed the monic polynomial of smallest possible degree that can be written as a multiple of $a(x)$ plus a multiple of $b(x)$. As a result, $c(x)$ does indeed have all the properties we would expect from a greatest common divisor. $\qquad\square$

■ Examples

1. Before looking at the Euclidean Algorithm in $F[x]$, we looked briefly at the problem of finding the greatest common divisor of $70x^4 - 70x^2$ and $30x^4 - 30x^3$ in $\mathbb{Q}[x]$. We now use the Euclidean Algorithm to complete the problem. First divide $70x^4 - 70x^2$ by $30x^4 - 30x^3$ to obtain

(7) $$70x^4 - 70x^2 = \left(\frac{7}{3}\right) \cdot (30x^4 - 30x^3) + (70x^3 - 70x^2).$$

Next, we divide $30x^4 - 30x^3$ by $70x^3 - 70x^2$ to obtain

$$30x^4 - 30x^3 = \left(\frac{3}{7}x\right) \cdot (70x^3 - 70x^2) + 0.$$

Since $70x^3 - 70x^2$ is the last nonzero remainder, it would be the greatest common divisor but for the fact that it is not monic. However, multiplying $70x^3 - 70x^2$ by $\frac{1}{70}$, we see that $x^3 - x^2 = gcd(70x^4 - 70x^2, 30x^4 - 30x^3)$.

In order to write $x^3 - x^2$ as a multiple of $70x^4 - 70x^2$ plus a multiple of $30x^4 - 30x^3$, we will go back through our equations and first write $70x^3 - 70x^2$ in this form. Then we will multiply our equation by $\frac{1}{70}$ to complete the problem. In particular, it follows from (7) that

$$70x^3 - 70x^2 = 1 \cdot \left(70x^4 - 70x^2\right) + \left(-\frac{7}{3}\right) \cdot \left(30x^4 - 30x^3\right).$$

Multiplying this equation by $\frac{1}{70}$ yields

$$x^3 - x^2 = \frac{1}{70} \cdot \left(70x^4 - 70x^2\right) + \left(-\frac{1}{30}\right) \cdot \left(30x^4 - 30x^3\right),$$

as desired.

2. Let us now compute $gcd(x^4 - x^2 - 2, x^4 + x^2 - 6)$ in $\mathbb{Q}[x]$. First, we divide $x^4 - x^2 - 2$ by $x^4 + x^2 - 6$ to obtain

(8) $$x^4 - x^2 - 2 = 1 \cdot \left(x^4 + x^2 - 6\right) + \left(-2x^2 + 4\right).$$

Next, we divide $x^4 + x^2 - 6$ by $-2x^2 + 4$ to obtain

$$x^4 + x^2 - 6 = \left(-\frac{1}{2}x^2 - \frac{3}{2}\right) \cdot \left(-2x^2 + 4\right) + 0.$$

Therefore, all that remains to find the greatest common divisor is to multiply $-2x^2 + 4$ by an appropriate number to obtain a monic polynomial. Thus, $x^2 - 2 = gcd(x^4 - x^2 - 2, x^4 + x^2 - 6)$.

To write $x^2 - 2$ as a multiple of $x^4 - x^2 - 2$ plus a multiple of $x^4 + x^2 - 6$, we can rewrite (8) as

$$-2x^2 + 4 = 1 \cdot \left(x^4 - x^2 - 2\right) + (-1) \cdot \left(x^4 + x^2 - 6\right).$$

Multiplying this equation by $-\frac{1}{2}$ gives us

$$x^2 - 2 = \left(-\frac{1}{2}\right) \cdot \left(x^4 - x^2 - 2\right) + \left(\frac{1}{2}\right) \cdot \left(x^4 + x^2 - 6\right),$$

as desired.

3. For another example, let us find the greatest common divisor of $x^3 + x^2 + x$ and $[2]_5 x^3 + [2]_5 x^2$ in $\mathbb{Z}_5[x]$. First, we divide $x^3 + x^2 + x$ by $[2]_5 x^3 + [2]_5 x^2$ to obtain

$$x^3 + x^2 + x = [3]_5 \cdot \left([2]_5 x^3 + [2]_5 x^2\right) + x.$$

Dividing $[2]_5 x^3 + [2]_5 x^2$ by x gives us

$$[2]_5 x^3 + [2]_5 x^2 = \left([2]_5 x^2 + [2]_5 x\right) \cdot x + 0.$$

Since x is already monic, $x = gcd\left(x^3 + x^2 + x, [2]_5 x^3 + [2]_5 x^2\right)$. It is also easy to see that we can write

$$x = [1]_5 \cdot \left(x^3 + x^2 + x\right) + [2]_5 \cdot \left([2]_5 x^3 + [2]_5 x^2\right).$$

∎

As you probably expected, if $a(x), b(x) \in F[x]$ with $gcd(a(x), b(x)) = 1$, then we say that $a(x)$ and $b(x)$ are **relatively prime**. You should convince yourself that $a(x)$ and $b(x)$ are relatively prime if and only if there is no irreducible polynomial $f(x)$ that is a divisor of both $a(x)$ and $b(x)$. It now follows that if $p(x), q(x) \in F[x]$ are irreducible, then either they are relatively prime or there is some nonzero $\alpha \in F$ such that $q(x) = \alpha p(x)$. Thus, if $p(x), q(x)$ are monic and irreducible, then either they are relatively prime or are equal. We can now return to the path that will lead to a proof of the second part of Theorem 12.4.

Lemma 12.14. *If $a(x), b(x), f(x)$ are nonzero polynomials in $F[x]$ such that $f(x) \mid (a(x) \cdot b(x))$ and $gcd(a(x), f(x)) = 1$, then $f(x) \mid b(x)$.*

Proof. Since $a(x)$ and $f(x)$ are relatively prime, we can write 1 as a multiple of $a(x)$ plus a multiple of $f(x)$. Therefore, there exist polynomials $r(x)$ and $s(x)$ such that

$$1 = r(x) \cdot a(x) + s(x) \cdot f(x).$$

Multiplying this equation by $b(x)$ results in

$$b(x) = b(x) \cdot (r(x) \cdot a(x)) + b(x) \cdot (s(x) \cdot f(x)) =$$
$$r(x) \cdot (a(x) \cdot b(x)) + (b(x) \cdot s(x)) \cdot f(x).$$

Having written $b(x)$ as a multiple of $a(x) \cdot b(x)$ plus a multiple of $f(x)$, Lemma 12.12 implies that $f(x) \mid b(x)$. □

The next lemma is essentially Corollary 3.10 modified for $F[x]$. Note that it is now the monic, irreducible polynomials in $F[x]$ that assume the role played by prime numbers in Corollary 3.10.

Lemma 12.15. *Let $p(x), q_1(x), q_2(x), \ldots, q_n(x)$ be monic, irreducible polynomials in $F[x]$ (which are not necessarily distinct). If*

$$p(x) \mid (q_1(x) \cdot q_2(x) \cdots q_n),$$

then $p(x)$ is equal to one of the $q_i(x)$'s.

Proof. We let T be the set of positive integers n such that whenever a monic, irreducible polynomial $p(x)$ divides the product $q_1(x) \cdot q_2(x) \cdots q_n(x)$ of n monic, irreducible polynomials, then $p(x)$ is equal to one of the $q_i(x)$'s. We need to show that $T = \mathbb{N}$, and we will proceed by using Mathematical Induction. First, we need to show that T contains 1. So let us consider the case where $p(x)$ and $q_1(x)$ are monic, irreducible polynomials such that $p(x) \mid q_1(x)$. This immediately implies that $p(x) = q_1(x)$, so $1 \in T$.

Next, we consider the case where T contains some positive integer k. We need to show that T also contains $k+1$. Therefore, suppose that we are now in the situation where $p(x), q_1(x), q_2(x), \ldots, q_k(x), q_{k+1}(x)$ are monic, irreducible polynomials such that

$$(9) \qquad p(x) \mid (q_1(x) \cdot q_2(x) \cdots q_k(x) \cdot q_{k+1}(x)).$$

We need to show that $p(x)$ is equal to one of the $q_i(x)$'s.

There are two possibilities: either $p(x) = q_{k+1}(x)$ or $p(x) \neq q_{k+1}(x)$. In the first case, we are done. In the second case, let

$$b(x) = q_1(x) \cdot q_2(x) \cdots q_k(x),$$

then (9) becomes $p(x) \mid (b(x) \cdot q_{k+1}(x))$. However, in this case, $p(x)$ and $q_{k+1}(x)$ are relatively prime. Therefore, we can apply Lemma 12.14 to assert that $p(x) \mid b(x)$. But $b(x)$ is a product of k monic, irreducible polynomials and T contains k. Therefore, $p(x)$ is indeed equal to one of the $q_i(x)$'s that appear in $b(x)$, and we are also done in this case. $\qquad\square$

We can now complete the proof of the main result of this chapter.

Proof of the final part of Theorem 12.4—uniqueness of the factorization into irreducible polynomials. We will proceed using the Second Version of Mathematical Induction. Let T denote those natural numbers n such that every $f(x) \in F[x]$ of degree n can be written uniquely, up to order, as an element of F times monic, irreducible polynomials in $F[x]$. Our goal is to show that $T = \mathbb{N}$. Mathematical Induction asserts that it will be enough for us to show that T contains 1 and that whenever T contains the set of numbers $\{1, 2, \ldots, k\}$, then it also contains the number $k+1$.

If $f(x) \in \alpha x + \beta \in F[x]$, then certainly $f(x) = \alpha\left(x + \frac{\beta}{\alpha}\right)$ is one way to express $f(x)$ as an element of F times monic, irreducible polynomials in $F[x]$. If $f(x) = \gamma p_1(x) \cdots p_n(x)$ is another factorization of $f(x)$, then since the leading coefficient of $f(x)$ is α and each $p_i(x)$ is monic, we can see that $\alpha = \gamma$. Since each $p_i(x)$ has degree at least 1 and $f(x)$ has degree 1, it is clear that $n = 1$. As a result,

$$\alpha\left(x + \frac{\beta}{\alpha}\right) = f(x) = \alpha p_1(x),$$

which immediately tells us that $p_1(x) = x + \frac{\beta}{\alpha}$. Thus, our factorization of $f(x)$ as $f(x) = \alpha\left(x + \frac{\beta}{\alpha}\right)$ is unique and T does contain 1.

Now suppose that T contains the set of numbers $\{1, 2, \ldots, k\}$; we need to show that T contains $k + 1$. Therefore, we may assume that every polynomial whose degree is at least 1 and is less than $k + 1$ can be written uniquely, up to order as an element of F times monic, irreducible polynomials in $F[x]$. Our job is to show that the same is true for every polynomial of degree $k + 1$. Therefore, let us suppose that $f(x) \in F[x]$ has degree $k + 1$ and that

$$f(x) = \alpha p_1(x) \cdot p_2(x) \cdots p_n(x) \quad \text{and} \quad f(x) = \beta q_1(x) \cdot q_2(x) \cdots q_m(x)$$

are two ways of writing $f(x)$ as an element of F times monic, irreducible polynomials in $F[x]$. To show that these two factorizations of $f(x)$ are identical, up to order, we need to show that there is a reordering of the $q_j(x)$'s such that $\alpha = \beta$, $n = m$, and $p_i(x) = q_i(x)$, for all $i \leq n$.

Since $p_1(x)$ is a divisor of $f(x)$, $p_1(x)$ must also be a divisor of $q_1(x) \cdot q_2(x) \cdots q_m(x)$. By Lemma 12.15, $p_1(x)$ must be equal to $q_k(x)$, for some $k \leq m$. We can now reorder the $q_j(x)$'s such that $p_1(x) = q_1(x)$. If we let $g(x) = \frac{f(x)}{p_1(x)}$, we now have

(10) $$g(x) = \alpha p_2(x) \cdot p_3(x) \cdots p_n(x) \quad \text{and} \quad g(x) = \beta q_2(x) \cdot q_3(x) \cdots q_m(x).$$

One possibility is that $g(x)$ has degree 0. In this case, the previous equations involving $g(x)$ reduce to $g(x) = \alpha$ and $g(x) = \beta$. Thus, $\alpha = \beta$ and it follows that $f(x) = \alpha p_1(x)$ is the only way to write $f(x)$ as an element of F times monic, irreducible polynomials in $F[x]$.

The only other possibility is that $g(x)$ has degree at least 1. In this case, $deg(g(x)) \in T$, so $g(x)$ can be written uniquely, up to order, as an element of F times monic, irreducible polynomials in $F[x]$. In light of this, it follows that we can reorder the $q_j(x)$'s in (10) so that $\alpha = \beta$, $n - 1 = m - 1$, and $p_i(x) = q_i(x)$, whenever $2 \leq i \leq n$. However, having already shown that $p_1(x) = q_1(x)$, it now follows immediately that $\alpha = \beta$, $n = m$, and $p_i(x) = q_i(x)$, for all $i \leq n$. Thus, we have succeeded in showing that the factorization of $f(x)$ is unique, up to order. Hence, $k + 1 \in T$, as desired. $\square$

As we will see in the next example, the unique factorization of polynomials can be used to help us find minimum polynomials.

■ Example

Let us find the minimum polynomial for $\sqrt{2} + \sqrt{7}$ over $\mathbb{Q}$. If we let $\alpha = \sqrt{2} + \sqrt{7}$, then squaring both sides yields

$$\alpha^2 = 9 + 2\sqrt{14}.$$

This implies that

$$\alpha^2 - 9 = 2\sqrt{14}$$

and squaring again results in

$$\alpha^4 - 18\alpha^2 + 81 = 56.$$

Thus, we can see that $\alpha^4 - 18\alpha^2 + 25 = 0$, and it follows that α is a root of $x^4 - 18x^2 + 25 \in \mathbb{Q}[x]$.

Whereas finding $x^4 - 18x^2 + 25$ was fairly easy, we still need to show that it is irreducible, and that is more difficult. One's first impulse is probably to try to use Eisenstein's Criterion, but clearly it does not apply in this case. However, the Rational Root Test does give us some information. Since none of $\pm 1, \pm 5, \pm 25$ are roots of $x^4 - 18x^2 + 25$, we know that our polynomial has no linear factors in $\mathbb{Q}[x]$. Thus, if $x^4 - 18x^2 + 25$ is not irreducible in $\mathbb{Q}[x]$, then it must be the product of two irreducible quadratics in $\mathbb{Q}[x]$. If we were still in Chapter 9, we might try to finish this problem using Gauss' Lemma and some brute-force computations. But we are now in a position to use Theorem 12.4.

In light of our work in Chapter 5 on the relationship between automorphisms and roots of polynomials, your intuition probably tells you that $\pm\sqrt{2}\pm\sqrt{7}$ are the four roots in $\mathbb{C}$ of $x^4 - 18x^2 + 25$. To see this, let $\alpha = \sqrt{2} - \sqrt{7}$ and then perform the identical operations we had performed when we had let $\alpha = \sqrt{2} + \sqrt{7}$. In this case, we now obtain $\alpha^2 = 9 - 2\sqrt{14}$, $\alpha^2 - 9 = -2\sqrt{14}$, $\alpha^4 - 18\alpha^2 + 81 = 56$, and finally $\alpha^4 - 18\alpha^2 + 25 = 0$. Thus $\sqrt{2} - \sqrt{7}$ is also a root of $x^4 - 18x^2 + 25$. But since all the exponents of x in $x^4 - 18x^2 + 25$ are even, γ is a root of $x^4 - 18x^2 + 25$ if and only if $-\gamma$ is a root. Thus, all four of $\pm\sqrt{2}\pm\sqrt{7}$ are roots of $x^4 - 18x^2 + 25$.

As a result, up to order, the *only* way to factor $x^4 - 18x^2 + 25$ in $\mathbb{C}[x]$ into monic, irreducible polynomials is as

$$x^4 - 18x^2 + 25 = \left(x - \left(\sqrt{2} + \sqrt{7}\right)\right) \cdot \left(x - \left(\sqrt{2} - \sqrt{7}\right)\right) \cdot$$

(11)
$$\left(x - \left(-\sqrt{2} + \sqrt{7}\right)\right) \cdot \left(x - \left(-\sqrt{2} - \sqrt{7}\right)\right).$$

Now, suppose $x^4 - 18x^2 + 25$ can indeed be factored into two monic, irreducible quadratics in $\mathbb{Q}[x]$. Then the uniqueness portion of Theorem 12.4 asserts that when we factor these two quadratics in $\mathbb{C}[x]$, we must obtain the same four linear factors as in (11). Thus, one of these irreducible quadratics must be a multiple of $x - (\sqrt{2} + \sqrt{7})$ in $\mathbb{C}[x]$. Hence, the product of $x - (\sqrt{2} + \sqrt{7})$, and at least one of the other three linear factors of $x^4 - 18x^2 + 25$ in $\mathbb{C}[x]$ must belong to $\mathbb{Q}[x]$. However, we can easily check that

$$\left(x - \left(\sqrt{2} + \sqrt{7}\right)\right) \cdot \left(x - \left(\sqrt{2} - \sqrt{7}\right)\right) = x^2 - 2\sqrt{2}x - 5 \notin \mathbb{Q}[x],$$

$$\left(x - \left(\sqrt{2} + \sqrt{7}\right)\right) \cdot \left(x - \left(-\sqrt{2} + \sqrt{7}\right)\right) = x^2 - 2\sqrt{7}x + 5 \notin \mathbb{Q}[x],$$

$$\left(x - \left(\sqrt{2} + \sqrt{7}\right)\right) \cdot \left(x - \left(-\sqrt{2} - \sqrt{7}\right)\right) = x^2 - \left(9 + 2\sqrt{14}\right) \notin \mathbb{Q}[x].$$

Thus, $x^4 - 18x^2 + 25$ has no quadratic factors in $\mathbb{Q}[x]$, so it is irreducible in $\mathbb{Q}[x]$. Therefore, Lemma 12.9 tells us that $x^4 - 18x^2 + 25$ is indeed the minimum polynomial for $\sqrt{2} + \sqrt{7}$ over $\mathbb{Q}$. ∎

Exercises for Sections 12.3 and 12.4

For exercises 1–9, in $\mathbb{Q}[x]$, let

$$f(x) = 10x^2 \left(x^2 + 1\right)^2 \left(x^2 + 4x + 10\right)^3 \left(x^3 - 7\right)^5 \left(x^4 + 5\right)^2 \left(x^3 + 3x - 6\right),$$

$$g(x) = -4(x+3)^2 \left(x^2 + 1\right)^3 \left(x^2 + 4x + 10\right)^4 \left(x^3 - 7\right)^3 \left(x^3 + 3x - 6\right)^2 \left(x^6 + 10x^5 - 15\right)^3,$$

$$h(x) = 8x^3(x+3) \left(x^2 + 4x + 10\right)^2 \left(x^3 - 7\right)^2 \left(x^4 + 5\right)^2 \left(x^2 - 8\right)^4 \left(x^6 + 10x^5 - 15\right)^2.$$

In exercises 1–4, you can use unique factorization instead of the Euclidean Algorithm to compute greatest common divisors.

1. Find $gcd(f(x), g(x))$.

2. Find $gcd(g(x), h(x))$.

3. Find $gcd(f(x), h(x))$.

4. Having defined the greatest common divisor of two polynomials, we can easily extend this definition to three polynomials as $gcd(a(x), b(x), c(x)) = gcd(gcd(a(x), b(x)), c(x))$. Using this definition, find $gcd(f(x), g(x), h(x))$.

For exercises 5–9, please also read the following:

In order to add rational functions, such as $\frac{5x}{(x+1)^2(x+3)^5} + \frac{7x-5}{(x+1)(x+3)^7}$, we usually begin by computing the least common multiple of the denominators. More formally, given nonzero polynomials $a_1(x), a_2(x), \ldots, a_n(x)$, we can let the least common multiple, denoted as $lcm(a_1(x), a_2(x), \ldots, a_n(x))$, be the monic polynomial of smallest degree that is a multiple of every $a_i(x)$.

5. If α is the product of the leading coefficients of nonzero polynomials $a(x)$ and $b(x)$, show that $lcm(a(x), b(x)) = \frac{\alpha^{-1}a(x)b(x)}{gcd(a(x),\, b(x))}$.

6. Find $lcm(f(x), g(x))$.

7. Find $lcm(g(x), h(x))$.

8. Find $lcm(f(x), h(x))$.

9. Find $lcm(f(x), g(x), h(x))$.

For exercises 10–17, you may first need to read the instructions for exercises 1–9. In $\mathbb{Z}_5[x]$, let

$$j(x) = [3]_5(x+[2]_5)^4(x+[4]_5)^2\left(x^2+[2]_5\right)\left(x^2+[3]_5\right)^5\left(x^2+x+[2]_5\right)^2,$$

$$k(x) = (x+[1]_5)^3(x+[2]_5)^3\left(x^2+[3]_5\right)^2\left(x^2+x+[1]_5\right)^2\left(x^2+x+[2]_5\right)^3,$$

$$l(x) = [4]_5x^3(x+[4]_5)\left(x^2+[3]_5\right)^3\left(x^2+x+[2]_5\right).$$

10. Find $gcd(j(x), k(x))$.

11. Find $gcd(k(x), l(x))$.

12. Find $gcd(j(x), l(x))$.

13. Find $gcd(j(x), k(x), l(x))$.

14. Find $lcm(j(x), k(x))$.

15. Find $lcm(k(x), l(x))$.

16. Find $lcm(j(x), l(x))$.

17. Find $lcm(j(x), k(x), l(x))$.

18. If $a(x), b(x)$ are nonzero polynomials and $c(x) = gcd(a(x), b(x))$, show that $gcd\left(\frac{a(x)}{c(x)}, \frac{b(x)}{c(x)}\right) = 1$.

19. Suppose $a(x), b(x)$ are nonzero polynomials such that $a(x) \pm b(x)$ are also nonzero. If $c(x) = gcd(a(x), b(x))$, find $gcd(a(x)+b(x), a(x)-b(x))$.

20. If $gcd(f(x), g(x)) = 1$ and $m, n \in \mathbb{N}$, show that $gcd(f(x)^m, g(x)^n) = 1$.

21. Prove or provide a counterexample to the following statement: if $c(x) = gcd(f(x), g(x))$, then $c(x)^2 = gcd\left(f(x)^2, g(x)^2\right)$.

22. Prove or provide a counterexample to the following statement: if $c(x) = gcd(f(x), g(x))$ and $n \in \mathbb{N}$, then $c(x)^n = gcd(f(x)^n, g(x)^n)$.

23. Prove or provide a counterexample to the following statement: if $c(x) = gcd(f(x), g(x))$, then $c(x)^2 = gcd\left(f(x)^2, g(x)^3\right)$.

24. Prove or provide a counterexample to the following statement: if $c(x) = gcd(f(x), g(x))$, then $c(x)^3 = gcd\left(f(x)^2, g(x)^3\right)$.

In exercises 25–32, find the minimum polynomial over $\mathbb{Q}$ of the given number. Recall, it is necessary to show that the polynomial you find is irreducible in $\mathbb{Q}[x]$.

25. $\sqrt{7}$

26. $3\left(5^{\frac{1}{7}}\right)$

27. $5 - 6i$

28. $-11 + 4\sqrt{13}$

29. $\sqrt{3} + \sqrt{5}$

30. $5\sqrt{6} + 2i$

31. $3\sqrt{17} - 2\sqrt{37}$

32. $i\left(19^{\frac{1}{6}}\right)$

In exercises 33–38, find the minimum polynomial over $\mathbb{Q}(\sqrt{2}) = \{a + b\sqrt{2} \mid a, b \in \mathbb{Q}\}$ of the given number. Recall that it is necessary to show that the polynomial you find is irreducible in $\mathbb{Q}(\sqrt{2})[x]$.

33. 19

34. $-5 + 2\sqrt{2}$

35. $\sqrt{2} + \sqrt{3}$

36. $7 - 4i$

37. $\sqrt{3} + \sqrt{5}$

38. $5\sqrt{2} - 7i\sqrt{3}$

In exercises 39–44, find the minimum polynomial over $\mathbb{Q}(i) = \{a + bi \mid a, b \in \mathbb{Q}\}$ of the given number. Recall that it is necessary to show that the polynomial you find is irreducible in $\mathbb{Q}(i)[x]$.

39. $\sqrt{3}$

40. $5 - 4i$

41. $2\sqrt{11} - 7i$

42. $i\sqrt{6}$

43. $\sqrt{3} + \sqrt{5}$

44. $i\sqrt{2}$

45. (a) Find $a(x), b(x) \in \mathbb{Q}[x]$ such that $a(x) \cdot (2x^5 + 6x^4 + 2x^3 + 5x^2 + 15x + 5)$
$+ b(x) \cdot (3x^3 + 2x^2 - 18x - 7) = gcd(2x^5 + 6x^4 + 2x^3 + 5x^2 + 15x + 5,$
$3x^3 + 2x^2 - 18x - 7)$.

   (b) Find $c(x), d(x) \in \mathbb{Q}[x]$ such that $c(x) \cdot (2x^5 + 6x^4 + 2x^3 + 5x^2 + 15x +$
$5) + d(x) \cdot (3x^3 + 2x^2 - 18x - 7) = 5x^4 + 15x^3 + 5x^2$.

   (c) Do there exist $e(x), f(x) \in \mathbb{Q}[x]$ such that $e(x) \cdot (2x^5 + 6x^4 + 2x^3 +$
$5x^2 + 15x + 5) + f(x) \cdot (3x^3 + 2x^2 - 18x - 7) = 7x^2 - 11$?

   (d) Describe all $m(x) \in \mathbb{Q}[x]$ such that there exist $g(x), h(x) \in \mathbb{Q}[x]$ such that
$g(x) \cdot (2x^5 + 6x^4 + 2x^3 + 5x^2 + 15x + 5) + h(x) \cdot (3x^3 + 2x^2 - 18x - 7) = m(x)$.

46. (a) Find $a(x), b(x) \in \mathbb{Q}[x]$ such that $a(x) \cdot (2x^3 - 6x + 1) + b(x) \cdot (x^2 +$
$5x - 7) = gcd(2x^3 - 6x + 1, x^2 + 5x - 7)$.

   (b) Find $c(x), d(x) \in \mathbb{Q}[x]$ such that $c(x) \cdot (2x^3 - 6x + 1) + d(x) \cdot (x^2 +$
$5x - 7) = 3x - 15$.

   (c) Find $s(x), t(x) \in \mathbb{R}[x]$ such that $s(x) \cdot (2x^3 - 6x + 1) + t(x) \cdot (x^2 + 5x - 7) = \pi x + e$.

   (d) Find $v(x), w(x) \in \mathbb{C}[x]$ such that $v(x) \cdot (2x^3 - 6x + 1) + w(x) \cdot (x^2 +$
$5x - 7) = (10 + 21i)x + (10 + 25i)$.

47. (a) Find $a(x), b(x) \in \mathbb{Z}_5[x]$ such that $a(x) \cdot (x^3 + x^2 + x + [1]_5) + b(x) \cdot$
$(x^2 + x + [3]_5) = gcd(x^3 + x^2 + x + [1]_5, x^2 + x + [3]_5)$.

   (b) Find $c(x), d(x) \in \mathbb{Z}_5[x]$ such that $c(x) \cdot (x^3 + x^2 + x + [1]_5) + d(x) \cdot$
$(x^2 + x + [3]_5) = [4]_5 x + [3]_5$.

   (c) Do there exist $e(x), f(x) \in \mathbb{Z}_5[x]$ such that $e(x) \cdot (x^3 + x^2 + x + [1]_5) + f(x) \cdot$
$(x^2 + x + [3]_5) = x + [1]_5$?

   (d) Describe all $m(x) \in \mathbb{Z}_5[x]$ such that there exist $g(x), h(x) \in \mathbb{Z}_5[x]$ such that
$g(x) \cdot (x^3 + x^2 + x + [1]_5) + h(x) \cdot (x^2 + x + [3]_5) = m(x)$.

48. (a) Find $a(x), b(x) \in \mathbb{Z}_3[x]$ such that $a(x) \cdot (x^3 + [2]_3 x) + b(x) \cdot (x^3 + [2]_3) =$
$gcd(x^3 + [2]_3 x, x^3 + [2]_3)$.

   (b) Find $c(x), d(x) \in \mathbb{Z}_3[x]$ such that $c(x) \cdot (x^3 + [2]_3 x) + d(x) \cdot (x^3 + [2]_3) =$
$[2]_3 x^2 + x$.

   (c) Do there exist $e(x), f(x) \in \mathbb{Z}_3[x]$ such that $e(x) \cdot (x^3 + [2]_3 x) + f(x) \cdot$
$(x^3 + [2]_3) = x^2 + x$?

(d)   Describe all $m(x) \in \mathbb{Z}_3[x]$ such that there exist $g(x), h(x) \in \mathbb{Z}_5[x]$ such that
$g(x) \cdot (x^3 + [2]_3 x) + h(x) \cdot (x^3 + [2]_3) = m(x)$.

49.   (a)   Find $a(x), b(x) \in \mathbb{R}[x]$ such that $a(x) \cdot (x^2 + 1) + b(x) \cdot (x + 1) = 1$.

(b)   Find $c(x), d(x) \in \mathbb{R}[x]$ such that $c(x) \cdot (x^2 + 1) + d(x) \cdot (x + 1) = 5x^2 - 3x - 2$.

(c)   Adapt your answer from part (b) to find a real number α and some $g(x) \in \mathbb{R}[x]$ of degree at most 1 such that $\alpha \cdot (x^2 + 1) + g(x) \cdot (x + 1) = 5x^2 - 3x - 2$.

(d)   Show that α and $g(x)$ satisfy the equation $\frac{\alpha}{x+1} + \frac{g(x)}{x^2+1} = \frac{5x^2-3x-2}{(x+1)(x^2+1)}$.

(e)   In calculus, a typical integration problem you might come across is $\int \frac{5x^2-3x-2}{(x+1)(x^2+1)} dx$. Explain why your answer from part (d) would be useful in solving this problem. In light of this, hopefully you can begin to see that the Euclidean Algorithm and Greatest Common Divisors in $\mathbb{R}[x]$ play a fundamental role in justifying the use of partial fraction decompositions in calculus.

12.5 Formal Derivatives and Multiple Roots

In light of Theorem 12.4, if $f(x) \in F[x]$, then we can always write $f(x)$ uniquely, up to order, as

(12) $$f(x) = \alpha(x - \alpha_1)^{n_1}(x - \alpha_2)^{n_2} \cdots (x - \alpha_t)^{n_t} g_1(x)^{m_1} g_2(x)^{m_2} \cdots g_s(x)^{m_s},$$

where $\alpha \in F, \alpha_1, \ldots, \alpha_t$ are distinct elements of F, and the $g_i(x)$'s are distinct monic, irreducible polynomials in $F[x]$ of degree at least 2. Observe that $\alpha_1, \alpha_2, \ldots, \alpha_t$ are the only roots of $f(x)$ in F. In this situation, we say that α_i **is a root of** $f(x)$ **of multiplicity** n_i.

■ Examples

1.   Let

$$f(x) = x^3(x - 5)^2(x + 3)^4(x + 10)\left(x^2 + 2x + 2\right)\left(x^2 - 2\right)^3 \in \mathbb{Q}[x].$$

Since our polynomial is completely factored in $\mathbb{Q}[x]$, we can say that 0 is a root of multiplicity 3, 5 is a root of multiplicity 2, -3 is a root of multiplicity 4, and -10 is a root of multiplicity 1.

If we look at this same polynomial in $\mathbb{R}[x]$, it is not as yet completely factored since $x^2 - 2$ is reducible over $\mathbb{R}$. The complete factorization of $f(x)$ in $\mathbb{R}[x]$ is

$$f(x) = x^3(x - 5)^2(x + 3)^4(x + 10)\left(x - \sqrt{2}\right)^3\left(x + \sqrt{2}\right)^3(x^2 + 2x + 2).$$

In $\mathbb{R}$, we now say that 0 is a root of multiplicity 3, 5 is a root of multiplicity 2, -3 is a root of multiplicity 4, -10 is a root of multiplicity 1, $\sqrt{2}$ is a root of multiplicity 3, and $-\sqrt{2}$ is a root of multiplicity 3.

If we move to $\mathbb{C}[x]$, then $f(x)$ is still not completely factored as $x^2 + 2x + 2$ is reducible over $\mathbb{C}$. In fact, the complete factorization of $f(x)$ in $\mathbb{C}[x]$ is

$$f(x) = x^3 (x-5)^2 (x+3)^4 (x+10)\left(x-\sqrt{2}\right)^3 \left(x+\sqrt{2}\right)^3 (x-(-1+i))(x-(-1-i)).$$

Thus, in $\mathbb{C}$, we say that 0 is a root of multiplicity 3, 5 is a root of multiplicity 2, -3 is a root of multiplicity 4, -10 is a root of multiplicity 1, $\sqrt{2}$ is a root of multiplicity 3, $-\sqrt{2}$ is a root of multiplicity 3, $-1+i$ is a root of multiplicity 1, and $-1-i$ is a root of multiplicity 1.

2. Consider $g(x) = x^3 + [1]_3$ in $\mathbb{Z}_2[3]$. Since $(x+[1]_3)^3 = x^3 + [1]_3$ in $\mathbb{Z}_2[3]$, the complete factorization of $g(x)$ is

$$g(x) = (x+[1]_3)^3 = (x-[2]_3)^3.$$

Thus, $[2]_3$ is a root of multiplicity 3.

■

Depending upon the situation, if $f(x) \in F[x]$ is factored as in (12), then there are two ways to count the number of roots of $f(x)$ in F. The simplest way is merely to say that $\alpha_1, \alpha_1, \ldots, \alpha_t$ are the roots of $f(x)$, so $f(x)$ has t roots. However, in some other cases, we count a root as many times as its multiplicity. For example, given $g(x) = (x-1)^5(x+3)^8 \in \mathbb{Q}[x]$, we consider 1 as counting as five roots of $g(x)$ and -3 as counting as eight roots of $g(x)$. If we count roots in this fashion and if $f(x)$ is factored as in (12), then the number of roots of $f(x)$ in F is $n_1 + n_2 + \cdots + n_t$. When counting roots in this manner, we say that we are counting roots **including multiplicities**. Observe that when $f(x)$ is factored as in (12), it is certainly the case that $deg(f(x)) \geq n_1 + n_2 + \cdots + n_t$. As a result, we have actually already proven

Corollary 12.16. *If $f(x) \in F[x]$ has degree $n \geq 1$, then the number of roots in F, including multiplicities, is at most n.*

There are many ways to state the Fundamental Theorem of Algebra. Our Theorem 6.12 is one version but, in other books, you may come across different versions of this wonderful result. All of these versions follow easily from Theorem 6.12 and the work in the chapter. For convenience, we list them all in one place as

Theorem 12.17—Versions of the Fundamental Theorem of Algebra

1. *Every $f(x) \in \mathbb{C}[x]$ of degree $n \geq 1$ has a root in $\mathbb{C}$.*

2. *Every $f(x) \in \mathbb{R}[x]$ of degree $n \geq 1$ has a root in $\mathbb{C}$.*

3. *Every $f(x) \in \mathbb{C}[x]$ of degree $n \geq 1$ can be written in the form*

$$f(x) = \alpha(x - \alpha_1)(x - \alpha_2) \cdots (x - \alpha_n),$$

where $\alpha, \alpha_1, \ldots, \alpha_n \in \mathbb{C}$.

4. *Every $f(x) \in \mathbb{R}[x]$ of degree $n \geq 1$ can be written as a product of linear polynomials and irreducible quadratic polynomials in $\mathbb{R}[x]$.*

5. *Every $f(x) \in \mathbb{R}[x]$ of degree $n \geq 3$ is reducible in $\mathbb{R}$.*

Proof. Part (1) is Theorem 6.12 and, since $\mathbb{R}[x] \subseteq \mathbb{C}[x]$, part (2) is merely a special case of Theorem 6.12. Recall that Theorem 6.12 implies that the only irreducible polynomials in $\mathbb{C}[x]$ have degree 1. This fact, along with Theorem 12.4, implies part (3). Part (4) is Corollary 12.8 and part (5) follows immediately from Corollary 12.8. $\square$

If $f(x), g(x)$ are already in completely factored form in $F[x]$, then it is easy to find $gcd(f(x), g(x))$. Similar to the case for greatest common divisors in $\mathbb{Z}$, the exponent of a monic, irreducible polynomial $p(x)$ in $gcd(f(x), g(x))$ is the smaller of the exponents to which $p(x)$ appears in the factorizations of $f(x)$ and $g(x)$. For example, if

$$f(x) = 6x^2(x+7)^4(x^2-3)^5(x^2+1)^2 \quad \text{and} \quad g(x) = 10x(x+7)^8(x^2-3)^6,$$

then $gcd(f(x), g(x)) = x(x+7)^4(x^2-3)^5$.

We have seen many examples where $f(x) \in F[x]$, $F \subseteq K$ are fields, and the complete factorization of $f(x)$ is very different in $F[x]$ than it is in $K[x]$. In light of this, it is natural to wonder if the greatest common divisor of two polynomials $f(x), g(x) \in F[x]$ changes when we move to a larger field K. The next result shows that although the factorization of $f(x)$ and $g(x)$ can change when we move to a larger field, the greatest common divisor remains the same.

Lemma 12.18. *If $F \subseteq K$ are fields and $f(x), g(x) \in F[x]$, then the greatest common divisor of $f(x)$ and $g(x)$ in $F[x]$ is the same as it is in $K[x]$.*

Proof. Let $c(x) = gcd(f(x), g(x))$ in $F[x]$ and $d(x) = gcd(f(x), g(x))$ in $K[x]$; we need to show that $c(x) = d(x)$. Since $F \subseteq K$, it follows that in $K[x]$ both $c(x)$ and $d(x)$ can be written as a multiple of $f(x)$ plus a multiple of $g(x)$. In addition, since $c(x)$ and $d(x)$ both divide $f(x)$ and $g(x)$ in $K[x]$, Lemma 12.12 implies that $c(x) \mid d(x)$ and $d(x) \mid c(x)$ in $K[x]$. Thus

$c(x) = \gamma d(x)$, for some $\gamma \in K$. But since $c(x)$ and $d(x)$ are both monic, we immediately see that $\gamma = 1$ and $c(x) = d(x)$, as desired. $\qquad\qquad\qquad\qquad\qquad\qquad\qquad\qquad\qquad\square$

We will conclude this chapter by dealing with a problem we mentioned in Chapter 1. A root of a polynomial is called a multiple root if its multiplicity is greater than one. In calculus, you probably noticed that the multiple roots of a polynomial $f(x) \in \mathbb{R}[x]$ were precisely the roots of $f(x)$ that were also roots of its derivative, $f'(x)$. As we shall see, this fact follows from an algorithm that tells us when a polynomial $f(x) \in F[x]$ has multiple roots in a field K containing F.

It is quite surprising that there is a simple algorithm for determining whether a polynomial has multiple roots. As we have seen, it can be extremely difficult to find the roots of a polynomial. In fact, Galois has shown that using only basic algebraic operations, there are many polynomials where it is impossible to find the roots. Thus, it is indeed a surprise that there is an algorithm, that essentially consists of computing a greatest common divisor, that will tell us if a polynomial has multiple roots.

Since we will be dealing with fields more general than $\mathbb{R}$ and $\mathbb{C}$, we will need to look at derivatives in a formal, algebraic way as opposed to as the limit of difference quotients. This leads us to

Definition 12.19. *If* $f(x) = a_n x^n + a_{n-1} x^{n-1} + \cdots + a_2 x^2 + a_1 x + a_0 \in F[x]$, *then the formal derivative* $f'(x)$ *of* $f(x)$ *is defined as*

$$f'(x) = n a_n x^{n-1} + (n-1) a_{n-1} x^{n-2} + \cdots + 2 a_2 x + a_1.$$

Although the formal derivative of a polynomial looks like the derivative from calculus, we need to be careful about some of the notation in Definition 12.19. For example, if $g(x) = a_m x^m \in F[x]$, then $g'(x) = m a_m x^{m-1}$. However, whereas $a \in F$, m is not necessarily an element of F, we need to clarify what we mean by the product $m a_m$.

If $a \in F$ and n is a positive integer, the term na will be a shorthand for the sum

$$a + a + \cdots + a$$

of n copies of a. As a result, it is certainly the case that $na \in F$. It is now easy to see that if $f(x) \in F[x]$, then $f'(x) \in F[x]$.

For example, consider $g(x) = [2]_5 x^5 + [3]_5 \in \mathbb{Z}_5[x]$. Since the sum of five copies of $[2]_5$ is equal to $[0]_5$ in $\mathbb{Z}_5$, it follows that $g'(x) = [0]_5$ despite the fact that $g(x)$ has degree 5. This example indicates that fields split into two very different classes.

The more familiar situation for us is when a field F has the property that, for every $n \in \mathbb{N}$, the sum of n copies of the multiplicative identity is not equal to the additive identity. This is certainly the case for $\mathbb{Q}$, $\mathbb{R}$, and $\mathbb{C}$. On the other hand, fields like $\mathbb{Z}_2$, $\mathbb{Z}_3$, $\mathbb{Z}_5$, and $\mathbb{Z}_p$, for any prime p, have the property that a sum of copies of the multiplicative identity does equal the additive identity. In this situation, let n be the smallest positive integer where this occurs. Observe that in our examples above, n is always a prime. This is not a coincidence.

Suppose $n = a \cdot b$, where $1 \le a, b \le n$. If we let A and B denote, respectively, the sum of a copies and b copies of the multiplicative identity, then the distributive law tells us that $A \cdot B$ is the sum of n copies of the multiplicative identity. Thus, $A \cdot B$ is equal to the additive identity. However, since we are working in a field, either A or B must be equal to the additive identity. By the minimality of n, either $a = n$ or $b = n$. As a result, n cannot be written as a product of smaller positive integers, so n is indeed prime. We can summarize these observations in

Definition 12.20. *Let F be a field. If a sum of copies of the multiplicative identity of F equals the additive identity, let p be the smallest number of copies where this occurs. In this case, p is a prime number and we say that F has characteristic p. On the other hand, if no sum of copies of the multiplicative identity equals the additive identity, we say that F has characteristic 0.*

At the moment, it is not clear why the formal derivative is useful to us. However, it turns out that the formal derivative satisfies the familiar product rule and that is precisely what we need to obtain our result on multiple roots. Since the formal derivative is defined algebraically, and not in terms of limits, the proof of the product rule will be purely algebraic.

Lemma 12.21. *If $f(x), g(x) \in F[x]$, then*

$$(f(x) \cdot g(x))' = f'(x) \cdot g(x) + f(x) \cdot g'(x).$$

Proof. Suppose $f(x) = a_n x^n + a_{n-1} x^{n-1} + \cdots + a_2 x^2 + a_1 x + a_0$ and $g(x) = b_m x^m + b_{m-1} x^{m-1} + \cdots + b_2 x^2 + b_1 x + b_0$, then

$$f'(x) = (na_n)x^{n-1} + (n-1)(a_{n-1})x^{n-2} + \cdots + (2a_2)x + a_1$$

and

$$g'(x) = (mb_m)x^{m-1} + (m-1)(b_{m-1})x^{m-2} + \cdots + (2b_2)x + b_1.$$

Therefore, if $t \ge 1$, then the distributive law tells us that the coefficient of x^{t-1} in $f'(x) \cdot g(x)$ is

$$(ta_t)(b_0) + (t-1)(a_{t-1})(b_1) + \cdots + (2a_2)(b_{t-2}) + (a_1)(b_{t-1})$$

and the coefficient of x^{t-1} in $f(x) \cdot g'(x)$ is

$$(a_{t-1})(b_1) + (a_{t-2})(2b_2) + \cdots + (a_2)(t-2)(b_{t-2}) + (a_1)(t-1)(b_{t-1}) + (a_0)(t)(b_t).$$

Adding these two expressions, we can see that the coefficient of $t-1$ in $f'(x) \cdot g(x) + f(x) \cdot g'(x)$ is

$$t(a_t \cdot b_0) + t(a_{t-1} \cdot b_1) + t(a_{t-2} \cdot b_2) + \cdots + t(a_2 \cdot b_{t-2}) + t(a_1 \cdot b_{t-1}) + t(a_0 \cdot b_t).$$

On the other hand, the coefficient of x^t in $f(x) \cdot g(x)$ is

$$a_t \cdot b_0 + a_{t-1} \cdot b_1 + a_{t-2} \cdot b_2 + \cdots + a_2 \cdot b_{t-2} + a_1 \cdot b_{t-1} + a_0 \cdot b_t.$$

Therefore, the coefficient of x^{t-1} in $(f(x) \cdot g(x))'$ is

$$t(a_t \cdot b_0 + a_{t-1} \cdot b_1 + a_{t-2} \cdot b_2 + \cdots + a_2 \cdot b_{t-2} + a_1 \cdot b_{t-1} + a_0 \cdot b_t).$$

As we can see, for every $t \geq 1$, the coefficient of x^{t-1} in $f'(x) \cdot g(x) + f(x) \cdot g'(x)$ and $(f(x) \cdot g(x))'$ is the same. Hence, $(f(x) \cdot g(x))' = f'(x) \cdot g(x) + f(x) \cdot g'(x)$, as desired. $\quad\square$

We can now prove the main result of this section.

Proposition 12.22. *If $f(x) \in F[x]$, then the multiple roots of $f(x)$ in a field K containing F are precisely the roots in K of $\gcd(f(x), f'(x))$. In particular, if $\gcd(f(x), f'(x)) = 1$ in $F[x]$, then $f(x)$ does not have multiple roots in any field containing F.*

Before proving Proposition 12.22, we will examine some examples.

■ Examples

1. Let $f(x) = x^3 + 4x^2 + 5x + 2 \in \mathbb{Q}[x]$; then $f'(x) = 3x^2 + 8x + 5$ and we need to compute $\gcd(x^3 + 4x^2 + 5x + 2, 3x^2 + 8x + 5)$. We now apply the Euclidean Algorithm. First, we divide $x^3 + 4x^2 + 5x + 2$ by $3x^2 + 8x + 5$ and obtain

    $$x^3 + 4x^2 + 5x + 2 = \left(\frac{1}{3}x + \frac{4}{9}\right) \cdot (3x^2 + 8x + 5) + \left(-\frac{2}{9}x - \frac{2}{9}\right).$$

    Next we divide $3x^2 + 8x + 5$ by $-\frac{2}{9}x - \frac{2}{9}$, and we have

    $$3x^2 + 8x + 5 = \left(-\frac{27}{2}x - \frac{45}{2}\right) \cdot \left(-\frac{2}{9}x - \frac{2}{9}\right) + 0.$$

    Our last nonzero remainder is $-\frac{2}{9}x - \frac{2}{9}$, but it is not monic. However, if we multiply it by $-\frac{9}{2}$, we obtain $x + 1$. Thus, $x + 1 = \gcd(x^3 + 4x^2 + 5x + 2, 3x^2 + 8x + 5)$. Since the only root of $x + 1$ is -1, Proposition 12.22 asserts that -1 is the only multiple root of $x^3 + 4x^2 + 5x + 2$ in any field containing $\mathbb{Q}$.

2. Let $g(x) = x^5 + 5x^4 + 8x^3 + 40x^2 + 16x + 80 \in \mathbb{Q}[x]$; then $g'(x) = 5x^4 + 20x^3 + 24x^2 + 80x + 16$ and we need to find $gcd(x^5 + 5x^4 + 8x^3 + 40x^2 + 16x + 80, 5x^4 + 20x^3 + 24x^2 + 80x + 16)$. First, we divide $x^5 + 5x^4 + 8x^3 + 40x^2 + 16x + 80$ by $5x^4 + 20x^3 + 24x^2 + 80x + 16$ and obtain

$$x^5 + 5x^4 + 8x^3 + 40x^2 + 16x + 80 =$$

$$\left(\frac{1}{5}x + \frac{1}{5}\right) \cdot (5x^4 + 20x^3 + 24x^2 + 80x + 16) + \left(-\frac{4}{5}x^3 + \frac{96}{5}x^2 - \frac{16}{5}x + \frac{384}{5}\right).$$

Next, divide $5x^4 + 20x^3 + 24x^2 + 80x + 16$ by $-\frac{4}{5}x^3 + \frac{96}{5}x^2 - \frac{16}{5}x + \frac{384}{5}$, and we have

$$5x^4 + 20x^3 + 24x^2 + 80x + 16 =$$

$$\left(-\frac{25}{4}x - 175\right) \cdot \left(-\frac{4}{5}x^3 + \frac{96}{5}x^2 - \frac{16}{5}x + \frac{384}{5}\right) + (3364x^2 + 13456).$$

Next, divide $-\frac{4}{5}x^3 + \frac{96}{5}x^2 - \frac{16}{5}x + \frac{384}{5}$ by $3364x^2 + 13456$ and we have

$$-\frac{4}{5}x^3 + \frac{96}{5}x^2 - \frac{16}{5}x + \frac{384}{5} = \left(-\frac{1}{4205}x + \frac{24}{4205}\right) \cdot (3364x^2 + 13456) + 0.$$

Our last nonzero remainder is $3364x^2 + 13456$, but it is not monic. If we multiply it by $\frac{1}{3364}$, we obtain $x^2 + 4$. Thus,

$$x^2 + 4 = gcd\left(x^5 + 5x^4 + 8x^3 + 40x^2 + 16x + 80, 5x^4 + 20x^3 + 24x^2 + 80x + 16\right).$$

The polynomial $x^2 + 4$ has no roots in $\mathbb{R}$ but does have roots $\pm 2i$ in $\mathbb{C}$. As a result, Proposition 12.22 tells us that $x^5 + 5x^4 + 8x^3 + 40x^2 + 16x + 80$ has no multiple roots in $\mathbb{R}$ but does have two multiple roots in $\mathbb{C}$.

In fact, we can say quite a bit more about $x^5 + 5x^4 + 8x^3 + 40x^2 + 16x + 80$. When we look at the complete factorization of $x^5 + 5x^4 + 8x^3 + 40x^2 + 16x + 80$ in $\mathbb{C}[x]$ the exponents of both $x - 2i$ and $x + 2i$ must be at least 2. Thus,

$$x^5 + 5x^4 + 8x^3 + 40x^2 + 16x + 80 = (x - 2i)^2(x + 2i)^2 s(x) = \left(x^2 + 4\right)^2 s(x),$$

for some $s(x) \in \mathbb{R}[x]$. By looking at the leading and trailing terms of $x^5 + 5x^4 + 8x^3 + 40x^2 + 16x + 80$ and $(x^2 + 4)^2 = x^4 + 8x^2 + 16$, it is easy to see that $s(x) = x + 5$. Thus,

$$x^5 + 5x^4 + 8x^3 + 40x^2 + 16x + 80 = \left(x^2 + 4\right)^2 (x + 5)$$

is the complete factorization of $x^5 + 5x^4 + 8x^3 + 40x^2 + 16x + 80$ in $\mathbb{R}[x]$.

3. Let α be a nonzero real number. It follows from Theorem 6.8 that the nth roots of α in $\mathbb{C}$ are all different. We can also illustrate this point using Proposition 12.22. If we

let $h(x) = x^n - \alpha \in \mathbb{C}[x]$, then $h'(x) = nx^{n-1}$. It is easy to see that $h(x)$ and $h'(x)$ are relatively prime, so $x^n - \alpha$ has no multiple roots in $\mathbb{C}$.

∎

We can now prove Proposition 12.22. The proof will incorporate and integrate many of the ideas from this chapter.

Proof of Proposition 12.22. In one direction, suppose K is a field containing F and $\alpha \in K$ is a root of $gcd(f(x), f'(x))$. Corollary 12.10 asserts that $x - \alpha$ is a factor of both $f(x)$ and $f'(x)$ in $K[x]$ and we now have

$$(13) \qquad\qquad f(x) = (x - \alpha) \cdot p(x),$$

for some $p(x) \in K[x]$. Taking the formal derivative and applying the product rule to the previous equation gives us

$$f'(x) = (x - \alpha)' \cdot p(x) + (x - \alpha) \cdot p'(x) = p(x) + (x - \alpha) \cdot p'(x).$$

Since $x - \alpha$ is also a factor of $f'(x)$, replacing x by α in the previous equation tells us that $p(\alpha) = 0$. Using Corollary 12.10, we can see that $x - \alpha$ is also a factor of $p(x)$ in $K[x]$. Therefore, $p(x) = (x - \alpha) \cdot q(x)$, for some $q(x) \in K[x]$. Plugging this into equation (13) results in

$$f(x) = (x - \alpha) \cdot p(x) = (x - \alpha) \cdot ((x - \alpha) \cdot q(x)) = (x - \alpha)^2 \cdot q(x).$$

Therefore, α is a multiple root of $f(x)$ in K.

In the other direction, suppose K is a field containing F and $\alpha \in K$ is a multiple root of $f(x)$. Therefore, we now have

$$(14) \qquad\qquad f(x) = (x - \alpha)^2 \cdot q(x),$$

where $q(x) \in K[x]$. Since

$$(x - \alpha)^2 = x^2 - 2\alpha x + \alpha^2,$$

it follows that

$$\left((x - \alpha)^2\right)' = 2x^2 - 2\alpha x = 2(x - \alpha).$$

Applying the product rule to equation (14) and substituting the previous equation, we obtain

$$f'(x) = \left((x - \alpha)^2\right)' \cdot q(x) + (x - \alpha)^2 \cdot q'(x) = 2(x - \alpha) \cdot q(x) + (x - \alpha)^2 \cdot q'(x).$$

Replacing x by α in the previous equation shows that $f'(\alpha) = 0$. Corollary 12.10 asserts that $x - \alpha$ is a factor of $f'(x)$, so it is a common factor of $f(x)$ and $f'(x)$ in $K[x]$. By Theorem 12.13(b), $x - \alpha$ is a factor of $gcd(f(x), f'(x))$, so α is a root of $gcd(f(x), f'(x))$. Observe that although we are doing computations in K and $K[x]$, Lemma 12.18 already told us that the greatest common divisor of $f(x)$ and $f'(x)$ remains the same regardless of whether we are working in $F[x]$ or $K[x]$.

Finally, if $gcd(f(x), f'(x)) = 1$, then $gcd(f(x), f'(x))$ has no roots in any field K containing F. Hence, in this situation, $f(x)$ does not have multiple roots in any field containing F. $\qquad\square$

If a field F has characteristic 0 and a is a nonzero element of F then, for any $n \in \mathbb{N}$, we have

$$ na = a + a + \cdots + a = 1 \cdot a + 1 \cdot a + \cdots + 1 \cdot a = (1 + 1 + \cdots + 1) \cdot a, $$

which is a product of two nonzero elements of F. As a result, $na \neq 0$ and we will make use of this fact in the following application of Proposition 12.22.

Corollary 12.23. *If $f(x) \in F[x]$ is irreducible in $F[x]$, where F has characteristic 0, then $f(x)$ has no multiple roots in any field K containing F.*

Proof. If the leading term of $f(x)$ is $a_n x^n$, then the leading term of $f'(x)$ is $na_n x^{n-1}$. Since F has characteristic 0, $na_n \neq 0$, hence $f'(x)$ is a polynomial of degree $n - 1$.

Since $f(x)$ is irreducible, the only monic polynomial of degree less than n than divides $f(x)$ is the constant polynomial 1. If $c(x) = gcd(f(x), f'(x))$, then $c(x)$ is a monic polynomial that divides both $f(x)$ and $f'(x)$. Since $c(x)$ divides $f'(x)$, the degree of $c(x)$ cannot exceed $n - 1$. In particular, $c(x)$ has smaller degree than $f(x)$. But since $c(x)$ also divides $f(x)$, it follows that $c(x) = 1$. Hence, $gcd(f(x), f'(x)) = 1$, and Proposition 12.22 tells us that $f(x)$ does not have multiple roots in any field containing F. $\qquad\square$

Specializing Corollary 12.23 to the more familiar situation of polynomials with rational coefficients, we immediately have

Corollary 12.24. *If $f(x) \in \mathbb{Q}[x]$ is irreducible in $\mathbb{Q}[x]$, then $f(x)$ has no multiple roots in $\mathbb{C}$.*

Observe that although Galois' work tells us that we cannot find the roots of $x^5 - 6x + 3$, $x^5 - 4x + 2$, and $x^5 - 8x + 6$ using our standard algebraic tools, Corollary 12.24 immediately tells us that none of these polynomials has a multiple root in $\mathbb{C}$.

Whereas Corollary 12.24 is quite concrete, Corollary 12.23 looks more abstract. In particular, you might wonder if the hypothesis that F have characteristic 0 is needed. We will now show that it is.

■ Example

Let $R = \mathbb{Z}_2[t^2]$ be the ring of polynomials with coefficients in $\mathbb{Z}_2$ using the variable t^2. Therefore, R consists of those polynomials where the exponent of t is always even.

In calculus, you dealt with the field $\mathbb{R}(x)$ of rational functions. It consisted of those functions that could be written as a quotient of polynomials in $\mathbb{R}[x]$. The construction of $\mathbb{R}(x)$ from $\mathbb{R}[x]$ is done using equivalence relations and is almost identical to the construction of $\mathbb{Q}$ from $\mathbb{Z}$.

In the same way, we can construct the field $F = \mathbb{Z}_2(t^2)$ from $R = \mathbb{Z}_2[t^2]$. Examples of elements of F are

$$\frac{[1]_2 t^2}{[1]_2 t^4 + [1]_2}, \quad \frac{[1]_2 t^4 + [1]_2}{[1]_2 t^6 + [1]_2 t^4}, \quad \frac{[1] t^6 + [1]_2 t^4}{[1]_2}.$$

Observe that if $\alpha \in F$, then α can be represented in many different ways in the form $\alpha = \frac{a(t)}{b(t)}$, where $a(t), b(t) \in \mathbb{Z}_2[t^2]$. Note that $deg(a(t)) - deg(b(t))$ must be even. If we now consider α^2, then $\alpha^2 = \frac{a(t)^2}{b(t)^2}$ and $deg(a(t)^2) - deg(b(t)^2)$ must be a multiple of 4. This tells us that if $\alpha \in F$, then α^2 cannot possibly to equal to t^2.

The upshot of all this is that the polynomial $x^2 - t^2 \in F[x]$ does not have any roots in F. Since $x^2 - t^2$ has degree 2, this tells us that $x^2 - t^2$ is irreducible in $F[x]$. Now, let $K = \mathbb{Z}_2(t)$ be the field consisting of all quotients of polynomials from $\mathbb{Z}_2[t]$. It is easy to see that K is a field that contains F. Furthermore, if we work in $K[x]$, we can factor $x^2 - t^2$ as

$$x^2 - t^2 = (x - t)^2.$$

Thus, $x^2 - t^2$ has multiple roots in K despite the fact that it is irreducible in F.

■

This example can be modified to the fields $F = \mathbb{Z}_p(t^p)$ and $K = \mathbb{Z}_p(t)$, where p is a prime. In this case, K contains F and both have characteristic p. The polynomial $x^p - t^p \in F[x]$ is irreducible in $F[x]$, yet it factors over $K[x]$ as

$$x^p - t^p = (x - t)^p.$$

In the exercises, we will verify that $x^p - t^p$ is indeed irreducible in $F[x]$. Thus, we will have shown that, for every prime p, there exists a field F of characteristic p and a polynomial $f(x) \in F[x]$ such that $f(x)$ is irreducible in $F[x]$ and has multiple roots in a field K containing F. This further illustrates the need for our field to have characteristic 0 in Corollary 12.23.

Exercises for Section 12.5

In exercises 1–14, find all multiple roots in $\mathbb{C}$ of the given polynomial.

1. $x^3 + 3x^2 - 9x + 5$

2. $x^3 + 3x^2 - 9x + 6$

3. $x^3 + 6x^2 + 15x + 12$

4. $x^3 + 7x^2 + 16x + 12$

5. $x^4 + 2x^3 + 8x^2 + 14x + 7$

6. $x^4 + 2x^3 + 8x^2 + 14x + 6$

7. $x^5 + 2x^4 - 6x^3 - 12x^2 + 9x + 18$

8. $x^5 + 2x^4 - 6x^3 - 12x^2 + 8x + 18$

9. $x^4 - 6x + 1$

10. $x^4 + x^2 - 6$

11. $x^5 - 3x^4 + 10x^3 - 30x^2 + 25x - 75$

12. $x^5 - 3x^4 + 9x^3 - 30x^2 + 24x - 75$

13. $x^4 - 3x^3 + 3x^2 + 12x - 12$

14. $x^4 - 4x^3 + x^2 + 12x - 12$

15. Does $x^3 + x \in \mathbb{Z}_2[x]$ have any roots in $\mathbb{Z}_2$ that are also roots of its derivative?

16. Does $x^3 + x \in \mathbb{Z}_3[x]$ have any roots in $\mathbb{Z}_3$ which are also roots of its derivative?

17. Does $x^3 + [2]_3x^2 + x \in \mathbb{Z}_3[x]$ have any roots in $\mathbb{Z}_3$ which are also roots of its derivative?

18. Let F be a field of characteristic $p \neq 0$. Suppose $f(x) \in F[x]$ is irreducible in $F[x]$ and has degree n, where n is not a multiple of p. Show that $f(x)$ does not have multiple roots in any field containing F.

19. Let $F \subseteq K$ be fields of characteristic $p \neq 0$ and suppose $a \in K$ such that $a \notin F$ and $a^p \in F$.
    (a) Show that if n is a positive integer less than p, then the polynomial $(x - a)^n$ belongs to $K[x]$ but does not belong to $F[x]$.

    (b) Completely factor $x^p - a^p$ in $K[x]$.

    (c) If $b(x) \in K[x]$ is a monic divisor of $x^p - a^p$ in $K[x]$, show that $b(x) = (x - a)^n$, for some integer n, where $0 \leq n \leq p$.

(d) Show that $x^p - a^p$ is irreducible in $F[x]$.

(e) Conclude that $x^p - a^p$ is an irreducible polynomial in $F[x]$ with multiple roots in K.

20. Let F be a field with a finite number of elements and let p denote the characteristic of F. (Observe that F cannot have characteristic 0.) Show that the function $\sigma : F \to F$ defined as $\sigma(r) = r^p$, for all $r \in F$, is an automorphism of the field F.

21. Give an example of a field F of characteristic $p \neq 0$ such that the function $\sigma : F \to F$ defined as $\sigma(r) = r^p$, for all $r \in F$, is not an automorphism of the field F because it fails to be surjective.

CHAPTER 13

Difference Functions and Partial Fractions

In this chapter, we make repeated use of Mathematical Induction and apply several ideas from Chapter 12 to address some problems that arise in courses in precalculus and calculus.

In precalculus courses, we are taught to recognize linear functions by applying the following test: A function is linear if and only if a fixed change in the variable always results in a fixed changed in the function.

Let us apply this test to the following table:

x:	-7	-3	1	5	9	13	17	21	25
$f(x)$:	30	23	16	9	2	-5	-12	-19	-26

Observe that throughout this table, as the variable continues to increase by 4, the function continues to decrease by 7. Thus, this function passes the test, and it is then easy to determine that this table is produced by the linear function $-\frac{7}{4}x + \frac{71}{4}$.

On the other hand, let us now apply this test to the following table:

x:	-9	-7	-5	-3	-1	1	3	5	7
$g(x)$:	-51	-16	1	6	5	4	9	26	61

Observe that whereas the variable continues to increase by 2 throughout the table, the change in the variable is not fixed. Thus, this function fails the test, and this table cannot be produced by a linear function. In this chapter, we will present a new test that is easy to apply that will allow us to determine when a table of values is produced by a polynomial. Then, if the table is indeed produced by a polynomial, we will find the smallest-degree polynomial that can produce the table. By applying this new test, we will show that the preceding table for $g(x)$ is produced by a cubic polynomial, and we will find that polynomial.

Next, we will revisit the types of formulas you were often asked to verify after first being introduced to Mathematical Induction. Typically, students are asked to use Mathematical

Induction to verify formulas such as

$$2^2 + 4^2 + 6^2 + \cdots + (2n-2)^2 + (2n)^2 = \frac{(2n)(n+1)(2n+1)}{3}.$$

Whereas it can be useful to verify formulas of this type, it is more important and far more interesting to actually derive them. We will show how to use our results on polynomials and tables of values to derive many formulas like the preceding one.

We will conclude this chapter by examining the algorithm we are introduced to in calculus that allows us to decompose rational functions into the sum of a polynomial and partial fractions. Recall that this algorithm is extremely useful in finding integrals of rational functions and is also often used to compute the sum of an infinite series. However, in calculus courses, it is never explained why this algorithm works. Ultimately, this algorithm produces a system of linear equations. Therefore, it is tempting to believe that the underlying reason that this algorithm works is based on some facts from linear algebra. However, it may come as somewhat of a surprise that the keys to partial fraction decomposition are actually the division algorithm and the Euclidean Algorithm in the polynomial ring $\mathbb{R}[x]$.

13.1 Difference Functions

Let us begin by looking at the following table:

x:	0	1	2	3	4	5	6	7	8
$f(x)$:	93	15	−9	21	105	243	435	681	981

It is quite natural to wonder if the preceding table can be produced by a polynomial. As it turns out, this table can be produced by a polynomial, and the smallest-degree polynomial that does the trick is $27x^2 - 105x + 93$. This example helps to motivate a two-part question. First, is there an easy algorithm to determine if a table of values was produced by a polynomial? Next, if the table was produced by a polynomial, how can we find a polynomial that does the job? The key to solving these problems will be to introduce a class of functions we will refer to as **difference functions**.

Definition 13.1. *Given a function $f(x)$, define $f_{(1)}(x)$, the* **first difference function of** $f(x)$, *as $f_{(1)}(x) = f(x+1) - f(x)$. If $n \geq 1$, define $f_{(n+1)}(x) = f_{(n)}(x+1) - f_{(n)}(x)$, and we call $f_{(n)}(x)$ the nth* **difference function of** $f(x)$.

To familiarize ourselves with difference functions, we begin with a few straightforward computations.

■ Examples

1. If $f(x) = 5$, then $f_{(1)}(x) = f(x+1) - f(x) = 5 - 5 = 0$.

2. If $g(x) = 4x - 2$, then

$$g_{(1)}(x) = g(x+1) - g(x) = (4(x+1) - 2) - (4x - 2) = (4x + 2) - (4x - 2) = 4.$$

Therefore,

$$g_{(2)}(x) = g_{(1)}(x+1) - g_{(1)}(x) = 4 - 4 = 0.$$

Thus, $g_{(n)}(x) = 0$, for all $n \geq 2$.

3. If $h(x) = 3x^2 + 2x + 5$, then

$$h_{(1)}(x) = h(x+1) - h(x) = (3(x+1)^2 + 2(x+1) + 5) - (3x^2 + 2x + 5) =$$
$$(3x^2 + 8x + 10) - (3x^2 + 2x + 5) = 6x + 5.$$

Furthermore,

$$h_{(2)}(x) = h_{(1)}(x+1) - h_{(1)}(x) = (6(x+1) + 5) - (6x + 5) =$$
$$(6x + 11) - (6x + 5) = 6.$$

Now observe that $h_{(n)}(x) = 0$, for all $n \geq 3$.

4. If $j(x) = 3^x$, then

$$j_{(1)}(x) = j(x+1) - j(x) = 3^{x+1} - 3^x = (3-1) \cdot 3^x = 2 \cdot 3^x.$$

Thus, $j_{(1)}(x)$ is also an exponential function, and it is now easy to see that $j_{(n)}(x)$ is an exponential function, for all n.

■

The goal of this section is to show that if the nth difference function $f_{(n)}(x)$ is a nonzero constant, then $f(x)$ is a polynomial of degree n. As we saw in calculus, if the nth derivative of a function is a nonzero constant, then the function is actually a polynomial of degree n. At first, it might seem odd that we can prove a result for difference functions that is so similar to one in calculus. However, if we reexamine the definition of the first difference function, we see that

$$f_{(1)}(x) = f(x+1) - f(x) = \frac{f(x+1) - f(x)}{1}.$$

Observe that $\frac{f(x+1)-f(x)}{1}$ is somewhat similar to $\lim_{h \to 0} \frac{f(x+h)-f(x)}{h}$, which is the definition of the derivative of $f(x)$. If we now look back at the preceding examples, we can see the formulas we obtained for difference functions are indeed quite similar to the analogous formulas for derivatives.

Definition 13.2. *If $n \in \mathbb{N}$, let $G_n(x) = (x+1)^n - x^n$.*

Observe that for any $n \geq 1$, $G_n(x)$ is the first difference function of the function $f(x) = x^n$. Once again, we will do some straightforward computations.

■ Examples

$$G_1(x) = (x+1) - x = 1$$
$$G_2(x) = (x+1)^2 - x^2 = 2x + 1$$
$$G_3(x) = (x+1)^3 - x^3 = 3x^2 + 3x + 1$$
$$G_4(x) = (x+1)^4 - x^4 = 4x^3 + 6x^2 + 4x + 1$$

■

In the preceding examples, we can see that the leading term of $G_n(x)$ is nx^{n-1}, which is the derivative of x^n. In light of this, the following lemma should not be too surprising.

Lemma 13.3

(a) $G_n(x) = (x+1)^n - x^n$ has degree $n-1$, for all $n \in \mathbb{N}$.

(b) If $f(x) \in \mathbb{R}[x]$ has degree n, where $n \geq 1$, then its first difference function $f_{(1)}(x)$ has degree $n-1$.

(c) If $f(x) \in \mathbb{R}[x]$ has degree n, then its nth difference function $f_{(n)}(x)$ is a nonzero constant, and its mth difference function $f_{(m)}(x)$ is equal to 0, for all $m > n$.

Proof. For part (a), since $(x+1)^n = x^n + nx^{n-1} + $ terms of lower degree, we have

$$G_n(x) = (x+1)^n - x^n = (x^n + nx^{n-1} + \text{terms of lower degree}) - x^n =$$

$$nx^{n-1} + \text{terms of lower degree}.$$

Thus, $G_n(x)$ has degree $n-1$.

For part (b), if $f(x) \in \mathbb{R}[x]$ has degree n, then

$$f(x) = \alpha_n x^n + \alpha_{n-1} x^{n-1} + \cdots + \alpha_1 x + \alpha_0,$$

where $\alpha_i \in \mathbb{R}$ and $\alpha_n \neq 0$. Then

$$f_{(1)}(x) = f(x+1) - f(x) =$$

$$(\alpha_n(x+1)^n + \alpha_{n-1}(x+1)^{n-1} + \cdots + \alpha_1(x+1) + \alpha_0) -$$

$$(\alpha_n x^n + \alpha_{n-1} x^{n-1} + \cdots + \alpha_1 x + \alpha_0) =$$

$$\alpha_n((x+1)^n - x^n) + \alpha_{n-1}((x+1)^{n-1} - x^{n-1}) + \cdots + \alpha_1((x+1) - x) =$$

$$\alpha_n G_n(x) + \alpha_{n-1} G_{n-1}(x) + \cdots + \alpha_1 G_1(x).$$

Since each $G_i(x)$ has degree $i - 1$ and $\alpha_n \neq 0$, we see that $f_{(1)}(x)$ has the same degree as $G_n(x)$. Thus, $f_{(1)}(x)$ has degree $n - 1$.

We know that taking the first derivative of a polynomial drops the degree by 1. In light of part (b), we now also know that this is the case when taking the first difference function of a polynomial. Therefore, if $f(x) \in \mathbb{R}[x]$ has degree n, the degree of $f_{(n)}(x)$ must be n less than n. Thus, $f_{(n)}(x)$ is a nonzero constant. Furthermore, since the first difference function of a constant function is 0, it immediately follows that if $m > n$, then $f_{(m)}(x) = 0$, thereby proving part (c).   □

Recall that our goal is to show that if $f_{(n)}(x)$ is a nonzero constant, then $f(x)$ is a polynomial of degree n. Although we are still quite far from this goal, we can observe that Lemma 13.3(c) tells us the converse is indeed true. Let us briefly return to the table of values we saw at the beginning of this section. If we insert the values of $f_{(1)}(x) = f(x+1) - f(x)$ into the table, we obtain

x:	0	1	2	3	4	5	6	7	8
$f(x)$:	93	15	−9	21	105	243	435	681	981
$f_{(1)}(x)$:	−78	−24	30	84	138	192	246	300	

Therefore, Lemma 13.3(c) tells us that since $f_{(1)}(x)$ is not constant, $f(x)$ cannot be linear. We will return to this example later in this section.

When looking at tables of values, it often suffices to examine functions whose domain is not necessarily the entire set of real numbers but is instead a subset of the integers. Throughout this chapter, we will let $\mathbb{N}_0$ denote the set $\mathbb{N} \cup \{0\}$ and will often only concern ourselves with functions defined on $\mathbb{N}_0$. Recall from calculus that if two functions have the same first derivative, they need not be the same function, but they must differ by a constant. As a result, if two functions have the same first derivative and agree at a single point, then they must be the same function. Given the similarity between difference functions and derivatives, it is not surprising that the analogous result also holds for difference functions.

Lemma 13.4. *Let $f, g : \mathbb{N}_0 \to \mathbb{R}$ be two functions such that $f_{(1)}(x) = g_{(1)}(x)$, for all $x \in \mathbb{N}_0$, and $f(0) = g(0)$. Then $f(x) = g(x)$, for all $x \in \mathbb{N}_0$.*

Proof. If $f_{(1)}(x) = g_{(1)}(x)$, for all $x \in \mathbb{N}_0$, and $f(0) = g(0)$, then let

$$T = \{n \in \mathbb{N}_0 \mid f(n) = g(n)\}.$$

It suffices to show that $T = \mathbb{N}_0$, and we will proceed by using Mathematical Induction. Since $f(0) = g(0)$, it is certainly the case that $0 \in T$. Therefore, we must now show that if T contains some $k \in \mathbb{N}_0$, then T also contains $k + 1$. Observe that

$$f_{(1)}(k) = f(k+1) - f(k) \quad \text{and} \quad g_{(1)}(k) = g(k+1) - g(k).$$

The previous equations immediately imply that

(1) $f(k+1) = f_{(1)}(k) + f(k)$ and $g(k+1) = g_{(1)}(k) + g(k)$.

However, since $k \in T$, we know that $f(k) = g(k)$. Combining this with the fact that $f_{(1)}(k) = g_{(1)}(k)$, the equations in (1) tell us that $f(k+1) = g(k+1)$. Therefore, $k+1 \in T$, as desired.   □

The next lemma indicates that any polynomial in $\mathbb{R}[x]$ can be rewritten in terms of polynomials of the form $G_n(x)$.

Lemma 13.5. *If $f(x) \in \mathbb{R}[x]$ has degree n then there exist $\alpha_1, \ldots, \alpha_{n+1} \in \mathbb{R}$ with $\alpha_{n+1} \neq 0$ such that $f(x) = \alpha_1 G_1(x) + \cdots + \alpha_{n+1} G_{n+1}(x)$.*

Proof. We will proceed by using the Second Version of Mathematical Induction. We begin by letting

$$T = \{n \in \mathbb{N}_0 \,|\, \text{all polynomials of degree } n \text{ have the desired property}\}.$$

It will suffice to show that $T = \mathbb{N}_0$.

We first need to show that $0 \in T$. If $f(x) \in \mathbb{R}[x]$ has degree 0, then $f(x) = \alpha_1$, for some nonzero $\alpha_1 \in \mathbb{R}$. Since $G_1(x) = 1$, we immediately see that $f(x) = \alpha_1 G_1(x)$. Thus $0 \in T$.

Now suppose that $k \in \mathbb{N}_0$ such that every element of the set $\{0, \ldots, k\}$ belongs to T. To conclude our proof, we need to show that $k+1 \in T$. Therefore, we need to show that if $f(x) \in \mathbb{R}[x]$ has degree $k+1$, then there exist $\alpha_1, \ldots, \alpha_{k+1}, \alpha_{k+2} \in \mathbb{R}$ such that

$$f(x) = \alpha_1 G_1(x) + \cdots + \alpha_{k+1} G_{k+1}(x) + \alpha_{k+2} G_{k+2}(x).$$

Let a denote the leading coefficient of $f(x)$, and then let $\alpha_{k+2} = \frac{a}{k+2}$. Observe that α_{k+2} is certainly nonzero. Furthermore, since $G_{k+2}(x)$ has degree $k+1$ with leading coefficient $k+2$, it is clear that $\alpha_{k+2} G_{k+2}(x)$ has the same degree and leading coefficient as $f(x)$. If $f(x) = \alpha_{k+2} G_{k+2}(x)$, then we are done. On the other hand, if $f(x) \neq \alpha_{k+2} G_{k+2}(x)$, then $f(x) - \alpha_{k+2} G_{k+2}(x)$ is a nonzero element of $\mathbb{R}[x]$ of degree less than $k+1$. If we let t denote the degree of $f(x) - \alpha_{k+2} G_{k+2}(x)$, then $t \in \{0, \ldots, k\} \subseteq T$. Since $t \in T$, there exist $\alpha_1, \ldots, \alpha_{t+1} \in \mathbb{R}$ such that

$$f(x) - \alpha_{k+2} G_{k+2}(x) = \alpha_1 G_1(x) + \cdots + \alpha_{t+1} G_{t+1}(x).$$

Finally, if we let $\alpha_m = 0$, for all m such that $t+1 < m < k+2$, the previous equation implies that

$$f(x) = \alpha_1 G_1(x) + \cdots + \alpha_{k+1} G_{k+1}(x) + \alpha_{k+2} G_{k+2}(x).$$

Thus, $k+1 \in T$, thereby concluding the proof.   □

We can now state and prove the first main result of this chapter.

Theorem 13.6. *If* $f : \mathbb{N}_0 \to \mathbb{R}$ *is a function such that its nth difference function,* $f_{(n)}(x)$, *is a nonzero constant, then* $f(x)$ *is a polynomial of degree n.*

Proof. We proceed with a proof by induction. To this end, let

$$T = \{k \in \mathbb{N} \,|\, \text{every function whose } k\text{th difference function}$$

$$\text{is a nonzero constant is equal to a polynomial of degree } k\};$$

and we will use Mathematical Induction to show that $T = \mathbb{N}$. In order to show that $1 \in T$, let us suppose that $f_1(x) = a$, for all $x \in \mathbb{N}_0$, where a is a nonzero real number. At this point, we would expect that $f(x)$ is a linear function with slope a and y-intercept $f(0)$. Thus, it makes sense to try to prove that $f(x)$ is equal to the linear function $ax + f(0)$. Observe that the first difference functions of both $f(x)$ and $ax + f(0)$ is the constant function a. Furthermore, plugging $x = 0$ into both $f(x)$ and $ax + f(0)$ gives the value $f(0)$. Therefore, Lemma 13.4 tells us that it is indeed the case that $f(x) = ax + f(0)$. Thus, $1 \in T$.

Now let us suppose that k is a positive integer that belongs to T; we must show that $k + 1$ also belongs to T. Therefore, suppose that $f_{(k+1)}(x) = a$, for all $x \in \mathbb{N}_0$, where a is a nonzero real number. We need to show that $f(x)$ is equal to some polynomial of degree $k + 1$. Observe that if we let $g(x) = f_{(1)}(x)$, then it immediately follows that

$$g_{(k)}(x) = f_{(k+1)}(x) = a.$$

Since the kth difference function of $g(x)$ is a nonzero constant and $k \in T$, we know that $g(x)$ is equal to some polynomial of degree k. In light of Lemma 13.5, there exist $\alpha_1, \ldots, \alpha_{k+1} \in \mathbb{R}$ with $\alpha_{k+1} \neq 0$ such that

$$g(x) = \alpha_{k+1} G_{k+1}(x) + \alpha_k G_k(x) + \cdots + \alpha_1 G_1(x).$$

Next, let

$$F(x) = \alpha_{k+1} x^{k+1} + \alpha_k x^k + \cdots + \alpha_1 x.$$

Then

$$F_{(1)}(x) = F(x + 1) - F(x) =$$

$$(\alpha_{k+1}(x+1)^{k+1} + \alpha_k(x+1)^k + \cdots + \alpha_1(x+1)) - (\alpha_{k+1} x^{k+1} + \alpha_k x^k + \cdots + \alpha_1 x) =$$

$$\alpha_{k+1}((x+1)^{k+1} - x^{k+1}) + \alpha_k((x+1)^{k+1} - x^k) + \cdots + \alpha_1((x+1) - x) =$$

$$\alpha_{k+1} G_{k+1}(x) + \alpha_k G_k(x) + \cdots + \alpha_1 G_1(x) = g(x) = f_{(1)}(x).$$

As a result, $f(x)$ and $F(x)$ have the same first difference function.

Now, let $H(x) = F(x) + f(0)$. Since $\alpha_{k+1} \neq 0$, $H(x)$ is a polynomial of degree $k+1$. It is easy to see that

$$H_{(1)}(x) = F_{(1)}(x) = f_{(1)}(x).$$

Furthermore, since $F(0) = 0$, we have

$$H(0) = F(0) + f(0) = 0 + f(0) = f(0).$$

Thus, $H(x)$ and $f(x)$ have the same first difference function and also have the same value at 0. Therefore, Lemma 13.4 tells us that $f(x) = H(x)$ and so, $f(x)$ is a polynomial of degree $k+1$. Hence, $k+1 \in T$, as desired. □

We can now use Theorem 13.6 to show that the table of values at the beginning of this section is not only produced by a polynomial but is produced by a quadratic. Having already inserted $f_{(1)}(x)$ onto the table and seen that $f(x)$ is not linear, we now insert $f_{(2)}(x) = f_{(1)}(x+1) - f_{(1)}(x)$ onto the table.

x:	0	1	2	3	4	5	6	7	8
$f(x)$:	93	15	−9	21	105	243	435	681	981
$f_{(1)}(x)$:	−78	−24	30	84	138	192	246	300	
$f_{(2)}(x)$:	54	54	54	54	54	54	54		

Since $f_{(2)}(x)$ is a nonzero constant, Theorem 13.6 asserts that our table can be produced by a polynomial of degree 2. Thus, $f(x) = ax^2 + bx + c$, for some $a, b, c \in \mathbb{R}$. To find the values of a, b, c, we will plug three of the values of x from our table into $f(x)$. Any three choices of x will work, and if we use the values $x = 0, 1, 2$ we obtain

$$93 = f(0) = a(0)^2 + b(0) + c = c$$
$$15 = f(1) = a(1)^2 + b(1) + c = a + b + c$$
$$-9 = f(2) = a(2)^2 + b(2) + c = 4a + b + c.$$

Therefore, we obtain the three linear equations

$$c = 93$$
$$a + b + c = 15$$
$$4a + 2b + c = -9.$$

Solving these three equations in three unknowns, we see that $a = 27, b = -105, c = 93$. Thus, $f(x) = 27x^2 - 105x + 93$ produces the table. Also note that $f(x)$ is the only quadratic polynomial that produces this table. To see this, let us suppose that $g(x)$ is also a quadratic polynomial that produces this table, and we will show that $f(x) = g(x)$. To accomplish this, let

$F(x) = f(x) - g(x)$; it will then suffice to show that $F(x) = 0$, for all x. It is certainly the case that $F(x)$ is a polynomial whose degree cannot exceed 2 with the additional properties that $F(0) = F(1) = F(2) = 0$. However, Corollary 12.16 asserts that if a polynomial $p(x)$ has three distinct roots, then either $p(x)$ has degree at least 3 or $p(x)$ is always equal to 0. As a result, $F(x)$ is always equal to 0, so $F(x) = f(x) - g(x) = 0$ and so $f(x) = g(x)$. Thus, $f(x) = 27x^2 - 105x + 93$ is the only quadratic polynomial that produces this table of values.

At first glance, Theorem 13.6 may not appear to handle as general a situation as was described in Chapter 1. Note that Theorem 13.6 only deals with functions whose domain is the set $\mathbb{N}_0 = \{0, 1, 2, 3, \ldots\}$. Furthermore, Theorem 13.6 only dealt with the case where x continually increased by 1, whereas in Chapter 1, we discussed the situation where x continually changed by a fixed real number a. Fortunately, it is easy to generalize Theorem 13.6 to this more general situation. Therefore, we will now state a more general version of Theorem 13.6 that, as we will see, has the advantage that it can be used more often.

Corollary 13.7. *Let $f(x)$ be a real-valued function whose domain is the set $\{b, b+a, b+ 2a, b+3a, b+4a, \ldots\}$, where $a, b \in \mathbb{R}$ and $a \neq 0$. In this case, let $f_{(1)}(x) = f(x+a) - f(x)$ and $f_{(n+1)}(x) = f_{(n)}(x+a) - f_{(n)}(x)$, for $n \geq 1$. If the function $f_{(n)}(x)$ is a nonzero constant, then $f(x)$ is a polynomial of degree n.*

Proof. Let $g(x) = f(b + a \cdot x)$; therefore, the domain of $g(x)$ is $\mathbb{N}_0$. If we define the difference functions of $g(x)$ as in Definition 13.1, and if the difference functions of $f(x)$ are as in the statement of this corollary, we have

$$g_{(1)}(x) = g(x+1) - g(x) = f(b+a(x+1)) - f(b+a \cdot x) =$$
$$f((b+a \cdot x)+a) - f(b+a \cdot x) = f_{(1)}(b+a \cdot x).$$

Repeating the same argument shows us that $g_{(2)}(x) = f_{(2)}(b + a \cdot x)$. It is now easy to see the fact, which we could have proved formally using Mathematical Induction, that $g_{(n)}(x) = f_{(n)}(b + a \cdot x)$, for all $n \geq 1$. Since the nth difference function of $f(x)$ is a nonzero constant, the same is now true of $g(x)$. Therefore, we can apply Theorem 13.6 to conclude that $g(x)$ is a polynomial of degree n. Thus, $f(b + a \cdot x)$ is also polynomial in x of degree n. If we make the substitution $y = b + a \cdot x$, then x gets replaced by $\frac{y-b}{a}$, and we see that $f(y)$ is a polynomial in y whose degree in y is also n. However, this is the same as saying that $f(x)$ is also a polynomial of degree n in x. $\square$

We can now use Corollary 13.7 to revisit the table that appeared at the beginning of this chapter:

x:	-9	-7	-5	-3	-1	1	3	5	7
$g(x)$:	-51	-16	1	6	5	4	9	26	61

Since the variable increases by 2 through the table, we can let $g_{(1)}(x) = g(x+2) - g(x)$ and $g_{(n+1)}(x) = g_{(n)}(x+2) - g_{(n)}(x)$, for $n \geq 1$. Adding $g_{(1)}(x)$ to the table, we obtain

x:	-9	-7	-5	-3	-1	1	3	5	7
$g(x)$:	-51	-16	1	6	5	4	9	26	61
$g_{(1)}(x)$:	35	17	5	-1	-1	5	17	35	

Since $g_{(1)}(x)$ is not a constant, we know that the table cannot be produced by a linear function. Therefore, we next add $g_{(2)}(x)$ to the table to obtain

x:	-9	-7	-5	-3	-1	1	3	5	7
$g(x)$:	-51	-16	1	6	5	4	9	26	61
$g_{(1)}(x)$:	35	17	5	-1	-1	5	17	35	
$g_{(2)}(x)$:	-18	-12	-6	0	6	12	18		

Observe that $g_{(2)}(x)$ is also not a constant. Thus, the table cannot be produced by a quadratic function and we add $g_{(3)}(x)$ to the table.

x:	-9	-7	-5	-3	-1	1	3	5	7
$g(x)$:	-51	-16	1	6	5	4	9	26	61
$g_{(1)}(x)$:	35	17	5	-1	-1	5	17	35	
$g_{(2)}(x)$:	-18	-12	-6	0	6	12	18		
$g_{(3)}(x)$:	6	6	6	6	6	6			

Since $g_{(3)}(x)$ is a nonzero constant, Corollary 13.7 now asserts that the table can be produced by a cubic polynomial. Thus, $g(x) = ax^3 + bx^2 + cx + d$. By plugging in any four values of x from the table, we will obtain four linear equations that will allow us to find a, b, c, d. In order to deal with smaller numbers, we will let $x = -3, -1, 1, 3$ and then obtain

$$6 = a(-3)^3 + b(-3)^2 + c(-3) + d = -27a + 9b - 3c + d,$$
$$5 = a(-1)^3 + b(-1)^2 + c(-1) + d = -a + b - c + d,$$
$$4 = a(1)^3 + b(1)^2 + c(1) + d = a + b + c + d,$$
$$9 = a(3)^3 + b(3)^2 + c(3) + d = 27a + 9b + 3c + d.$$

Therefore, we have obtained the four equations

$$-27a + 9b - 3c + d = 6$$
$$-a + b - c + d = 5$$
$$a + b + c + d = 4$$
$$27a + 9b + 3c + d = 9.$$

Solving these equations yields $a = \frac{1}{8}, b = \frac{3}{8}, c = -\frac{5}{8}, d = \frac{33}{8}$. Hence, the table can be produced by the cubic polynomial $\frac{1}{8}x^3 + \frac{3}{8}x^2 - \frac{5}{8}x + \frac{33}{8}$. It then follows, by Corollary 12.16, that this is the only cubic polynomial that produces the table.

In high school algebra courses, we learned that two data points determine a line. Earlier in this section, we used three data points to find a quadratic and four data points to find a cubic. We will now generalize these and prove that given any n data points, there always exists a unique polynomial of degree less than n that agrees with all n data points.

Proposition 13.8. *Let $\alpha_1, \ldots, \alpha_n, \beta_1, \ldots, \beta_n \in \mathbb{R}$, where all the α_i are different, Then there exists a unique $f(x) \in \mathbb{R}[x]$ of degree less than n such that $f(\alpha_1) = \beta_1, f(\alpha_2) = \beta_2, \ldots, f(\alpha_n) = \beta_n$.*

Proof. For each $i \leq n$, let

$$H_i(x) = (x - \alpha_1)(x - \alpha_2) \cdots (x - \alpha_{i-1})(x - \alpha_{i+1}) \cdots (x - \alpha_n).$$

In other words, $H_i(x)$ is the product of all the $x - \alpha_j$ except for $x - \alpha_i$. Next, let $\gamma_i = H(\alpha_i)$; since all the α_j are different, we see that $\gamma_i \neq 0$. Also note that if $i \neq j$, then $H_i(\alpha_j) = 0$.

For each $i \leq n$, consider the polynomial $\beta_i \gamma_i^{-1} H_i(x)$. It is easy to see that $\beta_i \gamma_i^{-1} H_i(\alpha_i) = \beta_i$ and $\beta_i \gamma_i^{-1} H_i(\alpha_j) = 0$, for all $j \neq i$. Therefore, if we let

$$f(x) = \beta_1 \gamma_1^{-1} H_1(x) + \beta_2 \gamma_2^{-1} H_2(x) + \cdots + \beta_n \gamma_n^{-1} H_n(x),$$

then

$$f(\alpha_i) = \beta_1 \gamma_1^{-1} H_1(\alpha_i) + \cdots + \beta_{i-1} \gamma_{i-1}^{-1} H_{i-1}(\alpha_i) +$$

$$\beta_i \gamma_i^{-1} H_i(\alpha_i) + \beta_{i+1} \gamma_{i+1}^{-1} H_{i+1}(\alpha_i) + \cdots + \beta_n \gamma_n^{-1} H_n(\alpha_i) =$$

$$0 + \cdots 0 + \beta_i + 0 + \cdots + 0 = \beta_i$$

for all $i \leq n$. Furthermore, since each $H_i(x)$ has degree $n - 1$, the degree of $f(x)$ is less than n.

All that remains is to show that $f(x)$ is the only polynomial of degree less than n such that $f(\alpha_i) = \beta_i$, for all $i \leq n$. Let $g(x)$ also be a polynomial of degree less than n such that $g(\alpha_i) = \beta_i$, for all $i \leq n$. We need to show that $f(x) = g(x)$. If we let $F(x) = f(x) - g(x)$, then $F(\alpha_i) = 0$, for all $i \leq n$. Therefore, $F(x)$ has at least n distinct roots and Corollary 12.16 asserts that $F(x)$ must either be 0 or have degree at least n. But since the degrees of $f(x)$ and $g(x)$ are less than n, the degree of $F(x) = f(x) - g(x)$ cannot possible be as large as n. Hence, $F(x) = f(x) - g(x) = 0$ and $f(x) = g(x)$, as desired. $\square$

We now have two different ways to find a polynomial that agrees with n data points. One is to construct the polynomial as in the proof of Proposition 13.8, whereas the other is to set up a system of linear equations as in the example preceding Proposition 13.8. We now provide an example illustrating both techniques.

■ Example

Find a polynomial $f(x)$ of degree less than 3 such that $f(-1) = 2$, $f(1) = 6$, $f(4) = 87$. Applying the technique used in the proof of Proposition 13.8, let

$$f(x) = 2\gamma_1^{-1}(x-1)(x-4) + 6\gamma_2^{-1}(x+1)(x-4) + 87\gamma_3^{-1}(x+1)(x-1),$$

where

$$\gamma_1 = ((-1) - 1)((-1) - 4) = 10, \quad \gamma_2 = (1+1)(1-4) = -6,$$

$$\gamma_3 = (4+1)(4-1) = 15.$$

Plugging the value of each γ_i into the preceding equation for $f(x)$, we obtain

$$f(x) = \frac{1}{5}(x-1)(x-4) - (x+1)(x-4) + \frac{29}{5}(x+1)(x-1).$$

Converting $f(x)$ to the more familiar form of a quadratic gives us $f(x) = 5x^2 + 2x - 1$.

In our second technique, we use the fact that $f(x) = ax^2 + bx + c$, for some $a, b, c \in \mathbb{R}$. Plugging $x = -1, 1, 4$ into $f(x)$, we obtain

$$2 = f(-1) = a(-1)^2 + b(-1) + c = a - b + c,$$

$$6 = f(1) = a(1)^2 + b(1) + c = a + b + c,$$

$$87 = f(4) = a(4)^2 + b(4) + c = 16a + 4b + c.$$

Solving the three linear equations

$$a - b + c = 2,$$

$$a + b + c = 6,$$

$$16a + 4b + c = 87,$$

gives us $a = 5$, $b = 2$, and $c = -1$. Thus, once again, $f(x) = 5x^2 + 2x - 1$. ■

In the preceding example, the second technique produced three linear equations in three unknowns. More generally, if we wished to find a polynomial of degree less than n that took on n prescribed values, this technique would produce n linear equations in n unknowns. Although a system of linear equations can have no solutions or an infinite number of solutions, Proposition 13.8 guarantees that the system we obtain in this situation does have a unique solution. Therefore, Proposition 13.8 is useful even when we use a technique for finding the polynomial other than the technique that appears in its proof.

13.2 Polynomials and Mathematical Induction

As an application of Corollary 13.7, we will show how to derive many of the formulas we were asked to verify when learning about Mathematical Induction. Several examples of these types of formulas are:

$$1+2+3+\cdots(n-1)+n = \frac{n(n+1)}{2},$$

$$1^2+2^3+3^2+\cdots(n-1)^2+n^2 = \frac{n(n+1)(2n+1)}{6},$$

$$1^3+2^3+3^3+\cdots(n-1)^3+n^3 = \frac{n^2(n+1)^2}{4},$$

$$1+3+5+\cdots+(2n-3)+(2n-1) = n^2$$

Whereas Mathematical Induction is a wonderful tool for proving the validity of these formulas, it does not address the much more important question of where these formulas come from. After all, the real excitement and challenge is in *finding* these formulas.

Observe that in the first formula, we are adding terms obtained from plugging natural numbers into the linear function $p(x) = x$ and the formula on the right side is a quadratic. In the second formula, we are adding terms obtained from plugging natural numbers into the quadratic function $p(x) = x^2$ and the formula on the right side is a cubic. Similarly, for the third formula, we are adding terms obtained from plugging natural numbers into the cubic $p(x) = x^3$ and the right side is a quartic. At this point, it is reasonable to ask the following question: If $p(x)$ is a polynomial of degree t, is the formula for $p(1) + p(2) + p(3) + \cdots + p(n-1) + p(n)$ always a polynomial of degree $t+1$? We can now use Corollary 13.7 to answer this question in the affirmative.

Corollary 13.9. *If $p(x) \in \mathbb{R}[x]$ is a polynomial of degree t and $k \in \mathbb{N}_0$, define the function f as*

$$f(n) = p(k) + p(k+1) + p(k+2) + p(k+3) + \cdots + p(n-1) + p(n),$$

where $n \in \mathbb{N}_0$ and $n \geq k$. Then f is a polynomial of degree $t+1$.

Proof. Observe that

$$f_{(1)}(n) = f(n+1) - f(n) =$$
$$(p(k) + p(k+1) + p(k+2) + \cdots + p(n-1) + p(n) + p(n+1)) -$$
$$(p(k) + p(k+1) + p(k+2) + \cdots + p(n-1) + p(n)) = p(n+1).$$

Since $p(x)$ has degree t, it follows that $f_{(1)}(n) = p(n+1)$ also has degree t. However, since the first difference function of f is a polynomial of degree t, the same calculations as in Lemma 13.3 tells us that the $t+1$st difference function of f is a nonzero constant. Corollary 13.7 now asserts that f is indeed a polynomial of degree $t+1$.   □

We will now work through several examples.

■ Example

Let us find a formula for the function

$$f(n) = 5+8+11+\cdots+(3n-1)+(3n+2).$$

Since f is obtained by adding values of the linear function $p(x) = 3x+2$ starting with $p(1)$, Corollary 13.9 tells that f is a quadratic. Therefore, there exist $a, b, c \in \mathbb{R}$ such that

(2) $f(n) = an^2 + bn + c.$

As in Proposition 13.8, to find a, b, c, we will need three values of $f(n)$. Since we know that $f(1) = 5$, $f(2) = 5+8 = 13$, $f(3) = 5+8+11 = 24$, we can plug $n = 1, 2, 3$ into (2) to obtain

$$5 = a+b+c$$
$$13 = 4a+2b+c$$
$$24 = 9a+3b+c.$$

We have reduced this problem down to solving a system of three linear equations in three unknowns. It is now easy to determine that the unique solution is $a = \frac{3}{2}, b = \frac{7}{2}, c = 0$. As a result, we have derived the formula

$$5+8+11+\cdots+(3n-1)+(3n+2) = \frac{3n^2+7n}{2}.$$

■

Hopefully, Corollary 13.9 and the preceding example have helped to take the mystery out of some of the formulas that appear when we are learning to use Mathematical Induction. As we use Corollary 13.9 to derive various formulas, the last step is always to solve a system of linear equations. These systems of equations can sometimes be rather messy and quite difficult to solve by hand. However, as we can soon see, by using difference functions, we can always reduce these problems down to a system of linear equations which is extremely easy to solve. In the next example we will derive the appropriate formula in two ways. The second technique will illustrate how using difference functions can greatly simplify the computations.

■ Example

Find a formula for the function

$$g(n) = 1^2 + 3^2 + 5^2 + \cdots + (2n-3)^2 + (2n-1)^2.$$

Since $g(n)$ is obtained by adding values of the quadratic $p(x) = (2x-1)^2$ starting with $p(1)$, Corollary 13.9 tells us that $g(n)$ is a cubic. As a result, there exist $a, b, c, d \in \mathbb{R}$ such that

(3) $$g(n) = an^3 + bn^2 + cn + d.$$

Using the same technique as in the previous example, by looking at four values of $g(n)$, we will obtain four linear equations in a, b, c, d. Since $g(1) = 1^2 = 1$, $g(2) = 1^2 + 3^2 = 10$, $g(3) = 1^2 + 3^2 + 5^2 = 35$, and $g(4) = 1^2 + 3^2 + 5^2 + 7^2 = 84$, we can now plug $n = 1, 2, 3, 4$ into (3) to obtain

$$1 = a + b + c + d$$
$$10 = 8a + 4b + 2c + d$$
$$35 = 27a + 9b + 3c + d$$
$$84 = 64a + 16b + 4c + d.$$

Solving these four equations requires some work, but eventually you will obtain the solution $a = \frac{4}{3}, b = 0, c = -\frac{1}{3}, d = 0$. Thus,

$$1^2 + 3^2 + 5^2 + \cdots + (2n-3)^2 + (2n-1)^2 = \frac{4}{3}n^3 - \frac{1}{3}n.$$

However, we can use difference functions to produce four linear equations involving a, b, c, d that are much easier to solve. In light of Corollary 13.9, we know that $g(n) = an^3 + bn^2 + cn + d$. Taking the first difference function of $g(n)$, we obtain

$$g_{(1)}(n) = g(n+1) - g(n) =$$
$$(a(n+1)^3 + b(n+1)^2 + c(n+1) + d) - (an^3 + bn^2 + cn + d) =$$
$$a((n+1)^3 - n^3) + b((n+1)^2 - n^2) + c((n+1) - n) =$$
$$a((n^3 + 3n^2 + 3n + 1) - n^3) + b((n^2 + 2n + 1) - n^2) + c((n+1) - n) =$$

(4) $$(3a)n^2 + (3a + 2b)n + (a + b + c).$$

On the other hand, as we saw in the proof of Corollary 13.9, $g_{(1)}(n) = p(n+1)$. Since $p(x) = (2x-1)^2$, we obtain

(5) $$g_{(1)}(n) = (2(n+1) - 1)^2 = (2n+1)^2 = 4n^2 + 4n + 1.$$

By comparing the coefficients of the various powers of n in equations (4) and (5), we obtain the linear equations

$$3a = 4$$

$$3a + 2b = 4$$

$$a + b + c = 1.$$

Observe that this system of linear equations is extremely easy to solve. The first equation tells us that $a = \frac{4}{3}$. Plugging the value of a into the second equation yields $b = 0$. Then plugging the values of a and b into the third equation gives us $c = -\frac{1}{3}$. Finally, letting $n = 1$ in the formula $g(n) = an^3 + bn^2 + cn + d$ results in the equation

$$a + b + c + d = 1.$$

But having already found a, b, c, this equation immediately tells us that $d = 0$. Therefore, we have once again—this time with much less work—derived the formula

$$1^2 + 3^2 + 5^2 + \cdots + (2n - 3)^2 + (2n - 1)^2 = \frac{4}{3}n^3 - \frac{1}{3}n.$$

∎

As the degrees of the polynomials involved increase, so does the amount of work saved by using difference functions. This will become clearer in the next example.

▪ Example

Find a formula for

$$F(n) = 1^4 + 2^4 + 3^4 + \cdots + n^4.$$

In light of Corollary 13.9, we know that F will be a polynomial of degree 5. Thus, we need to find $a, b, c, d, e, f \in \mathbb{R}$ such that

(6) $$F(n) = 1^4 + 2^4 + 3^4 + \cdots + n^4 = an^5 + bn^4 + cn^3 + dn^2 + en + f.$$

Taking the first difference function of F, we obtain

$$F_{(n)} = F(n + 1) - F(n) =$$

$$(a(n + 1)^5 + b(n + 1)^4 + c(n + 1)^3 + d(n + 1)^2 + e(n + 1) + f) -$$

$$(an^5 + bn^4 + cn^3 + dn^2 + en + f) =$$

$$a((n + 1)^5 - n^5) + b((n + 1)^4 - n^4) + c((n + 1)^3 - n^3) + d((n + 1)^2 - n^2) +$$

$$e((n + 1) - n) = a(5n^4 + 10n^3 + 10n^2 + 5n + 1) + b(4n^3 + 6n^2 + 4n + 1) +$$

$$c(3n^2 + 3n + 1) + d(2n + 1) + e = (5a)n^4 + (10a + 4b)n^3 + (10a + 6b + 3c)n^2 +$$

(7) $$(5a + 4b + 3c + 2d)n + (a + b + c + d + e).$$

On the other hand, since $g_{(1)}(n) = p(n+1)$ and $p(x) = x^4$, we obtain

(8) $$g_{(1)}(n) = (n+1)^4 = n^4 + 4n^3 + 6n^2 + 4n + 1.$$

By comparing the coefficients of the various powers of n in equations (7) and (8), we obtain the linear equations

$$5a = 1$$
$$10a + 4b = 4$$
$$10a + 6b + 3c = 6$$
$$5a + 4b + 3c + 2d = 4$$
$$a + b + c + d + e = 1.$$

Going in order from the top linear equation to the bottom, we can now easily determine, in sequence, the values of a, b, c, d, e. It follows that $a = \frac{1}{5}$, $b = \frac{1}{2}$, $c = \frac{1}{3}$, $d = 0$, and $e = -\frac{1}{30}$. Finally, by letting $n = 1$ in equation (6) we see that

$$a + b + c + d + e + f = 1.$$

Having already found the values of a, b, c, d, e, this equation immediately tells us that $f = 0$. Thus,

$$1^4 + 2^4 + 3^4 + \cdots + n^4 = \frac{1}{5}n^5 + \frac{1}{2}n^4 + \frac{1}{3}n^3 - \frac{1}{30}n.$$

On the other hand, if we attempted to solve this problem by letting $n = 1, 2, 3, 4, 5, 6$, in equation (6), then we would obtain the following system of linear equations:

$$a + b + c + d + e + f = 1$$
$$32a + 16b + 8c + 4d + 2e + f = 17$$
$$243a + 81b + 27c + 9d + 3e + f = 98$$
$$1024a + 256b + 64c + 16d + 4e + f = 354$$
$$3125a + 625b + 125c + 25d + 5e + f = 979$$
$$7776a + 1296b + 216c + 36d + 6e + f = 2275.$$

Certainly, this system of linear equations is much more complicated than one obtained using difference functions. Thus, Corollary 13.9 and difference functions do provide us with a fairly easy way of deriving many important formulas.

■

Exercises for Sections 13.1 and 13.2

1. (a) Use difference functions to determine the smallest degree of a polynomial that can produce the following table.

   (b) Then find the polynomial of smallest possible degree that does produce the following table.

x:	-6	-5	-4	-3	-2	-1	0	1	2
$f(x)$:	83	72	61	50	39	28	17	6	-5

2. (a) Use difference functions to determine the smallest degree of a polynomial that can produce the following table.

   (b) Then find the polynomial of smallest possible degree that does produce the following table.

x:	-22	-14	-6	2	10	18	26	34	42
$g(x)$:	-26.5	-24.5	-22.5	-20.5	-18.5	-16.5	-14.5	-12.5	-10.5

3. (a) Use difference functions to determine the smallest degree of a polynomial that can produce the following table.

   (b) Then find the polynomial of smallest possible degree that does produce the following table.

x:	3	4	5	6	7	8	9	10	11
$h(x)$:	24	24	22	18	12	4	-6	-18	-32

4. (a) Use difference functions to determine the smallest degree of a polynomial that can produce the following table.

   (b) Then find the polynomial of smallest possible degree that does produce the following table.

x:	-8	-5	-2	1	4	7	10	13	16
$p(x)$:	289	163	73	19	1	19	73	163	289

5. (a) Use difference functions to determine the smallest degree of a polynomial that can produce the following table.

   (b) Then find the polynomial of smallest possible degree that does produce the following table.

x:	-10	-6	-2	2	6	10	14	18	22
$q(x)$:	293	129	29	-7	21	113	269	489	773

6. (a) Use difference functions to determine the smallest degree of a polynomial that can produce the following table.

(b) Then find the polynomial of smallest possible degree that does produce the following table.

x:	−23	−18	−13	−8	−3	2	7	12	17
f(x):	−1461	−871	−431	−141	−1	−11	−171	−481	−941

7. (a) Use difference functions to determine the smallest degree of a polynomial that can produce the following table.

   (b) Then find the polynomial of smallest possible degree that does produce the following table.

x:	−2	−1	0	1	2	3	4	5	6
g(x):	−19	−2	1	2	13	46	113	226	397

8. (a) Use difference functions to determine the smallest degree of a polynomial that can produce the following table.

   (b) Then find the polynomial of smallest possible degree that does produce the following table.

x:	−19	−14	−9	−4	1	6	11	16	21
h(x):	−6944	−2809	−774	−89	−4	231	1366	4151	9336

9. (a) Use difference functions to determine the smallest degree of a polynomial that can produce the following table.

   (b) Then find the polynomial of smallest possible degree that does produce the following table.

x:	−8	−5	−2	1	4	7	10	13	16
p(x):	−1205	−314	−17	10	91	550	1711	3898	7435

10. (a) Use difference functions to determine the smallest degree of a polynomial that can produce the following table.

    (b) Then find the polynomial of smallest possible degree that does produce the following table.

x:	−7	−5	−3	−1	1	3	5	7	9
q(x):	362	154	66	50	58	42	−46	−254	−630

11. If $F(n) = 1 + 5 + 9 + \cdots + (4n - 3)$, write $F(n)$ as a polynomial.

12. If $G(n) = 8 + 13 + 18 + \cdots + (5n - 7)$, write $G(n)$ as a polynomial.

13. If $F(n) = 1 \cdot 2 + 2 \cdot 3 + 3 \cdot 4 + \cdots + n(n + 1)$, write $F(n)$ as a polynomial.

14. If $G(n) = 1 \cdot 2 + 3 \cdot 4 + 5 \cdot 6 + \cdots + (2n - 1)(2n)$, write $G(n)$ as a polynomial.

15. If $W(n) = 1 \cdot 1 + 2 \cdot 3 + 3 \cdot 5 + \cdots + (n)(2n - 1)$, write $W(n)$ as a polynomial.

16. If $T(n) = 1 \cdot 2 \cdot 3 + 2 \cdot 3 \cdot 4 + 3 \cdot 4 \cdot 5 + \cdots + (n)(n+1)(n+2)$, write $T(n)$ as a polynomial.

17. If $S(n) = 1 \cdot 2 \cdot 3 + 3 \cdot 4 \cdot 5 + 5 \cdot 6 \cdot 7 + \cdots + (2n-1)(2n)(2n+1)$, write $S(n)$ as a polynomial.

18. If $H(n) = 1 \cdot 2 \cdot 3 + 4 \cdot 5 \cdot 6 + 7 \cdot 8 \cdot 9 + \cdots + (3n-2)(3n-1)(3n)$, write $H(n)$ as a polynomial.

19. If $F(n) = \sum_{i=2}^{n}(i^2 + 4i)$, write $F(n)$ as a polynomial.

20. If $G(n) = \sum_{i=1}^{n}(3i^2 - 2i)$, write $G(n)$ as a polynomial.

21. If $H(n) = \sum_{i=1}^{n}(i^3 + 5i)$, write $H(n)$ as a polynomial.

22. If $S(n) = \sum_{i=4}^{n}(2i^3 - 7i + 5)$, write $S(n)$ as a polynomial.

23. If $D(n) = 1^4 + 2^4 + 3^4 + \cdots + n^4$, write $D(n)$ as a polynomial.

24. If $T(n) = 1^5 + 2^5 + 3^5 + \cdots + n^5$, write $T(n)$ as a polynomial.

25. If $W(n) = 1^6 + 2^6 + 3^6 + \cdots + n^6$, write $W(n)$ as a polynomial.

26. On an 8×8 checkerboard, there are 64 squares each with dimension 1×1. The checkerboard also contains 2×2 squares, 3×3 squares, all the way up to an 8×8 square. What is the total number of squares on the checkerboard?

27. Repeat exercise 26, but instead suppose you had a 10×10 checkerboard. In this case, what is the total number of squares on the checkerboard?

28. We will now generalize exercises 26 and 27. If you have an $n \times n$ checkerboard, what is the total number of squares on the checkerboard?

In exercises 29–38, we will examine additional similarities between difference functions and derivatives. Throughout these exercises, we will make frequent use of the definition of difference functions that appeared in Definition 13.1.

29. If $f_{(n)}(x) = 0$, where $n \in \mathbb{N}$, show that there exist $\alpha_0, \alpha_1, \ldots, \alpha_{n-1} \in \mathbb{R}$ such that $f(x) = \alpha_{n-1}x^{n-1} + \cdots + \alpha_1 x + \alpha_0$.

30. Show that if $f(x), g(x)$ are functions such that $f_{(n)}(x) = g_{(n)}(x)$, where $n \in \mathbb{N}$, then there exist $\alpha_0, \alpha_1, \ldots, \alpha_{n-1} \in \mathbb{R}$ such that $f(x) = g(x) + \alpha_{n-1}x^{n-1} + \cdots + \alpha_1 x + \alpha_0$.

31. Let $f(x) = \alpha a^x$, where $\alpha, a \in \mathbb{R}$, $\alpha \neq 0$, $a > 0$, and $a \neq 1$.
    (a) Show that $f_{(1)}(x) = \alpha(a-1)a^x$.
    (b) Show that, for all $n \in \mathbb{N}$, $f_{(n)}(x) = \alpha(a-1)^n a^x$.

32. Use exercise 31 to find the value of $a \in \mathbb{R}$ such that $f(x) = \alpha a^x$ has the property that $f_{(1)}(x) = f(x)$.

33. Let $f(x)$ be a function such that $f_{(1)}(x) = f(x)$ and let $g(x) = \frac{f(x)}{2^x}$.

    (a) Show that $g_{(1)}(x) = 0$.

    (b) Use Lemma 13.4 or exercise 29 to conclude that $f(x) = \alpha 2^x$, for some $\alpha \in \mathbb{R}$.

34. (a) Show that if $f(x) = \sin(x)$, then $f_{(1)}(x) = \alpha \sin(x) + \beta \cos(x)$, for some $\alpha, \beta \in \mathbb{R}$ and determine the exact values of α, β.

    (b) Show that if $g(x) = \cos(x)$, then $g_{(1)}(x) = \gamma \sin(x) + \delta \cos(x)$, for some $\gamma, \delta \in \mathbb{R}$ and determine the exact values of γ, δ.

    (c) If $A, B \in \mathbb{R}$, $n \in \mathbb{N}$, and $F(x) = A \sin(x) + B \cos(x)$, show that there exist $a, b \in \mathbb{R}$ such that $F_{(n)}(x) = a \sin(x) + b \cos(x)$.

In exercises 35–38, you will need to find functions that produce a given difference function. This is analogous to finding antiderivatives in calculus.

35. (a) Use exercises 30 and 31 to find all functions $f(x)$ such that $f_{(1)}(x) = 10^x$.

    (b) Use part (a) to find the unique function $f(x)$ such that $f_{(1)}(x) = 10^x$ and $f(0) = 3$.

36. (a) Use exercises 30 and 31 to find all functions $g(x)$ such that $g_{(1)}(x) = e^x$.

    (b) Use part (a) to find the unique function $g(x)$ such that $g_{(1)}(x) = e^x$ and $g(0) = 7$.

37. (a) Find all functions $f(x)$ such that $f_{(2)}(x) = 3^x$.

    (b) Find the unique function $f(x)$ such that $f_{(2)}(x) = 3^x$, $f(0) = 10$, and $f_{(1)}(0) = 21$.

38. (a) Find all functions $g(x)$ such that $g_{(2)}(x) = 15^x$.

    (b) Find the unique function $g(x)$ such that $g_{(2)}(x) = 15^x$, $g(0) = 19$, and $g_{(1)}(0) = 51$.

In exercises 39–42, we will make reference to the functions $G_n(x)$ that were defined in Definition 13.2.

39. (a) Find $\alpha, \beta \in \mathbb{R}$ such that $\alpha G_2(x) + \beta G_1(x) = 6x - 10$.

    (b) Use part (a) to find all functions $f(x)$ such that $f_{(1)}(x) = 6x - 10$.

    (c) Find the unique function $f(x)$ such that $f_{(1)}(x) = 6x - 10$ and $f(0) = 28$.

40. (a) Find $\alpha, \beta \in \mathbb{R}$ such that $\alpha G_2(x) + \beta G_1(x) = 9x - 16$.

    (b) Use part (a) to find all functions $f(x)$ such that $f_{(1)}(x) = 9x - 16$.

    (c) Find the unique function $f(x)$ such that $f_{(1)}(x) = 9x - 16$ and $f(0) = -11$.

41. (a) Find $\alpha, \beta, \gamma \in \mathbb{R}$ such that $\alpha G_3(x) + \beta G_2(x) + \gamma G_1(x) = x^2$.

   (b) Use part (a) to find all functions $f(x)$ such that $f_{(1)}(x) = x^2$.

   (c) Find the unique function $f(x)$ such that $f_{(1)}(x) = x^2$ and $f(0) = 1987$.

42. (a) Find $\alpha, \beta, \gamma \in \mathbb{R}$ such that $\alpha G_3(x) + \beta G_2(x) + \gamma G_1(x) = 2x^2 - 9x + 7$.

   (b) Use part (a) to find all functions $f(x)$ such that $f_{(1)}(x) = 2x^2 - 9x + 7$.

   (c) Find the unique function $f(x)$ such that $f_{(1)}(x) = 2x^2 - 9x + 7$ and $f(0) = -44$.

Some of our previous exercises dealt with finding formulas for the sums of values of polynomials. In exercises 43 and 44, we will use difference functions to study sums of values of functions involving factorials and exponential.

43. Let $S(n) = 1 \cdot 1! + 2 \cdot 2! + 3 \cdot 3! + \cdots + n \cdot n!$.
   (a) Find the formula for $S_{(1)}(n)$.

   (b) You might suspect that the formula for $S(n)$ involves factorials that are larger than $n!$. In light of this, you might conjecture that the formula for $S(n)$ involves $(n+1)!$. Therefore, in an attempt to find a formula for $S(n)$, let $T(n) = (n+1)!$ and find the formula for $T_{(1)}(n)$.

   (c) Compare the formulas for $S_{(1)}(n)$ and $T_{(1)}(n)$.

   (d) Use part (c) along with exercise 30 and the fact that $T(1) = 2$ to find the formula for $S(n)$.

44. (a) Let $F(n) = 1 + a + a^2 + \cdots + a^n$, where $a \in \mathbb{R}$ and $a \neq 1$. Find $F_{(1)}(n)$.

   (b) You might suspect that the formula for $F(n)$ involves a^{n+1}. Therefore, let $G(n) = a^{n+1}$ and compare $F_{(1)}(n)$ and $G_{(1)}(n)$.

   (c) Use part (b) to show that $F(n) = \frac{1}{a-1}a^{n+1} + \alpha$, for some $\alpha \in \mathbb{R}$.

   (d) Use the fact that $F(0) = 1$, to find α from part (c).

   (e) Write down the formula you have derived for $F(n)$ and then compare it to the formula $1 + a + a^2 + \cdots + a^n = \frac{1 - a^{n+1}}{1 - a}$ you have likely seen in precalculus and calculus courses.

In exercises 45–54, we will let

$$F(n) = p(0) + p(1)a + p(2)a^2 + \cdots + p(n)a^n,$$

where $n \geq 0$, $a \in \mathbb{R}$, $a \neq 1$, and p is a polynomial of degree t. We will use difference functions to find a formula for $F(n)$. To do this, for every $n \in \mathbb{N}_0$, let $H_n(x) = a(x+1)^n - x^n$. Observe that the functions $H_n(x)$ are similar to the functions $G_n(x)$ defined in Definition 13.2.

45. Show that $H_n(x)$ has degree n, for all $n \in \mathbb{N}_0$.

46. Show that for every polynomial $f(x) \in \mathbb{R}[x]$ of degree n, there exist $\alpha_0, \alpha_1, \ldots, \alpha_n$ such that $f(x) = \alpha_0 H_0(x) + \alpha_1 H_1(x) + \cdots + \alpha_n H_n(x)$. (Hint: Look at the proof of Lemma 13.5.)

47. Use exercise 46 to show that for every polynomial $f(x) \in \mathbb{R}[x]$ of degree n, there exists $h(x) \in \mathbb{R}[x]$ of degree n such that $f(x) = ah(x+1) - h(x)$.

48. Show that $F_{(1)}(n) = p(n+1)a^{n+1}$.

49. In light of exercise 48, one might guess that $F(n)$ is of the form $g(n)a^{n+1}$, for some polynomial g. Therefore, let $T(n) = g(n)a^{n+1}$, where g is a polynomial, and then compute $T_{(1)}(n)$.

50. Use exercise 47 to show that there exists some $h(x) \in \mathbb{R}[x]$ such that $T(x) = h(x)a^{n+1}$ has the property that $T_{(1)}(x) = F_{(1)}(n)$.

51. Use exercises 30 and 47 to show that there exists a polynomial $h(x)$ of degree t and some $\alpha \in \mathbb{R}$ such that $p(0) + p(1)a + p(2)a^2 + \cdots + p(n)a^n = h(n)a^{n+1} + \alpha$.

52. Use exercise 51 to find a formula for the sum $0 \cdot 2^0 + 1 \cdot 2^1 + 2 \cdot 2^2 + \cdots + n \cdot 2^n$. Observe that by letting $n = 0, 1, 2$, you can generate three linear equations that will enable you to find the coefficients of h and the value of α in the previous exercise.

53. Use exercise 51 to find a formula for the sum $0 \cdot 5^0 + 1 \cdot 5^1 + 2 \cdot 5^2 + \cdots + n \cdot 5^n$.

54. Generalize the formulas you found in the previous two exercises by finding the formula for the sum $a + 2a^2 + 3a^3 + \cdots + na^n$. Your final answer should be in terms of a and n.

In exercises 55–58, we will determine all $p(x) \in \mathbb{C}[x]$ that have the following properties:

(i) $p(\alpha) \in \mathbb{R}$, for all $\alpha \in \mathbb{R}$, and

(ii) $p(\beta) \notin \mathbb{R}$, for all $\beta \in \mathbb{C}$ such that $\beta \notin \mathbb{R}$.

55. Let $F \subseteq L$ be fields, where both F and L are infinite sets. Suppose $p(x) \in L[x]$ has the property that $p(\alpha) \in F$, for all $\alpha \in F$. Prove that all the coefficients of the polynomial $p(x)$ must belong to the smaller field F. (Hint: Think about the ideas behind Proposition 13.8 and its proof.)

56. (a) Let $p(x) \in \mathbb{R}[x]$ have a positive leading coefficient. Show that there exists a real number T such that the polynomial $p(x) + T$ has at most one real root and no multiple roots in $\mathbb{R}$. (Hint: It might help to first look at this problem graphically.)

    (b) Use part (b) to show that every $p(x) \in \mathbb{R}[x]$ has the property that there exists a real number T such that $p(x) + T$ has at most one real root and no multiple roots in $\mathbb{R}$.

57. Let $p(x) \in \mathbb{R}[x]$ have degree at least two. Use exercise 56 to show that there exist some $\alpha \in \mathbb{C}$ such that $\alpha \notin \mathbb{R}$ and $p(\alpha) \in \mathbb{R}$.

58. Use exercises 55 and 57 to find all $p(x) \in \mathbb{C}[x]$ that have the properties described before exercise 55.

13.3 Partial Fraction Decomposition

We now turn our attention to the second main topic of this chapter: the partial fraction decomposition of rational functions. When studying techniques of integration in calculus, a typical problem we might run across is

$$\int \frac{4x^4 - 17x^3 - 6x^2 - 26x - 17}{x^3 - 5x^2 + x - 5}\, dx.$$

In an attempt to find a function whose derivative is $\frac{4x^4 - 17x^3 - 6x^2 - 26x - 17}{x^3 - 5x^2 + x - 5}$, we would proceed as follows:

Step I Use the division algorithm to divide $4x^4 - 17x^3 - 6x^2 - 26x - 17$ by $x^3 - 5x^2 + x - 5$ to obtain a quotient of $4x + 3$ and a remainder of $5x^2 - 9x - 2$. We can divide the equation

$$4x^4 - 17x^3 - 6x^2 - 26x - 17 = (4x+3)(x^3 - 5x^2 + x - 5) + (5x^2 - 9x - 2),$$

by $x^3 - 5x^2 + x - 5$ to obtain

$$\frac{4x^4 - 17x^3 - 6x^2 - 26x - 17}{x^3 - 5x^2 + x - 5} = (4x+3) + \frac{5x^2 - 9x - 2}{x^3 - 5x^2 + x - 5}.$$

Since $4x+3$ is easy to integrate, the problem now reduces to integrating $\frac{5x^2-9x-2}{x^3-5x^2+x-5}$. Observe that whereas the degree of $4x^4 - 17x^3 - 6x^2 - 26x - 17$ exceeds the degree of the denominator, the degree of $5x^2 - 9x - 2$ is less than the degree of the denominator. This fact will be essential for performing the partial fraction decomposition.

Step II Completely factor the denominator of $\frac{5x^2-9x-2}{x^3-5x^2+x-5}$ into irreducible polynomials in $\mathbb{R}[x]$. Recall that the Fundamental Theorem of Algebra tells us that the only irreducible polynomials in $\mathbb{R}[x]$ are all linear polynomials and those quadratic polynomials that have no real roots. We can then factor $x^3 - 5x^2 + x - 5$ as $(x-5)(x^2+1)$.

Step III Write $\frac{5x^2-9x-2}{x^3-5x^2+x-5}$ as a sum of partial fraction using the denominators obtained in Step II. Therefore, we need to find real numbers a, b, c such that

$$\frac{5x^2-9x-2}{x^3-5x^2+x-5} = \frac{a}{x-5} + \frac{bx+c}{x^2+1}.$$

Next, we multiply both sides of this equation by $x^3-5x^2+x-5 = (x-5)(x^2+1)$ and obtain

$$5x^2-9x-2 = a(x^2+1)+(bx+c)(x-5) = (ax^2+a)+(bx^2+(-5b+c)x-5c) =$$

$$(9) \qquad\qquad (a+b)x^2+(-5b+c)x+(a-5c).$$

The only way the polynomials in (9) can be equal is if the coefficients of x^2, $x^1 = x$, and $x^0 = 1$ are all equal. By setting the corresponding coefficients in (9) equal to each other, we obtain the following system of linear equations:

$$a+b = 5$$

$$-5b+c = -9$$

$$a-5c = -2.$$

The solution to this system of equations is $a = 3$, $b = 2$, $c = 1$. Thus

$$\frac{5x^2-9x-2}{x^3-5x^2+x-5} = \frac{3}{x-5} + \frac{2x+1}{x^2+1}.$$

Step IV Having decomposed $\frac{4x^4-17x^3-6x^2-26x-17}{x^3-5x^2+x-5}$ as

$$\frac{4x^4-17x^3-6x^2-26x-17}{x^3-5x^2+x-5} = (4x+3) + \frac{5x^2-9x-2}{x^3-5x^2+x-5} =$$

$$(4x+3) + \frac{3}{x-5} + \frac{2x+1}{x^2+1},$$

we now have

$$\int \frac{4x^4-17x^3-6x^2-26x-17}{x^3-5x^2+x-5}\,dx =$$

$$\int (4x+3)\,dx + \int \frac{3}{x-5}\,dx + \int \frac{2x+1}{x^2+1}\,dx =$$

$$2x^2+3x+3\ln|x-5|+\ln|x^2+1|+\arctan(x)+C.$$

Recall that there are two types of partial fractions. The simpler type is of the form $\frac{a}{(mx+b)^n}$, where $a, m, b \in \mathbb{R}$, $n \in \mathbb{N}$, and both a and m are nonzero. Examples of this type of partial fraction are

$$\frac{7}{x-2}, \quad \frac{\sqrt{2}}{x^3}, \quad \frac{-5}{4x-3}, \quad \text{and} \quad \frac{\pi}{(2x+7)^8}.$$

The second type of partial fraction is of the form $\frac{\alpha x+\beta}{(ax^2+bx+c)^n}$, where $\alpha, \beta, a, b, c \in \mathbb{R}$, $n \in \mathbb{N}$, at least one of α, β is nonzero, and the quadratic ax^2+bx+c has no real roots. Examples of this type of partial fraction are

$$\frac{3x-8}{x^2+4}, \quad \frac{\sqrt{2}}{(2x^2+1)^4}, \quad \frac{11x+76}{x^2+1}, \quad \text{and} \quad \frac{\pi-7}{(3x^2+7x+55)^9}.$$

Observe that whereas $\frac{3x-8}{x^2+4}$ is a partial fraction, $\frac{3x-8}{x^2-4}$ is not a partial fraction as the quadratic in the denominator does have real roots.

The partial fraction decomposition of rational functions asserts that every rational function can be written as the sum of a polynomial and partial fractions. When we look back at the previous example, we see that Step IV is entirely based on calculus and takes place after having already decomposed $\frac{4x^4-17x^3-6x^2-26x-17}{x^3-5x^2+x-5}$.

The actual decomposition of $\frac{4x^4-17x^3-6x^2-26x-17}{x^3-5x^2+x-5}$ takes place in Steps I, II, and III. Observe that Step I is a straightforward application of the division algorithm for polynomials. Step II is the version of the Fundamental Theorem of Algebra, which states that all irreducible polynomials in $\mathbb{R}[x]$ must have degree 1 or 2. Thus, the real issue is why does Step III work? More precisely, why does setting things up the way we do in Step III always result in a system of linear equations that has a solution?

It turns out that the answer lies in properties of polynomials that can be derived from the division algorithm. In order to prove this, we will once again examine and exploit the similarities between the positive integers and polynomials with coefficients in a field. We begin with a lemma about rational numbers which follows from the Euclidean Algorithm.

Lemma 13.10. *Let $\frac{a}{b}$ be a nonzero rational number, where $b = d_1 d_2 \cdots d_n$, such that each d_i is a positive integer with d_i and d_j relatively prime whenever $i \neq j$. Then*

$$\frac{a}{b} = A + \frac{c_1}{d_1} + \frac{c_2}{d_2} + \cdots + \frac{c_n}{d_n},$$

where $A, c_1, c_2, \ldots, c_n \in \mathbb{Z}$ and $0 \leq c_i < d_i$, for all i.

Intuition. In some sense, Lemma 13.10 undoes the addition and subtraction of fractions. When we add and subtract fractions, we first obtain a common denominator as in

$$\frac{4}{7} + \frac{5}{9} = \frac{36 + 35}{63} = \frac{71}{63}.$$

However, Lemma 13.10 asserts that we can do the opposite. For example, it says that $\frac{451}{63}$ can be decomposed as

$$\frac{451}{63} = A + \frac{c_1}{7} + \frac{c_2}{9},$$

where $A, c_1, c_2 \in \mathbb{Z}$, $0 \le c_1 < 7$, and $0 \le c_2 < 9$.

Let's see how to perform this decomposition. Since 7 and 9 are relatively prime, the division algorithm tells us that there exist $\alpha_1, \alpha_2 \in \mathbb{Z}$ such that

$$1 = 9\alpha_1 + 7\alpha_2.$$

There are many possible choices for α_1, α_2. One such choice is $\alpha_1 = -3$ and $\alpha_2 = 4$, which gives us

$$1 = 9(-3) + 7(4).$$

Multiplying both sides by 451 yields

$$451 = 9(-1353) + 7(1804).$$

Then dividing both sides by 63 gives us

$$\frac{451}{63} = \frac{-1353}{7} + \frac{1804}{9}$$

Observe that neither -1353 nor 1804 lies within the proper range. However the division algorithm tells us that

$$-1353 = (-194) \cdot 7 + 5 \quad \text{and} \quad 1804 = (200) \cdot 9 + 4.$$

Therefore we can replace $\frac{-1353}{7}$ by $-194 + \frac{5}{7}$ and also replace $\frac{1804}{9}$ by $200 + \frac{4}{9}$. As a result, we now have

$$\frac{451}{63} = \frac{-1353}{7} + \frac{1804}{9} = \left(-194 + \frac{5}{7}\right) + \left(200 + \frac{4}{9}\right) =$$

$$(-194 + 200) + \frac{5}{7} + \frac{4}{9} = 6 + \frac{5}{7} + \frac{4}{9}.$$

Thus, $A = 6$, $c_1 = 5$, and $c_2 = 4$ yields the desired decomposition. In this example, 63 was factored into the product of two relative prime numbers. More generally, if the number of relatively prime factors exceeds two, we can prove that an appropriate decomposition exists by applying Mathematical Induction to the number of factors of the denominator.

We can now prove Lemma 13.10.

Proof. We will proceed by Mathematical Induction and begin by letting

$$T = \{n \in \mathbb{N} \,|\, \text{a decomposition exists whenever the denominator}$$

$$\text{is a product of } n \text{ factors that are relatively prime to each other}\}.$$

We must first show that $1 \in T$. Therefore, we will examine the rational number $\frac{a}{b}$, where $b = d_1$. If use the division algorithm to divide a by d_1 and then let A denote the quotient and c_1 the remainder, we obtain

$$a = A \cdot d_1 + c_1,$$

where $A, c_1 \in \mathbb{Z}$ and $0 \le c_1 < d_1$. Dividing this equation by $b = d_1$ results in

$$\frac{a}{b} = A + \frac{c_1}{d_1}.$$

Thus, we have decomposed $\frac{a}{b}$ and so, $1 \in T$.

To conclude the proof, it now suffices to show that if a natural number k belongs to T, then so does $k + 1$. We need to examine the rational number $\frac{a}{b}$, where $b = d_1 d_2 \cdots d_k d_{k+1}$, such that d_i and d_j are relative prime whenever $i \ne j$. Let $e = d_1 d_2 \cdots d_k$; observe that e and d_{k+1} are relative prime. Therefore, the Euclidean Algorithm implies that there exist $\alpha_1, \alpha_2 \in \mathbb{Z}$ such that

$$1 = \alpha_1 \cdot d_{k+1} + \alpha_2 \cdot e.$$

If we multiply this equation by a and let $\beta_1 = a\alpha_1$ and $\beta_2 = a\alpha_2$, we obtain

$$a = (a\alpha_1) \cdot d_{k+1} + (a\alpha_2) \cdot e = \beta_1 \cdot d_{k+1} + \beta_2 \cdot e.$$

Since $b = e \cdot d_{k+1}$, dividing this equation by b yields

$$\frac{a}{b} = \frac{\beta_1}{e} + \frac{\beta_2}{d_{k+1}}. \tag{10}$$

Using the fact that $k \in T$, we can decompose $\frac{\beta_1}{e}$ as

$$\frac{\beta_1}{e} = A_1 + \frac{c_1}{d_1} + \frac{c_2}{d_2} + \cdots \frac{c_k}{d_k}, \tag{11}$$

where $A_1, c_1, c_2, \ldots, c_k \in \mathbb{Z}$ and $0 \le c_i < d_i$, for all i. Furthermore, since $1 \in T$, we can rewrite $\frac{\beta_2}{d_{k+1}}$ as

$$\frac{\beta_2}{d_{k+1}} = A_2 + \frac{c_{k+1}}{d_{k+1}},$$

where $A_2, c_{k+1} \in \mathbb{Z}$ and $0 \le c_{k+1} < d_{k+1}$. If we substitute both equation (11) and the previous equation back into equation (10) and then let $A = A_1 + A_2$, we obtain

$$\frac{a}{b} = \frac{\beta_1}{e} + \frac{\beta_2}{d_{k+1}} = \left(A_1 + \frac{c_1}{d_1} + \frac{c_2}{d_2} + \cdots + \frac{c_k}{d_k}\right) + \left(A_2 + \frac{c_{k+1}}{d_{k+1}}\right) =$$

$$(A_1 + A_2) + \frac{c_1}{d_1} + \frac{c_2}{d_2} + \cdots + \frac{c_k}{d_k} + \frac{c_{k+1}}{d_{k+1}} = A + \frac{c_1}{d_1} + \frac{c_2}{d_2} + \cdots + \frac{c_k}{d_k} + \frac{c_{k+1}}{d_{k+1}}.$$

Since $A, c_1, c_2, \ldots, c_{k+1} \in \mathbb{Z}$ and $0 \le c_i < d_i$, for all i, we have successfully decomposed $\frac{a}{b}$. Thus, $k + 1 \in T$, thereby concluding the proof. $\qquad\square$

The only tools used to prove Lemma 13.10 are the division algorithm, the Euclidean Algorithm, and Mathematical Induction. In light of this, it should come as no surprise that by using the division algorithm and Euclidean Algorithm for polynomials over a field, we can easily adapt the proof of Lemma 13.10 to prove an analogous result for decomposing rational functions. We will keep the notation in Lemma 13.11 as close as possible to the notation in Lemma 13.10 to make the similarities between positive integers and polynomials as transparent as possible.

Lemma 13.11. *Let F be a field and let $a(x), b(x)$ be nonzero elements of the polynomial ring $F[x]$. Suppose $b(x) = d_1(x)d_2(x) \cdots d_n(x)$, where each $d_i(x) \in F[x]$ has degree at least one and $d_i(x), d_j(x)$ are relatively prime whenever $i \ne j$. Then*

$$\frac{a(x)}{b(x)} = A(x) + \frac{c_1(x)}{d_1(x)} + \frac{c_2(x)}{d_2(x)} + \cdots + \frac{c_n(x)}{d_n(x)},$$

where $A(x), c_1(x), c_2(x), \ldots, c_n(x) \in F[x]$ and, for all i, either $c_i(x) = 0$ or has degree less than the degree of $d_i(x)$.

Proof. All the key ideas of this proof are contained in the proof of Lemma 13.10. It will simply be a matter of making some minor modifications. However, it is worth noting that although our ultimate goal is to work with polynomials in $\mathbb{R}[x]$, this lemma does hold for polynomials over any field.

We will proceed by Mathematical Induction and begin by letting

$$T = \{n \in \mathbb{N} \mid a \text{ decomposition exists whenever}$$

$$b(x) \text{ is a product of } n \text{ factors that are relatively prime to each other}\}.$$

We must first show that $1 \in T$. Therefore, we begin with $\frac{a(x)}{b(x)}$, where $b(x) = d_1(x)$. Using the division algorithm in $F[x]$ to divide $a(x)$ by $d_1(x)$ and then letting $A(x)$ denote the quotient and $c_1(x)$ the remainder, we obtain

$$a(x) = A(x) \cdot d_1(x) + c_1(x),$$

where $A(x), c_1(x) \in F[x]$ and either $c_1(x) = 0$ or $c_1(x)$ has smaller degree than $d_1(x)$. Dividing this equation by $b(x) = d_1(x)$ results in

$$\frac{a(x)}{b(x)} = A(x) + \frac{c_1(x)}{d_1(x)}.$$

Thus, we have decomposed $\frac{a(x)}{b(x)}$ and so, $1 \in T$.

It now suffices to show that if k belongs to T, then so does $k + 1$. We need to examine $\frac{a(x)}{b(x)}$, where $b(x) = d_1(x)d_2(x) \cdots d_k(x)d_{k+1}(x)$, where $d_i(x), d_j(x)$ are relative prime whenever $i \neq j$. Let $e(x) = d_1(x)d_2(x) \cdots d_k(x)$; observe that $e(x)$ and $d_{k+1}(x)$ are relative prime in $F[x]$. Therefore, the Euclidean Algorithm in $F[x]$ implies that there exist $\alpha_1(x), \alpha_2(x) \in F[x]$ such that

$$1 = \alpha_1(x) \cdot d_{k+1}(x) + \alpha_2(x) \cdot e(x).$$

If we multiply this equation by $a(x)$ and let $\beta_1(x) = a(x)\alpha_1(x)$ and $\beta_2(x) = a\alpha_2(x)$, we obtain

$$a(x) = \beta_1(x) \cdot d_{k+1}(x) + \beta_2(x) \cdot e(x).$$

Since $b(x) = e(x) \cdot d_{k+1}(x)$, dividing this equation by $b(x)$ yields

(12)
$$\frac{a(x)}{b(x)} = \frac{\beta_1(x)}{e(x)} + \frac{\beta_2(x)}{d_{k+1}(x)}.$$

However, $k \in T$, therefore we can decompose $\frac{\beta_1(x)}{e(x)}$ as

(13)
$$\frac{\beta_1(x)}{e(x)} = A_1(x) + \frac{c_1(x)}{d_1(x)} + \frac{c_2(x)}{d_2(x)} + \cdots \frac{c_k(x)}{d_k(x)},$$

where $A_1(x), c_1(x), c_2(x), \ldots, c_k(x) \in F[x]$ and, for all i, either $c_i(x) = 0$ or $c_i(x)$ has smaller degree than $d_i(x)$. Furthermore, since $1 \in T$, we can rewrite $\frac{\beta_2(x)}{d_{k+1}(x)}$ as

$$\frac{\beta_2(x)}{d_{k+1}(x)} = A_2(x) + \frac{c_{k+1}(x)}{d_{k+1}(x)},$$

where $A_2(x), c_{k+1}(x) \in F[x]$ and $c_{k+1}(x) = 0$ or $c_{k+1}(x)$ has smaller degree than $d_{k+1}(x)$. Substituting both equation (13) and the previous equation back into equation (12) and then letting $A(x) = A_1(x) + A_2(x)$, we obtain

$$\frac{a(x)}{b(x)} = \frac{\beta_1(x)}{e(x)} + \frac{\beta_2}{d_{k+1}(x)} =$$

$$\left(A_1(x) + \frac{c_1(x)}{d_1(x)} + \frac{c_2(x)}{d_2(x)} + \cdots \frac{c_k(x)}{d_k(x)} \right) + \left(A_2(x) + \frac{c_{k+1}(x)}{d_{k+1}(x)} \right) =$$

$$A(x) + \frac{c_1(x)}{d_1(x)} + \frac{c_2(x)}{d_2(x)} + \cdots \frac{c_k(x)}{d_k(x)} + \frac{c_{k+1}(x)}{d_{k+1}(x)}.$$

Since $A(x), c_1(x), c_2(x), \ldots, c_{k+1}(x) \in F[x]$ and, for all i, either $c_i(x) = 0$ or $c_i(x)$ has smaller degree than $d_i(x)$, we have successfully decomposed $\frac{a(x)}{b(x)}$. Thus, $k + 1 \in T$, thereby concluding the proof $\qquad \square$

The next big piece of the puzzle relies on a fact about positive integers that you have known since grade school but whose analog for polynomials might well be unfamiliar to you. Almost every computation you have ever seen with positive integers has probably taken place in base 10. Recall that in base 10, the real meaning of the number 493 is

$$4 \cdot 10^2 + 9 \cdot 10^1 + 3 \cdot 10^0.$$

In computer science, frequent use is made of base 2 as well as base 16. However, any positive integer greater than 1 is a perfectly valid base for representing the positive integers. For example, since

$$493 = 1 \cdot 256 + 1 \cdot 128 + 1 \cdot 64 + 1 \cdot 32 + 1 \cdot 8 + 1 \cdot 4 + 1 \cdot 1 =$$

$$1 \cdot 2^8 + 1 \cdot 2^7 + 1 \cdot 2^6 + 1 \cdot 2^5 + 0 \cdot 2^4 + 1 \cdot 2^3 + 1 \cdot 2^2 + 0 \cdot 2^1 + 1 \cdot 2^0,$$

the representation of 493 in base 2 is 111101101. Similarly, if we wish to look at 493 in base 5, then we have

$$493 = 3 \cdot 125 + 4 \cdot 25 + 3 \cdot 5 + 3 \cdot 1 = 3 \cdot 5^3 + 4 \cdot 5^2 + 3 \cdot 5^1 + 3 \cdot 5^0.$$

Thus, in base 5, the representation of 493 is 3433.

Although we may not think of it this way, when we write polynomials we are expressing them using a base of x. After all, the polynomial $6x^3 - 7x^2 - 5x + 40$ really means $6 \cdot x^3 + (-7) \cdot x^2 + (-5) \cdot x^1 + 40 \cdot x^0$. Just as any positive integer can be used as a base to represent the positive integers, an analogous fact holds for polynomials. In particular, any polynomial of degree at least one can be used as a base to represent all polynomials. This idea is probably less foreign to you than realize. In calculus, you represented many functions by Taylor polynomials. If we look at the Taylor polynomial for $6x^3 - 7x^2 - 5x + 40$ centered at $x = 1$, we obtain $34 - (x - 1) + 11(x - 1)^2 + 6(x - 1)^3$. Reordering terms, we can rewrite this as

$$6x^3 - 7x^2 - 5x + 40 = 6 \cdot (x - 1)^3 + 11 \cdot (x - 1)^2 + (-1) \cdot (x - 1)^1 + 34 \cdot (x - 1)^0.$$

Thus, we have written the polynomial $6x^3 - 7x^2 - 5x + 40$ using the base $x - 1$. Similarly, if we want to write $6x^3 - 7x^2 - 5x + 40$ using the base $x + 5$, we would find the Taylor polynomial for $6x^3 - 7x^2 - 5x + 40$ centered at -5. We would then obtain

$$6x^3 - 7x^2 - 5x + 40 = 6 \cdot (x + 5)^3 + (-97) \cdot (x + 5)^2 + 515 \cdot (x + 5)^1 + (-860) \cdot (x + 5)^0.$$

At first, it seems unclear whether we can also represent polynomials using any base that has degree greater than one. If we begin to experiment, we see that we can represent $6x^3 - 7x^2 - 5x + 40$ using x^2 as the base as

$$6x^3 - 7x^2 - 5x + 40 = (6x - 7) \cdot (x^2)^1 + (-5x + 40) \cdot (x^2)^0.$$

Similarly, if we wanted to represent $6x^3 - 7x^2 - 5x + 40$ using $x^2 + 1$ as the base, experimentation would eventually tell us that

$$6x^3 - 7x^2 - 5x + 40 = (6x - 7) \cdot (x^2 + 1)^1 + (-11x + 47) \cdot (x^2 + 1)^0.$$

Observe that when we write a number in base 10, the allowable nonzero coefficients of the powers of 10 are the positive integers less than 10. Similarly, in base 5, the allowable nonzero coefficients of the powers of 5 are the positive integers less than 5. When we look at the analogous situation for polynomials, when one uses a base of x or $x - 1$ or $x + 5$, one would expect that the allowable nonzero coefficients are the nonzero polynomials of degree less than one. Similarly, when using a base of x^2 or $x^2 + 1$, one would expect that the allowable nonzero coefficients are the nonzero polynomials of degree less than two.

At this point it is reasonable to conjecture that when trying to represent all polynomials using the polynomial $p(x)$ as a base, then the allowable nonzero coefficients are all polynomials whose degrees are less than the degree of $p(x)$. However, you might be wondering how we find the coefficients of the powers of $p(x)^n$. If you fully understand the situation for positive integers, then you should not be too surprised that we use the division algorithm to find the coefficients. We express this more formally in

Lemma 13.12. *Let F be a field and let $p(x) \in F[x]$ have degree $m \geq 1$. Then every nonzero element of $F[x]$ can be expressed using $p(x)$ as the base. More precisely, if $a(x)$ is a nonzero element of $F[x]$, then there exists $t \geq 0$ and $b_0(x), b_1(x), \ldots, b_t(x) \in F[x]$ such that*

$$a(x) = b_t(x) \cdot p(x)^t + b_{t-1}(x) \cdot p(x)^{t-1} + \cdots + b_1(x) \cdot p(x)^1 + b_0(x) \cdot p(x)^0,$$

where each $b_i(x)$ is either 0 or has degree less than m.

Proof. Given $p(x)$, we will proceed using the Second Version of Mathematical Induction. To this end, we let

$$T = \{n \in \mathbb{N}_0 \mid \text{every element of } F[x] \text{ of degree } n \text{ can be written using}$$

$$p(x) \text{ as the base}\}.$$

First, we need to show that $0 \in T$. However, if $a(x)$ has degree 0, then by letting $t = 0$ and $b_0(x) = a(x)$, we have

$$a(x) = a(x) \cdot 1 = b_0(x) \cdot p(x)^0.$$

Since $b_0(x)$ has smaller degree than $p(x)$, we have written $a(x)$ using $p(x)$ as the base. Thus, $0 \in T$, as desired.

Next, let us suppose that $k \geq 0$ has the property that $\{0, 1, \ldots, k\} \subseteq T$. It suffices to show that $k + 1 \in T$, so we may now assume that $a(x) \in F[x]$ has degree $k + 1$; If $k + 1 < m$, then we can represent $a(x)$ using $p(x)$ as the base by letting $b_0(x) = a(x)$ to give us

$$a(x) = a(x) \cdot 1 = b_0(x) \cdot p(x)^0,$$

as desired.

Therefore, for the remainder of the proof, we may assume that $k + 1 \geq m$. Using the division algorithm in $\mathbb{Z}$, we can divide $k + 1$ by m. Letting t denote the quotient, we obtain

$$k + 1 = tm + r,$$

where $0 \leq r < m$. Now, if we use the division algorithm in $F[x]$ to divide $a(x)$ by $p(x)^t$ and let $b_t(x)$ denote the quotient, we have

(14) $$a(x) = b_t(x) \cdot p(x)^t + r(x),$$

where $r(x) = 0$ or has degree less than the degree of $p(x)^t$.

Since $a(x)$ has degree $k + 1$ and $p(x)^t$ has degree tm, it follows the degree of $b_t(x)$ is $(k + 1) - tm = r < m$. Thus, $b_t(x)$ has smaller degree than $p(x)$. If $r(x) = 0$, then

$$a(x) = b_t(x) \cdot p(x)^t$$

is a representation of $a(x)$ using $p(x)$ as a base. On the other hand, if $r(x) \neq 0$, then the degree of $r(x)$ is less than $k + 1$. Therefore, the degree of $r(x)$ belongs to the set $\{0, 1, \ldots, k\}$, which is a subset of T. As a result, $r(x)$ can be written using $p(x)$ as a base. Furthermore, since the degree of $r(x)$ is also less than tm, it follows that whenever a term of the form $p(x)^j$ appears in the representation of $r(x)$, the exponent j must be less than t. Therefore, $r(x)$ can be written as

$$r(x) = b_{t-1}(x) \cdot p(x)^{t-1} + \cdots + b_1(x) \cdot p(x)^1 + b_0(x) \cdot p(x)^0,$$

where each $b_i(x)$ is either 0 or has degree less than m. Substituting this representation for $r(x)$ into equation (14), we obtain

$$a(x) = b_t(x) \cdot p(x)^t + b_{t-1}(x) \cdot p(x)^{t-1} + \cdots + b_1(x) \cdot p(x)^1 + b_0(x) \cdot p(x)^0,$$

where each $b_i(x)$ is either 0 or has degree less than m. Therefore, $a(x)$ can be represented using $p(x)$ as a base, so $k+1 \in T$, as desired. $\qquad\square$

Recall that the primary goal of this section is to prove the validity of the partial fraction decomposition technique that is shown but not proven in calculus. That result will easily follow from the main result of this section which we now have all the pieces to prove.

Theorem 13.13. *Let F be a field and let $a(x), b(x)$ be nonzero elements of the polynomial ring $F[x]$. Suppose $b(x) = p_1(x)^{m_1} p_2(x)^{m_2} \cdots p_n(x)^{m_n}$ where each $p_i(x) \in F[x]$ has degree at least one, each $m_i \geq 1$, and $p_i(x), p_j(x)$ are relatively prime whenever $i \neq j$. Then*

$$\frac{a(x)}{b(x)} = A(x) + \frac{b_{1,1}(x)}{p_1(x)^1} + \cdots + \frac{b_{1,m_1}(x)}{p_1(x)^{m_1}} + \cdots + \frac{b_{n,1}(x)}{p_n(x)^1} + \cdots + \frac{b_{n,m_n}(x)}{p_n(x)^{m_n}}$$

where $A(x), b_{i,j}(x) \in F[x]$ and, for all i, j, either $b_{i,j}(x) = 0$ or has degree less than the degree of $p_i(x)$.

Proof. For every $1 \leq i \leq n$, let $d_i(x) = p_i(x)^{m_i}$. Then each $d_i(x)$ has degree at least one and $d_i(x), d_j(x)$ are relatively prime whenever $i \neq j$. Therefore, we can apply Lemma 13.11 to assert that

$$(15) \qquad \frac{a(x)}{b(x)} = A(x) + \frac{c_1(x)}{d_1(x)} + \frac{c_2(x)}{d_2(x)} + \cdots + \frac{c_n(x)}{d_n(x)},$$

where $A(x), c_1(x), c_2(x), \ldots, c_n(x) \in F[x]$ and, for all i, either $c_i(x) = 0$ or has degree less than the degree of $d_i(x)$.

To conclude the proof, it remains to show that for every $1 \leq i \leq n$, we decompose $\frac{c_i(x)}{d_i(x)}$ as

$$\frac{c_i(x)}{d_i(x)} = \frac{b_{i,1}(x)}{p_i(x)^1} + \cdots + \frac{b_{i,m_i}(x)}{p_i(x)^{m_i}},$$

where, for every j, $b_{i,j}(x) = 0$ or has degree less than the degree of $p_i(x)$. In light of Lemma 13.12, we can write $c_i(x)$ using $p_i(x)$ as the base. As in the proof of Lemma 13.12, since the degree of $c_i(x)$ is less than the degree of $p_i(x)^{m_i}$, no term in the representation of $c_i(x)$ appears where the exponent of $p_i(x)$ exceeds $m_i - 1$. Therefore,

$$c_i(x) = b_{i,1}(x) \cdot p_i(x)^{m_i-1} + \cdots + b_{i,m_i}(x) \cdot p_i(x)^0,$$

where, for each j, either $b_{i,j}(x) = 0$ or has degree less than the degree of $p_i(x)$.

Dividing the previous equation by $d_i(x) = p_i(x)^{m_i}$ yields

(16)
$$\frac{c_i(x)}{d_i(x)} = \frac{b_{i,1}(x)}{p_i(x)^1} + \cdots + \frac{b_{i,m_i}(x)}{p_i(x)^{m_i}}.$$

Finally, for every $1 \le i \le n$, replace $\frac{c_i(x)}{d_i(x)}$ in equation (15) by the decomposition in equation (16), to obtain

$$\frac{a(x)}{b(x)} = A(x) + \frac{b_{1,1}(x)}{p_1(x)^1} + \cdots + \frac{b_{1,m_1}(x)}{p_1(x)^{m_1}} + \cdots + \frac{b_{n,1}(x)}{p_n(x)^1} + \cdots + \frac{b_{n,m_n}(x)}{p_n(x)^{m_n}},$$

where $A(x), b_{i,j}(x) \in F[x]$ and, for all i, j, either $b_{i,j}(x) = 0$ or has degree less than the degree of $p_i(x)$. Thus, $\frac{a(x)}{b(x)}$ can indeed be decomposed in the desired manner. $\qquad\square$

We can now easily prove the validity of the partial fraction decomposition technique in calculus by specializing Theorem 13.13 to the case where all our polynomials belong to $\mathbb{R}[x]$.

Corollary 13.14. *Let $a(x), b(x) \in \mathbb{R}[x]$, where $a(x) \ne 0$ and $b(x)$ has degree at least one. Suppose*

$$b(x) = p_1(x)^{m_1} p_2(x)^{m_2} \cdots p_s(x)^{m_s} q_1(x)^{n_1} q_2(x)^{n_2} \cdots q_t(x)^{n_t}$$

is a factorization of $b(x)$ into irreducible polynomials in $\mathbb{R}[x]$, where each $p_i(x)$ has degree one, each $q_j(x)$ has degree two, and every $m_i, n_j \ge 1$. Then there exist $A(x) \in \mathbb{R}[x]$ and $a_{i,j}, b_{i,j}, c_{i,j} \in \mathbb{R}$ such that

$$\frac{a(x)}{b(x)} = A(x) + \frac{a_{1,1}}{p_1(x)^1} + \cdots + \frac{a_{1,m_1}}{p_1(x)^{m_1}} + \cdots + \frac{a_{s,1}}{p_s(x)^1} + \cdots + \frac{a_{s,m_s}}{p_s(x)^{m_s}} +$$
$$\frac{b_{1,1}x + c_{1,1}}{q_1(x)^1} + \cdots + \frac{b_{1,n_1}x + c_{1,n_1}}{q_1(x)^{n_1}} + \cdots + \frac{b_{t,1}x + c_{t,1}}{q_t(x)^1} + \cdots + \frac{b_{t,n_t}x + c_{t,n_t}}{q_t(x)^{n_t}}.$$

Furthermore, $A(x) = 0$ if and only if the degree of $b(x)$ exceeds the degree of $a(x)$.

The proof of Corollary 13.14 follows easily from Theorem 13.13. However, it is easy to get lost in a sea of notation. So, before proving the corollary, let us examine what it says about a specific example.

■ Example

Consider the rational function

$$\frac{9x^4 - 5x^2 + 2x - 6}{x^2(2x-1)^3(5x+2)^1(x^2+1)^3(2x^2+3)^1(x^2+3x+17)^2}.$$

Since the denominator has greater degree than the numerator, Corollary 13.14 asserts there exist $a_{i,j}, b_{i,j}, c_{i,j} \in \mathbb{R}$ such that

$$\frac{9x^4 - 5x^2 + 2x - 6}{x^2(2x-1)^3(5x+2)^1(x^2+1)^3(2x^2+3)^1(x^2+3x+17)^2} =$$

$$\frac{a_{1,1}}{x^1} + \frac{a_{1,2}}{x^2} + \frac{a_{2,1}}{(2x-1)^1} + \frac{a_{2,2}}{(2x-1)^2} + \frac{a_{2,3}}{(2x-1)^3} + \frac{a_{3,1}}{(5x+2)^1} +$$

$$\frac{b_{1,1}x + c_{1,1}}{(x^2+1)^1} + \frac{b_{1,2}x + c_{1,2}}{(x^2+1)^2} + \frac{b_{1,3}x + c_{1,3}}{(x^2+1)^3} + \frac{b_{2,1}x + c_{2,1}}{(2x^2+3)^1} +$$

$$\frac{b_{3,1}x + c_{3,1}}{(x^2+3x+17)^1} + \frac{b_{3,2}x + c_{3,2}}{(x^2+3x+17)^2}.$$

It is a rather daunting and time-consuming task to actually find the 18 real numbers that appear in the numerators of the right-hand side of the previous equation. Most likely, you would begin by multiplying both sides of the equation by $x^2(2x-1)^3(5x+2)^1$ $(x^2+1)^3(2x^2+3)^1(x^2+3x+17)^2$. This yields an equality of polynomials where the largest exponent appearing is 17. Comparing the coefficients of $x^0, x^1, \ldots, x^{17}$ yields 18 linear equations in 18 unknowns. In general, a system of linear equations need not have a solution. But Corollary 13.14, which relied on the division algorithm and Euclidean Algorithm in $\mathbb{R}[x]$, guarantees that there is indeed a solution.

∎

Proof of Corollary 13.14. Theorem 13.13 provides us with a decomposition of $\frac{a(x)}{b(x)}$, but we need to show that this decomposition is of the form described in Corollary 13.14 Since each $p_i(x)$ in the factorization of $b(x)$ is linear, every term in the decomposition of $\frac{a(x)}{b(x)}$ that has a power of $p_i(x)$ in the denominator must have a real number in the numerator. Similarly, since every $q_j(x)$ in the factorization of $b(x)$ is a quadratic, every term in the decomposition of $\frac{a(x)}{b(x)}$ that has a power of $q_j(x)$ in the denominator must have a numerator of the form $\alpha x + \beta$, where $\alpha, \beta \in \mathbb{R}$. Thus, in our situation, the decomposition described in Theorem 13.13 has the desired form.

Finally, in the decomposition described in this corollary, let

$$g(x) = \frac{a_{1,1}}{p_1(x)^1} + \cdots + \frac{a_{1,m_1}}{p_1(x)^{m_1}} + \cdots + \frac{a_{s,1}}{p_s(x)^1} + \cdots + \frac{a_{s,m_s}}{p_s(x)^{m_s}} +$$

$$\frac{b_{1,1}x + c_{1,1}}{q_1(x)^1} + \cdots + \frac{b_{1,n_1}x + c_{1,n_1}}{q_1(x)^{n_1}} + \cdots + \frac{b_{t,1}x + c_{t,1}}{q_t(x)^1} + \cdots + \frac{b_{t,n_t}x + c_{t,n_t}}{q_t(x)^{n_t}}.$$

Therefore,

$$\frac{a(x)}{b(x)} = A(x) + g(x),$$

and if we multiply both sides of this equation by $b(x)$, we obtain

$$a(x) = b(x) \cdot A(x) + b(x) \cdot g(x).$$

Observe that when we multiply $g(x)$ by $b(x)$, we obtain a sum of polynomials, all of which have degree less than the degree of $b(x)$. If $A(x) = 0$, then the degree of the right-hand side of the previous equation is smaller than the degree of $b(x)$. Thus, the degree of $a(x)$ is smaller than the degree of $b(x)$. Hence, in this case, the degree of $b(x)$ exceeds the degree of $a(x)$.

On the other hand, if $A(x) \neq 0$, then the degree of the right-hand side of the previous equation is at least as large as the degree of $b(x)$. Hence, the degree of $a(x)$ is at least as large as the degree of $b(x)$. Thus, in this case, the degree of $b(x)$ does not exceed the degree of $a(x)$. Combining these observations, we can see that $A(x) = 0$ if and only if the degree of $b(x)$ exceeds the degree of $a(x)$. $\square$

Exercises for Section 13.3

1. Write the number 315 in base 2.

2. Write the number 315 in base 3.

3. Write the number 315 in base 5.

4. Write the number 315 in base 8.

5. Write the number 721 in base 2.

6. Write the number 721 in base 3.

7. Write the number 721 in base 5.

8. Write the number 721 in base 8.

9. Write the number 2009 in base 2.

10. Write the number 2009 in base 3.

11. Write the number 2009 in base 5.

12. Write the number 2009 in base 8.

13. Write the polynomial $7x^3 + 4x^2 + 9x - 6$ using $x - 2$ as the base.

14. Write the polynomial $7x^3 + 4x^2 + 9x - 6$ using $x + 5$ as the base.

15. Write the polynomial $7x^3 + 4x^2 + 9x - 6$ using x^2 as the base.

16. Write the polynomial $7x^3 + 4x^2 + 9x - 6$ using $x^2 + 4$ as the base.

17. Write the polynomial $2x^5 - 3x^4 + 6x + 3$ using $x + 1$ as the base.

18. Write the polynomial $2x^5 - 3x^4 + 6x + 3$ using $x - 8$ as the base.

19. Write the polynomial $2x^5 - 3x^4 + 6x + 3$ using x^2 as the base.

20. Write the polynomial $2x^5 - 3x^4 + 6x + 3$ using $x^2 + 1$ as the base.

21. Write the polynomial $2x^5 - 3x^4 + 6x + 3$ using x^3 as the base.

22. Write the polynomial $2x^5 - 3x^4 + 6x + 3$ using $x^3 + 5$ as the base.

23. Write the polynomial $3x^{11} - 5x^9 + 2x^8 - 14x^5 - x^4 + 7x^3 - 11x^2 + 5x + 1$ using x^2 as the base.

24. Write the polynomial $3x^{11} - 5x^9 + 2x^8 - 14x^5 - x^4 + 7x^3 - 11x^2 + 5x + 1$ using x^3 as the base.

25. Write the polynomial $3x^{11} - 5x^9 + 2x^8 - 14x^5 - x^4 + 7x^3 - 11x^2 + 5x + 1$ using x^4 as the base.

26. Write the polynomial $3x^{11} - 5x^9 + 2x^8 - 14x^5 - x^4 + 7x^3 - 11x^2 + 5x + 1$ using x^5 as the base.

27. Find integers A, B, C such that $\frac{211}{36} = A + \frac{B}{4} + \frac{C}{9}$, where $0 \leq B \leq 3, 0 \leq C \leq 8$.

28. Find integers A, B, C such that $\frac{714}{50} = A + \frac{B}{2} + \frac{C}{25}$, where $0 \leq B \leq 1, 0 \leq C \leq 24$.

29. Find integers A, B, C, D such that $\frac{451}{30} = A + \frac{B}{2} + \frac{C}{3} + \frac{D}{5}$, where $0 \leq B \leq 1, 0 \leq C \leq 2, 0 \leq D \leq 4$.

30. Find integers A, B, C, D such that $\frac{4397}{700} = A + \frac{B}{4} + \frac{C}{25} + \frac{D}{7}$, where $0 \leq B \leq 3, 0 \leq C \leq 24, 0 \leq D \leq 6$.

31. (a) Find integers A, B such that $\frac{1}{18} = \frac{A}{2} + \frac{B}{9}$. (Many different answers are possible.)

    (b) Suppose you have a large bowl and two small cups that hold $\frac{1}{2}$ and $\frac{1}{9}$ cups of water each, respectively. Explain how you could use the bowl and cups so the bowl will contain exactly $\frac{1}{18}$ of a cup of water.

32. (a) Find integers A, B such that $\frac{1}{60} = \frac{A}{4} + \frac{B}{15}$. (Many different answers are possible.)

    (b) Suppose you have a large bowl and two small cups that hold $\frac{1}{4}$ and $\frac{1}{15}$ cups of water each, respectively. Explain how you could use the bowl and cups so the bowl will contain exactly $\frac{1}{60}$ of a cup of water.

In exercises 33–42, decompose the rational function into a sum of a polynomial and rational functions in $\mathbb{R}[x]$ as described in Corollary 13.14.

33. $\frac{-x+10}{(x-1)(x+2)}$

34. $\frac{-x^3+31x^2+7x-19}{(x-1)^2(x+2)^2}$

35. $\frac{3x^3+15x^2+13x+20}{x(x+5)}$

36. $\frac{4x^3+2x^2-13x+10}{x^2(x+5)}$

37. $\frac{2x^2-5x-3}{(x+1)(x^2+1)}$

38. $\frac{13x^3+37x^2+64x+5}{(x^2+4)(x+1)^2}$

39. $\frac{4x^4-10x^3+40x^2-91x+63}{x(x^2+9)}$

40. $\frac{16x^5-3x^4+287x^3-68x^2+1285x-405}{x(x^2+9)^2}$

41. $\frac{10x^3+36x^2+19x+81}{(x^2+1)(x^2+4)}$

42. $\frac{4x^5-28x^4+84x^3-786x^2-676x-476}{(x+1)(x-8)(x^2+25)}$

In exercises 43–46, let $F(x) = \frac{5x^3+4x^2+5x-6}{(x^2-2)(x^2+1)}$.

43. Find $a_1, a_2, a_3, a_4 \in \mathbb{Q}$ such that $F(x) = \frac{a_1x+a_2}{x^2-2} + \frac{a_3x+a_4}{x^2+1}$.

44. Find $b_1, b_2, b_3, b_4 \in \mathbb{R}$ such that $F(x) = \frac{b_1}{x-\sqrt{2}} + \frac{b_2}{x+\sqrt{2}} + \frac{b_3x+b_4}{x^2+1}$.

45. Find $c_1, c_2, c_3, c_4 \in \mathbb{Q}(i)$ such that $F(x) = \frac{c_1x+c_2}{x^2-2} + \frac{c_3}{x-i} + \frac{c_4}{x+i}$.

46. Find $d_1, d_2, d_3, d_4 \in \mathbb{C}$ such that $F(x) = \frac{d_1}{x-\sqrt{2}} + \frac{d_2}{x+\sqrt{2}} + \frac{d_3}{x-i} + \frac{d_4}{x+i}$.

In exercises 47–50, let $G(x) = \frac{3x^3-3x^2-9x+23}{(x^2-3)(x^2+4)}$.

47. Find $a_1, a_2, a_3, a_4 \in \mathbb{Q}$ such that $G(x) = \frac{a_1x+a_2}{x^2-3} + \frac{a_3x+a_4}{x^2+4}$.

48. Find $b_1, b_2, b_3, b_4 \in \mathbb{R}$ such that $G(x) = \frac{b_1}{x-\sqrt{3}} + \frac{b_2}{x+\sqrt{3}} + \frac{b_3x+b_4}{x^2+4}$.

49. Find $c_1, c_2, c_3, c_4 \in \mathbb{Q}(i)$ such that $G(x) = \frac{c_1x+c_2}{x^2-3} + \frac{c_3}{x-2i} + \frac{c_4}{x+2i}$.

50. Find $d_1, d_2, d_3, d_4 \in \mathbb{C}$ such that $G(x) = \frac{d_1}{x-\sqrt{3}} + \frac{d_2}{x+\sqrt{3}} + \frac{d_3}{x-2i} + \frac{d_4}{x+2i}$.

In exercises 51–54, let $H(x) = \frac{x^2+69}{(x^2-11)(x^2+9)}$.

51. Find $a_1, a_2, a_3, a_4 \in \mathbb{Q}$ such that $H(x) = \frac{a_1x+a_2}{x^2-11} + \frac{a_3x+a_4}{x^2+9}$.

52. Find $b_1, b_2, b_3, b_4 \in \mathbb{R}$ such that $H(x) = \frac{b_1}{x-\sqrt{11}} + \frac{b_2}{x+\sqrt{11}} + \frac{b_3x+b_4}{x^2+9}$.

53. Find $c_1, c_2, c_3, c_4 \in \mathbb{Q}(i)$ such that $H(x) = \frac{c_1x+c_2}{x^2-11} + \frac{c_3}{x-3i} + \frac{c_4}{x+3i}$.

54. Find $d_1, d_2, d_3, d_4 \in \mathbb{C}$ such that $H(x) = \frac{d_1}{x-\sqrt{11}} + \frac{d_2}{x+\sqrt{11}} + \frac{d_3}{x-3i} + \frac{d_4}{x+3i}$.

55. Show that the decomposition in Theorem 13.13 is unique. This requires showing that no other $A(x), b_{i,j}(x) \in F[x]$ are possible subject to the condition that for all i, j, either $b_{i,j}(x) = 0$ or has degree less than the degree of $p_i(x)$.

56. Show that the partial fraction decomposition in Corollary 13.14 is unique. This requires showing that no other $A(x) \in \mathbb{R}[x]$ and $a_{i,j}, b_{i,j}, c_{i,j} \in \mathbb{R}$ are possible.

CHAPTER 14

An Introduction to Linear Algebra and Vector Spaces

In the next few chapters, we will often look at "chains" of fields such as

$$\mathbb{R} \subseteq \mathbb{C}, \mathbb{Q} \subseteq \mathbb{Q}(\sqrt{2}), \text{ and } \mathbb{Q} \subseteq \mathbb{Q}(\sqrt{3}) \subseteq \mathbb{Q}(\sqrt{3}, i).$$

In order to better understand these chains, we need to introduce a new concept that deals with the "relative" size of a field compared to a smaller one. Since all the fields in these chains are infinite sets, it is not at all clear what this new concept means. To develop an understanding of this concept, we will need to look at many examples and introduce some new ideas and terms.

14.1 Examples, Examples, Examples, and a Definition

We begin this section with

■ Examples

1. $\mathbb{R} \subseteq \mathbb{C}$; every $\alpha \in \mathbb{C}$ can be written as $\alpha = a + bi$, where $a, b \in \mathbb{R}$. Therefore, every element of $\mathbb{C}$ can be described using two elements of $\mathbb{R}$. So, in some sense, we can think of $\mathbb{C}$ as being two times as large as $\mathbb{R}$.

2. $\mathbb{Q} \subseteq \mathbb{Q}(\sqrt{2})$; every $\alpha \in \mathbb{Q}(\sqrt{2})$ can be written as $\alpha = a + b\sqrt{2}$, where $a, b \in \mathbb{Q}$. Since every element of $\mathbb{Q}(\sqrt{2})$ can be described using two elements of $\mathbb{Q}$, we can think of $\mathbb{Q}(\sqrt{2})$ as being two times as large as $\mathbb{Q}$.

3. $\mathbb{Q} \subseteq \mathbb{Q}(\sqrt{3}) \subseteq \mathbb{Q}(\sqrt{3}, i)$; using this chain of fields, there are three chances for us to compare the relative sizes of fields, as we can compare $\mathbb{Q}(\sqrt{3}, i)$ to $\mathbb{Q}$, $\mathbb{Q}(\sqrt{3})$ to $\mathbb{Q}$,

and $\mathbb{Q}(\sqrt{3}, i)$ to $\mathbb{Q}(\sqrt{3})$. Every $\alpha \in \mathbb{Q}(\sqrt{3}, i)$ can be written as

$$\alpha = a + b\sqrt{3} + ci + di\sqrt{3} = \left(a + b\sqrt{3}\right) + \left(c + d\sqrt{3}\right)i,$$

where $a, b, c, d \in \mathbb{Q}$. Therefore, it takes four elements of $\mathbb{Q}$ to describe elements of $\mathbb{Q}(\sqrt{3}, i)$ and two elements of $\mathbb{Q}$ to describe elements of $\mathbb{Q}(\sqrt{3})$. Finally, the two elements $a + b\sqrt{3}$ and $c + d\sqrt{3}$ of $\mathbb{Q}(\sqrt{3})$ can be used to describe elements of $\mathbb{Q}(\sqrt{3}, i)$. Therefore, we can think of $\mathbb{Q}(\sqrt{3}, i)$ as being four times as large as $\mathbb{Q}$ and two times as large as $\mathbb{Q}(\sqrt{3})$. We can also think of $\mathbb{Q}(\sqrt{3})$ as being two times as large as $\mathbb{Q}$. ∎

As we develop and formalize the idea of relative size, we are led to concept of **dimension**. Since dimension is the most important concept in linear algebra, it will be the primary focus of this chapter.

Throughout this chapter, we will frequently refer to sets such as the polynomials in $\mathbb{Q}[x]$ of degree less than 3 or the polynomials in $\mathbb{R}[x]$ of degree less than 5. Technically, neither of these sets contains the polynomial $f(x) = 0$, since we did not assign a degree to this particular polynomial. However, for every $n \in \mathbb{N}$, we will adopt the convention that the polynomial $f(x) = 0$ is considered to belong to the set of polynomials of degree less than n.

Let us begin by looking at two large collections of examples. In each collection, we will try to find algebraic properties that are common to all of the examples.

■ Examples—Collection I

1. All polynomials in $\mathbb{Q}[x]$ of degree less than $2 = \{a + bx \mid a, b \in \mathbb{Q}\}$

2. All polynomials in $\mathbb{Q}[x]$ of degree less than $3 = \{a + bx + cx^2 \mid a, b, c \in \mathbb{Q}\}$

3. $\mathbb{Q}[x]$

4. All polynomials in $\mathbb{Q}[x]$ of degree less than 3 with 0 constant term $= \{ax + bx^2 \mid a, b \in \mathbb{Q}\}$

5. All ordered pairs of elements of $\mathbb{Q} = \{(a, b) \mid a, b \in \mathbb{Q}\}$

6. All ordered four-tuples of elements of $\mathbb{Q} = \{(a, b, c, d) \mid a, b, c, d \in \mathbb{Q}\}$

7. $\mathbb{Q}(i) = \{a + bi \mid a, b \in \mathbb{Q}\}$

8. $\mathbb{Q}(\sqrt{2}) = \{a + b\sqrt{2} \mid a, b \in \mathbb{Q}\}$

9. All ordered pairs of elements of $\mathbb{Q}$ such that the sum of the components is $0 = \{(a, b) \mid a, b \in \mathbb{Q} \text{ and } a + b = 0\}$

10. All four-tuples of elements of $\mathbb{Q}$ such that the first component plus twice the second component plus five times the third component minus three times the fourth component is $0 = \{(a, b, c, d) \mid a, b, c, d \in \mathbb{Q} \text{ and } a + 2b + 5c - 3d = 0\}$

When we look at examples 1, 2, 3, 4, 7, and 8, it should be clear that given any two elements from one of these sets, these elements can be added to produce a third element of the set. For example, $2 - 5x + 8x^2$ and $7 + 2x - 11x^2$ are typical elements from the set in 2 and their sum, $9 - 3x - 3x^2$, is also an element of the set in 2. For a more general example, if $a_1 x + b_1 x^2$ and $a_2 x + b_2 x^2$ are elements from the set in 4, then their sum, $(a_1 + a_2)x + (b_1 + b_2)x^2$, is also an element from the set in 4.

To deal with examples 5, 6, 9, and 10, we must review addition in these sets. Given ordered pairs, we add them componentwise and

$$(a, b) + (c, d) = (a + c, b + d).$$

More generally, given ordered n-tuples, where $n \in \mathbb{N}$, we again add componentwise and

$$(a_1, a_2, \ldots, a_n) + (b_1, b_2, \ldots, b_n) = (a_1 + b_1, a_2 + b_2, \ldots, a_n + b_n).$$

Having said this, it should now be easy to see that we can always add elements in examples 5 and 6. However, the situation for examples 9 and 10 is a little more complicated.

In example 9, we can always add two ordered pairs and obtain a third ordered pair. However, we need to make sure that the element we obtained satisfies the extra condition that the sum of the two components is 0. To check this, suppose (a, b) and (c, d) are two elements from example 9. Thus,

(1) $$a + b = 0 \text{ and } c + d = 0.$$

Adding our two elements we obtain $(a + c, b + d)$. To check if the sum of the two components in this ordered pair is 0, we use the information from equation (1) to obtain

$$(a + c) + (b + d) = (a + b) + (c + d) = 0 + 0 = 0,$$

as desired. Thus the sum of any two elements from the set in example (9) remains in that set.

In example 10, suppose (a_1, b_1, c_1, d_1) and (a_2, b_2, c_2, d_2) are two elements from this example. Then

(2) $a_1 + 2b_1 + 5c_1 - 3d_1 = 0$ and $a_2 + 2b_2 + 5c_2 - 3d_2 = 0.$

Adding these two elements, we obtain $(a_1 + a_2, b_1 + b_2, c_1 + c_2, d_1 + d_2)$. To check if this element belongs to the set in 10, we need to check that the first component plus twice the second component plus five times the third component minus three times the fourth component is 0. Using the information in equation (2), we obtain

$$(a_1 + a_2) + 2(b_1 + b_2) + 5(c_1 + c_2) - 3(d_1 + d_2) =$$
$$(a_1 + 2b_1 + 5c_1 - 3d_1) + (a_2 + 2b_2 + 5c_2 - 3d_2) =$$
$$0 + 0 = 0,$$

as desired. As a result, whenever we add elements from the set in example 10, we remain in that set.

It is not hard to see that addition in all ten of our examples is associative and commutative. Also note that every example has an additive identity element. In examples 1–4, the polynomial $f(x) = 0$ is the additive identity. This helps to illustrate why it is convenient to consider $f(x) = 0$ to be an element of the set of polynomials of degree less than n, regardless of the value of n. For examples 5 and 9, the ordered pair $(0, 0)$ is the additive identity and, in examples 6 and 10, the ordered 4-tuple $(0, 0, 0, 0)$ is the additive identity. In these examples, not only is it important that $(0, 0)$ and $(0, 0, 0, 0)$ behave like an identity, but it is also important that they belong to the given set. So, in example 10, before we can state that $(0, 0, 0, 0)$ is the additive identity, we first need to convince ourselves that it satisfies the condition that the first component plus twice the second component plus five times the third component minus three times the fourth component is 0. Finally, in examples 7 and 8, the number 0 is the additive identity.

Furthermore, in each example, it is also not hard to see that every element has an additive inverse. In particular, in example 2, the additive inverse of $a + bx + cx^2$ is $-a - bx - cx^2$, and in example 10, the additive inverse of (a, b, c, d) is $(-a, -b, -c, -d)$. At this point, we need to make a comment similar to the preceding one about checking that an element belongs to the set before declaring that it is the additive identity. In example 10, before we can say $(-a, -b, -c, -d)$ is the additive inverse of (a, b, c, d), we first need to convince ourselves that $(-a, -b, -c, -d)$ also satisfies the condition that the first component plus twice the second component plus five times the third component minus three times the fourth component is 0. Therefore, as you go about convincing yourself that every element has an additive inverse in the other eight examples, remember that the additive inverse needs to be an element of the set in the example.

To summarize, all ten of our examples are groups under addition and the addition is commutative. We can now examine the situation regarding multiplication. As before, when dealing with n-tuples, multiplication will be done componentwise. Therefore, given $(a_1, a_2, \ldots, a_n)$ and $(b_1, b_2, \ldots, b_n)$, we have

$$(a_1, a_2, \ldots, a_n) \cdot (b_1, b_2, \ldots, b_n) = (a_1 \cdot b_1, a_2 \cdot b_2, \ldots, a_n \cdot b_n).$$

Observe that examples (3),(5),(6),(7),(8) are all commutative rings. But things are quite different in examples (1),(2),(4),(9),(10). The polynomial x belongs to the set in example 1, but the product $x \cdot x = x^2$ does not. Similarly, the polynomial x^2 belongs to the sets in examples 2 and 4, but the product $x^2 \cdot x^2 = x^4$ does not. Note that the ordered pair $(1, -1)$ belongs to the set in example 9, but the product $(1, -1) \cdot (1, -1) = (1, 1)$ does not satisfy the condition that the sum of the components is 0. Similarly, the 4-tuple $(2, -1, 0, 0)$ belongs to the set in example 10, but the product $(2, -1, 0, 0) \cdot (2, -1, 0, 0) = (4, 2, 0, 0)$ does not satisfy the extra condition placed on the components. Thus, in five of our ten examples, our sets are not closed under multiplication.

Before going any further, we should stop and note that some subtleties arise as we develop the concept of dimension. It is easy to see that every element in the set in example 9 can be written using two elements of $\mathbb{Q}$. However, if (a, b) belongs to this set, a and b also satisfy the condition $a + b = 0$. Therefore, $a = -b$, and we can now write (a, b) as $(-b, b)$. As a result, although every element of this set consists of two components, we really only need a single element of $\mathbb{Q}$ in order to describe it. In this example, the number 2 describes the ordered pair $(-2, 2)$, whereas the number $-\frac{3}{5}$ describes the element $\left(\frac{3}{5}, -\frac{3}{5}\right)$. Therefore, at the moment, it is unclear whether the relative size of this set compared to $\mathbb{Q}$ should be 1 or 2. However, later in this chapter, it will become clear that the relative size is 1.

Similarly, every element in example 10 is of the form (a, b, c, d), where $a, b, c, d \in \mathbb{Q}$ and can be written using four elements of $\mathbb{Q}$. However, since $a + 2b + 5c - 3d = 0$, we see that $a = -2b - 5c + 3d$. Therefore, we can now write (a, b, c, d) as $(-2b - 5c + 3d, b, c, d)$. Thus, every element can now be written using only the three elements b, c, d of $\mathbb{Q}$. So, at this point, it remains unclear whether the relative size of this set compared to $\mathbb{Q}$ is 3 or 4 or perhaps some entirely different number.

Although ordinary multiplication is not possible in five of our ten examples, as we develop the concept of dimension, it will suffice to look at a somewhat different and weaker form of multiplication. The sets in examples 3, 7, and 8 are all commutative rings that contain $\mathbb{Q}$, so we can always multiply elements of these sets by elements of $\mathbb{Q}$. However, notice that although the sets in examples 1, 2, and 4 are not rings, we can always multiply elements of these sets by elements of $\mathbb{Q}$. For example, in 4, observe that $5 \cdot (3x - 8x^2) = 15x - 40x^2$. When dealing with n-tuples in 5, 6, 9, and 10, we can also multiple by elements of $\mathbb{Q}$ by doing the multiplication componentwise. In other words, given $\alpha \in \mathbb{Q}$ and n-tuple $(a_1, a_2, \ldots, a_n)$, we have

$$\alpha \cdot (a_1, a_2, \ldots, a_n) = (\alpha a_1, \alpha a_2, \ldots, \alpha a_n).$$

An important aspect of this type of multiplication is that when we multiply an element from the sets in 9 or 10 by an element in $\mathbb{Q}$, we remain in the set. To check this for example 9, suppose $\alpha \in \mathbb{Q}$ and suppose (a, b) is an ordered pair such that $a + b = 0$. Then $\alpha \cdot (a, b) = (\alpha a, \alpha b)$. To see that this element still belongs to the set in 9, observe that

$$\alpha a + \alpha b = \alpha(a + b) = \alpha \cdot 0 = 0,$$

as required. Similarly, if $\alpha \in \mathbb{Q}$ and if (a, b, c, d) belongs to the set in 10 then $a + 2b + 5c - 3d = 0$ and $\alpha \cdot (a, b, c, d) = (\alpha a, \alpha b, \alpha c, \alpha d)$. To check that $(\alpha a, \alpha b, \alpha c, \alpha d)$ belongs to the set in 10, we need to check that the first component plus twice the second component plus five times the third component minus three times the fourth component is 0. Observe that

$$\alpha a + 2(\alpha b) + 5(\alpha c) - 3(\alpha d) = \alpha(a + 2b + 5c - 3d) = \alpha \cdot 0 = 0,$$

as required.

As we summarize the properties shared by all ten of our examples, we can now say that all ten are

(a)   commutative groups under addition and

(b)   allow multiplication by elements of $\mathbb{Q}$.

Let us now look at a second collection of examples.

■ Examples—Collection II

1.   All polynomials in $\mathbb{R}[x]$ of degree less than $2 = \{a + bx | a, b \in \mathbb{R}\}$

2.   All polynomials in $\mathbb{R}[x]$ of degree less than $3 = \{a + bx + cx^2 | a, b, c \in \mathbb{R}\}$

3. $\mathbb{R}[x]$

4.   All polynomials in $\mathbb{R}[x]$ of degree less than 3 with 0 constant term
     $= \{ax + bx^2 | a, b \in \mathbb{R}\}$

5.   All ordered pairs of elements of $\mathbb{R} = \{(a, b) | a, b \in \mathbb{R}\}$

6.   All ordered four-tuples of elements of $\mathbb{R} = \{(a, b, c, d) | a, b, c, d \in \mathbb{R}\}$

7. $\mathbb{C} = \{a + bi | a, b \in \mathbb{R}\}$

8. All ordered pairs of elements of $\mathbb{R}$ such that the sum of the components is $0 = \{(a, b)|a, b \in \mathbb{R} \text{ and } a + b = 0\}$

9. All four-tuples of elements of $\mathbb{R}$ such that the first component plus twice the second component plus five times the third component minus three times the fourth component is $0 = \{(a, b, c, d)|a, b, c, d \in \mathbb{R} \text{ and } a + 2b + 5c - 3d = 0\}$

10. $\{f : \mathbb{R} \to \mathbb{R}|f \text{ is continuous everywhere}\}$

11. $\{f : \mathbb{R} \to \mathbb{R}|f \text{ is differentiable everywhere}\}$

12. $\{f : \mathbb{R} \to \mathbb{R}|f \text{ is continuous everywhere and } f(5) = 0\}$

Many of these 12 examples are similar to examples from our first collection. Once again observe that all 12 sets in these examples are commutative groups under addition. Also note, in some of these examples, that it is not always possible to multiply two elements and remain in the set. In particular, this is the situation in 1, 2, 4, 8, and 9. However, in all 12 examples, it is always possible to multiply an element from one of our sets by a real number and still stay in the set. In our first collection of examples, all of our sets were commutative groups under addition in which we were also allowed to multiply by elements of $\mathbb{Q}$. In our second collection of examples, all of our sets are commutative groups under addition in which we are also allowed to multiply by elements of $\mathbb{R}$. Thus, as we attempt to formalize things, we need to realize that we are looking at two interconnected algebraic objects: a set V that is a commutative group under addition and a set F that is a field. The connection between V and F is that we can multiply elements of V by elements of F to obtain elements of V. By collecting the properties possessed by V and F, we are led to

Definition 14.1. *A set V is called a vector space over a field F if V is a commutative group under addition and elements of V can be multiplied by elements of F to produce elements of V such that, for all $\alpha, \beta \in F$ and $v, w \in V$, we have*

1. $\alpha \cdot (v + w) = \alpha \cdot v + \alpha \cdot w,$

2. $(\alpha + \beta) \cdot v = \alpha \cdot v + \beta \cdot v,$

3. $\alpha \cdot (\beta \cdot v) = (\alpha\beta) \cdot v,$ *and*

4. $1 \cdot v = v,$ *where 1 is the multiplicative identity in F.*

It is important to understand the meaning of properties 1–4 in Definition 14.1. Properties 1 and 2 are types of distributive laws. For example, suppose

$$V = \{(a, b)|a, b \in \mathbb{R}\} \text{ and } F = \mathbb{R}.$$

To illustrate property (1), we have

$$3 \cdot ((4, -7) + (-8, 2)) = 3 \cdot (-4, -5) = (-12, -15)$$

and

$$3 \cdot (4, -7) + 3 \cdot (-8, 2) = (12, -21) + (-24, 6) = (-12, -15).$$

Thus,

$$3 \cdot ((4, -7) + (-8, 2)) = 3 \cdot (4, -7) + 3 \cdot (-8, 2).$$

To illustrate property 2, observe that

$$(5+9) \cdot (11, -5) = 14 \cdot (11, -5) = (154, -70)$$

and

$$5 \cdot (11, -5) + 9 \cdot (11, -5) = (55, -25) + (99, -45) = (154, -70).$$

Thus,

$$(5+9) \cdot (11, -5) = 5 \cdot (11, -5) + 9 \cdot (11, -5).$$

Property 3 is a type of associative law. Remember that since F is a field, we are allowed to multiply elements of F. Also, we are allowed to multiply an element of V by an element of F to obtain an element of V. However, we are not permitted to multiply two elements of V. When we analyze the terms that appear in property 3, keep in mind that $\alpha\beta \in F$, whereas $\beta \cdot v, \alpha \cdot (\beta \cdot v), (\alpha\beta) \cdot v \in V$. To illustrate this property, using V and F as previously, we have

$$-2 \cdot (7 \cdot (-4, 1)) = -2 \cdot (-28, 7) = (56, -14)$$

and

$$(-2 \cdot 7) \cdot (-4, 1) = -14 \cdot (-4, 1) = (56, -14).$$

Thus,

$$-2 \cdot (7 \cdot (-4, 1)) = (-2 \cdot 7) \cdot (-4, 1).$$

Observe that a second associative law of the form $\alpha \cdot (vw) = (\alpha \cdot v)w$ would not make any sense as we are not permitted to multiply elements of V by elements of V.

The easiest property to understand is 4. It asserts that the multiplicative identity of F continues to behave like a multiplicative identity, even when we multiply it by elements of V.

If V is a vector space over a field F, we may simply say that V is a vector space over F. Sometimes we are even more informal and do not mention F at all and merely say that V is a vector space. We call the elements of V **vectors** and the elements of F **scalars**. Multiplication of an element of V by an element of F is called **scalar multiplication**. Observe that both V and F have an additive identity, and, depending on the situation, this could cause some confusion. If the additive identity of V is of the form $(0, 0)$ or $(0, 0, 0, 0)$, then it looks nothing like the additive identity of F and no confusion arises. But, in situations where there might be some confusion recognizing which elements are vectors and which are scalars, vectors may be written with an arrow, such as $\vec{v}$.

In Lemma 5.14(c), we showed that multiplication by 0 in a ring always gives 0 as the answer. As we think about the analogous situation for vector spaces, we note that we can multiply vectors by the scalar 0 and can also multiply the vector $\vec{0}$ by scalars. We would expect that, in both cases, we would obtain $\vec{0}$ as the answer. The next lemma shows that this is the case and also provides us with some basic facts that will be useful when performing various computations within a vector space. When reading the following proofs, make sure you understand why all the various equalities hold. In virtually every case, equality will be a consequence of the associative and distributive laws along with the properties of additive identities and inverses.

Lemma 14.2. *If V is a vector space over a field F, then*

(a) $0 \cdot v = \vec{0}$, *for all $v \in V$.*

(b) $\alpha \cdot \vec{0} = \vec{0}$, *for all $\alpha \in F$.*

(c) $(-\alpha) \cdot v$ *is the additive inverse of $\alpha \cdot v$, for all $\alpha \in F$ and $v \in V$. In particular $(-1) \cdot v$ is the additive inverse of v.*

(d) *If $\alpha, \beta \in F$ and $v \in V$ such that $\alpha \cdot v = \beta \cdot v$, then either $\alpha = \beta$ or $v = \vec{0}$. In particular, if $\alpha \cdot v = \vec{0}$, then either $\alpha = 0$ or $v = \vec{0}$.*

Proof. For part (a), if $v \in V$, we have

$$0 \cdot v = 0 \cdot v + \vec{0} = 0 \cdot v + (0 \cdot v - 0 \cdot v) =$$
$$(0 \cdot v + 0 \cdot v) - 0 \cdot v = (0 + 0) \cdot v - 0 \cdot v = 0 \cdot v - 0 \cdot v = \vec{0}.$$

For part (b), if $\alpha \in F$, we have

$$\alpha \cdot \vec{0} = \alpha \cdot \vec{0} + \vec{0} = \alpha \cdot \vec{0} + (\alpha \cdot \vec{0} - \alpha \cdot \vec{0}) =$$
$$(\alpha \cdot \vec{0} + \alpha \cdot \vec{0}) - \alpha \cdot \vec{0} = \alpha \cdot (\vec{0} + \vec{0}) - \alpha \cdot \vec{0} = \alpha \cdot \vec{0} - \alpha \cdot \vec{0} = \vec{0}.$$

For part (c), if $v \in V$, we can use part (a) to obtain

$$\alpha \cdot v + (-\alpha) \cdot v = (\alpha - \alpha) \cdot v = 0 \cdot v = \vec{0}.$$

Since addition is commutative, we also know that $(-\alpha) \cdot v + \alpha \cdot v = \vec{0}$. Therefore, when we add $(-\alpha) \cdot v$ to $\alpha \cdot v$, we obtain the additive identity of V. Thus $(-\alpha) \cdot v$ is the additive inverse of $\alpha \cdot v$. If we specialize to the situation where $\alpha = 1$, we can see that $(-1) \cdot v$ is the additive inverse of $1 \cdot v = v$.

For part (d), suppose $\alpha \cdot v = \beta \cdot v$; we need to show that if $\alpha \neq \beta$ then $v = \vec{0}$. It now follows from part (c) that $(\alpha - \beta) \cdot v = \vec{0}$. If $\alpha \neq \beta$, then $\alpha - \beta$ is invertible, and if we multiply the equation

$$\vec{0} = (\alpha - \beta) \cdot v$$

by $(\alpha - \beta)^{-1}$ and use part (b), we obtain

$$\vec{0} = (\alpha - \beta)^{-1} \cdot \vec{0} = (\alpha - \beta)^{-1}((\alpha - \beta) \cdot v) = ((\alpha - \beta)^{-1}(\alpha - \beta)) \cdot v = 1 \cdot v = v.$$

Thus, if $\alpha \neq \beta$, then $v = \vec{0}$, as desired.

Finally, if $\alpha \cdot v = \vec{0}$ then, using part (a), we have $\alpha \cdot v = 0 \cdot v$. It now follows from our previous argument that either $\alpha = 0$ or $v = \vec{0}$. $\qquad \square$

Having looked at many examples of sets which are vector spaces, it is also instructive to look at some sets which are not vector spaces.

■ Examples—Some Sets That Are *Not* Vector Spaces over $\mathbb{Q}$

1. Let $V_1 = \{(a, b) | a, b \in \mathbb{Q}$ and $a \cdot b \geq 0\}$; observe that

    $$(2, 1), (-1, -2) \in V_1 \quad \text{but} \quad (2, 1) + (-1, -2) = (1, -1) \notin V_1.$$

    Thus, V_1 is not closed under addition and is therefore not a vector space.

2. Let $V_2 = \{a + bx \in \mathbb{Q}[x] | a^2 = b^2\}$; then

    $$1 + x, 1 - x \in V_1 \quad \text{but} \quad (1 + x) + (1 - x) = 2 \notin V_2.$$

    Thus, V_2 is also not closed under addition, so it is also not a vector space.

3. Let $V_3 = \{(a, b) | a, b \in \mathbb{Q}$ and $a + b = 1\}$; then $(1, 0) \in V_3$ but $2 \cdot (1, 0) = (2, 0) \notin V_3$. Therefore, scalar multiplication is not possible in V_3. You should also check that V_3 is not closed under addition.

4. Let $V_4 = \{(a, b)|a, b \in \mathbb{Z}\}$; although V_4 is a group under addition, scalar multiplication is not possible as

$$(1, 1) \in V_4 \text{ and } \frac{1}{2} \cdot (1, 1) = \left(\frac{1}{2}, \frac{1}{2}\right) \notin V_4.$$

∎

Suppose $V = \{(a, b)|a, b \in \mathbb{Q}\}$ and $W = \{(a, b)|a, b \in \mathbb{R}\}$; then certainly V is a vector space over $\mathbb{Q}$ and W is a vector space over $\mathbb{R}$. Note that W is also a vector space over $\mathbb{Q}$. More generally, if a set is a vector space over a field F, then it is also a vector space over any field contained in F. However, a similar result does not hold when we move to a field that is larger than F. In particular, V is not a vector space over $\mathbb{R}$. Observe that $(0, 1) \in V$ but $\sqrt{2} \cdot (0, 1) = (0, \sqrt{2}) \notin V$, so scalar multiplication is not possible in V.

We now need to focus our attention on two examples of particular importance.

∎ Examples—Two Fundamental Examples

1. If F is a field, let

$$F^n = \{(a_1, a_2, \ldots, a_n)|a_i \in F\}.$$

F^n is a vector space over F where both addition and scalar multiplication are done componentwise.

2. If $F \subseteq K$ are fields, then K is a vector space over F. To see this, we first note that K is certainly a group under addition. Furthermore, since K is a field, K satisfies the distributive laws, its multiplication is associative, and it has a multiplicative identity. Therefore, properties 1–4 of Definition 14.1 hold for all elements of K. However, since $F \subseteq K$, scalar multiplication of elements of K by elements of F is merely a special case of multiplication in K. Thus, scalar multiplication automatically inherits properties 1–4 of Definition 14.1.

∎

In Chapter 5, we remarked that if R is a commutative ring, then R might be a field but it need not be a field. Similarly, if V is a vector space over F, then V might also be a commutative ring or field but it need not be. For example, $\mathbb{Q}[x]$ is not only a vector space over $\mathbb{Q}$ but is also a commutative ring. On the other hand, $V = \{ax + bx^2|a, b \in \mathbb{Q}\}$ is a vector space over $\mathbb{Q}$ but is not a commutative ring. When dealing with fields $F \subseteq K$, in some applications it is useful to temporarily ignore the fact that K is a field and to simply think of K as being a vector space over F.

Exercises for Section 14.1

In exercises 1–14, you will be doing computations in

$$\mathbb{Q}^4 = \{(a, b, c, d)|a, b, c, d \in \mathbb{Q}\},$$

which is a vector space over $\mathbb{Q}$. In $\mathbb{Q}^4$, let

$$v_1 = (5, 0, 0, 6), \quad v_2 = (1, 2, 0, 0), \quad v_3 = (0, 30, 0, -18),$$
$$v_4 = (0, 6, 4, 0), \quad v_5 = (1, 0, 5, 2).$$

1. Compute $\frac{3}{5}v_2$.

2. Compute $-\frac{8}{3}v_3$.

3. Compute $3v_2 - 6v_4$.

4. Compute $2v_3 - \frac{5}{3}v_4$.

5. Compute $-v_1 + 4v_2 - \frac{1}{2}v_5$.

6. Compute $3v_2 - \frac{3}{4}v_4 + 7v_5$.

7. Compute $-5v_2 - 7v_3 + v_4 - 6v_5$.

8. Compute $-2v_1 + \frac{5}{2}v_2 - 6v_3 + v_5$.

9. Compute $4v_1 - v_2 + 8v_3 - 5v_4 + 11v_5$.

10. Compute $-3v_1 - 7v_2 + v_3 + v_4 + 9v_5$.

11. Can you find $\alpha, \beta \in \mathbb{Q}$ such that $v_3 = \alpha \cdot v_1 + \beta \cdot v_2$?

12. Can you find $\alpha, \beta \in \mathbb{Q}$ such that $v_1 = \alpha \cdot v_2 + \beta \cdot v_3$?

13. Can you find $\alpha, \beta \in \mathbb{Q}$ such that $v_5 = \alpha \cdot v_1 + \beta \cdot v_4$?

14. Show that if $\alpha, \beta, \gamma, \delta \in \mathbb{Q}$ such that $\alpha \cdot v_2 + \beta \cdot v_3 + \gamma \cdot v_4 + \delta \cdot v_5 = (0, 0, 0, 0)$, then $\alpha = \beta = \gamma = \delta = 0$.

In exercises 15–30, you will be doing computations in $\mathbb{R}[x]$, which is a vector space over $\mathbb{R}$. In $\mathbb{R}[x]$, let

$$w_1 = 3x^2 - 1, \quad w_2 = 4x + 5, \quad w_3 = x^3 - 2x,$$
$$w_4 = 2x^3 + 5, \quad w_5 = 9x - 2.$$

15. Compute $\sqrt{2}w_2$.

16. Compute $\frac{1}{2}w_4$.

17. Compute $7w_1 + \pi w_3$.

18. Compute $-3w_3 + \frac{3}{4}w_5$.

19. Compute $11w_1 - \frac{6}{5}w_2 + 8w_4$.

20. Compute $-\sqrt{7}w_3 - 6w_4 + 5w_5$.

21. Compute $4w_1 - \frac{2}{3}w_3 + 3w_4 - 7w_5$.

22. Compute $6w_1 + w_3 - \frac{1}{3}w_4 - 4w_5$.

23. Compute $2w_1 + 3w_2 - 5w_3 - w_4 + 7w_5$.

24. Compute $-9w_1 - 6w_2 + 8w_3 - w_4 + 2w_5$.

25. Can you find $\alpha, \beta \in \mathbb{R}$ such that $w_1 = \alpha \cdot w_2 + \beta \cdot w_3$?

26. Can you find $\alpha, \beta \in \mathbb{R}$ such that $w_4 = \alpha \cdot w_2 + \beta \cdot w_3$?

27. Can you find $\alpha, \beta \in \mathbb{R}$ such that $1 = \alpha \cdot w_2 + \beta \cdot w_5$?

28. Can you find $\alpha, \beta \in \mathbb{R}$ such that $x = \alpha \cdot w_2 + \beta \cdot w_5$?

29. Can you find $\alpha, \beta, \gamma \in \mathbb{R}$ such that $x^3 = \alpha \cdot w_2 + \beta \cdot w_4 + \gamma \cdot w_5$?

30. Show that if $\alpha, \beta, \gamma, \delta \in \mathbb{R}$ such that $\alpha \cdot w_1 + \beta \cdot w_2 + \gamma \cdot w_3 + \delta \cdot w_5 = 0$, then $\alpha = \beta = \gamma = \delta = 0$.

In exercises 31–42, determine if the given set is a vector space over $\mathbb{Q}$. If it is not a vector space, briefly explain why it fails to be a vector space.

31. $\{f(x) \in \mathbb{Q}[x]\,|\, f(10) = 0\}$.

32. $\{f(x) \in \mathbb{Q}[x]\,|\, f(10) > 0\}$.

33. $\{f(x) \in \mathbb{Q}[x]\,|\, f(1) = f(2)\}$.

34. $\{f(x) \in \mathbb{Q}[x]\,|\, f(x) \text{ has a root in } \mathbb{Q}\}$.

35. $\{f(x) \in \mathbb{Q}[x]\,|\, f(0) \in \mathbb{Z}\}$.

36. $\{g(x) \in \mathbb{R}[x]\,|\, g(0) \in \mathbb{Q}\}$.

37. $\{g(x) \in \mathbb{R}[x]\,|\, g'(0) \in \mathbb{Q}\}$.

38. $\{g(x) \in \mathbb{R}[x]\,|\, g(i) = 0\}$.

39. $\{g(x) \in \mathbb{R}[x]\,|\, g(i) \in \mathbb{Q}\}$.

40. $\{g(x) \in \mathbb{R}[x]\,|\, g(i) \in \mathbb{Z}\}$.

41. $\mathbb{Q}(\sqrt{2})$

42. $\{r \in \mathbb{Q}(\sqrt{2}) \mid r^2 \in \mathbb{Q}\}$.

In exercises 43–52, determine if the given subset of $\mathbb{R}^4$ is a vector space over $\mathbb{R}$. If it is not a vector space, briefly explain why it fails to be a vector space.

43. $\{(a, b, c, d) \in \mathbb{R}^4 \mid a = b \text{ and } c = d\}$.

44. $\{(a, b, c, d) \in \mathbb{R}^4 \mid a^2 = b^2\}$.

45. $\{(a, b, c, d) \in \mathbb{R}^4 \mid b = c = 0\}$.

46. $\{(a, b, c, d) \in \mathbb{R}^4 \mid a = 2b - 3c\}$.

47. $\{(a, b, c, d) \in \mathbb{R}^4 \mid ab \geq 0\}$.

48. $\{(a, b, c, d) \in \mathbb{R}^4 \mid a + b < c + d\}$.

49. $\{(a, b, c, d) \in \mathbb{R}^4 \mid 2a + 4c = 5b + d\}$.

50. $\{(a, b, c, d) \in \mathbb{R}^4 \mid b = c \text{ and } d = 0\}$.

51. $\{(a, b, c, d) \in \mathbb{R}^4 \mid a^2 + b^2 + c^2 + d^2 > 0\}$.

52. $\{(a, b, c, d) \in \mathbb{R}^4 \mid a \in \mathbb{Q}\}$.

14.2 Spanning Sets and Linear Independence

Now that we have worked with quite a few examples of vector spaces, we can start introducing the concepts that will lead us to an understanding of dimension.

Definition 14.3. *If V is a vector space over a field F and if $v_1, v_2, \ldots, v_n \in V$, then a linear combination of $v_1, v_2, \ldots, v_n$ is any element of V that can be written in the form $\alpha_1 \cdot v_1 + \alpha_2 \cdot v_2 + \cdots + \alpha_n \cdot v_n$, where each $\alpha_i \in F$.*

■ Examples

1. $V = \mathbb{Q}[x]$ is a vector space over $\mathbb{Q}$ and let us examine $1, x, x^2 \in \mathbb{Q}[x]$. The elements

$$3 \cdot 1 + 4 \cdot x + (-5) \cdot x^2 = -5x^2 + 4x + 3,$$

$$0 \cdot 1 + 0 \cdot x + 2 \cdot x^2 = 2x^2,$$

$$\frac{1}{3} \cdot 1 + 2 \cdot x + 0 \cdot x^2 = 2x + \frac{1}{3},$$

$$0 \cdot 1 + 0 \cdot x + 0 \cdot x^2 = 0$$

are all linear combinations of $1, x, x^2$. However, observe that $x^3, 2x^4 - 1$, and $x^5 - 6x + 1$ are not linear combinations of $1, x, x^2$.

2. $V = \mathbb{Q}^3$ is a vector space over $\mathbb{Q}$ and let us examine $(1, 2, 3), (-1, 0, 2) \in \mathbb{Q}^3$. The elements

$$2 \cdot (1, 2, 3) + 5 \cdot (-1, 0, 2) = (-3, 4, 16),$$

$$(-4) \cdot (1, 2, 3) + 3 \cdot (-1, 0, 2) = (-7, -8, -6),$$

$$0 \cdot (1, 2, 3) + (-8) \cdot (-1, 0, 2) = (8, 0, -16)$$

are all linear combinations of $(1, 2, 3), (-1, 0, 2)$.

Suppose we wish to determine if $(1, 0, 1)$ is a linear combination of $(1, 2, 3)$ and $(-1, 0, 2)$. This would mean that there exist $\alpha, \beta \in \mathbb{Q}$ such that

(3) $$\alpha \cdot (1, 2, 3) + \beta \cdot (-1, 0, 2) = (1, 0, 1).$$

However,

(4) $$\alpha \cdot (1, 2, 3) + \beta \cdot (-1, 0, 2) = (\alpha - \beta, 2\alpha, 3\alpha + 2\beta).$$

In light of equations (3) and (4), α and β would simultaneously need to satisfy the equations

(5) $$\alpha - \beta = 1, \quad 2\alpha = 0, \quad 3\alpha + 2\beta = 1.$$

The only α, β which satisfy the first two equations in (5) are $\alpha = 0, \beta = -1$. But these values of α, β do not satisfy the third equation in (5). Thus, there are no $\alpha, \beta \in \mathbb{Q}$ that simultaneously satisfy the three equations in (5). Therefore, $(1, 0, 1)$ is not a linear combination of $(1, 2, 3)$ and $(-1, 0, 2)$.

Now suppose we wish to determine if $(10, 12, 12)$ is a linear combination of $(1, 2, 3)$ and $(-1, 0, 2)$. Using (4), we see that we need to find $\alpha, \beta \in \mathbb{Q}$ such that

(6) $$\alpha - \beta = 10, \quad 2\alpha = 12, \quad 3\alpha + 2\beta = 10.$$

If we solve the first two equations in (6), we see that $\alpha = 6, \beta = -4$ is a solution. These values of α, β are also a solution to the third equation in (6). Thus,

$$(10, 12, 10) = 6 \cdot (1, 2, 3) + (-4) \cdot (-1, 0, 2)$$

and so, $(10, 12, 10)$ is a linear combination of $(1, 2, 3)$ and $(-1, 0, 2)$.

The calculations done in the previous example certainly suggest that there is a connection between linear equations and our work on vector spaces.

Definition 14.4. *If V is a vector space over F and if S is a finite subset of V then the span of S, written as $span(S)$, is the set of all linear combinations of elements of S. If $span(S) = V$, then we say that S spans V or S is a spanning set of V.*

■ Examples

1. Let $V = \mathbb{Q}[x]$; this is a vector space over $\mathbb{Q}$, and we will look at the span of various subsets of V.

    (a) If $S_1 = \{1, x\}$, then $span(S_1)$ consists of those elements of V that can be written in the form $\alpha \cdot 1 + \beta \cdot x$, where $\alpha, \beta \in \mathbb{Q}$. Thus, $span(S_1)$ is the set of all polynomials in $\mathbb{Q}[x]$ of degree less than 2. In particular, $x^3 \notin span(S_1)$.

    (b) If $S_2 = \{x, x^2\}$, then $span(S_2)$ consists of those elements of V that can be written in the form $\alpha \cdot x + \beta \cdot x^2$, where $\alpha, \beta \in \mathbb{Q}$. Thus, $span(S_2)$ is the set of all polynomials in $\mathbb{Q}[x]$ of degree less than 3 whose constant term is 0. As a result, neither $x^2 + 1$ nor x^3 belongs to $span(S_2)$.

    (c) If $S_3 = \{1, x^2, x^4, x^6, x^8\}$, then $span(S_3)$ consists of all polynomials in $\mathbb{Q}[x]$ of degree less than 9 such that only terms with an even exponent appear. For example, $7x^8 - 4x^6 + \frac{2}{7}x^2 - \frac{11}{3} \in span(S_3)$, whereas x^{10} and $5x^6 - 8x^3$ do not belong to $span(S_3)$.

2. Let $W = \mathbb{R}^3$; this is a vector space over $\mathbb{R}$, and we will look at the span of various subsets of W.

    (a) If $S_1 = \{(1, 0, 0), (0, 1, 0), (0, 0, 1)\}$, then S_1 spans W. To see this, observe that if $(\alpha, \beta, \gamma) \in \mathbb{R}^3$, then

    $$(\alpha, \beta, \gamma) = \alpha \cdot (1, 0, 0) + \beta \cdot (0, 1, 0) + \gamma \cdot (0, 0, 1) \in span(S_1).$$

    (b) If $S_2 = \{(0, 1, 0), (0, 0, 1)\}$, then $span(S_2)$ consists of all those elements in W whose first component is 0. To see this, note that

    $$(0, \alpha, \beta) = \alpha \cdot (0, 1, 0) + \beta \cdot (0, 0, 1) \in span(S_2).$$

    (c) If $S_3 = \{(6, 6, 6)\}$, then $span(S_3)$ consists of all those elements in W where all three components are equal. Observe that

    $$(\alpha, \alpha, \alpha) = \frac{\alpha}{6} \cdot (6, 6, 6) \in span(S_3).$$

    (d) Let $S_4 = \{(1, 2, 0), (2, 1, 0), (0, 0, 1)\}$; we claim that S_4 spans W. In example (a), we showed that the set $S_1 = \{(1, 0, 0), (0, 1, 0), (0, 0, 1)\}$ spans W. We will now show that all three members of S_1 belong to $span(S_4)$. For the vector $(0, 0, 1)$, there is nothing to prove. To see that the first two elements of S_1 belong to $span(S_4)$, observe that

$$(1, 0, 0) = -\frac{1}{3} \cdot (1, 2, 0) + \frac{2}{3} \cdot (2, 1, 0) \text{ and } (0, 1, 0) = \frac{2}{3} \cdot (1, 2, 0) + \frac{-1}{3} \cdot (2, 1, 0).$$

Having shown that $S_1 \subseteq span(S_4)$, we can see that if $(\alpha, \beta, \gamma) \in \mathbb{R}^3$, then

$$(\alpha, \beta, \gamma) = \alpha \cdot (1, 0, 0) + \beta \cdot (0, 1, 0) + \gamma (0, 0, 1) =$$

$$\alpha \cdot \left(-\frac{1}{3} \cdot (1, 2, 0) + \frac{2}{3} \cdot (2, 1, 0) \right) + \beta \cdot \left(\frac{2}{3} \cdot (1, 2, 0) + \frac{-1}{3} \cdot (2, 1, 0) \right) + \gamma \cdot (0, 0, 1) =$$

$$\frac{-\alpha + 2\beta}{3} \cdot (1, 2, 0) + \frac{2\alpha - \beta}{3} \cdot (2, 1, 0) + \gamma \cdot (0, 0, 1) \in span(S_4).$$

∎

This example illustrates the useful fact that in order to show that a set T spans W, it suffices to show that $span(T)$ contains some other spanning set S. We record this as

Lemma 14.5. *Let V be a vector space over F and let S and T be finite subsets of V. If $S \subseteq span(T)$, then $span(S) \subseteq span(T)$. Therefore, if $S \subseteq span(T)$ and S spans V, then T also spans V.*

Proof. Suppose $S \subseteq span(T)$ and let $s \in span(S)$, we need to show that $s \in span(T)$. Since $s \in span(S)$, there exist $v_i \in S$ and $\alpha_i \in F$ such that

$$(7) \qquad s = \alpha_1 \cdot v_1 + \alpha_2 \cdot v_2 + \cdots + \alpha_n \cdot v_n.$$

However, each $v_i \in span(T)$, so there exist $w_j \in T$ and $\beta_{ij} \in F$ such that

$$(8) \qquad v_i = \beta_{i1} \cdot w_1 + \beta_{i2} \cdot w_2 + \cdots + \beta_{im} \cdot w_m,$$

for all $i \leq n$. Therefore, in (7), we can replace each v_i by a linear combination of the w_j's from (8) to obtain

$$s = \alpha_1 \cdot (\beta_{11} \cdot w_1 + \beta_{12} \cdot w_2 + \cdots + \beta_{1m} \cdot w_m) + \alpha_2 \cdot (\beta_{21} \cdot w_1 + \beta_{22} \cdot w_2 + \cdots + \beta_{2m} \cdot w_m) +$$

$$\cdots + \alpha_n \cdot (\beta_{n1} \cdot w_1 + \beta_{n2} \cdot w_2 + \cdots + \beta_{nm} \cdot w_m) =$$

$$(\alpha_1 \beta_{11} + \alpha_2 \beta_{21} + \cdots + \alpha_n \beta_{n1}) \cdot w_1 + (\alpha_1 \beta_{12} + \alpha_2 \beta_{22} + \cdots + \alpha_n \beta_{n2}) \cdot w_2 +$$

$$\cdots + (\alpha_1 \beta_{1m} + \alpha_2 \beta_{2m} + \cdots + \alpha_n \beta_{nm}) \cdot w_m \in span(T).$$

Thus, $span(S) \subseteq span(T)$.

It now easily follows that if $S \subseteq span(T)$ and S spans V, then $V = span(S) \subseteq span(T)$. Hence, T spans V. $\qquad\qquad\qquad\qquad\qquad\qquad\qquad\qquad\qquad\qquad\qquad\qquad\qquad\qquad\quad$ $\square$

If we consider the vector space $\mathbb{Q}^2$ over $\mathbb{Q}$, it is easy to see that $S = \{(1, 0), (1, 1), (0, 1)\}$ spans $\mathbb{Q}^2$. In this case, elements of $\mathbb{Q}^2$ can be written in several different ways as linear combinations of elements of S. For example, we have

$$(4, 2) = 2 \cdot (1, 0) + 2 \cdot (1, 1) + 0 \cdot (0, 1),$$
$$(4, 2) = 0 \cdot (1, 0) + 4 \cdot (1, 1) + (-2) \cdot (0, 1),$$
$$(4, 2) = 4 \cdot (1, 0) + 0 \cdot (1, 1) + 2 \cdot (0, 1).$$

In fact, not only can $\mathbb{Q}^2$ be spanned by S, but it can also be spanned by any subset of S that contains two elements. Thus, S is a spanning set for V, but it is not the smallest possible spanning set for V. It turns out that given a vector space V over a field F, we will be interested in sets that not only span V but have the additional property that no smaller subset spans V. To help us better understand this idea, we have

Lemma 14.6. *If V is a vector space over a field F and $S = \{v_1, v_2, \ldots, v_n\}$ is a finite subset of V, then the following statements are equivalent:*

(a) *No element of S can be written as a linear combination of the other elements of S.*

(b) *For every $w \in span(S)$, there is only one way to express w as a linear combination of the elements of S.*

(c) *The only way to express $\vec{0}$ as a linear combination of the elements of S is*
$$\vec{0} = 0 \cdot v_1 + 0 \cdot v_2 + \cdots + 0 \cdot v_n.$$

Proof. Let us first suppose that statement (a) is true; we will prove that statement (b) also holds. If $w \in span(S)$, then there exist $\alpha_i \in F$ such that

$$w = \alpha_1 \cdot v_1 + \alpha_2 \cdot v_2 + \cdots + \alpha_n \cdot v_n.$$

To show that this is the only way to express w as a linear combination of the elements of S, we must show that if there exist $\beta_i \in F$ such that

$$w = \beta_1 \cdot v_1 + \beta_2 \cdot v_2 + \cdots + \beta_n \cdot v_n,$$

then $\alpha_i = \beta_i$, for all $i \leq n$. We will proceed with a proof by contradiction, so let us assume that there is some subscript j such that $\alpha_j \neq \beta_j$. Using the fact that

$$\alpha_1 \cdot v_1 + \alpha_2 \cdot v_2 + \cdots + \alpha_n \cdot v_n = \beta_1 \cdot v_1 + \beta_2 \cdot v_2 + \cdots + \beta_n \cdot v_n,$$

we can subtract various elements from both sides of this equation so that the v_j term will be on the left side of equation and all the other v_i's will be on the right side. After doing this, we obtain

$$(\alpha_j - \beta_j) \cdot v_j = (\beta_1 - \alpha_1) \cdot v_1 + \cdots + (\beta_{j-1} - \alpha_{j-1}) \cdot v_{j-1}$$

(9)
$$+ (\beta_{j+1} - \alpha_{j+1}) \cdot v_{j+1} + \cdots + (\beta_n - \alpha_n) \cdot v_n.$$

Since $\alpha_j \neq \beta_j$, the element $\alpha_j - \beta_j$ has a multiplicative inverse in F. If we multiply both sides of (9) by $(\alpha_j - \beta_j)^{-1}$ and let $\gamma_i = (\alpha_j - \beta_j)^{-1}(\beta_i - \alpha_i)$, for $i \leq n$, we obtain

$$v_j = \gamma_1 \cdot v_1 + \cdots + \gamma_{j-1} \cdot v_{j-1} + \gamma_{j+1} \cdot v_{j+1} + \cdots + \gamma_n \cdot v_n.$$

But this is a contradiction, as we have written v_j as a linear combination of the other elements of S. Therefore, it must be the case that $\alpha_i = \beta_i$, for all $i \leq n$, and there is indeed only one way to express w as a linear combination of the elements of S.

If statement (b) is true, it is quite easy to show that statement (c) is also true. Lemma 14.2(c) told us that $0 \cdot v = \vec{0}$, for all $v \in V$. Therefore, it immediately follows that

$$\vec{0} = 0 \cdot v_1 + 0 \cdot v_2 + \cdots + 0 \cdot v_n.$$

As a result, we have certainly illustrated one way to express $\vec{0}$ as a linear combination of the elements of S. However, statement (b) asserts that this must be the only one way to express $\vec{0}$ as a linear combination of the elements of S.

Finally, let us suppose statement (c) is true, and we will prove that statement (a) is true. By way of contradiction, let us suppose that there is some $v_j \in S$ that can be written as a linear combination of the others. Therefore, there exist $\alpha_i \in F$ such that

$$v_j = \alpha_1 \cdot v_1 + \cdots + \alpha_{j-1} \cdot v_{j-1} + \alpha_{j+1} \cdot v_{j+1} + \cdots + \alpha_n \cdot v_n.$$

If we subtract v_j from both sides and apply Lemma 14.2(c), we obtain

$$\vec{0} = \alpha_1 \cdot v_1 + \cdots + \alpha_{j-1} \cdot v_{j-1} + (-1) \cdot v_j + \alpha_{j+1} \cdot v_{j+1} + \cdots + \alpha_n \cdot v_n$$

However, this equation is a second way to express $\vec{0}$ as a linear combination of the elements of S. This contradicts statement (c), thereby concluding the proof. $\square$

We now give a name to those sets which satisfy the conditions in the previous lemma.

Definition 14.7. *Let V be a vector space over a field F. We say that a set of vectors $\{v_1, v_2, \ldots, v_n\}$ is linearly independent if it satisfies any of the conditions in Lemma 14.6. If the set $\{v_1, v_2, \ldots, v_n\}$ is not linearly independent, we say that it is linearly dependent.*

To convince yourself that you understand Definition 14.7, you should try to prove that if a set S is linearly independent, then every nonempty subset of S is also linearly independent.

If you are given a set of vectors, it is often quite tedious to determine whether or not they are linearly independent. After looking at the set for awhile, you might get a sense of whether one of the vectors can be written as a linear combination of the others. If you succeed in showing that one of the vectors can be written this way, then you have shown that they are linearly dependent. On the other hand, you may feel that there is only one way to express $\vec{0}$ as a linear combination of the vectors in the set. If you show this, then you will have shown that they are linearly independent. Regardless of whether the vectors ultimately turn out to be linearly independent or linearly dependent, the computation involved once again indicates the link between vector spaces and linear equations.

■ Examples

$\mathbb{Q}[x]$ is a vector space over $\mathbb{Q}$; we will look at various subsets of $\mathbb{Q}[x]$ and will determine if they are linearly independent.

1. Let $S_1 = \{4x^2 + x, 3x + 1, 5\}$, we want to check if there exists a second linear combination of the elements of S_1 which is equal to 0. To this end, suppose $\alpha, \beta, \gamma \in \mathbb{Q}$ such that

$$0 = \alpha \cdot (4x^2 + x) + \beta \cdot (3x + 1) + \gamma \cdot (5).$$

The right side of the previous equation is equal to $4\alpha x^2 + (\alpha + 3\beta)x + (\beta + 5\gamma)$. Recall that the only way a polynomial can be equal to the zero polynomial is for all its coefficients to be equal to 0. Hence, α, β, γ must simultaneously be solutions of

$$4\alpha = 0, \quad \alpha + 3\beta = 0, \quad \beta + 5\gamma = 0.$$

It is not hard to see that the only solution to these equations is $\alpha = \beta = \gamma = 0$. Thus, there is only one way to express 0 as a linear combination of the vectors in S_1, so these vectors are indeed linearly independent.

2. Let $S_2 = \{x^2 + 1, 4, 2x^2\}$; if you noticed that

$$x^2 + 1 = \frac{1}{2} \cdot (2x^2) + \frac{1}{4} \cdot (4),$$

then it is clear that the vectors in S_2 are linearly dependent. If you did not notice this, then you could check if there is a second way to express 0 as a linear combination of the elements of S_2. Therefore, let us now suppose that $\alpha, \beta, \gamma \in \mathbb{Q}$ such that

$$0 = \alpha \cdot (x^2 + 1) + \beta \cdot (4) + \gamma \cdot (2x^2).$$

The polynomial on the right side is equal to $(\alpha + 2\gamma)x^2 + (\alpha + 4\beta)$. As a result, α, β, γ must be solutions of

$$\alpha + 2\gamma = 0, \quad \alpha + 4\beta = 0.$$

There are many solutions to these equations, in particular $\alpha = -4$, $\beta = 1$, $\gamma = 2$ is one of them. Thus, the vectors in S_2 are linearly dependent.

3. Let $S_3 = \{x, x^3, x^5\}$; if a linear combination of these vectors was equal to 0, then there exist α, β, $\gamma \in \mathbb{Q}$ such that

$$0 = \alpha \cdot x + \beta \cdot x^3 + \gamma \cdot x^5.$$

However, this immediately implies that $\alpha = \beta = \gamma = 0$. Therefore, there is only one way to express 0 as a linear combination of elements of S_3, so the elements in S_3 are linearly independent.

4. Let $S_4 = \{x + 1, x^3 + 1, x^5 + 1\}$; to check if these vectors are linearly independent, suppose there exist α, β, $\gamma \in \mathbb{Q}$ such that

$$0 = \alpha \cdot (x + 1) + \beta \cdot (x^3 + 1) + \gamma \cdot (x^5 + 1).$$

As a result,

$$\gamma x^5 + \beta x^3 + (\alpha + \beta + \gamma) = 0.$$

This immediately implies that $\alpha = \beta = \gamma = 0$. Thus, the elements in S_4 are linearly independent.

To obtain more experience with linearly independent sets and spanning sets, we will look at various subsets of $\mathbb{R}^2$.

■ Examples—Linearly Independent and Spanning Sets in $\mathbb{R}^2$

The set $\mathbb{R}^2$ is a vector space over $\mathbb{R}$. We will now list various subsets of $\mathbb{R}^2$ that are linearly independent and some that are spanning sets and will then make some observations.

Linearly Independent Sets	Spanning Sets
$\{(0, 1), (1, 0)\}$	$\{(0, 1), (1, 0)\}$
$\{(1, 2)\}$	$\{(0, 1), (1, 1), (1, 0)\}$
$\{(3, 4), (5, -1)\}$	$\{(2, 4), (3, 1), (5, -1)\}$
$\{(1, 2), (-2, 0)\}$	$\{(1, 2), (-2, 0)\}$

(Continued)

Linearly Independent Sets	Spanning Sets
$\{(0, 1)\}$	$\{(2, 1), (3, 1), (5, 0), (7, 3)\}$
$\{(1, 1)\}$	$\{(4, 5), (6, 7), (8, 5)\}$
$\{(-2, 6), (-2, 5)\}$	$\{(7, 4), (-8, 11)\}$

We can now make the following observations:

1. The number of elements in each of the linearly independent sets is less than or equal to the number of elements in each of the spanning sets.

2. The sets that are on both lists contain exactly two elements.

∎

When we intuitively think of the relative size of $\mathbb{R}^2$ compared to $\mathbb{R}$, we think of $\mathbb{R}^2$ as being two times as large as $\mathbb{R}$. At the same time, every subset of $\mathbb{R}^2$ in our example that is both linearly independent and spans $\mathbb{R}^2$ has two elements. This is no coincidence.

Similarly, if you look at various subsets of the vector space $\mathbb{Q}^3$ over $\mathbb{Q}$, then you will observe that every linearly independent subset has at most 3 elements and every spanning set has at least 3 elements. Therefore, those subsets of $\mathbb{Q}^3$ that are both linearly independent and span $\mathbb{Q}^3$ have exactly 3 elements. Again note that the number 3 agrees with our intuitive notion of what the relative size of $\mathbb{Q}^3$ is compared to $\mathbb{Q}$.

14.3 Basis and Dimension

Having looked at quite a few examples, we can now take a more general view of things. Suppose V is a vector space over a field F, and let $S = \{v_1, v_2, \ldots, v_n\}$ be a subset of V, which is both linearly independent and spans V. If $(\alpha_1, \alpha_2, \ldots, \alpha_n)$ is an n-tuple of elements of F, then we can associate to this n-tuple the element

$$w = \alpha_1 \cdot v_1 + \alpha_2 \cdot v_2 + \cdots + \alpha_n \cdot v_n \in V.$$

Since S spans V, every element of V can be obtained in this way from S and an n-tuple of elements of F. Furthermore, since S is linearly independent, Lemma 14.6(b) asserts that different n-tuples of elements of F always produce different elements of V. As a result, using S, there is now a one-to-one correspondence between the elements of V and n-tuples of elements of F. Having now seen that every element of V can be expressed uniquely using n elements of F, it makes sense to say that the relative size of V compared to F is equal to the number of elements in S. This appears to have successfully formalized the idea of relative size. However, one problem remains. If we believe that the relative size of V compared to F should be the number of elements in a set that is both linearly independent and spans V, we need to deal with the fact that V can have many different subsets which are both linearly independent and span V. Therefore, for the idea of relative size to truly makes sense, we must deal with the following.

Question: If a vector space V has various subsets that are simultaneously linearly independent and span V, do all these subsets have the same size?

This is a fundamental question, and its answer will appear in Corollary 14.10. In order to prove Corollary 14.10, we first need to prove Theorem 14.8. It asserts that, in a vector space, the number of elements in a spanning set cannot be smaller than the number of elements in a linearly independent set. Once we prove Theorem 14.8, it will be easy to prove Corollary 14.10. The combination of Theorem 14.8 and Corollary 14.10 is the foundation behind much of linear algebra. Not surprisingly, they are the main results of this section, and they will allow us to completely formalize and understand the concept of dimension.

Theorem 14.8. *Let V be a vector space over a field F. If $\{v_1, v_2, \ldots, v_s\}$ spans V and if $\{w_1, w_2, \ldots w_r\}$ is a linearly independent subset of V, then $s \geq r$.*

Once we prove Theorem 14.8, it will be easy to prove that any two subsets of V that are both linearly independent and spanning must have the same size. However, to prove Theorem 14.8, we first need

Lemma 14.9. *Let V be a vector space over F and let $e_1, e_2, \ldots, e_n, e, f_1, f_2, \ldots, f_m$ be elements of V such that*

(a) *the set $\{e_1, e_2, \ldots, e_n, e\}$ is linearly independent and*

(b) *the set $\{e_1, e_2, \ldots, e_n, f_1, f_2, \ldots, f_m\}$ spans V.*

Then there exists some $k \leq m$ such that the set

$$T = \{e_1, e_2, \ldots, e_n, e, f_1, f_2, \ldots, f_{k-1}, f_{k+1}, \ldots, f_m\}$$

also spans V. In other words, the element e can replace one of the f_j's in the set in (b) and the new set will still span V.

Proof. Since the set $\{e_1, e_2, \ldots, e_n, f_1, f_2, \ldots, f_m\}$ spans V, we know that there exist $\alpha_i, \beta_j \in F$ such that

(10) $$e = \alpha_1 \cdot e_1 + \alpha_2 \cdot e_2 + \cdots + \alpha_n \cdot e_n + \beta_1 \cdot f_1 + \beta_2 \cdot f_2 + \cdots + \beta_m \cdot f_m.$$

If all the β_j's in (10) were 0, equation (10) would reduce to

$$e = \alpha_1 \cdot e_1 + \alpha_2 \cdot e_2 + \cdots + \alpha_n \cdot e_n.$$

However, this contradicts the fact that the set $\{e_1, e_2, \ldots, e_n, e\}$ is linearly independent. Therefore, there exists some subscript $k \leq m$ such that $\beta_k \neq 0$. If we subtract e and $\beta_k \cdot f_k$ from both sides of (10) and apply Lemma 14.2(c), we obtain

$$-\beta_k \cdot f_k = \alpha_1 \cdot e_1 + \alpha_2 \cdot e_2 + \cdots + \alpha_n \cdot e_n + (-1) \cdot e +$$
$$\beta_1 \cdot f_1 + \beta_2 \cdot f_2 + \cdots + \beta_{k-1} \cdot f_{k-1} + \beta_{k+1} \cdot f_{k+1} + \cdots + \beta_m \cdot f_m.$$

Let $T = \{e_1, e_2, \ldots, e_n, e, f_1, f_2, \ldots, f_{k-1}, f_{k+1}, \ldots, f_m\}$; we would like to show that $f_k \in span(T)$. But when we multiply both sides of our previous equation by $(-\beta_k)^{-1}$, it becomes clear that f_k belongs to $span(T)$. Therefore, $span(T)$ contains every element of the original spanning set and Lemma 14.5 now asserts that T must also span V. □

Lemma 14.9 is the key to the proof of Theorem 14.8 as it allows us to replace, one element at a time, the elements of the spanning set $\{v_1, v_2, \ldots, v_s\}$ by elements of the linearly independent set $\{w_1, w_2, \ldots w_r\}$.

Proof of Theorem 14.8. By way of contradiction, let us assume that $s < r$. Since $\{w_1\}$ is a linearly independent set and $\{v_1, v_2, \ldots, v_s\}$ is a spanning set, Lemma 14.9 asserts that we can replace one of the v_j's by w_1 and still have a spanning set. By reordering the v_j's, we may assume that we can replace v_1 by w_1. Therefore, we may now assume that $\{w_1, v_2, \ldots, v_s\}$ is a spanning set.

But now, $\{w_1, w_2\}$ is a linearly independent set and $\{w_1, v_2, \ldots, v_s\}$ is a spanning set. Again applying Lemma 14.9, we can replace one of the remaining v_j's by w_2 and still have a spanning set. By reordering the remaining v_j's, we may assume that we can replace v_2 by w_2. Thus, $\{w_1, w_2, v_3, \ldots, v_s\}$ is a spanning set. Observe that we are now in the case where $\{w_1, w_2, w_3\}$ is a linearly independent set and $\{w_1, w_2, v_3, \ldots, v_s\}$ is a spanning set. Note that we can again apply Lemma 14.9 and reorder the remaining v_j's so that $\{w_1, w_2, w_3, v_4, \ldots, v_s\}$ is a spanning set.

Since we are assuming that $s < r$, we can continue to apply Lemma 14.9 and replace elements of the spanning set $\{v_1, v_2, \ldots, v_s\}$ by elements of the linearly independent set $\{w_1, w_2, \ldots w_r\}$ until we have exhausted all the v_j's. When we have reached the point that we have replaced every v_j, we see that the set $\{w_1, w_2, \ldots w_s\}$ is also a spanning set. But since $s < r$, we can talk about the element w_{s+1}. Since $\{w_1, w_2, \ldots w_s\}$ is a spanning set, w_{s+1} must be in the span of these other w_i's. However, by Lemma 14.6(a), no element of a linearly independent set can be in the span of the others and this contradicts the fact that $\{w_1, w_2, \ldots w_s, w_{s+1}\}$ is linearly independent. As a result, it is impossible for s to be less than r and $s \geq r$, as desired. □

We can now easily prove

Corollary 14.10.   *Let V be a vector space over a field F and let S and T be finite subsets of V, both of which are linearly independent and span V. Then S and T have the same number of elements.*

Proof. Let n be the number of elements in S and m the number of elements in T. Since S is a spanning set and T is linearly independent, Theorem 14.8 tells us that $n \geq m$. Changing our perspective, it is also the case that S is linearly independent and T is a spanning set. Now Theorem 14.8 tells us that $m \geq n$. Combining these two facts, we see that $n = m$, as desired. $\square$

In light of Corollary 14.10, all subsets of a vector space V that are linearly independent and span V must have the same size. Therefore, it now makes perfect sense to consider the relative size of a vector space V compared to the field F to be the number of elements in a subset of V that is both linearly independent and spanning. Clearly, sets that are both linearly independent and spanning are quite important, and this motivates the following:

Definition 14.11.   *Let V be a vector space over a field F. If the set $S = \{v_1, v_2, \ldots, v_n\}$ is both linearly independent and spans V, then we call S a **basis** of V. We call the number of elements in S the **dimension** of V over F.*

Given a vector space V over a field F, it does not have a unique basis. However, Corollary 14.10 tells us that every basis of V over F has the same number of elements. Thus, we have achieved our goal of showing that the concept of dimension is well defined. For example, observe that in the vector space $\mathbb{R}^2$ over $\mathbb{R}$, each of the sets

$$\{(1, 0), (0, 1)\}, \quad \{(2, 2), (3, 1)\}, \quad \{(-1, 3), (7, 4)\}$$

is a basis. Corollary 14.10 tells us that *every* basis of $\mathbb{R}^2$ over $\mathbb{R}$ has 2 elements in it. Thus, the dimension of $\mathbb{R}^2$ over $\mathbb{R}$ is 2.

■ Examples—Some Vector Spaces and Their Dimensions

In each of these examples, we will write down one basis for each vector space. However, keep in mind that for each of these examples, there are an infinite number of choices you could make for a basis. But always remember that every basis for a given vector space must have the same number of elements.

1. If F is a field, then F^n is a vector space over F. The set

$$\{(1, 0, 0, \ldots, 0), (0, 1, 0, \ldots, 0), (0, 0, 1, \ldots, 0), \ldots, (0, 0, 0, \ldots, 1)\}$$

is a basis. This set has n elements, thus F^n has dimension n over F. For example, $\{(1, 0), (0, 1)\}$ is a basis for $\mathbb{R}^2$ over $\mathbb{R}$, hence $\mathbb{R}^2$ has dimension 2 over $\mathbb{R}$. Similarly, $\{(1, 0, 0), (0, 1, 0), (0, 0, 1)\}$ is a basis for $\mathbb{Q}^3$ over $\mathbb{Q}$, so $\mathbb{Q}^3$ has dimension 3 over $\mathbb{Q}$.

2. $\mathbb{C}$ is a vector space over $\mathbb{R}$ and $\{1, i\}$ is a basis. Therefore, $\mathbb{C}$ has dimension 2 over $\mathbb{R}$.

3. $\mathbb{Q}(\sqrt{3}) = \{\alpha + \beta\sqrt{3} | \alpha, \beta \in \mathbb{Q}\}$ is a vector space over $\mathbb{Q}$ and $\{1, \sqrt{3}\}$ is certainly a spanning set. But to be a basis, it also needs to be linearly independent. To this end, suppose there exist $\alpha, \beta \in \mathbb{Q}$ such that $\alpha \cdot 1 + \beta \cdot \sqrt{3} = 0$. If $\beta \neq 0$, then $\sqrt{3} = -\frac{\alpha}{\beta} \in \mathbb{Q}$, which contradicts the fact that $\sqrt{3}$ is not rational. Therefore, $\beta = 0$, and it is now easy to see that $\alpha = 0$. Thus, both α and β must be 0, and the set $\{1, \sqrt{3}\}$ is indeed linearly independent. Hence, $\mathbb{Q}(\sqrt{3})$ has dimension 2 over $\mathbb{Q}$.

4. If F is a field, let $P_n(x)$ denote all the polynomials in $F[x]$ of degree less than n. Recall that we have adopted the convention that $P_n(x)$ contains the polynomial $f(x) = 0$. Then $P_n(x)$ is a vector space over F, and it is not hard to check that the set $\{1, x, x^2, \ldots, x^{n-1}\}$ is a basis. Thus, $P_n(x)$ has dimension n over F.

5. Let V be the polynomials in $P_n(x)$ with constant term 0. In this case, $\{x, x^2, \ldots, x^{n-1}\}$ is a basis for V over F, so V has dimension $n - 1$ over F.

6. If $n \geq 2$, let W be the polynomials in $P_n(x)$ that have 1 as a root. Every element of W must simultaneously have degree less than n and also have $x - 1$ as a factor. Therefore, if $f(x) \in W$, then $f(x) = q(x)(x - 1)$, where $q(x) = \alpha_0 + \alpha_1 x + \cdots + \alpha_{n-2}x^{n-2}$. We can now rewrite the equation $f(x) = q(x)(x - 1)$ as

$$f(x) = \alpha_0(x - 1) + \alpha_1 x(x - 1) + \cdots + \alpha_{n-2}x^{n-2}(x - 1).$$

Therefore, the set $\{(x - 1), x(x - 1), x^2(x - 1), \ldots, x^{n-2}(x - 1)\}$ spans W. Observe that this set is also linearly independent, for if

$$\alpha_0(x - 1) + \alpha_1 x(x - 1) + \cdots + \alpha_{n-2}x^{n-2}(x - 1) = 0,$$

then

$$(\alpha_0 + \alpha_1 x + \cdots + \alpha_{n-2}x^{n-2})(x - 1) = 0,$$

which implies that

$$\alpha_0 + \alpha_1 x + \cdots + \alpha_{n-2}x^{n-2} = 0.$$

Thus, each $\alpha_i = 0$, so $\{(x - 1), x(x - 1), x^2(x - 1), \ldots, x^{n-2}(x - 1)\}$ is also linearly independent. Since $\{(x - 1), x(x - 1), x^2(x - 1), \ldots, x^{n-2}(x - 1)\}$ has $n - 1$ elements, W has dimension $n - 1$ over F.

7. Earlier in this chapter, we looked at the vector space

$$V = \{(a, b) | a, b \in \mathbb{Q} \quad \text{and} \quad a + b = 0\}$$

and wondered if its dimension over $\mathbb{Q}$ should be 1 or 2. The set $\{(-1, 1)\}$ is a spanning set for V over $\mathbb{Q}$, and it also follows from Lemma 14.2(d) that any subset of a vector space consisting of only one element is linearly independent provided the element is not $\vec{0}$. Thus, the set $\{(-1, 1)\}$ is a basis for V over $\mathbb{Q}$ and V has dimension 1 over $\mathbb{Q}$.

8. Earlier in this chapter, we also examined the vector space

$$W = \{(a, b, c, d) | a, b, c, d \in \mathbb{Q} \ \text{ and } \ a + 2b + 5c - 3d = 0\}.$$

At the time, it was unclear if the dimension was 3, or 4, or some entirely different number. You can now check that $\{(-2, 1, 0, 0), (-5, 0, 1, 0), (3, 0, 0, 1)\}$ is a basis for W over $\mathbb{Q}$. Thus, W has dimension 3 over $\mathbb{Q}$.

Throughout this chapter, we have seen that linearly independent sets need not be spanning sets, and spanning sets need not be linearly independent. However, we will now show that if a vector space V has dimension n, then a set with n elements is linearly independent if and only if it spans V. In light of this, if you already know a vector space has dimension n, then to show that a set with n elements is a basis, you only need to verify that the set is either linearly independent or spans V. We record this as

Lemma 14.12. *Let V be a vector space of dimension n over the field F. Then a set $S = \{v_1, v_2, \ldots, v_n\}$ is linearly independent if and only if it spans V.*

Proof. In one direction, let us suppose that S is linearly independent. We need to show that every $w \in V$ belongs to $span(S)$. Certainly, if $w \in S$, then there is nothing to prove. On the other hand, if $w \notin S$, then the set $S \cup \{w\}$ contains $n + 1$ elements. However, since every basis of V contains n elements, V certainly contains subsets of size n that are spanning sets. Therefore, Theorem 14.10 implies that the set $S \cup \{w\}$ is not linearly independent. As a result, there exist $\alpha_i, \alpha \in F$, not all of which are 0, such that

(11)
$$\alpha_1 \cdot v_1 + \alpha_2 \cdot v_2 + \cdots + \alpha_n \cdot v_n + \alpha \cdot w = 0.$$

Observe that α cannot be 0, otherwise S would be linearly dependent. Since $\alpha \neq 0$, we can subtract $\alpha \cdot w$ from both sides of (14.11) and then multiply both sides by $-\alpha^{-1}$ to see that $w \in span(S)$. Thus, S does span V.

In the other direction, let us suppose that S spans V and we will show that S is linearly independent using a proof by contradiction. If S is not linearly independent, then Lemma 14.6 tells us that some $v_j \in S$ is in the span of the remaining elements. By reordering the elements of S, we may assume that v_1 is in the span of the set $T = \{v_2, v_3, \ldots, v_n\}$. However, since $v_1 \in span(T)$, T has the property that $span(T)$ contains the spanning set S and Lemma 14.5

asserts that T also spans V. Every basis of V is a linearly independent set with n elements, yet T is a spanning set with only $n-1$ elements. This contradicts Theorem 14.8, thus S is linearly independent. $\qquad\square$

In light of Lemma 14.12, if we already know that a vector space has dimension n, then it is often quite easy to determine whether a particular subset with n elements is a basis.

■ Examples

1. Let $P_4(x)$ be the polynomials of degree less than 4 is $\mathbb{Q}[x]$ and let $S = \{1, x+1, x^2+1, x^3+1\}$. Then

$$x = (x+1) - 1, \quad x^2 = (x^2+1) - 1, \quad x^3 = (x^3+1) - 1 \in span(S).$$

    Therefore, $span(S)$ contains the spanning set $\{1, x, x^2, x^3\}$. By Lemma 14.5, S spans $P_4(x)$. However, S has 4 elements and $P_4(x)$ has dimension 4 over $\mathbb{Q}$, so Lemma 14.12 tells us that S is a basis of $P_4(x)$.

2. Consider the subset $\{(2,0), (-4,3)\}$ of $\mathbb{R}^2$. If $\alpha, \beta \in \mathbb{R}$ have the property that $\alpha \cdot (2,0) + \beta \cdot (-4,3) = (0,0)$, then α, β must satisfy the equations

$$2\alpha - 4\beta = 0 \quad \text{and} \quad 3\beta = 0.$$

    It is easy to see that $\alpha = \beta = 0$ is the only solution, so $\{(2,0), (-4,3)\}$ is linearly independent. Since $\mathbb{R}^2$ has dimension 2 over $\mathbb{R}$ and our set has 2 elements, Lemma 14.12 implies that $\{(2,0), (-4,3)\}$ is a basis of $\mathbb{R}^2$.

There certainly exist vector spaces V whose dimension over F is not finite. For example, consider the vector space $\mathbb{Q}[x]$ over $\mathbb{Q}$. If we let $S = \{p_1(x), p_2(x), \ldots, p_n(x)\}$ be any finite subset of $\mathbb{Q}[x]$, then we can let m denote the largest exponent that occurs in any of the polynomials in S. Since the exponent x^{m+1} does not appear in any of the polynomials in S, it is clear that $x^{m+1} \notin span(S)$. Thus, S does not span $\mathbb{Q}[x]$. This tells us that no finite subset of $\mathbb{Q}[x]$ spans $\mathbb{Q}[x]$, so $\mathbb{Q}[x]$ cannot contain a finite subset that is both linearly independent and spans. In cases like this, we say that our vector space V is *infinite dimensional* over F. In the next result, we will see that the existence of a finite subset that spans is not only necessary but also sufficient for a vector space to have finite dimension.

Lemma 14.13. *Let V be a vector space over a field F. Then the dimension of V over F is finite if and only if V contains a spanning set which is finite.*

Proof. Clearly if the dimension of V is finite, then any basis of V is also a finite spanning set. In the other direction, suppose $S = \{v_1, v_2, \ldots, v_m\}$ is a finite set that spans V. The set S certainly contains subsets that are linearly independent, so we can let T be a subset of S of the largest possible size that is linearly independent. Our goal is to show that T also spans V, so it suffices to show that $span(T)$ contains S. By reordering the elements of S, we may assume that $T = \{v_1, v_2, \ldots, v_k\}$, where $k \le m$. If $T = S$, then certainly $span(T)$ contains S. On the other hand, if $T \ne S$, let $v \in S$ such that $v \notin T$. Observe that the set $T \cup \{v\}$ is a subset of S that is larger than T. By our choice of T, the set $T \cup \{v\}$ must be linearly dependent. As a result, there exist $\alpha_i, \alpha \in F$, not all of which are 0, such that

$$\alpha_1 \cdot v_1 + \alpha_2 \cdot v_2 + \cdots + \alpha_k \cdot v_k + \alpha \cdot v = \vec{0}.$$

Since T is linearly independent, it must be the case that $\alpha \ne 0$. Therefore, we can solve for v and see that v can be written as a linear combination of the elements of S. Thus, $v \in span(T)$, so $V = span(S) = span(T)$. Therefore, T is a basis of V, so V has dimension $k \le m$. $\square$

Observe that the proof of Lemma 14.13 really shows that if T is a linearly independent subset of a vector space such that any set that contains T and is not equal to T is linearly dependent, then T is actually a basis. This idea will appear again several times in the next section.

Exercises for Sections 14.2 and 14.3

In exercises 1–12, you will be working in $\mathbb{R}^4$, which is a vector space over $\mathbb{R}$. In each exercise, determine if the given set is

(a) a spanning set for $\mathbb{R}^4$,

(b) linearly independent over $\mathbb{R}$,

(c) both a spanning set and linearly independent, or

(d) neither a spanning set nor linearly independent.

Briefly explain your answer.

1. $\{(1, 2, 0, 0), \ (3, 0, 1, 0), \ (0, 0, 0, 1)\}$

2. $\{(4, 1, 3, 0), \ (0, 0, 1, 1), \ (4, 1, 5, 2)\}$

3. $\{(0, 1, 1, 1), \ (1, 2, 0, 0), \ (2, 0, -4, -4)\}$

4. $\{(1, 1, 1, 1), \ (0, 1, 2, 3), \ (0, 0, 1, 1)\}$

5. $\{(1, 0, 0, 0), \ (1, 1, 0, 0), \ (1, 1, 1, 0), \ (1, 1, 1, 1)\}$

6. $\{(2, 2, 0, 0),\ (-1, -1, -1, -1),\ (1, 2, 3, 4),\ (0, 0, 1, 1)\}$

7. $\{(4, 1, 0, 0),\ (-12, -3, 1, 0),\ (0, 0, 1, 0),\ (0, 0, 0, 1)\}$

8. $\{(1, 2, 2, 2),\ (0, 1, 1, 1),\ (0, 0, 1, 1),\ (0, 0, 2, 1)\}$

9. $\{(3, 0, 1, 1),\ (1, 0, 2, 2),\ (-1, 0, 1, -1),\ (2, 0, 3, 4),\ (4, 0, 1, 1)\}$

10. $\{(1, 0, 0, 0),\ (0, 2, 0, 0),\ (0, 0, 4, 4),\ (0, 0, 1, 2),\ (0, 0, -1, 0)\}$

11. $\{(2, 1, 2, 1),\ (0, 1, 0, 1),\ (0, 0, 0, 1), (1, 1, 1, 1),\ (0, 3, 0, 1)\}$

12. $\{(-1, 0, 1, 2),\ (0, 1, 0, 1),\ (2, 1, 0, 0),\ (0, 0, 1, 2),\ (0, 0, 1, -1)\}$

In exercises 13–24, you will be working in $P_3(x)$, the set of polynomials of degree less than 3 with coefficients in $\mathbb{Q}$. $P_3(x)$ is a vector space over $\mathbb{Q}$ and, in each exercise, determine if the given set is

(a) a spanning set for $P_3(x)$,

(b) linearly independent over $\mathbb{Q}$,

(c) both a spanning set and linearly independent, or

(d) neither a spanning set nor linearly independent.

Briefly explain your answer.

13. $\{x + 1,\ 2x^2 - 5\}$

14. $\{-x + 14,\ 2x - 28\}$

15. $\{7x + 11,\ 7x - 11\}$

16. $\{5x - 35,\ -2x + 14\}$

17. $\{1,\ x,\ x^2\}$

18. $\{x + 1,\ x^2 + 1,\ x^2 - x\}$

19. $\{x + 1,\ x + 2,\ x^2 + 3\}$

20. $\{2,\ x^2 - 3x + 4,\ 2x^2 - 6x\}$

21. $\{x^2 + x,\ x^2 + x + 1,\ x^2 + x + 2,\ x^2 + x + 3\}$

22. $\{x^2 - 1,\ x^2 + 1,\ x + 1,\ x - 1\}$

23. $\{2x^2 + 1,\ 4x^2 + x + 2,\ 3x,\ 10x^2 - 21x + 5\}$

24. $\{x - 5,\ x^2 - 25,\ x^2 - 6x + 5,\ x^2 - 10x + 25\}$

In exercises 25–30, you are given a subset of $\mathbb{R}^4$, which is also a vector space over $\mathbb{R}$. In each exercise, find the dimension and a basis. Keep in mind that there are many choices for a basis.

25. $\{(a, b, c, d) \in \mathbb{R}^4 | b = d = 0\}$

26. $\{(a, b, c, d) \in \mathbb{R}^4 | a = b = c = d\}$

27. $\{(a, b, c, d) \in \mathbb{R}^4 | a + b = 0 \text{ and } c - d = 0\}$

28. $\{(a, b, c, d) \in \mathbb{R}^4 | b - 2c + 3d = 0\}$

29. $\{(a, b, c, d) \in \mathbb{R}^4 | 2a - 3b + c - 8d = 0\}$

30. $\{(a, b, c, d) \in \mathbb{R}^4 | a = 0, b = 2c, \text{ and } 2b - 5c + 8d = 0\}$

In exercises 31–36, $P_4(x)$ will denote the polynomials of degree less than 4 with coefficients in $\mathbb{Q}$. In each exercise, you will be given a subset of $P_4(x)$ that is also a vector space over $\mathbb{Q}$ and you will need to find the dimension and a basis. Keep in mind that there are many choices for a basis.

31. $\{f(x) \in P_4(x) | f(0) = 0\}$

32. $\{f(x) \in P_4(x) | f(5) = f(10)\}$

33. $\{f(x) \in P_4(x) | f(1) = f(2)\}$

34. $\{f(x) \in P_4(x) | f(-1) = f(0) = f(1) = 0\}$

35. $\{f(x) \in P_4(x) | f(0) = f(1) = f(2)\}$

36. $\{f(x) \in P_4(x) | f(0) = f(1), \ f(2) = f(3), \text{ and } f(4) = f(5)\}$

In exercises 37–38, let $S = \{v_1, v_2, \ldots, v_n\} \subset V$, where V is a vector space over a field F.

37. Show that $span(S)$ is also a vector space over F.

38. Show that the dimension of $span(S)$ over F is at most n and the dimension is exactly n if and only if S is linearly independent.

In Section 14.1, we indicated that if $F \subseteq K$ are fields then K is always a vector space over F. In exercises 39–46, we examine the more general situation where $F \subseteq R$, F is a field, and R is a commutative ring.

39. Let $R = \{(a, b) | a, b \in \mathbb{Q}\}$ and let $F = \{(a, 0) | a \in \mathbb{Q}\}$. Observe that R is a commutative ring, where both addition and multiplication are done componentwise, and F is a field. Since R is a commutative ring, the first three properties of Definition 14.1 are satisfied. Prove that R is not a vector space over F by showing that property (4) of Definition 14.1 does not hold.

40. Let $R = \{(a, b)|a, b \in \mathbb{Q}\}$ and let $K = \{(a, a)|a \in \mathbb{Q}\}$. Similar to the situation in exercise 39, R is a commutative ring, K is field, and the first three properties of Definition 14.2 are automatically satisfied. Prove that R is a vector space over K by showing that property (4) of Definition 14.1 does hold.

41. Let $F \subseteq R$, where F is a field and R is a commutative ring. If we let e denote the multiplicative identity of F, show that R is a vector space over F if and only if e is also the multiplicative identity of R. Then examine how this result applies to the examples in exercises 39 and 40.

42. Suppose $F \subseteq R$, where F is a field and R is an integral domain. Show that R is a vector space over F.

43. Let $F \subseteq R$, where F is a field, R is a commutative ring, and R is a vector space over F. Suppose $r \in R$ and $p(t) = x^n + a_{n-1}x^{n-1} + \cdots + a_1 x + a_0 \in F[t]$ have the properties that $p(r) = 0$ and $p(t)$ is the monic polynomial of smallest degree in $F[t]$ which has r as a root. (Note that when plugging r into $p(t)$ or any other element of $F[t]$, the computations are done within the ring R.)

    (a) Show that if $a_0 = 0$, then r is a zero divisor in R.

    (b) Show that if $a_0 \neq 0$, then r has a multiplicative inverse in R.

44. Let $F \subseteq R$, where F is a field, R is a commutative ring, and R is a vector space of dimension n over F. Show that for every $r \in R$, there exists a monic $p(t) \in F[t]$ of degree at most n such that $p(r) = 0$.

45. In Chapter 7, we showed that if a ring is finite, then every nonzero element is either invertible or a zero divisor. We also saw that this does not hold for rings in general. Prove that if $F \subseteq R$, where F is a field, R is a commutative ring, and R is a finite dimensional vector space over F, then every nonzero element of R is either invertible or a zero divisor.

46. Suppose $F \subseteq R$, where F is a field and R is an integral domain. In light of exercise 42, we know that R is a vector space over F. Show that if R is finite dimensional over F, then R is a field.

47. Let $F \subseteq K$ be fields. Suppose $r \in K$ such that r is the root of some nonzero polynomial in $F[t]$ and let n denote the degree of the minimum polynomial for r over F. Show that the set $\{1, r, r^2, \ldots, r^{n-1}\}$ is linearly independent over F.

48. Use exercise 47 to show that the set $\{1, 2^{\frac{1}{7}}, 2^{\frac{2}{7}}, 2^{\frac{3}{7}}, 2^{\frac{4}{7}}, 2^{\frac{5}{7}}, 2^{\frac{6}{7}}\}$ is a subset of $\mathbb{R}$ which is linearly independent over $\mathbb{Q}$.

49. Let $r = \sqrt{2} + \sqrt{3}$.

    (a) Show that the minimum polynomial for r over $\mathbb{Q}$ has degree 4.

(b) Compute r^2 and r^3 and then consider the sets $S = \{1, r, r^2, r^3\}$ and $T = \{1, \sqrt{2}, \sqrt{3}, \sqrt{6}\}$, both of which are subsets of $\mathbb{R}$. Show that S and T have the same span over $\mathbb{Q}$.

(c) Use exercises 37 and 38 to show that if $span(T)$ represents the span of T over $\mathbb{Q}$, then $span(T)$ has dimension 4 over $\mathbb{Q}$.

50. If $\mathbb{Q}(\sqrt{2}) = \{a + b\sqrt{2} \mid a, b \in \mathbb{Q}\}$, use exercise 49 to show that $\sqrt{3} \notin \mathbb{Q}(\sqrt{2})$ and $\sqrt{6} \notin \mathbb{Q}(\sqrt{2})$.

51. If $\mathbb{Q}(\sqrt{3}) = \{a + b\sqrt{3} \mid a, b \in \mathbb{Q}\}$, use exercise 49 to show that $\sqrt{2} \notin \mathbb{Q}(\sqrt{3})$ and $\sqrt{6} \notin \mathbb{Q}(\sqrt{3})$.

52. If $\mathbb{Q}(\sqrt{6}) = \{a + b\sqrt{6} \mid a, b \in \mathbb{Q}\}$, use exercise 49 to show that $\sqrt{2} \notin \mathbb{Q}(\sqrt{6})$ and $\sqrt{3} \notin \mathbb{Q}(\sqrt{6})$.

53. Let $s = \sqrt{5} + \sqrt{7}$.

(a) Show that the minimum polynomial for s over $\mathbb{Q}$ has degree 4.

(b) Compute s^2 and s^3 and then consider the sets $S = \{1, s, s^2, s^3\}$ and $T = \{1, \sqrt{5}, \sqrt{7}, \sqrt{35}\}$, both of which are subsets of $\mathbb{R}$. Show that S and T have the same span over $\mathbb{Q}$.

(c) Use exercises 37 and 38 to show that if $span(T)$ represents the span of T over $\mathbb{Q}$, then $span(T)$ has dimension 4 over $\mathbb{Q}$.

54. If $\mathbb{Q}(\sqrt{5}) = \{a + b\sqrt{5} \mid a, b \in \mathbb{Q}\}$, use exercise 53 to show that $\sqrt{7} \notin \mathbb{Q}(\sqrt{5})$ and $\sqrt{35} \notin \mathbb{Q}(\sqrt{5})$.

55. If $\mathbb{Q}(\sqrt{7}) = \{a + b\sqrt{7} \mid a, b \in \mathbb{Q}\}$, use exercise 53 to show that $\sqrt{5} \notin \mathbb{Q}(\sqrt{7})$ and $\sqrt{35} \notin \mathbb{Q}(\sqrt{7})$.

56. If $\mathbb{Q}(\sqrt{35}) = \{a + b\sqrt{35} \mid a, b \in \mathbb{Q}\}$, use exercise 53 to show that $\sqrt{5} \notin \mathbb{Q}(\sqrt{35})$ and $\sqrt{7} \notin \mathbb{Q}(\sqrt{35})$.

57. Show that the set $S = \{2^{\frac{i}{12}} \mid 0 \le i \le 11\}$, which is a subset of $\mathbb{R}$, is linearly independent over $\mathbb{Q}$.

58. Let K be a field such that $\mathbb{Q} \subseteq K \subseteq \mathbb{R}$ and K is finite dimensional over $\mathbb{Q}$. Use exercise 57 to show that if K contains both $2^{\frac{1}{3}}$ and $2^{\frac{1}{4}}$, then the dimension of K over $\mathbb{Q}$ must be at least 12.

59. Show that the set $S = \{3^{\frac{i}{35}} \mid 0 \le i \le 34\}$, which is a subset of $\mathbb{R}$, is linearly independent over $\mathbb{Q}$.

60. Let L be a field such that $\mathbb{Q} \subseteq L \subseteq \mathbb{R}$ and L is finite dimensional over $\mathbb{Q}$. Use exercise 59 to show that if L contains both $3^{\frac{1}{5}}$ and $3^{\frac{1}{7}}$, then the dimension of L over $\mathbb{Q}$ must be at least 35.

61. Let V be a vector space of dimension n over the finite field F, where F has q elements. How many elements does V have?

14.4 Subspaces and Linear Equations

We can now begin to look more closely at the connection between vector spaces and linear equations.

Definition 14.14. *If V is a vector space over a field F, then a subset U of V is called a subspace if*

(a) $u_1 + u_2 \in U$, *for all $u_1, u_2 \in U$ and*

(b) $\alpha \cdot u \in U$, *for all $\alpha \in F$ and $u \in U$.*

Observe that if U is a subspace of V, then property (a) from Definition 14.14 tells us that we can always add elements of U and remain in U. Furthermore, property (b) from Definition 14.14 tells us that

$$0 \cdot u = \vec{0}, \ (-1) \cdot u = -u \in U,$$

for all $u \in U$. Therefore, U contains the additive identity of V and also contains the additive inverse of each of its own elements. Since addition in U inherits being associative and commutative from V, U is also a commutative group under addition. Furthermore, property (b) from Definition 14.14 tells us that we can always multiply an element of U by a scalar and remain in U. Scalar multiplication in U inherits properties 1–4 of Definition 14.1 from V, so U now satisfies all the properties of a vector space. As a result, any subspace of V is also a vector space over F.

■ Examples

1. If V is a vector space over F, then the sets $\{\vec{0}\}$ and V are both subspaces of V. Since $\{\vec{0}\}$ does not contain any nonzero vectors, we consider it to have dimension zero over F.

2. If S is any finite subset of V, then $span(S)$ is a subspace of V.

3. If we let $P_n(x)$ be the polynomials of degree less than n in $\mathbb{R}[x]$, then we have

$$P_1(x) \subseteq P_2(x) \subseteq P_3(x) \subseteq P_4(x) \subseteq \cdots$$

> Each $P_n(x)$ is a subspace of $\mathbb{R}[x]$. In fact, if $m \leq n$ then $P_m(x)$ is a subspace of $P_n(x)$.

∎

One would suspect that the dimension of a subspace cannot exceed the dimension of the original vector space. In fact, we have

Proposition 14.15. *Let V be a vector space over a field F.*

(a) *If V has dimension n over F and if U is a subspace of V, then the dimension of U is at most n. Furthermore, $U = V$ if and only if U also has dimension n over F.*

(b) *If $U_1, U_2, \ldots, U_m$ are subspaces of V, then the intersection $U_1 \cap U_2 \cap \cdots \cap U_m$ is also a subspace of V.*

Proof. For part (a), since V has dimension n, any linearly independent subset of U also has at most n elements. Since U certainly contains subsets which are linearly independent, we can let S be a linearly independent subset of U which is as large as possible. The argument used in the proof of Lemma 14.13 shows that S is a basis of U. Since S can have at most n elements, the dimension of U is at most n.

For the second piece of part (a), if $U = V$, then clearly U also has dimension n. On the other hand, if U has dimension n, then any basis S of U is a linearly independent subset of V with n elements. By Lemma 14.12, S spans V, so $U = span(S) = V$, as desired.

For part (b), since each U_i is a subspace, it follows that if $\alpha \in F$ and

$$u_1, u_2 \in U_1 \cap U_2 \cap \cdots \cap U_m,$$

then $u_1 + u_2$ and $\alpha \cdot u_1$ belong to each of the U_i's. Thus,

$$u_1 + u_2, \alpha \cdot u_1 \in U_1 \cap U_2 \cap \cdots \cap U_m$$

and we see that $U_1 \cap U_2 \cap \cdots \cap U_m$ satisfies the two conditions of Definition 14.14. As a result, $U_1 \cap U_2 \cap \cdots \cap U_m$ is also a subspace. □

In the next lemma, if $(\alpha_1, \alpha_2, \ldots, \alpha_n) \in F^n$, we define the function

$$T : F^n \to F$$

as

$$T(\vec{x}) = \alpha_1 \cdot x_1 + \alpha_2 \cdot x_2 + \cdots \alpha_n \cdot x_n,$$

for all $\vec{x} = (x_1, x_2, \ldots, x_n) \in F^n$. For example, using the element $(3, -2) \in \mathbb{R}^2$, we can define the function $T : \mathbb{R}^2 \to \mathbb{R}$ as $T((x_1, x_2)) = 3x_1 - 2x_2$, for all $(x_1, x_2) \in \mathbb{R}^2$.

Lemma 14.16. *If* $(\alpha_1, \alpha_2, \ldots, \alpha_n) \in F^n$, *then*

$$T(\vec{x} + \vec{y}) = T(\vec{x}) + T(\vec{y}) \text{ and } T(\alpha \cdot \vec{x}) = \alpha \cdot T(\vec{x}),$$

for all $\vec{x} = (x_1, x_2, \ldots, x_n), \vec{y} = (y_1, y_2, \ldots y_n) \in F^n$ *and* $\alpha \in F$. *Furthermore, the set* $U = \{\vec{x} \in F^n | T(\vec{x}) = 0\}$ *is a subspace of* F^n.

Proof. Observe that

$$T(\vec{x} + \vec{y}) = T((x_1 + y_1, x_2 + y_2, \ldots, x_n + y_n)) =$$

$$\alpha_1(x_1 + y_1) + \alpha_2(x_2 + y_2) + \cdots \alpha_n(x_n + y_n) =$$

$$(\alpha_1 \cdot x_1 + \alpha_2 \cdot x_2 + \cdots + \alpha_n \cdot x_n) + (\alpha_1 \cdot y_1 + \alpha_2 \cdot y_2 + \cdots + \alpha_n \cdot y_n) = T(\vec{x}) + T(\vec{y}).$$

Similarly, we also have

$$T(\alpha \cdot \vec{x}) = T((\alpha x_1, \alpha x_2, \ldots \alpha x_n)) = \alpha_1(\alpha x_1) + \alpha_2(\alpha x_2) + \cdots + \alpha_n(\alpha x_n) =$$

$$\alpha(\alpha_1 \cdot x_1 + \alpha_2 \cdot x_2 + \cdots + \alpha_n \cdot x_n) = \alpha \cdot T(\vec{x}).$$

To see that U is a subspace, if $\vec{x}, \vec{y} \in U$ and $\alpha \in F$, it follows that

$$T(\vec{x} + \vec{y}) = T(\vec{x}) + T(\vec{y}) = 0 + 0 = 0$$

and

$$T(\alpha \cdot \vec{x}) = \alpha \cdot T(\vec{x}) = \alpha \cdot 0 = 0.$$

Thus, $\vec{x} + \vec{y}, \alpha \cdot \vec{x} \in U$ and U is indeed a subspace of F^n. □

■ Examples

1. Let $T : \mathbb{R}^2 \to \mathbb{R}$ be the function $T((x_1, x_2)) = 4x_1 + x_2$, for all $(x_1, x_2) \in \mathbb{R}^2$. Then $U = \{(x_1, x_2) \in \mathbb{R}^2 | 4x_1 + x_2 = 0\}$ is a subspace of $\mathbb{R}^2$. Since $U \neq \mathbb{R}^2$, the dimension of U over $\mathbb{R}$ is less than 2. On the other hand, since $(1, -4) \in U$, the dimension of U over $\mathbb{R}$ is at least 1. Combining these two facts, U has dimension 1 over $\mathbb{R}$ and the set $\{(1, -4)\}$ is a basis.

2. Let $T : \mathbb{R}^3 \to \mathbb{R}$ be the function $T((x_1, x_2, x_3)) = x_1 + 2x_2 - 4x_3$, for all $(x_1, x_2, x_3) \in \mathbb{R}^3$. Then $U = \{(x_1, x_2, x_3) \in \mathbb{R}^3 | x_1 + 2x_2 - 4x_3 = 0\}$ is a subspace of $\mathbb{R}^3$. Since $U \neq \mathbb{R}^3$, the dimension of U over $\mathbb{R}$ is less than 3. On the other hand, the set $\{(4, 0, 1), (-2, 1, 0)\}$ is a linearly independent subset of U. Thus, the dimension of U over $\mathbb{R}$ is at least 2. Combining these facts, U has dimension 2 over $\mathbb{R}$ and $\{(4, 0, 1), (-2, 1, 0)\}$ is a basis. ■

We can begin applying the theory developed in this chapter to linear equations.

Corollary 14.17. *Let U be the set of n-tuples $\vec{x} = (x_1, x_2, \ldots, x_n) \in F^n$ such that*

$$\alpha_{11} \cdot x_1 + \alpha_{12} \cdot x_2 + \cdots + \alpha_{1n} \cdot x_n = 0,$$
$$\alpha_{21} \cdot x_1 + \alpha_{22} \cdot x_2 + \cdots + \alpha_{2n} \cdot x_n = 0,$$
$$\cdots$$
$$\cdots$$
$$\alpha_{m1} \cdot x_1 + \alpha_{m2} \cdot x_2 + \cdots + \alpha_{mn} \cdot x_n = 0,$$

where $\alpha_{ij} \in F$ and $1 \le i \le m, 1 \le j \le n$. Then U is a subspace of F^n. Furthermore, if the field F is infinite, then U is either an infinite set or consists only of the single element $\vec{0} = (0, 0, \ldots, 0)$.

Proof. For every $i \le m$, let $T_i : F^n \to F$ be the function

$$T_i(\vec{x}) = \alpha_{i1} x_1 + \alpha_{i2} x_2 + \cdots + \alpha_{in} x_n,$$

for all $\vec{x} \in F^n$. In addition, let

$$U_i = \left\{ \vec{x} \in F^n | T_i(\vec{x}) = 0 \right\}.$$

Then, by Lemma 14.16, each U_i is a subspace of F^n. Therefore, by Proposition 14.15, $U_1 \cap U_2 \cap \cdots \cap U_m$ is also a subspace of F^n. However, $U = U_1 \cap U_2 \cap \cdots \cap U_m$, so U is also a subspace of F^n.

Finally, suppose U contains an element $\vec{x} \ne \vec{0}$. By Lemma 14.2(d), $\alpha \cdot \vec{x}$ is a different element of F^n, for each different $\alpha \in F$. Thus, if F is infinite, then $\{\alpha \cdot \vec{x} | \alpha \in F\}$ is a subset of U that is also infinite. As a result, if F is infinite and U contains an element other than $\vec{0}$, then U must also be infinite. $\qquad\square$

We can now consider an example where our field is finite.

■ Examples

Let $T : \mathbb{Z}_5^3 \to \mathbb{Z}_5$ be the function

$$T((x_1, x_2, x_3)) = [3]_5 x_1 + [2]_5 x_2 + [1]_5 x_3,$$

where $(x_1, x_2, x_3) \in \mathbb{Z}_5^3$. Then

$$U = \left\{ (x_1, x_2, x_3) \in \mathbb{Z}_5^3 | [3]_5 x_1 + [2]_5 x_2 + [1]_5 x_3 = [0]_5 \right\}$$

is a subspace of $\mathbb{Z}_5{}^3$. Observe that the set $\{([1]_5, [1]_5, [0]_5), ([1]_5, [0]_5, [2]_5)\}$ is a basis for U. Therefore, every element of U can be written uniquely as

$$\alpha \cdot ([1]_5, [1]_5, [0]_5) + \beta \cdot ([1]_5, [0]_5, [2]_5),$$

where $\alpha, \beta \in \mathbb{Z}_5$. Since there are $5 \cdot 5 = 25$ different pairs of elements (α, β), where $\alpha, \beta \in \mathbb{Z}_5$, we see that U contains exactly 25 elements.

∎

The collection of equations displayed in Corollary 14.17 is often referred to as a **system of linear equations**. In these equations, $x_1, x_2, \ldots, x_n$ are referred to as **variables** or as **unknowns**. Therefore, we say that Corollary 14.17 deals with a system of m linear equations in n unknowns. If the right side of every equation in a system of linear equations is 0, we say that the system is **homogeneous**.

Given a system of homogeneous linear equations, it is clear that $\vec{0}$ is always a solution. However, a system of homogeneous linear equations need not have any solutions other than $\vec{0}$. For example, the system of equations

$$x_1 + x_2 = 0 \quad \text{and} \quad x_1 + 2x_2 = 0$$

certainly has no solution other than $\vec{0} = (0, 0)$. This system consists of two equations and two unknowns. An important application of Theorem 14.8 is that if the number of unknowns exceeds the number of equations, then a system of homogeneous linear equations must have a solution other than $\vec{0}$. Deriving this fact from Theorem 14.8 will require an interesting change in perspective, as we will temporarily need to think of the x_i's as scalars and not as unknowns.

Theorem 14.18. *Consider the following system of m homogeneous linear equation in n unknowns:*

$$\alpha_{11} \cdot x_1 + \alpha_{12} \cdot x_2 + \cdots + \alpha_{1n} \cdot x_n = 0,$$
$$\alpha_{21} \cdot x_1 + \alpha_{22} \cdot x_2 + \cdots + \alpha_{2n} \cdot x_n = 0,$$
$$\cdots$$
$$\cdots$$
$$\alpha_{m1} \cdot x_1 + \alpha_{m2} \cdot x_2 + \cdots + \alpha_{mn} \cdot x_n = 0,$$

where $\alpha_{ij} \in F$ and $1 \leq i \leq m, 1 \leq j \leq n$. If the number of unknowns exceeds the number of equations, then there exist solutions other than $\vec{0}$ in F^n.

Proof. Given $\alpha_{ij} \in F$, where $1 \le i \le m, 1 \le j \le n$, let

$$\vec{a}_1 = (\alpha_{11}, \alpha_{21}, \ldots, \alpha_{m1}), \vec{a}_2 = (\alpha_{12}, \alpha_{22}, \ldots, \alpha_{m2}), \ldots,$$

$$\vec{a}_n = (\alpha_{1n}, \alpha_{2n}, \ldots, \alpha_{mn}).$$

Since each $\vec{a}_j$ is an m-tuple of elements of F, we can consider $\{\vec{a}_1, \vec{a}_2, \ldots, \vec{a}_n\}$ to be a set of n vectors belonging to the vector space F^m.

Since F^m has dimension m over F and $n > m$, Theorem 14.8 asserts that the set $\{\vec{a}_1, \vec{a}_2, \ldots, \vec{a}_n\}$ must be linearly dependent. Therefore, there exist $x_1, x_2, \ldots x_n \in F$, not all of which are 0, such that

(12) $$x_1 \cdot \vec{a}_1 + x_2 \cdot \vec{a}_2 + \cdots + x_n \cdot \vec{a}_n = \vec{0}.$$

It is important to keep in mind that both sides of equation (12) belong to F^m. Therefore, we can rewrite equation (12) in a form that explicitly shows the m components of both sides of equation (12). Since this is an essential part of the proof, you should be careful to check that you understand where all these components come from. Equation (12) can now be rewritten as

$$(\alpha_{11} \cdot x_1 + \alpha_{12} \cdot x_2 + \cdots + \alpha_{1n} \cdot x_n, \alpha_{21} \cdot x_1 + \alpha_{22} \cdot x_2 + \cdots + \alpha_{2n} \cdot x_n, \ldots,$$

$$\alpha_{m1} \cdot x_1 + \alpha_{m2} \cdot x_2 + \cdots + \alpha_{mn} \cdot x_n) = (0, 0, \ldots, 0).$$

Since each component in the preceding equation is equal to 0, it is now clear that $\vec{x} = (x_1, x_2, \ldots, x_n)$ is a solution to all m homogeneous linear equations. Furthermore, since at least one of $x_1, x_2, \ldots x_n$ is not zero, $\vec{x} \ne \vec{0}$ in F^n. $\square$

We now turn our attention to systems of linear equations that are not necessarily homogeneous. Consider the following system of linear equations:

$$\alpha_{11} \cdot x_1 + \alpha_{12} \cdot x_2 + \cdots + \alpha_{1n} \cdot x_n = \beta_1,$$

$$\alpha_{21} \cdot x_1 + \alpha_{22} \cdot x_2 + \cdots + \alpha_{2n} \cdot x_n = \beta_2,$$

$$\cdots$$

$$\cdots$$

$$\alpha_{m1} \cdot x_1 + \alpha_{m2} \cdot x_2 + \cdots + \alpha_{mn} \cdot x_n = \beta_m,$$

where $\alpha_{ij} \in F$, $\beta_i \in F$, and $1 \le i \le m$, $1 \le j \le n$. In this case, we refer to the homogeneous linear equations

$$\alpha_{11} \cdot x_1 + \alpha_{12} \cdot x_2 + \cdots + \alpha_{1n} \cdot x_n = 0,$$
$$\alpha_{21} \cdot x_1 + \alpha_{22} \cdot x_2 + \cdots + \alpha_{2n} \cdot x_n = 0,$$
$$\cdots$$
$$\cdots$$
$$\alpha_{m1} \cdot x_1 + \alpha_{m2} \cdot x_2 + \cdots + \alpha_{mn} \cdot x_n = 0$$

as the **corresponding system of homogeneous linear equations**. In our next result, we will show that in order to find all the solutions of a system of linear equations, it suffices to find one solution provided you can also find all the solutions of the corresponding system of homogeneous linear equations.

Corollary 14.19.   *Let W be the set of n-tuples $\vec{x} = (x_1, x_2, \ldots, x_n)$ such that*

$$\alpha_{11} \cdot x_1 + \alpha_{12} \cdot x_2 + \cdots + \alpha_{1n} \cdot x_n = \beta_1,$$
$$\alpha_{21} \cdot x_1 + \alpha_{22} \cdot x_2 + \cdots + \alpha_{2n} \cdot x_n = \beta_2,$$
$$\cdots$$
$$\cdots$$
$$\alpha_{m1} \cdot x_1 + \alpha_{m2} \cdot x_2 + \cdots + \alpha_{mn} \cdot x_n = \beta_m,$$

where $\alpha_{ij} \in F$, $\beta_i \in F$, and $1 \le i \le m$, $1 \le j \le n$. If $\vec{y} \in W$ and U is the set of solutions of the corresponding homogeneous linear equations, then every element in W is of the form $\vec{y} + \vec{u}$, where $\vec{u} \in U$. Conversely, if $\vec{y} \in W$ and $\vec{u} \in U$, then $\vec{y} + \vec{u} \in W$. Furthermore, if the field F is infinite and W contains more than one element, then W is an infinite set.

Proof. As in the proof of Corollary 14.17, for every $i \le m$, we let $T_i : F^n \to F$ be the function

$$T_i(\vec{x}) = \alpha_{i1} x_1 + \alpha_{i2} x_2 + \cdots + \alpha_{in} x_n,$$

for all $\vec{x} \in F^n$. Observe that $\vec{x} \in W$ if and only if $T_i(\vec{x}) = \beta_i$, for all $i \le m$.

Now suppose $\vec{y}$ is a fixed element of W. If $\vec{x} \in W$, by applying both parts of Lemma 14.16, we see that

$$T_i(\vec{x} - \vec{y}) = T_i(\vec{x} + (-1)y) = T_i(\vec{x}) + T_i((-1)y) = T_i(\vec{x}) - T_i(\vec{y}) = \beta_i - \beta_i = 0,$$

for all $i \le m$. This means that $\vec{x} - \vec{y} \in U$, so $\vec{x} - \vec{y} = \vec{u}$, for some $\vec{u} \in U$. Thus, $\vec{x} = \vec{y} + \vec{u}$.

Conversely, if $\vec{y} \in W$ and $\vec{u} \in U$, then it follows from Lemma 14.16 that

$$T_i(\vec{y} + \vec{u}) = T_i(\vec{y}) + T_i(\vec{u}) = \beta_i + 0 = \beta_i,$$

for all $i \le m$. Thus, $\vec{y} + \vec{u} \in W$.

In addition, if W has more than one element, let $\vec{x}, \vec{y} \in W$ such that $\vec{x} \ne \vec{y}$. Then our preceding argument shows that $\vec{u} = \vec{x} - \vec{y}$ is a nonzero element of U. Using Corollary 14.17, if F is infinite, then so is the set U. Since F^n is a group under addition, a now familiar argument tells us that if $\vec{y} \in W$ and $\vec{u}_1, \vec{u}_2 \in U$ such that $\vec{u}_1 \ne \vec{u}_2$, then $\vec{y} + \vec{u}_1 \ne \vec{y} + \vec{u}_2$. Therefore, if U is an infinite set, so is W. $\qquad\square$

■ Examples

1. In $\mathbb{R}$, consider the linear equation $2x_1 + 3x_2 = 7$. In this case, the corresponding homogeneous linear equation is $2x_1 + 3x_2 = 0$. The set of solutions of this homogeneous linear equation is a subspace of $\mathbb{R}^2$ of dimension less than 2. Since $(3, -2)$ is a solution to this equation, the solutions of this equation are a 1-dimensional subspace of $\mathbb{R}^2$ with basis $\{(3, -2)\}$. It is easy to see that $(2, 1)$ is a solution of $2x_1 + 3x_2 = 7$. Therefore, Corollary 14.19 tells us that all solutions of $2x_1 + 3x_2 = 7$ are of the form

    $$(2, 1) + \alpha \cdot (3, -2) = (2 + 3\alpha, 1 - 2\alpha),$$

    where $\alpha \in \mathbb{R}$.

2. In $\mathbb{Q}$, consider the linear equation $5x_1 - 2x_2 + 4x_3 = 9$. The set of solutions of the corresponding homogeneous linear equation $5x_1 - 2x_2 + 4x_3 = 0$ must be a subspace of $\mathbb{Q}^3$ of dimension less than 3. On the other hand, $(2, 5, 0)$ and $(0, 2, 1)$ are two linearly independent solutions of $5x_1 - 2x_2 + 4x_3 = 0$. Thus, the solutions of $5x_1 - 2x_2 + 4x_3 = 0$ are a 2-dimensional subspace of $\mathbb{Q}^3$ with basis $\{(2, 5, 0),$ $(0, 2, 1)\}$. Furthermore, $(1, 0, 1)$ is a solution of $5x_1 - 2x_2 + 4x_3 = 9$. Therefore, Corollary 14.19 asserts that all solutions of $5x_1 - 2x_2 + 4x_3 = 9$ are of the form

    $$(1, 0, 1) + \alpha \cdot (2, 5, 0) + \beta \cdot (0, 2, 1) = (1 + 2\alpha, 5\alpha + 2\beta, 1 + \beta),$$

    where $\alpha, \beta \in \mathbb{Q}$.

Exercises for Section 14.4

In exercises 1–20, you will be examining various subsets of $\mathbb{R}[x]$. In each case, determine if the set is a subspace of $\mathbb{R}[x]$ over $\mathbb{R}$. If the set is not a subspace, provide a reason. If the set is a subspace and is also finite dimensional, compute the dimension.

1. Polynomials in $\mathbb{R}[x]$ of degree less than 10.

2. Polynomials in $\mathbb{R}[x]$ of degree less than 10 with a constant term of 0.

3. Polynomials in $\mathbb{R}[x]$ of odd degree.

4. Polynomials in $\mathbb{R}[x]$ of even degree.

5. Polynomials in $\mathbb{R}[x]$ where the coefficient of every even power of x is 0.

6. Polynomials in $\mathbb{R}[x]$ of degree less than 8 where the coefficient of every even power of x is 0.

7. Polynomials in $\mathbb{R}[x]$ of degree less than 7 that are also multiples of x^5.

8. Polynomials in $\mathbb{R}[x]$ of degree 1.

9. Polynomials in $\mathbb{R}[x]$ which have no real roots.

10. Polynomials in $\mathbb{R}[x]$ of degree less than 5 that have no real roots.

11. Polynomials in $\mathbb{R}[x]$ where the coefficient of x^3 is 0.

12. Polynomials in $\mathbb{R}[x]$ of degree less than 5 where the coefficient of x^3 is 0.

13. Polynomials in $\mathbb{R}[x]$ where the coefficient of x^3 is not 0.

14. Polynomials in $\mathbb{R}[x]$ where the coefficients of x and x^2 are different or are both equal to 0.

15. Polynomials in $\mathbb{R}[x]$ where the coefficients of x, x^2, and x^3 are all the same.

16. Polynomials in $\mathbb{R}[x]$ of degree less than 7 where the coefficients of x, x^2, and x^3 are all the same.

17. Polynomials in $\mathbb{R}[x]$ that are multiples of $x^2 + 1$.

18. Polynomials in $\mathbb{R}[x]$ that are multiples of $x^2 + 1$ and have degree less than 20.

19. Polynomials in $\mathbb{R}[x]$ that are monic.

20. Polynomials in $\mathbb{R}[x]$ that have more than one real root.

21. Let V be a vector space over a field F of dimension n. If $m \in \mathbb{N}$ and $m < n$, show that V contains a subspace of dimension m over F.

In exercises 22–26, we generalize some of the ideas from Lemma 14.16. Let V, W be vector spaces over a field F and let $T : V \rightarrow W$ be a function such that $T(v_1 + v_2) = T(v_1) + T(v_2)$ and $T(\alpha \cdot v_1) = \alpha \cdot T(v_1)$, for all $v_1, v_2 \in V$ and $\alpha \in F$. We call T a **linear transformation**. Since V and W are groups under addition, observe that T is a homomorphism of groups that also preserves scalar multiplication.

22. If $v_1, \ldots, v_n \in V$ and $\alpha_1, \ldots, \alpha_n \in F$, show that $T(\alpha_1 \cdot v_1 + \cdots + \alpha_n \cdot v_n) = \alpha_1 \cdot T(v_1) + \cdots + \alpha_n \cdot T(v_n)$.

23. Let $Ker(T) = \{v \in V | T(v) = 0\}$; show that $Ker(T)$ is a subspace of V.

24. Let $Im(T) = \{w \in W |$ there exists $v \in V$ such that $w = T(v)\}$; show that $Im(T)$ is a subspace of W.

25. Suppose $w_1, w_2, \ldots, w_n \in Im(T)$ are linearly independent in W and let $v_1, v_2, \ldots, v_n \in V$ such that $T(v_i) = w_i$, for $1 \leq i \leq n$. Show that the set $\{v_1, v_2, \ldots, v_n\}$ is linearly independent in V.

26. Suppose V has dimension n over F.
    (a) Show that $Im(T)$ has dimension less than or equal to n.

    (b) If we let m denote the dimension of $Im(T)$, show that there exists a linearly independent subset $\{v_1, v_2, \ldots, v_m\}$ of V such that $\{T(v_1), T(v_2), \ldots, T(v_m)\}$ is a basis for $Im(T)$.

    (c) Show that $Ker(T)$ has dimension less than or equal to n.

    (d) If we let l denote the dimension of $Ker(T)$ and let $\{u_1, u_2, \ldots, u_l\}$ be a basis for $Ker(T)$, show that $\{v_1, v_2, \ldots, v_m, u_1, u_2, \ldots, u_l\}$ is a basis for V.

    (e) Conclude that if V is finite dimensional, then the dimension of V is equal to the dimension of $Im(T)$ plus the dimension of $Ker(T)$.

In exercises 53 and 54 following Section 8.1, you saw that a group cannot be the union of two proper subgroups. Since every vector space is also a group under addition, it immediately follows that a vector space cannot be the union of two proper subspaces. In exercises 27–30, we will prove the much stronger result that a vector space cannot be the union of any finite number of proper subspaces, provided the underlying field is infinite.

27. Let V be a vector space over an infinite field F and let $U_1, U_2, \ldots, U_n$ be subspaces of V. Suppose $u, v \in V$ such that $u + \alpha \cdot v \in U_1 \cup U_2 \cup \cdots \cup U_n$, for all $\alpha \in F$. Show that there is some $i \leq n$ such that both u and v belong to U_i.

28. Let $U_1, U_2, \ldots, U_n, U_{n+1}$ be subspaces of a vector space V over an infinite field F. If $u, v \in V$ such that $u \notin U_1 \cup U_2 \cup \cdots \cup U_n$ and $v \notin U_{n+1}$, show that there is some $\alpha \in F$ such that $u + \alpha \cdot v \notin U_1 \cup U_2 \cup \cdots \cup U_n \cup U_{n+1}$.

29. Let V be a vector space over an infinite field F and let $U_1, U_2, \ldots, U_n$ be subspaces of V, none of which is equal to V. Show that $V \neq U_1 \cup U_2 \cup \cdots \cup U_n$. If you try to prove this using Mathematical Induction, you should first take a look at exercises 27 and 28.

30. Let V be a vector space over an infinite field F and let $U_1, U_2, \ldots, U_n$ be subspaces of V. If $U_1 \cup U_2 \cup \cdots \cup U_n$ is a subspace of V, show that there is some $m \leq n$ such that U_m contains every U_i, where $i \leq n$.

31. The solutions of

$$3x + 8y + 2z = 0$$

are a subspace of $\mathbb{R}^3$. Find the dimension and a basis for this subspace.

32. The solutions of

$$5x - 6y + 5z = 0$$

are a subspace of $\mathbb{R}^3$. Find the dimension and a basis for this subspace.

33. The solutions of

$$4x - 5y + 2z = 0$$
$$7x + 10y - 3z = 0$$

are a subspace of $\mathbb{R}^3$. Find the dimension and a basis for this subspace.

34. The solutions of

$$2x + 3y + 2z = 0$$
$$-4x + 9y - z = 0$$

are a subspace of $\mathbb{R}^3$. Find the dimension and a basis for this subspace.

35. The solutions of

$$-x + 3y + 2z + 11w = 0$$

are a subspace of $\mathbb{R}^4$. Find the dimension and a basis for this subspace.

36. The solutions of

$$5x - 8y + 10z - 7w = 0$$

are a subspace of $\mathbb{R}^4$. Find the dimension and a basis for this subspace.

37. The solutions of

$$11x - 5y + 6z + 2w = 0$$
$$6x - 4z + 5w = 0$$

are a subspace of $\mathbb{R}^4$. Find the dimension and a basis for this subspace.

38. The solutions of

$$-7x + 4y + 18z - 5w = 0$$
$$4x - y + 9w = 0$$

are a subspace of $\mathbb{R}^4$. Find the dimension and a basis for this subspace.

39. The solutions of

$$x - 4y + 2z + 3w = 0$$
$$6x + 2y - z + 5w = 0$$
$$-5y + 7w = 0$$

are a subspace of $\mathbb{R}^4$. Find the dimension and a basis for this subspace.

40. The solutions of

$$7x + 4y - 8z - 10w = 0$$
$$-4x + 3y + 5z + w = 0$$
$$-x + 11y - 14z + 2w = 0$$

are a subspace of $\mathbb{R}^4$. Find the dimension and a basis for this subspace.

41. Find a solution of

$$3x + 8y + 2z = 3$$

in $\mathbb{R}^3$. Then use your answer from exercise 31 to find all solutions of this equation.

42. Find a solution of

$$5x - 6y + 5z = -12$$

in $\mathbb{R}^3$. Then use your answer from exercise 32 to find all solutions of this equation.

43. Find a solution of

$$4x - 5y + 2z = -90$$
$$7x + 10y - 3z = 7$$

in $\mathbb{R}^3$. Then use your answer from exercise 33 to find all solutions of these equations.

44. Find a solution of

$$2x + 3y + 2z = 5$$
$$-4x + 9y - z = 50$$

in $\mathbb{R}^3$. Then use your answer from exercise 34 to find all solutions of these equations.

45. Find a solution of

$$-x + 3y + 2z + 11w = 32$$

in $\mathbb{R}^4$. Then use your answer from exercise 35 to find all solutions of this equation.

46. Find a solution of

$$5x - 8y + 10z - 7w = 73$$

in $\mathbb{R}^4$. Then use your answer from exercise 36 to find all solutions of this equation.

47. Find a solution of

$$11x - 5y + 6z + 2w = 14$$
$$6x - 4z + 5w = -6$$

in $\mathbb{R}^4$. Then use your answer from exercise 37 to find all solutions of these equations.

48. Find a solution of

$$-7x + 4y + 18z - 5w = -60$$
$$4x - y + 9w = 86$$

in $\mathbb{R}^4$. Then use your answer from exercise 38 to find all solutions of these equations.

49. Find a solution of

$$x - 4y + 2z + 3w = 5$$
$$6x + 2y - z + 5w = 69$$
$$-5y + 7w = 29$$

in $\mathbb{R}^4$. Then use your answer from exercise 39 to find all solutions of these equations.

50. Find a solution of

$$7x + 4y - 8z - 10w = 31$$
$$-4x + 3y + 5z + w = -9$$
$$-x + 11y - 14z + 2w = 149$$

in $\mathbb{R}^4$. Then use your answer from exercise 40 to find all solutions of these equations.

CHAPTER 15

Degrees and Galois Groups of Field Extensions

In Chapter 16, we will prove that $60°$ angles cannot be trisected with a ruler and compass. To do this, for every real number that can be constructed, we will associate a field K. It will turn out that K is a finite dimensional vector space over $\mathbb{Q}$, and we will need to determine the dimension of K over $\mathbb{Q}$.

Then, in Chapter 17, we will prove the insolvability of the quintic. In this situation, to every polynomial $f(x) \in \mathbb{Q}[x]$, we will associate a field L such that L is also finite dimensional over $\mathbb{Q}$. However, not only will we be interested in the dimension of L over $\mathbb{Q}$, but we will be even more concerned with the structure of the Galois group $Gal(L/\mathbb{Q})$.

As we can see, fundamental and concrete questions in abstract algebra can often be reduced to questions involving fields, dimensions of vector spaces, and groups of automorphisms. Since the dimension of a field compared to a subfield is often related to the degree of a polynomial, we adopt the following terminology:

Definition 15.1. *If $F \subseteq K$ are fields, we call K a field extension of F. The dimension of K as a vector space over F is often referred to as the degree of K over F and denoted as $[K : F]$. Furthermore, if $[K : F]$ is finite, we say that K is a finite extension of F.*

In light of our goals and the fact that we now often use the word *degree* in place of *dimension*, it is clear that we need to turn our attention to the degrees and Galois groups of field extensions.

15.1 Degrees of Field Extensions

Most of the fields we will deal with are related to roots of polynomials in $\mathbb{Q}[x]$. Having proven the Fundamental Theorem of Algebra in Chapter 6, we know that all these roots belong to $\mathbb{C}$. Therefore, virtually all of our fields K will have the property that $\mathbb{Q} \subseteq K \subseteq \mathbb{C}$. Proofs of the Fundamental Theorem of Algebra that rely almost entirely on algebra require an enormous amount of algebraic machinery. Therefore, in most textbooks, a proof of the insolvability of the quintic appears before a proof of the Fundamental Theorem of Algebra. That approach makes developing the Galois Theorem needed to prove the insolvability of the quintic very

573

abstract. However, we have the Fundamental Theorem of Algebra at our disposal and therefore know that the relevant fields lie between $\mathbb{Q}$ and $\mathbb{C}$. This will help in making our discussion of the Galois Theorem and the insolvability of the quintic relatively concrete.

In order to be a field, a set and its two operations need to satisfy nine axioms. Therefore, checking if a set is a field can sometimes be quite tedious. That can make it difficult to generate a large collection of examples. However, the next result will make it quite easy to generate examples of fields that lie between $\mathbb{Q}$ and $\mathbb{C}$.

Lemma 15.2. *Let $\mathbb{Q} \subseteq K \subseteq L \subseteq \mathbb{C}$ be such that K is a field and L is a finite dimensional vector space over K. Then L is a field if and only if it is closed under multiplication.*

Proof. One direction is clear, for if L is a field, then it must be closed under multiplication. The heart of this result is the other direction. By virtue of being a vector space, L automatically satisfies all the field axioms that deal solely with addition. Furthermore, since L is a subset of $\mathbb{C}$, multiplication in L satisfies the associative and commutative laws. For the same reason, the distributive laws are also satisfied. In addition, since L contains $\mathbb{Q}$, it contains a multiplicative identity. Therefore, even without using the fact that L is finite dimensional over K, we have succeeded in showing that L is a commutative ring.

At this point, in order to be a field, all that remains is to show that every nonzero element of L has a multiplicative inverse. It is easy to jump the gun here. Since L is contained in $\mathbb{C}$ and $\mathbb{C}$ is a field, it is indeed the case that every nonzero element of L has a multiplicative inverse that is contained in $\mathbb{C}$. But that is not enough. We need that the multiplicative inverse belongs to L. This is where we use that L is finite dimensional over K.

Suppose $b \in L$ is nonzero; we know $b^{-1} \in \mathbb{C}$, and we need to show $b^{-1} \in L$. Let n denote the dimension of L over K and then consider the elements $1, b, b^2, \ldots, b^n \in L$. Since any collection of more than n elements of L must be linearly independent over K, there exist $\alpha_i \in K$, not all of which are zero, such that

$$\alpha_0 + \alpha_1 b + \cdots + \alpha_{n-1} b^{n-1} + \alpha_n b^n = 0.$$

Let s be the smallest integer such that $\alpha_s \neq 0$. Note that $s \geq 0$ and $s < n$. For every i, we can let $\gamma_i = -\alpha_s^{-1} \alpha_i$, then multiplying the preceding equation by $\alpha_s^{-1}(b^{-1})^{s+1}$ results in

$$b^{-1} - \gamma_{s+1} - \gamma_{s+2} b - \cdots - \gamma_n b^{n-s-1} = 0.$$

This can be rewritten as

$$b^{-1} = \gamma_{s+1} + \gamma_{s+2} b + \cdots + \gamma_n b^{n-s-1},$$

so b^{-1} is a linear combination over K of elements of L. Hence, $b^{-1} \in L$, as desired. $\square$

Observe that in order to use Lemma 15.2, we do not need to know the dimension of L, but we merely need to know that it is finite dimensional over K. Lemma 14.13 indicated that for L to be finite dimensional, it suffices to have a finite spanning set. Thus, in many of our applications of Lemma 15.2, L will initially be described as the span over K of some finite subset of $\mathbb{C}$. In order to show that L is a field, it will suffice to show that L is closed under multiplication. But since L is the span of a finite set, this will merely require that the product of any two elements of the spanning set also belongs to L. This is typically a fairly straightforward computation. Thus, it will be easy to generate many examples of fields. Also, in many of our applications of Lemma 15.2, we will be in the special case where $K = \mathbb{Q}$.

■ Examples

1. Let $\mathbb{Q}(\sqrt{3}) = \{\alpha + \beta\sqrt{3} \,|\, \alpha, \beta \in \mathbb{Q}\}$. Since $\mathbb{Q}(\sqrt{3})$ is the span over $\mathbb{Q}$ of the elements $1, \sqrt{3} \in \mathbb{C}$, it is clear that $\mathbb{Q}(\sqrt{3})$ is a vector space over $\mathbb{Q}$ of dimension at most 2. Our spanning set $\{1, \sqrt{3}\}$ has only two elements, and the product of any of them remains in $\mathbb{Q}(\sqrt{3})$. Therefore, $\mathbb{Q}(\sqrt{3})$ is indeed a field.

   Before leaving this example, we should mention that although the proof of Lemma 15.2 indicates that every nonzero element of $\mathbb{Q}(\sqrt{3})$ has a multiplicative inverse in $\mathbb{Q}(\sqrt{3})$, the proof does not provide a particularly efficient way of finding it. In fact, to find the inverse of $\alpha + \beta\sqrt{3}$, it is more helpful to note that

$$\left(\alpha + \beta\sqrt{3}\right)\left(\alpha - \beta\sqrt{3}\right) = \alpha^2 - 3\beta^2.$$

   Since $\sqrt{3} \notin \mathbb{Q}$, you can check that if $\alpha + \beta\sqrt{3} \neq 0$, then $0 \neq \alpha^2 - 3\beta^2 \in \mathbb{Q}$. Dividing the previous equation by $\alpha^2 - 3\beta^2$ now results in

$$\left(\alpha + \beta\sqrt{3}\right)\left(\frac{\alpha}{\alpha^2 - 3\beta^2} - \frac{\beta}{\alpha^2 - 3\beta^2}\sqrt{3}\right) = 1.$$

   Therefore, $\frac{\alpha}{\alpha^2 - 3\beta^2} - \frac{\beta}{\alpha^2 - 3\beta^2}\sqrt{3} \in \mathbb{Q}(\sqrt{3})$ is the multiplicative inverse of $\alpha + \beta\sqrt{3}$.

2. Let $\mathbb{Q}(\sqrt{3}, i) = \{a + bi \,|\, a, b \in \mathbb{Q}(\sqrt{3})\}$. Observe that $\mathbb{Q}(\sqrt{3}, i)$ is the span of the set $\{1, i\}$ over the field $\mathbb{Q}(\sqrt{3})$ from the previous example. Thus, $\mathbb{Q}(\sqrt{3}, i)$ is a vector space over $\mathbb{Q}(\sqrt{3})$ of dimension at most 2. However, it is also quite easy to see that the product of any elements of the spanning set $\{1, i\}$ lies in $\mathbb{Q}(\sqrt{3}, i)$. Thus, $\mathbb{Q}(\sqrt{3}, i)$ is indeed a field. At this point, there are several ways to interpret the meaning of the symbol $\mathbb{Q}(\sqrt{3}, i)$. However, when we arrive at Theorem 15.4, we will see that all possible interpretations yield the same set.

3. Let $\mathbb{Q}(7^{\frac{1}{3}}) = \{\alpha + \beta 7^{\frac{1}{3}} + \gamma 7^{\frac{2}{3}} \,|\, \alpha, \beta, \gamma \in \mathbb{Q}\}$. Then $\mathbb{Q}(7^{\frac{1}{3}})$ is the span over $\mathbb{Q}$ of the finite set $\{1, 7^{\frac{1}{3}}, 7^{\frac{2}{3}}\}$. Since the product of any two elements from the spanning set belongs

to $\mathbb{Q}(7^{\frac{1}{3}})$, we see that $\mathbb{Q}(7^{\frac{1}{3}})$ is a field. Note that the set $V = \{\alpha + \beta 7^{\frac{1}{3}} | \alpha, \beta \in \mathbb{Q}\}$ is not a field, as $7^{\frac{1}{3}}$ belongs to the spanning set but $7^{\frac{1}{3}} \cdot 7^{\frac{1}{3}} = 7^{\frac{2}{3}} \notin V$. We should point out that $7^{\frac{2}{3}}$ certainly does not look like an element of V, but it does requires some work to verify that $7^{\frac{2}{3}}$ cannot be written in the form $\alpha + \beta 7^{\frac{1}{3}}$, where $\alpha, \beta \in \mathbb{Q}$. However, for the moment, we will accept that $7^{\frac{2}{3}} \notin V$, as we will soon prove a more general result that easily covers this situation.

4. Let $\mathbb{Q}(\sqrt{2}+\sqrt{5}) = \{a + b(\sqrt{2}+\sqrt{5}) + c(\sqrt{2}+\sqrt{5})^2 + d(\sqrt{2}+\sqrt{5})^3 | a, b, c, d \in \mathbb{Q}\}$. Since $\mathbb{Q}(\sqrt{2}+\sqrt{5})$ is the span over $\mathbb{Q}$ of the finite set

$$\left\{ 1, \sqrt{2}+\sqrt{5}, (\sqrt{2}+\sqrt{5})^2, (\sqrt{2}+\sqrt{5})^3 \right\},$$

to see that $\mathbb{Q}(\sqrt{2}+\sqrt{5})$ is a field, all we need to do is to show that the product of any elements of the set $\{1, \sqrt{2}+\sqrt{5}, (\sqrt{2}+\sqrt{5})^2, (\sqrt{2}+\sqrt{5})^3\}$ remains in $\mathbb{Q}(\sqrt{2}+\sqrt{5})$.

Before doing the necessary computations, you might wonder why the largest exponent of $\sqrt{2}+\sqrt{5}$ that appears in our spanning set is 3. Observe that if we let $w = \sqrt{2}+\sqrt{5}$, then $w^2 = 7 + 2\sqrt{10}$, so $(w^2 - 7)^2 = 40$. We can rewrite this fact as

$$w^4 = 14w^2 - 9.$$

In other words,

$$\left(\sqrt{2}+\sqrt{5}\right)^4 = 14\left(\sqrt{2}+\sqrt{5}\right)^2 - 9,$$

which tells us that $(\sqrt{2}+\sqrt{5})^4 \in \mathbb{Q}(\sqrt{2}+\sqrt{5})$. Furthermore, our preceding work also tells us that

$$w^5 = 14w^3 - 9w.$$

Going one step further, we obtain

$$w^6 = 14w^4 - 9w^2 = 14(w^2 - 9) - 9w^2.$$

These equations tell us that $(\sqrt{2}+\sqrt{5})^5$, $(\sqrt{2}+\sqrt{5})^6 \in \mathbb{Q}(\sqrt{2}+\sqrt{5})$. Thus, the product of any elements from our spanning set is in $\mathbb{Q}(\sqrt{2}+\sqrt{5})$, so $\mathbb{Q}(\sqrt{2}+\sqrt{5})$ is a field. After doing these computations, we can see the reason that 3 was largest power of $\sqrt{2}+\sqrt{5}$ that needed to be in the spanning set followed from the fact that $\sqrt{2}+\sqrt{5}$ satisfied a polynomial of degree 4 in $\mathbb{Q}[x]$. This enabled us to write larger powers of $\sqrt{2}+\sqrt{5}$ as linear combinations of elements from our spanning set. As we shall see, in Theorem 15.3, this is a special case of a very important and more general fact.

5. Let $\mathbb{Q}(\text{cis}(\frac{\pi}{5})) = \{\sum_{j=0}^{9} a_j(\text{cis}(\frac{\pi}{5}))^j | a_j \in \mathbb{Q}\}$. The set $\mathbb{Q}(\text{cis}(\frac{\pi}{5}))$ is spanned over $\mathbb{Q}$ by the finite set $\{(\text{cis}(\frac{\pi}{5}))^j | 0 \le j \le 9\}$. By DeMoivre's Theorem, $(\text{cis}(\frac{\pi}{5}))^{10} = 1$. Note that the product of any two elements of our spanning set is of the form $(\text{cis}(\frac{\pi}{5}))^n$, where $n \ge 0$. Applying the division algorithm, if r is the remainder after n is divided by 10, then $0 \le r \le 9$ and we have

$$\left(\text{cis}\left(\frac{\pi}{5}\right)\right)^n = \left(\text{cis}\left(\frac{\pi}{5}\right)\right)^r \in \mathbb{Q}\left(\text{cis}\left(\frac{\pi}{5}\right)\right).$$

Therefore, the product of any elements of our spanning set does belong to $\mathbb{Q}(\text{cis}(\frac{\pi}{5}))$, so $\mathbb{Q}(\text{cis}(\frac{\pi}{5}))$ is a field. ∎

In the preceding five examples, we used Lemma 15.2 to show that various sets were fields. At that point, we were not terribly concerned with computing the degrees of these fields over smaller fields. In particular, in the last example we were content to merely say that $\mathbb{Q}(\text{cis}(\frac{\pi}{5}))$ is a field, and its degree over $\mathbb{Q}$ is at most 10. It turns out that the degree of this extension is significantly smaller than 10. Computing degrees of field extensions can often require a long series of computations. However, we now present a result that, in many cases, will make it much easier to find the degree of a field extension. It will be stated for fields more general than those that lie between $\mathbb{Q}$ and $\mathbb{C}$, but the increased generality should not make things more difficult.

Theorem 15.3. *Let $K \subseteq L$ be fields and let $p(x)$ be an irreducible polynomial of degree n in $K[x]$. If $\theta \in L$ is a root of $p(x)$, let*

$$K(\theta) = \left\{ \alpha_0 + \alpha_1\theta + \alpha_2\theta^2 + \cdots + \alpha_{n-1}\theta^{n-1} | \alpha_0, \alpha_1, \ldots, \alpha_{n-1} \in K \right\}.$$

Then $K(\theta)$ is a field extension of K such that $\{1, \theta, \theta^2, \ldots, \theta^{n-1}\}$ is a basis and $[K(\theta) : K] = n$.

Proof. One way to think $K(\theta)$ is as the set consisting of 0 as well as all elements of L that can be obtained by plugging θ into an element of $K[x]$ of degree less than n. Certainly, $K(\theta)$ is the span over K of the finite set $\{1, \theta, \ldots, \theta^{n-1}\}$. In light of Lemma 15.2, to show that $K(\theta)$ is a field, it suffices to show that the product of any two elements of our spanning set belongs to $K(\theta)$. The product of any two elements of $\{1, \theta, \ldots, \theta^{n-1}\}$ is of the form θ^m, for some $m \ge 0$, and this leads us to an examination of the element $x^m \in K[x]$.

The Division Algorithm in $K[x]$ (Theorem 12.6) tells us that there exist $q(x), r(x) \in K[x]$ such that

$$x^m = q(x) \cdot p(x) + r(x) \text{ with } deg(r(x)) < n \text{ or } r(x) = 0.$$

If we replace x by θ then, inside the field L, we have

$$\theta^m = q(\theta) \cdot p(\theta) + r(\theta) = q(\theta) \cdot 0 + r(\theta) = r(\theta).$$

Since either $r(x)$ has degree less than n or $r(x) = 0$, it follows that $r(\theta) \in K(\theta)$. Thus, $\theta^m = r(\theta) \in K(\theta)$ and $K(\theta)$ is indeed a field.

If we look back at the previous paragraph, we can see that the division algorithm actually shows that $K(\theta)$ as the set of *all* elements of L obtained by plugging θ into elements of $K[x]$. Until now, it has not been necessary to assume that $p(x)$ is irreducible. However, we will need this assumption to show that $[K(\theta) : K] = n$. Since the set $\{1, \theta, \ldots, \theta^{n-1}\}$ spans $K(\theta)$ over K, it only remains to show that this set is linearly independent over K. To this end, suppose $\alpha_i \in K$ such that

(1) $$\alpha_0 + \alpha_1 \theta + \alpha_2 \theta^2 + \cdots + \alpha_{n-1} \theta^{n-1} = 0.$$

We need to show that $\alpha_i = 0$, for $0 \le i \le n - 1$. Let

$$r(x) = \alpha_0 + \alpha_1 x + \alpha_2 x^2 + \cdots + \alpha_{n-1} x^{n-1} \in K[x];$$

if $r(x) \ne 0$, then it follows from (1) that $r(x)$ is an element of $K[x]$ of degree smaller than n that has θ as a root. However, since $p(x)$ is irreducible and has degree n, Lemma 12.9 asserts that there are no nonzero polynomials in $K[x]$ of degree less than n that have θ as a root. Thus, $r(x) = 0$. As a result, all the coefficients of $r(x)$ are 0, so every $\alpha_i = 0$. Thus, the set $\{1, \theta, \ldots, \theta^{n-1}\}$ is indeed linearly independent over K. $\square$

To this point, we have been defining objects such as $\mathbb{Q}(\sqrt{3})$, $\mathbb{Q}(i)$, $\mathbb{Q}(7^{\frac{1}{3}})$, and $\mathbb{Q}(\sqrt{2} + \sqrt{5})$ on a case by case basis. However, Theorem 15.3 leads us to a single definition that describes all of these objects. If $K \subseteq L$ are fields and $\alpha \in L$ is a root of some irreducible $p(x) \in K[x]$ of degree n, then $K(\theta)$ is the field

$$\left\{ \alpha_0 + \alpha_1 \theta + \alpha_2 \theta^2 + \cdots + \alpha_{n-1} \theta^{n-1} \mid \alpha_0, \alpha_1, \ldots, \alpha_{n-1} \in K \right\}$$

■ Examples

1. Since $\sqrt{3}$ is a root of $x^2 - 3$ and $x^2 - 3$ is irreducible in $\mathbb{Q}[x]$ and has degree 2, Theorem 15.3 immediately asserts that the field $\mathbb{Q}(\sqrt{3})$ has basis $\{1, \sqrt{3}\}$ over $\mathbb{Q}$ and $[\mathbb{Q}(\sqrt{3}) : \mathbb{Q}] = 2$.

2. Since i is a root of $x^2 + 1$ and this polynomial is irreducible and of degree 2 in $\mathbb{Q}[x]$, it follows that $\mathbb{Q}(i)$ has basis $\{1, i\}$ over $\mathbb{Q}$ and $[\mathbb{Q}(i) : \mathbb{Q}] = 2$. Similarly, $x^2 + 1$ is irreducible and of degree 2 in $\mathbb{R}[x]$, so $\mathbb{R}(i)$ has basis $\{1, i\}$ over $\mathbb{R}$ and $[\mathbb{R}(i) : \mathbb{R}] = 2$. On the other hand, $x^2 + 1$ is not irreducible in $\mathbb{C}[x]$. In fact, the monic, irreducible

polynomial in $\mathbb{C}[x]$ that has i as a root is $x - i$. Since this polynomial has degree 1, $[\mathbb{C}(i) : \mathbb{C}] = 1$. Observe that this is simply another way of saying that $\mathbb{C}(i) = \mathbb{C}$.

3. The number $7^{\frac{1}{3}}$ is a root of $x^3 - 7$ which is irreducible and of degree 3 in $\mathbb{Q}[x]$. Therefore, the field $\mathbb{Q}(7^{\frac{1}{3}})$ has basis $\{1, 7^{\frac{1}{3}}, 7^{\frac{2}{3}}\}$ over $\mathbb{Q}$ and $[\mathbb{Q}(7^{\frac{1}{3}}) : \mathbb{Q}] = 3$. In particular, it is impossible to write $7^{\frac{2}{3}}$ as a linear combination of the other two basis elements. Thus, Theorem 15.3 immediately allows us to say that $7^{\frac{2}{3}}$ cannot be written in the form $\alpha + \beta 7^{\frac{1}{3}}$, where $\alpha, \beta \in \mathbb{Q}$. Recall that this was an issue that we did not resolve when it first arose in the third example following Lemma 15.2.

4. Earlier in this chapter, we saw that the number $\sqrt{2} + \sqrt{5}$ is a root of the polynomial $x^4 - 14x^2 + 9 \in \mathbb{Q}[x]$. Therefore, $\mathbb{Q}(\sqrt{2} + \sqrt{5})$ is a field and $[\mathbb{Q}(\sqrt{2} + \sqrt{5}) : \mathbb{Q}] \le 4$. However, to determine if 4 is indeed the degree of this extension, we need to check if $x^4 - 14x^2 + 9$ is the minimum polynomial for $\sqrt{2} + \sqrt{5}$ over $\mathbb{Q}$. Remember that this is the same as determining if $x^4 - 14x^2 + 9$ is irreducible in $\mathbb{Q}[x]$.

    If we look back at the calculations which produced the polynomial $x^4 - 14x^2 + 9$ and replace $\sqrt{2} + \sqrt{5}$ by $\sqrt{2} - \sqrt{5}$, we can see that $\sqrt{2} - \sqrt{5}$ is also a root of $x^4 - 14x^2 + 9$. Furthermore, since $x^4 - 14x^2 + 9$ only involves even powers of x, the additive inverses of these two roots must also be roots of $x^4 - 14x^2 + 9$. Thus $\sqrt{2} + \sqrt{5}, \sqrt{2} - \sqrt{5}, -\sqrt{2} - \sqrt{5}, -\sqrt{2} + \sqrt{5}$ are the four roots of $x^4 - 14x^2 + 9$. Since we now know the four roots of $x^4 - 14x^2 + 9$ in $\mathbb{R}$ and $\mathbb{C}$, it must factor in both $\mathbb{R}[x]$ and $\mathbb{C}[x]$ as,

$$x^4 - 14x^2 + 9 =$$

(2) $\left(x - \left(\sqrt{2} + \sqrt{5}\right)\right)\left(x - \left(\sqrt{2} - \sqrt{5}\right)\right)\left(x - \left(-\sqrt{2} - \sqrt{5}\right)\right)\left(x - \left(-\sqrt{2} + \sqrt{5}\right)\right).$

If $x^4 - 14x^2 + 9$ were reducible in $\mathbb{Q}[x]$, then it would either have a linear factor or be a product of two irreducible quadratics. However, the rational root test shows that $x^4 - 14x^2 + 9$ has no rational roots, so it has no linear factors in $\mathbb{Q}[x]$. The Unique Factorization Theorem (Theorem 12.4) asserts that $x^4 - 14x^2 + 9$ factors uniquely in both $\mathbb{R}[x]$ and $\mathbb{C}[x]$. Therefore, if $x^4 - 14x^2 + 9$ were a product of two irreducible quadratics in $\mathbb{Q}[x]$, then after factoring these quadratics in $\mathbb{R}[x]$ and $\mathbb{C}[x]$, we would obtain the identical factorization as in equation (2). Therefore, if $x^4 - 14x^2 + 9$ was a product of two quadratics in $\mathbb{Q}[x]$, we could then regroup the four linear factors in equation (2) so that they could be multiplied to produce two quadratic factors in $\mathbb{Q}[x]$. However, you should check that whenever we multiply $x - (\sqrt{2} + \sqrt{5})$ by any of the three other linear factors, the quadratic polynomial we obtain contains a coefficient which does not belong to $\mathbb{Q}$. Thus, $x^4 - 14x^2 + 9$ is irreducible in $\mathbb{Q}[x]$ and so, $[\mathbb{Q}(\sqrt{2} + \sqrt{5}) : \mathbb{Q}] = 4$.

5. Earlier, we examined $\mathbb{Q}(\text{cis}(\frac{\pi}{5}))$ and saw that it was a field whose dimension over $\mathbb{Q}$ was at most 10 because $\{(\text{cis}(\frac{\pi}{5}))^j | 0 \le j \le 9\}$ was a spanning set. To determine the exact degree of this extension, we need to find the minimum polynomial in $\mathbb{Q}[x]$ satisfied by $\text{cis}(\frac{\pi}{5})$. Certainly $\text{cis}(\frac{\pi}{5})$ is a root of $x^{10} - 1$, and since $x^{10} - 1$ is a difference of squares, we have

$$x^{10} - 1 = (x^5 - 1)(x^5 + 1).$$

Since $\text{cis}\left(\frac{\pi}{5}\right)$ is a root of $x^5 + 1$, we now know that $[\mathbb{Q}(\text{cis}(\frac{\pi}{5})) : \mathbb{Q}] \le 5$. However, -1 is a root of $x^5 + 1$ and we have

$$x^5 + 1 = (x + 1)(x^4 - x^3 + x^2 - x + 1).$$

Therefore, $\text{cis}\left(\frac{\pi}{5}\right)$ is a root of $x^4 - x^3 + x^2 - x + 1$, so $[\mathbb{Q}(\text{cis}(\frac{\pi}{5})) : \mathbb{Q}] \le 4$. We now need to determine if $x^4 - x^3 + x^2 - x + 1$ is irreducible over $\mathbb{Q}$.

Suppose $p(x), q(x) \in \mathbb{Q}[x]$ such that

$$x^4 - x^3 + x^2 - x + 1 = p(x) \cdot q(x).$$

Observe that by replacing x by $-x$ in the previous equation, we obtain

$$x^4 + x^3 + x^2 + x + 1 = p(-x) \cdot q(-x).$$

However, in the last example at the end of Section 9.4, we used a variation of Eisenstein's Criterion to show that $x^4 + x^3 + x^2 + x + 1$ is irreducible in $\mathbb{Q}[x]$. As a result, either $p(-x)$ or $q(-x)$ has degree 4. It easily follows that either $p(x)$ or $q(x)$ also has degree 4. Therefore, $x^4 - x^3 + x^2 - x + 1$ cannot be written as a product of polynomials of smaller degree in $\mathbb{Q}[x]$, so $x^4 - x^3 + x^2 - x + 1$ is irreducible in $\mathbb{Q}[x]$.   ■

These examples indicate that determining the degrees of some extensions of the form $\mathbb{Q}(\theta)$ can be handled in seconds using Theorem 15.3, yet others require not only Theorem 15.3 but also a series of computations. When dealing with field extensions and other topics in abstract algebra, we can often come up with very quick answers to questions by invoking powerful theorems. This is quite satisfying and yields short and elegant solutions. However, there will always be problems where fancy algebraic machinery will take us only so far, and to complete these problems, we will need to roll up our sleeves and do lots and lots of old-fashioned paper and pencil computations.

We now fully understand exactly which fields are represented by the symbols

$$\mathbb{Q}\left(\sqrt{3}\right), \ \mathbb{Q}(i), \ \mathbb{R}(i), \ \mathbb{Q}\left(7^{\frac{1}{3}}\right), \ \mathbb{Q}\left(\sqrt{2} + \sqrt{5}\right), \ \mathbb{Q}\left(\text{cis}\left(\frac{\pi}{5}\right)\right).$$

We can now turn attention to determining what we mean by the symbols

$$\mathbb{Q}\left(\sqrt{3}, i\right), \ \mathbb{Q}\left(\sqrt{2}, \sqrt{3}, \sqrt{5}\right), \ \mathbb{Q}\left(7^{\frac{1}{3}}, \operatorname{cis}\left(\frac{\pi}{5}\right)\right).$$

The fields

$$\mathbb{Q}\left(\sqrt{2}\right), \ \mathbb{Q}\left(2^{\frac{1}{4}}\right), \ \mathbb{Q}\left(2^{\frac{1}{6}}\right), \ \mathbb{Q}\left(2^{\frac{1}{100}}\right), \ \mathbb{R}$$

all have the property that they lie between $\mathbb{Q}$ and $\mathbb{C}$ and also contain $\sqrt{2}$. However, any field that lies between $\mathbb{Q}$ and $\mathbb{C}$ and contains $\sqrt{2}$ must contain every element of the form $a + b\sqrt{2}$, where $a, b \in \mathbb{Q}$. Therefore, $\mathbb{Q}(\sqrt{2})$ is clearly the *smallest* field that lies between $\mathbb{Q}$ and $\mathbb{C}$ and contains $\sqrt{2}$.

More generally, if $K \subseteq L$ are fields and $\theta \in L$ is the root of some nonzero element of $K[x]$, then any field that lies between K and L and also contains θ, must contain θ^n, for all $n \in \mathbb{N}$. Therefore, such a field contains K as well as the entire spanning set for $K(\theta)$ over K. As a result, although there are likely many fields that lie between K and L and contain θ, every one of them must contain $K(\theta)$. Therefore, $K(\theta)$ is the *smallest* field that lies between K and L and also contains θ.

Using the previous paragraph as a guide, we can look at $\mathbb{Q}(\sqrt{3}, i)$. It should represent the smallest field that lies between $\mathbb{Q}$ and $\mathbb{C}$ and contains both $\sqrt{3}$ and i. If we want to view things abstractly, we could simply say that $\mathbb{Q}(\sqrt{3}, i)$ is the intersection of all fields that lie between $\mathbb{Q}$ and $\mathbb{C}$ and contain both $\sqrt{3}$ and i. We would then need to check that this intersection is indeed a field in its own right. Although this is a legitimate way to view this problem, to a certain degree, it is quite unsatisfying. Not only do we want to know that a smallest field exists that lies between $\mathbb{Q}$ and $\mathbb{C}$ and contains both $\sqrt{3}$ and i, but we would also like to explicitly describe its elements in terms of $\mathbb{Q}$, $\sqrt{3}$, and i.

To this end, let V be the span over $\mathbb{Q}$ of all elements of the form $(\sqrt{3})^n i^m$, where $n, m \geq 0$. Observe that not only is V a vector space over $\mathbb{Q}$, but it is also closed under multiplication. In addition, since $\sqrt{3}$ is a root of $x^2 - 3$, $(\sqrt{3})^n$ is always equal to an element in the span over $\mathbb{Q}$ of the set $\{1, \sqrt{3}\}$. Similarly, i is a root of $x^2 + 1$, so i^m is always equal to an element in the span over $\mathbb{Q}$ of the set $\{1, i\}$. As a result, the set V is equal the span over $\mathbb{Q}$ of the finite set $\{1, \sqrt{3}, i, \sqrt{3}i\}$. By Lemma 15.2, we know that V is a field. Observe that every field that lies between $\mathbb{Q}$ and $\mathbb{C}$ and also contains $\sqrt{3}$ and i must contain V, so V is the smallest field with these properties. Furthermore, we can describe every element of V in terms of $\mathbb{Q}$, $\sqrt{3}$, and i. Note that the only issue we have not as yet settled is whether the degree of V over $\mathbb{Q}$ is 4 or something smaller. This set V is what we mean by the symbol $\mathbb{Q}(\sqrt{3}, i)$. We are now in a position to generalize this example.

Theorem 15.4. *Let $K \subseteq L$ be fields and let $\theta_1, \theta_2, \ldots, \theta_n \in L$ such that each θ_i is the root of some nonzero $p_i(x) \in K[x]$ of degree m_i in $K[x]$. If we let*

$$K(\theta_1, \theta_2, \ldots, \theta_n) = \left\{ \sum a_{(j_1, j_2, \ldots, j_n)} \theta_1^{j_1} \theta_2^{j_2} \cdots \theta_n^{j_n} \, \big| \, a_{(j_1, j_2, \ldots, j_n)} \in K \text{ and each } j_i < m_i \right\},$$

then $K(\theta_1, \theta_2, \ldots, \theta_n)$ is the smallest field that lies between K and L and also contains each θ_i. Furthermore, the degree of $K(\theta_1, \theta_2, \ldots, \theta_n)$ over K is at most $m_1 \cdot m_2 \cdots m_n$.

Proof. Let W denote the span over K of all elements of L of the form $\theta_1^{l_1} \theta_2^{l_2} \cdots \theta_n^{l_n}$, where no restriction is placed on how large the various l_i can be. Since our spanning set for W is closed under multiplication, it is clear that W is closed under multiplication. Also observe that $K(\theta_1, \theta_2, \ldots, \theta_n)$ is the span over K of all elements of the form $\theta_1^{j_1} \theta_2^{j_2} \cdots \theta_n^{j_n}$, where each j_i is bounded by m_i. Therefore, $K(\theta_1, \theta_2, \ldots, \theta_n)$ is a subset of W and has a spanning set with at most $m_1 \cdot m_2 \cdots m_n$ distinct elements, so the dimension of $K(\theta_1, \theta_2, \ldots, \theta_n)$ over K is at most $m_1 \cdot m_2 \cdots m_n$. Also observe that $K(\theta_1, \theta_2, \ldots, \theta_n)$ must be contained in any field that lies between K and L and also contains each θ_i. Therefore, it suffices to show that $K(\theta_1, \theta_2, \ldots, \theta_n)$ is a field.

Since $K(\theta_1, \theta_2, \ldots, \theta_n)$ is finite dimensional over K and W is closed under multiplication, Lemma 15.2 asserts that to conclude the proof, we only need to show that $K(\theta_1, \theta_2, \ldots, \theta_n) = W$. Therefore, we need to show that every element of the form $\theta_1^{l_1} \theta_2^{l_2} \cdots \theta_n^{l_n}$ belongs to $K(\theta_1, \theta_2, \ldots, \theta_n)$, even if no restrictions are placed on the l_i. We now make use of some ideas that appeared in the proof of Theorem 15.3. For each l_i, we will apply the division algorithm in $K[x]$ and divide x^{l_i} by $p_i(x)$ to obtain

$$x^{l_i} = q_i(x) \cdot p_i(x) + r_i(x),$$

where $q_i(x), r_i(x) \in K[x]$ and either $r_i(x) = 0$ or has degree less than m_i. Substituting θ_i into the preceding polynomials, we can see that in L we have

$$\theta_i^{l_i} = q_i(\theta_i) \cdot p_i(\theta_i) + r_i(\theta_i) = q_i(\theta_i) \cdot 0 + r_i(\theta_i) = r_i(\theta_i).$$

Therefore, in the expression $\theta_1^{l_1} \theta_2^{l_2} \cdots \theta_n^{l_n}$, we can replace each $\theta_i^{l_i}$ by $r_i(\theta_i)$ to obtain

$$\theta_1^{l_1} \theta_2^{l_2} \cdots \theta_n^{l_n} = r_1(\theta_1) r_2(\theta_2) \cdots r_n(\theta_n).$$

Each $r_i(\theta)$ belongs to the span over K of the set $\{1, \theta_i, \theta_i^2, \ldots, \theta_i^{m_i - i}\}$. Therefore, after repeated applications of the distributive law, the product $r_1(\theta_1) r_2(\theta_2) \cdots r_n(\theta_n)$ belongs to the span over K of elements of L of the form $\theta_1^{j_1} \theta_2^{j_2} \cdots \theta_n^{j_n}$, where each j_i is less than m_i. Thus,

$$\theta_1^{l_1} \theta_2^{l_2} \cdots \theta_n^{l_n} = r_1(\theta_1) r_2(\theta_2) \cdots r_n(\theta_n) \in K(\theta_1, \theta_2, \ldots, \theta_n).$$

Hence, $W = K(\theta_1, \theta_2, \ldots, \theta_n)$, thereby concluding the proof. $\qquad \square$

Let us now take a look at the field $\mathbb{Q}(\sqrt{3}, i, \sqrt{2}, 7^{\frac{1}{3}}, 11^{\frac{1}{7}})$. Theorem 15.4 tells us what a spanning set over $\mathbb{Q}$ for this field looks like. It then becomes clear that this field is the same as the fields

$$\mathbb{Q}\left(i, \sqrt{2}, \sqrt{3}, 7^{\frac{1}{3}}, 11^{\frac{1}{7}}\right), \ \mathbb{Q}\left(11^{\frac{1}{7}}, \sqrt{3}, i, \sqrt{2}, 7^{\frac{1}{3}}\right),$$

$$\mathbb{Q}\left(\sqrt{3}, i, 7^{\frac{1}{3}}, 11^{\frac{1}{7}}, \sqrt{2}\right), \ \text{and} \ \mathbb{Q}\left(\sqrt{3}, i, 7^{\frac{1}{3}}, \sqrt{2}, 11^{\frac{1}{7}}\right).$$

More generally, when we look at the field $K(\theta_1, \theta_2, \ldots, \theta_n)$, no matter how we permute the order of the θ_i, we continue to obtain the same field. This holds because multiplication in L is commutative, so Theorem 15.4 continues to produce the same spanning sets regardless of the order of the θ_i.

We can now look at another field similar to $\mathbb{Q}(\sqrt{3}, i, \sqrt{2}, 7^{\frac{1}{3}}, 11^{\frac{1}{7}})$ and examine $\mathbb{Q}(\sqrt{3}, \sqrt{2})$ $(i)(7^{\frac{1}{3}}, 11^{\frac{1}{7}})$. Observe that this field is built in steps. First, starting with $\mathbb{Q}$, we use Theorem 15.4 to construct the smallest field contained in $\mathbb{C}$, which also contains $\sqrt{3}$ and $\sqrt{2}$, thereby obtaining $\mathbb{Q}(\sqrt{3}, \sqrt{2})$. Next, starting with $\mathbb{Q}(\sqrt{3}, \sqrt{2})$, we use Theorems 15.3 or 15.4 to construct the smallest field contained in $\mathbb{C}$ and also contains i to give us $\mathbb{Q}(\sqrt{3}, \sqrt{2})(i)$. Finally, we begin with $\mathbb{Q}(\sqrt{3}, \sqrt{2})(i)$ and apply Theorem 15.4 to construct the smallest field contained in $\mathbb{C}$, which also contains $7^{\frac{1}{3}}$ and $11^{\frac{1}{7}}$ to obtain the field $\mathbb{Q}(\sqrt{3}, \sqrt{2})(i)(7^{\frac{1}{3}}, 11^{\frac{1}{7}})$. As we use Theorem 15.3 and 15.4 to construct $\mathbb{Q}(\sqrt{3}, \sqrt{2})(i)(7^{\frac{1}{3}}, 11^{\frac{1}{7}})$, we can see that it has the same spanning set over $\mathbb{Q}$ as does $\mathbb{Q}(\sqrt{3}, i, \sqrt{2}, 7^{\frac{1}{3}}, 11^{\frac{1}{7}})$. Thus, $\mathbb{Q}(\sqrt{3}, \sqrt{2})(i)(7^{\frac{1}{3}}, 11^{\frac{1}{7}}) = \mathbb{Q}(\sqrt{3}, i, \sqrt{2}, 7^{\frac{1}{3}}, 11^{\frac{1}{7}})$.

More generally, if $K \subseteq L$ are fields and $\theta_1, \ldots, \theta_5 \in L$ are all roots of nonzero elements of $K[x]$, then Theorems 15.3 and 15.4 tell us that the fields $K(\theta_1, \theta_2, \theta_3, \theta_4, \theta_5)$ and $K(\theta_1, \theta_2)$ $(\theta_3)(\theta_4, \theta_5)$ are the same because they have the same spanning set over K. However, there is a problem that arises that can be handled by extending a concept introduced in Chapter 11.

Definition 15.5. *Let $K \subseteq L$ be fields. An element $\theta \in L$ is said to be algebraic over K if it is the root of some nonzero $p(x) \in K[x]$.*

Recall, in Chapter 11, we said that an element of $\mathbb{R}$ was algebraic if it satisfied some nonzero $p(x) \in Q[x]$. Using the terminology of Definition 15.5, Theorem 15.4 tells us that the construction of the field $K(\theta_1, \ldots, \theta_n)$ requires that each $\theta_i \in L$ is algebraic over K. We just remarked that the fields $K(\theta_1, \theta_2, \theta_3, \theta_4, \theta_5)$ and $K(\theta_1, \theta_2)(\theta_3)(\theta_4, \theta_5)$ are the same provided each θ_i is algebraic over K. However, if we look carefully at the construction of $K(\theta_1, \theta_2)$ $(\theta_3)(\theta_4, \theta_5)$ with Theorems 15.3 and 15.4, we need that θ_1, θ_2 are algebraic over K, θ_3 is algebraic over $K(\theta_1, \theta_2)$, and θ_4, θ_5 are algebraic over $K(\theta_1, \theta_2)(\theta_3)$. On the surface, it appears that θ_3 could be algebraic over $K(\theta_1, \theta_2)$ without being algebraic over K. Similarly, θ_4, θ_5 could be algebraic over $K(\theta_1, \theta_2)(\theta_3)$ without being algebraic over K. This raises the question whether it is possible to select $\theta_i \in L$ such that $K(\theta_1, \theta_2)(\theta_3)(\theta_4, \theta_5)$ can be constructed with Theorem 15.4 but $K(\theta_1, \theta_2, \theta_3, \theta_4, \theta_5)$ cannot. This question will be answered by Theorem 15.9. However, we first need to prove a result that will not only enable us to prove Theorem 15.9 but also to compute degrees of field extensions such as $[\mathbb{Q}(\sqrt{3}, i, \sqrt{2}, 7^{\frac{1}{5}}, 11^{\frac{1}{7}}) : \mathbb{Q}]$.

Theorem 15.6. *Let $F \subseteq K \subseteq L$ be fields. If $[L : K]$ and $[K : F]$ are both finite, then* $[L : F] = [L : K] \cdot [K : F]$.

Proof. Since L is a finite dimensional vector space over K, let $v_1, v_2, \ldots, v_n \in L$ be a basis for L over K. Similarly, K is a finite dimensional vector space over F and we can let $w_2, w_2, \ldots, w_m \in K$ be a basis for K over F. Next, consider all elements of L of the form $w_i v_j$, where $1 \leq i \leq m, 1 \leq j \leq n$. To complete the proof, it suffices to show that these nm elements of L are indeed linear independent and a spanning set for L over F.

To this end, let $v \in L$. Since the v_j span L over K, there exist $\beta_j \in K$ such that

$$v = \beta_1 v_1 + \beta_2 v_2 + \cdots \beta_n v_n.$$

On the other hand, since $\beta_j \in K$, for all j, there exist $\alpha_{ij} \in F$ such that

$$\beta_j = \alpha_{1j} w_1 + \alpha_{2j} w_2 + \cdots + \alpha_{mj} w_m.$$

Combining the previous two equations, we have

$$v = \beta_1 v_1 + \beta_2 v_2 + \cdots \beta_n v_n = (\alpha_{11} w_1 + \alpha_{21} w_2 + \cdots + \alpha_{m1} w_m) v_1 +$$

$$(\alpha_{12} w_1 + \alpha_{22} w_2 + \cdots + \alpha_{m2} w_m) v_2 + \cdots + (\alpha_{1n} w_1 + \alpha_{2n} w_2 + \cdots + \alpha_{mn} w_m) v_n =$$

$$\sum_{i \leq m; j \leq n} \alpha_{ij} (w_i v_j).$$

Thus, the elements of the form $w_i v_j$ do span L over F.

Next, suppose $\alpha_{ij} \in F$ such that $\sum_{i \leq m; j \leq n} \alpha_{ij} (w_i v_j) = 0$. To show that the $w_i v_j$ are linearly independent, we need to show that each $\alpha_{ij} = 0$. Observe that

$$0 = \sum_{i \leq m; j \leq n} \alpha_{ij} (w_i v_j) = \left(\sum_{i \leq m} \alpha_{i1} w_i \right) v_1 + \left(\sum_{i \leq m} \alpha_{i2} w_i \right) v_2 + \cdots + \left(\sum_{i \leq m} \alpha_{in} w_i \right) v_n.$$

The v_j are linearly independent over K and, for all j, $\sum_{i \le m} \alpha_{ij} w_i \in K$. Therefore, for all j,

$$\alpha_{1j} w_1 + \alpha_{2j} w_2 + \cdots + \alpha_{mj} w_m = 0.$$

However, the w_i are linearly independent over F and every $\alpha_{ij} \in F$. Therefore, for all i, j, we have $\alpha_{ij} = 0$. Thus, the nm elements $w_i v_j$ are both linearly independent and span L over F. Hence, $[L : F] = nm = [L : K][K : F]$, as desired. $\qquad\square$

Theorem 15.6 will also be the key piece of the puzzle, in the next chapter, when we prove that $60°$ angles cannot be constructed. At this point, it is natural to wonder if, when given fields $K \subseteq L$, how many fields E are there such that $K \subseteq E \subseteq L$. We will now show that if $[L : K]$ is prime, then no field can lie properly between K and L.

Corollary 15.7. *Let $K \subseteq L$ be fields.*

(a) *If $[L : K] = 1$, then $L = K$.*

(b) *If $[L : K] = p$, where p is prime, and E is a field such that $K \subseteq E \subseteq L$, then either $E = L$ or $E = K$.*

Proof. Part (a) should come as little surprise as the concept of dimension reflects the relative size of a vector space over a field. Therefore, if $[L : K] = 1$, we would intuitively think of L as being the same size as K, so L and K should be equal. For a formal proof, let β be any nonzero element of K. Since $[L : K] = 1$, the set $\{\beta\}$ is subset of L with is linearly independent over K and has the same size as the dimension of L over K. By Lemma 14.12, $\{\beta\}$ is also a spanning set for L over K. Hence, if $l \in L$, there exists $\alpha \in K$ such that $\alpha \cdot \beta = l$. Since both α and β belong to K, so does l. Thus, $L = K$, as desired.

For part (b), we begin by looking at E as a vector space over F. Since E is a subspace of L, Proposition 14.15 tells us that $[E : F]$ is finite. Next, we look at L as a vector space over E. Any basis for L over F is certainly a spanning set for L over E. By Lemma 14.13, $[L : E]$ is also finite. Therefore, we can now apply Theorem 15.6 to assert that

$$p = [L : K] = [L : E] \cdot [E : F].$$

Since p is prime, either $[L : E] = 1$ or $[E : K] = 1$. However, part (a) now tells us that either $E = L$ or $E = K$, thereby concluding the proof. $\qquad\square$

Observe that Corollary 15.7(b) immediately tells us that there are no fields that lie properly between $\mathbb{R}$ and $\mathbb{C}$. Similarly, there are no fields that lie properly between $\mathbb{Q}$ and $\mathbb{Q}(\sqrt{2})$, nor any that lie properly between $\mathbb{Q}$ and $\mathbb{Q}(6^{\frac{1}{13}})$. We can now use Theorem 15.6 to reexamine some of the field extensions introduced earlier in this section.

■ Examples

1. Let us compute $[\mathbb{Q}(\sqrt{3}, i) : \mathbb{Q}]$. Since $\mathbb{Q}(\sqrt{3}, i) = \mathbb{Q}(\sqrt{3})(i)$, we can look at this problem as

$$\mathbb{Q} \subseteq \mathbb{Q}(\sqrt{3}) \subseteq \mathbb{Q}(\sqrt{3})(i).$$

Theorem 15.6 now asserts that

$$\left[\mathbb{Q}(\sqrt{3}, i) : \mathbb{Q}\right] = \left[\mathbb{Q}(\sqrt{3})(i) : \mathbb{Q}\right] = \left[\mathbb{Q}(\sqrt{3})(i) : \mathbb{Q}(\sqrt{3})\right]\left[\mathbb{Q}(\sqrt{3}) : \mathbb{Q}\right].$$

Since $\sqrt{3}$ is a root of $x^2 - 3$ and $x^2 - 3$ is irreducible in $\mathbb{Q}[x]$, $[\mathbb{Q}(\sqrt{3}) : \mathbb{Q}] = 2$. Note that $\mathbb{Q}(\sqrt{3}) \subseteq \mathbb{R}$, therefore $i \notin \mathbb{Q}(\sqrt{3})$. Therefore, i is a root of $x^2 + 1$ and $x^2 + 1$ is irreducible in $\mathbb{Q}(\sqrt{3})[x]$. Thus, $[\mathbb{Q}(\sqrt{3})(i) : \mathbb{Q}(\sqrt{3})] = 2$. As a result, $[\mathbb{Q}(\sqrt{3}, i) : \mathbb{Q}] = 2 \cdot 2 = 4$.

At this point, one could ask if we could have also looked at this problem as

$$(3) \qquad \left[\mathbb{Q}(\sqrt{3}, i) : \mathbb{Q}\right] = \left[\mathbb{Q}(i)(\sqrt{3}) : \mathbb{Q}\right] = \left[\mathbb{Q}(i)(\sqrt{3}) : \mathbb{Q}(i)\right]\left[\mathbb{Q}(i) : \mathbb{Q}\right].$$

If we took this approach, we would immediately obtain that $[\mathbb{Q}(i) : \mathbb{Q}] = 2$. However, it takes some work to show that $\sqrt{3} \notin \mathbb{Q}(i)$. Therefore, the first approach is the easier one. On the other hand, note that since the first approach told us that $[\mathbb{Q}(\sqrt{3}, i) : \mathbb{Q}] = 4$, we could combine that with equation (3) to see that $[\mathbb{Q}(i)(\sqrt{3}) : \mathbb{Q}(i)] = 2$. Therefore, we can use Theorem 15.6 to quickly establish that $\sqrt{3} \notin \mathbb{Q}(i)$.

2. We will now compute $[\mathbb{Q}(7^{\frac{1}{3}}, \mathrm{cis}(\frac{\pi}{5})) : \mathbb{Q}]$. Two approaches are to consider this problem as either

$$\left[\mathbb{Q}(7^{\frac{1}{3}}, \mathrm{cis}(\frac{\pi}{5})) : \mathbb{Q}\right] = \left[\mathbb{Q}(7^{\frac{1}{3}})(\mathrm{cis}(\frac{\pi}{5})) : \mathbb{Q}\right] =$$

$$\left[\mathbb{Q}(7^{\frac{1}{3}})(\mathrm{cis}(\frac{\pi}{5})) : \mathbb{Q}(7^{\frac{1}{3}})\right]\left[\mathbb{Q}(7^{\frac{1}{3}}) : \mathbb{Q}\right]$$

or as

$$\left[\mathbb{Q}(7^{\frac{1}{3}}, \mathrm{cis}(\frac{\pi}{5})) : \mathbb{Q}\right] = \left[\mathbb{Q}(\mathrm{cis}(\frac{\pi}{5}))(7^{\frac{1}{3}}) : \mathbb{Q}\right] =$$

$$\left[\mathbb{Q}(\mathrm{cis}(\frac{\pi}{5}))(7^{\frac{1}{3}}) : \mathbb{Q}(\mathrm{cis}(\frac{\pi}{5}))\right]\left[\mathbb{Q}(\mathrm{cis}(\frac{\pi}{5})) : \mathbb{Q}\right].$$

As we will see, the easiest solution to this problem combines both approaches. Since $7^{\frac{1}{3}}$ is a root of $x^3 - 7$ and $\text{cis}(\frac{\pi}{5})$ is a root of $x^4 - x^3 + x^2 - x + 1$ and both of these polynomials are irreducible in $\mathbb{Q}[x]$, we know that

$$\left[\mathbb{Q}(7^{\frac{1}{3}}) : \mathbb{Q}\right] = 3 \text{ and } \left[\mathbb{Q}(\text{cis}(\frac{\pi}{5})) : \mathbb{Q}\right] = 4.$$

Furthermore, $x^3 - 7$ might not be irreducible in $\mathbb{Q}(\text{cis}(\frac{\pi}{5}))[x]$ and $x^4 - x^3 + x^2 - x + 1$ might not be irreducible in $\mathbb{Q}(7^{\frac{1}{3}})[x]$, so we can also say that

$$\left[\mathbb{Q}(\text{cis}(\frac{\pi}{5}))(7^{\frac{1}{3}}) : \mathbb{Q}(\text{cis}(\frac{\pi}{5}))\right] \leq 3 \text{ and } \left[\mathbb{Q}(7^{\frac{1}{3}})(\text{cis}(\frac{\pi}{5})) : \mathbb{Q}(7^{\frac{1}{3}})\right] \leq 4.$$

Summarizing, we know the following:

(a) $[\mathbb{Q}(7^{\frac{1}{3}}, \text{cis}(\frac{\pi}{5})) : \mathbb{Q}] \leq 12$.

(b) $[\mathbb{Q}(7^{\frac{1}{3}}, \text{cis}(\frac{\pi}{5})) : \mathbb{Q}]$ is a multiple of 4.

(c) $[\mathbb{Q}(7^{\frac{1}{3}}, \text{cis}(\frac{\pi}{5})) : \mathbb{Q}]$ is a multiple of 3.

Since $[\mathbb{Q}(7^{\frac{1}{3}}, \text{cis}(\frac{\pi}{5})) : \mathbb{Q}]$ is divisible by the relatively prime integers 3 and 4, it must be divisible by their product, $12 = 3 \cdot 4$. Combined with the fact that $[\mathbb{Q}(7^{\frac{1}{3}}, \text{cis}(\frac{\pi}{5})) : \mathbb{Q}] \leq 12$, we can now conclude that $[\mathbb{Q}(7^{\frac{1}{3}}, \text{cis}(\frac{\pi}{5})) : \mathbb{Q}] = 12$. Observe that at this point, we could now use Theorem 15.6 to assert that $x^3 - 7$ is irreducible in $\mathbb{Q}(\text{cis}(\frac{\pi}{5}))[x]$ and $x^4 - x^3 + x^2 - x + 1$ is irreducible in $\mathbb{Q}(7^{\frac{1}{3}})[x]$.

3.   We now turn our attention to $[\mathbb{Q}(\sqrt{3}, i, \sqrt{2}, 7^{\frac{1}{5}}, 11^{\frac{1}{7}}) : \mathbb{Q}]$. Solving this problem will require applying Theorem 15.6 several times. First we will consider

$$\left[\mathbb{Q}(\sqrt{3}, i, \sqrt{2}) : \mathbb{Q}\right], \left[\mathbb{Q}(7^{\frac{1}{5}}) : \mathbb{Q}\right], \text{ and } \left[\mathbb{Q}(11^{\frac{1}{7}}) : \mathbb{Q}\right].$$

We know that $[\mathbb{Q}(\sqrt{3}) : \mathbb{Q}] = 2$. You should do the computations necessary to convince yourself that $\sqrt{2} \notin \mathbb{Q}(\sqrt{3})$. Combined with the fact that $\sqrt{2}$ is a root of $x^2 - 2$, it follows that $[\mathbb{Q}(\sqrt{3})(\sqrt{2}) : \mathbb{Q}(\sqrt{3})] = 2$. We can now use Theorem 15.6 to see that $[\mathbb{Q}(\sqrt{3}, \sqrt{2}) : \mathbb{Q}] = 4$. Note that $\mathbb{Q}(\sqrt{3}, \sqrt{2}) \subseteq \mathbb{R}$, so $i \notin \mathbb{Q}(\sqrt{3}, \sqrt{2})$. When we combine this with the fact that i is a root of $x^2 + 1$, we see that $[\mathbb{Q}(\sqrt{3}, \sqrt{2})(i) : \mathbb{Q}(\sqrt{3}, \sqrt{2})] = 2$. Theorem 15.6 now tells us that $[\mathbb{Q}(\sqrt{3}, i, \sqrt{2}) : \mathbb{Q}] = 8$.

Observe that $7^{\frac{1}{5}}$ is a root of $x^5 - 7$ and $11^{\frac{1}{7}}$ is a root of $x^7 - 11$, and both of these polynomials are irreducible in $\mathbb{Q}[x]$ using Eisenstein's Criterion. Therefore, $[\mathbb{Q}(7^{\frac{1}{5}}) : \mathbb{Q}] = 5$ and $[\mathbb{Q}(11^{\frac{1}{7}}) : \mathbb{Q}] = 7$.

Theorem 15.6 tells us that $[\mathbb{Q}(\sqrt{3}, i, \sqrt{2}, 7^{\frac{1}{5}}, 11^{\frac{1}{7}}) : \mathbb{Q}]$ must be divisible by $[\mathbb{Q}(\sqrt{3}, i, \sqrt{2} : \mathbb{Q}], [\mathbb{Q}(7^{\frac{1}{5}}) : \mathbb{Q}]$, and $[\mathbb{Q}(11^{\frac{1}{7}}) : \mathbb{Q}]$. In light of our preceding work, $[\mathbb{Q}(\sqrt{3}, i, \sqrt{2}, 7^{\frac{1}{5}}, 11^{\frac{1}{7}}) : \mathbb{Q}]$ is divisible by 8, 5, and 7. Since any pair of these numbers is relatively prime, it follows that $[\mathbb{Q}(\sqrt{3}, i, \sqrt{2}, 7^{\frac{1}{5}}, 11^{\frac{1}{7}}) : \mathbb{Q}]$ is divisible by the product $8 \cdot 5 \cdot 7 = 280$.

Now let's take a slightly different approach. Using Theorem 15.6, we have

$$\left[\mathbb{Q}(\sqrt{3}, i, \sqrt{2}, 7^{\frac{1}{5}}, 11^{\frac{1}{7}}) : \mathbb{Q}\right] =$$

$$\left[\mathbb{Q}(\sqrt{3}, i, \sqrt{2}, 7^{\frac{1}{5}})(11^{\frac{1}{7}}) : \mathbb{Q}(\sqrt{3}, i, \sqrt{2}, 7^{\frac{1}{5}})\right]\left[\mathbb{Q}(\sqrt{3}, i, \sqrt{2})(7^{\frac{1}{5}}) : \mathbb{Q}(\sqrt{3}, i, \sqrt{2})\right]$$

$$\left[\mathbb{Q}(\sqrt{3}, i, \sqrt{2}) : \mathbb{Q}\right].$$

Since $11^{\frac{1}{7}}$ is a root of $x^7 - 11$ and $x^7 - 11$ may or may not be irreducible in $\mathbb{Q}(\sqrt{3}, i, \sqrt{2}, 7^{\frac{1}{5}})[x]$, we know that

$$\left[\mathbb{Q}(\sqrt{3}, i, \sqrt{2}, 7^{\frac{1}{5}})(11^{\frac{1}{7}}) : \mathbb{Q}(\sqrt{3}, i, \sqrt{2}, 7^{\frac{1}{5}})\right] \le 7.$$

Similarly, $7^{\frac{1}{5}}$ is a root of $x^5 - 7$ and $x^5 - 7$ may or may not be irreducible in $\mathbb{Q}(\sqrt{3}, i, \sqrt{2})[x]$, so

$$\left[\mathbb{Q}(\sqrt{3}, i, \sqrt{2})(7^{\frac{1}{5}}) : \mathbb{Q}(\sqrt{3}, i, \sqrt{2})\right] \le 5.$$

In addition, we already showed that $[\mathbb{Q}(\sqrt{3}, i, \sqrt{2}) : \mathbb{Q}] = 8$. Therefore, Theorem 15.6 now tells us that

$$\left[\mathbb{Q}(\sqrt{3}, i, \sqrt{2}, 7^{\frac{1}{5}}, 11^{\frac{1}{7}}) : \mathbb{Q}\right] =$$

$$\left[\mathbb{Q}(\sqrt{3}, i, \sqrt{2}, 7^{\frac{1}{5}})(11^{\frac{1}{7}}) : \mathbb{Q}(\sqrt{3}, i, \sqrt{2}, 7^{\frac{1}{5}})\right]\left[\mathbb{Q}(\sqrt{3}, i, \sqrt{2})(7^{\frac{1}{5}}) : \mathbb{Q}(\sqrt{3}, i, \sqrt{2})\right]$$

$$\left[\mathbb{Q}(\sqrt{3}, i, \sqrt{2}) : \mathbb{Q}\right] \le 7 \cdot 5 \cdot 8 = 280.$$

Our preceding work now tells us that not only is $[\mathbb{Q}(\sqrt{3}, i, \sqrt{2}, 7^{\frac{1}{5}}, 11^{\frac{1}{7}}) : \mathbb{Q}] \le 280$, but it is also divisible by 280. Combining these facts, we now know that $[\mathbb{Q}(\sqrt{3}, i, \sqrt{2}, 7^{\frac{1}{5}},$

$11^{\frac{1}{7}}) : \mathbb{Q}] = 280$. Once again, we could now go back and use Theorem 15.6 to show that $x^7 - 11$ is irreducible in $\mathbb{Q}(\sqrt{3}, i, \sqrt{2}, 7^{\frac{1}{5}})[x]$ and $x^5 - 7$ is irreducible in $\mathbb{Q}(\sqrt{3}, i, \sqrt{2})[x]$.

∎

If $K \subseteq L$ are fields such that $[L : K]$ is finite, it follows that every $\theta \in L$ is algebraic over K. To see this, suppose $[L : K] = n$ and consider the elements $1, \theta, \ldots, \theta^n$. Observe that this set must be linearly dependent over K, so there exist $\alpha_i \in K$ such that

$$\alpha_0 + \alpha_1 \theta + \cdots + \alpha_n \theta^n = 0,$$

with at least one $\alpha_i \neq 0$. If we let

$$p(x) = \alpha_0 + \alpha_1 x + \cdots + \alpha_n x^n,$$

then $p(x)$ is a nonzero element of $K[x]$ that has θ as a root. Therefore, θ is algebraic over K. This motivates

Definition 15.8. *If $K \subseteq L$ are fields, we say that L is algebraic over K if every element of L is algebraic over K.*

We saw that if $[L : K]$ is finite, then L is algebraic over K. However, the converse does not hold. For example, consider the field $L = \mathbb{Q}(2^{\frac{1}{2}}, 2^{\frac{1}{3}}, 2^{\frac{1}{4}}, 2^{\frac{1}{5}}, \ldots)$. Then L is the smallest field that lies between $\mathbb{Q}$ and $\mathbb{C}$ and contains $2^{\frac{1}{n}}$, for every $n \in \mathbb{N}$. If $\theta \in L$, then θ can be written using only a finite number of roots of 2. Therefore, there exists some $m \in \mathbb{N}$ such that $\theta \in \mathbb{Q}(2^{\frac{1}{2}}, 2^{\frac{1}{3}}, \ldots, 2^{\frac{1}{m}})$. Using Theorem 15.4, it follows that $[\mathbb{Q}(2^{\frac{1}{2}}, 2^{\frac{1}{3}}, \ldots, 2^{\frac{1}{m}}) : \mathbb{Q}]$ is finite. Since $\mathbb{Q}(\theta)$ is a subfield of $\mathbb{Q}(2^{\frac{1}{2}}, 2^{\frac{1}{3}}, \ldots, 2^{\frac{1}{m}})$, it must also be a finite extension $\mathbb{Q}$. Thus, θ is algebraic over $\mathbb{Q}$ and so, L is algebraic over $\mathbb{Q}$.

If $[L : \mathbb{Q}]$ were finite, then $[L : \mathbb{Q}] = t$, for some $t \in \mathbb{N}$. However, L contains the field $\mathbb{Q}(2^{\frac{1}{t+1}})$. Observe that $2^{\frac{1}{t+1}}$ is a root of the polynomial $x^{t+1} - 2$, and, by Eisenstein's Criterion, this polynomial is irreducible in $\mathbb{Q}[x]$. Therefore, Theorem 15.3 asserts that $[\mathbb{Q}(2^{\frac{1}{t+1}}) : \mathbb{Q}] = t + 1$. Thus, $\mathbb{Q}(2^{\frac{1}{t+1}})$ is a subfield of L whose dimension over $\mathbb{Q}$ exceeds that of L, a contradiction. As a result, L is an algebraic extension of $\mathbb{Q}$ that is not a finite extension of $\mathbb{Q}$. However, even though algebraic extensions need not be finite extensions, there is a result on algebraic extensions that is somewhat analogous to Theorem 15.6.

Theorem 15.9. *Let $F \subseteq K \subseteq L$ be fields such that L is algebraic over K and K is algebraic over F. Then L is algebraic over F.*

Proof. Let $\theta \in L$; since L is algebraic over K, there exists some nonzero

$$p(x) = \alpha_0 + \alpha_1 x + \cdots + \alpha_n x^n \in K[x]$$

such that $p(\theta) = 0$. Since K is algebraic over F, each α_i is algebraic over F. Therefore, by Theorem 15.4, we know that $F(\alpha_0, \alpha_1, \ldots, \alpha_n)$ is a finite dimensional field extension of F. However, by Theorem 15.6, we have

$$[F(\alpha_0, \ldots, \alpha_n, \theta) : F] = [F(\alpha_0, \ldots, \alpha_n, \theta) : F(\alpha_0, \alpha_1, \ldots, \alpha_n)][F(\alpha_0, \alpha_1, \ldots, \alpha_n) : F].$$

Observe that $p(x) \in F(\alpha_0, \alpha_1, \ldots, \alpha_n)[x]$ and θ is a root of $p(x)$, so Theorem 15.3 tells us that $[F(\alpha_0, \ldots, \alpha_n, \theta) : F(\alpha_0, \alpha_1, \ldots, \alpha_n)]$ is finite. Since we already know that $[F(\alpha_0, \alpha_1, \ldots, \alpha_n) : F]$ is finite, the preceding equation tells us that $[F(\alpha_0, \ldots, \alpha_n, \theta) : F]$ is finite. Since θ is contained in a finite extension of F, θ is algebraic over F. Thus, every element of L is algebraic over F, as desired. $\qquad\square$

Observe that Theorem 15.9 immediately resolves the problem we faced before Definition 15.5 regarding extensions of the form $K(\theta_1, \theta_2)(\theta_3)(\theta_4, \theta_5)$. Indeed, if $K(\theta_1, \theta_2)$ is an algebraic extension of K and θ_3 is algebraic over $K(\theta_1, \theta_2)$, then Theorem 15.9 implies that θ_3 is algebraic over K. Similarly, if $K(\theta_1, \theta_2)(\theta_3)$ is algebraic over K and θ_4, θ_5 are algebraic over $K(\theta_1, \theta_2)(\theta_3)$, then Theorem 15.9 asserts that θ_4, θ_5 are also algebraic over K. Therefore, if $K \subseteq L$ and $\theta_i \in L$ such that we can construct $K(\theta_1, \theta_2)(\theta_3)(\theta_4, \theta_5)$ using Theorem 15.4, then each θ_i is algebraic over K. As a result, we can also use Theorem 15.4 to construct $K(\theta_1, \theta_2, \theta_3, \theta_4, \theta_5)$ and $K(\theta_1, \theta_2)(\theta_3)(\theta_4, \theta_5) = K(\theta_1, \theta_2, \theta_3, \theta_4, \theta_5)$. A more general way to look at this is that when constructing fields that lie between K and L and contain various $\theta_i \in L$, Theorems 15.4 and 15.9 give us complete freedom regarding both the ordering and grouping of the θ_i.

Exercises for Section 15.1

In exercises 1–8, we examine fields F where $\mathbb{Q} \subseteq F \subseteq Q(\sqrt{7}, 5^{\frac{1}{3}})$.

1. Compute $[\mathbb{Q}(\sqrt{7}) : \mathbb{Q}]$.

2. Compute $[\mathbb{Q}(5^{\frac{1}{3}}) : \mathbb{Q}]$.

3. Compute $[\mathbb{Q}(\sqrt{7}, 5^{\frac{1}{3}}) : \mathbb{Q}]$.

4. Compute $[\mathbb{Q}(\sqrt{7}, 5^{\frac{1}{3}}) : \mathbb{Q}(\sqrt{7})]$.

5. Compute $[\mathbb{Q}(\sqrt{7}, 5^{\frac{1}{3}}) : \mathbb{Q}(5^{\frac{1}{3}})]$.

6. If K is a field such that $\mathbb{Q} \subseteq K \subseteq \mathbb{Q}(\sqrt{7}, 5^{\frac{1}{3}})$, what are the only possible values for $[K : \mathbb{Q}]$?

7. If $\alpha \in Q(\sqrt{7})$ and $\alpha \notin \mathbb{Q}$, find $[\mathbb{Q}(\sqrt{7}) : \mathbb{Q}(\alpha)]$.

8. If $\beta \in Q(5^{\frac{1}{3}})$ and $\beta \notin \mathbb{Q}$, find $[\mathbb{Q}(5^{\frac{1}{3}}) : \mathbb{Q}(\beta)]$.

In exercises 9–27, we examine fields K where $\mathbb{Q} \subseteq K \subseteq Q(14^{\frac{1}{4}}, 31^{\frac{1}{5}}, i)$.

9. Compute $[\mathbb{Q}(14^{\frac{1}{4}}) : \mathbb{Q}]$.

10. Compute $[\mathbb{Q}(31^{\frac{1}{5}}) : \mathbb{Q}]$.

11. Compute $[\mathbb{Q}(i) : \mathbb{Q}]$.

12. Compute $[\mathbb{Q}(14^{\frac{1}{4}}, 31^{\frac{1}{5}}) : \mathbb{Q}]$.

13. Compute $[\mathbb{Q}(14^{\frac{1}{4}}, i) : \mathbb{Q}]$.

14. Compute $[\mathbb{Q}(31^{\frac{1}{5}}, i) : \mathbb{Q}]$.

15. Compute $[\mathbb{Q}(14^{\frac{1}{4}}, 31^{\frac{1}{5}}, i) : \mathbb{Q}]$.

16. Compute $[\mathbb{Q}(14^{\frac{1}{4}}, 31^{\frac{1}{5}}) : \mathbb{Q}(14^{\frac{1}{4}})]$.

17. Compute $[\mathbb{Q}(14^{\frac{1}{4}}, i) : \mathbb{Q}(14^{\frac{1}{4}})]$.

18. Compute $[\mathbb{Q}(14^{\frac{1}{4}}, 31^{\frac{1}{5}}) : \mathbb{Q}(31^{\frac{1}{5}})]$.

19. Compute $[\mathbb{Q}(31^{\frac{1}{5}}, i) : \mathbb{Q}(31^{\frac{1}{5}})]$.

20. Compute $[\mathbb{Q}(14^{\frac{1}{4}}, i) : \mathbb{Q}(i)]$.

21. Compute $[\mathbb{Q}(31^{\frac{1}{5}}, i) : \mathbb{Q}(i)]$.

22. Compute $[\mathbb{Q}(14^{\frac{1}{4}}, 31^{\frac{1}{5}}, i) : \mathbb{Q}(14^{\frac{1}{4}}, 31^{\frac{1}{5}})]$.

23. Compute $[\mathbb{Q}(14^{\frac{1}{4}}, 31^{\frac{1}{5}}, i) : \mathbb{Q}(14^{\frac{1}{4}}, i)]$.

24. Compute $[\mathbb{Q}(14^{\frac{1}{4}}, 31^{\frac{1}{5}}, i) : \mathbb{Q}(31^{\frac{1}{5}}, i)]$.

25. If K is a field such that $\mathbb{Q} \subseteq K \subseteq \mathbb{Q}(14^{\frac{1}{4}}, 31^{\frac{1}{5}}, i)$, what are the only possible values for $[K : \mathbb{Q}]$?

26. If $\alpha \in \mathbb{Q}(31^{\frac{1}{5}})$ and $\alpha \notin \mathbb{Q}$, find $[\mathbb{Q}(\alpha) : \mathbb{Q}]$.

27. Find $\beta \in \mathbb{Q}(14^{\frac{1}{4}})$ such that $[\mathbb{Q}(\beta) : \mathbb{Q}] = 2$.

28. If $F, K_1, K_2, \ldots, K_n, L$ are fields, such that $F \subseteq K_i \subseteq L$, for all i, show that $K = \bigcap_{i=1}^{n} K_i$ is a field such that $F \subseteq K \subseteq L$.

In exercises 29–34, let $a, b \in \mathbb{C}$ such that $[\mathbb{Q}(a) : \mathbb{Q}] = m$ and $[\mathbb{Q}(b) : \mathbb{Q}] = n$, where m and n are relatively prime.

29. Compute $[\mathbb{Q}(a, b) : \mathbb{Q}]$.

30. Compute $[\mathbb{Q}(a, b) : \mathbb{Q}(a)]$ and then check if this is equal to $[\mathbb{Q}(b) : \mathbb{Q}]$.

31. Compute $[\mathbb{Q}(a, b) : \mathbb{Q}(b)]$ and then check if this is equal to $[\mathbb{Q}(a) : \mathbb{Q}]$.

32. If $c \in \mathbb{Q}(a)$ and $d \in \mathbb{Q}(b)$, show that $[\mathbb{Q}(c) : \mathbb{Q}]$ and $[\mathbb{Q}(d) : \mathbb{Q}]$ are relatively prime.

33. If $c \in \mathbb{Q}(a)$ and $d \in \mathbb{Q}(b)$, show that $[\mathbb{Q}(c, d) : \mathbb{Q}] = [\mathbb{Q}(c) : \mathbb{Q}] \cdot [\mathbb{Q}(d) : \mathbb{Q}]$.

34. Show that $\mathbb{Q}(a) \cap \mathbb{Q}(b) = \mathbb{Q}$.

In exercises 35–44, we examine fields L where $\mathbb{Q} \subseteq L \subseteq Q(2^{\frac{1}{6}}, 2^{\frac{1}{10}})$. If $\alpha, \beta \in \mathbb{C}$ are algebraic over $\mathbb{Q}$, then $[\mathbb{Q}(\alpha, \beta) : \mathbb{Q}(\alpha)] \leq [\mathbb{Q}(\beta) : \mathbb{Q}]$ and $[\mathbb{Q}(\alpha, \beta) : \mathbb{Q}(\beta)] \leq [\mathbb{Q}(\alpha) : \mathbb{Q}]$. In many of the preceding exercises, equality occurred. However, as we shall see in the following exercises, equality does not always occur.

35. Show $\mathbb{Q}(2^{\frac{1}{6}}, 2^{\frac{1}{10}}) = \mathbb{Q}(2^{\frac{1}{30}})$.

36. Compute $[\mathbb{Q}(2^{\frac{1}{6}}, 2^{\frac{1}{10}}) : \mathbb{Q}]$. If necessary, use exercise 35.

37. Compute $[\mathbb{Q}(2^{\frac{1}{6}}) : \mathbb{Q}]$.

38. Compute $[\mathbb{Q}(2^{\frac{1}{10}}) : \mathbb{Q}]$.

39. Compute $[\mathbb{Q}(2^{\frac{1}{6}}, 2^{\frac{1}{10}}) : \mathbb{Q}(2^{\frac{1}{6}})]$ and check if this is equal to $[\mathbb{Q}(2^{\frac{1}{10}}) : \mathbb{Q}]$.

40. Find the minimum polynomial for $2^{\frac{1}{10}}$ over $\mathbb{Q}$ and over $\mathbb{Q}(2^{\frac{1}{6}})$.

41. Compute $[\mathbb{Q}(2^{\frac{1}{6}}, 2^{\frac{1}{10}}) : \mathbb{Q}(2^{\frac{1}{10}})]$ and check if this is equal to $[\mathbb{Q}(2^{\frac{1}{6}}) : \mathbb{Q}]$.

42. Find the minimum polynomial for $2^{\frac{1}{6}}$ over $\mathbb{Q}$ and over $\mathbb{Q}(2^{\frac{1}{10}})$.

43. Compute $[\mathbb{Q}(2^{\frac{1}{6}}, 2^{\frac{1}{10}}) : \mathbb{Q}(2^{\frac{1}{2}})]$.

44. Show that $\mathbb{Q}(2^{\frac{1}{6}}) \cap \mathbb{Q}(2^{\frac{1}{10}}) = \mathbb{Q}(2^{\frac{1}{2}})$.

In exercises 45–54, we examine fields E where $\mathbb{Q} \subseteq E \subseteq Q(7^{\frac{1}{6}}, 7^{\frac{1}{10}})$.

45. Show $\mathbb{Q}(7^{\frac{1}{6}}, 7^{\frac{1}{15}}) = \mathbb{Q}(7^{\frac{1}{30}})$.

46. Compute $[\mathbb{Q}(7^{\frac{1}{6}}, 7^{\frac{1}{15}}) : \mathbb{Q}]$. If necessary, use exercise 45.

47. Compute $[\mathbb{Q}(7^{\frac{1}{6}}) : \mathbb{Q}]$.

48. Compute $[\mathbb{Q}(7^{\frac{1}{15}}) : \mathbb{Q}]$.

49. Compute $[\mathbb{Q}(7^{\frac{1}{6}}, 7^{\frac{1}{15}}) : \mathbb{Q}(7^{\frac{1}{6}})]$ and check if this is equal to $[\mathbb{Q}(7^{\frac{1}{15}}) : \mathbb{Q}]$.

50. Find the minimum polynomial for $7^{\frac{1}{15}}$ over $\mathbb{Q}$ and over $\mathbb{Q}(7^{\frac{1}{6}})$.

51. Compute $[\mathbb{Q}(7^{\frac{1}{6}}, 7^{\frac{1}{15}}) : \mathbb{Q}(7^{\frac{1}{15}})]$ and check if this is equal to $[\mathbb{Q}(7^{\frac{1}{6}}) : \mathbb{Q}]$.

52. Find the minimum polynomial for $7^{\frac{1}{6}}$ over $\mathbb{Q}$ and over $\mathbb{Q}(7^{\frac{1}{15}})$.

53. Compute $[\mathbb{Q}(7^{\frac{1}{6}}, 7^{\frac{1}{15}}) : \mathbb{Q}(7^{\frac{1}{3}})]$.

54. Show that $\mathbb{Q}(7^{\frac{1}{6}}) \cap \mathbb{Q}(7^{\frac{1}{15}}) = \mathbb{Q}(7^{\frac{1}{3}})$.

In exercises 55–68, we examine fields K where $\mathbb{Q} \subseteq K \subseteq Q(\sqrt{2}, \sqrt{3}, i)$.

55. Compute $[\mathbb{Q}(\sqrt{2}) : \mathbb{Q}]$.

56. Compute $[\mathbb{Q}(\sqrt{3}) : \mathbb{Q}]$.

57. Compute $[\mathbb{Q}(i) : \mathbb{Q}]$.

58. Compute $[\mathbb{Q}(\sqrt{2}, \sqrt{3}) : \mathbb{Q}]$.

59. Compute $[\mathbb{Q}(\sqrt{2}, i) : \mathbb{Q}]$.

60. Compute $[\mathbb{Q}(\sqrt{3}, i) : \mathbb{Q}]$.

61. Compute $[\mathbb{Q}(\sqrt{2}, \sqrt{3}, i) : \mathbb{Q}]$.

62. Compute both $[\mathbb{Q}(\sqrt{6}) : \mathbb{Q}]$ and $[\mathbb{Q}(\sqrt{2}, \sqrt{3}) : \mathbb{Q}(\sqrt{6})]$.

63. Compute both $[\mathbb{Q}(i\sqrt{2}) : \mathbb{Q}]$ and $[\mathbb{Q}(\sqrt{2}, i) : \mathbb{Q}(i\sqrt{2})]$.

64. Compute both $[\mathbb{Q}(i\sqrt{3}) : \mathbb{Q}]$ and $[\mathbb{Q}(\sqrt{3}, i) : \mathbb{Q}(i\sqrt{3})]$.

65. Compute both $[\mathbb{Q}(\sqrt{2} + \sqrt{3}) : \mathbb{Q}]$ and $[\mathbb{Q}(\sqrt{2}, \sqrt{3}) : \mathbb{Q}(\sqrt{2} + \sqrt{3})]$.

66. Compute both $[\mathbb{Q}(\sqrt{2} + \sqrt{i}) : \mathbb{Q}]$ and $[\mathbb{Q}(\sqrt{2}, i) : \mathbb{Q}(\sqrt{2} + i)]$.

67. Compute both $[\mathbb{Q}(\sqrt{3} + i) : \mathbb{Q}]$ and $[\mathbb{Q}(\sqrt{3}, i) : \mathbb{Q}(\sqrt{3} + i)]$.

68. Compute $[\mathbb{Q}(\sqrt{2} + \sqrt{3} + i) : \mathbb{Q}]$.

69. If $K \subseteq L$ are fields, show that $a \in L$ is algebraic over K if and only if a^2 is algebraic over K.

70. Let $K \subseteq L$ be fields where $a \in L$ and $f(x) \in K[x]$ has degree at least one. In a generalization of exercise 69, show that a is algebraic over K if and only if $f(a)$ is algebraic over K.

71. If $K \subseteq L$ are fields, show that the elements of L that are algebraic over K are a field.

72. If $K \subseteq L$ are fields and $a \in L$ is algebraic over K and $b \in L$ is not algebraic over K, show that $a + b$ is not algebraic over K.

15.2 Simple Extensions

In the previous section, we began by looking at fields of the form $\mathbb{Q}(\sqrt{3})$, $\mathbb{Q}(i)$, and $\mathbb{Q}(7^{\frac{1}{3}})$ and then more complicated fields like $\mathbb{Q}(\sqrt{3}, i, \sqrt{2}, 7^{\frac{1}{3}}, 11^{\frac{1}{7}})$. It certainly appears the more elements we attach to a field, the more complicated things become. However, it turns out that

$$\mathbb{Q}\left(\sqrt{3}, i\right) = \mathbb{Q}\left(\sqrt{3} + i\right) \quad \text{and} \quad \mathbb{Q}\left(\sqrt{2}, \sqrt{3}\right) = \mathbb{Q}\left(\sqrt{2} + \sqrt{3}\right).$$

Therefore, we can sometimes take a field that is generated by more than one element and simplify the situation by generating it by a single element. It is natural to ask, how often we can do this? To answer this, we begin with

Definition 15.10. *If $K \subseteq L$ are fields with $[L : K]$ finite, we say that L is a simple extension if there exists some $\theta \in L$ such that $L = K(\theta)$.*

The goal of this section is to show that whenever $\mathbb{Q} \subseteq K \subseteq L \subseteq \mathbb{C}$ are fields such that $[L : K]$ is finite, then L is a simple extension of K. It is not hard to prove slightly more general results, but for the applications we have in mind, it suffices to look at fields that lie between $\mathbb{Q}$ and $\mathbb{C}$. We first need a technical lemma about linear functions.

Lemma 15.11. *Let A, B be finite subsets of $\mathbb{C}$ and let $\alpha, \beta \in \mathbb{C}$, where $\alpha \notin A$. Then there exists a nonzero integer m such that the linear function*

$$l(x) = m(x - \alpha) + \beta$$

has the property that $l(a) \notin B$, for all $a \in A$.

Proof. Since A and B are finite sets, there are only a finite number of elements of $\mathbb{C}$ that are of the form $\frac{b-\beta}{a-\alpha}$, where $a \in A$ and $b \in B$. Therefore, we can choose some nonzero $m \in \mathbb{Z}$ that is not of this form. We claim that the function $l(x) = m(x - \alpha) + \beta$ has the desired property. By way of contradiction, suppose $a \in A$ such that $l(a) = b \in B$. Then we have

$$b = m(a - \alpha) + \beta,$$

and solving for m yields $m = \frac{b-\beta}{a-\alpha}$, which contradicts our choice of m. Thus, $l(x)$ has the desired property. $\square$

We can now prove

Theorem 15.12. *Let $\mathbb{Q} \subseteq K \subseteq L \subseteq \mathbb{C}$ be such that $[L : K]$ is finite. Then there exists some $\theta \in L$ such that $L = K(\theta)$.*

Proof. Since $[L : K]$ is finite, there exists a finite set $\{\theta_1, \ldots, \theta_n\}$ that is the basis for L over K. In particular, this says that $L = K(\theta_1, \ldots, \theta_n)$. Therefore, if $[L : K]$ is finite, there always exists some $n \in \mathbb{N}$ such that $L = K(\theta_1, \ldots, \theta_n)$. Therefore, it will now suffice to use Mathematical Induction to show that whenever $L = K(\theta_1, \ldots, \theta_n)$, there exists $\theta \in L$ such that $L = K(\theta)$.

The case where $n = 1$ is obvious, so we need to show that if the result holds for some $k \in \mathbb{N}$, then it also holds for $k+1$. Therefore, let us consider the field $L = K(\theta_1, \ldots, \theta_k, \theta_{k+1})$. Observe that if we let $K_1 = K(\theta_1, \ldots, \theta_k)$, then the fact that the result holds for k tells us that there exists $\alpha \in K_1 \subseteq L$ such that $K_1 = K(\alpha)$. For convenience, we can let $\beta = \theta_{k+1}$ and using this notation, we now have

$$L = K(\theta_1, \ldots, \theta_k, \theta_{k+1}) = K(\theta_1, \ldots, \theta_k)(\theta_{k+1}) = K(\alpha)(\beta) = K(\alpha, \beta).$$

Since α, β are algebraic over K, we can let $f(x), g(x)$ be, respectively, the minimum polynomials for α and β in $K[x]$. Next, we can apply the Fundamental Theorem of Algebra to assert that both $f(x)$ and $g(x)$ can be factored into linear factors in $\mathbb{C}[x]$. Now let A be all the roots of $f(x)$ other than α and let B be all the roots of $g(x)$.

Using A, B, α, β from the previous paragraph, we can let $l(x) = m(x - \alpha) + \beta$ be the linear function described in Lemma 15.11. We can rewrite $l(x)$ as $mx + (\beta - m\alpha)$, so $l(x) \in K(\beta - m\alpha)[x]$. If we let $h(x) = g(l(x))$, then since $f(x), g(x) \in K[x] \subseteq K(\beta - m\alpha)[x]$, we can see that the polynomials $f(x)$ and $h(x)$ both belong to the polynomial ring $K(\beta - m\alpha)[x]$.

Observe that

$$h(\alpha) = g(l(\alpha)) = g(m(\alpha - \alpha) + \beta) = g(\beta) = 0.$$

Therefore, α is a root of both $h(x)$ and $f(x)$. On the other hand, suppose a is any root of $f(x)$ other than α. Then, by our choice of $l(x)$, $l(a)$ is not a root of $g(x)$. Hence, computing in $\mathbb{C}$, we have

$$h(a) = g(l(a)) \neq 0.$$

Thus, α is the only root that $f(x)$ and $h(x)$ have in common.

Computing in the polynomial ring $K(\beta - m\alpha)[x]$, we can let $c(x)$ denote the greatest common divisor of $f(x)$ and $h(x)$. Recall that Lemma 12.18 tells us that $c(x)$ is the greatest common divisor of $f(x)$ and $h(x)$, even if we work in the larger polynomial ring $\mathbb{C}[x]$. Observe that the roots in $\mathbb{C}$ of $c(x)$ are precisely the roots that $f(x)$ and $h(x)$ have in common. As a result, α is the only root of $c(x)$. Furthermore, since $f(x)$ is irreducible in $K[x]$, Corollary 12.23 tells us that $f(x)$ does not have any multiple roots in $\mathbb{C}[x]$. Since any multiple root of $c(x)$ would also be a multiple root of $f(x)$, it follows that α is a root of $c(x)$ of multiplicity one, so

$$c(x) = x - \alpha \in K(\beta - m\alpha)[x].$$

Every coefficient of $c(x)$ must belong to $K(\beta - m\alpha)$, so $\alpha \in K(\beta - m\alpha)$, which immediately tells us that

$$\beta = (\beta - m\alpha) + m \cdot \alpha \in K(\beta - m\alpha).$$

Since $\alpha, \beta \in K(\beta - m\alpha)$, we have

$$K(\beta - m\alpha) = K(\alpha, \beta) = L.$$

If we now let $\theta = \beta - m\alpha$, then $L = K(\theta)$, as desired. $\qquad\square$

The combination of Lemma 15.11 and Theorem 15.12 not only tells us that finite extensions $K \subseteq L$ that lie between $\mathbb{Q}$ and $\mathbb{C}$ are simple extensions, it also explains how to find an appropriate θ such that $L = K(\theta)$.

■ Examples

1. Consider $\mathbb{Q}(\sqrt{3}, i)$; we can begin by letting $\alpha = i$ and $\beta = \sqrt{3}$. According to the arguments in Lemma 15.11 and Theorem 15.12, we only need to avoid nonzero integers that are of the form $\frac{b - \sqrt{3}}{a - i}$, where a is a root of $x^2 + 1$ and b is a root of $x^2 - 3$. Since this restriction does not exclude any nonzero integers, it follows that $\mathbb{Q}(\sqrt{3} - mi) = \mathbb{Q}(\sqrt{3}, i)$, for every nonzero integer m. In particular, $\mathbb{Q}(\sqrt{3} + i) = \mathbb{Q}(\sqrt{3}, i)$.

2. Let us now examine $\mathbb{Q}(\sqrt{2}, \sqrt{3})$; we can begin by taking $\alpha = \sqrt{3}$ and $\beta = \sqrt{2}$. In this example, the nonzero integers we need to avoid are those of the form $\frac{b - \sqrt{2}}{a - \sqrt{3}}$, where a is a root of $x^2 - 3$ and b is a root of $x^2 - 2$. Once again, this restriction does not exclude any nonzero integers, thus $\mathbb{Q}(\sqrt{2} - m\sqrt{3}) = \mathbb{Q}(\sqrt{2}, \sqrt{3})$, for every nonzero integer m. In particular, $\mathbb{Q}(\sqrt{2} + \sqrt{3}) = \mathbb{Q}(\sqrt{2}, \sqrt{3})$.

3. For a more involved example, let us combine the previous examples and consider $\mathbb{Q}(\sqrt{2}, \sqrt{3}, i)$. From our work in the previous example, we know that $\mathbb{Q}(\sqrt{2}, \sqrt{3}, i) = \mathbb{Q}(\sqrt{2} + \sqrt{3}, i)$. This time, let $\alpha = i$ and $\beta = \sqrt{2} + \sqrt{3}$. The only nonzero integers we need to avoid are those of the form $\frac{b - (\sqrt{2} + \sqrt{3})}{a - i}$, where a is a root of $x^2 + 1$ and b is a root of the minimum polynomial for $\sqrt{2} + \sqrt{3}$. Observe that $\sqrt{2} + \sqrt{3}$ is a root of $x^4 - 10x^2 + 1$, and it follows from Theorem 9.15 that $x^4 - 10x^2 + 1$ is the minimum polynomial for $\sqrt{2} + \sqrt{3}$ in $\mathbb{Q}[x]$. The four roots of $x^4 - 10x^2 + 1$ are easily seen to be $\pm\sqrt{2} \pm \sqrt{3}$. Therefore, when we are looking at objects of the form $\frac{b - (\sqrt{2} + \sqrt{3})}{a - i}$, b is always a real number, and the only choice for a is $-i$. In particular, the nonzero values of $\frac{b - (\sqrt{2} + \sqrt{3})}{a - i}$ are never real numbers. Thus,

$\mathbb{Q}(\sqrt{2}, \sqrt{3}, i) = \mathbb{Q}(\sqrt{2} + \sqrt{3} - mi)$, for every nonzero integer m. In particular, $\mathbb{Q}(\sqrt{2}, \sqrt{3}, i) = \mathbb{Q}(\sqrt{2} + \sqrt{3} + i)$.

4. The preceding examples might be somewhat misleading, as you might be tempted to believe that whenever $\mathbb{Q}(\alpha, \beta)$ is a finite extension of $\mathbb{Q}$, then $\mathbb{Q}(\alpha, \beta) = \mathbb{Q}(\alpha + \beta)$. We now provide an example that shows that this is not the case. Consider $\mathbb{Q}(\sqrt{2} + i, \sqrt{3} - i)$; if we let $\alpha = \sqrt{2} + i$ and $\beta = \sqrt{3} - i$, then $\alpha + \beta = \sqrt{2} + \sqrt{3}$. However, $\mathbb{Q}(\sqrt{2} + \sqrt{3}) \subseteq \mathbb{R}$, whereas $\mathbb{Q}(\sqrt{2} + i, \sqrt{3} - i)$ contains elements of $\mathbb{C}$ which do not belong to $\mathbb{R}$. Thus,

$$\mathbb{Q}(\sqrt{2} + i, \sqrt{3} - i) \neq \mathbb{Q}(\sqrt{2} + \sqrt{3}).$$

Observe that $\alpha + \beta$ corresponds to the value of $\beta - m\alpha$ when $m = -1$ and if we examine the roots of the minimum polynomials for $\sqrt{2} + i, \sqrt{3} - i$, we can see that the only nonzero integer excluded by Lemma 15.11 and Theorem 15.12 is $m = -1$. Therefore,

$$\mathbb{Q}(\sqrt{2} + i, \sqrt{3} - i) = \mathbb{Q}((\sqrt{3} - i) - m(\sqrt{2} + i)),$$

for any integer m other than 0 and -1. In particular, if we let $m = 1$, we obtain

$$\mathbb{Q}(\sqrt{2} + i, \sqrt{3} - i) = \mathbb{Q}((\sqrt{3} - i) - (\sqrt{2} + i)).$$

∎

Since Theorem 15.12 allows us to view various finite extensions as simple extensions, it greatly assists us in obtaining our first result comparing the size of a Galois group to the degree of a field extension. Recall that Galois groups were defined back in Definition 5.17. Also in Chapter 5, Corollary 5.13 generalized the familiar fact that roots of polynomials in $\mathbb{R}[x]$ occur in conjugate pairs in $\mathbb{C}$. In the following proof, we will use Corollary 5.13 as it tells us that if $g \in Gal(L/K)$ and if $\theta \in L$ is the root of some $p(x) \in K[x]$, then $g(\theta)$ is also a root of $p(x)$.

Theorem 15.13. *Let $\mathbb{Q} \subseteq K \subseteq L \subseteq \mathbb{C}$ be fields such that $[L : K]$ is finite. Then $|Gal(L/K)| \leq [L : K]$.*

Proof. If $g \in Gal(L/K)$, we need to see how g can behave on the various elements of L. Since $[L : K]$ is finite, Theorem 15.12 asserts that $L = K(\theta)$, for some $\theta \in L$. Let $n = [L : K]$; therefore, if $p(x)$ is the minimum polynomial in $K[x]$ for θ then, by Theorem 15.3, we know that $p(x)$ has degree n. If $v \in L$, Theorem 15.3 also tells us that

$$v = \alpha_0 + \alpha_1\theta + \cdots + \alpha_{n-1}\theta^{n-1},$$

where every $\alpha_i \in K$. Since g is an automorphism that is the identity map on K, we have

$$g(v) = g(\alpha_0 + \alpha_1\theta + \cdots + \alpha_{n-1}\theta^{n-1}) = g(\alpha_0) + g(\alpha_1\theta) + \cdots + g(\alpha_{n-1}\theta^{n-1}) =$$
$$g(\alpha_0) + g(\alpha_1)g(\theta) + \cdots + g(\alpha_{n-1})g(\theta^{n-1}) = \alpha_0 + \alpha_1 g(\theta) + \cdots + \alpha_{n-1} g(\theta^{n-1}) =$$
$$\alpha_0 + \alpha_1 g(\theta) + \cdots + \alpha_{n-1} g(\theta)^{n-1}.$$

As a result, if we knew $g(\theta)$, then we would immediately know $g(v)$. In other words, the behavior of g on θ completely determines its behavior on every element of L. By Corollary 5.13, $g(\theta)$ must also be a root of $p(x)$. However, $p(x)$ has at most m roots $\theta_1, \ldots, \theta_m$ in L, where $m \leq n$. Therefore, the only possible values for $g(\theta)$ are $\theta_1, \ldots, \theta_m$. At this point, it is not clear that each θ_i actually results in an element of $Gal(L/K)$. However, it is clear that there are at most $m \leq n$ choices for $g(\theta)$, hence

$$|Gal(L/K)| \leq m \leq n = [L : K],$$

as desired. $\square$

■ Examples

1. Since $[\mathbb{C} : \mathbb{R}] = 2$, Theorem 15.13 tells us that $Gal(\mathbb{C}/\mathbb{R}) \leq 2$. However, since we know that the identity map and complex conjugation are two different elements of $|Gal(\mathbb{C}/\mathbb{R})| \leq 2$, we know that

    $$|Gal(\mathbb{C}/\mathbb{R})| = 2 = [\mathbb{C} : \mathbb{R}].$$

2. We know that $[\mathbb{Q}(\sqrt{2}) : \mathbb{Q}] = 2$; therefore, Theorem 15.13 tells us that $|Gal(\mathbb{Q}(\sqrt{2})/\mathbb{Q})| \leq 2$. On the other hand, in Chapter 5 we showed that the identity map and the function defined as $\sigma(a + b\sqrt{2}) = a - b\sqrt{2}$, for all $a, b \in \mathbb{Q}$, are both elements of $Gal(\mathbb{Q}(\sqrt{2})/\mathbb{Q})$. Combining these observations, we have

    $$|Gal(\mathbb{Q}(\sqrt{2})/\mathbb{Q})| = 2 = [\mathbb{Q}(\sqrt{2}) : \mathbb{Q}].$$

    In the two preceding examples, we had $|Gal(L/K)| = [L : K]$. It turns out that when examining the relationship between groups, fields, and polynomials, this is the ideal situation. However, the following example, which we first saw in Chapter 5, indicates that this ideal situation does not always occur.

3. The degree of $\mathbb{Q}(2^{\frac{1}{3}})$ over $\mathbb{Q}$ is 3. In addition, if $g \in Gal(\mathbb{Q}(2^{\frac{1}{3}})/\mathbb{Q})$, then g sends $2^{\frac{1}{3}}$ to a root of $x^3 - 2$ that belongs to $\mathbb{Q}(2^{\frac{1}{3}})$. However, the other two roots of $x^3 - 2$ are not real numbers and $\mathbb{Q}(2^{\frac{1}{3}}) \subseteq \mathbb{R}$. Therefore, $2^{\frac{1}{3}}$ is the only root of $x^2 - 3$ that

belongs to $\mathbb{Q}(2^{\frac{1}{3}})$. Hence $g(2^{\frac{1}{3}}) = 2^{\frac{1}{3}}$ and g is the identity map on all of $\mathbb{Q}(2^{\frac{1}{3}})$. As a result,

$$\left| Gal(\mathbb{Q}(2^{\frac{1}{3}})/\mathbb{Q}) \right| = 1 < 3 = \left[\mathbb{Q}(2^{\frac{1}{3}}) : \mathbb{Q} \right].$$

4. Next we consider $\mathbb{Q}(\sqrt{3}, i)$; since $[\mathbb{Q}(\sqrt{3}, i) : \mathbb{Q}] = 4$, Theorem 15.13 tells us that $|Gal(\mathbb{Q}(\sqrt{3}, i)/\mathbb{Q})| \leq 4$. In Chapter 5, we saw that there were four candidates g_1, g_2, g_3, g_4 for elements of $Gal(\mathbb{Q}(\sqrt{3}, i)/\mathbb{Q})$, and they can each be described in terms of their behavior on $\sqrt{3}$ and i as follows:

$$g_1(\sqrt{3}) = \sqrt{3}, \ g_1(i) = i, \qquad g_2(\sqrt{3}) = -\sqrt{3}, \ g_2(i) = i,$$
$$g_3(\sqrt{3}) = \sqrt{3}, \ g_3(i) = -i, \quad g_4(\sqrt{3}) = -\sqrt{3}, \ g_4(i) = -i.$$

It turns out that each of the preceding four candidates does indeed yield different elements of $Gal(\mathbb{Q}(\sqrt{3}, i)/\mathbb{Q})$, so

$$\left| Gal(\mathbb{Q}(\sqrt{3}, i)/\mathbb{Q}) \right| = 4 = \left[\mathbb{Q}(\sqrt{3}, i) : \mathbb{Q} \right].$$

∎

In the last example, we stated but did not prove that g_1, g_2, g_3, g_4 were all elements of the Galois group. We omitted the proof because, at this point, the computations needed would be very messy. But it raises the question, are there some conditions on K and L that guarantee that $|Gal(L/K)| = [L : K]$? In particular, if we knew that the previous example satisfied these conditions, then we would automatically know that $|Gal(\mathbb{Q}(\sqrt{3}, i)/\mathbb{Q})| = 4 = [\mathbb{Q}(\sqrt{3}, i) : \mathbb{Q}] = 4$. Thus, we would immediately know, without doing messy computations, that all four of the g_i are elements of $Gal(\mathbb{Q}(\sqrt{3}, i)/\mathbb{Q})$. More generally, if we are trying to describe the elements of $Gal(L/K)$, then our job becomes much easier if we already know the size of $Gal(L/K)$. In light of this, the goal of the next section will be to find a natural condition that will indeed guarantee that $|Gal(L/K)| = [L : K]$.

15.3 Splitting Fields and Their Galois Groups

In the previous section, we saw that if $\mathbb{Q} \subseteq K \subseteq L \subseteq \mathbb{C}$, with $[L : K]$ finite, then $|Gal(L/K)| \leq [L : K]$. We then examined some examples where $|Gal(L/K)| = [L : K]$ and others where $|Gal(L/K)| < [L : K]$. This raises the question whether there is some natural condition we can place on L that will guarantee that $|Gal(L/K)| = [L : K]$.

Recall

$$\left| Gal(\mathbb{Q}(\sqrt{2})/\mathbb{Q}) \right| = 2 = \left[\mathbb{Q}(\sqrt{2}) : \mathbb{Q} \right]$$

and

(4)
$$\left| Gal(\mathbb{Q}(2^{\frac{1}{3}})/\mathbb{Q}) \right| = 1 < 3 = \left[\mathbb{Q}(2^{\frac{1}{3}}) : \mathbb{Q} \right].$$

We need to examine what it is about $\mathbb{Q}(\sqrt{2})$ and $\mathbb{Q}(2^{\frac{1}{3}})$ that caused the first part of (4) to be an equality and the second part to be an inequality. Observe that $\mathbb{Q}(\sqrt{2})$ is obtained from $\mathbb{Q}$ by adding on one of the roots of the polynomial $x^2 - 2 \in \mathbb{Q}[x]$. However, it turns out that $\mathbb{Q}(\sqrt{2})$ actually contains all the roots of $x^2 - 2$. On the other hand, $\mathbb{Q}(2^{\frac{1}{3}})$ is obtained from $\mathbb{Q}$ by adding on one root of $x^3 - 2$, but it does not contain the other two roots of $x^3 - 2$. This is no coincidence, and it motivates

Definition 15.14.   *Let $\mathbb{Q} \subseteq K \subseteq \mathbb{C}$ be fields and let $f(x) \in K[x]$. If $\theta_1, \ldots, \theta_m \in \mathbb{C}$ are the roots of $f(x)$, we call the field $K(\theta_1, \ldots, \theta_m)$ the splitting field for $f(x)$ over K.*

Let us now briefly revisit the four examples from the end of the previous section.

■ Examples

1. $\mathbb{C}$ is an extension of $\mathbb{R}$ and is the splitting field over $\mathbb{R}$ of the polynomial $x^2 + 1 \in \mathbb{R}[x]$. In fact, the quadratic formula tells us that if $f(x)$ is any irreducible quadratic in $\mathbb{R}[x]$, then $\mathbb{C}$ is the splitting field for $f(x)$ over $\mathbb{R}$. Furthermore, the Fundamental Theorem of Algebra tells us that if $g(x) \in \mathbb{R}[x]$, then the splitting field for $g(x)$ over $\mathbb{R}$ is either $\mathbb{R}$ or $\mathbb{C}$ depending on whether or not all the roots of $g(x)$ are real.

2. $\mathbb{Q}(\sqrt{2})$ is an extension of $\mathbb{Q}$ and is the splitting field over $\mathbb{Q}$ of $x^2 - 2 \in \mathbb{Q}[x]$. Using the quadratic formula, we can see that $\mathbb{Q}(\sqrt{2})$ is the splitting field over $\mathbb{Q}$ of every polynomial of the form $x^2 - 2ax + (a^2 - 2b^2)$, where $a, b \in \mathbb{Q}$ and $b \neq 0$. Therefore it is also the splitting field over $\mathbb{Q}$ of the polynomials $x^2 - 2x - 1, x^2 - 18$, and $x^2 - 8x - 34$.

3. The field $\mathbb{Q}(2^{\frac{1}{3}})$ is *not* the splitting field of $x^3 - 2$ over $\mathbb{Q}$. If we let $\omega = \mathrm{cis}(\frac{2\pi}{3})$, then the three roots of $x^3 - 2$ are $2^{\frac{1}{3}}, 2^{\frac{1}{3}}\omega, 2^{\frac{1}{3}}\omega^2$. Therefore, $\mathbb{Q}(2^{\frac{1}{3}}, 2^{\frac{1}{3}}\omega, 2^{\frac{1}{3}}\omega^2)$ is the splitting field of $x^3 - 2$. Observe that this field is the same as the field $\mathbb{Q}(2^{\frac{1}{3}}, \omega)$. This is a larger field than $\mathbb{Q}(2^{\frac{1}{3}})$ as $\mathbb{Q}(2^{\frac{1}{3}}) \subseteq \mathbb{R}$, yet $\mathbb{Q}(2^{\frac{1}{3}}, \omega)$ contains ω, which is not an element of $\mathbb{R}$.

4. $\mathbb{Q}(\sqrt{3}, i)$ is the splitting field for $(x^2 - 3)(x^2 + 1) \in \mathbb{Q}[x]$ over $\mathbb{Q}$. Observe that $\mathbb{Q}(\sqrt{3}, i)$ is also the splitting field of $(x^2 - 3)(x^2 + 1)$ over both $\mathbb{Q}(i)$ and $\mathbb{Q}(\sqrt{3})$.

■

The last example illustrates the fact that if L is the splitting field over K of some $f(x) \in K[x]$, then, for any field E with $K \subseteq E \subseteq L$, L is also the splitting field for $f(x)$ over E. Before proving that being a splitting field guarantees equality between the degree of the field extension and the size of the Galois group, we need two lemmas.

Lemma 15.15. *Let $\mathbb{Q} \subseteq K \subseteq L \subseteq \mathbb{C}$ be fields such that L is the splitting field of some $f(x) \in K[x]$. If $\pi : L \to \mathbb{C}$ is an injective homomorphism which is the identity map on K, then $\pi(L) = L$.*

Proof. Let $\theta_1, \ldots, \theta_m \in \mathbb{C}$ be the roots of $f(x)$. Therefore, $L = K(\theta_1, \ldots, \theta_m)$ and, by Theorem 15.4, L is a finite extension of K. Applying Theorem 15.12, there exists some $\alpha \in L$ such that $L = K(\alpha)$. Next, let $p(x) \in K[x]$ be the minimum polynomial for α over K and let n be the degree of $p(x)$.

Since $\alpha \in K(\theta_1, \ldots, \theta_m)$, Theorem 15.4 asserts that α can be written as a linear combination over K of elements of the form $\theta_1{}^{j_1}\theta_2{}^{j_2} \cdots \theta_m{}^{j_m}$, where each $j_t \geq 0$. We would like to apply Corollary 5.13 to π, but Corollary 5.13 is stated in terms of automorphisms, and we haven't yet shown that π is an automorphism of L. However, if you look back at the proofs of Proposition 5.11 and Corollary 5.13, it is enough for π to be an injective homomorphism from L to $\mathbb{C}$ for every step in these proofs to apply to π. Therefore, by applying the conclusion of Corollary 5.13, we can conclude that $\pi(\theta_t)$ is also a root of $f(x)$. In particular, this says that $\pi(\theta_t) = \theta_s$, for some s. Hence, $\pi(\theta_t) \in L = K(\alpha)$. But this immediately implies that $\pi(\alpha)$ is a linear combination over K of elements of the form $\pi(\theta_1)^{j_1}\pi(\theta_2)^{j_2} \cdots \pi(\theta_m)^{j_m}$, all of which belong to $L = K(\alpha)$. As a result, $\pi(\alpha) \in K(\alpha)$.

Since π is a homomorphism that is the identity on K, we have $\pi(K(\alpha)) = K(\pi(\alpha))$. When we combine this with the fact that $\pi(\alpha) \in K(\alpha)$, we see that

$$\pi(L) = \pi(K(\alpha)) = K(\pi(\alpha)) \subseteq K(\alpha) = L.$$

The set $\pi(L)$ is now a vector space over K which is contained in L. If we let $\beta = \pi(\alpha)$, then, since the conclusion of Corollary 5.13 applies to π, β is also a root of the polynomial $p(x)$. As a result, $[K(\beta) : K]$ and $[K(\alpha) : K]$ are equal as they are both equal to the degree of $p(x)$. Hence, $\pi(L) = \pi(K(\alpha)) = K(\beta)$ is a subspace of $L = K(\alpha)$ yet has the same dimension over K as L does. Proposition 14.15 now asserts that $\pi(L) = L$. □

We continue with

Lemma 15.16. *Let $\mathbb{Q} \subseteq K \subseteq \mathbb{C}$ be fields and let $p(x) \in K[x]$ be irreducible of degree n. If $\theta_1, \theta_2 \in \mathbb{C}$ are roots of $p(x)$, define the function $\pi : K(\theta_1) \to K(\theta_2)$ as*

$$\pi \left(\alpha_0 + \alpha_1\theta_1 + \alpha_2{\theta_1}^2 + \cdots + \alpha_{n-1}{\theta_1}^{n-1} \right) = \alpha_0 + \alpha_1\theta_2 + \alpha_2{\theta_2}^2 + \cdots + \alpha_{n-1}{\theta_2}^{n-1},$$

where each $\alpha_i \in K$. Then π is a bijective homomorphism of fields that is the identity map on K.

Except for showing that the function π preserves multiplication, the proof of Lemma 15.16 is fairly straightforward. Before reading the proof, you should try to prove that π is bijective and preserves addition. In the proof, we will repeatedly use the fact that for every element $a \in K(\theta_1)$, there exists a unique polynomial in $K[x]$, which we will denote as $f_a(x)$, such that $a = f_a(\theta_1)$ and either $deg(f_a(x)) < n$ or $f_a(x) = 0$. Clearly, for every $c \in K(\theta_2)$, there also exists a unique polynomial $f_c(x) \in K[x]$ with the same properties such that $c = f_c(\theta_2)$. Representing elements of $K(\theta_1)$ and $K(\theta_2)$ as values of polynomials might appear to make parts of the proof of Lemma 15.16 more complicated than need be, but it will be good practice for the part of the proof where we show that π preserves multiplication.

Proof of Lemma 15.16. By Theorem 15.3, for $i = 1, 2$, $K(\theta_i)$ is a field extension of K with basis $\{1, \theta_i, {\theta_i}^2, \ldots, {\theta_i}^{n-1}\}$. If $c \in K(\theta_2)$, let $f_c(x) \in K[x]$ be the polynomial described above such that $c = f_c(\theta_2)$. Observe that

$$c = f_c(\theta_2) = \pi(f_c(\theta_1)),$$

hence π is surjective.

Next, if $a, b \in K(\theta_1)$, let $f_a(x), f_b(x) \in K[x]$ be the polynomials just described such that $a = f_a(\theta_1)$ and $b = f_b(\theta_1)$. If $\pi(a) = \pi(b)$, then we have

$$f_a(\theta_2) = \pi(f_a(\theta_1)) = \pi(a) = \pi(b) = \pi(f_b(\theta_1)) = f_b(\theta_2).$$

Therefore, θ_2 is a root of $f_a(x) - f_b(x)$. Observe that $f_a(x) - f_b(x)$ either has degree less than the degree of $p(x)$ or is equal to 0. Since $p(x)$ has the smallest degree among all nonzero polynomials with θ_2 as a root, it follows that $f_a(x) - f_b(x) = 0$. Thus, $f_a(x) = f_b(x)$, which implies that

$$a = f_a(\theta_1) = f_b(\theta_1) = b.$$

Thus, π is also injective, so π is a bijection. It is also easy to see that if $a \in K$, then the polynomial $f_a(x) \in K[x]$ such that $a = f_a(\theta_1)$ is a constant polynomial. Therefore, $\pi(a) = a$, so π is the identity on K.

To show that π preserves addition, let $a, b \in K(\theta_1)$ and once again let $f_a(x)$, $f_b(x) \in K[x]$ be the polynomials just described such that $a = f_a(\theta_1)$ and $b = f_b(\theta_1)$. We now have

$$\pi(a+b) = \pi(f_a(\theta_1) + f_b(\theta_1)) = f_a(\theta_2) + f_b(\theta_2)$$

$$= \pi(f_a(\theta_1)) + \pi(f_b(\theta_1)) = \pi(a) + \pi(b).$$

Finally, to show that π preserves multiplication, let $a, b, f_a(x), f_b(x)$ be as in the previous paragraph. Applying the division algorithm in $K[x]$, there exist $q(x), r(x) \in K[x]$, where $deg(r(x)) < n$ or $r(x) = 0$, such that

$$f_a(x) \cdot f_b(x) = q(x) \cdot p(x) + r(x).$$

As a result

$$\pi(ab) = \pi(f_a(\theta_1) f_b(\theta_1)) = \pi(q(\theta_1) p(\theta_1) + r(\theta_1)) = \pi(r(\theta_1)) =$$

$$r(\theta_2) = q(\theta_2) p(\theta_2) + r(\theta_2) = f_a(\theta_2) f_b(\theta_2) = \pi(f_a(\theta_1))\pi(f_b(\theta_1)) = \pi(a)\pi(b).$$

Thus, π preserves multiplication, thereby concluding the proof. $\qquad\square$

We can now prove the main result of this section.

Theorem 15.17. *Let $\mathbb{Q} \subseteq K \subseteq L \subseteq \mathbb{C}$ be such that L is a splitting field of some $f(x) \in K[x]$. Then $|Gal(L/K)| = [L : K]$.*

Proof. By Theorem 15.12, $L = K(\alpha)$, for some $\alpha \in L$. If we let $p(x)$ be the minimum polynomial for α over $K[x]$ and let n denote the degree of $p(x)$, then Corollary 12.23 tells us that $p(x)$ has n distinct roots in $\mathbb{C}$ that we can denote as $\beta_1, \ldots, \beta_n$. Since Theorem 15.3 tells us that $[L : K] = n$, it will suffice to show that there is exactly one element of $Gal(L/K)$ corresponding to each β_i.

By Lemma 15.16, for each $i \le n$, there is a bijective homomorphism from $K(\alpha)$ to $K(\beta_i)$ that is the identity on K. In particular, π_i sends L to $\mathbb{C}$, so Lemma 15.15 tells us that $\pi(L) = L$. Thus, each π_i is indeed an element of $Gal(L/K)$. On the other hand, all of the π_i's are different, as they give different values when we plug in α. Hence, $Gal(L/K)$ has at least n different elements. But having already shown that $|Gal(L/K)| \le [L : K]$, we can now conclude that when L is a splitting field over K, we have $|Gal(L/K)| = [L : K]$. $\qquad\square$

It is useful to consider what role Theorem 15.17 actually plays in helping us to determine the structure of $Gal(L/K)$. In the proof of Theorem 15.17, we used the fact that $L = K(\alpha)$ to determine the size of $Gal(L/K)$ by looking at how the various π_i act on α. Although this is extremely helpful in computing the size of $Gal(L/K)$, it is not particularly useful in determining the product of elements of $Gal(L/K)$. For example, suppose π_1, π_2 are as in the proof of Theorem 15.17, and we wish to determine if π_1 and π_2 commute. Observe that

$$\pi_1\pi_2(\alpha) = \pi_1(\beta_2) \quad \text{and} \quad \pi_2\pi_1(\alpha) = \pi_2(\beta_1).$$

Unfortunately, without additional information about the field, there is no way of knowing if $\pi_1(\beta_2)$ and $\pi_2(\beta_1)$ are equal. Thus, there are some limitations on how we can apply Theorem 15.17. But depending on the situation, using Theorem 15.17 to determine the size of $Gal(L/K)$, can be very helpful. To see this, we revisit the last example from the previous section.

■ Example

When we tried to describe the elements of $Gal(\mathbb{Q}(\sqrt{3}, i)/\mathbb{Q})$, Corollary 5.13 indicated that there were four possible functions g_1, g_2, g_3, g_4 that could be elements of $Gal(\mathbb{Q}(\sqrt{3}, i)/\mathbb{Q})$. Since these four functions are the identity on $\mathbb{Q}$, their behavior on $Gal(\mathbb{Q}(\sqrt{3}, i)/\mathbb{Q})$ is completely described by their action on $\sqrt{3}$ and i:

$$g_1(\sqrt{3}) = \sqrt{3}, \; g_1(i) = i, \quad g_2(\sqrt{3}) = -\sqrt{3}, \; g_2(i) = i,$$

$$g_3(\sqrt{3}) = \sqrt{3}, \; g_3(i) = -i, \quad g_4(\sqrt{3}) = -\sqrt{3}, \; g_4(i) = -i.$$

However, as we concluded the previous section, it was unclear which of these four functions actually belong to $Gal(\mathbb{Q}(\sqrt{3}, i)/\mathbb{Q})$. We could have checked that each one was indeed an automorphism of $Gal(\mathbb{Q}(\sqrt{3}, i)/\mathbb{Q})$, but that would be very tedious and time consuming.

However, it is clear that $\mathbb{Q}(\sqrt{3}, i)$ is the splitting field over $\mathbb{Q}$ of the polynomial $(x^2 - 3)(x^2 + 1)$. Therefore, we can now use Theorem 15.17 to make quick work of this problem. Indeed, Theorem 15.17 tells us that

$$\left| Gal(\mathbb{Q}(\sqrt{3}, i)/\mathbb{Q}) \right| = \left[\mathbb{Q}(\sqrt{3}, i) : \mathbb{Q} \right] = 4.$$

Since $|Gal(\mathbb{Q}(\sqrt{3}, i)/\mathbb{Q})| = 4$, it is clear that all four of g_1, g_2, g_3, g_4 are elements of $Gal(\mathbb{Q}(\sqrt{3}, i)/\mathbb{Q})$.

Note that having expressed the four elements of $Gal(\mathbb{Q}(\sqrt{3}, i)/\mathbb{Q})$ in terms of their action of $\sqrt{3}$ and i, it is easy to check that $Gal(\mathbb{Q}(\sqrt{3}, i)/\mathbb{Q})$ is abelian and has the property that the square of each of its elements is the identity map.

For the remainder of this section, we will look at additional applications of Theorem 15.17. But first we need the following application of Eisenstein's Criterion.

Lemma 15.18. *If p is prime, then the polynomial $x^{p-1} + \cdots + x + 1$ is irreducible in $\mathbb{Q}[x]$.*

Proof. Suppose

$$x^{p-1} + \cdots + x + 1 = \left(\alpha_s x^s + \cdots + \alpha_1 x + \alpha_0\right)\left(\beta_t x^t + \cdots + \beta_1 x + \beta_0\right)$$

in $\mathbb{Q}[x]$. We need to show that either $s = 0$ or $t = 0$. First, multiply both sides of the previous equation by $x - 1$ to obtain

$$x^p - 1 = \left(\alpha_s x^s + \cdots + \alpha_1 x + \alpha_0\right)\left(\beta_t x^t + \cdots + \beta_1 x + \beta_0\right)(x - 1).$$

Next, replace x by $y + 1$ to give us

$$(y+1)^p - 1 = \left(\alpha_s(y+1)^s + \cdots + \alpha_1(y+1) + \alpha_0\right)\left(\beta_t(y+1)^t + \cdots + \beta_1(y+1) + \beta_0\right)(y).$$

After expanding out the $(y+1)^p$ term on the left-hand side and then dividing both sides by y, we have

$$y^{p-1} + py^{p-2} + \frac{p(p-1)}{2}y^{p-3} + \cdots \frac{p(p-1)}{2}y + p =$$
$$(\alpha_s(y+1)^s + \cdots + \alpha_1(y+1) + \alpha_0)(\beta_t(y+1)^t + \cdots + \beta_1(y+1) + \beta_0).$$

If we expand all the terms on the right-hand side of the equation of the form $(y+1)^i$, for some i, then the right-hand side is seen to be a product of polynomials in $\mathbb{Q}[y]$ of degrees s and t. However, every coefficient on the left-hand side, with the exception of the leading coefficient, is a multiple of the prime p. Furthermore, the constant term on the left-hand side is not a multiple of p^2. Therefore, we can apply Eisenstein's Criterion to assert that the polynomial on the left-hand side is irreducible in $\mathbb{Q}[y]$. Looking at the product on the right-hand side, it must now be the case that either $s = 0$ or $t = 0$, as desired. $\quad\square$

We can now examine the splitting field of $x^p - 1$ over $\mathbb{Q}$.

Corollary 15.19. *Let L be the splitting field over $\mathbb{Q}$ of $x^p - 1$, where p is prime. Then*

1. $[L : \mathbb{Q}] = |Gal(L/\mathbb{Q})| = p - 1$.

2. *If $\omega = \mathrm{cis}(\frac{2\pi}{p})$, then $Gal(L/\mathbb{Q}) = \{g_1, \ldots, g_{p-1}\}$, where $L = \mathbb{Q}(\omega)$ and $g_i(\omega) = \omega^i$, for $1 \leq i \leq p - 1$.*

3. *$Gal(L/\mathbb{Q})$ is abelian.*

Proof. By DeMoivre's Theorem, if $\omega = \mathrm{cis}(\frac{2\pi}{p})$, then the p roots of $x^p - 1$ are of the form ω^i, for $0 \leq i \leq p - 1$. Since all powers of ω belong to $\mathbb{Q}(\omega)$, it is clear the $L = \mathbb{Q}(\omega)$.

Since $x^p - 1 = (x - 1)(x^{p-1} + \cdots + x + 1)$, it follows that all the roots of $x^{p-1} + \cdots + x + 1$ are of the form ω^i, where $1 \le i \le p - 1$. Lemma 15.18 tells us that $x^{p-1} + \cdots + x + 1$ is irreducible in $\mathbb{Q}[x]$. Therefore, Theorems 15.3 and 15.17 combine to tell us that

$$[L : \mathbb{Q}] = p - 1 = |Gal(L/\mathbb{Q})|.$$

Since $L = \mathbb{Q}(\omega)$, every element of $Gal(L/\mathbb{Q})$ is determined by its action on ω and Corollary 5.13 tells us that ω must get sent to another root of $x^{p-1} + \cdots + x + 1$. Therefore, the only possible candidates to be elements of $Gal(L/\mathbb{Q})$ are the $p - 1$ functions g_i which act on ω as follows

$$g_i(\omega) = \omega^i,$$

for $1 \le i \le p - 1$. However, since there are only $p - 1$ candidates and $Gal(L/\mathbb{Q})$ has exactly $p - 1$ elements, it follows that $Gal(L/\mathbb{Q}) = \{g_1, \ldots, g_{p-1}\}$.

Finally, to see that $Gal(L/\mathbb{Q})$ is abelian, let $g_i, g_j \in Gal(L/\mathbb{Q})$. We now have

$$g_i g_j(\omega) = g_i(\omega^j) = g_i(\omega)^j = (\omega^i)^j = \omega^{ij} =$$
$$(\omega^j)^i = g_j(\omega)^i = g_j(\omega^i) = g_j g_i(\omega).$$

Therefore $g_i g_j$ and $g_j g_i$ agree on ω and so, $g_i g_j = g_j g_i$. Thus, $Gal(L/\mathbb{Q})$ is abelian. $\qquad\square$

■ Examples

1. If $\omega_1 = \text{cis}(\frac{2\pi}{5})$, then $\mathbb{Q}(\omega_1)$ is the splitting field of $x^4 + x^3 + x^2 + x + 1$ over $\mathbb{Q}$ and $Gal(\mathbb{Q}(\omega_1)/\mathbb{Q})$ is an abelian group with 4 elements.

2. If $\omega_2 = \text{cis}(\frac{2\pi}{23})$, then $\mathbb{Q}(\omega_2)$ is the splitting field of $x^{22} + x^{21} + \cdots + x + 1$ over $\mathbb{Q}$ and $Gal(\mathbb{Q}(\omega_2)/\mathbb{Q})$ is an abelian group with 22 elements.

■

We now prove that another collection of Galois groups is abelian.

Corollary 15.20. *Suppose $n \in \mathbb{N}$, $a \in \mathbb{Q}$, and let $\omega = \text{cis}(\frac{2\pi}{n})$. If L is the splitting field of $x^n - a$ over $\mathbb{Q}(\omega)$, then $Gal(L/\mathbb{Q}(\omega))$ is abelian.*

Proof. Let γ be a root of $x^n - a$. Then $\mathbb{Q}(\omega)(\gamma)$ must be contained in L. However, the n roots of $x^n - a$ are $\gamma, \gamma\omega, \ldots, \gamma\omega^{n-1}$, and they all belong to $\mathbb{Q}(\omega)(\gamma)$. Hence, $L = \mathbb{Q}(\omega)(\gamma)$. As a result, the elements of $Gal(L/\mathbb{Q}(\omega))$ are completely determined by their action on γ. However, Corollary 5.13 tells us that every element of $Gal(L/K)$ must send γ to another root of $x^n - a$.

As a result, if $g, h \in Gal(L/\mathbb{Q}(\omega))$, then there exist i, j such that

$$g(\gamma) = \gamma\omega^i \quad \text{and} \quad h(\gamma) = \gamma\omega^j.$$

Remember that g, h are the identity on $\mathbb{Q}(\omega)$, so we now have

$$gh(\gamma) = g(\gamma\omega^j) = g(\gamma)g(\omega^j) = (\gamma\omega^i)(\omega^j) =$$
$$\gamma\omega^{i+j} = (\gamma\omega^j)(\omega^i) = h(\gamma)h(\omega^i) = h(\gamma\omega^i) = hg(\gamma).$$

Since gh and hg agree on γ, we see that $gh = hg$, hence $Gal(L/K)$ is abelian. $\qquad\square$

■ Examples

1. Let L be the splitting field of $x^4 - 5$ over $\mathbb{Q}(i)$. Then $Gal(L/\mathbb{Q}(i))$ is abelian using Corollary 15.20 with $n = 4$.

2. Let $\omega = \operatorname{cis}(\frac{2\pi}{7})$ and let L be the splitting field of $x^7 - 12$ over $\mathbb{Q}(\omega)$. Then $Gal(L/\mathbb{Q}(\omega))$ is abelian using Corollary 15.20 with $n = 7$. ■

In some earlier examples, we knew the size of $Gal(L/K)$ but did not know if it was abelian. In the two preceding examples, we were not concerned with the size of the Galois group, but we do know they are abelian. It certainly is *not* the case that Galois groups of splitting fields are always abelian, as we see in

Corollary 15.21. *Let $p \geq 3$ be a prime and let $a \in \mathbb{Q}$ such that $x^p - a$ is irreducible in $\mathbb{Q}[x]$. If L is the splitting field of $x^p - a$ over $\mathbb{Q}$, then $Gal(L/\mathbb{Q})$ is a nonabelian group with $p(p-1)$ elements.*

Proof. Let γ be a root of $x^p - a$ and let $\omega = \operatorname{cis}(\frac{2\pi}{p})$; the p roots of $x^p - a$ are $\gamma, \gamma\omega, \dots,$ $\gamma\omega^{p-1}$. Observe that all p of these roots belong to the field $\mathbb{Q}(\gamma, \omega)$. On the other hand, since $\omega = \frac{\gamma\omega}{\gamma}$, it follows that both γ and ω belong to L. Thus, $L = \mathbb{Q}(\gamma, \omega)$.

Since γ is a root of $x^p - a$, the minimum polynomial for γ in $\mathbb{Q}(\omega)[x]$ has degree at most p. Thus, $[\mathbb{Q}(\gamma, \omega) : \mathbb{Q}(\omega)] \leq p$. In addition, by Lemma 15.18, $[\mathbb{Q}(\omega) : \mathbb{Q}] = p - 1$. Therefore,

$$[L : \mathbb{Q}] = [\mathbb{Q}(\gamma, \omega) : \mathbb{Q}] = [\mathbb{Q}(\gamma, \omega) : \mathbb{Q}(\omega)] \cdot [\mathbb{Q}(\omega) : \mathbb{Q}] =$$
$$[\mathbb{Q}(\gamma, \omega) : \mathbb{Q}(\omega)] \cdot (p - 1) \leq p(p - 1).$$

On the other hand, since $x^p - a$ is irreducible in $\mathbb{Q}[x]$,

$$[L : \mathbb{Q}] = [\mathbb{Q}(\gamma, \omega) : \mathbb{Q}] = [\mathbb{Q}(\gamma, \omega) : \mathbb{Q}(\gamma)] \cdot [\mathbb{Q}(\gamma) : \mathbb{Q}] = [\mathbb{Q}(\gamma, \omega) : \mathbb{Q}(\gamma)] \cdot p.$$

As a result, $[L : \mathbb{Q}]$ is divisible by both p and $p-1$. However, p and $p-1$ are relatively prime, so $[L : \mathbb{Q}]$ is divisible by $p(p-1)$. Combining this with the fact that $[L : \mathbb{Q}] \leq p(p-1)$, we now know that $[L : \mathbb{Q}] = p(p-1)$. In light of Theorem 15.17, we have $|Gal(L/\mathbb{Q})| = p(p-1)$.

If $g \in Gal(L/\mathbb{Q})$, then g is determined by its action on γ and ω. Corollary 5.13 tells us that g must send γ to one of the p roots of $x^p - a$ and also must send ω to one of the $p-1$ roots of $x^{p-1} + \cdots + x + 1$. Therefore, the only possible candidates to be elements of $Gal(L/K)$ are the functions $g_{i,j}$, where

$$g_{i,j}(\gamma) = \gamma\omega^i \quad \text{and} \quad g_{i,j}(\omega) = \omega^j,$$

for $0 \leq i \leq p-1, 1 \leq j \leq p-1$. However, since $|Gal(L/\mathbb{Q})| = p(p-1)$, all $p(p-1)$ of our candidates are indeed elements of $Gal(L/K)$.

To show that $Gal(L/K)$ is not abelian, it suffices to find two elements that do not commute. To this end, consider $g_{1,1}$ and $g_{1,2}$; then we have

$$g_{1,1}g_{1,2}(\gamma) = g_{1,1}(\gamma\omega) = g_{1,1}(\gamma)g_{1,1}(\omega) = (\gamma\omega)\omega = \gamma\omega^2,$$

whereas

$$g_{1,2}g_{1,1}(\gamma) = g_{1,2}(\gamma\omega) = g_{1,2}(\gamma)g_{1,2}(\omega) = (\gamma\omega)\omega^2 = \gamma\omega^3.$$

Since the products $g_{1,1}g_{1,2}$ and $g_{1,2}g_{1,1}$ behave differently on γ, we know that $g_{1,1}g_{1,2} \neq g_{1,2}g_{1,1}$. Hence, $Gal(L/K)$ is not abelian. □

■ Examples

1. Let L be the splitting field of $x^3 - 20$ over $\mathbb{Q}$. This polynomial is irreducible in $\mathbb{Q}[x]$ using Eisenstein's Criterion with the prime 5. Corollary 15.21 now applies with the prime $p = 3$. Therefore, $Gal(L/\mathbb{Q})$ is a group that is nonabelian and has six elements.

2. Now let L be the splitting field of $x^{53} - 22$ over $\mathbb{Q}$. Using Eisenstein's Criterion with the prime 2 or 11, we can see that the polynomial is irreducible in $\mathbb{Q}[x]$. Corollary 15.21 now asserts, with the prime $p = 53$, that $Gal(L/\mathbb{Q})$ is a group with $2756 = 53 \cdot 52$ elements that is not abelian. ■

Although the Galois groups in the previous examples are not abelian, in some sense, they are not far from being abelian. In Section 8.3, we formalized this notion when we introduced the concept of a group being solvable. In fact, we will soon show that the Galois groups in Corollary 15.21 are solvable. However, we need

Definition 15.22. *If G is a group and $g, h \in G$, then let $\langle g, h \rangle = ghg^{-1}h^{-1}$ and we call this element of G the commutator of g and h.*

If we let e denote the identity of G, then you should check that g and h commute if and only if $\langle g, h \rangle = e$. More generally, observe that a group G is abelian if and only if all commutators of elements of G belong to the set $\{e\}$. Now suppose $\mathbb{Q} \subseteq K \subseteq L \subseteq \mathbb{C}$ are fields. If $g \in Gal(L/K)$, then g is an automorphism of L, which is the identity map on K. Since K contains $\mathbb{Q}$, g is also the identity map on $\mathbb{Q}$. Thus, $g \in Gal(L/\mathbb{Q})$ and it follows that

$$Gal(L/\mathbb{Q}) \supseteq Gal(L/K).$$

We now look at a situation somewhat more general than the one in Corollary 15.20.

Corollary 15.23. *Let $n \in \mathbb{N}$, $a \in \mathbb{Q}$, and let L and K be the splitting fields over $\mathbb{Q}$ of $x^n - a$ and $x^n - 1$, respectively. Then*

1. $\mathbb{Q} \subseteq K \subseteq L \subseteq \mathbb{C}$ *and* $Gal(L/\mathbb{Q}) \supseteq Gal(L/K)$.

2. *if $g, h \in Gal(L/\mathbb{Q})$, then $\langle g, h \rangle \in Gal(L/K)$.*

3. $Gal(L/K)$ *is abelian.*

Before proving Corollary 15.22, let us think about what it actually means. We have the chain of groups

$$Gal(L/\mathbb{Q}) \supseteq Gal(L/K) \supseteq \{e\},$$

with the properties that

 (i) $\langle g, h \rangle \in Gal(L/K)$, for all $g, h \in Gal(L/\mathbb{Q})$, and

(ii) $\langle g, h \rangle \in \{e\}$, for all $g, h \in Gal(L/K)$.

By Proposition 8.33, we can see that although $Gal(L/\mathbb{Q})$ might not be abelian, it must be solvable.

We would certainly consider $x^n - a$ to be an example of a polynomial in $\mathbb{Q}[x]$ whose roots can be found using an algebraic algorithm like the quadratic formula. The fact that the Galois group of its splitting field over $\mathbb{Q}$ is solvable is not a coincidence. In fact, it is at the heart of what goes on when we prove the insolvability of the quintic in Chapter 17. We will show that if a polynomial in $\mathbb{Q}[x]$ is solvable by radicals, then the Galois group over $\mathbb{Q}$ of its splitting field is a solvable group. Then, we will produce polynomials in $\mathbb{Q}[x]$ of degree 5 whose splitting fields over $\mathbb{Q}$ have Galois groups that are *not* solvable. Thus, Corollary 15.23 foreshadows the path we will take in proving the insolvability of the quintic.

Proof of Corollary 15.23. For (1), as in the proof of Corollary 15.20, let γ be a root of $x^n - a$ and let $\omega = \text{cis}(\frac{2\pi}{n})$. Then $K = \mathbb{Q}(\omega)$ and $L = \mathbb{Q}(\omega, \gamma)$, thus $\mathbb{Q} \subseteq K \subseteq L \subseteq \mathbb{C}$ and $Gal(L/\mathbb{Q}) \supseteq Gal(L/K)$.

For (2), if $g, h \in Gal(L/\mathbb{Q})$, then $g^{-1}, h^{-1} \in Gal(L/\mathbb{Q})$ and Corollary 5.13 asserts that $g^{-1}(\omega)$ and $h^{-1}(\omega)$ must also be roots of $\frac{x^n-1}{x-1} = x^{n-1} + \cdots + x + 1$. Therefore, $g^{-1}(\omega) = \omega^i$ and $h^{-1}(\omega) = \omega^j$, where $1 \leq i, j \leq n - 1$. Observe that $g(\omega^i) = \omega$ and $h(\omega^j) = \omega$, and we now have

$$\langle g, h \rangle(\omega) = ghg^{-1}h^{-1}(\omega) = ghg^{-1}(\omega^j) = gh(g^{-1}(\omega)^j) =$$
$$gh((\omega^i)^j) = gh((\omega^j)^i) = g((h(\omega^j))^i) = g(\omega^i) = \omega.$$

Therefore, $\langle g, h \rangle$ is an element of $Gal(L/\mathbb{Q})$ that is also the identity map on ω. Hence, $\langle g, h \rangle$ is the identity on all of $K = \mathbb{Q}(\omega)$, so $\langle g, h \rangle \in Gal(L/K)$.

Since (3) is actually Corollary 15.20, the proof is complete. $\square$

■ Example

As in the example preceding Definition 15.22, let L be the splitting field of $x^{53} - 22$ over $\mathbb{Q}$. Then $Gal(L/\mathbb{Q})$ is a group with $2756 = 53 \cdot 52$ elements that is not abelian. However, if we let $\omega = \text{cis}(\frac{2\pi}{53})$, then we have the chain of groups

$$Gal(L/\mathbb{Q}) \supseteq Gal(L/\mathbb{Q}(\omega)) \supseteq \{e\}.$$

Since the commutator of any two elements of $Gal(L/\mathbb{Q})$ is an element of $Gal(L/\mathbb{Q}(\omega))$, and the commutator of any two elements of $Gal(L/\mathbb{Q}(\omega))$ belongs to $\{e\}$, Proposition 8.33 tells us that $Gal(L/\mathbb{Q})$ is solvable.

■

Given fields $F \subseteq K \subseteq L$, Theorem 15.6 revealed the relationship between $[L : K]$, $[K : F]$, and $[L : K]$. We conclude this chapter by revealing a relationship between the Galois groups $Gal(L/K)$, $Gal(K/F)$, and $Gal(L/F)$. This relationship will be essential in proving the insolvability of the quintic.

Corollary 15.24. *Let $\mathbb{Q} \subseteq F \subseteq K \subseteq L \subseteq \mathbb{C}$ be fields such that K is the splitting field over F of some $f(x) \in F[x]$. For every $g \in Gal(L/F)$, let g_K denote the restriction of g to K. Then the function*

$$\psi : Gal(L/F) \to Gal(K/F)$$

defined as $\psi(g) = g_K$, for all $g \in Gal(L/F)$ is a homomorphism of groups. Furthermore, the kernel of ψ is the group $Gal(L/K)$.

Proof. The first thing we need to check is that the values of ψ do indeed belong to $Gal(K/F)$. To this end, observe that if $g \in Gal(L/F)$, then g_K is an injective homomorphism from K to $\mathbb{C}$, which is the identity map on F. Since K is a splitting field over F of some polynomial in $F[x]$, Lemma 15.15 tells us that $g_K(K) = K$. Therefore, g_K is an automorphism of K, so $g_K \in Gal(K/F)$. Thus, ψ does send elements of $Gal(L/F)$ to $Gal(K/F)$.

Next, we need to show that ψ is a homomorphism of groups. If $g, h \in Gal(L/F)$, then $\psi(gh)$ and $\psi(g)\psi(h)$ are both elements of $Gal(K/F)$. In fact, if $\alpha \in K$, then the way we compute the values of both $\psi(gh)$ and $\psi(g)\psi(h)$ is to plug α into the composition gh. Since $\psi(gh)$ and $\psi(g)\psi(h)$ agree on all elements of K, $\psi(gh) = \psi(g)\psi(h)$. Thus, ψ is a homomorphism.

Finally, if $g \in Gal(L/F)$, then g belongs to the kernel of ψ precisely if g_K is the identity map on K. But the automorphisms of L that are the identity map on K are, by the definition of the Galois group, the elements of the group $Gal(L/K)$. Thus, $Gal(L/K)$ is indeed the kernel of ψ. $\square$

■ Example

Once again, we consider the case where L is the splitting field of $x^{53} - 22$ over $\mathbb{Q}$. If we let $\omega = \text{cis}(\frac{2\pi}{53})$, then $\mathbb{Q}(\omega)$ is the splitting field of $x^{53} - 1$ over $\mathbb{Q}$ and we can examine the chain of fields $\mathbb{Q} \subseteq \mathbb{Q}(\omega) \subseteq L$. By Lemma 15.15, the restriction of every $g \in Gal(L/\mathbb{Q})$ to $\mathbb{Q}(\omega)$ yields an element of $Gal(\mathbb{Q}(\omega)/\mathbb{Q})$. The elements of $Gal(L/\mathbb{Q})$ that act like the identity on $\mathbb{Q}(\omega)$ are precisely the elements of $Gal(L/\mathbb{Q}(\omega))$.

■

Back in Definition 5.12, given a commutative ring R with automorphism σ, we introduced the set $R^\sigma = \{r \in R | \sigma(r) = r\}$. Having discussed Galois groups and extension fields throughout this chapter, there are some natural ways to extend this concept. Given fields $K \subseteq L$, if H is a subgroup of $Gal(L/K)$, we can look at the elements of L that are fixed by all elements of H. In the opposite direction, given a field M lying between K and L, we can look at the elements of $Gal(L/K)$ that act as the identity on M. To be more precise, we have

Definition 15.25. *Let $K \subseteq L$ be fields and let $G = Gal(L/K)$. If H is a subgroup of G, we let $L^H = \{l \in L | h(l) = l,$ for every $h \in H\}$ and call this set the fixed subfield of L under H. In the opposite direction, if M is a field such that $K \subseteq M \subseteq L$, we let $G_M = \{g \in G | g(m) = m,$ for every $m \in M\}$.*

Although we refer to L^H as a fixed subfield, we have not yet shown that it is actually a field. In the other direction, although we have not shown that G_M is a subgroup of G, you might guess that it is. This leads us to

Proposition 15.26.   *Let $K \subseteq L$ be fields and let $G = Gal(L/K)$. If H is a subgroup of G, then L_H is a field such that $K \subseteq L^H \subseteq L$. Furthermore, if M is a field such that $K \subseteq M \subseteq L$, then G_M is a subgroup of G.*

Proof. In order to show that L^H is a field, we first need to use Proposition 5.15 to show that L^H is a commutative ring. Every element of H certainly acts as the identity map on K, hence $K \subseteq L^H$. Since $0, 1, -1 \in K$, it follows that $0, 1, -1 \in L^H$ and part (b) of Proposition 5.15 is satisfied. Therefore, to satisfy part (a) of Proposition 5.15 and to complete the proof that L^H is a commutative ring, it suffices to show that L^H is closed under addition and multiplication.

Suppose $x, y \in L^H$, we must show that $x + y, x \cdot y \in L^H$. If $h \in H$, we know that $h(x) = x$ and $h(y) = y$. Since h is an automorphism, it follows that

$$h(x + y) = h(x) + h(y) = x + y$$

and

$$h(x \cdot y) = h(x) \cdot h(y) = x \cdot y.$$

Since the previous equations hold for all $h \in H$, we have $x + y, x \cdot y \in L^H$.

To finish showing that L^H is a field, we must show that if x is a nonzero element of L^H, then $x^{-1} \in L^H$. Since $x \in L$, we know that $x^{-1} \in L$. However, we still need to show that if $h \in H$, then $h(x^{-1}) = x^{-1}$. Since $h(x) = x$ and h is an automorphism, we have

$$1 = h(1) = h(x \cdot x^{-1}) = h(x) \cdot h(x^{-1}) = x \cdot h(x^{-1}).$$

Multiplying the far ends of the previous equation by x^{-1} results in $x^{-1} = h(x^{-1})$. Thus, $x^{-1} \in L^H$ and L^H is indeed a field.

In the other direction, in order to prove that G_M is a subgroup of G, Proposition 8.6 asserts that it suffices to show that if $g, h \in G_M$, then we also have $gh^{-1} \in G_M$. To show that $gh^{-1} \in G_M$, we need to verify that $gh^{-1}(m) = m$, for all $m \in M$. If $m \in M$, we have $g(m) = m$ and $h(m) = m$. Observe that the last equation also tells us that $h^{-1}(m) = m$. We now have

$$gh^{-1}(m) = g(h^{-1}(m)) = g(m) = m,$$

thereby concluding the proof.   □

We examine the correspondence from Definition 15.25 in the following example.

■ Example

Let L be the splitting field of $(x^2 - 30)(x^2 + 1)$ over $\mathbb{Q}$. It is easy to see that $L = \mathbb{Q}(\sqrt{30}, i)$. Observe that

$$[L : \mathbb{Q}] = \left[\mathbb{Q}(\sqrt{30}, i) : \mathbb{Q}\right] = \left[\mathbb{Q}(\sqrt{30}, i) : \mathbb{Q}(\sqrt{30})\right] \cdot \left[\mathbb{Q}(\sqrt{30}) : \mathbb{Q}\right].$$

Since $\mathbb{Q}(\sqrt{30}) \subseteq \mathbb{R}$ and $i \notin \mathbb{R}$, it follows that $i \notin \mathbb{Q}(\sqrt{30})$ and $[\mathbb{Q}(\sqrt{30}, i) : \mathbb{Q}(\sqrt{30})] = 2$. Furthermore, we know that $[\mathbb{Q}(\sqrt{30}) : \mathbb{Q}] = 2$, and our observations now combine to tell us that $[L : \mathbb{Q}] = 4$. In light of Theorem 15.17, we also know that $|Gal(L/\mathbb{Q})| = 4$.

The behavior of each $g \in Gal(L/\mathbb{Q})$ is completely determined by its action on $\sqrt{30}$ and i. Corollary 5.13 now asserts that if $g \in Gal(L/\mathbb{Q})$ then $g(\sqrt{30})$ must be a root of $x^2 - 30$ and $g(i)$ must be a root of $x^2 + 1$. This leaves us with four candidates to be elements of $Gal(L/\mathbb{Q})$ and they are g_1, g_2, g_3, g_4, where

$$g_1(\sqrt{30}) = \sqrt{30} \quad \text{and} \quad g_1(i) = i,$$

$$g_2(\sqrt{30}) = -\sqrt{30} \quad \text{and} \quad g_2(i) = i,$$

$$g_3(\sqrt{30}) = \sqrt{30} \quad \text{and} \quad g_3(i) = -i,$$

$$g_4(\sqrt{30}) = -\sqrt{30} \quad \text{and} \quad g_4(i) = -i.$$

Since $|Gal(L/\mathbb{Q})| = 4$, all four of the preceding candidates are indeed elements of $Gal(L/\mathbb{Q})$, so $Gal(L/\mathbb{Q}) = \{g_1, g_2, g_3, g_4\}$. By examining how compositions of the various g_j act on $\sqrt{30}$ and i, we obtain the following table for $Gal(L/\mathbb{Q})$.

$\circ$	g_1	g_2	g_3	g_4
g_1	g_1	g_2	g_3	g_4
g_2	g_2	g_1	g_4	g_3
g_3	g_3	g_4	g_1	g_2
g_4	g_4	g_3	g_2	g_1

If we let $G = Gal(L/\mathbb{Q})$, then a brief examination of the table for G indicates that G has exactly five subgroups and they are

$$H_1 = \{g_1\}, \quad H_2 = \{g_1, g_2\}, \quad H_3 = \{g_1, g_3\}, \quad H_4 = \{g_1, g_4\}, \quad H_5 = G.$$

The set $\{1, \sqrt{30}, i, i\sqrt{30}\}$ is a basis for L over $\mathbb{Q}$, so every element of L can be expressed uniquely as

$$a + b\sqrt{30} + ci + di\sqrt{30},$$

where $a, b, c, d \in \mathbb{Q}$. By examining the action of the automorphisms in G on elements of this type, you can see which elements of L are fixed by all the automorphisms in the five different subgroups of G. Doing the required computations, you should find that

$$L^{H_1} = L, \quad L^{H_2} = \{a + ci | a, c \in \mathbb{Q}\} = \mathbb{Q}(i),$$

$$L^{H_3} = \left\{a + b\sqrt{30} | a, b \in \mathbb{Q}\right\} = \mathbb{Q}(\sqrt{30}),$$

$$L^{H_4} = \left\{a + di\sqrt{30} | a, d \in \mathbb{Q}\right\} = \mathbb{Q}(i\sqrt{30}),$$

$$L^{H_5} = \mathbb{Q}.$$

It certainly appears that it is not easy to find all the fields that lie between $\mathbb{Q}$ and L. This is an issue we will deal with later, but, for now, let us concern ourselves with the five fields

$$M_1 = L, \quad M_2 = \mathbb{Q}(i), \quad M_3 = \mathbb{Q}(\sqrt{30}), \quad M_4 = \mathbb{Q}(i\sqrt{30}), \quad M_5 = \mathbb{Q}.$$

By checking which automorphisms from G act as the identity map on the various M_i, it follows that

$$G_{M_1} = \{g_1\} = H_1, \quad G_{M_2} = \{g_1, g_2\} = H_2, \quad G_{M_3} = \{g_1, g_3\} = H_3,$$

$$G_{M_4} = \{g_1, g_4\} = H_4, \quad G_{M_5} = \{g_1, g_2, g_3, g_4\} = H_5 = G.$$

In this example, it turns out that the correspondence that goes from H to L^H is actually a bijection from the set of subgroups of $Gal(L/\mathbb{Q})$ to the set of fields that lie between $\mathbb{Q}$ and L. This is no coincidence, as a fundamental result in Galois theory states that if $\mathbb{Q} \subseteq K \subseteq L \subseteq \mathbb{C}$ are fields such that L is the splitting field of some $f(x) \in K[x]$, then the correspondence from H to L^H described in Definition 15.25 is indeed a bijection between the subgroups of $Gal(L/K)$ and the fields that lie between K and L. Furthermore, the other correspondence in Definition 15.25 from M to G_M is the inverse of the bijection from H to L^H.

■

Observe that if $\mathbb{Q} \subseteq K \subseteq L \subseteq \mathbb{C}$ are fields, then there is no obvious way to go about finding all the fields that lie between K and L. On the other hand, if L is a finite extension of K, then $Gal(L/K)$ is a finite group and it might not be too difficult to find all of its subgroups. In the case that L is the splitting field over K of some $f(x) \in K[x]$, then the beautiful correspondence in Definition 15.25 from H to L^H indicates that knowing all the subgroups of $Gal(L/K)$

enables you to find all the fields that lie between K and L. Therefore, in our previous example, the correspondence from H to L^H actually tells us that there are exactly five fields that lie between $\mathbb{Q}$ and $\mathbb{Q}(\sqrt{30}, i)$.

We will not prove this fundamental result as the proof requires developing more Galois theory than is needed to prove the insolvability of the quintic. However, in the exercises, you will get a great deal of practice doing computations and working with this important correspondence between subgroups of $Gal(L/K)$ and fields that lie between K and L.

Exercises for Sections 15.2 and 15.3

In exercises 1–25, we let L be the splitting field over $\mathbb{Q}$ of $x^3 - 10$.

1. Show that $L = \mathbb{Q}(10^{\frac{1}{3}}, \omega)$, where $\omega = \text{cis}(\frac{2\pi}{3})$.

2. Show that $[L : \mathbb{Q}] = 6$ and determine the size of $Gal(L/\mathbb{Q})$.

3. If $g \in Gal(L/\mathbb{Q})$, show that $g(10^{\frac{1}{3}})$ is equal to either $10^{\frac{1}{3}}$, $10^{\frac{1}{3}}\omega$, or $10^{\frac{1}{3}}\omega^2$.

4. If $g \in Gal(L/\mathbb{Q})$, show that $g(\omega)$ is equal to either ω or ω^2.

5. Show that the set $\{1, 10^{\frac{1}{3}}, 10^{\frac{2}{3}}, \omega, 10^{\frac{1}{3}}\omega, 10^{\frac{2}{3}}\omega\}$ is a basis for L over $\mathbb{Q}$ and then explain why the action of every $g \in Gal(L/\mathbb{Q})$ is determined by its behavior on $10^{\frac{1}{3}}$ and ω.

6. If we define $g_{i,j}$ as $g_{i,j}(10^{\frac{1}{3}}) = 10^{\frac{1}{3}}\omega^i$ and $g_{i,j}(\omega) = \omega^j$, for $0 \leq i \leq 2$ and $1 \leq j \leq 2$, show that the six functions of the form $g_{i,j}$ make up all the elements of $Gal(L/\mathbb{Q})$.

Exercises 7–25 will examine the correspondences discussed in Definition 15.25 between the subgroups of $Gal(L/\mathbb{Q})$ and the fields that lie between $\mathbb{Q}$ and L. When performing computations with the various subgroups and fields involved with these correspondences, you should feel free to use the following table for $Gal(L/\mathbb{Q})$, as well as the basis for L over $\mathbb{Q}$ provided in exercise 5.

$\circ$	$g_{0,1}$	$g_{0,2}$	$g_{1,1}$	$g_{1,2}$	$g_{2,1}$	$g_{2,2}$
$g_{0,1}$	$g_{0,1}$	$g_{0,2}$	$g_{1,1}$	$g_{1,2}$	$g_{2,1}$	$g_{2,2}$
$g_{0,2}$	$g_{0,2}$	$g_{0,1}$	$g_{2,2}$	$g_{2,1}$	$g_{1,2}$	$g_{1,1}$
$g_{1,1}$	$g_{1,1}$	$g_{1,2}$	$g_{2,1}$	$g_{2,2}$	$g_{0,1}$	$g_{0,2}$
$g_{1,2}$	$g_{1,2}$	$g_{1,1}$	$g_{0,2}$	$g_{0,1}$	$g_{2,2}$	$g_{2,1}$
$g_{2,1}$	$g_{2,1}$	$g_{2,2}$	$g_{0,1}$	$g_{0,2}$	$g_{1,1}$	$g_{1,2}$
$g_{2,2}$	$g_{2,2}$	$g_{2,1}$	$g_{1,2}$	$g_{1,1}$	$g_{0,2}$	$g_{0,1}$

7. If H_1 is the subgroup $\{g_{0,1}\}$, find the elements of the field L^{H_1}.

8. Compute $[L : L^{H_1}]$ and $[L^{H_1} : \mathbb{Q}]$.

9. If H_2 is the subgroup $\{g_{0,1}, g_{0,2}\}$, find the elements of the field L^{H_2}.

10. Compute $[L : L^{H_2}]$ and $[L^{H_2} : \mathbb{Q}]$.

11. If H_3 is the subgroup $\{g_{0,1}, g_{1,2}\}$, find the elements of the field L^{H_3}.

12. Compute $[L : L^{H_3}]$ and $[L^{H_3} : \mathbb{Q}]$.

13. If H_4 is the subgroup $\{g_{0,1}, g_{2,2}\}$, find the elements of the field L^{H_4}.

14. Compute $[L : L^{H_4}]$ and $[L^{H_4} : \mathbb{Q}]$.

15. If H_5 is the subgroup $\{g_{0,1}, g_{1,1}, g_{2,1}\}$, find the elements of the field L^{H_5}.

16. Compute $[L : L^{H_5}]$ and $[L^{H_5} : \mathbb{Q}]$.

17. If H_6 is all of $Gal(L/\mathbb{Q})$, find the elements of the field L^{H_6}.

18. Compute $[L : L^{H_6}]$ and $[L^{H_6} : \mathbb{Q}]$.

19. If $M_1 = \mathbb{Q}$, find the elements of the subgroup G_{M_1}.

20. If $M_2 = \mathbb{Q}(\omega)$, find the elements of the subgroup G_{M_2}.

21. If $M_3 = \mathbb{Q}(10^{\frac{1}{3}}\omega)$, find the elements of the subgroup G_{M_3}.

22. If $M_4 = \mathbb{Q}(10^{\frac{1}{3}}\omega^2)$, find the elements of the subgroup G_{M_4}.

23. If $M_5 = \mathbb{Q}(10^{\frac{1}{3}})$, find the elements of the subgroup G_{M_5}.

24. If $M_6 = L$, find the elements of the subgroup G_{M_6}.

25. Is $Gal(L/\mathbb{Q})$ abelian? If not, is it solvable?

In exercises 26–47, we let L be the splitting field over $\mathbb{Q}$ of $x^4 + 1$.

26. Show that $L = \mathbb{Q}(\omega)$, where $\omega = \text{cis}(\frac{\pi}{4})$.

27. Show that $[L : \mathbb{Q}] = 4$ and determine the size of $Gal(L/\mathbb{Q})$.

28. If $g \in Gal(L/\mathbb{Q})$, show that $g(\omega)$ is equal to either ω, ω^3, $-\omega$, or $-\omega^3$.

29. Show that the set $\{1, \omega, \omega^2, \omega^3\}$ is a basis for L over $\mathbb{Q}$ and then explain why the action of every $g \in Gal(L/\mathbb{Q})$ is determined by its behavior on ω.

30. If we define e, g, h, k as $e(\omega) = \omega$, $g(\omega) = \omega^3$, $h(\omega) = -\omega$, and $k(\omega) = -\omega^3$, show that the four functions of the form e, g, h, k make up all the elements of $Gal(L/\mathbb{Q})$.

Exercises 31–47 will examine the correspondences discussed in Definition 15.25 between the subgroups of $Gal(L/\mathbb{Q})$ and the fields that lie between $\mathbb{Q}$ and L. When performing computations with the various subgroups and fields involved with these correspondences, you should feel free to use the table below for $Gal(L/\mathbb{Q})$ as well as the basis for L over $\mathbb{Q}$ provided in exercise 29.

$\circ$	e	g	h	k
e	e	g	h	k
g	g	e	k	h
h	h	k	e	g
k	k	h	g	e

31. If H_1 is the subgroup $\{e\}$, find the elements of the field L^{H_1}.

32. Compute $[L : L^{H_1}]$ and $[L^{H_1} : \mathbb{Q}]$.

33. If H_2 is the subgroup $\{e, g\}$, find the elements of the field L^{H_2}.

34. Compute $[L : L^{H_2}]$ and $[L^{H_2} : \mathbb{Q}]$.

35. If H_3 is the subgroup $\{e, h\}$, find the elements of the field L^{H_3}.

36. Compute $[L : L^{H_3}]$ and $[L^{H_3} : \mathbb{Q}]$.

37. If H_4 is the subgroup $\{e, k\}$, find the elements of the field L^{H_4}.

38. Compute $[L : L^{H_4}]$ and $[L^{H_4} : \mathbb{Q}]$.

39. If H_5 is all of $Gal(L/\mathbb{Q})$, find the elements of the field L^{H_5}.

40. Compute $[L : L^{H_5}]$ and $[L^{H_5} : \mathbb{Q}]$.

41. Show that L contains i, $\sqrt{2}$, and $i\sqrt{2}$.

42. Determine the values of the automorphisms g, h, k on i, $\sqrt{2}$, and $i\sqrt{2}$.

43. If $M_1 = \mathbb{Q}$, find the elements of the subgroup G_{M_1}.

44. If $M_2 = \mathbb{Q}(i\sqrt{2})$, find the elements of the subgroup G_{M_2}.

45. If $M_3 = \mathbb{Q}(i)$, find the elements of the subgroup G_{M_3}.

46. If $M_4 = \mathbb{Q}(\sqrt{2})$, find the elements of the subgroup G_{M_4}.

47. If $M_5 = L$, find the elements of the subgroup G_{M_5}.

In exercises 48–102, we let L be the splitting field over $\mathbb{Q}$ of $(x^2 - 5)(x^2 - 7)(x^2 + 1)$.

48. Show that $L = \mathbb{Q}(\sqrt{5}, \sqrt{7}, i)$.

49. Show that $[L : \mathbb{Q}] = 8$ and determine the size of $Gal(L/\mathbb{Q})$.

50. If $g \in Gal(L/\mathbb{Q})$, show that $g(\sqrt{5})$ is equal to either $\sqrt{5}$ or $-\sqrt{5}$.

51. If $g \in Gal(L/\mathbb{Q})$, show that $g(\sqrt{7})$ is equal to either $\sqrt{7}$ or $-\sqrt{7}$.

52. If $g \in Gal(L/\mathbb{Q})$, show that $g(i)$ is equal to either i or $-i$.

53. Show that the set $\{1, \sqrt{5}, \sqrt{7}, \sqrt{35}, i, i\sqrt{5}, i\sqrt{7}, i\sqrt{35}\}$ is a basis for L over $\mathbb{Q}$, and then explain why the action of every $g \in Gal(L/\mathbb{Q})$ is determined by its behavior on $\sqrt{5}, \sqrt{7}$, and i.

54. If we define $g_{j,k,l}$ as $g_{j,k,l}(\sqrt{5}) = (-1)^j \sqrt{5}$, $g_{j,k,l}(\sqrt{7}) = (-1)^k \sqrt{7}$, and $g_{j,k,l}(i) = (-1)^l i$, for $0 \le j, k, l \le 1$, show that the eight functions of the form $g_{j,k,l}$ make up all the elements of $Gal(L/\mathbb{Q})$.

Exercises 55–102 will examine the correspondences discussed in Definition 15.25 between the subgroups of $Gal(L/\mathbb{Q})$ and the fields that lie between $\mathbb{Q}$ and L. When performing computations with the various subgroups and fields involved with these correspondences, you should feel free to use the table below for $Gal(L/\mathbb{Q})$ as well as the basis for L over $\mathbb{Q}$ provided in exercise 53.

$\circ$	$g_{0,0,0}$	$g_{0,0,1}$	$g_{0,1,0}$	$g_{0,1,1}$	$g_{1,0,0}$	$g_{1,0,1}$	$g_{1,1,0}$	$g_{1,1,1}$
$g_{0,0,0}$	$g_{0,0,0}$	$g_{0,0,1}$	$g_{0,1,0}$	$g_{0,1,1}$	$g_{1,0,0}$	$g_{1,0,1}$	$g_{1,1,0}$	$g_{1,1,1}$
$g_{0,0,1}$	$g_{0,0,1}$	$g_{0,0,0}$	$g_{0,1,1}$	$g_{0,1,0}$	$g_{1,0,1}$	$g_{1,0,0}$	$g_{1,1,1}$	$g_{1,1,0}$
$g_{0,1,0}$	$g_{0,1,0}$	$g_{0,1,1}$	$g_{0,0,0}$	$g_{0,0,1}$	$g_{1,1,0}$	$g_{1,1,1}$	$g_{1,0,0}$	$g_{1,0,1}$
$g_{0,1,1}$	$g_{0,1,1}$	$g_{0,1,0}$	$g_{0,0,1}$	$g_{0,0,0}$	$g_{1,1,1}$	$g_{1,1,0}$	$g_{1,0,1}$	$g_{1,0,0}$
$g_{1,0,0}$	$g_{1,0,0}$	$g_{1,0,1}$	$g_{1,1,0}$	$g_{1,1,1}$	$g_{0,0,0}$	$g_{0,0,1}$	$g_{0,1,0}$	$g_{0,1,1}$
$g_{1,0,1}$	$g_{1,0,1}$	$g_{1,0,0}$	$g_{1,1,1}$	$g_{1,1,0}$	$g_{0,0,1}$	$g_{0,0,0}$	$g_{0,1,1}$	$g_{0,1,0}$
$g_{1,1,0}$	$g_{1,1,0}$	$g_{1,1,1}$	$g_{1,0,0}$	$g_{1,0,1}$	$g_{0,1,0}$	$g_{0,1,1}$	$g_{0,0,0}$	$g_{0,0,1}$
$g_{1,1,1}$	$g_{1,1,1}$	$g_{1,1,0}$	$g_{1,0,1}$	$g_{1,0,0}$	$g_{0,1,1}$	$g_{0,1,0}$	$g_{0,0,1}$	$g_{0,0,0}$

55. If H_1 is the subgroup $\{g_{0,0,0}\}$, find the elements of the field L^{H_1}.

56. Compute $[L : L^{H_1}]$ and $[L^{H_1} : \mathbb{Q}]$.

57. If H_2 is the subgroup $\{g_{0,0,0}, g_{0,0,1}\}$, find the elements of the field L^{H_2}.

58. Compute $[L : L^{H_2}]$ and $[L^{H_2} : \mathbb{Q}]$.

59. If H_3 is the subgroup $\{g_{0,0,0}, g_{0,1,0}\}$, find the elements of the field L^{H_3}.

60. Compute $[L : L^{H_3}]$ and $[L^{H_3} : \mathbb{Q}]$.

61. If H_4 is the subgroup $\{g_{0,0,0}, g_{0,1,1}\}$, find the elements of the field L^{H_4}.

62. Compute $[L : L^{H_4}]$ and $[L^{H_4} : \mathbb{Q}]$.

63. If H_5 is the subgroup $\{g_{0,0,0}, g_{1,0,0}\}$, find the elements of the field L^{H_5}.

64. Compute $[L : L^{H_5}]$ and $[L^{H_5} : \mathbb{Q}]$.

65. If H_6 is the subgroup $\{g_{0,0,0}, g_{1,0,1}\}$, find the elements of the field L^{H_6}.

66. Compute $[L : L^{H_6}]$ and $[L^{H_6} : \mathbb{Q}]$.

67. If H_7 is the subgroup $\{g_{0,0,0}, g_{1,1,0}\}$, find the elements of the field L^{H_7}.

68. Compute $[L : L^{H_7}]$ and $[L^{H_7} : \mathbb{Q}]$.

69. If H_8 is the subgroup $\{g_{0,0,0}, g_{1,1,1}\}$, find the elements of the field L^{H_8}.

70. Compute $[L : L^{H_8}]$ and $[L^{H_8} : \mathbb{Q}]$.

71. If H_9 is the subgroup $\{g_{0,0,0}, g_{0,0,1}, g_{0,1,0}, g_{0,1,1}\}$, find the elements of the field L^{H_9}.

72. Compute $[L : L^{H_9}]$ and $[L^{H_9} : \mathbb{Q}]$.

73. If H_{10} is the subgroup $\{g_{0,0,0}, g_{0,0,1}, g_{1,0,0}, g_{1,0,1}\}$, find the elements of the field $L^{H_{10}}$.

74. Compute $[L : L^{H_{10}}]$ and $[L^{H_{10}} : \mathbb{Q}]$.

75. If H_{11} is the subgroup $\{g_{0,0,0}, g_{0,0,1}, g_{1,1,0}, g_{1,1,1}\}$, find the elements of the field $L^{H_{11}}$.

76. Compute $[L : L^{H_{11}}]$ and $[L^{H_{11}} : \mathbb{Q}]$.

77. If H_{12} is the subgroup $\{g_{0,0,0}, g_{0,1,0}, g_{1,0,0}, g_{1,1,0}\}$, find the elements of the field $L^{H_{12}}$.

78. Compute $[L : L^{H_{12}}]$ and $[L^{H_{12}} : \mathbb{Q}]$.

79. If H_{13} is the subgroup $\{g_{0,0,0}, g_{0,1,0}, g_{1,0,1}, g_{1,1,1}\}$, find the elements of the field $L^{H_{13}}$.

80. Compute $[L : L^{H_{13}}]$ and $[L^{H_{13}} : \mathbb{Q}]$.

81. If H_{14} is the subgroup $\{g_{0,0,0}, g_{0,1,1}, g_{1,0,0}, g_{1,1,1}\}$, find the elements of the field $L^{H_{14}}$.

82. Compute $[L : L^{H_{14}}]$ and $[L^{H_{14}} : \mathbb{Q}]$.

83. If H_{15} is the subgroup $\{g_{0,0,0}, g_{0,1,1}, g_{1,0,1}, g_{1,1,0}\}$, find the elements of the field $L^{H_{15}}$.

84. Compute $[L : L^{H_{15}}]$ and $[L^{H_{15}} : \mathbb{Q}]$.

85. If H_{16} is all of $Gal(L/\mathbb{Q})$, find the elements of the field $L^{H_{16}}$.

86. Compute $[L : L^{H_{16}}]$ and $[L^{H_{16}} : \mathbb{Q}]$.

87. If $M_1 = \mathbb{Q}$, find the elements of the subgroup G_{M_1}.

88. If $M_2 = \mathbb{Q}(i\sqrt{35})$, find the elements of the subgroup G_{M_2}.

89. If $M_3 = \mathbb{Q}(i\sqrt{7})$, find the elements of the subgroup G_{M_3}.

90. If $M_4 = \mathbb{Q}(i\sqrt{5})$, find the elements of the subgroup G_{M_4}.

91. If $M_5 = \mathbb{Q}(i)$, find the elements of the subgroup G_{M_5}.

92. If $M_6 = \mathbb{Q}(\sqrt{35})$, find the elements of the subgroup G_{M_6}.

93. If $M_7 = \mathbb{Q}(\sqrt{7})$, find the elements of the subgroup G_{M_7}.

94. If $M_8 = \mathbb{Q}(\sqrt{5})$, find the elements of the subgroup G_{M_8}.

95. If $M_9 = \mathbb{Q}(i\sqrt{5}, i\sqrt{7})$, find the elements of the subgroup G_{M_9}.

96. If $M_{10} = \mathbb{Q}(\sqrt{35}, i)$, find the elements of the subgroup $G_{M_{10}}$.

97. If $M_{11} = \mathbb{Q}(\sqrt{7}, i\sqrt{5})$, find the elements of the subgroup $G_{M_{11}}$.

98. If $M_{12} = \mathbb{Q}(\sqrt{7}, i)$, find the elements of the subgroup $G_{M_{12}}$.

99. If $M_{13} = \mathbb{Q}(\sqrt{5}, i\sqrt{7})$, find the elements of the subgroup $G_{M_{13}}$.

100. If $M_{14} = \mathbb{Q}(\sqrt{5}, i)$, find the elements of the subgroup $G_{M_{14}}$.

101. If $M_{15} = \mathbb{Q}(\sqrt{5}, \sqrt{7})$, find the elements of the subgroup $G_{M_{15}}$.

102. If $M_{16} = L$, find the elements of the subgroup $G_{M_{16}}$.

For exercises 103–105, please read the following:

In Theorem 15.12, we showed that if $\mathbb{Q} \subseteq K \subseteq L \subseteq \mathbb{C}$ are fields such that $[L : K]$ is finite, then $L = K(\theta)$, for some $\theta \in L$. In exercises 103–105, we will show that there exist fields $K \subseteq L$, with $[L : K]$ finite, such that L is *not* a simple extension of K. Thus, the hypothesis in Theorem 15.12 that K contain $\mathbb{Q}$ is needed. In doing these exercises, you might want to first review the final example from Section 12.5 as well as exercises 19 and 21 after Section 12.5.

103. Let $K = \mathbb{Z}_p(t_1{}^p, t_2{}^p)$, $E = \mathbb{Z}_p(t_1{}^p, t_2)$, and $L = \mathbb{Z}_p(t_1, t_2)$. Observe that K, E, L are fields with $K \subseteq E \subseteq L$.
    (a) Compute $[E : K]$ and $[L : E]$.

    (b) Use part (a) to find $[L : K]$.

104. (a) Show that if $\alpha \in \mathbb{Z}_p[t_1, t_2]$, then $\alpha^p \in \mathbb{Z}_p[t_1{}^p, t_2{}^p]$.

    (b) If $K \subseteq L$ are as in exercise 103, use part (a) of this exercise to show that if $\beta \in L$ then $\beta^p \in K$.

    (c) Show that if $\theta \in L$, then $[K(\theta) : K]$ is equal to 1 or p.

105. Use exercises 103 and 104 to show that if $\theta \in L$, then $L \neq K(\theta)$. This shows that although L is a finite extension of K, it is not a simple extension of K.

For exercise 106, please read the following:

Exercises 1–25, 26–47, and 48–102, dealt with a series of questions about the splitting fields over $\mathbb{Q}$ of the polynomials $x^3 - 10$, $x^4 + 1$, and $(x^2 - 5)(x^2 - 7)(x^2 + 1)$, respectively. For additional practice with splitting fields, Galois groups, and fixed fields, in exercise 106, we will examine the splitting field of $x^4 - 2$ over $\mathbb{Q}$. In this exercise, the Galois group will be a group we have seen before, but we have not as yet seen it arise as the Galois group of a splitting field over $\mathbb{Q}$. Before beginning this exercise, keep in mind that it has many, many parts and could easily have been split into over 40 separate exercises.

106. Let L be the splitting field of $x^4 - 2$ over $\mathbb{Q}$.

   (a)   Show that $L = \mathbb{Q}(2^{\frac{1}{4}}, i)$ and then find $[L : \mathbb{Q}]$ and $|Gal(L/\mathbb{Q})|$.

   (b)   Describe the behavior of each element of $Gal(L/\mathbb{Q})$ on $2^{\frac{1}{4}}$ and i and then find the complete table for $Gal(L/\mathbb{Q})$.

   (c)   To what familiar group is $Gal(L/\mathbb{Q})$ isomorphic?

   (d)   Use the table for $Gal(L/\mathbb{Q})$ to find all subgroups of $Gal(L/\mathbb{Q})$.

   (e)   For every subgroup H of $Gal(L/\mathbb{Q})$, find the fixed field L^H.

   (f)   For each subgroup H of $Gal(L/\mathbb{Q})$, check if H is a normal subgroup of $Gal(L/\mathbb{Q})$ and also check if the corresponding fixed field L^H is the splitting field of some polynomial over $\mathbb{Q}$.

   (g)   Was there any relationship in part (f) between when H is normal and when L^H is a splitting field?

CHAPTER 16

Geometric Constructions

In the final two chapters, we settle the two great "impossibility" problems introduced in Chapter 1—namely, trisecting angles with a ruler and compass and finding a formula for the roots of fifth-degree polynomials. For both problems, we will be confronted with the difficult issue of how we go about proving that something cannot be done.

Proving that angles cannot be trisected with a ruler and compass will be done in three steps:

I.   Formally define what it means for a real number to be constructible with ruler and compass.

II. Show that the degree of a field extension over $\mathbb{Q}$ of a field generated by a constructible real number must be 2^n, where $n \geq 0$ is an integer.

III. Show that if angles could be trisected, then numbers can be constructed that do *not* generate extensions of $\mathbb{Q}$ of degree 2^n.

16.1 Constructible Points and Constructible Real Numbers

In Chapter 2, we showed that depending on our definition of the word *ruler*, it is possible to trisect angles. Therefore, in this chapter, we need to be very precise about the meaning of the terms we use when proving that various geometric objects *cannot* be constructed. Not only must we be very careful about what we mean by *ruler* and *compass*, but we will also need to carefully describe exactly how the ruler and compass are allowed to be used.

We begin with the xy-plane and designate the points $(0, 0)$ and $(1, 0)$ on the x-axis.

This is done for two reasons. First, the only time we will be allowed to use either the ruler or compass is when we draw a line or circle with points that have previously been constructed. Therefore, in order to get started, we need two initial points, $(0, 0)$ and $(1, 0)$, that we already consider to be constructed. Second, as we construct additional points in the plane beyond $(0, 0)$ and $(1, 0)$, we will need to determine the x and y coordinates of these points. By beginning with the points $(0, 0)$ and $(1, 0)$, we associate the number 1 to the distance between these two particular points, and this will be used to determine the coordinates of other points as they are constructed.

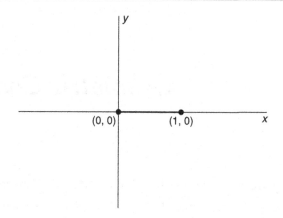

Allowable Moves: Let $P_0 = (0, 0)$ and $P_1 = (1, 0)$ be the initial constructible points.

1. The ruler may *only* be used to draw a line connecting two points that have already been constructed.

2. The compass may *only* be used to draw a circle using two points that have already been constructed, where one of the points is the center of the circle and the other lies on the circumference of the circle.

We consider a point in the plane to be constructible if it is the intersection of

(a)   two lines,

(b)   two circles, or

(c)   one line and one circle,

both of which have been drawn according to the allowable moves.

More formally, we record this as

Definition 16.1. *A point P in the xy-plane is constructible if there exists a sequence of points*

$$P_0 = (0, 0), \; P_1 = (1, 0), \; P_2, \ldots, P_n = P,$$

such that every P_i, for $2 \leq i \leq n$, is the intersection of either two lines, two circles, or one line and one circle that have been constructed using only the points $P_0, P_1, \ldots, P_{i-1}$.

This immediately leads us to

Definition 16.2. *A real number α is constructible if it is the x or y coordinate of a constructible point.*

Observe that our list of allowable moves eliminates any ambiguity about how the ruler can be used. All markings on the ruler are to be ignored, so the ruler can only be used like a straightedge. In particular, the "trisection" done in Chapter 2 uses a move that is clearly not permitted.

When you use a compass, you might occasionally draw a circle and then move the compass to another location to draw a new circle using the same radius. If you have done this, then you have allowed the compass to keep the same radius as it moved from one location to the next, and, in this case, we say that you have used a *rigid* compass. However, we do *not* assume that our compass is rigid, so preserving the radius when the compass is moved is not an allowable move.

On the other hand, it is now reasonable to ask whether a *sequence* of allowable moves will allow us to move the compass from one constructible point to another while preserving the radius. If the answer is yes, then our allowable moves will enable us to perform all the constructions that can be done with a rigid compass. In order to show that the answer is indeed yes and to also develop a sense of which real numbers are constructible, we begin with four basic constructions.

Basic construction #1—Constructing a perpendicular line from a point on a line
Given the line connecting points A and B, we will construct the line perpendicular to this one at the point A.

(a) Draw the circle with center A having B on the circumference and let C denote the other point where this circle intersects the line connecting A and B.

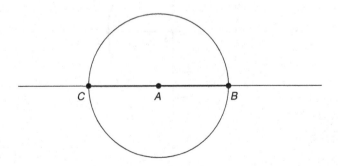

(b) Draw the circle centered at B with C on the circumference and also draw the circle centered at C with B on the circumference. Let D be either of the intersection points of these two circles.

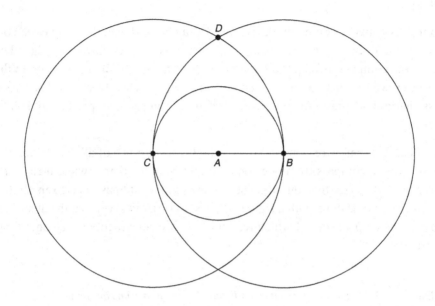

(c) Draw the line connecting D and A.

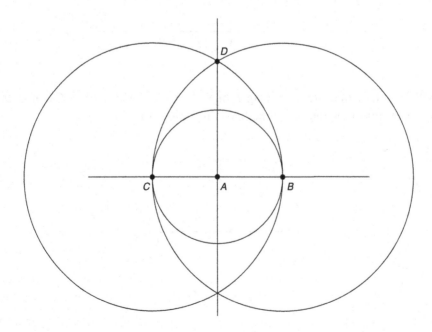

Observe that BCD is an equilateral triangle and triangles ABD and ACD are congruent. The line connecting A and D is perpendicular to the line connecting A and B at the point A. In fact, the line connecting A and D is the perpendicular bisector of the line segment BC.   □

Basic construction #2—Dropping a perpendicular line to a line from a point off the line
Given a point C, not on the line connecting points A and B, we will construct the line through C that is perpendicular to the line connecting A and B.

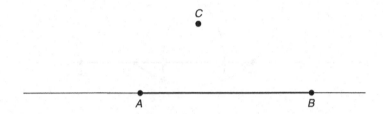

(a)   Draw the circle with center C having A on the circumference and then let D denote the point, other than A, where this circle intersects the line connecting A and B.

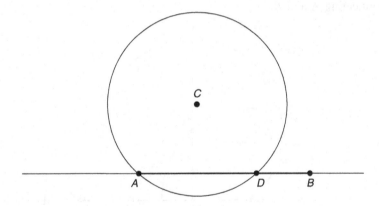

(b)   Draw the circles centered at A and at D that both have C on the circumference and then let E denote the point, other than C, where these two circles intersect.

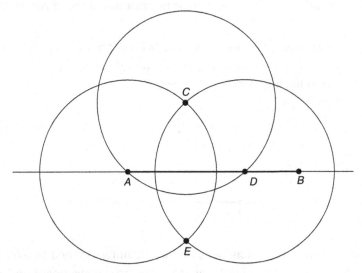

(c)   Draw the line connecting the points C and E. It passes through C and is perpendicular to the line connecting A and B.

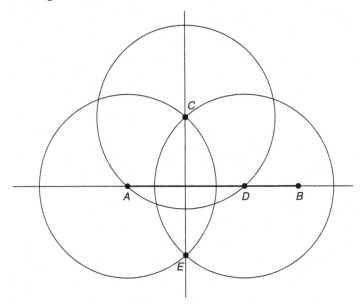

Observe that $ACDE$ is a rhombus, triangles ACE and DCE are congruent, and the line connecting C and E is the perpendicular bisector of the line segment AD.   □

In part (a) of the previous construction, we let D denote the second point where the circle centered at C intersected the line connecting A and B. You should convince yourself that it is possible that there is no second intersection point, but this happens if and only if the line connecting C and A is perpendicular to the original line.

Basic construction #3—Constructing a line parallel to a given line through a point off the line
Given a point C, not on the line connecting points A and B, we will construct the line through C that is parallel to the line connecting A and B.

(a) Using basic construction #2, drop a perpendicular line from C to the line connecting A and B and let D denote the point where these two lines intersect.

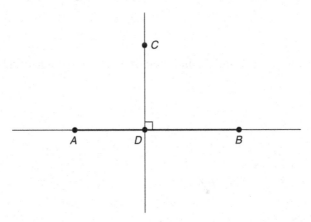

(b) Using basic construction #1, construct the line perpendicular to the line connecting C and D at C.

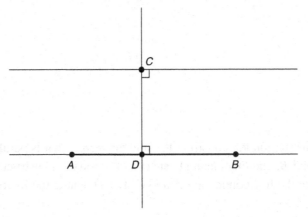

Observe that the line constructed in part (b) and the line connecting A and B are both perpendicular to the line connecting C and D. Therefore, the line constructed in part (b) is parallel to the original line.   □

Basic construction #4—Transporting the distance between two constructible points to use as the radius of a circle centered at a third constructible point

Given points A, B, C, we will draw a circle centered at C whose radius is equal to the distance between A and B.

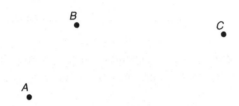

(a)   Draw the line connecting A and B, and then draw the line connecting A and C.

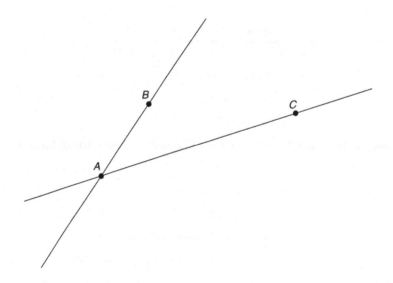

(b)   Using basic construction #3, construct the line through C that is parallel to the line connecting A and B, and then, using basic construction #3, construct the line through B that is parallel to the line connecting A and C. Let D denote the intersection of these two lines.

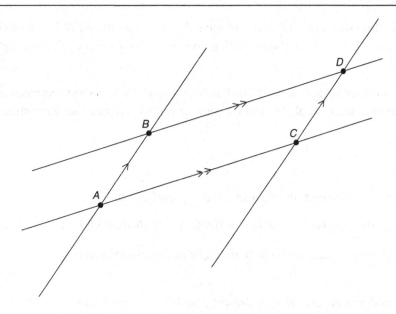

(c)   Draw the circle centered at C with D on the circumference.

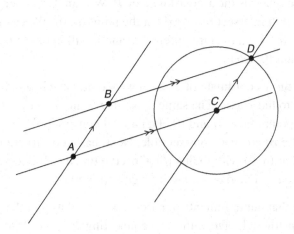

Observe that $ABDC$ is a parallelogram. Therefore, line segments AB and CD have the same length. As a result, the radius of the circle centered at C with D on the circumference is equal to the distance between A and B.   □

The previous construction assumes that C does not lie on the line connecting A and B. You should convince yourself that if A, B, C do lie on the same line, then you can construct a point

B' such that the line segments AB and AB' have the same length and B' is not on the line connecting A and C. Therefore, the preceding construction will still work after replacing the point B with B'.

Since each of our four basic constructions has been shown to be a short sequence of allowable moves, we can now freely apply the basic constructions to generate many constructible numbers.

Theorem 16.3

(a)   If $\alpha, \beta \in \mathbb{R}$ are constructible, then so are $\alpha + \beta$ and $\alpha - \beta$.

(b)   If $\alpha, \beta \in \mathbb{R}$ are constructible, with $\alpha > 0$ and $\beta > 0$, then $\alpha\beta$ and $\frac{\alpha}{\beta}$ are both constructible.

(c)   If $\alpha \in \mathbb{R}$ is constructible and $\alpha > 0$, then $\sqrt{\alpha}$ is also constructible.

Proof. We should first observe that by drawing the line through the points $(0, 0)$ and $(1, 0)$ and then constructing the line perpendicular to this one at $(0, 0)$, our allowable moves and basic constructions have allowed us to draw both the x- and y-axes.

For part (a), since α is constructible, it is a coordinate of a constructible point P. For the moment, let us assume that α is the x coordinate of P. We can drop a perpendicular from P to the x-axis, and it will intersect the x-axis at the point $(\alpha, 0)$. We can now draw the circle centered at $(0, 0)$ with $(\alpha, 0)$ on the circumference, and it will intersect the x- and y-axes at the points $(\pm\alpha, 0)$ and $(0, \pm\alpha)$.

Observe that if α was the y coordinate of P, then we would have begun by first dropping a perpendicular from P to the y-axis. The same reasoning as in the previous paragraph would once again show that points $(\pm\alpha, 0)$ and $(0, \pm\alpha)$ are all constructible. It now follows that the points $(\pm\beta, 0)$ and $(0, \pm\beta)$ are also constructible. Next, basic construction #4 allows us to draw a circle centered at $(\alpha, 0)$ with radius $|\beta|$. Observe that it will intersect the x-axis at the points $(\alpha \pm \beta, 0)$. Thus, $\alpha + \beta$ and $\alpha - \beta$ are both constructible.

For part (b), we know that our arguments in part (a) have told us that the points $A = (\beta, 0)$ and $B = (0, \alpha)$ are constructible. Draw the line connecting $P_1 = (1, 0)$ to B, and then draw the line parallel to this one through A. If we let C denote the point where the line drawn through A intersects the y-axis, then triangles $P_0 P_1 B$ and $P_0 A C$ are similar. As a result, the corresponding sides are in proportion and if we let γ be the y coordinate of C, we have

$$\frac{\gamma}{\alpha} = \frac{\beta}{1}.$$

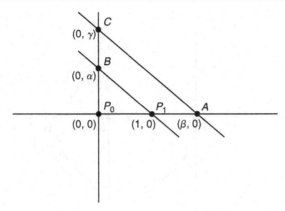

Therefore, $\gamma = \alpha\beta$ and since γ has been shown to be constructible, we can conclude that $\alpha\beta$ is constructible.

Next, we let A and B be as in the previous construction, but we now draw the line connecting A and B and then draw the line parallel to this one through P_1. If we let D be the point where the line drawn through P_1 intersects the y-axis, then the triangles P_0P_1D and P_0AB are similar. Since the corresponding sides are in proportion, when we let δ be the y coordinate of D, we have

$$\frac{\delta}{\alpha} = \frac{1}{\beta}.$$

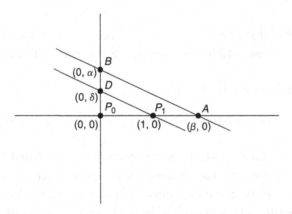

Therefore, $\delta = \frac{\alpha}{\beta}$ and since δ has been shown to be constructible, we see that $\frac{\alpha}{\beta}$ is constructible.

For part (c), if we let $A = (\alpha, 0)$ and $B = (\alpha + 1, 0)$, then A and B are both constructible. Recall that our first basic construction actually showed how to construct the perpendicular bisector of a line segment. If we let C be the point where the perpendicular bisector of the line segment P_0B intersects the x-axis, then $C = \left(\frac{\alpha+1}{2}, 0\right)$. Next, draw the circle centered at C with P_0 on the circumference. Then draw the line perpendicular to the x-axis at A and let D denote the point in the first quadrant where this line intersects the circle centered at C.

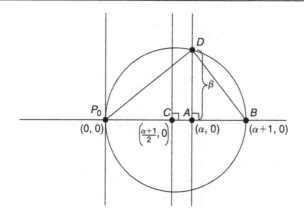

Observe that $\angle P_0DB$ is a right angle, so the triangles AP_0D and ADB are similar to $\angle AP_0D = \angle ADB$ and $\angle ADP_0 = \angle ABD$. If we let β be the y coordinate of D, then β is the length of line segment AD, α is the length of the line segment P_0A, and 1 is the length of the line segment AB.

Since the lengths of the corresponding sides of similar triangles are in proportion, we have

$$\frac{\beta}{\alpha} = \frac{1}{\beta}.$$

This immediately implies that $\beta^2 = \alpha$, so $\beta = \sqrt{\alpha}$. Since β is the y coordinate of a constructible point, it is constructible. As a result, $\sqrt{\alpha}$ is constructible, as required. $\qquad\square$

In light of Theorem 16.3, we easily obtain

Corollary 16.4. *The set of constructible numbers is a field lying between $\mathbb{Q}$ and $\mathbb{R}$.*

Proof. Since $P_0 = (0, 0)$ and $P_1 = (1, 0)$ are constructible points, 0 and 1 are constructible numbers. Theorem 16.3(a) asserts that beginning with 0, whenever we add or subtract copies of 1, we continue to obtain constructible numbers. Therefore, every element of $\mathbb{Z}$ is constructible. Every positive element of $\mathbb{Q}$ is the quotient of elements of $\mathbb{N}$, so Theorem 16.3(b) tells us that every positive rational number is constructible. However, Theorem 16.3(a) also tells us that the additive inverse of every constructible number is constructible, so every element of $\mathbb{Q}$ is constructible. Since the set of constructible numbers was defined as a subset of $\mathbb{R}$, it follows that this set lies between $\mathbb{Q}$ and $\mathbb{R}$.

The set of constructible numbers is contained in the field $\mathbb{R}$, so in order to show that it is a subfield of $\mathbb{R}$, it suffices to show that it is a group under addition and its nonzero elements are a group under multiplication. In light of Proposition 8.6(a), to show that the constructible numbers are a group under addition, we only need to show that $\alpha - \beta$ is constructible whenever α and β are. However, Theorem 16.3(a) certainly tells us that this is indeed the case.

If α, β are nonzero constructible numbers, then Theorem 16.3(b) shows that $\frac{|\alpha|}{|\beta|}$ is constructible. Thus, using Theorem 16.3(a), we know that $\frac{|\alpha|}{|\beta|}$ and $-\frac{|\alpha|}{|\beta|}$ are both constructible. Since $\frac{\alpha}{\beta}$ is equal to either $\frac{|\alpha|}{|\beta|}$ or $-\frac{|\alpha|}{|\beta|}$, it follows that $\frac{\alpha}{\beta}$ is certainly constructible. Therefore, Proposition 8.6(a) now applies to tell us that the nonzero constructible numbers are a group under multiplication and the set of constructible numbers is indeed a subfield of $\mathbb{R}$. $\qquad\square$

Theorem 16.3 and Corollary 16.4 enable us to easily generate constructible numbers.

■ Examples

The following numbers are all constructible:

$$3, \ \sqrt{2}, \ \sqrt{5}, \ \sqrt{2}+\sqrt{5}, \ \frac{\sqrt{2}+\sqrt{5}}{\sqrt{2}-\sqrt{5}}, \ \sqrt{\sqrt{2}+\sqrt{5}}, \ \sqrt{\sqrt{2}+\sqrt{5}+\sqrt{3}}$$

■

It takes some work, but it can be shown that the degrees of the minimum polynomials over $\mathbb{Q}$ of the preceding seven numbers are, respectively, $1, 2, 2, 4, 4, 8, 16$. Observe that each of these degrees is a power of 2. It certainly leads one to conjecture that if $\alpha \in \mathbb{R}$ is constructible, then $[\mathbb{Q}(\alpha) : \mathbb{Q}] = 2^n$, where $n \geq 0$ is an integer. In order to prove this, we need to start translating some of our observations about constructions into facts about field extensions.

Lemma 16.5. *Let K be a field such that $\mathbb{Q} \subseteq K \subseteq \mathbb{R}$ and suppose (α_1, β_1) and (α_2, β_2) are constructible points such that $\alpha_1, \alpha_2, \beta_1, \beta_2 \in K$.*

(a) *If l is the line connecting (α_1, β_1) and (α_2, β_2), then l has equation $ax + by = c$, where $a, b, c \in K$.*

(b) *If C is the circle centered at (α_1, β_1) with (α_2, β_2) on the circumference, then C has equation $(x - a)^2 + (y - b)^2 = c$, where $a, b, c \in K$.*

Proof. For part (a), if l is vertical, then it has equation $x = \alpha_1$, which is certainly of the desired form. On the other hand, if l is not vertical, then $\alpha_2 \neq \alpha_1$ and the point-slope formula tells us that l has equation

$$(y - \beta_1) = \left(\frac{\beta_2 - \beta_1}{\alpha_2 - \alpha_1}\right)(x - \alpha_1).$$

This equation can be rewritten as

$$(\beta_2 - \beta_1)x + (\alpha_1 - \alpha_2)y = \alpha_1\beta_2 - \alpha_2\beta_1$$

and since $\beta_2 - \beta_1, \alpha_1 - \alpha_2, \alpha_1\beta_2 - \alpha_2\beta_1 \in K$, this equation is of the desired form.

The distance between (α_1, β_1) and (α_2, β_2) is $\sqrt{(\alpha_2 - \alpha_1)^2 + (\beta_2 - \beta_1)^2}$, so the equation of C is

$$(x - \alpha_1)^2 + (y - \beta_1)^2 = (\alpha_2 - \alpha_1)^2 + (\beta_2 - \beta_1)^2.$$

Since $\alpha_1, \beta_1, (\alpha_2 - \alpha_1)^2, (\beta_2 - \beta_1)^2 \in K$, this equation is also of the desired form. $\square$

We can now prove the main result of this section.

Theorem 16.6. *If $\alpha \in \mathbb{R}$ is constructible, then $[\mathbb{Q}(\alpha) : \mathbb{Q}] = 2^l$, where $l \geq 0$ is an integer.*

Proof. If α is constructible, then Definitions 16.1 and 16.2 tell us that α is one of the coordinates of a point P such that there is a sequence of points

$$P_0 = (0, 0), \ P_1 = (1, 0), \ P_2, \ldots, P_n = P$$

and, for $2 \leq i \leq n$, every P_i is the intersection of either two lines, two circles, or one line and one circle that have been constructed using only the points $P_0, P_1, \ldots, P_{i-1}$.

We now use the sequence of points to construct a sequence of fields

$$K_1 = \mathbb{Q}, \ K_2 = K_1(\alpha_2, \beta_2), \ K_3 = K_2(\alpha_3, \beta_3), \ldots, K_n = K_{n-1}(\alpha_n, \beta_n)$$

where, for $2 \leq i \leq n$, $P_i = (\alpha_i, \beta_i)$. Observe that, for $2 \leq i \leq n$, K_i is the smallest field that contains $\mathbb{Q}$ as well as the coordinates of the points $P_0, P_1, \ldots, P_{i-1}, P_i$.

Using Theorem 14.6 and the fact that $K_1 = \mathbb{Q}$, we have

(1) $$[K_n : \mathbb{Q}] = [K_n : K_{n-1}] \cdot [K_{n-1} : K_{n-2}] \cdots [K_3 : K_2] \cdot [K_2 : K_1].$$

We now claim that, for $2 \leq t \leq n$, $[K_t : K_{t-1}]$ is equal to 1 or 2. Note that K_t is the smallest field that contains K_{t-1}, α_t, and β_t. In addition, P_t is the intersection of two lines, a line and a circle, or two circles, where the lines and circles are drawn using points whose coordinates belong to K_{t-1}. Our proof now branches out into three cases.

Case I—Two Lines: By Lemma 16.5, P_t is the intersection of two lines with equations $a_1x + b_1y = c_1$ and $a_2x + b_2y = c_2$, where $a_1, a_2, b_1, b_2, c_1, c_2 \in K_{t-1}$. Since the two lines are different, at most one of them can be vertical. If one of them is vertical then, without loss of generality, we may assume it is the first. Therefore, $b_1 = 0$, $a_1 \neq 0$, and we can rewrite the equation of the first line as $x = \frac{c_1}{a_1}$. Since the second line is not vertical, we know $b_2 \neq 0$, and plugging $x = \frac{c_1}{a_1}$ into this equation gives us

$$y = \frac{a_1 c_2 - a_2 c_1}{a_1 b_2}.$$

Therefore,

$$P_t = \left(\frac{c_1}{a_1}, \frac{a_1 c_2 - a_2 c_1}{a_1 b_2} \right).$$

As a result, the coordinates of P_t are obtained from elements of the field K_{t-1} by applying operations which always produce elements of the same field. Therefore, both coordinates of P_t belong to K_{t-1}, so $K_t = K_{t-1}$ and so $[K_t : K_{t-1}] = 1$.

Next, suppose that neither line is vertical. Therefore, both b_1 and b_2 are nonzero. Since the two lines have different slopes, it follows that $-\frac{a_1}{b_1} \neq -\frac{a_2}{b_2}$ which immediately implies that $a_2 b_1 - a_1 b_2 \neq 0$. As a result, when we solve the simultaneous equations $a_1 x + b_1 y = c_1$ and $a_2 x + b_2 y = c_2$, we obtain

$$x = \frac{b_1 c_2 - b_2 c_1}{a_2 b_1 - a_1 b_2}, \quad y = \frac{a_2 c_1 - a_1 c_2}{a_2 b_1 - a_1 b_2}.$$

Therefore,

$$P_t = \left(\frac{b_1 c_2 - b_2 c_1}{a_2 b_1 - a_1 b_2}, \frac{a_2 c_1 - a_1 c_2}{a_2 b_1 - a_1 b_2} \right).$$

Using the same argument as in the previous paragraph, the coordinates of P_t are obtained by applying operations to elements of K_{t-1} that produce elements of K_{t-1}. Once again, the coordinates of P_t belong to K_{t-1}, so $K_t = K_{t-1}$ and $[K_t : K_{t-1}] = 1$.

Case II—One Line and One Circle: By Lemma 16.5, P_t is the intersection of a line with equation $a_1 x + b_1 y = c_1$ and a circle with equation $(x - a_2)^2 + (y - b_2)^2 = c_2$, where $a_1, a_2, b_1, b_2, c_1, c_2 \in K_{t-1}$. If the line is vertical, then $b_1 = 0$, $a_1 \neq 0$, and we can rewrite its equation as $x = \frac{c_1}{a_1}$. Plugging this into the equation of the circle, we obtain

$$\left(\frac{c_1}{a_1} - a_2 \right)^2 + (y - b_2)^2 = c_2.$$

Since $P_t = (\alpha_t, \beta_t)$, the previous equation tells us that β_t satisfies a quadratic polynomial with coefficients in K_{t-1}. As a result,

$$[K_{t-1}(\beta_t) : K_{t-1}] = 1 \text{ or } 2.$$

Since $\alpha_t = \frac{c_1}{a_1} \in K_{t-1} \subseteq K_{t-1}(\beta_t)$, we have

$$[K_{t-1}(\alpha_t, \beta_t) : K_{t-1}(\beta_t)] = 1,$$

which implies that

$$[K_t : K_{t-1}] = [K_{t-1}(\alpha_t, \beta_t) : K_{t-1}(\beta_t)] \cdot [K_{t-1}(\beta_t) : K_{t-1}] = 1 \text{ or } 2.$$

If the line is not vertical, then $b_1 \neq 0$, and we can rewrite its equation as

$$y = -\frac{a_1}{b_1}x + \frac{c_1}{b_1}.$$

Plugging this into the equation of the circle results in

$$(x - a_2)^2 + \left(\left(-\frac{a_1}{b_1}x + \frac{c_1}{b_1} \right) - b_2 \right)^2 = c_2.$$

The preceding equation is a quadratic polynomial in x with leading coefficient $1 + \left(\frac{a_1}{b_1} \right)^2$. Since $P_t = (\alpha_t, \beta_t)$, we see that α_t satisfies a quadratic polynomial with coefficients in K_{t-1}. As a result,

$$[K_{t-1}(\alpha_t) : K_{t-1}] = 1 \text{ or } 2.$$

Furthermore, since the point (α_t, β_t) satisfies the equation $y = -\frac{a_1}{b_1}x + \frac{c_1}{b_1}$, we have

$$\beta_t = -\frac{a_1}{b_1}\alpha_t + \frac{c_1}{b_1} \in K_{t-1}(\alpha_t).$$

It now follows that $[K_{t-1}(\alpha_t, \beta_t) : K_{t-1}(\alpha_t)] = 1$, hence

$$[K_t : K_{t-1}] = [K_{t-1}(\alpha_t, \beta_t) : K_{t-1}(\alpha_t)] \cdot [K_{t-1}(\alpha_t) : K_{t-1}] = 1 \text{ or } 2.$$

Case III—Two Circles: By Lemma 16.5, $P_t = (\alpha_t, \beta_t)$ is one of the intersection points of circles with equations $(x - a_1)^2 + (y - b_1)^2 = c_1$ and $(x - a_2)^2 + (y - b_2)^2 = c_2$, where $a_1, a_2, b_1, b_2, c_1, c_2 \in K_{t-1}$. Therefore, the ordered pair (α_t, β_t) is also a solution of the following equation that is obtained by subtracting the equations of the two circles:

$$((x - a_1)^2 + (y - b_1)^2) - ((x - a_2)^2 + (y - b_2)^2) = c_1 - c_2.$$

This can be rewritten as

$$(2a_2 - 2a_1)x + (a_1^2 - a_2^2) + (2b_2 - 2b_1)y + (b_1^2 - b_2^2) = c_1 - c_2,$$

which simplifies to

$$(2a_2 - 2a_1)x + (2b_2 - 2b_1)y = (c_1 - c_2) - \left(a_1{}^2 - a_2{}^2\right) - \left(b_1{}^2 - b_2{}^2\right).$$

Observe that this last equation is of the form $ax + by = c$, where $a, b, c \in K_{t-1}$. Therefore, the point $P_t = (\alpha_t, \beta_t)$ is an intersection point of a line and a circle of the form we examined in Case II. Hence, the conclusions in Case II also apply in this case, and we once again have

$$[K_t : K_{t-1}] = [K_{t-1}(\alpha_t, \beta_t) : K_{t-1}] = 1 \text{ or } 2.$$

In all three cases, we have succeeded in showing that $[K_t : K_{t-1}]$ is equal to 1 or 2, for $2 \le t \le n$. Therefore, when we look back at equation (1), every term that is being multiplied on the right-hand side of

$$[K_n : \mathbb{Q}] = [K_n : K_{n-1}] \cdot [K_{n-1} : K_{n-2}] \cdots [K_3 : K_2] \cdot [K_2 : K_1]$$

is either 1 or 2. Therefore, $[K_n : \mathbb{Q}] = 2^m$, for some integer $m \ge 0$.

Finally, if α is constructible, then α is one of the coordinates of the final point P_n of the sequence of points

$$P_0 = (0, 0), \ P_1 = (1, 0), \ P_2, \ldots, P_n = P$$

introduced at the beginning of this proof. Since the field K_n contains both coordinates of P_n, we know that $\alpha \in K_n$, and Theorem 14.6 asserts that

$$2^m = [K_n : \mathbb{Q}] = [K_n : \mathbb{Q}(\alpha)] \cdot [\mathbb{Q}(\alpha) : \mathbb{Q}].$$

As a result, $[\mathbb{Q}(\alpha) : \mathbb{Q}]$ is a divisor of 2^m, so $[\mathbb{Q}(\alpha) : \mathbb{Q}] = 2^l$, for some integer $l \ge 0$. $\square$

16.2 The Impossibility of Trisecting Angles

We now have all the tools needed to prove that angles cannot be trisected. But first, we need to put this problem in perspective. If α is the distance between two constructible points, then basic construction #4 asserts that we can construct a circle with radius α centered at the origin. Since the point $(\alpha, 0)$ is an intersection point of this circle and the x-axis, it follows that α is constructible. Theorem 16.3(b) now indicates that

$$\frac{\alpha}{2}, \frac{\alpha}{3}, \frac{\alpha}{4}, \ldots, \frac{\alpha}{n}, \ldots$$

are also constructible. This tells us that any constructible line segment can be bisected, trisected, or, more generally, divided into n equal pieces, for any $n \in \mathbb{N}$.

It is then natural to turn our attention to angles. If an angle has been constructed, it is not hard to bisect it. To see this, if we are given an angle that has been constructed, let A denote its vertex.

Basic construction #4 allows us to draw the circle with radius 1 centered at A. Let B and C denote the points where this circle intersects the two rays forming our angle. Next, draw the circles with radius 1 centered at B and at C and let D denote the intersection point, other than A, of these circles. Draw the line connecting A and D and observe that triangles ABD and ACD are congruent. Hence, angles BAD and CAD are also congruent.

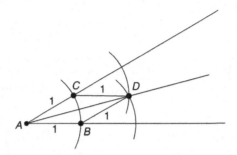

Since each of these angles is equal to half of our original angle, the line connecting A and D has bisected our original angle.

Having shown that we can split a line segment into n equal parts, for any $n \in \mathbb{N}$, and then having shown that angles can be bisected, it is natural to try to trisect angles. Our work in Section 16.1 has enabled us to produce lists of numbers which are constructible as well as lists of numbers which are not constructible. For example, Theorem 16.3 tells us that every number on the following list is constructible:

$$\sqrt{2}, \ \sqrt{3}, \ \sqrt{5}, \ \sqrt{6}, \ \sqrt{7}, \ \sqrt[4]{2}, \ \sqrt[4]{3}, \ \sqrt[4]{5}, \ \sqrt[4]{6}, \ \sqrt[4]{7}, \ \sqrt[8]{2}, \ \sqrt[8]{3}, \ \sqrt[8]{5}, \ \sqrt[8]{6}, \ \sqrt[8]{7}.$$

On the other hand, every number of the list

$$\sqrt[3]{2}, \ \sqrt[3]{3}, \ \sqrt[3]{5}, \ \sqrt[3]{6}, \ \sqrt[3]{7}, \ \sqrt[5]{2}, \ \sqrt[5]{3}, \ \sqrt[5]{5}, \ \sqrt[5]{6}, \ \sqrt[5]{7}, \ \sqrt[6]{2}, \ \sqrt[6]{3}, \ \sqrt[6]{5}, \ \sqrt[6]{6}, \ \sqrt[6]{7}$$

satisfies an irreducible polynomial in $\mathbb{Q}[x]$ of degree 3, 5, or 6. Thus, Theorem 16.6 tells us that every number of this list cannot be constructed with a ruler and compass. In fact, it will be Theorem 16.6 that is the key to the proof that angles cannot be trisected.

Theorem 16.7. *There are angles that cannot be trisected with ruler and compass. In particular, 60° angles cannot be trisected.*

Proof. We begin by drawing the circle with center $P_0 = (0, 0)$ with point $P_1 = (1, 0)$ on its circumference and then drawing the circle with center P_1 that has P_0 on its circumference. If we let A be the point in the first quadrant where these circles intersect, then we can see that triangle AP_0P_1 is an equilateral triangle. Hence, $\angle AP_0P_1 = 60°$.

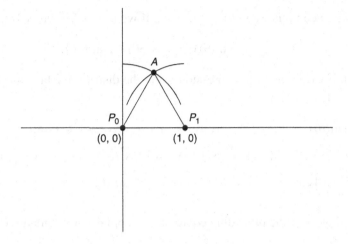

Next, by way of contradiction, let us suppose there was a sequence of allowable moves that could trisect angles. Then, in particular, we could trisect $\angle A P_0 P_1$ to construct a 20° angle with vertex at P_0 and one ray starting at P_0 and continuing through P_1. The other ray would begin at P_0 and would intersect the circle centered at P_0 with radius 1 at some constructible point B in the first quadrant. Therefore, $\angle B P_0 P_1 = 20°$. By the definition of the cosine function, the x coordinate of B is $\cos(20°)$.

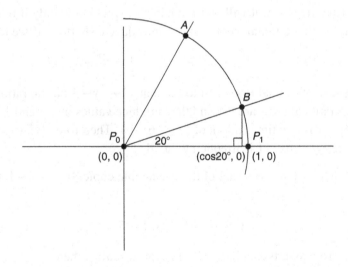

As a result, if a sequence of allowable moves enabled us to trisect angles, then the number $\cos(20°)$ would be constructible. Thanks to Theorem 16.6, this observation will soon lead to a contradiction.

There are two ways to compute $(\cos(\theta) + i\sin(\theta))^3$. If we use DeMoivre's Theorem, we obtain

$$(\cos(\theta) + i\sin(\theta))^3 = \cos(3\theta) + i\sin(3\theta).$$

On the other hand, if we expand the expression using the distributive law and use the identity $\sin^2(\theta) + \cos^2(\theta) = 1$, we obtain

$$(\cos(\theta) + i\sin(\theta))^3 = (\cos^3(\theta) - 3\cos(\theta)\sin^2(\theta)) + i(3\cos^2(\theta)\sin(\theta) - \sin^3(\theta)) =$$

$$(\cos^3(\theta) - 3\cos(\theta)(1 - \cos^2(\theta))) + i(3\cos^2(\theta)\sin(\theta) - \sin^3(\theta)) =$$

$$(4\cos^3(\theta) - 3\cos(\theta)) + i(3\cos^2(\theta)\sin(\theta) - \sin^3(\theta)).$$

Comparing the real parts of the preceding equations, we obtain the triple-angle formula

$$\cos(3\theta) = 4\cos^3(\theta) - 3\cos(\theta).$$

Since $\cos(60°) = \frac{1}{2}$, if we let $\theta = 20°$ and also let $a = \cos(20°)$, we can see that a satisfies

$$\frac{1}{2} = 4a^3 - 3a.$$

After multiplying this by 2 and subtracting 1, we can see that a is a root of

$$8x^3 - 6x - 1 \in \mathbb{Q}[x].$$

Since $8x^3 - 6x - 1$ has degree 3, it will be irreducible in $\mathbb{Q}[x]$ if and only if it has no rational roots. Before applying the Rational Root Test, we introduce a shortcut. If we let $y = 2x$, then

$$8x^3 - 6x - 1 = (2x)^3 - 3(2x) - 1 = y^3 - 3y - 1.$$

It is very easy to use the Rational Root Test to see that $y^3 - 3y - 1$ has no rational roots as the only candidates for rational roots are ± 1 and they produce values of -3 and 1. Since $y = 2x$, any root of $y^3 - 3y - 1$ is two times a root of $8x^3 - 6x - 1$. Therefore, $8x^3 - 6x - 1$ also has no rational roots, hence $8x^3 - 6x - 1$ is irreducible in $\mathbb{Q}[x]$.

We have just seen that $\cos(20°)$ is a root of the irreducible cubic $8x^3 - 6x - 1 \in \mathbb{Q}[x]$. Therefore,

$$[\mathbb{Q}(\cos(20°)) : \mathbb{Q}] = 3.$$

However, Theorem 16.6 asserts that if $\cos(20°)$ is constructible, then

$$[\mathbb{Q}(\cos(20°)) : \mathbb{Q}] = 2^l,$$

for some integer $l \geq 0$. This is a contradiction, as $3 \neq 2^l$, for any integer l. As a result, $20°$ angles cannot be constructed, so angles cannot be trisected with a ruler and compass. □

We conclude this chapter with a brief discussion of two other famous constructibility problems: the "Doubling of the Cube" and the "Squaring of the Circle." A cube with volume 1 has sides of length 1. The Doubling of the Cube problem asks whether it is possible to construct a line segment whose length would produce a cube whose volume was 2. Certainly, a cube of volume 2 would have sides of length $2^{\frac{1}{3}}$. However, $2^{\frac{1}{3}}$ is a root of the polynomial $x^3 - 2$, which is irreducible over $\mathbb{Q}$. Therefore,

$$[\mathbb{Q}(2^{\frac{1}{3}}) : \mathbb{Q}] = 3,$$

so Theorem 16.6 tells us that $2^{\frac{1}{3}}$ is not constructible. Therefore, the cube cannot be doubled with a ruler and compass.

The area of a circle of radius 1 is π. The Squaring of the Circle problem asks whether it is possible to construct a line segment whose length would produce a square whose area is π. Once again, Theorem 16.6 plays a fundamental role. However, parts of this problem require ideas not covered in this course. If a square has area equal to π, then the length of its sides must be $\sqrt{\pi}$. Therefore, if we could square the circle, then Theorem 16.6 would imply that $[\mathbb{Q}(\sqrt{\pi}) : \mathbb{Q}]$ was equal to 2^n, for some integer $n \geq 0$.

It turns out that π is transcendental. This is another way of saying that π is not algebraic, so it is not a root of any nonconstant polynomial with rational coefficients. However, the proof that π is transcendental is quite long and would take us very far afield. But once you know that π is transcendental, it is easy to show that $\sqrt{\pi}$ is also transcendental. Thus, at this point, Theorem 16.6 would tell you that we cannot square the circle with ruler and compass.

Exercises for Sections 16.1 and 16.2

1. If $\alpha, \beta \in \mathbb{R}$ such that α is constructible and β is not constructible, prove that $\alpha + \beta$ is not constructible.

2. If $\alpha, \beta \in \mathbb{R}$ such that α is a nonzero constructible number and β is not constructible, prove that $\alpha \cdot \beta$ is not constructible.

3. If $n°$ and $m°$ angles can be constructed, explain why $(n+m)°$, $(n-m)°$, and $\left(\frac{n}{2}\right)°$ angles can also be constructed.

4. Show that $36°$ angles can be constructed. You may want to look at exercise 18 following Section 11.1.

5. Show that $3°$ angles can be constructed.

6. If $n \in \mathbb{N}$, show that $n°$ angles can be constructed if and only if n is a multiple of 3.

In exercises 7–18, determine whether the given number is constructible and explain your answer. Exercises 1 and 2 may be useful is solving these problems.

7. $17^{\frac{1}{8}}$

8. $23^{\frac{1}{6}}$

9. $17^{\frac{1}{8}} + 23^{\frac{1}{6}}$

10. $5 - 2\left(17^{\frac{1}{8}}\right)$

11. $\dfrac{7 + 9\left(17^{\frac{1}{8}}\right)}{4 + \sqrt{2}}$

12. $\dfrac{3\left(23^{\frac{1}{6}}\right)}{\sqrt{19 - 2\sqrt{3}}}$

13. $\dfrac{\sqrt{7 - 4\sqrt{3}}}{4^{\frac{1}{3}} - 9}$

14. $\left(\dfrac{5^{\frac{1}{4}} + \sqrt{29}}{4 - \sqrt{5}}\right)^{\frac{1}{8}}$

15. $\dfrac{3^{\frac{1}{4}} - 4^{\frac{1}{3}}}{\sqrt{2} + \sqrt{3}}$

16. $4^{\frac{1}{3}} + 11^{\frac{1}{5}}$

17. $\left(\dfrac{5^{\frac{1}{4}} + 7^{\frac{1}{8}} + 11^{\frac{1}{16}}}{13^{\frac{1}{32}} + 5}\right)^{\frac{1}{4}}$

18. $2^{\frac{1}{7}} - 19^{\frac{1}{11}}$

19. If $\mathbb{D}$ is the field of constructible numbers, show that $\mathbb{D}$ is not a finite extension of $\mathbb{Q}$.

20. If $\mathbb{D}$ is the field of constructible numbers, show that $[\mathbb{D}(a) : \mathbb{D}] \neq 2$, for every $a \in \mathbb{R}$.

CHAPTER 17

Insolvability of the Quintic

In our final chapter, we pull together much of the mathematical machinery developed in the first 16 chapters to give a proof of Galois's classic result on the insolvability of the quintic. In Chapter 10, we saw that all polynomials of degree $1, 2, 3, 4$ can be solved by radicals. This means that we can find the roots of these polynomials by adding, subtracting, multiplying, dividing, and taking nth roots, for various $n \in \mathbb{N}$, of combinations of the coefficients.

Back in Chapter 1, we wondered how we go about proving that something cannot be done. More precisely, how does one prove that polynomials of degree greater than or equal to 5 cannot be solved by radicals? Applying the tools we have developed, we do this in three steps:

I.   Formally define, in terms of field extensions, what it means for a polynomial to be solvable by radicals.

II. Prove that the Galois group of the splitting field of a polynomial that is solvable by radicals is a solvable group.

III. Exhibit polynomials such that the Galois groups of their splitting fields are not solvable groups.

17.1 Radical Extensions and Their Galois Groups

The first step in proving the insolvability of the quintic is to formally define, in terms of field extensions, exactly what it means for a polynomial to be solvable by radicals. To see what the definition should be, we will look at several examples and then examine what they have in common. Throughout this chapter, for $m \in \mathbb{N}$, we will let $\omega_m = \operatorname{cis}\left(\frac{2\pi}{m}\right)$.

■ Examples

1. Let $f(x) = x^2 - 11 \in \mathbb{Q}[x]$. The roots of $f(x)$ are $\pm\sqrt{11}$ and they both belong to the field $L_1 = \mathbb{Q}(\sqrt{11})$.

2. Let $g(x) = x^3 - 5 \in \mathbb{Q}[x]$. The roots of $g(x)$ are

$$\sqrt[3]{5}, \ \omega_3\sqrt[3]{5}, \ \omega_3{}^2\sqrt[3]{5}$$

and they all belong to the field $L_2 = \mathbb{Q}(\omega_3, \sqrt[3]{5})$.

3. Let $h(x) = x^6 - 15x^4 - 8x^3 + 75x^2 - 120x - 109 \in \mathbb{Q}[x]$. The roots of $h(x)$ are

$$\pm\sqrt{5} + \omega_3{}^i \sqrt[3]{4},$$

for $0 \leq i \leq 2$, and they all belong to the field $L_3 = \mathbb{Q}(\omega_6, \sqrt{5}, \sqrt[3]{4})$.

4. Let $\beta = \sqrt[4]{6 - 20(\sqrt[5]{7})}$; β is the root of some polynomial in $\mathbb{Q}[x]$ and β belongs to the field $L_4 = \mathbb{Q}\left(\omega_{20}, \sqrt[5]{7}, \sqrt[4]{6 - 20(\sqrt[5]{7})}\right)$.

5. Let $\gamma = 14\sqrt[3]{28} + \sqrt{3\sqrt{7} + 4\sqrt[7]{13}}$; γ is the root of some polynomial in $\mathbb{Q}[x]$ and γ belongs to the field

$$L_5 = \mathbb{Q}\left(\omega_{252}, \sqrt[3]{28}, \sqrt{7}, \sqrt[7]{13}, \sqrt[6]{3\sqrt{7} + 4\sqrt[7]{13}}\right).$$

∎

Observe that another way to look at the fields that arose in our five examples are

1. $L_1 = \mathbb{Q}(a_1)$, where $a_1{}^2 \in \mathbb{Q}$;

2. $L_2 = \mathbb{Q}(\omega_3, b_1)$, where $b_1{}^3 \in \mathbb{Q}(\omega_3)$;

3. $L_3 = \mathbb{Q}(\omega_6, c_1, c_2)$, where $c_1{}^2 \in \mathbb{Q}(\omega_6)$ and $c_2{}^3 \in \mathbb{Q}(\omega_6, c_1)$;

4. $L_4 = \mathbb{Q}(\omega_{20}, d_1, d_2)$, where $d_1{}^5 \in \mathbb{Q}(\omega_{20})$ and $d_2{}^4 \in \mathbb{Q}(\omega_{20}, d_1)$;

5. $L_5 = \mathbb{Q}(\omega_{252}, e_1, e_2, e_3, e_4)$, where $e_1{}^3 \in \mathbb{Q}(\omega_{252})$, $e_2{}^2 \in \mathbb{Q}(\omega_{252}, e_1)$, $e_3{}^7 \in \mathbb{Q}(\omega_{252}, e_1, e_2)$, and $e_4{}^6 \in \mathbb{Q}(\omega_{252}, e_1, e_2, e_3)$.

In fact, if we want L_1 to look more like L_2, L_3, L_4, and L_5, we can express L_1 as

$$L_1 = \mathbb{Q}(\omega_2, a_1), \quad \text{where} \quad a_1{}^2 \in \mathbb{Q}(\omega_2),$$

since $\omega_2 = -1 \in \mathbb{Q}$.

In all five examples, the roots of our polynomials belong to fields of the form

$$\mathbb{Q}(\omega_m, \alpha_1, \ldots, \alpha_{n-1}, \alpha_n),$$

where $\alpha_1{}^{m_1} \in \mathbb{Q}(\omega_m), \alpha_2{}^{m_2} \in \mathbb{Q}(\omega_m, \alpha_1), \alpha_3{}^{m_3} \in \mathbb{Q}(\omega_m, \alpha_1, \alpha_2), \ldots, \alpha_n{}^{m_n} \in \mathbb{Q}(\omega_m, \alpha_1, \ldots, \alpha_{n-1})$, and $m = m_1 \cdot m_2 \cdots m_n$.

It certainly appears that fields of the type described above will play an important role in the proof of the insolvability of the quintic. This motivates

Definition 17.1. *A field L is called a radical extension, if there exist $\alpha_i \in \mathbb{C}$ and $m_i \in \mathbb{N}$ such that*

(a) $L = \mathbb{Q}(\omega_m, \alpha_1, \ldots, \alpha_{n-1}, \alpha_n),$

(b) $\omega_m = \operatorname{cis}\left(\frac{2\pi}{m}\right),$ *where* $m = m_1 \cdot m_2 \cdots m_n,$

(c) $\alpha_1{}^{m_1} \in \mathbb{Q}(\omega_m), \alpha_2{}^{m_2} \in \mathbb{Q}(\omega_m, \alpha_1), \alpha_3{}^{m_3} \in \mathbb{Q}(\omega_m, \alpha_1, \alpha_2), \ldots, \alpha_n{}^{m_n} \in \mathbb{Q}(\omega_m,$
$\alpha_1, \ldots, \alpha_{n-1}).$

Expressions that arose in our examples, such as

$$\sqrt[4]{6 - 20(\sqrt[5]{7})} \quad \text{and} \quad 14\sqrt[3]{28} + \sqrt[6]{3\sqrt{7} + 4\sqrt[7]{13}},$$

are exactly the types of elements of $\mathbb{C}$ we would expect to see as roots of polynomials in which the roots are obtained by adding, subtracting, multiplying, dividing, and taking nth roots, for $n \in \mathbb{N}$, of combinations of the coefficients. Since expressions like the preceding ones belong to radical extensions, it seems reasonable to say that if $\alpha \in \mathbb{C}$ is the root of some $f(x) \in \mathbb{Q}[x]$ that is solvable by radicals, then α belongs to some radical extension. Therefore, it appears that the proper definition of what it means for $f(x) \in \mathbb{Q}[x]$ to be solvable by radicals is that all of its roots belong to some radical extension. We record this as

Definition 17.2. *A polynomial $f(x) \in \mathbb{Q}[x]$ is solvable by radicals if there exists a radical extension L such that $L \supseteq K \supseteq \mathbb{Q}$, where K is the splitting field of $f(x)$ over $\mathbb{Q}$.*

Having succeeded in giving a formal definition, in terms of field extensions, what it means for a polynomial to be solvable by radicals, we now need to examine properties of radical extensions. If L is a radical extension, then $L = \mathbb{Q}(\omega_m, \alpha_1, \ldots, \alpha_{n-1}, \alpha_n)$ and repeated applications of Theorem 15.6 tell us that

$$[L : \mathbb{Q}] = [\mathbb{Q}(\omega_m, \alpha_1, \ldots, \alpha_{n-1}, \alpha_n) : \mathbb{Q}(\omega_m, \alpha_1, \ldots, \alpha_{n-1})] \cdots$$

$$[\mathbb{Q}(\omega_m, \alpha_1, \alpha_2) : \mathbb{Q}(\omega_m, \alpha_1)] \cdot [\mathbb{Q}(\omega_m, \alpha_1) : \mathbb{Q}(\omega_m)] \cdot [\mathbb{Q}(\omega_m) : \mathbb{Q}].$$

Since $\alpha_i{}^{m_i} \in \mathbb{Q}(\omega_m, \alpha_1, \ldots, \alpha_{i-1})$, for $2 \leq i \leq n$, we know that

$$[\mathbb{Q}(\omega_m, \alpha_1, \ldots, \alpha_{i-1}, \alpha_i) : \mathbb{Q}(\omega_m, \alpha_1, \ldots, \alpha_{i-1})] \leq m_i.$$

In addition

$$[\mathbb{Q}(\omega_m, \alpha_1) : \mathbb{Q}(\omega_m)] \leq m_1 \quad \text{and} \quad [\mathbb{Q}(\omega_m) : \mathbb{Q}] \leq m.$$

Combining these facts, it follows that

$$[L : \mathbb{Q}] \leq m_n \cdot m_{n-1} \cdots m_2 \cdot m_1 \cdot m = m^2.$$

As a result, all radical extensions are finite extensions of $\mathbb{Q}$. Thus, every element of a radical extension is algebraic over $\mathbb{Q}$. This is the reason why, in examples (4) and (5), we could say that

$$\sqrt[4]{6 - 20(\sqrt[5]{7})} \quad \text{and} \quad 14\sqrt[3]{28} + \sqrt[6]{3\sqrt{7} + 4\sqrt[7]{13}},$$

were roots of polynomials in $\mathbb{Q}[x]$, even though we did not explicitly mention polynomials that had them as roots.

One thing that might strike you as odd about radical extensions is the inclusion of $\omega_m = \text{cis}\left(\frac{2\pi}{m}\right)$. There are two reasons for including ω_m. The first, which is certainly not apparent at this time, is that it will simplify the computation of Galois groups of radical extensions.

The second reason concerns expressions of the form $\sqrt[4]{6 - 20(\sqrt[5]{7})}$. This expression represents an element of $\mathbb{C}$ that, when raised to the fourth power, is equal to $6 - 20(\sqrt[5]{7})$. Since there are no real numbers whose fourth power is $6 - 20(\sqrt[5]{7})$, there is some ambiguity as to which element of $\mathbb{C}$ is meant by the expression $\sqrt[4]{6 - 20(\sqrt[5]{7})}$.

DeMoivre's Theorem asserts that there are four different elements of $\mathbb{C}$ that, when raised to the fourth power, give us $6 - 20(\sqrt[5]{7})$. Furthermore, if β is one of them, then the other three are $\omega_4\beta$, $\omega_4^2\beta$, and $\omega_4^3\beta$. However, since our radical extension contains ω_4, once it contains one element of $\mathbb{C}$ whose fourth power is $6 - 20(\sqrt[5]{7})$, it automatically contains all four. Therefore, any ambiguity about which fourth root of $6 - 20(\sqrt[5]{7})$ is contained in a radical extension is no longer a concern.

As a result, by letting $\omega_m = \text{cis}\left(\frac{2\pi}{m}\right)$, where $m = m_1 \cdot m_2 \cdots m_n$, our radical extension contains all m_i of the m_ith roots of 1, for $1 \leq i \leq n$. Therefore, whenever our radical extension contains one m_ith root of an element of $\mathbb{C}$, it will automatically contain all m_i of them.

It is now time to examine the Galois groups of radical extensions. To simplify the notation used with radical extensions, if

$$L = \mathbb{Q}(\omega_m, \alpha_1, \ldots, \alpha_{n-1}, \alpha_n)$$

is a radical extension, let

$$L_i = \mathbb{Q}(\omega_m, \alpha_1, \ldots, \alpha_{i-1}, \alpha_i),$$

for $1 \leq i \leq n$, and let $L_0 = \mathbb{Q}(\omega_m)$.

We now have the chain of fields

$$L = L_n \supseteq L_{n-1} \supseteq \cdots \supseteq L_2 \supseteq L_1 \supseteq L_0 \supseteq \mathbb{Q}.$$

We would like to use this chain of fields to produce a chain of groups. To do this, we need

Lemma 17.3. *If $F \subseteq K \subseteq L$ are fields, then $Gal(L/K)$ is a subgroup of $Gal(L/F)$.*

Proof. Every element of $Gal(L/K)$ is an automorphism of the field L that is the identity map on the field K. Since K contains F, every element of $Gal(L/K)$ is certainly the identity map on the smaller field F, so $Gal(L/K)$ is a subset of $Gal(L/F)$. Since $Gal(L/K)$ is already a group, we know that if $g, h \in Gal(L/K)$, then $gh^{-1} \in Gal(L/K)$. Proposition 8.6(a) now tells us that $Gal(L/K)$ is a subgroup of $Gal(L/F)$. $\square$

■ Example

We revisit an example from Chapter 15. Let $\mathbb{Q}(\sqrt{3}, i)$; then $Gal(Q(\sqrt{3}, i)/\mathbb{Q})$ is a group of order 4. Since $Gal(Q(\sqrt{3}, i)/\mathbb{Q})$ consists of automorphisms that are the identity on $\mathbb{Q}$, its elements can be completely described in terms of their behavior on $\sqrt{3}$ and i as follows:

$$g_1(\sqrt{3}) = \sqrt{3}, \ g_1(i) = i, \quad g_2(\sqrt{3}) = -\sqrt{3}, \ g_2(i) = i,$$

$$g_3(\sqrt{3}) = \sqrt{3}, \ g_3(i) = -i, \quad g_4(\sqrt{3}) = -\sqrt{3}, \ g_4(i) = -i.$$

Since $\mathbb{Q}(i)$, $\mathbb{Q}(\sqrt{3})$, and $\mathbb{Q}(i\sqrt{3})$ are all fields containing $\mathbb{Q}$ and contained in $\mathbb{Q}(\sqrt{3}, i)$, Lemma 17.3 tells us that

$$Gal(\mathbb{Q}(\sqrt{3}, i)/\mathbb{Q}(i)), \ \ Gal(\mathbb{Q}(\sqrt{3}, i)/\mathbb{Q}(\sqrt{3})), \ \ \text{and} \ \ Gal(\mathbb{Q}(\sqrt{3}, i)/\mathbb{Q}(i\sqrt{3}))$$

are all subgroups of $Gal(Q(\sqrt{3}, i)/\mathbb{Q})$.

It is now not hard to see that

$$Gal(\mathbb{Q}(\sqrt{3}, i)/\mathbb{Q}(i)) = \{g_1, g_2\} \subseteq Gal(Q(\sqrt{3}, i)/\mathbb{Q}),$$

$$Gal(\mathbb{Q}(\sqrt{3}, i)/\mathbb{Q}(\sqrt{3})) = \{g_1, g_3\} \subseteq Gal(Q(\sqrt{3}, i)/\mathbb{Q}),$$

$$Gal(\mathbb{Q}(\sqrt{3}, i)/\mathbb{Q}(i\sqrt{3})) = \{g_1, g_4\} \subseteq Gal(Q(\sqrt{3}, i)/\mathbb{Q}).$$

Using the notation for the chain of fields that we introduced before Lemma 17.3, if we let $G = Gal(L/\mathbb{Q})$, then Lemma 17.3 tells us that $Gal(L/L_i)$ is a subgroup of G, for $0 \leq i \leq n$. In fact, we obtain the chain of subgroups

$$G = Gal(L/\mathbb{Q}) \supseteq Gal(L/L_0) \supseteq Gal(L/L_1) \supseteq \cdots \supseteq Gal(L/L_{n-1}) \supseteq$$

$$Gal(L/L_n) = \{e\}.$$

Therefore, if we wish to prove that $Gal(L/\mathbb{Q})$ is a solvable group, it suffices to show that this chain of subgroups satisfies the condition in Proposition 8.33. Parts of the proof of Proposition 17.4 should look familiar as many of the ideas in the proof have already appeared in the proofs of Corollaries 15.19, 15.20, and 15.23.

Proposition 17.4. *If L is a radical extension, then $Gal(L/\mathbb{Q})$ is a solvable group.*

Proof. Since L is a radical extension, there exist $\alpha_i \in \mathbb{C}$ and $m_i \in \mathbb{N}$ such that

$$L = \mathbb{Q}(\omega_m, \alpha_1, \ldots, \alpha_{n-1}, \alpha_n),$$

where $\alpha_1{}^{m_1} \in \mathbb{Q}(\omega_m)$, $\alpha_2{}^{m_2} \in \mathbb{Q}(\omega_m, \alpha_1)$, $\alpha_3{}^{m_3} \in \mathbb{Q}(\omega_m, \alpha_1, \alpha_2)$, $\ldots$, $\alpha_n{}^{m_n} \in \mathbb{Q}(\omega_m, \alpha_1, \ldots, \alpha_{n-1})$, and $m = m_1 \cdot m_2 \cdots m_n$.

If we let $L_i = \mathbb{Q}(\omega_m, \alpha_1, \ldots, \alpha_{i-1}, \alpha_i)$, for $1 \leq i \leq n$, and let $L_0 = \mathbb{Q}(\omega_m)$, then we have the chain of fields

$$L = L_n \supseteq L_{n-1} \supseteq \cdots \supseteq L_2 \supseteq L_1 \supseteq L_0 \supseteq \mathbb{Q}.$$

Next, if we let $G = Gal(L/\mathbb{Q})$, then Lemma 17.6 turns the chain of fields into the following chain of subgroups:

$$G = Gal(L/\mathbb{Q}) \supseteq Gal(L/L_0) \supseteq Gal(L/L_1) \supseteq \cdots \supseteq Gal(L/L_{n-1}) \supseteq$$
$$Gal(L/L_n) = \{e\}.$$

According to Proposition 8.33, in order to show that $Gal(L/\mathbb{Q})$ is solvable, it now suffices to show that

(a) $ghg^{-1}h^{-1} \in Gal(L/L_0)$, for all $g, h \in Gal(L/\mathbb{Q})$, and

(b) $ghg^{-1}h^{-1} \in Gal(L/L_{i+1})$, for all $g, h \in Gal(L/L_i)$, where $0 \leq i \leq n-1$.

For part (a), let $g, h \in Gal(L/\mathbb{Q})$; since $L_0 = \mathbb{Q}(\omega_m)$, in order to show that $ghg^{-1}h^{-1} \in Gal(L/L_0)$, we need to show that $ghg^{-1}h^{-1}$ is the identity on ω_m. The polynomial $x^m - 1$ has coefficients in $\mathbb{Q}$, therefore Corollary 5.13 tells us g^{-1} and h^{-1} must both send roots of $x^m - 1$ to roots of $x^m - 1$. Since every root of $x^m - 1$ is a power of ω_m, if we simplify the notation by letting ω take the place of ω_m, there exist $i, j \in \mathbb{N}$ such that

$$g^{-1}(\omega) = \omega^i \quad \text{and} \quad h^{-1}(\omega) = \omega^j.$$

Therefore, $g(\omega^i) = \omega$ and $h(\omega^j) = \omega$, and since g, h, g^{-1}, h^{-1} are all automorphisms, we have

$$(ghg^{-1}h^{-1})(\omega) = (ghg^{-1})(h^{-1}(\omega)) = (ghg^{-1})(\omega^j) =$$
$$(gh)(g^{-1}(\omega^j)) = (gh)((g^{-1}(\omega))^j) = (gh)((\omega^i)^j) = (gh)(\omega^{ij}) =$$
$$g(h(\omega^{ij})) = g((h(\omega^j))^i) = g(\omega^i) = \omega.$$

Therefore, $ghg^{-1}h^{-1}(\omega_m) = \omega_m$ and it is indeed the case that if $g, h \in Gal(L/\mathbb{Q})$, then $ghg^{-1}h^{-1} \in Gal(L/L_0)$.

For part (b), if $0 \leq i \leq n-1$ and $g, h \in Gal(L/L_i)$, we need to show that $ghg^{-1}h^{-1} \in Gal(L/L_{i+1})$. Recall that $L_{i+1} = L_i(\alpha_{i+1})$, where

$$\alpha_{i+1}{}^{m_{i+1}} = a,$$

for some $a \in L_i$. Since α_{i+1} is a root of $x^{m_{i+1}} - a \in L_i[x]$, both g^{-1} and h^{-1} must send α_{i+1} to another root of $x^{m_{i+1}} - a$. However, every root of $x^{m_{i+1}} - a$ is equal to α_{i+1} times a root of $x^{m_{i+1}} - 1$. On the other hand, every root of $x^{m_{i+1}} - 1$ is a power of ω_m. If we once again simplify the notation by letting $\omega = \omega_m$, there exists $j, k \in \mathbb{N}$ such that

$$g^{-1}(\alpha_{i+1}) = \omega^j \alpha_{i+1} \quad \text{and} \quad h^{-1}(\alpha_{i+1}) = \omega^k \alpha_{i+1}.$$

Since g and h are the identity on L_i, they are the identity on all powers of ω. Combining this with the previous equation and the fact that g, h, g^{-1}, h^{-1} are all automorphisms, we have

$$(ghg^{-1}h^{-1})(\alpha_{i+1}) = (ghg^{-1})(h^{-1}(\alpha_{i+1})) = (ghg^{-1})(\omega^k \alpha_{i+1}) =$$

$$(gh)(g^{-1}(\omega^k \alpha_{i+1})) = (gh)(g^{-1}(\omega^k)g^{-1}(\alpha_{i+1})) = (gh)(\omega^k g^{-1}(\alpha_{i+1})) =$$

$$(gh)(\omega^k \omega^j \alpha_{i+1}) = (gh)(\omega^{j+k} \alpha_{i+1}) = g(h(\omega^{j+k} \alpha_{i+1})) =$$

$$g(h(\omega^j)h(\omega^k \alpha_{i+1})) = g(\omega^j \alpha_{i+1}) = \alpha_{i+1}.$$

Since $ghg^{-1}h^{-1}$ has been shown to be the identity on L_i and α_{i+1}, it follows that it is the identity on $L_{i+1} = L_i(\alpha_{i+1})$. Thus, $ghg^{-1}h^{-1}$ is an automorphism of L which is the identity on L_{i+1}, so $ghg^{-1}h^{-1}$ does indeed belong to $Gal(L/L_{i+1})$, as required. $\qquad\square$

Proving that Galois groups of radical extensions are solvable is a huge step towards the proof of the insolvability of the quintic. If $f(x) \in \mathbb{Q}[x]$ is solvable by radicals, we would like to show that $Gal(K/\mathbb{Q})$ is solvable, where K is the splitting field of $f(x)$. We know that there exists a radical extension L such that $L \supseteq K \supseteq \mathbb{Q}$, and Proposition 17.4 tells us that $Gal(L/\mathbb{Q})$ is solvable. But this does not yet tell us that $Gal(K/\mathbb{Q})$ is solvable. Although Corollary 8.40 is not stated in these exact terms, if you look back at the proof of this result, you will see that the proof tells us that subgroups as well as images under homomorphisms of solvable groups are solvable. However, $Gal(K/\mathbb{Q})$ has not been shown to be either a subgroup or the image under a homomorphism of $Gal(L/\mathbb{Q})$.

On the other hand, if L could be replaced by a radical extension M of $\mathbb{Q}$ that not only contained K but was also a splitting field over $\mathbb{Q}$, then we would eventually be able to show that $Gal(K/\mathbb{Q})$ is the image under a homomorphism of the solvable group $Gal(M/\mathbb{Q})$. The next two lemmas are rather long and technical, but they are exactly what we need to show that there exists a radical extension M with the desired properties.

Lemma 17.5. *If $L_1 = \mathbb{Q}(\beta_1)$, $L_2 = \mathbb{Q}(\beta_2)$, ..., $L_t = \mathbb{Q}(\beta_t)$ are radical extensions, then there exists $m \in \mathbb{N}$ such that $\mathbb{Q}(\omega_m, \beta_1, ..., \beta_t)$ is a radical extension.*

Proof. Since L_i is a radical extension, for $1 \le i \le t$, we have

$$L_i = \mathbb{Q}(\beta_i) = \mathbb{Q}(\omega_{m_i}, \alpha_{i,1}, \alpha_{i,2}, ..., \alpha_{i,n_i}).$$

where $\alpha_{i,j} \in \mathbb{C}$, $\alpha_{i,1}{}^{m_{i,1}} \in \mathbb{Q}(\omega_{m_i})$, $\alpha_{i,j}{}^{m_{i,j}} \in \mathbb{Q}(\omega_{m_i}, \alpha_{i,1}, ..., \alpha_{i,j-1})$, for $2 \le j \le n_i$, and $m_i = m_{i,1} \cdot m_{i,2} \cdots m_{i,n_i}$.

We now make two simple observations. The first is that if $F \subseteq K$ are fields and $\gamma \in \mathbb{C}$ has the property that $\gamma^k \in F$, then it is clear that $\gamma^k \in K$. Next, if $n, a, b \in \mathbb{N}$ such that $n = a \cdot b$, then

$$\omega_n{}^a = \left(\operatorname{cis}\left(\frac{2\pi}{n} \right) \right)^a = \operatorname{cis}\left(\frac{a \cdot 2\pi}{n} \right) = \operatorname{cis}\left(\frac{2\pi}{b} \right) = \omega_b.$$

Thus, $\omega_b \in \mathbb{Q}(\omega_n)$ and $\mathbb{Q}(\omega_b) \subseteq \mathbb{Q}(\omega_n)$.

In light of these observations, consider the field

$$L = \mathbb{Q}(\omega_m, \alpha_{1,1}, ..., \alpha_{1,n_1}, \alpha_{2,1}, ..., \alpha_{2,n_2}, ..., \alpha_{t,1}, ..., \alpha_{t,n_t}),$$

where $m = m_1 \cdot m_2 \cdots m_t$.

We now need to show that L is a radical extension. Since m is a multiple of each $m_{i,j}$, where $1 \le i \le t$ and $1 \le j \le n_i$, we know that $\mathbb{Q}(\omega_{m_{i,j}}) \subseteq \mathbb{Q}(\omega_m)$. As a result, $\alpha_{1,1}{}^{m_{1,1}} \in \mathbb{Q}(\omega_m)$ and $\alpha_{i,j}{}^{m_{i,j}}$ belongs to the field generated over $\mathbb{Q}$ by all the elements to the left of $\alpha_{i,j}$ in the definition of L. For example,

$$\alpha_{3,2}{}^{m_{3,2}} \in \mathbb{Q}(\omega_{m_3}, \alpha_{3,1}) \subseteq \mathbb{Q}(\omega_m, \alpha_{1,1}, ..., \alpha_{1,n_1}, \alpha_{2,1}, ..., \alpha_{2,n_2}, \alpha_{3,1}).$$

Since m is the product of all the $m_{i,j}$, we see that L is indeed a radical extension.

Next, since $\beta_i \in \mathbb{Q}(\omega_{m_i}, \alpha_{i,1}, \alpha_{i,2}, ..., \alpha_{i,n_i}) \subseteq L$, for all i, it follows that

$$\mathbb{Q}(\omega_m, \beta_1, ..., \beta_t) \subseteq L.$$

On the other hand, since each $\alpha_{i,j} \in \mathbb{Q}(\beta_i)$, we can also see that every generator of L over $\mathbb{Q}$ belongs to $\mathbb{Q}(\omega_m, \beta_1, ..., \beta_t)$. Therefore,

$$L \subseteq \mathbb{Q}(\omega_m, \beta_1, ..., \beta_t).$$

As a result, $L = \mathbb{Q}(\omega_m, \beta_1, ..., \beta_t)$, hence $\mathbb{Q}(\omega_m, \beta_1, ..., \beta_t)$ is a radical extension. $\qquad\square$

We can now prove the extremely useful

Lemma 17.6. *If $f(x) \in \mathbb{Q}[x]$ is solvable by radicals, then there exists a radical extension M such that $M \supseteq K \supseteq \mathbb{Q}$, where M is the splitting field of some $q(x) \in \mathbb{Q}[x]$ over $\mathbb{Q}$ and K is the splitting field of $f(x)$ over $\mathbb{Q}$.*

Proof. Since $f(x)$ is solvable by radicals, K is contained in a radical extension $L = \mathbb{Q}(\omega_m, \alpha_1, \ldots, \alpha_{n-1}, \alpha_n)$ that satisfies the properties in Definition 17.1. In light of Theorem 15.12, there exists $\beta \in \mathbb{C}$ such that $L = \mathbb{Q}(\beta)$. Next, let $h(x) \in \mathbb{Q}[x]$ be the minimum polynomial for β over $\mathbb{C}$, and also let t denote the degree of $h(x)$. If $\gamma \in \mathbb{C}$ is another root of $h(x)$, then Lemma 17.6 asserts that the map

$$\pi : \mathbb{Q}(\beta) \to \mathbb{Q}(\gamma)$$

defined as

$$\pi(a_0 + a_1\beta + \cdots + a_{t-1}\beta^{t-1}) = a_0 + a_1\gamma + \cdots + a_{t-1}\gamma^{t-1},$$

for all $a_i \in \mathbb{Q}$, is a bijective homomorphism of fields that is the identity on $\mathbb{Q}$. We need to show that $\mathbb{Q}(\gamma)$ is also a radical extension.

To this end, since π is a bijective homomorphism which is the identity on $\mathbb{Q}$, we have

$$\mathbb{Q}(\gamma) = \mathbb{Q}(\pi(\beta)) = \pi(\mathbb{Q}(\beta)) = \pi(\mathbb{Q}(\omega_m, \alpha_1, \ldots, \alpha_{n-1}, \alpha_n)) =$$

$$\mathbb{Q}(\pi(\omega_m), \pi(\alpha_1), \ldots, \pi(\alpha_{n-1}), \pi(\alpha_n)).$$

The terms of the form $\omega_m{}^j$, for $0 \leq j \leq m - 1$, are the m distinct roots of $x^m - 1$. Since π is an injective homomorphism, observe that the terms of the form $\pi(\omega_m)^j$ are also the m distinct roots of $x^m - 1$. As a result, $\pi(\omega_m)$ is a power of ω_m, and, conversely, ω_m is also a power of $\pi(\omega_m)$. Therefore, a field extension of $\mathbb{Q}$ contains ω_m if and only if it contains $\pi(\omega_m)$. In light of this and our preceding work, we now have

$$\mathbb{Q}(\gamma) = \mathbb{Q}(\omega_m, \pi(\alpha_1), \ldots, \pi(\alpha_{n-1}), \pi(\alpha_n)).$$

By repeatedly using the fact that π is a bijective homomorphism that is the identity on $\mathbb{Q}$, we obtain $\pi(\alpha_1)^{m_1} \in \mathbb{Q}(\omega_m)$, $\pi(\alpha_2)^{m_2} \in \mathbb{Q}(\omega_m, \pi(\alpha_1))$, $\pi(\alpha_3)^{m_3} \in \mathbb{Q}(\omega_m, \pi(\alpha_1), \pi(\alpha_2))$, $\ldots$, $\pi(\alpha_n)^{m_n} \in \mathbb{Q}(\omega_m, \pi(\alpha_1), \ldots, \pi(\alpha_{n-1}))$. Thus, $\mathbb{Q}(\gamma)$ satisfies all the properties in Definition 17.1 and is indeed a radical extension.

If we let $\beta_1, \ldots, \beta_t \in \mathbb{C}$ be the roots of $h(x)$, our previous argument shows that each $\mathbb{Q}(\beta_i)$ is a radical extension. Therefore, Lemma 17.5 tells us that there exists $l \in \mathbb{N}$ such that $\mathbb{Q}(\omega_l, \beta_1, \ldots, \beta_t)$ is also a radical extension. If we let $M = \mathbb{Q}(\omega_l, \beta_1, \ldots, \beta_t)$, we observe that M also contains the l roots of $x^l - 1$. As a result, M is the splitting field over $\mathbb{Q}$ of the polynomial $(x^l - 1)h(x) \in \mathbb{Q}[x]$. Thus, in addition to containing both L and K, M is a radical extension that is the splitting field over $\mathbb{Q}$ of $q(x) = (x^l - 1)h(x) \in \mathbb{Q}[x]$, as required. $\square$

The pieces are now in place to prove the main result of this section. The idea behind the proof will be to compare the Galois groups $Gal(M/\mathbb{Q})$ and $Gal(K/\mathbb{Q})$, where $M \supseteq K \supseteq \mathbb{Q}$ and K and M are both splitting fields over $\mathbb{Q}$ of elements of $\mathbb{Q}[x]$. In this situation, we will show that $Gal(K/\mathbb{Q})$ is the image under a group homomorphism of $Gal(M/\mathbb{Q})$. The proof is one of the more abstract ones in this book, so we will first look at two examples that illustrate this situation.

■ Examples

1. Let $M = \mathbb{Q}(\sqrt[3]{2}, \omega_3)$ and $K = \mathbb{Q}(\omega_3)$, where $\omega_3 = \text{cis}\left(\frac{2\pi}{3}\right)$. Then we have $M \supseteq K \supseteq \mathbb{Q}$, where K is the splitting field over $\mathbb{Q}$ of $x^3 - 1$ and M is the splitting field over $\mathbb{Q}$ of $x^3 - 2$. Corollary 15.21 indicated that $Gal(M/\mathbb{Q})$ is a nonabelian group of order 6, so it must be isomorphic to S_3. We will use the same notation we used in Section 8.1 when we provided a multiplication table for S_3. Recall that since the elements of $Gal(M/\mathbb{Q})$ are the identity on $\mathbb{Q}$, they are completely determined by their behavior on $\sqrt[3]{2}$ and ω_3.

$$e(\sqrt[3]{2}) = \sqrt[3]{2} \text{ and } e(\omega_3) = \omega_3; \qquad f(\sqrt[3]{2}) = \sqrt[3]{2} \text{ and } e(\omega_3) = \omega_3{}^2;$$
$$g(\sqrt[3]{2}) = \sqrt[3]{2} \cdot \omega_3 \text{ and } g(\omega_3) = \omega_3{}^2; \qquad h(\sqrt[3]{2}) = \sqrt[3]{2} \cdot \omega_3{}^2 \text{ and } h(\omega_3) = \omega_3{}^2;$$
$$j(\sqrt[3]{2}) = \sqrt[3]{2} \cdot \omega_3{}^2 \text{ and } j(\omega_3) = \omega_3; \qquad k(\sqrt[3]{2}) = \sqrt[3]{2} \cdot \omega_3 \text{ and } k(\omega_3) = \omega_3;$$

Observe that if $x \in Gal(M/\mathbb{Q})$, then

$$x(\omega_3) \in \{\omega_3, \omega_3{}^3\} \subset \mathbb{Q}(\omega_3) = K.$$

However, since K is a splitting field, this is no surprise. Indeed, Lemma 15.15 guarantees that $x(K) = K$, for any $x \in Gal(M/\mathbb{Q})$. Since the automorphisms in $Gal(M/\mathbb{Q})$ send elements of K to K, we can consider the function

$$\phi : Gal(M/\mathbb{Q}) \to Gal(K/\mathbb{Q})$$

defined as $\phi(x) = \bar{x}$, where $\bar{x}$ denotes the restriction of x to the smaller field K. Recall that Corollary 15.24 guarantees that ϕ is a group homomorphism. We now have

$$\bar{e}(\omega_3) = \bar{j}(\omega_3) = \bar{k}(\omega_3) = \omega_3$$

and

$$\bar{f}(\omega_3) = \bar{g}(\omega_3) = \bar{h}(\omega_3) = \omega_3{}^2.$$

The preceding equations show that the image of ϕ contains both elements of $Gal(K/\mathbb{Q})$. In addition, we can see the images under ϕ of the three elements e, j, k are the identity element in $Gal(K/\mathbb{Q})$. Therefore, the set $\{e, j, k\}$ is the kernel of ϕ.

Another way to look at this is that the kernel of ϕ consists of the elements of $Gal(M/\mathbb{Q})$ which are the identity of K, so the kernel of ϕ is equal to $Gal(M/K)$. Putting all these pieces together, Theorem 8.39(a)—Isomorphism Theorem for Groups tells us that

$$Gal(M/\mathbb{Q})/Gal(M/K) \approx Gal(K/\mathbb{Q}).$$

2. Let $M = \mathbb{Q}(\sqrt{2}, \sqrt{3})$ and $K = \mathbb{Q}(\sqrt{2})$; then $M \supseteq K \supseteq \mathbb{Q}$, where K is the splitting field over $\mathbb{Q}$ of $x^2 - 2$ and M is the splitting field over $\mathbb{Q}$ of $(x^2 - 2)(x^2 - 3)$. It is easy to see that $Gal(M/\mathbb{Q})$ is a group of order 4 that contains no element of order 4, so $Gal(M/\mathbb{Q})$ is isomorphic to $C_2 \times C_2$. We can now represent the four elements of $Gal(M/\mathbb{Q})$ by looking at their behavior on $\sqrt{2}$ and $\sqrt{3}$.

$$e(\sqrt{2}) = \sqrt{2} \;\; \text{and} \;\; e(\sqrt{3}) = \sqrt{3}; \qquad f(\sqrt{2}) = \sqrt{2} \;\; \text{and} \;\; f(\sqrt{3}) = -\sqrt{3};$$

$$g(\sqrt{2}) = -\sqrt{2} \;\; \text{and} \;\; g(\sqrt{3}) = \sqrt{3}; \qquad h(\sqrt{2}) = -\sqrt{2} \;\; \text{and} \;\; h(-\sqrt{3}) = -\sqrt{3}.$$

Observe that if $x \in Gal(M/\mathbb{Q})$, then

$$x(\sqrt{2}) \in \{\sqrt{2}, -\sqrt{2}\} \subset \mathbb{Q}(\sqrt{2}) = K.$$

As in the previous example, this is no surprise as Lemma 15.15 guarantees that automorphisms in $Gal(M/\mathbb{Q})$ send elements of K to K. As before, we can define the function

$$\phi : Gal(M/\mathbb{Q}) \to Gal(K/\mathbb{Q})$$

defined as $\phi(x) = \bar{x}$, where $\bar{x}$ denotes the restriction of x to the smaller field K. Corollary 15.24 asserts that ϕ is a group homomorphism, and we can observe that

$$\bar{e}(\sqrt{2}) = \bar{f}(\sqrt{2}) = \sqrt{2}$$

and

$$\bar{g}(\sqrt{2}) = \bar{h}(\sqrt{2}) = -\sqrt{2}.$$

We can see that the image of ϕ contains both elements of $Gal(K/\mathbb{Q})$ and the kernel of ϕ consists of the set $\{e, f\}$. Note that the kernel of ϕ consists of the same elements as $Gal(M/K)$, so Theorem 8.39(a)—Isomorphism Theorem for Groups tells us that

$$Gal(M/\mathbb{Q})/Gal(M/K) \approx Gal(K/\mathbb{Q}).$$

∎

We can now state and prove Galois's beautiful result on the insolvability of polynomials by radicals.

Theorem 17.7.   *If $f(x) \in \mathbb{Q}[x]$ is solvable by radicals and K is the splitting field of $f(x)$ over $\mathbb{Q}$, then $Gal(K/\mathbb{Q})$ is a solvable group.*

Proof. Definition 17.2 and Lemma 17.6 tell us that there exists a radical extension M such that $M \supseteq K \supseteq \mathbb{Q}$ and M is the splitting field over $\mathbb{Q}$ of some $q(x) \in \mathbb{Q}[x]$. Observe that since M is the smallest field that contains $\mathbb{Q}$ and all the roots of $q(x)$, it is also the smallest field that contains K and the roots of $q(x)$. Therefore, M is also the splitting field of $q(x)$ over K.

When we combine the preceding observation with the fact that K and M are splitting fields over $\mathbb{Q}$, Theorem 15.17 tells us that

(1) $|Gal(M/K)| = [M : K], |Gal(K/\mathbb{Q})| = [K : \mathbb{Q}]$,   and $|Gal(M/\mathbb{Q})| = [M : \mathbb{Q}]$.

Since K is a splitting field, Lemma 15.15 asserts that for every $g \in Gal(M/\mathbb{Q})$, we have $g(K) = K$. As a result, we can define the function

$$\phi : Gal(M/\mathbb{Q}) \rightarrow Gal(K/\mathbb{Q})$$

as

$$\phi(g) = \overline{g},$$

where $\overline{g}$ is the restriction of g to the smaller field K.

Corollary 15.24 tells us that ϕ is a homomorphism and $Gal(M/K)$ is the kernel of ϕ. Applying Theorem 8.39(a)—Isomorphism Theorem for Groups, we now have

$$Gal(M/\mathbb{Q})/Gal(M/K) \approx Im(\phi).$$

Our goal is to show that $Gal(K/\mathbb{Q})$ is solvable. Proposition 17.4 told us that $Gal(M/\mathbb{Q})$ is solvable, and the proof of Corollary 8.40 indicates that the image under a homomorphism of a solvable group is solvable. As a result, we do know that $Im(\phi)$ is a solvable group. Therefore, we will be done if we can show that $Im(\phi)$ is all of $Gal(K/\mathbb{Q})$. Theorem 15.6 tells us that

$$[M : \mathbb{Q}] = [M : K] \cdot [K : \mathbb{Q}]$$

and Lagrange's Theorem tells us that

$$|Im(\phi)| = \frac{|Gal(M/\mathbb{Q})|}{|Gal(M/K)|}.$$

When we combine this with equation (1), we have

$$|Im(\phi)| = \frac{|Gal(M/\mathbb{Q})|}{|Gal(M/K)|} = \frac{[M : \mathbb{Q}]}{[M : K]} = [K : \mathbb{Q}] = |Gal(K/\mathbb{Q})|.$$

Since $Im(\phi) \subseteq Gal(K/\mathbb{Q})$ and $|Im(\phi)| = |Gal(K/\mathbb{Q})|$, it now follows that $Im(\phi) = Gal(K/\mathbb{Q})$. Having already shown that $Im(\phi)$ is solvable, we know now that $Gal(K/\mathbb{Q})$ is solvable, as required. $\qquad\square$

17.2 A Proof of the Insolvability of the Quintic

At the beginning of this chapter, we described the three steps needed to prove the insolvability of the quintic. In Section 17.1, we worked through the first two steps. For the final step, we now present a technique for producing polynomials which are not solvable by radicals.

Theorem 17.8. *Suppose $f(x) \in \mathbb{Q}[x]$ has degree p, is irreducible in $\mathbb{Q}[x]$, and has exactly $p-2$ real roots, where p is prime. If we let K be the splitting field of $f(x)$ over $\mathbb{Q}$, then $Gal(K/\mathbb{Q})$ is the symmetric group S_p.*

Proof. Since $f(x)$ is irreducible over $\mathbb{Q}[x]$, the Fundamental Theorem of Algebra and Corollary 10.24 assert that $f(x)$ has p distinct roots in $\mathbb{C}$, which we will call $\alpha_1, \alpha_2, \ldots, \alpha_p$. Corollary 5.13 tells us that every element of $Gal(L/K)$ sends roots of $f(x)$ to roots of $f(x)$. In addition, since $K = \mathbb{Q}(\alpha_1, \alpha_2, \ldots, \alpha_p)$, every element of $Gal(K/\mathbb{Q})$ is a different bijection of the set of roots of $f(x)$. Therefore, $Gal(K/\mathbb{Q})$ is a subgroup of S_p, and we need to show that $Gal(K/\mathbb{Q})$ is all of S_p.

Since $f(x)$ has exactly $p-2$ real roots, we can order the p roots so that α_1 and α_2 do not belong to $\mathbb{R}$ and the other $p-2$ roots do belong to $\mathbb{R}$. Lemma 15.15 asserts that since K is a splitting field, complex conjugation restricts to an automorphism g of K. Therefore, g belongs to $Gal(K/\mathbb{Q})$, and we have

$$g(\alpha_1) = \alpha_2, \quad g(\alpha_2) = \alpha_1, \quad \text{and} \quad g(\alpha_i) = \alpha_i,$$

for $3 \leq i \leq p$. We can see that g is the transposition $(12) \in S_p$.

Since $f(x)$ has degree p, $[\mathbb{Q}(\alpha_1) : \mathbb{Q}] = p$. However, Theorem 15.6 tells us that

$$[K : \mathbb{Q}] = [K : \mathbb{Q}(\alpha_1)] \cdot [\mathbb{Q}(\alpha_1) : \mathbb{Q}],$$

so $[K : \mathbb{Q}]$ is a multiple of p. On the other hand, since K is a splitting field, Theorem 15.17 asserts that

$$|Gal(K/\mathbb{Q})| = [K : \mathbb{Q}].$$

As a result, $|Gal(K/\mathbb{Q})|$ is a multiple of p, therefore Sylow's Theorem tells us that $Gal(K/\mathbb{Q})$ contains a subgroup of order p, and it immediately follows that $Gal(K/\mathbb{Q})$ contains an element of order p.

At this point, we have succeeded in showing that $Gal(K/\mathbb{Q})$ is a subgroup of S_p that contains both a transposition and an element of order p. Theorem 8.51 now tells us that $Gal(K/\mathbb{Q})$ is indeed equal to all of S_p. □

It has been a very long and interesting journey. We have been introduced to many algebraic objects and concepts such as automorphisms, solvable groups, symmetric groups, Sylow's Theorem, irreducibility criteria, splitting fields, and Galois groups. By applying these and various other algebraic ideas, we have achieved one of the primary goals of this course, the ability to produce fifth-degree polynomials that are insolvable by radicals.

Corollary 17.9. *If p is a prime and $n \geq 2$ is an integer, then the polynomial $x^5 - npx + p$ is not solvable by radicals.*

Proof. We will begin by showing that $f(x) = x^5 - npx + p$ satisfies the conditions of Theorem 17.8. First, using the prime p, Eisenstein's Criterion tells us that $f(x)$ is irreducible over $\mathbb{Q}$. Next, the derivative of $f(x)$ is $5x^4 - np$ and the only real roots of $f'(x)$ are $\pm\sqrt[4]{\frac{np}{5}}$. Since $f'(x)$ has only two real roots, Rolle's Theorem asserts that $f(x)$ has at most three real roots.

We would now like to show that $f(x)$ changes sign three times, for that would allow us to use the Intermediate Value Theorem to assert that $f(x)$ has at least three real roots. When x is a negative real number with a large absolute value, $f(x) < 0$. However, $f(0) = p > 0$, so there exists at least one negative root. Next, observe that

$$f(1) = 1 - np + p = 1 - p(n-1) < 0,$$

so $f(x)$ also has a root between 0 and 1. Finally, for large values of x, $f(x)$ is again positive, so there is a third sign change and a third real root occurring when $x > 1$.

As a result, $f(x)$ is irreducible over $\mathbb{Q}$, has degree 5, and has exactly 3 real roots. We can now apply Theorem 17.8, and it tells us that if K is the splitting field of $x^5 - npx + p$ over $\mathbb{Q}$, then $Gal(K/\mathbb{Q}) = S_5$. Theorem 8.53 asserts that S_5 is not solvable, and it now follows from Theorem 17.7 that $x^5 - npx + p$ is not solvable by radicals. □

By plugging values of n and p into $x^5 - npx + p$, we can easily produce examples of fifth-degree polynomials that are not solvable by radicals.

■ Examples

In light of Corollary 17.9, the following quintics are insolvable by radicals:

$$x^5 - 4x + 2, \quad x^5 - 6x + 2, \quad x^5 - 8x + 2, \quad x^5 - 10x + 2, \quad x^5 - 12x + 2,$$

$$x^5 - 6x + 3, \quad x^5 - 9x + 3, \quad x^5 - 12x + 3, \quad x^5 - 15x + 3, \quad x^5 - 18x + 3,$$

$$x^5 - 10x + 5, \quad x^5 - 15x + 5, \quad x^5 - 20x + 5, \quad x^5 - 25x + 5, \quad x^5 - 30x + 5,$$
$$x^5 - 14x + 7, \quad x^5 - 21x + 7, \quad x^5 - 28x + 7, \quad x^5 - 35x + 7, \quad x^5 - 42x + 7.$$

∎

We used Theorem 17.8 to prove Corollary 17.9 and this corollary allowed us to produce an infinite number of polynomials of degree five not solvable by radicals. However, we do not need to stop with quintics. Theorem 17.8 can also be used to produce polynomials of higher degree not solvable by radicals. In fact, our next corollary allows us to produce an infinite number of seventh-degree polynomials that are insolvable by radicals.

Corollary 17.10. *If p is a prime then the polynomial* $x^7 - 5px^5 + 7px + p$ *is not solvable by radicals.*

Proof. If we let $g(x) = x^7 - 5px^5 + 7px + p$, then Eisenstein's Criterion with the prime p implies that $g(x)$ is irreducible over $\mathbb{Q}$. Observe that

$$g''(x) = 42x^5 - 100px^3 = x^3(42x^2 - 100p).$$

Therefore, $g''(x)$ has only three distinct real roots, so Rolle's Theorem tells us that $g'(x)$ has at most four distinct real roots. Another application of Rolle's Theorem now tells us that $g(x)$ has at most five distinct real roots. However, since $g(x)$ is irreducible, Corollary 10.24 tells us that all of its roots are distinct. Therefore, $g(x)$ has at most five real roots.

On the other hand, consider the following values of $g(x)$:

$$g(-3) = -2187 + 1195p > 0, \quad g(-1) = -1 - p < 0,$$
$$g(0) = p > 0, \quad g(2) = 128 - 145p < 0.$$

Certainly, if x is negative with a large absolute value, we know that $g(x) < 0$. Therefore the Intermediate Value Theorem asserts that there is a real root when $x < -3$. In addition, we can see that $g(x)$ changes sign between -3 and -1, between -1 and 0, and between 0 and 2. Therefore, the Intermediate Value Theorem says that there are at least three more real roots between -3 and 2. In addition, when x is large, we know that $g(x) > 0$. Therefore, there is yet another sign change and at least one more real root when $x > 2$. As a result, $g(x)$ has at least five real roots, and the preceding paragraph showed that $g(x)$ had at most five real roots. Thus, $g(x)$ has exactly five real roots.

The preceding arguments show that $g(x)$ is irreducible over $\mathbb{Q}$, has degree 7, and has exactly 5 real roots. Theorem 17.8 now asserts that if K is the splitting field of $x^7 - 5px^5 + 7px + p$ over $\mathbb{Q}$, then $Gal(K/\mathbb{Q}) = S_7$. Theorem 8.53 asserts that S_7 is not solvable, and it now follows from Theorem 17.7 that $x^7 - 5px^5 + 7px + p$ is not solvable by radicals. □

By plugging primes into $x^7 - 5px^5 + 7px + p$, we can now produce an infinite number of polynomials of degree seven that are insolvable by radicals.

■ Examples

It follows from Corollary 17.10 that the following seventh-degree polynomials are insolvable by radicals:

$$x^7 - 10x^5 + 14x + 2, \quad x^7 - 15x^5 + 21x + 3, \quad x^7 - 25x^5 + 35x + 5,$$
$$x^7 - 35x^5 + 49x + 7, \quad x^7 - 55x^5 + 77x + 11, \quad x^7 - 65x^5 + 91x + 13.$$

■

Exercises for Sections 17.1 and 17.2

In exercises 1–6, use the notation from Definition 17.1 and find $m \in \mathbb{N}$ and $\alpha_1, \ldots, \alpha_n \in \mathbb{C}$ such that the given element belongs to the radical extension $\mathbb{Q}(\omega_m, \alpha_1, \ldots, \alpha_n)$.

1. $6\sqrt{2} - 8\sqrt{3}$

2. $4\sqrt{5} + 7\sqrt{19} - 14\sqrt{23}$

3. $2\sqrt[3]{6} - 3\sqrt[4]{11}$

4. $6\sqrt[3]{10} + 5\sqrt[6]{3} - 18\sqrt{13}$

5. $\sqrt[4]{6 + \sqrt[7]{23}}$

6. $\sqrt{8\sqrt[3]{19} - 4\sqrt[7]{51} + \sqrt[5]{24}}$

Before doing exercises 7–14 or 15–22, please read the following paragraph:

In the proof of the insolvability of the quintic, a key step was the fact that if $M \supseteq K \supseteq \mathbb{Q}$ are fields such that both M and K are splitting fields of polynomials in $\mathbb{Q}[x]$, then the homomorphism

$$\phi : Gal(M/\mathbb{Q}) \to Gal(K/\mathbb{Q})$$

described in Corollary 15.24 is a surjection. In exercises 7–14 and 15–22, we provide examples where M is not a splitting field and ϕ is not surjective. This explains why it was necessary to replace the field L in Proposition 17.4 by the splitting field M in Theorem 17.7.

In exercises 7–14, let $K = \mathbb{Q}(2^{\frac{1}{3}}, \text{cis}\left(\frac{2\pi}{3}\right))$ and $M = \mathbb{Q}(2^{\frac{1}{9}}, \text{cis}\left(\frac{2\pi}{3}\right))$. Observe that $M \supseteq K \supseteq \mathbb{Q}$.

7. Show that K is the splitting field over $\mathbb{Q}$ of some $f(x) \in \mathbb{Q}[x]$.

8. Show that M is the splitting field over K of some $g(x) \in K[x]$.

9. Show that if $g \in Gal(M/\mathbb{Q})$, then $g(2^{\frac{1}{3}}) = 2^{\frac{1}{3}}$.

10. Show that the homomorphism $\phi : Gal(M/\mathbb{Q}) \to Gal(K/\mathbb{Q})$ from Corollary 15.24 is not a surjection.

11. Show that M is not the splitting field over $\mathbb{Q}$ of any $h(x) \in \mathbb{Q}[x]$.

12. Determine $[K : \mathbb{Q}]$, $[M : \mathbb{Q}]$, and $[M : K]$.

13. Determine $|Gal(K/\mathbb{Q})|$, $|Gal(M/\mathbb{Q})|$, and $|Gal(M/K)|$.

14. Determine $|Im(\phi)|$ and compare it to $|Gal(K/\mathbb{Q})|$.

In exercises 15–22, let $K = \mathbb{Q}\left(6^{\frac{1}{5}}, cis\left(\frac{2\pi}{5}\right)\right)$ and $M = \mathbb{Q}\left(6^{\frac{1}{25}}, cis\left(\frac{2\pi}{5}\right)\right)$. Observe that $M \supseteq K \supseteq \mathbb{Q}$.

15. Show that K is the splitting field over $\mathbb{Q}$ of some $f(x) \in \mathbb{Q}[x]$.

16. Show that M is the splitting field over K of some $g(x) \in K[x]$.

17. Show that if $g \in Gal(M/\mathbb{Q})$, then $g\left(6^{\frac{1}{5}}\right) = 6^{\frac{1}{5}}$.

18. Show that the homomorphism $\phi : Gal(M/\mathbb{Q}) \to Gal(K/\mathbb{Q})$ from Corollary 15.24 is not a surjection.

19. Show that M is not the splitting field over $\mathbb{Q}$ of any $h(x) \in \mathbb{Q}[x]$.

20. Determine $[K : \mathbb{Q}]$, $[M : \mathbb{Q}]$, and $[M : K]$.

21. Determine $|Gal(K/\mathbb{Q})|$, $|Gal(M/\mathbb{Q})|$, and $|Gal(M/K)|$.

22. Determine $|Im(\phi)|$ and compare it to $|Gal(K/\mathbb{Q})|$.

In exercises 23–30, you will need to manipulate the polynomials from Corollaries 17.9 and 17.10 to produce additional infinite classes of polynomials that are not solvable by radicals.

23. Show that if p is a prime and $n \geq 2$ is an integer, then $x^5 - npx - p$ is not solvable by radicals.

24. Show that if p is a prime and $n \geq 2$ is an integer, then $p^4 x^5 - npx + 1$ is not solvable by radicals.

25. Show that if p is a prime and $n \geq 2$ is an integer, then $px^5 - npx^4 + 1$ is not solvable by radicals.

26. Show that if p is a prime and $n \geq 2$ is an integer, then $x^5 - npx^4 + p^4$ is not solvable by radicals.

27. Show that if p is prime, then $x^7 - 5px^5 + 7px - p$ is not solvable by radicals.

28. Show that if p is prime, then $p^6 x^7 - 5p^5 x^5 + 7px + 1$ is not solvable by radicals.

29. Show that if p is prime, then $px^7 + 7px^6 - 5px^2 + 1$ is not solvable by radicals.

30. Show that if p is prime, then $x^7 + 7px^6 - 5p^5 x^2 + p^6$ is not solvable by radicals.

In exercises 31–34, we look at some polynomials that, despite having degree at least 5, are solvable by radicals. When doing these exercises you may use the fact, which was shown in Chapter 10, that polynomials of degree at most 4 are solvable by radicals.

31. If $a, b, c, d \in \mathbb{Q}$ with $a \neq 0$, show that $ax^6 + bx^4 + cx^2 + d$ is solvable by radicals.

32. If $a, b, c, d, e \in \mathbb{Q}$ with $a \neq 0$, show that $ax^8 + bx^6 + cx^4 + dx^2 + e$ is solvable by radicals.

33. If $a, b, c \in \mathbb{Q}$ with $a \neq 0$, show that $ax^6 + bx^3 + c$ is solvable by radicals.

34. If $a, b, c, d \in \mathbb{Q}$ with $a \neq 0$, show that $ax^9 + bx^6 + cx^3 + d$ is solvable by radicals.

In exercises 35–40, we examine fields of the form $\mathbb{Q}(\sqrt{p_1}, \sqrt{p_2}, \ldots, \sqrt{p_n})$, where $p_1, p_2, \ldots, p_n$ are distinct primes. This will generalize the work we have done on fields of the form $\mathbb{Q}(\sqrt{2})$, $\mathbb{Q}(\sqrt{7}, \sqrt{11})$, and $\mathbb{Q}(\sqrt{2}, \sqrt{3}, \sqrt{5})$. In these exercises, $p_1, p_2, \ldots, p_n$ will be distinct primes, and $d_1, d_2, \ldots, d_{2^n}$ will be the 2^n positive integers of the form $p_1^{i_1} \cdot p_2^{i_2} \cdots p_n^{i_n}$, where each i_j is either 0 or 1. We will also let $T = \left\{ \sqrt{d_1}, \sqrt{d_2}, \ldots, \sqrt{d_{2^n}} \right\}$.

35. If the set T is linearly independent over $\mathbb{Q}$, show that

$$\left[\mathbb{Q}(\sqrt{p_1}, \sqrt{p_2}, \ldots, \sqrt{p_n}) : \mathbb{Q} \right] = 2^n.$$

36. If the set T is linearly independent over $\mathbb{Q}$, show that

$$\left| Gal(\mathbb{Q}(\sqrt{p_1}, \sqrt{p_2}, \ldots, \sqrt{p_n})/\mathbb{Q}) \right| = 2^n.$$

37. Suppose the set T is linearly independent over $\mathbb{Q}$ and also suppose that $v \in \mathbb{Q}(\sqrt{p_1}, \sqrt{p_2}, \ldots, \sqrt{p_n})$ is a linear combination of at least two elements of T. Show that there exists $g \in Gal(\mathbb{Q}(\sqrt{p_1}, \sqrt{p_2}, \ldots, \sqrt{p_n})/\mathbb{Q})$ such that $g(v) \neq v$ and $g(v) \neq -v$.

38. If p is a prime that does not belong to the set $\{p_1, p_2, \ldots, p_n\}$, and if the set T is linearly independent over $\mathbb{Q}$, show that $\sqrt{p} \notin \mathbb{Q}(\sqrt{p_1}, \sqrt{p_2}, \ldots, \sqrt{p_n})$.

39. Use Mathematical Induction and exercises 35–38 to show that if $p_1, p_2, \ldots, p_n$ are distinct primes, then

$$[\mathbb{Q}(\sqrt{p_1}, \sqrt{p_2}, \ldots, \sqrt{p_n}) : \mathbb{Q}] = 2^n.$$

40. If $p_1, p_2, \ldots, p_n$ are distinct primes, show that

$$Gal(\mathbb{Q}(\sqrt{p_1}, \sqrt{p_2}, \ldots, \sqrt{p_n})/\mathbb{Q}) \approx C_2 \times C_2 \times \cdots \times C_2,$$

the direct product of n copies of the cyclic group of order 2.

17.3 Kronecker's Theorem

Included in the subtitle of this book is the phrase "from the integers to the insolvability of the quintic." We conclude this book by coming full circle and returning to the integers. We will look at some properties of the integers that can be generalized and applied to more abstract settings.

The proof we presented of the insolvability of the quintic made heavy use of the Fundamental Theorem of Algebra. The fact that every nonconstant $p(x) \in \mathbb{C}[x]$ has a root in $\mathbb{C}$ meant that when we looked at the roots of $f(x) \in \mathbb{Q}[x]$, we could look at fields K such that $\mathbb{Q} \subseteq K \subseteq \mathbb{C}$ and K was a finite extension of $\mathbb{Q}$. Knowing that the splitting field of $f(x)$ lived between $\mathbb{Q}$ and $\mathbb{C}$ greatly simplified the work needed to compute $Gal(K/\mathbb{Q})$.

On the other hand, it is very common in abstract algebra books to not use the Fundamental Theorem of Algebra when proving the insolvability of the quintic. Using that approach, proofs of facts about splitting fields and Galois groups become much more abstract and technical. An important piece of the puzzle becomes the proof that if K is a field and if $f(x) \in K[x]$ has degree at least 1, then there exists a field L such that L is a finite extension of K and also contains a root of $f(x)$. The main goal of this section is to prove this fact, which is known as Kronecker's Theorem. Its proof will reveal interesting properties of rings as well as show us how the integers can be useful in other contexts.

As we saw in Chapter 8, if G is a group and H is a subgroup, then the set of left (or right) cosets forms a group if and only if H is normal. If R is a ring and S is a subgroup under addition then, since R is an abelian group under addition, S is certainly a normal subgroup. Therefore, the set of cosets, R/S, is a group under addition. It is reasonable to ask if R/S is also a ring.

Since addition in R/S is defined by adding the names of the cosets, in trying to multiply cosets, it would make sense to define coset multiplication as

$$(a+S)(b+S) = ab + S.$$

In R, we can both add and multiply, but since S is a subgroup under addition, we will express cosets using additive notation. In order to check if the multiplication of cosets is well defined, we need to determine if changing the names of the cosets can change the product when we multiply.

To this end, suppose $a \in R$ and $b \in S$. Since $b + S = 0 + S$, if the multiplication of cosets was well defined, then we would have

$$ab + S = (a+S)(b+S) = (a+S)(0+S) = 0 + S$$

and

$$ba + S = (b+S)(a+S) = (0+S)(a+S) = 0 + S.$$

Therefore, if coset multiplication is well defined, it follows that

$$ab + S = 0 + S = ba + S,$$

so $ab, ba \in S$. As we will soon see, this condition is not only necessary, but is also sufficient. This motivates

Definition 17.11. *If R is a ring, a subgroup I under addition is called an ideal if $ab, ba \in I$, for all $a \in R, b \in I$.*

Since the definition of a ring included the commutativity of multiplication, requiring both ab and ba to belong to I is redundant. However, the reason we do this is the same reason that the definition of a ring includes two distributive laws. Namely, if we drop the assumption of commutativity and study noncommutative rings, we need both ab and ba to belong to I for the multiplication of cosets to be well defined. In this section, we will primarily concern ourselves with the ideals of $\mathbb{Z}$ and of $F[x]$, for fields F, and will determine the ideals that arise in these cases.

■ Examples

1. The Ideals of $\mathbb{Z}$

    For any ring R, the sets $\{0\}$ and R are always ideals. If $I \neq \{0\}$ is an ideal of $\mathbb{Z}$, suppose $a \neq 0$ belongs to I. Then exactly one of a or $-a$ is positive, so I contains at least one positive integer. Therefore, the Well Ordering Principle asserts that there is a smallest positive integer n belonging to I. If we let $n\mathbb{Z} = \{nb \,|\, b \in \mathbb{Z}\}$, we claim that $I = n\mathbb{Z}$.

    Since I is an ideal of $\mathbb{Z}$ and $n \in I$, it follows that $n\mathbb{Z} \subseteq I$. In the other direction, if $m \in I$, the division algorithm tells us that there exists $q, r \in \mathbb{Z}$ such that

    $$m = q \cdot n + r \quad \text{and} \quad 0 \leq r < n.$$

    Observe that both m and $q \cdot n$ belong to I, so

    $$r = m - q \cdot n \in I.$$

    If $r \neq 0$, then r would be an element of I that is both positive and smaller than n. But this contradicts the minimality of n, so $r = 0$ and so,

    $$m = q \cdot n \in n\mathbb{Z}.$$

    As a result, $I \subseteq n\mathbb{Z}$ and so, $I = n\mathbb{Z}$.

    If R is a ring and $r \in R$, we will let $(r) = \{r \cdot s \,|\, s \in R\}$. Using this notation, the preceding argument tells us that every nonzero ideal of $\mathbb{Z}$ is of the form (n), for some $n \in \mathbb{N}$. Observe that the set $\{0\}$ is equal to (0), so every ideal of $\mathbb{Z}$ is of the form (n), for some $n \in \mathbb{Z}$. For example, (2) consists of all even integers, (100) consists of all multiples of 100, and (1) is the entire set $\mathbb{Z}$.

∎

We can now exploit similarities between $\mathbb{Z}$ and $F[x]$ to describe all ideals of $F[x]$.

∎ Examples

2. The Ideals of $F[x]$, where F is a field.

    One of the similarities between the rings $\mathbb{Z}$ and $F[x]$ is that they both have a division algorithm. We will exploit this to find all ideals of $F[x]$. If $I \neq 0$ is an ideal of $F[x]$, we will first consider the case where I contains a nonzero constant $a \in F$. In this situation, for any $f(x) \in F[x]$,

    $$f(x) = (f(x)a^{-1}) \cdot a \in I,$$

    so I is equal to all of $F[x]$.

    On the other hand, suppose I contains no nonzero constants. Then every nonzero element of $F[x]$ has positive degree, and the Well Ordering Principle asserts that there is a smallest positive degree n from among all elements of I. Now let $p(x) \in I$ have degree n; we claim $I = (p(x))$.

    Since I is an ideal, it is certainly the case that $(p(x)) \subseteq I$. On the other hand, if $f(x) \in I$, the division algorithm tells us that there exist $q(x), r(x) \in F[x]$ such that

    $$f(x) = q(x) \cdot p(x) + r(x)$$

    and $r(x) = 0$ or has degree smaller than n.

    Since $q(x) \cdot p(x) \in I$, we now have

    $$r(x) = f(x) - q(x) \cdot p(x),$$

    so $r(x) \in I$. It is impossible for $r(x)$ to have positive degree, for that would contradict the minimality of the degree of $p(x)$. Hence, $r(x) = 0$ and

    $$f(x) = q(x) \cdot p(x) \in (p(x)).$$

    As a result, $I \subseteq (p(x))$, so $I = (p(x))$.

    For example, (x) equals the set of polynomials with 0 constant term, $(x - 5)$ is all multiples of $x - 5$ and therefore consists of all polynomials with 5 as a root, and (1) is equal to all of $F[x]$.

∎

We can now prove

Theorem 17.12. *If R is a ring and I is a subgroup under addition, we define the multiplication of cosets as*

$$(a + I)(b + I) = ab + I,$$

for all $a, b \in R$. Then multiplication is well defined and R/I forms a ring if and only if I is an ideal.

Proof. Our earlier discussion showed that if multiplication is well defined, then I must be an ideal. Now we will show that if I is an ideal, then multiplication is well defined. In particular, we need to show that changing the names of the cosets does not change the answer when we multiply. To this end, suppose $a, b, c, d \in R$ such that

$$a + I = b + I \quad \text{and} \quad c + I = d + I,$$

we must show that

$$(a + I)(c + I) = (b + I)(d + I).$$

We know that $a - b, c - d \in I$, so there exist $x, y \in I$ such that $a = b + x$ and $c = d + y$. Thus

$$ac = (b + x)(d + y) = bd + xd + by + xy.$$

Since I is an ideal, $xd, by, xy \in I$, hence $ac - bd \in I$. This tells us that $ac + I = bd + I$.

The preceding calculations and the definition of coset multiplication combine to tell us that

$$(a + I)(c + I) = ac + I = bd + I = (b + I)(d + I),$$

as desired. Thus the multiplication of cosets is well defined.

Since I is a normal subgroup of R under addition, Corollary 8.22 verified that R/I is a group. Next we need to show that associativity under multiplication, commutativity under addition and multiplication, and the distributive laws are inherited by R/I from R. These properties are verified, one at a time, in the following four equations as

$$(a + I)((b + I)(c + I)) = (a + I)(bc + I) = a(bc) + I = (ab)c + I =$$
$$(ab + I)(c + I) = ((a + I)(b + I))(c + I);$$
$$(a + I) + (b + I) = (a + b) + I = (b + a) + I = (b + I) + (a + I);$$
$$(a + I)(b + I) = (ab) + I = (ba) + I = (b + I)(a + I);$$
$$(a + I)((b + I) + (c + I)) = (a + I)((b + c) + I) = a(b + c) + I = (ab + ac) + I =$$
$$(ab + I) + (ac + I) = ((a + I)(b + I)) + ((a + I)(c + I)),$$

for all $a, b, c \in R$.

Observe that since multiplication in R/I is commutative, we only needed to verify one distributive law. On the other hand, if we were dealing with noncommutative rings, we would need to verify the other distributive law for R/I to be a ring.

Finally,

$$(a+I)(1+I) = a \cdot 1 + I = a + I = 1 \cdot a + I = (1+I)(a+I),$$

so $1+I$ is the identity element of R/I under multiplication. Thus, we have verified that by virtue of inheriting properties from R, R/I does indeed satisfy all the axioms of a ring. □

When G was a group with normal subgroup H, we called groups of the form G/H quotient groups. Similarly, if R is a ring with ideal I, we call rings of the form R/I *quotient rings*.

It turns out that we are already quite familiar with rings of the form R/I when $R = \mathbb{Z}$. To see this, if $n > 1$ and $I = (n)$, then the cosets $a+I$ and $b+I$ are equal in $R/I = \mathbb{Z}/(n)$ if and only if $a-b$ is a multiple of n. But that is precisely the condition used to define $\mathbb{Z}_n$. As a result, the cosets in $\mathbb{Z}/(n)$ are the same as the equivalence classes in $\mathbb{Z}_n$. Therefore, rings of the form $\mathbb{Z}_n$ are merely the special case of quotient rings R/I, where $R = \mathbb{Z}$ and $I = (n)$.

Recall that in $\mathbb{Z}_n$, every equivalence class has an infinite number of names. However, the division algorithm told us that each equivalence class contains exactly one element from the set $\{0, 1, \dots, n-1\}$. Thus, as a convenience, we tend to express elements of $\mathbb{Z}_n$ as

$$[0]_n, [1]_n, \dots, [n-1]_n.$$

We will now carry this type of notation to a somewhat similar situation.

In a previous example, we saw that if $R = F[x]$, where F is a field, and if I is an ideal of R, then $I = (f(x))$, for some $f(x) \in F[x]$. As a result, when we form the quotient ring $R/I = F[x]/(f(x))$, the cosets $a(x) + I$ and $b(x) + I$ are equal if and only if $a(x) - b(x)$ is a multiple of $f(x)$. Since $F[x]$ has a division algorithm similar to the one for $\mathbb{Z}$, there will be some parallels between the cosets in $F[x]/(f(x))$ and those in $\mathbb{Z}/(n)$.

Let us consider the case where $f(x) \in F[x]$ has degree $n \geq 1$ and we will examine the coset

$$a(x) + I,$$

where $a(x) \in F[x]$ and $I = (f(x))$. The division algorithm tells us that there exist $q(x), r(x) \in F[x]$ such that

$$a(x) = q(x) \cdot f(x) + r(x),$$

where $r(x) = 0$ or $deg(r(x)) < n$. Since $a(x) - r(x) = q(x) \cdot f(x) \in (f(x))$, it follows that

$$a(x) + I = r(x) + I.$$

Furthermore, we can write $r(x)$ as $\alpha_0 + \alpha_1 x + \cdots + \alpha_{n-1} x^{n-1}$, where each $\alpha_i \in F$.

In light of the preceding, just as every equivalence class in $\mathbb{Z}_n$ contains exactly one element from the set $\{0, 1, \ldots, n-1\}$, we can see that every coset in $F[x]/(f(x))$ contains exactly one element of the form

$$\alpha_0 + \alpha_1 x + \cdots + \alpha_{n-1} x^{n-1},$$

where each $\alpha_i \in F$. As was the case for $\mathbb{Z}_n$, there are an infinite number of names for each equivalence class in $F[x]/(f(x))$. However, as a convenience, we can express each coset in $F[x]/(f(x))$ as

$$[\alpha_0 + \alpha_1 x + \cdots + \alpha_{n-1} x^{n-1}]_{f(x)},$$

where each $\alpha_i \in F$.

Let us now look at the set

$$[F]_{f(x)} = \{[\alpha]_{f(x)} \mid \alpha \in F\} \subseteq F[x]/(f(x)).$$

Observe that the set $[F]_{f(x)}$ is essentially a copy of the field F, only the elements of F look slightly different in this context. The advantage of introducing the set $[F]_{f(x)}$, is that we can now view F as a subset of $F[x]/(f(x))$. By expressing elements $\alpha \in F$ as $[\alpha]_{f(x)}$, we can now consider $F[x]/(f(x))$ to be a commutative ring which contains the field F. We will now examine some examples of rings of the form $F[x]/(f(x))$.

▪ Examples

1. Let $R = \mathbb{Q}[x]$ and $I = (x^2 - 2)$, then every element in $R/I = \mathbb{Q}[x]/(x^2 - 2)$ can be written in the form $[a + bx]_{x^2-2}$, where $a, b \in \mathbb{Q}$.

   Addition in $\mathbb{Q}[x]/(x^2 - 2)$ is quite straightforward, and we have

   $$[a + bx]_{x^2-2} + [c + dx]_{x^2-2} = [(a+c) + (b+d)x]_{x^2-2},$$

   where $a, b, c, d \in \mathbb{Q}$.

   In $\mathbb{Q}[x]/(x^2 - 2)$, x^2 and 2 belong to the same coset as their difference is $x^2 - 2$. Thus, in $\mathbb{Q}[x]/(x^2 - 2)$, any multiple of x^2 can be replaced by the corresponding multiple of 2. As a result, we have

   $$[a + bx]_{x^2-2} \cdot [c + dx]_{x^2-2} = [(a+bx)(c+dx)]_{x^2-2} = [ac + (ad + bc)x + (bd)x^2]_{x^2-2}$$

   $$= [ac + (ad + bc)x + (bd)(2)]_{x^2-2} = [(ac + 2bd) + (ad + bc)x]_{x^2-2}.$$

Summarizing this, if $a, b, c, d \in \mathbb{Q}$, then

$$[a+bx]_{x^2-2} \cdot [c+dx]_{x^2-2} = [(ac+2bd)+(ad+bc)x]_{x^2-2}.$$

For example, in $\mathbb{Q}[x]/(x^2-2)$, we have

$$[2-3x]_{x^2-2} \cdot [4+x]_{x^2-2} = [12-10x-3x^2]_{x^2-2} = [6-10x]_{x^2-2},$$

$$[1+2x]_{x^2-2} \cdot [3-5x]_{x^2-2} = [3+x-10x^2]_{x^2-2} = [-17+x]_{x^2-2},$$

$$[4+x]_{x^2-2} \cdot \left[\frac{2}{7} - \frac{1}{14}x\right]_{x^2-2} = \left[\frac{8}{7} - \frac{1}{14}x^2\right]_{x^2-2} = [1]_{x^2-2}.$$

Observe that the last example indicates that

$$\left[\frac{2}{7} - \frac{1}{14}x\right]_{x^2-2}$$

is the multiplicative inverse of $[4+x]_{x^2-2}$. In fact, if

$$[a+bx]_{x^2-2} \neq 0,$$

then

$$\left[\frac{a}{a^2-2b^2} - \frac{b}{a^2-2b^2}x\right]_{x^2-2}$$

is the multiplicative inverse of $[a+bx]_{x^2-2}$. To check this, you first need to multiply $[a+bx]_{x^2-2}$ and $\left[\frac{a}{a^2-2b^2} - \frac{b}{a^2-2b^2}x\right]_{x^2-2}$ and see that you obtain $[1]_{x^2-2}$ as the answer. But you also need to check that if

$$[a+bx]_{x^2-2} \neq 0,$$

then the fact that $\sqrt{2}$ is not rational implies that $a^2-2b^2 \neq 0$.

As a result, $\mathbb{Q}[x]/(x^2-2)$ is not only a commutative ring but it is also a field. Therefore $\mathbb{Q}[x]/(x^2-2)$ is a field extension of $\mathbb{Q}$ of degree 2 as the set $\{[1]_{x^2-2}, [x]_{x^2-2}\}$ is a basis for $\mathbb{Q}[x]/(x^2-2)$ over $\mathbb{Q}$.

Next, consider the polynomial

$$p(T) = T^2 - 2 \in \mathbb{Q}[T].$$

Clearly, the field $\mathbb{Q}$ does not contain any roots of T^2-2. But what about the field $\mathbb{Q}[x]/(x^2-2)$? Recall, in this context, we write elements of $\mathbb{Q}$ in the form $[a]_{x^2-2}$, where $a \in \mathbb{Q}$. As a result, in this context, we should rewrite $p(T)$ as $T^2 - [2]_{x^2-2}$. When we plug $[x]_{x^2-2}$ into $p(T)$, we obtain

$$p([x]_{x^2-2}) = [x]_{x^2-2}{}^2 - [2]_{x^2-2} = [x^2]_{x^2-2} - [2]_{x^2-2} =$$

$$[x^2 - 2]_{x^2-2} = [0]_{x^2-2}.$$

Thus, not only is $\mathbb{Q}[x]/(x^2 - 2)$ a field extension of $\mathbb{Q}$, but it also contains a root of $p(T) = T^2 - 2 \in \mathbb{Q}[T]$.

2. Let $R = \mathbb{R}[x]$ and $I = (x^2 + 1)$, then we can write every element of $R/I = \mathbb{R}[x]/(x^2 + 1)$ in the form $[a + bx]_{x^2+1}$, where $a, b \in \mathbb{R}$.

It is easy to see that when we add elements in $\mathbb{R}[x]/(x^2 + 1)$ we have

$$[a + bx]_{x^2+1} + [c + dx]_{x^2+1} = [(a+c) + (b+d)x]_{x^2+1},$$

where $a, b, c, d \in \mathbb{R}$.

In $\mathbb{R}[x]/(x^2 + 1)$, x^2 and -1 belong to the same coset as their difference is $x^2 + 1$. Thus, in $\mathbb{R}[x]/(x^2 + 1)$, we have

$$[a + bx]_{x^2+1} \cdot [c + dx]_{x^2+1} = [(a+bx)(c+dx)]_{x^2+1} = [ac + (ad + bc)x + (bd)x^2]_{x^2+1}$$

$$= [ac + (ad + bc)x + (bd)(-1)]_{x^2+1} = [(ac - bd) + (ad + bc)x]_{x^2+1}.$$

Summarizing, if $a, b, c, d \in \mathbb{R}$, we have

$$[a + bx]_{x^2+1} \cdot [c + dx]_{x^2+1} = [(ac - bd) + (ad + bc)x]_{x^2+1}.$$

For example, in $\mathbb{R}[x]/(x^2 + 1)$, we have

$$[3 + 4x]_{x^2+1} \cdot [2 - x]_{x^2+1} = [6 + 5x - 4x^2]_{x^2+1} = [10 + 5x]_{x^2+1},$$

$$[1 + 2x]_{x^2+1} \cdot [5 - 7x]_{x^2+1} = [5 + 3x - 14x^4]_{x^2+1} = [19 + 3x]_{x^2+1},$$

$$[2 + 7x]_{x^2+1} \cdot \left[\frac{2}{53} - \frac{7}{53}x\right]_{x^2+1} = \left[\frac{4}{53} - \frac{49}{53}x^2\right]_{x^2+1} = [1]_{x^2+1}.$$

The last example illustrates that

$$\left[\frac{2}{53} - \frac{7}{53}x\right]_{x^2+1}$$

is the multiple inverse of $[2 + 7x]_{x^2+1}$. More generally, if $[a + bx]_{x^2+1} \neq 0$, then at least one of a or b is nonzero. As a result, $a^2 + b^2 \neq 0$, and it then is easy to check that

$$\left[\frac{a}{a^2 + b^2} - \frac{b}{a^2 + b^2}x\right]_{x^2+1}$$

is the multiplicative inverse of $[a+bx]_{x^2+1}$. Therefore, $\mathbb{R}[x]/(x^2+1)$ is a field extension of $\mathbb{R}$ of degree 2 and the set $\{[1]_{x^2+1}, [x]_{x^2+1}\}$ is a basis for $\mathbb{R}[x]/(x^2+1)$ over $\mathbb{R}$.

If we look at the polynomial

$$q(T) = T^2 + 1 \in \mathbb{R}[T],$$

then we know that $\mathbb{R}$ does not contain any roots of $T^2 + 1$. However, we will now consider the field $\mathbb{R}[x]/(x^2+1)$. In this context, we write elements of $\mathbb{R}$ in the form $[a]_{x^2+1}$, where $a \in \mathbb{R}$. We now rewrite $q(T)$ as $T^2 + [1]_{x^2+1}$, and when we plug $[x]_{x^2+1}$ into $q(T)$, we now obtain

$$q\left([x]_{x^2+1}\right) = [x]_{x^2+1}{}^2 + [1]_{x^2+1} = [x^2]_{x^2+1} + [1]_{x^2+1} =$$

$$[x^2 + 1]_{x^2+1} = [0]_{x^2+1}.$$

Thus, $\mathbb{R}[x]/(x^2+1)$ is a field extension of $\mathbb{R}$ that also contains a root of $q(T) = T^2 + 1 \in \mathbb{R}[T]$.

■

In the preceding two examples, $F[x]/(f(x))$ turned out to be a field that was a finite extension of F. The polynomials $p(T)$ and $q(T)$ in these examples were really the original polynomial $f(x)$, with the variable x replaced by the new variable T. In both cases, $f(T)$ did not have a root in F, but $[x]_{f(x)}$ was a root of $f(T)$ in $F[x]/(f(x))$. However, before going any further in this direction, we need to observe that not all rings of the form $F[x]/(f(x))$ are fields. To see this, we have

■ Example

Let $R = \mathbb{Q}[x]$ and $I = (x^2 - 9)$, then every element in $R/I = \mathbb{Q}[x]/(x^2-9)$ can be written in the form $[a+bx]_{x^2-9}$, where $a, b \in \mathbb{Q}$. Since x^2 and 9 belong to the same coset, if $a, b, c, d \in \mathbb{Q}$, we have

$$[a+bx]_{x^2-9} \cdot [c+dx]_{x^2-9} = [ac + (ad+bc)x + (bd)x^2]_{x^2-9} =$$

$$[(ac + 9bd) + (ad + bc)x^2]_{x^2-9}.$$

In particular,

$$[3+x]_{x^2-9} \cdot [-3+x]_{x^2-9} = [-9 + x^2]_{x^2-9} = [0]_{x^2-9}.$$

As a result, $[3+x]_{x^2-9}$ and $[-3+x]_{x^2-9}$ are nonzero elements of $\mathbb{Q}[x]/(x^2-9)$ whose product is $[0]_{x^2-9}$. Thus, $[3+x]_{x^2-9}$ and $[-3+x]_{x^2-9}$ are zero divisors, so $\mathbb{Q}[x]/(x^2-9)$ is not a field.

■

In our previous three examples, $\mathbb{Q}[x]/(x^2 - 2)$ and $\mathbb{R}[x]/(x^2 + 1)$ are fields, whereas $\mathbb{Q}[x]/(x^2 - 9)$ is not a field. If we wish to determine precisely when rings of the form $F[x]/(f(x))$ are fields, we will again exploit the parallels between $F[x]$ and $\mathbb{Z}$. Recall, that in Chapter 7 we used the Euclidean Algorithm to show that $\mathbb{Z}_n$ was a field if and only if n was prime. Observe that there is also a Euclidean Algorithm in $F[x]$, and irreducible polynomials play the role in $F[x]$ that primes play in $\mathbb{Z}$. This observation, along with a look back at our three examples of rings of the form $F[x]/(f(x))$, is evidence that $F[x]/(f(x))$ should be a field precisely when $f(x)$ is irreducible in $F[x]$. We record this as

Theorem 17.13. *Let $f(x) \in F[x]$ have degree at least one. Then the ring $F[x]/(f(x))$ is a field if and only if $f(x)$ is irreducible in $F[x]$.*

Proof. In one direction, suppose $f(x)$ is not irreducible in $F[x]$. Then we can write $f(x) = a(x) \cdot b(x)$, where both $a(x)$ and $b(x)$ are elements of $F[x]$ having smaller degree than $f(x)$. As a result, neither $a(x)$ nor $b(x)$ can be multiples of $f(x)$, so

$$[a(x)]_{f(x)} \quad \text{and} \quad [b(x)]_{f(x)}$$

are both nonzero elements of $F[x]/(f(x))$.

However,

$$[a(x)]_{f(x)} \cdot [b(x)]_{f(x)} = [a(x) \cdot b(x)]_{f(x)} = [f(x)]_{f(x)} = [0]_{f(x)}.$$

As a result, $[a(x)]_{f(x)}$ and $[b(x)]_{f(x)}$ are zero divisors in $F[x]/(f(x))$, so $F[x]/(f(x))$ is not a field.

In the other direction, suppose $f(x)$ is irreducible in $F[x]$. We need to show that if $a(x) \in F[x]$ such that $[a(x)]_{f(x)} \neq 0$ in $F[x]/(f(x))$, then $[a(x)]_{f(x)}$ is invertible in $F[x]/(f(x))$. Since $f(x)$ is irreducible in $F[x]$ and $a(x)$ is not a multiple of $f(x)$, it follows that $a(x)$ and $f(x)$ are relatively prime in $F[x]$. The Euclidean Algorithm in $F[x]$ asserts that there exist $r(x)$, $s(x) \in F[x]$ such that

$$r(x) \cdot a(x) + s(x) \cdot f(x) = 1.$$

Therefore, when we multiply in $F[x]/(f(x))$, we have

$$[r(x)]_{f(x)} \cdot [a(x)]_{f(x)} = [r(x) \cdot a(x)]_{f(x)} = [1 - s(x) \cdot f(x)]_{f(x)} = [1]_{f(x)}.$$

Thus, $[r(x)]_{f(x)}$ is the multiplicative inverse of $[a(x)]_{f(x)}$ in $F[x]/(f(x))$ and it is indeed the case that every nonzero element of $F[x]/(f(x))$ has a multiplicative inverse. Hence, $F[x]/(f(x))$ is a field. $\qquad\square$

Based on the discussion that preceded Theorem 17.13, we can now prove the main result of this section.

Theorem 17.14—Kronecker's Theorem. *If F is a field and $f(T) \in F[T]$ has degree $n \geq 1$, then there exists a field K containing F such that K contains a root of $f(x)$ and K has degree at most n over F.*

Proof. Since $f(T)$ has degree at least one, $f(T)$ can be written as a product of irreducible polynomials in $F[T]$. If we let $p(T) \in F[T]$ be an irreducible factor of $f(T)$ and if we let m denote the degree of $p(T)$, then clearly $m \leq n$ and any root of $p(T)$ is also a root of $f(T)$. Therefore, it suffices to find a field K containing F such that K contains a root of $p(T)$ and has degree at most m over F.

If we let $p(x)$ be the same polynomial as $p(T)$, except with the variable x, Theorem 17.13 tells us that the ring $F[x]/(p(x))$ is a field. We also know that every element of $F[x]/(p(x))$ can be written as

$$[\alpha_0 + \alpha_1 x + \cdots + \alpha_{m-1} x^{m-1}]_{p(x)},$$

where each $\alpha_i \in F$. Since elements of F are represented in $F[x]/(p(x))$ in the form $[\alpha]_{p(x)}$, where $\alpha \in F$, we now have

$$[\alpha_0 + \alpha_1 x + \cdots + \alpha_{m-1} x^{m-1}]_{p(x)} =$$

$$[\alpha_0]_{p(x)} \cdot [1]_{p(x)} + [\alpha_1]_{p(x)} \cdot [x]_{p(x)} + \cdots + [\alpha_{m-1}]_{p(x)} \cdot [x^{m-1}]_{p(x)}.$$

This last equation tells us that the set

$$\{[1]_{p(x)}, [x]_{p(x)}, \ldots, [x^{m-1}]_{p(x)}\}$$

spans $F[x]/(p(x))$ over F. Thus, the degree of $F[x]/(p(x))$ over F is at most m. Although it is not needed for this proof, it is not hard to see that the set $\{[1]_{p(x)}, [x]_{p(x)}, \ldots, [x^{m-1}]_{p(x)}\}$ is also linearly independent over F. Thus, the degree is $F[x]/(p(x))$ over F is exactly m.

In light of the preceding, it now suffices to show that $F[x]/(p(x))$ contains a root of $p(T)$. Based on our examples, $[x]_{p(x)}$ is the obvious candidate. Using our representation of elements of F as cosets in $F[x]/(f(x))$, we know that $p(T)$ can be written as

$$p(T) = [\beta_0]_{p(x)} + \cdots + [\beta_{m-1}]_{p(x)} \cdot T^{m-1} + [\beta_m]_{p(x)} \cdot T^m,$$

where each $\beta_i \in F$ and $\beta_m \neq 0$. Using the fact that $[p(x)]_{p(x)} = [0]_{p(x)}$ in $F[x]/(p(x))$, when we plug our candidate $[x]_{p(x)}$ into $p(T)$, we obtain

$$p([x]_{p(x)}) = [\beta_0]_{p(x)} + \cdots + [\beta_{m-1}]_{p(x)} \cdot ([x]_{p(x)})^{m-1} + [\beta_m]_{p(x)} \cdot ([x]_{p(x)})^m =$$

$$[\beta_0 + \cdots + \beta_{m-1} x^{m-1} + \beta_m x^m]_{p(x)} = [p(x)]_{p(x)} = [0]_{p(x)}.$$

Thus, $[x]_{p(x)}$ is indeed a root of $p(T)$, as required. $\qquad\square$

When we compare the fields $\mathbb{Q}[x]/(x^2 - 2)$ and $\mathbb{Q}(\sqrt{2})$, they appear to be essentially the same field. Elements of both fields look like polynomials of degree at most one with rational coefficients. The field $\mathbb{Q}[x]/(x^2 - 2)$ contains the element $[x]_{x^2-2}$, which has the interesting property that its square is equal to 2, whereas the element $\sqrt{2} \in \mathbb{Q}(\sqrt{2})$ plays an analogous role in $\mathbb{Q}(\sqrt{2})$. Similarly, the fields $\mathbb{R}[x]/(x^2 + 1)$ and $\mathbb{C}$ seem to be essentially the same field. The element $[x]_{x^2+1}$ plays the same role in $\mathbb{R}[x]/(x^2 + 1)$ that the element i plays in $\mathbb{C}$.

In Chapter 8, when two groups were essentially the same group, except that the elements might have different names, we said that the groups were isomorphic. More formally, we defined two groups to be isomorphic when there existed a homomorphism between them that was also a bijection. In Chapter 9, we introduced homomorphisms of rings and used this concept to prove various irreducibility criteria. In an attempt to formalize the notion of two different rings or fields being essentially the same algebraic object, we have

Definition 17.15.   *Rings R_1 and R_2 are said to be isomorphic if there exists a ring homomorphism $\phi : R_1 \to R_2$ which is also a bijection.*

It is natural to wonder if there are straightforward ways to show that $\mathbb{Q}[x]/(x^2 - 2)$ and $\mathbb{Q}(\sqrt{2})$ are isomorphic and also that $\mathbb{R}[x]/(x^2 + 1)$ and $\mathbb{C}$ are isomorphic. The tool that we will use will be the analog for rings of Theorem 8.39(a)—Isomorphism Theorem for Groups.

Theorem 17.16—Isomorphism Theorem for Rings.   *Let $\phi : R_1 \to R_2$ be a homomorphism of rings.*

(a)   *$Ker(\phi) = \{r \in R_1 \,|\, \phi(r_1) = 0\}$ is an ideal of R_1.*

(b)   *$Im(\phi) = \{\phi(r) \,|\, r \in R_1\}$ is a ring.*

(c)   *The rings $R_1/Ker(\phi)$ and $Im(\phi)$ are isomorphic.*

Proof. For part (a), since ϕ is a homomorphism of groups, Theorem 8.24(c) tells us that $Ker(\phi)$ is a subgroup under addition of R_1. In this situation, 0 represents the additive identity of R_2 and we know that multiplication by 0 in R_2 always results in 0. To check that $Ker(\phi)$ is an ideal of R_1, if we let $a \in Ker(\phi)$ and $r \in R_1$, we have

$$\phi(ar) = \phi(a)\phi(r) = 0 \cdot \phi(r) = 0$$

and

$$\phi(ra) = \phi(r)\phi(a) = \phi(r) \cdot 0 = 0.$$

Therefore, $ar, ra \in Ker(\phi)$, hence $Ker(\phi)$ is an ideal of R_1.

For part (b), since ϕ is a homomorphism of groups, we again cite Theorem 8.24(c), this time to assert that $Im(\phi)$ is a subgroup of R_2 under addition. Furthermore, since $Im(\phi)$ is a subset of the ring R_2, we know that the associative and commutative laws of multiplication and the distributive laws hold when dealing with elements of $Im(\phi)$. Therefore, in order to verify that $Im(\phi)$ is a ring, it suffices to show that $Im(\phi)$ is closed under multiplication and also contains a multiplicative identity.

Observe that if $r, s \in Im(\phi)$, then there exist $a, b \in R_1$ such that $r = \phi(a)$ and $s = \phi(b)$. Therefore,

$$rs = \phi(a)\phi(b) = \phi(ab) \in Im(\phi),$$

so $Im(\phi)$ is closed under multiplication.

If we let 1 denote the multiplicative identity of R_1, then $\phi(1)$ is the likely candidate to be the multiplicative identity of R_2. To check that it is, if $r \in Im(\phi)$, then $r = \phi(a)$, for some $a \in R_1$, and we have

$$r\phi(1) = \phi(a)\phi(1) = \phi(a \cdot 1) = \phi(a) = r$$

as well as

$$\phi(1)r = \phi(1)\phi(a) = \phi(1 \cdot a) = \phi(a) = r.$$

Thus, $r\phi(1) = r = \phi(1)r$. Therefore, $\phi(1)$ is the multiplicative identity of $Im(\phi)$ and $Im(\phi)$ is a ring.

Since R_2 is commutative, it was not necessary to show that both $r\phi(1)$ and $\phi(1)r$ were equal to r. However, by providing the extra details, we are actually providing a proof that also works for noncommutative rings.

For part (c), most of the proof is already taken care of by the proof of Theorem 8.39(a)—Isomorphism Theorem for Groups. In particular, that proof tells us that if we define

$$v : R_1/Ker(\phi) \rightarrow Im(\phi)$$

as

$$v(a + Ker(\phi)) = \phi(a),$$

for all $a \in R_1$, then v is an isomorphism between the groups $R_1/Ker(\phi)$ and $Im(\phi)$.

As a result, it now suffices to show that v preserves multiplication. To this end, if we are given two elements of $R_1/Ker(\phi)$, then we can represent them as $a + Ker(\phi)$ and $b + Ker(\phi)$, where $a, b \in R_1$. Since ϕ is a homomorphism of rings, we know that it preserves multiplication, so

$$v((a + Ker(\phi)) \cdot (b + Ker(\phi))) = v(ab + Ker(\phi)) = \phi(ab) =$$
$$\phi(a)\phi(b) = v(a + Ker(\phi)) \cdot v(b + Ker(\phi)),$$

as required. $\qquad\Box$

We will conclude this section by using the Isomorphism Theorem for Rings to further examine the quotient rings $\mathbb{Q}[x]/(x^2 - 2)$, $\mathbb{R}[x]/(x^2 + 1)$, and $\mathbb{Q}[x]/(x^2 - 9)$.

■ Examples

1. Let $\phi : \mathbb{Q}[x] \to \mathbb{Q}(\sqrt{2})$ be defined as $\phi(f(x)) = f(\sqrt{2})$, for all $f(x) \in \mathbb{Q}[x]$. Observe that if $f(x), g(x) \in \mathbb{Q}[x]$, we have

    $$\phi(f(x) + g(x)) = f(\sqrt{2}) + g(\sqrt{2}) = \phi(f(x)) + \phi(g(x))$$

    and

    $$\phi(f(x) \cdot g(x)) = f(\sqrt{2}) \cdot g(\sqrt{2}) = \phi(f(x)) \cdot \phi(g(x)).$$

    As a result, ϕ is a homomorphism of rings. If $a, b \in \mathbb{Q}$, then

    $$\phi(a + bx) = a + b\sqrt{2},$$

    so ϕ is certainly surjective.

    Corollary 12.7 tells us that the elements of $\mathbb{Q}[x]$ that have $\sqrt{2}$ as a root are precisely the multiples of $x^2 - 2$. Observe that this is equivalent to saying that $Ker(\phi) = (x^2 - 2)$. Since we know that $Im(\phi) = \mathbb{Q}(\sqrt{2})$ and $Ker(\phi) = (x^2 - 2)$, Theorem 17.16(c) asserts that

    $$\mathbb{Q}[x]/(x^2 - 2) \approx \mathbb{Q}(\sqrt{2}).$$

2. Let $\phi : \mathbb{R}[x] \to \mathbb{C}$ be defined as $\phi(f(x)) = f(i)$, for all $f(x) \in \mathbb{R}[x]$. Using a short argument almost identical to one in the previous example, we can see that ϕ is a homomorphism. Indeed, if $f(x), g(x) \in \mathbb{R}[x]$, we have

    $$\phi(f(x) + g(x)) = f(i) + g(i) = \phi(f(x)) + \phi(g(x))$$

    and

    $$\phi(f(x) \cdot g(x)) = f(i) \cdot g(i) = \phi(f(x)) \cdot \phi(g(x)).$$

    If $a, b \in \mathbb{R}$, then

    $$\phi(a + bx) = a + bi,$$

    so ϕ is surjective. Corollary 12.7 tells us that the elements of $\mathbb{R}[x]$ that have i as a root are precisely the multiples of $x^2 + 1$. This immediately tells us that $Ker(\phi) = (x^2 + 1)$, so Theorem 17.16(c) tells us that

    $$\mathbb{R}[x]/(x^2 + 1) \approx \mathbb{C}.$$

3. Let $\phi : \mathbb{Q}[x] \to \mathbb{Q} \times \mathbb{Q}$ be defined as $\phi(f(x)) = (f(3), f(-3))$, for all $f(x) \in \mathbb{Q}[x]$. The ring $\mathbb{Q} \times \mathbb{Q}$ consists of all ordered pairs of rational numbers where addition and multiplication are done componentwise. In particular, if $a, b, c, d \in \mathbb{Q}$, then

$$(a, b) + (c, d) = (a+c, b+d) \quad \text{and} \quad (a, b) \cdot (c, d) = (ac, bd).$$

Observe that in $\mathbb{Q} \times \mathbb{Q}$ we have $(1, 0) \cdot (0, 1) = (0, 0)$. Therefore, $\mathbb{Q} \times \mathbb{Q}$ has zero divisors and thus cannot be a field.

If $f(x), g(x) \in \mathbb{Q}[x]$, we have

$$\phi(f(x) + g(x)) = (f(3) + g(3), f(-3) + g(-3)) =$$

$$(f(3), f(-3)) + (g(3), g(-3)) = \phi(f(x)) + \phi(g(x))$$

and

$$\phi(f(x) \cdot g(x)) = (f(3) \cdot g(3), f(-3) \cdot g(-3)) =$$

$$(f(3), f(-3)) \cdot (g(-3), g(-3)) = \phi(f(x)) \cdot \phi(g(x)).$$

Thus, ϕ is a homomorphism of rings.

If $a, b \in \mathbb{Q}$ then, when we let $f(x) = \left(\frac{a-b}{6}\right)x + \frac{a+b}{2}$, it is easy check that $f(3) = a$ and $f(-3) = b$. Thus,

$$\phi(f(x)) = (f(3), f(-3)) = (a, b),$$

which shows that ϕ is surjective.

Next, if $f(x) \in \mathbb{Q}[x]$, it is not hard to see that the three conditions of having 3 and -3 as roots, being a multiple of $x^2 - 9$, and belonging to $Ker(\phi)$ are all equivalent. Therefore, $Ker(\phi) = (x^2 - 9)$, and Theorem 17.16(c) now asserts that

$$\mathbb{Q}[x]/(x^2 - 9) \approx \mathbb{Q} \times \mathbb{Q}.$$

In an earlier example, we showed that $\mathbb{Q}[x]/(x^2 - 9)$ was not a field because the cosets containing $3 + x$ and $-3 + x$ are nonzero, yet their product was the coset containing 0. Since ϕ is an isomorphism, the product of $\phi(3 + x)$ and $\phi(-3 + x)$ should now be the additive identity of $\mathbb{Q} \times \mathbb{Q}$, which is $(0, 0)$. Observe that this is indeed the case as

$$\phi(3 + x) \cdot \phi(-3 + x) = (6, 0) \cdot (0, -6) = (0, 0).$$

Exercises for Section 17.3

1. Let R be a commutative ring. If $r \in R$, show that the ideal $(r) = \{rs \mid s \in R\}$ is equal to all of R if and only if r has a multiplicative inverse in R.

2. Suppose n, m are nonzero integers such that the ideals $(n) = \{nb \mid b \in \mathbb{Z}\}$ and $(m) = \{mb \mid b \in \mathbb{Z}\}$ are equal. What can you say about the relationship between n and m?

3. Suppose $f(x), g(x)$ are nonzero elements of $F[x]$, where F is a field, such that the ideals $(f(x)) = \{f(x) \cdot h(x) \mid h(x) \in F[x]\}$ and $(g(x)) = \{g(x) \cdot h(x) \mid h(x) \in F[x]\}$ are equal. What can you say about the relationship between $f(x)$ and $g(x)$?

4. Using the notation of exercise 2, suppose n, m are nonzero integers such that $(n) \subseteq (m)$. What can you say about the relationship between n and m?

5. Using the notation of exercise 3, suppose $f(x), g(x)$ are nonzero elements of $F[x]$, where F is a field, such that $(f(x)) \subseteq (g(x))$. What can you say about the relationship between $f(x)$ and $g(x)$?

6. Let I, J be ideals of a commutative ring R. If $I + J = \{a + b \mid a \in I, b \in J\}$, show that $I + J$ is also an ideal of R.

7. Let I, J be ideals of a commutative ring R. Show that $I \cap J$ is also an ideal of R.

8. Using the notation of exercises 2 and 6, if n, m are nonzero integers and $c = gcd(n, m)$, show that $(n) + (m) = (c)$.

9. Let $f(x), g(x)$ be nonzero elements of $F[x]$, where F is a field. Using the notation of exercises 3 and 6, if $c(x) = gcd(f(x), g(x))$, show that $(f(x)) + (g(x)) = (c(x))$.

10. Using the notation of exercise 2, if n, m are nonzero integers and $d = lcm(n, m)$, show that $(n) \cap (m) = (d)$.

11. Using the notation of exercise 3, if $f(x), g(x)$ are nonzero elements of $F[x]$, where F is a field, and $d(x) = lcm(f(x), g(x))$, show that $(f(x)) \cap (g(x)) = (d(x))$.

Before doing exercise 12, please read the following:

In Chapter 12, we saw that a huge difference between $\mathbb{Z}[x]$ and $F[x]$, where F is a field, is that there is a division algorithm in $F[x]$ but not in $\mathbb{Z}[x]$. In this chapter, we saw that every ideal of $F[x]$ is of the form $(f(x)) = \{f(x) \cdot h(x) \mid h(x) \in F[x]\}$, for some $f(x) \in F[x]$. Since the proof of this relied on the division algorithm, it may not come as much of a surprise that there exist ideals of $\mathbb{Z}[x]$ not of the form $(g(x)) = \{g(x) \cdot k(x) \mid k(x) \in \mathbb{Z}[x]\}$.

12. In $\mathbb{Z}[x]$, using the notation of exercise 6, let $I = (x) + (2)$. We can also think of I as $\{f(x) \in \mathbb{Z}[x] \mid f(0) \text{ is an even integer}\}$. Show that there does not exist any $g(x) \in \mathbb{Z}[x]$ such that $I = (g(x))$.

In exercises 13–18, we will be working in the field $K = \mathbb{Q}[x]/(x^2 + 6x - 2)$. In your answers to these exercises, please write elements of K in the form $[a + bx]_{x^2+6x-2}$, where $a, b \in \mathbb{Q}$.

13. Compute $[8 - 5x]_{x^2+6x-2} \cdot [2 + 3x]_{x^2+6x-2}$.

14. Compute $[-6 + 7x]_{x^2+6x-2} \cdot [11 - 4x]_{x^2+6x-2}$.

15. Find the general formula for the product $[a + bx]_{x^2+6x-2} \cdot [c + dx]_{x^2+6x-2}$, where $a, b, c, d \in \mathbb{Q}$.

16. If $a, b \in \mathbb{Q}$ such that $[a + bx]_{x^2+6x-2}$ is not zero, find $([a + bx]_{x^2+6x-2})^{-1}$.

17. Show that both $[x]_{x^2+6x-2}$ and $[-6 - x]_{x^2+6x-2}$ are roots in K of $T^2 + 6T - 2 \in \mathbb{Q}[T]$.

18. Find two elements in K whose square is $[11]_{x^2+6x-2}$.

In exercises 19–24, we will be working in the field $L = \mathbb{Q}[x]/(x^3 - 7)$. In your answers to these exercises, please write elements of L in the form $[a + bx + cx^2]_{x^3-7}$, where $a, b, c \in \mathbb{Q}$.

19. Compute $[2 - 5x]_{x^3-7} \cdot [6 + 3x^2]_{x^3-7}$.

20. Compute $[5x - 7x^2]_{x^3-7} \cdot [-5 + 2x]_{x^3-7}$.

21. Compute $[1 + 4x - x^2]_{x^3-7} \cdot [8 - 7x + 5x^2]_{x^3-7}$.

22. Find the general formula for the product $[a + bx + cx^2]_{x^3-7} \cdot [d + ex + fx^2]_{x^3-7}$, where $a, b, c, d, e, f \in \mathbb{Q}$.

23. How many roots of $T^3 - 7 \in \mathbb{Q}[T]$ belong to L?

24. Show that $[-2 + x]_{x^3-7}$ is a root of $T^3 + 6T^2 + 12T + 1 \in \mathbb{Q}[T]$.

In exercises 25–30, we will be working in the commutative ring $R = \mathbb{Q}[x]/(x^2 - 25)$. In your answers to these exercises, please write elements of R in the form $[a + bx]_{x^2-25}$, where $a, b, c \in \mathbb{Q}$.

25. Compute $[3 + 2x]_{x^2-25} \cdot [-11 - 4x]_{x^2-25}$.

26. Find the general formula for the product $[a + bx]_{x^2-25} \cdot [c + dx]_{x^2-25}$, where $a, b, c, d \in \mathbb{Q}$.

27. Find an element $r \in R$ such that $[7 - 8x]_{x^2-25} \cdot r = [1]_{x^2-25}$.

28. Find a nonzero element $s \in R$ such that $[15 - 3x]_{x^2-25} \cdot s = [0]_{x^2-25}$.

29. Show that every nonzero element of R is either invertible or a zero divisor.

30. Show that $[a + bx]_{x^2-25}$ is invertible in R if and only if a is equal to neither $5b$ nor $-5b$.

31. Show that $\mathbb{Z}_2[x]/(x^2 + x + [1]_2)$ is a field with 4 elements.

32. Show that $\mathbb{Z}_2[x]/(x^3+x+[1]_2)$ is a field with 8 elements.

33. Show that $\mathbb{Z}_3[x]/(x^2+[1]_3)$ is a field with 9 elements.

34. Show that $\mathbb{Z}_3[x]/(x^3+[2]_3x+[1]_3)$ is a field with 27 elements.

In exercises 35–40, we examine finite fields. In exercises 35–37, you will show that if F is a finite field, then $|F| = p^n$, where p is prime and $n \in \mathbb{N}$. Conversely, in exercises 38–40, you will show that given a prime p and $n \in \mathbb{N}$, there exists a field F with p^n elements.

35. If F is a finite field, show that F has characteristic $p > 0$.

36. If F has characteristic $p > 0$, show that F contains a field E which is isomorphic to the field $\mathbb{Z}_p$.

37. If F is a finite field, show that F has p^n elements, where p is the characteristic and n is the dimension of F over the subfield isomorphic to $\mathbb{Z}_p$.

38. If L is any field and $f(x) \in L[x]$ has degree $n \geq 1$, show that there exists a field M that contains L and also contains n roots, counting multiplicities, of $f(x)$.

39. If p is a prime number and $n \in \mathbb{N}$, show that there exists a field K which contains $\mathbb{Z}_p$ and also contains p^n different roots of the polynomial $x^{p^n} - x \in \mathbb{Z}_p[x]$. (Technically, in $\mathbb{Z}_p[x]$, this polynomial is written as $[1]_p x^p + [p-1]_p x$.)

40. Let K be as in exercise 39 and let $F = \{r \in K \mid r^{p^n} = r\}$. Show that F is a field with exactly p^n elements.

41. Let $\phi : \mathbb{Q}[x] \to \mathbb{C}$ be the ring homomorphism defined as $\phi(f(x)) = f(3-2i)$, for all $f(x) \in \mathbb{Q}[x]$.

    (a) Describe, as simply as possible, the elements of $Ker(\phi)$.

    (b) Describe, as simply as possible, the elements of $Im(\phi)$.

    (c) Based on the Isomorphism Theorem for Rings and parts (a) and (b), which two rings can now be seen to be isomorphic?

42. Let $\phi : \mathbb{C}[x] \to \mathbb{C} \oplus \mathbb{C}$ be the ring homomorphism defined as $\phi(g(x)) = (g(5), g(-2))$, for all $g(x) \in \mathbb{C}[x]$.

    (a) Describe, as simply as possible, the elements of $Ker(\phi)$.

    (b) Describe, as simply as possible, the elements of $Im(\phi)$.

    (c) Based on the Isomorphism Theorem for Rings and parts (a) and (b), which two rings can now be seen to be isomorphic?

43. Let $\phi : \mathbb{Q}[x] \to \mathbb{C} \oplus \mathbb{C}$ be the ring homomorphism defined as $\phi(h(x)) = (h(\sqrt{3}), h(i))$, for all $h(x) \in \mathbb{Q}[x]$.

    (a) Describe, as simply as possible, the elements of $Ker(\phi)$.

    (b) Describe, as simply as possible, the elements of $Im(\phi)$.

    (c) Based on the Isomorphism Theorem for Rings and parts (a) and (b), which two rings can now be seen to be isomorphic?

44. Let $\phi : \mathbb{Q}[x] \to \mathbb{R} \oplus \mathbb{R} \oplus \mathbb{R}$ be the ring homomorphism defined as $\phi(f(x)) = (f(\sqrt{5}), f(11), f(-9))$, for all $f(x) \in \mathbb{Q}[x]$.

    (a) Describe, as simply as possible, the elements of $Ker(\phi)$.

    (b) Describe, as simply as possible, the elements of $Im(\phi)$.

    (c) Based on the Isomorphism Theorem for Rings and parts (a) and (b), which two rings can now be seen to be isomorphic?

In exercises 45–72, we will try to better understand fields of the form $\mathbb{Q}(\omega)$, where $\omega = \text{cis}\left(\frac{2\pi}{n}\right)$, for some $n \geq 1$. Our goals will be to

   (a) Find $[\mathbb{Q}(\omega) : \mathbb{Q}]$ and the minimum polynomial for ω over $\mathbb{Q}$.

   (b) Determine the structure of $Gal(\mathbb{Q}(\omega)/\mathbb{Q})$.

Throughout these exercises, n will be a fixed positive integer and ω will denote $\text{cis}\left(\frac{2\pi}{n}\right)$. We say that $\gamma \in \mathbb{C}$ is a **primitive** nth root of 1 if $\gamma^n = 1$ and $\gamma^d \neq 1$, for every positive integer d which is less than n. Observe that ω is always one of the primitive nth roots of 1. For any $m \in \mathbb{N}$, let W_m be the set of all primitive mth roots of 1. Next, we define the **nth cyclotomic polynomial** to be

$$\Psi_n(x) = \Pi_{\zeta \in W_n}(x - \zeta).$$

Thus, the roots of $\Psi_n(x)$ are precisely the primitive nth roots of 1. For example,

$$\Psi_1(x) = x - 1, \quad \Psi_2(x) = x - (-1) = x + 1,$$

and

$$\Psi_3(x) = \left(x - \text{cis}\left(\frac{2\pi}{3}\right)\right)\left(x - \text{cis}\left(\frac{4\pi}{3}\right)\right) = x^2 + x + 1.$$

We will also let $1 = d_1 < d_2 < \cdots < d_{t-1} < d_t = n$ be all the positive divisors of n.

45. In Chapter 7, we introduced the Euler ϕ-function. If $m_1, m_2, \ldots, m_{\phi(n)}$ are the $\phi(n)$ positive integers which are less than n and relatively prime to n, show that $\omega^{m_1}, \omega^{m_2}, \ldots, \omega^{m_{\phi(n)}}$ are the primitive nth roots of 1.

46. If $d_i \neq d_j$ are divisors of n, show that $W_{d_i} \cap W_{d_j} = \emptyset$.

47. Show that union of sets, $W_1 \cup W_{d_2} \cup \cdots \cup W_{d_{t-1}} \cup W_n$, is equal to set of all the roots in $\mathbb{C}$ of $x^n - 1$.

48. Use exercises 46 and 47 to show that

$$x^n - 1 = \Psi_1(x) \cdot \Psi_{d_2}(x) \cdots \Psi_{d_{t-1}}(x) \cdot \Psi_n(x).$$

49. Use the previous exercise and the Second Version of Mathematical Induction to prove that $\Psi_n(x) \in \mathbb{Q}[x]$, for every $n \in \mathbb{N}$.

50. Having already shown that $\Psi_n(x) \in \mathbb{Q}[x]$, now show that $\Psi_n(x) \in \mathbb{Z}[x]$. You might want to think about the ideas used toward the end of the proof of Gauss' Lemma.

51. If p is a prime number, show that $\Psi_p(x) = x^{p-1} + x^{p-2} + \cdots + x + 1$.

Before doing exercises 52–57, observe that in light of the formula in exercise 48, the easiest way to find $\Psi_n(x)$ is often to use the previously computed $\Psi_d(x)$, for divisors d of n.

52. Find $\Psi_4(x)$.

53. Find $\Psi_6(x)$.

54. Find $\Psi_8(x)$.

55. Find $\Psi_9(x)$.

56. Find $\Psi_{10}(x)$.

57. Find $\Psi_{12}(x)$.

58. Use Gauss' Lemma to show that there exist $f(x), g(x) \in \mathbb{Z}[x]$ such that $x^n - 1 = f(x) \cdot g(x)$, $f(x)$ is irreducible in $\mathbb{Q}[x]$, and $f(\omega) = 0$.

In all the remaining exercises, $f(x)$ and $g(x)$ will refer to the polynomials in exercise 58. In these exercises, $\gamma \in \mathbb{C}$ will denote a root of $f(x)$ and p will be a prime number that does not divide n.

59. Show that if γ^p is not a root of $f(x)$, then γ^p is a root of $g(x)$ and γ is a root of $g(x^p)$.

60. Assume once again that γ^p is not a root of $f(x)$ and then show that $g(x^p)$ is a multiple of $f(x)$.

61. Let $\rho : \mathbb{Z}[x] \to \mathbb{Z}_p[x]$ be the ring homomorphism used in the proof of Gauss' Lemma. Observe that

$$[1]_p x^n + [p-1]_p = \rho(x^n - 1) = \rho(f(x) \cdot g(x)) = \rho(f(x)) \cdot \rho(g(x)).$$

Show that $\rho(g(x))^p = \rho(g(x^p))$, and then show that if γ^p is not a root of $f(x)$, then $\rho(g(x))^p$ is a multiple of $\rho(f(x))$.

62. Kronecker's Theorem tells us that there exists a field K containing $\mathbb{Z}_p$ that contains a root r of $\rho(f(x))$. If γ^p is not a root of $f(x)$, show that r is also a root of $\rho(g(x))$.

63. Use Proposition 12.22 to show that $[1]_p x^n + [p-1]_p$ has no multiple roots in any field K containing $\mathbb{Z}_p[x]$.

64. Use exercises 62 and 63 to show that γ^p must be a root of $f(x)$.

65. If $d > 1$ is relatively prime to n, show that the prime factorization of d consists of primes that do not divide n.

66. Use exercises 64 and 65, to show every primitive nth root of 1 is a root of $f(x)$.

67. Use exercise 66 to show that the degree of $f(x)$ is at least as large as the degree of $\Psi_n(x)$.

68. Show that $\Psi_n(x)$ is irreducible in $\mathbb{Q}[x]$.

69. Show that $[\mathbb{Q}(\omega) : \mathbb{Q}] = \phi(n)$.

70. Show that $\mathbb{Q}(\omega)$ is the splitting field of $x^n - 1$ over $\mathbb{Q}$.

71. Show that $Gal(\mathbb{Q}(\omega)/\mathbb{Q}) \approx U(\mathbb{Z}_n)$ is the group of invertible elements in $\mathbb{Z}_n$.

72. Describe the behavior of each element of $Gal(\mathbb{Q}(\omega)/\mathbb{Q})$ on ω.

Bibliography

R.B.J.T. Allenby, *Rings, Fields and Groups*, Edward Arnold, London, 1991.

E. Artin, *Galois Theory*, Notre Dame University Press, Notre Dame, 1944.

M. Artin, *Algebra*, Prentice Hall, New Jersey, 1991.

J. Beachy and W. Blair, *Abstract Algebra*, Waveland Press, Illinois, 2006.

G.D. Birkhoff and S. MacLane, *A Survey of Modern Algebra*, Macmillan, New York, 1941.

M. Hall, *The Theory of Groups*, Macmillan, New York, 1959.

I.N. Herstein, *Noncommutative Rings*, Carus Mathematical Monographs No. 15, Mathematical Association of American, Washington, D.C., 1994.

I.N. Herstein, *Topics in Algebra*, John Wiley & Sons, New York, 1975.

I. Kaplansky, *Fields and Rings*, University of Chicago Press, Chicago, 1969.

T.Y. Lam, *A First Course in Noncommutative Rings*, Springer-Verlag, New York, 2001.

I. Niven, *Irrational Numbers*, Carus Mathematical Monographs No. 11, Mathematical Association of American, Distributed by John Wiley & Sons, New York, 1956.

I. Niven and H. Zuckerman, *An Introduction to the Theory of Numbers*, John Wiley & Sons, New York, 1960.

D.S. Passman, *A Course in Ring Theory*, Wadsworth and Brooks/Cole, California, 1991.

J. Rotman, *An Introduction to the Theory of Groups*, Springer-Verlag, New York, 1995.

L. Rowen, *Ring Theory: Student Edition*, Academic Press, Boston, 1991.

T. Shifrin, *Abstract Algebra: A Geometric Approach*, Prentice Hall, 1996.

I. Stewart, *Galois Theory*, Chapman & Hall/CRC, Florida, 2004.

Bibliography

Index

Geometry (*continued*)
nonabelian subgroups of S_n viewed in, 279–285
ruler/compass construction and, 6–7
Goldbach's conjecture (prime numbers), 62
Greatest common divisors, **76**
computing, 78–79
Euclidean Algorithm finding, 79–91, 460–470
in $F[x]$, 462–466
multiple roots of polynomials algorithm with, 477–479
of nonzero polynomials, 460
of polynomials moving to larger fields, 476–477
prime factorization and, 78
questions regarding, 77
unique factorization of polynomials for, 470
uniqueness of prime factorization and, 76–79
Green, B., 242–243
Group(s), **56**. *See also* Abelian groups; Cyclic groups; Factor groups; Finite groups; Solvable group(s); Subgroup(s); Symmetric group(s)
of bijective functions, 178, 267–271
centralizers showing structure of, 295
Class Equation and isomorphism in, 328–329
of commutative rings, 178
commutative rings under addition as, 266
commutative rings with multiplicative inverses as, 267
of cosets, 322
definition of, 265–266
dihedral, 283–286, 290
direct product and, 316
of fields, 179
fields under addition as, 266
with finite subgroups, 337–338
homomorphism of, 610–611
homomorphisms of, with injective functions, 309–311
identity element of, 271
importance of understanding, 182
in insolvability of the quintic, 56

invertible elements in commutative rings under multiplication as, 266–267
Invertible elements of rings, under multiplication, as, 238–239
isomorphism of, 287–288, 674
Isomorphism Theorem for, 656, 675
mathematical proofs and, 297–299
non commutative, 181–182
nonabelian, 270–271
nonempty subsets and, 272
subgroups of bijective function, 277–279
subsets of, 271–274
two "same," 286–287
Group theory, 4, 14
Sylow's Theorem and, 312
Groups of automorphisms, of commutative rings, 177–182

H

Homogeneous system of linear equations, **564**
corresponding, 566–567
unknowns and, 564–565
Homomorphism(s)
bijective, 602, 653
of groups, 610–611
groups with injective functions and, 309–311
injective, 601, 611
isomorphisms compared to, 307
normal subgroups and, 307–309
of rings, 370–374

I

Ideal(s), 370, **378**
Kronecker's Theorem and, 664–665
Identity element, 54–55
of groups, 271
Identity map, 53–54
Imaginary numbers, complex numbers as, 139
Imagination. *See* Creativity
Impedance, **151**
of parallel circuit, 152–154
of series circuit, 152
Impossibility of trisecting angles, 639
approach for, 623
Rational Root Test and, 642
with ruler and compass construction, 640–643
Including multiplicities, **475**

Indirect proof, 27
Infinite dimensional, 554–555
Infinite series. *See also* Convergent infinite series
prime numbers and, 63
real numbers as, 100
Injective functions, 48
composition of, 52–53
homomorphisms and groups with, 309–311
mathematical proof challenges for, 52
Injective homomorphisms, 601, 611
Insolvability of polynomials by radicals, 655–660
Insolvability of the quintic, **4**, 14–15, 182, 411, 453, 610, 645–683
Fundamental Theorem of Algebra and, 663
Galois groups of radical extensions and, 651–654
groups in, 56
mathematical proof of, 657–660
radical extensions and, 646
solvable groups and Galois' work on, 325
Integer solutions, **397**
Integers. *See also* Negative integers; Positive integer(s); Sum of squares of two integers
divisibility and, 61–62
importance of, 61
polynomials, similarities to, 29, 61, 437–444
prime numbers and, 61–64
real numbers as quotients of, 98–100
table of values and subsets of, 491
uniqueness of prime factorization and, 64–67
Integers modulo n. *See* $\mathbb{Z}_n$
Integral combinations, 79–84
Integral domain, polynomial multiplication of coefficients in, 367–369
Intermediate Value Theorem, 423
Fundamental Theorem of Algebra compared to, 220–221
Least Upper Bound Property proving, 107–108
mathematical proof of, 108–110
Rational Root Test and, 380
real numbers compared to rational numbers using, 105–106
roots of complex numbers and, 211